About the Author

Dr Chakrabarti P.K. graduated from the Presidency College (now Presidency University) under Calcutta University with honours in Physics and subsequently did his postgraduation in Physics from the same university.

He published a number of research papers in renowned International Journals on Transport properties in Liquids — a theoretical investigation, for which he was awarded PhD degree in Physics from the University of Calcutta.

Electronics which was a hobby at one time turned into passion in no time, Dr Chakrabarti made a number of electronic gadgets and obtained patents in some of them.

Chakrabarti Dr P.K. has established himself as an author of distinction with a number of good textbooks for BSc and MSc level, some of which deserve mentioning.

1. Geometrical and Physical Optics (using matrix method formalism)
2. Theory and experiments on Thermal Physics
3. Theory of Linear Circuits and Rudiments of Electronics
4. Advanced Textbook on Heat
5. Theory of Electricity and Magnetism
6. Triumph of Mechanics (in press)
7. Mechanics and General Properties of Matter
8. Mathematical Physics etc.

About the Book

P.K.C. (SXC, Kol)
Presents
Theory of Linear Circuits and Rudiments of Electronics

Based on P.K.C.'s lectures on electronics delivered from time to time at UG Level, SXC (Kol). It contains both analog and digital part and also discrete and integrated technology. The book contains a well blended combination of theoretical and experimental exposition of different topics in electronics.

The beauty of the book is that the whole course is first discussed theoretically in simple and elegant language, followed by appropriate circuitry with elaborate physical explanation. An electronic circuit, it is to be admitted, if looked at casually seems to be as arid as a desert but if you critically analyse the same and ask yourself why and how the active and passive elements of the circuit are arranged in it and find the answer yourself, there is nothing else to excel your pleasure. Read the book carefully and derive the pleasure hidden in this 'Treasure Island'.

THEORY OF LINEAR CIRCUITS
and
RUDIMENTS OF ELECTRONICS

Dr P. K. Chakrabarti MSc PhD
Dean, PG Studies, St. Xavier's College, Kolkata

New Central Book Agency (P) Ltd
LONDON
HYDERABAD ERNAKULAM BHUBANESWAR
NEW DELHI KOLKATA PUNE GUWAHATI

NCBA

REGD OFFICE

8/1 Chintamoni Das Lane, Kolkata 700 009, India
email: ncbapvtltd@eth.net

OVERSEAS

NCBA (UK) Ltd, 149 Park Avenue North, Northampton, NN3 2HY, UK
email: ncbauk@yahoo.com

EXPORT

NCBA Exports Pvt. Ltd, 212 Shahpur Jat, New Delhi 110 049
email: ncbaexports@ncbapvtltd.com

BRANCHES

212,Shahpur Jat
New Delhi 110 049
email: ncbadel@ncbapvtltd.com

House No. 3-1-315, 1st Floor
Nimboliadda, Kachiguda, Hyderabad 500 027
email: ncbahydb@ncbapvtltd.com

Shop Nos. 3 & 4, Vinayak Towers
681/B Budhwar Peth, Appa Balwant Chowk
Pune 411 002
email: ncbapune@ncbapvtltd.com

GSS Shopping Complex, 1st Floor
Opposite College Post Office, Convent Road
Ernakulam 682 035
email: ncbaernk@ncbapvtltd.com

Shop No. 15, Bharati Towers, Ground Floor
Forest Park, Bhubaneswar 751 001
email: ncbabub@ncbapvtltd.com

Radhamayee Complex (Second Floor)
Opp. SBI, Panbazar, Jaswanta Road
Guwahati 781 001
email: ncbaguwahati@ncbapvtltd.com

THEORY OF LINEAR CIRCUITS AND RUDIMENTS OF ELECTRONICS • Dr P. K. Chakrabarti

First Published: July 2015

PUBLISHER AND TYPESETTER
New Central Book Agency (P) Ltd
8/1 Chintamoni Das Lane, Kolkata 700 009

PRINTER
New Central Book Agency (P) Ltd
Web-Offset Division, Dhulagarh, Sankrail, Howrah

COVER DESIGNER
Soumen Paul

PROJECT TEAM
Dipan Roy

TECHNICAL EDITOR
Abhijit Sen, Prodip Baidy, Pradip Biswas and Prabhat Jas

ISBN 978 81 7381 962 9

Price: ₹ 435.00

Dedicated
To the loving memory of
MY PARENTS
who were and still are
my source of inspiration

Contents

Preface xix

PART I: NETWORK LAWS AND THEOREMS

Chapter 1: Linear Circuit Analysis 3–41

1.1 Introduction 3
1.2 What is an Ideal Voltage Source and Current Source 3
1.3 Laws and Theorems related to Circuits and Networks 4
1.4 Network Analysis: Different Techniques 9
1.5 Method of Supernode Analysis 12
1.6 Method of Mesh Analysis 14
1.7 Essence of the Method 14
1.8 Supermesh Analysis 17
1.9 Illustrative Examples 17
1.10 Numerical Problems 20
1.11 Solution of Mixed Numerical Problems 23
1.12 Problems on Mesh Analysis 30
1.13 Numerical Problems 32
1.14 Short Theory Questions with Answers 36
1.15 More Mixed Problems Solved (Taken from Univ. Exams.) 38

Chapter 2: Complex Impedance 42–66

2.1 Introduction 42
2.2 Essence of the Rotating Vector Method to Solve Alternating Current Problems 42
2.3 LR Circuit 43
2.4 CR Circuit 45
2.5 LCR Circuit 46
2.6 Parallel Combination of Circuit Elements 48
2.7 Series and Parallel Resonant Circuit 49
2.8 Parallel Resonant Circuit or Anti-resonant Circuit 52
2.9 Q-Factor and Its Effect on the Bandwidth $\Delta\omega$ of a Series Resonant Circuit 52

2.10 Parallel Resonance Circuit: Q-factor in such a Circuit 53
2.11 Power Factor — Wattless Component of Current 55
2.12 Numerical Problems 59
2.13 Exercises 65

Chapter 3: Inductively Coupled Circuits **67–85**
3.1 Introduction 67
3.2 Transformers 67
3.3 Mathematical Theory of the Transformer 68
3.4 Practical Transformer 72
3.5 Efficiency of a Transformer — All-day Efficiency 75
3.6 Transformer Losses and Their Remedy 76
3.7 LCR Coupled Circuits 76
3.8 Force on a Conductor Due to ac flowing in a Neighbouring Circuit 80
3.9 Numerical Problems 81
3.10 Exercises 85

Chapter 4: Rotating Magnetic Field **86–92**
4.1 Introduction 86
4.2 Rotating Magnetic Field by a Three-Phase ac Source 88
4.3 The Coupling Acting on a Small Coil Placed in a Rotating Magnetic Field 88
4.4 Application of the Rotating Magnetic Field 90
4.5 Construction of an Induction Motor 90
4.6 Numerical Problems 91
4.7 Exercises 91

Chapter 5: AC Measurements **93–107**
5.1 Introduction 93
5.2 Kirchhoff's Laws in ac Network 93
5.3 Laws of Series and Parallel Impedances 94
5.4 Alternating Current Bridges 94
5.5 Different Bridges for Determination of Different Circuit Elements 96
5.6 Wagner's Earthing Device for Accurate Measurement with ac Bridge 103
5.7 Effective Resistance at High-frequency Skin Effect 104
5.8 Exercises 107

PART II: SEMICONDUCTOR ELECTRONICS [ANALOG AND DIGITAL]

Chapter 1: Electron Ballistic **111–118**
1.1 What is Electronics 111

1.2 Electron Ballistics 111
1.3 Analog Electronics vs. Digital Electronics 115
1.4 Numerical Problems 115
1.5 Exercises 117

Chapter 2: Semiconductor Electronics **119–145**
2.1 Introduction 119
2.2 Semiconductors and Insulators 119
2.3 Extrinsic or Doped Semiconductor 120
2.4 Electron Distributioh and Fermi level in Semiconductors 121
2.5 *pn*-Junction: Semiconductor Diode 126
2.6 Rectification Action of a Junction Diode 130
2.7 Peak Inverse Voltage (PIV) 132
2.8 Bridge Rectifier 132
2.9 Diode as Clipper 133
2.10 Diode Used as Clamper 134
2.11 Diode Used as Voltage Multiplier 135
2.12 Diode Used as Damper 137
2.13 Diode Used as GATE — OR and AND GATES 137
2.14 Zener Diode 139
2.15 A Few Other Types of Diodes 141
2.16 Numerical Problems 143
2.17 Exercises 145

Chapter 3: Transistors — Basic Characteristics and as Amplifying Device **146–173**
3.1 What is a Transistor 146
3.2 Elements of a Transistor 146
3.3 Transistor Fabrication 146
3.4 Transistor Biasing and Mechanism of Current Flow 147
3.5 Different Current Components in a Transistor 148
3.6 Different Transistor Configurations and Gains of a Transistor 149
3.7 Transistor Characteristics 150
3.8 Distinct Regions of the Characteristic Curves of a Transistor 151
3.9 Calculation of Transistor Parameters from the Characteristics 152
3.10 Transistor Biasing and Thermal Stability of the Biasing Arrangement 153
3.11 Numerical Problems 157
3.12 Different Configurational Amplifier Circuits 159
3.13 Solved Numerical Problems 163
3.14 Exercises on CB Amplifier 166
3.15 Common Emitter Amplifier 166

3.16 Short Theory Questions with Answers 169
3.17 Numerical Problems 170
3.18 Exercises 171
3.19 Common Collector Amplifier 171
3.20 Exercises on CE, CB and CC Modes 172

Chapter 4: Transistor Modelling **174–190**
4.1 Introduction 174
4.2 Essential Requirements of Parameter Modelling 174
4.3 What are the Different Parametric Modelling 174
4.4 Hybrid Parameter Model Details and its Application 177
4.5 Analysis of the Transistor Amplifier in CE Mode 178
4.6 Numerical Problems 180
4.7 Short Theory Questions with Answers 187
4.8 Numerical Problems 188
4.9 Exercises 189

Chapter 5: Multistage Amplifier **191–216**
5.1 Introduction 191
5.2 RC Coupled Audio Frequency Amplifier 191
5.3 Transformer Coupled Amplifier 197
5.4 Short Theory Questions with Answers 201
5.5 Numerical Problems 204
5.6 Exercises 207
5.7 Amplifiers Revisited 208

Chapter 6: Oscillators **217–237**
6.1 Transistor Oscillators: Feedback in Amplifiers 217
6.2 Basic Components of an Oscillator 217
6.3 Commonly Used Oscillator Circuits 218
6.4 Operational Amplifier Version of the Oscillators (Op-Amp Version) 226
6.5 Multivibrator (Transistor Version) 230
6.6 Numerical Problems 236
6.7 Short Theory Questions with Answers 236
6.8 Exercises 237

Chapter 7: Field Effect Transistors **238–252**
7.1 Introduction 238
7.2 Junction Field Effect Transistor (JFET) 238
7.3 Short Theory Questions with Answers 249

7.4 Numerical Problems 251

Chapter 8: Overview of Field Effect Transistors (MOSFETs) and Prelude to IC Technology 253–264

8.1 JFET and MOSFET Compared 253
8.2 VLSI Design about the Transistors and Wires 254
8.3 General MOS Structure 254
8.4 Operation of a D-Mode MOSFET (DM MOS) 256
8.5 Enhancement Moded Operation 257
8.6 Exclusive Enhancement MOS: (EEMOS) 257
8.7 Characteristic Curves: Output Characteristics and Transfer Characteristics 258
8.8 The *n*-MOS Inverter 259
8.9 Formal Procedure for IC Fabrication 260
8.10 Summary of the *n*-MOS Depletion Transistor Fabrication Process 261
8.11 C-MOS Fabrication: Twin Tub Process 262
8.12 Electrical Behaviour of *n*-MOS Transistor Structure 262

Chapter 9: Integrated Circuits IC 265–271

9.1 What is an IC 265
9.2 Constructional Features of an IC 265
9.3 Fabrication of an IC 266
9.4 Formation of active and passive elements in IC: How a resistance is formed in an IC 267
9.5 Short Theory Questions and Answers 269

Chapter 10: Operational Amplifiers (Op-Amp) 272–303

10.1 What is an Operational Amplifier 272
10.2 Symbolic representation of an Op-Amp 273
10.3 Meaning of the terms Inverting and Non-inverting Terminals 273
10.4 The Heart of an Op-Amp is a Differential Amplifier 274
10.5 Basic Requirement of a Constant Current Source 274
10.6 Modes of Input Supply: Inverting and Non-inverting Input Terminals 275
10.7 Gain of the Op-Amp 276
10.8 What is the cause of the high gain of the Op-Amp? 278
10.9 Errors (Offset Voltage or Current) generated by the Op-Amp 278
10.10 Block Diagram of an Op-Amp 279
10.11 The Analysis of an Op-Amp IC 741 280
10.12 Power Supply for an Op-Amp 281
10.13 Measurement of Operational Amplifier Parameters 282
10.14 Applications of Op-Amp 285
10.15 Some Practical Project Circuits using Op-Amp 741 291
10.16 Short Theory Questions with Answers 292

10.17 Numerical Problems 294
10.18 What is a Logic Gate and Logic Function? 295
10.19 Binary Number System and Decimal Number System 296

Chapter 11: Regulated Power Supply (RPS) 304–328
11.1 Introduction 304
11.2 Short Theory Questions with Answers 311
11.3 Numerical Problems 311
11.4 Challenging Numerical Problems 314
11.5 Theory Questions Exercises 315
11.6 SMPS — revolution in Power Electronics 323

Chapter 12: IC 555 (Bipolar Version) and IC 7555 (C-MOS Version) Timer: IC 556 — the Dual Version of IC 555 329–338
12.1 Introduction 329
12.2 How does the flip-flop work? 329
12.3 How does the concept of time come in? At regular intervals of time, the flip-flop sets and resets? 330
12.4 Solved Theory and Numerical Problems 336
12.5 Theory Questions Exercises 338

Chapter 13: Power Control Devices and Circuits 339–362
13.1 Silicon Controlled Rectifier 339
13.2 Learning about an SCR 339
13.3 Operation of an SCR 340
13.4 Some important terms in relation to an SCR 342
13.5 Different facts about an SCR explained 342
13.6 How and when the SCR can be made to stop firing? 343
13.7 SCR Characteristics 345
13.8 Mathematical Viewpoint 346
13.9 Application of SCR 346
13.10 TRIAC (Triode ac switch) 348
13.11 DIAC (Diode ac switch) 349
13.12 Diac-Triac combination used in power control circuit — a fan regulator or a lamp dimmer circuit of practical importance 350
13.13 Shockley Diode 350
13.14 Different SCR Packages 351
13.15 Uni-junction transistor 352
13.16 Model of a UJT 352
13.17 Characteristic Curve of a UJT 353

13.18 Applications of UJT 354
13.19 Short Theory Questions with Answers 355
13.20 Solved Numerical Problems 357
13.21 Numerical Problems Exercises 360
13.22 Theory Questions 361
13.23 MCQ Type 361

PART III: DIGITAL ELECTRONICS

Chapter 1: Analog Circuits and Digital Circuits 365–366

Chapter 2: Oft-Quoted Number Systems: Data Representation and the Decimal and Binary Number Systems 367–377

2.1 Introduction 367
2.2 What are the different number systems and their conversions? 367
2.2.1 What do you mean by radix or base of a number system? 367
2.2.2 What is the meaning of positional number system and non-positional number system? 368
2.3 Details of decimal number system representation 368
2.4 Details of the binary number system 369
2.4.1 Scheme of conversion of decimal sys.em to binary system 369
2.4.2 Decimal fraction 369
2.4.3 Conversion of decimal fraction to binary system 370
2.4.4 Number in decimal system having integral part as well as fractional part 370
2.4.5 Conversion from decimal system to binary 370
2.5 Different transformation techniques for conversion from decimal to binary and vice versa 371
2.6 Conversion methods of binary to decimal 372
2.7 Short Theory Questions with Answers 373
2.8 True or False Statements 377
2.9 Questions Exercises 377

Chapter 3: Octal and Hexadecimal Number Systems — Their Conversion to Decimal and Binary Systems and vice versa 378–398

3.1 Introduction 378
3.2 Octal Number System 378
3.3 Conversion of Binary to Octal and vice versa, also its Decimal equivalent 378
3.4 How does a computer recognise the sign of a number? 393
3.5 What is the representation of a real number in a computer? 393
3.6 Floating Point Representation 394
3.7 Data representation to a computer through different types of coding 394
3.8 ASCII (ASKEE) Code 395

3.9 Image Representation — Bitmap Representation 395
3.10 Short Theory Questions with Answers 396
3.11 Short Theory Questions 398
3.12 Long Questions 398

Chapter 4: Boolean Algebra — Digital Gates 399–435
4.1 Introduction 399
4.2 More details about the basic gates 401
4.3 A few Boolean theorems with their implementing circuits: DeMorgan's theorems 408
4.4 Basic laws and basic gates 412
4.5 Boolean algebra laws 412
4.6 Boolean algebra revisited 414
4.6.1 What are the fundamental or basic gates? 414
4.6.2 Give the implementing circuits of (a) AND (b) OR (c) NOT 414
4.6.3 What are universal gates? Are NAND and NOR gates universal gates? Justify 417
4.7 From circuit representation to Boolean expression 419
4.8 All the Boolean theorems at a glance for ready reference 420
4.9 Short theory questions with answers 421
4.10 Short theory questions and answers 426
4.11 Solved Problems 429
4.12 Exercises 432
4.13 Problems Exercises 433
4.14 Theory Questions 434

Chapter 5: Boolean Function in Different Forms 436–442
5.1 Introduction 436
5.2 Maxterm and Minterm: Literals 437
5.3 Development of Boolean expression from the truth table representation 438
5.4 Simplification of a Boolean expression by MAP method: Minimisation of Boolean function 439

Chapter 6: Switching Circuit Re-visited 443–448
6.1 Short Theory Questions with Answers 446
6.2 Questions Exercises 448

Chapter 7: Combinational and Sequential Logical Circuits 449–472
7.1 Digital systems 449
7.2 Combinational digital systems 449
7.3 Choice of implementing circuit for realisation of Boolean expression 452
7.4 How to realise a Boolean expression by using universal gates? 452
7.5 Half-subtractor and full-subtractor 453

7.6 Multiplexer (MUX) 455
7.7 Demultiplexer 455
7.8 Encoder 456
7.9 Decoder 458
7.10 Sequential digital systems 459
7.11 Flip-flops (F-F) 460
7.12 Registers 464
7.13 Counters and their classifications 465
7.14 Down-counter 468
7.15 Short Theory Questions with Answers 468
7.16 Theory Questions Exercises 470
7.17 Some Assorted Problems from Different Chapters 470

PART IV: SIGNAL COMMUNICATION

Chapter 1: Analog Signal Communication 475–498
1.1 Communication Theory: Basic theory of signals and its communication 475
1.2 Dimension of a signal 477
1.3 Essentials of signal processing 478
1.4 Review of analog signal communication 480
1.5 Frequency spectrum of AM-wave 482
1.6 Significance of modulation index (m) 484
1.7 Power Transmitted along with AM-wave 485
1.8 Neither under-modulation nor over-modulation is desired in AM-transmission. Explain 486
1.9 Linear amplitude modulator 486
1.10 Merits and Demerits of AM-transmission 487
1.11 Frequency modulated wave transmission: FM-transmission 488
1.12 Actual FM-wave generation in practice 490
1.13 Solved Numerical Problems 491
1.14 Short Theory Questions with Answers 492
1.15 Review Questions 494
1.16 MCQs 494
1.17 Random signals, Random variables 495
1.18 Demodulation of radio signal 495
1.19 Radio communication system: Radio transmitter and receiver 496

Chapter 2: Digital Signal and Its Communication 499–525
2.1 Introduction 499
2.2 What is a Digital Signal? 499
2.3 Digital Signal Processing and Its Usefulness 499

2.4 Essential Schematics of Digital Communication System 500
2.5 Discrete Time Signal — It is the Result of Sampling 503
2.6 Illustrative Examples 504
2.7 Quantising and Encoding 506
2.8 The Practical AD Converter (ADC) 511
2.9 Digital Signal Converted to Analog Signal (DAC) 512
2.10 DAC — Digital to Analog Converter 512
2.11 IC Version of ADC and DAC 515
2.12 Signals and Response to Signals by Different Systems 516
2.13 Short Theory Questions with Answers 519
2.14 Solved Numerical Problems 522
2.15 Theory Questions Exercises 524

Chapter 3: Transformation Techniques for Signal Analysis 526–554
3.1 What are the Types of Transformations? 526
3.2 Z-transform and Its Applications 527
3.3 Z-transform of Some Standard Time-domain Functions 528
3.4 Worked-out Examples 532
3.5 Inverse Z-transform: $Z^{-1}\{G(z)\}$ 533
3.6 Solved Numerical Problems 537
3.7 Solved Numerical Problems 540
3.8 Short Theory Questions with Answers 544
3.9 Theory Questions Exercises 548
3.10 Numerical Problems Exercises 548
3.11 Objective type questions 552
3.12 Numerical Problems 553

PART V: SPECTRAL ANALYSIS

Chapter 1: Fourier Series and Its Application 557–574
1.1 Introduction 557
1.2 Fourier Series Representation of a Continuous Time Periodic Signals 557
1.3 FS for Continuous Time Periodic Function: Application to Continuous Time Signals 559
1.4 Importance of Fourier's Theorem 563
1.5 Evaluation of Riemann Zeta Function: Evaluation of the Sum of Some Standard Series 563
1.6 Short Theory Questions with Answers 565
1.7 A Few More Solved Problems 566
1.8 Exercises 569
1.9 Fourier Analysis of a Rectangular Pulse of length A_0 and width T 570
1.10 Fourier Transform for Continuous Time Function (CTFT) 571

Chapter 2: Fourier Transform and Its Application 575–592

2.1 Introduction: What is Fourier Transform? 575
2.2 Some Basic Properties of this Transform 575
2.3 Some Properties of sine and cosine FT 577
2.4 Solved Problems 577
2.5 Evaluation of Some Integrations by Fourier Transform 580
2.6 Application of Fourier Transform to Solve Differential Equation 582
2.6.1 Solution of ordinary differential equation 582
2.6.2 Solution of partial differential equation 582
2.7 Wave Equation 584
2.8 Solution of Poisson equation 585
2.9 Short Theory Questions with Answers 586
2.10 Exercises 587
2.11 Integral Equations 588
2.12 Discrete Fourier Series (DFS) 589
2.13 Fourier Transform for Discrete Time Aperiodic Signal (DTFT) 590
2.14 Short Theory Questions with Answers 591

PART VI: MICROELECTRONIC DEVICES

Chapter 1: Microcontroller 595–604

1.1 Introduction 595
1.2 Microcontroller: In quest of RD work with a microcontroller 596
1.3 Learning about a microcontroller 596
1.4 A microcontroller and a microprocessor compared 596
1.5 CPU is the brain of a microcontroller 597
1.6 Classification of microcontroller 598
1.7 Microcontroller vs. Microprocessor (Architectural differences) 599
1.8 Microcontroller 8051 — a member of MCS 51 family 600
1.9 Learning about the pin-configuration 601
1.10 Describing the functions of different pins 603

Chapter 2: Microprocessor 605–609

2.1 Introduction 605
2.2 A wonder gift from the micro-wonderland — a computer, a counting machine 606
2.3 What are the characteristics of a computer storage system? 608
2.4 Primary and secondary storages 608

Index 611-613

Preface

This book entitled, *Theory of Linear Circuits and Rudiments of Electronics* is basically a printed outcome of the lectures delivered by the author from time to time to his undergraduate students at St. Xavier's College, Kolkata.

The theory of linear circuits is itself of vast expanse, then comes the ever-expanding subject of electronics. It is a fantastic task to combine them in a single treatise, but the author hopes he has been successful in making an apt and beautiful blend of essential matters.

The portion on the theory of linear circuits has been developed from basic ideas to more complex designs. The portion on electronics — both analog and digital, discrete and integrated as well — has been written according to the best possible reasoning and long practical experience.

In this connection I would like to express my gratefulness to those eminent authors whose books I have freely consulted and made use of.

I would fail in my duties if I do not mention and remember my wife Mrs. Krishna Chakrabarti, who spared me most of the time while I was preparing the manuscript.

And, lastly, I am thankful to the management and staff of New Central Book Agency (P) Ltd, who are directly or indirectly associated with the publication of the book.

9 May 2015
56 Raja S.C. Mullick Road
Ekta Heights, Flat 3A3
Kolkata 700 032

Pradip Kumar Chakrabarti

Part I

Network Laws and Theorems

Chapter 1

Linear Circuit Analysis

1.1 Introduction

Circuit or Network refers to both electric and electronic components combined in such a way as to perform a specific task. A linear network is a combination of elements, which satisfy linearity relationship between voltage and current through them. We begin our discussion with an ideal voltage source and an ideal current source.

1.2 What is an Ideal Voltage Source?

An ideal voltage source is one which produces a voltage or p.d. of v_s volts not caring what is connected across its terminals [Fig. 1.1]. Actually, however, in case of real voltage source, the source voltage v_s and the terminal voltage v_t are different. It can be represented by an ideal voltage source with a resistance in series with it [Fig. 1.2]. So we can write

$$\boxed{v_t = v_s - i_s r_s}\,. \tag{1.1}$$

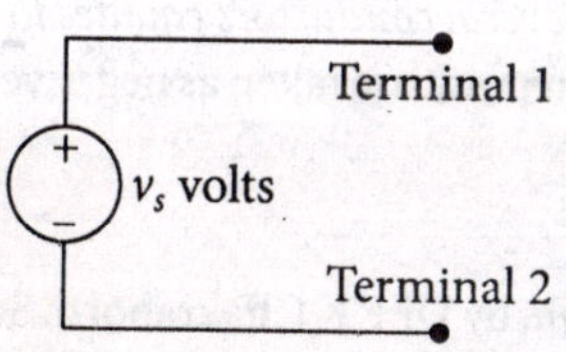

Fig. 1.1

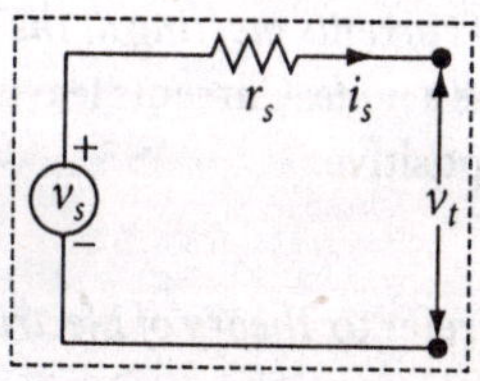

Fig. 1.2

The graphical representation of $v \sim i$ characteristic of a real voltage source [Fig. 1.3].

So it is evident that a voltage source with zero internal resistance ($r_s = 0$) behaves as an ideal voltage source.

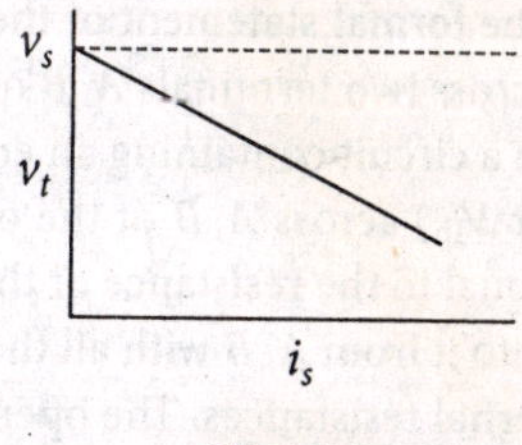

Fig. 1.3

Ideal current source

An ideal current source is a two-terminal element in which the current i_c remains constant not caring what is the voltage v_s across its terminals [Fig. 1.4].

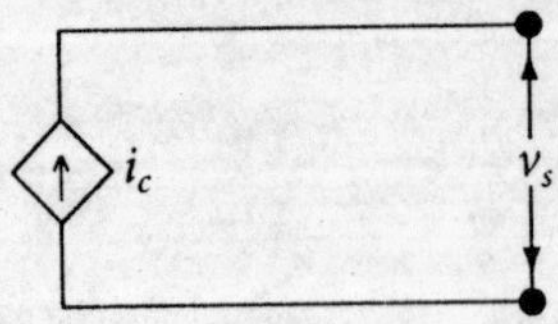

Fig. 1.4

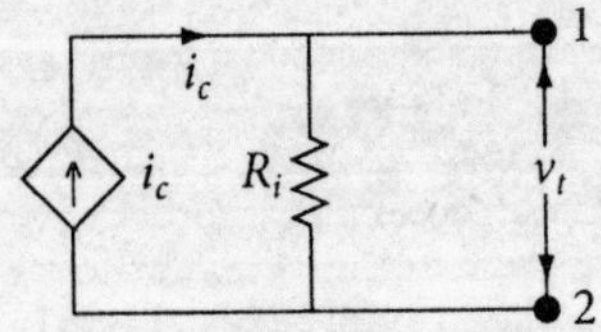

Fig. 1.5

A real current source may, however, be represented by an ideal current source with a resistance R_i connected across its terminals [Fig. 1.5].

Suppose, the terminals 1 and 2 are connected through a load z_l, then

$$i_L = i_c \times \frac{R_i}{R_i + z_l} = i_c \times \frac{1}{1 + (z_l/R_i)} = i_c, \text{ if } R_i \text{ is large compared to } z_l.$$

So a real current source behaves as an ideal current source if its internal resistance is very large.

1.3 Laws and Theorems related to Circuits and Networks

A circuit in its simple form contains an energy source, a load or sink and a connecting wire. A network, on the other hand, is a combination of a few simple units.

As a generalisation of Ohm's law, we have introduced two famous laws of electricity and based on these two laws again, we have formulated a few important theorems, which we will presently discuss.

Kirchhoff's voltage laws (KVL)

In a closed mesh, the algebraic sum of the rise and fall of voltage equates to zero. Rise of voltage is taken as positive, fall as negative.

So, $\Sigma R_j I_j = \Sigma v_j$. (KVL)

Kirchhoff's current law (KCL)

The algebraic sum of currents meeting at the function point of several conductors equates to zero. The junction point is referred to as a *node*. Currents leaving out of a junction point are taken as negative and those flowing into it are taken as positive.

Thus, $\Sigma I_j = 0$. (KCL)

For more details, refer to *Theory of Electricity and Magnetism*, by Dr P K Chakraborti, Scitech Publication, Chennai.

Thévenin's theorem

The formal statement of the Thévenin's theorem is given as follows: Current through a load resistance R_l across two terminals A, B of a network is equal to the current flowing in a circuit containing an emf equal to the open-circuited voltage (V_{oc} or V_{Th}) across A, B of the original network and a resistance in series equal to the resistance of the network as it appears when looked back into it from A, B with all the active sources being replaced by their internal resistances. The open-circuited voltage is also called *Thévenin's voltage* (V_{Th}) and the resistance in series the so-called Thévenin's resistance (R_{Th}).

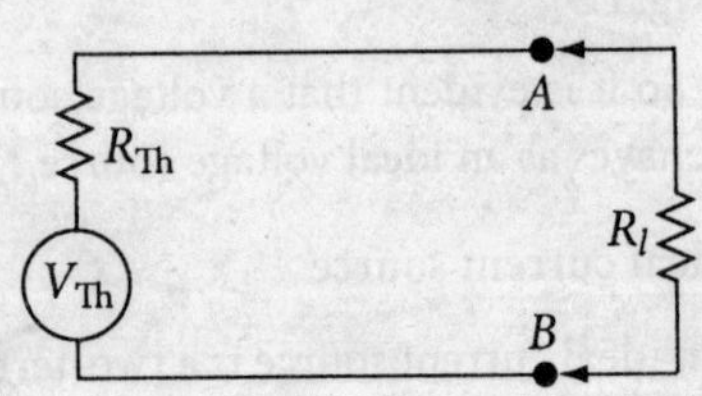

Fig. 1.6

Thévenin's equivalent circuit gives $I_L = \frac{V_{Th}}{(R_{Th}+R_l)}$.

Norton's theorem

This dual of the Thévenin's theorem is the replacement of a circuit by an equivalent current generator.

It states that a linear network containing active sources like emf and passive elements like resistances may be replaced by a constant current generator whose generated output is the short-circuited current (I_{sc}) through the load resistance shorted and a resistance R_N in parallel with the current generator. This current generator is the Norton's current generator ($I_{sc} = I_N$) and the resistance is the so-called Norton's resistance (R_N). Norton's equivalent circuit is as drawn in Fig. 1.7.

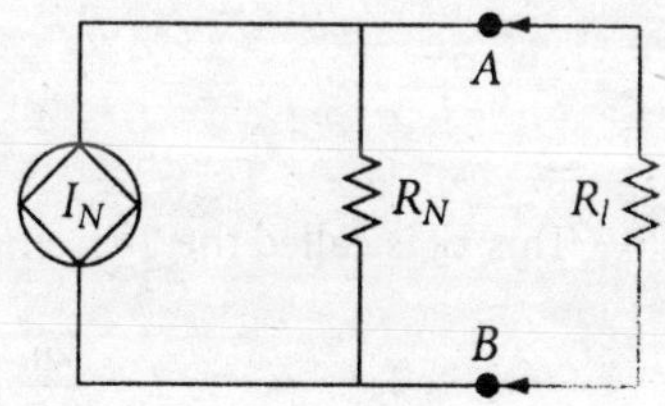

Fig. 1.7

Norton's equivalent gives $I_L = I_N \times \frac{R_N}{(R_N+R_l)}$.

Analytical basis of Thévenin's and Norton's theorem

Proof of Thévenin's theorem

Let there be a linear network having active sources and passive elements in the form of resistance. Between output terminals A, B an impedance z_l is connected. V_0 is the voltage across the load z_l (say), due to current I flowing through it.

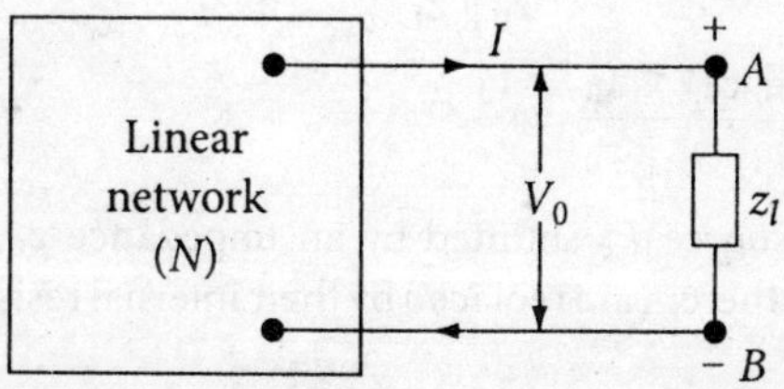

Fig. 1.8

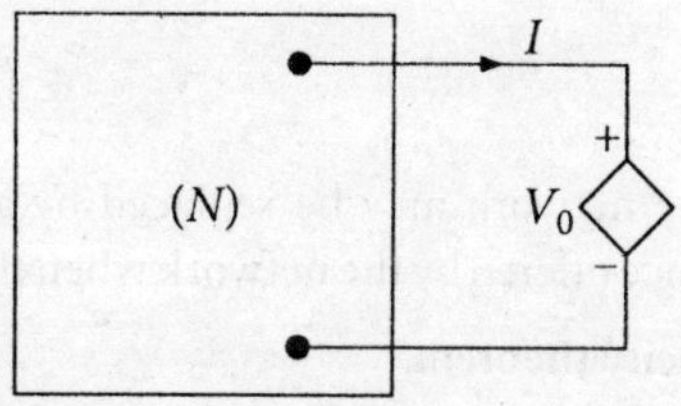

Fig. 1.9

Replacing the impedance by the voltage source V_0, we have the equivalent network [Fig. 1.9].

The voltage V_0 and the current I do not in any way alter the configuration of the network (N). So we can think of an external source V_0 and internal source or sources inside the network, (N). So current I may be found by superposition of two currents: (i) I_{sc}, called the *short-circuited current*, when the so-called external source V_0 is replaced by its internal resistance [Fig. 1.10] and (ii) current I_0 that flows in the external path due to V_0 when the internal sources are replaced by their internal resistances [Fig. 1.11]. The impedance offered by the network N becomes z_0 (say).

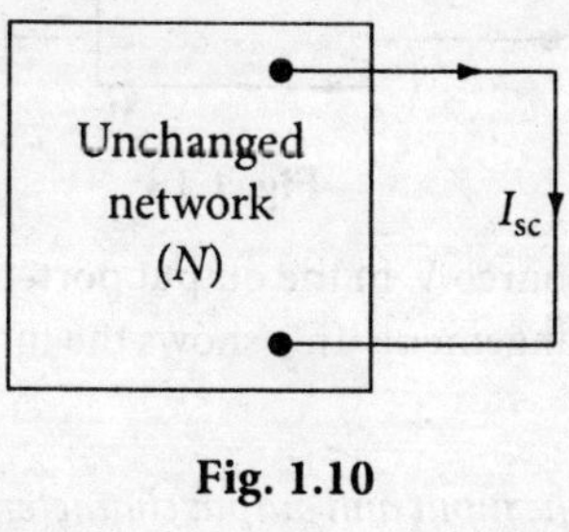

Fig. 1.10

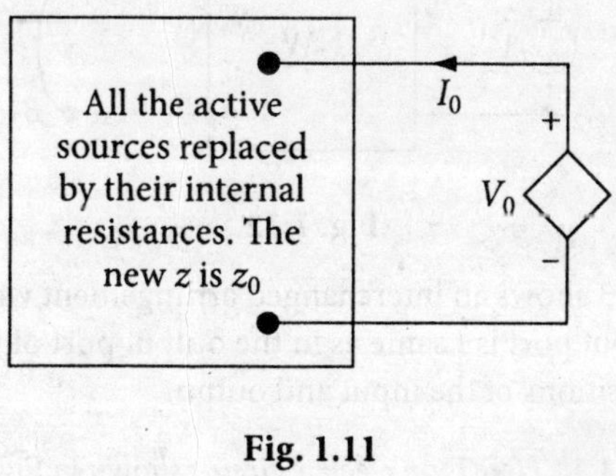

Fig. 1.11

Considering Figs. 1.9, 1.10 and 1.11, we can write $I = I_{sc} - I_0$.

Now,

$$I_0 = \frac{V_0}{z_0} \Rightarrow z_0 = \frac{V_0}{I_0}, \therefore I = I_{sc} - \frac{V_0}{z_0}.$$

If z_l is set equal to an infinite value, the terminals A, B are said to be open and V_0 becomes $V_{oc} = V_{Th}$, called *open-circuited voltage* or the *Thévenin's voltage.* In this case, then $I = 0$ and we would get from above

$$0 = I_{sc} - \frac{V_{Th}}{z_0} \Rightarrow I_{sc} = \frac{V_{Th}}{z_0}.$$

This z_0 is called the *Thévenin's impedance.*

$$\therefore I = \frac{V_{Th}}{z_0} - \frac{V_0}{z_0} \Rightarrow V_0 = V_{Th} - I z_0 \Rightarrow I z_l = V_{Th} - I z_0 \Rightarrow I = \frac{V_{Th}}{z_l + z_0}.$$

From the above equation we see that current through z_l is as if due to a source voltage V_{Th}, called *Thévenin's voltage* with $z_0 (= z_{Th})$ called the *Thévenin's impedance* in series [Fig. 1.6].

Proof of Norton's theorem

$$I = I_{sc} - \frac{V_0}{z_0}; \qquad I_{sc} \times (z_0 \parallel z_l) = V_0 \Rightarrow I_{sc} = \frac{V_0}{z_0 \parallel z_l} = \frac{V_0(z_0 + z_l)}{z_0 z_l}.$$

$$\therefore I_L = \frac{V_0(z_0 + z_l)}{z_0 z_l} - \frac{V_0}{z_0} = \frac{V_0}{z_l}.$$

So, the network may be replaced by a current source I_{sc} shunted by an impedance z_0, which is the impedance offered by the network when the sources therein are replaced by their internal resistance [Fig. 1.7].

Reciprocity theorem

The essence of this theorem is that the inputs and outputs of a network are always in balance.

Statement. If some inputs either from a current source or a voltage source be supplied to a network, the ratio of response (output) in any circuit element to the input is constant even when there is interchange in the positions of the input and output or in other words, the input and output positions are mutually interchangeable.

We have shown a network, where a voltage source is included in the input terminals A_1, A_2 [Fig. 1.12]. The output is manifested in the form of a current I in the output port.

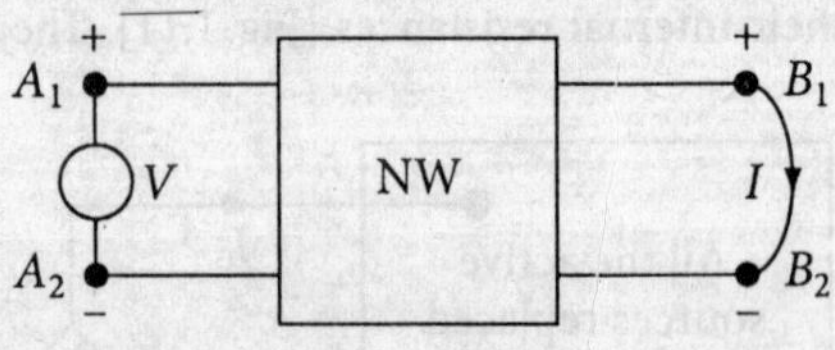

Fig. 1.12

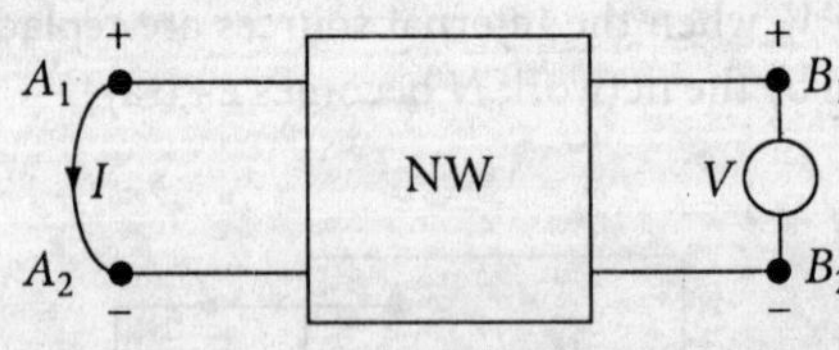

Fig. 1.13

Fig. 1.13 shows an interchanged arrangement with voltage source V in the output port, the current flow in the input port is I same as in the output port of the first arrangement. This shows the interchangeability in the positions of the input and output.

Example 1.1. *Here is an electric circuit shown in* Fig. 1.14 *with the input and output characteristic parameters defined. Show that Reciprocity theorem is obeyed.*

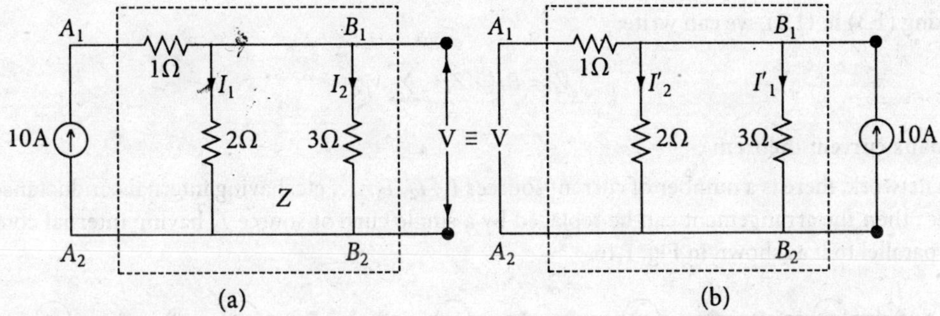

Fig. 1.14

In Fig. 1.14(a), we have

$$Z = \frac{3\times 2}{3+2} = 1.2\Omega; \quad V = 10\times 1.2 = 12\text{V}; \quad I_1 = 10\times\frac{3}{3+2} = 6\text{A}; \quad I_2 = 10\times\frac{2}{5} = 4\text{A}.$$

$\therefore\ V = 3\times 4 = 12\text{V}.$

In Fig. 1.14(b), we have

$$I_1' = 10\times\frac{2}{5} = 4\text{A}; \quad I_2' = 10\times\frac{3}{5} = 6\text{A}; \quad Z = \frac{3\times 2}{3+2} = 1.2\Omega.$$

V (between A_1 and A_2) $= 10\times 1.2 = 12\text{V}.$

$\therefore\ V = 6\times 2 = 12\text{V}.$

So, the direct arrangement and interchanged arrangement are identical. Hence, the Reciprocity theorem is verified.

Millman's theorem

This theorem has two forms, one is w.r.t. voltage sources, the other w.r.t. current sources. This aims to replace a number of voltage sources by a single voltage source or a number of current sources by a single current source.

Formally stated, this theorem stands as: In any network, if there be a number of voltage sources V_1, V_2, V_3, ..., etc. with internal resistances $R_1(/Z_1)$, $R_2(/Z_2)$, $R_3(/Z_3)$,..., etc. in series, and then all are connected in parallel, as shown in Fig. 1.15, then these voltage sources may be replaced by a single voltage source V_s with internal resistance $R_i(/Z_i)$ such that

$$V_s = \frac{V_1 g_1 + V_2 g_2 + \ldots V_n g_n}{(g_1+g_2+\cdots+g_n)} = \frac{\Sigma V_j g_j}{\Sigma g_j}, \tag{1.2}$$

where g_j is the conductance of the jth branch and the internal resistance $R_i(/Z_i)$ is given by

$$R_i(/Z_i) = \frac{1}{g_1+g_2+\cdots+g_n} = \frac{1}{\Sigma g_j}. \tag{1.3}$$

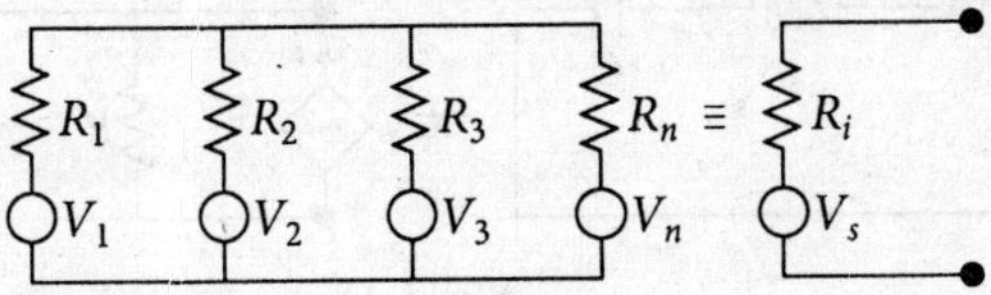

Fig. 1.15

Taking (1.3) in (1.2), we can write:

$$V_s = R_i(/Z_i)\cdot \sum V_j g_j. \tag{1.4}$$

Millman's current theorem

If in a network, there is a number of current sources $I_1, I_2, I_3, \ldots,$ etc. having internal conductances g_1, g_2, g_3, etc., then the arrangement can be replaced by a single current source I_s having internal conductance G_i in parallel to it as shown in Fig. 1.16.

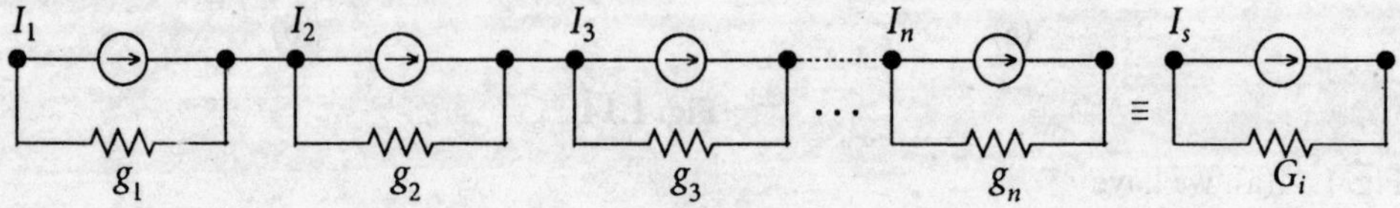

Fig. 1.16

The equivalent current source is given as

$$I_s = \frac{I_1R_1 + i_2R_2 + \cdots + I_nR_n}{(R_1 + R_2 + \ldots R_n)} = \frac{\Sigma I_jR_j}{\Sigma R_j} \text{ and } G_i = \frac{1}{(R_1 + R_2 + \cdots - R_n)} = \frac{1}{\Sigma R_j}.$$

Example 1.2. *Find the current I as shown in* Fig. 1.17 *by applying Millman's theorem.*

$$V_s = \frac{g_1V_1 + g_2V_2}{(g_1+g_2)} = \frac{\frac{1}{2}\times 10 + \frac{1}{5}\times 20}{\left(\frac{1}{2}+\frac{1}{5}\right)} = \frac{5+4}{\frac{7}{10}} = \frac{90}{7}\text{V}.$$

$$R_i = \frac{1}{g_1+g_2} = \frac{1}{\frac{1}{2}+\frac{1}{.5}} = \frac{10}{7}\Omega.$$

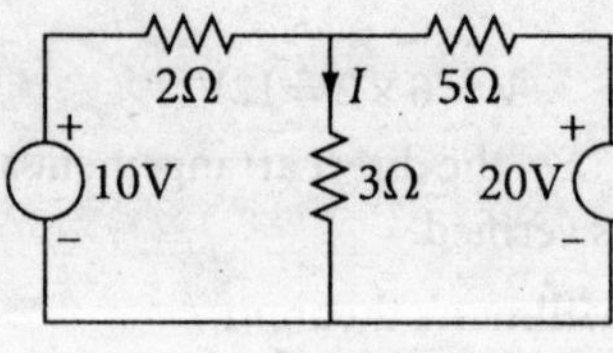

Fig. 1.17

$$\therefore\ I = \frac{V_s}{(R_i+3)} = \frac{\frac{90}{7}}{\frac{10}{7}+3} = \frac{90}{7}\times\frac{7}{31} = 2.9\text{A}.$$

Duals and Duality

In circuit theory we discover similarities between pairs of physical quantities like voltage and current, resistance and conductance, etc. to designate them as duals, and circuit relations can be written in pairs. This principle of writing circuit equations in terms of matching pairs is referred to as principle of duality.

So circuit equation may be written in terms of anyone of the duals. Thus,

$$V = iZ = i\left(R + j\omega L + \frac{1}{j\omega C}\right); \quad i = i_1 + i_2 + i_3; \quad i = YV = \left(G + \frac{1}{j\omega L_2} + j\omega C\right)V.$$

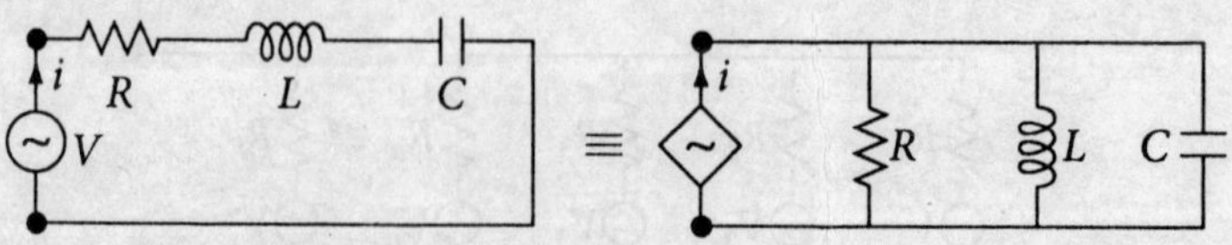

Fig. 1.18

These two relations are dual relations.

1.4 Network Analysis: Different Techniques

A circuit is also called a *network* and finding the element either the voltage or current is called *analysis.* If all the elements, voltages and currents are found, the analysis is said to be complete.

For the purpose of analysis we usually adopt two techniques, one is the nodal analysis, the other is the mesh analysis. There is still a third method of analysis and that is termed as Graph Theory of approach for network analysis. We will consider here only the first two methods.

Nodal analysis

In this technique of analysis we identify the different nodes. Nodes may be of three types (1) *simple node* (2) *compound node* (3) *supernode.*

1. *Simple node*: It is the join of two branches only.
2. *Compound node*: It is the join of more than two branches.
3. *Supernode*: If there is a voltage source between two nodes, they are said to constitute a supernode.

After identification of nodes, simple or compound, we assign voltages at the nodes w.r.t. the voltage of a common node, which is usually at zero potential. This node is also called *reference node.* Label the remaining $(N-1)$ nodal voltages. Apply KCL at these nodes assuming currents at these nodes to leave out of them. Thus, we get $(N-1)$ equations and solving them we would get the voltages at $(N-1)$ nodes. Thus, knowing $(N-1)$ voltages at the nodes other than that of the reference node, we can get the currents in the different branches of the circuit. Thus, the analysis of the network is complete. A network may contain an independent current source like I_1 or more than one independent current source like I_1, I_2, etc. Assuming currents to leave out of a node and then applying KCL, write down KCL equations and solve them.

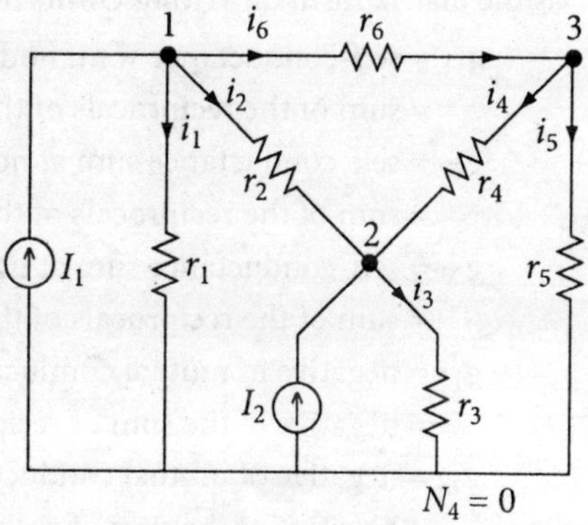

Fig. 1.19

Example 1.1. *Consider the network shown in* Fig. 1.19.

There are four nodes 1, 2, 3 of compound nature and another N_4 as one of reference. Let the voltages at these nodes be denoted by V_1, V_2, V_3 and $V_R(=0)$ respectively.

At node 1:

$$+I_1 - i_1 - i_2 - i_6 = 0 \Rightarrow I_1 = i_1 + i_2 + i_6 \Rightarrow I_1 = \frac{V_1 - 0}{r_1} + \frac{V_1 - V_2}{r_2} + \frac{V_1 - V_3}{r_6}$$

$$\Rightarrow V_1\left(\frac{1}{r_1} + \frac{1}{r_2} + \frac{1}{r_6}\right) + V_2\left(-\frac{1}{r_2}\right) + V_3\left(-\frac{1}{r_6}\right) = I_1 \Rightarrow V_1 g_{11} + V_2 g_{12} + V_3 g_{13} = I_1. \quad (1.5)$$

At node 2:

$$i_2 + (+i_4) + (I_2) - i_3 = 0 \Rightarrow -i_2 - i_4 + i_3 = I_2 \Rightarrow -\frac{V_1 - V_2}{r_2} - \frac{V_3 - V_2}{r_4} + \frac{V_2 - 0}{r_3} = +I_2$$

$$\Rightarrow -\frac{V_1}{r_2} + V_2\left(\frac{1}{r_2} + \frac{1}{r_4} + \frac{1}{r_3}\right) - \frac{V_3}{r_4} = I_2. \quad (1.6)$$

At node 3:

$$-i_6 + i_4 + i_5 = 0 \Rightarrow -\frac{V_1 - V_3}{r_6} + \frac{V_3 - V_2}{r_4} + \frac{V_3 - 0}{r_5} = 0 \Rightarrow -\frac{V_1}{r_6} + V_2\left(-\frac{1}{r_4}\right) + V_3\left(\frac{1}{r_4} + \frac{1}{r_5} + \frac{1}{r_6}\right). \quad (1.7)$$

Collecting the above three equations, we write,

$$V_1\left(\frac{1}{r_1}+\frac{1}{r_2}+\frac{1}{r_6}\right)+V_2\left(-\frac{1}{r_2}\right)+V_3\left(-\frac{1}{r_6}\right)=I_1$$

$$+V_1\left(-\frac{1}{r_2}\right)+V_2\left(\frac{1}{r_2}+\frac{1}{r_4}+\frac{1}{r_3}\right)+V_3\left(\frac{1}{-r_4}\right)=I_2$$

$$+V_1\left(-\frac{1}{r_6}\right)+V_2\left(-\frac{1}{r_4}\right)+V_3\left(\frac{1}{r_6}+\frac{1}{r_5}\right)=0$$

$$\Rightarrow g_{11}V_1+g_{12}V_2+g_{13}V_3=I_1 \tag{1.8}$$

$$g_{21}+g_{22}V_2+g_{23}V_3=I_2 \tag{1.9}$$

$$g_{31}V_1+g_{32}V_2+g_{33}V_3=0 \tag{1.10}$$

$$\Rightarrow \begin{bmatrix} g_{11} & g_{12} & g_{13} \\ g_{21} & g_{22} & g_{23} \\ g_{31} & g_{32} & g_{33} \end{bmatrix}\begin{bmatrix} V_1 \\ V_2 \\ V_3 \end{bmatrix}=\begin{bmatrix} I_1 \\ I_2 \\ I_3 \end{bmatrix}, \text{ in this case } I_3=0. \tag{1.11}$$

The g-matrix in (1.11) is called the *conductance matrix* and the compact form $[g][V]=[I]$ is referred to as the matrix form of writing Ohm's law. We have

g_{11} = self-conductance w.r.t. node 1:
= sum of the reciprocals of the resistances connected to node (1)

g_{22} = self-conductance sum at node 2:
= sum of the reciprocals of the resistance common to node (2)

g_{33} = self-conductance sum at node 3:
= sum of the reciprocals of the resistances common to node (3)

g_{12} = negative of mutual conductance sum between node (1) and node (2)
= negative of the sum of reciprocals of resistances common to both of nodes (1) and (2)

g_{13} = negative of mutual conductance sum between node 1 and 3:
= negative of the sum of reciprocals of the resistances common to node (1) and node (3).

Similarly,

g_{21} = negative of material conductance sum between nodes (2) and (1)
= g_{12}

g_{23} = negative of mutual conductance sum between nodes (2) and (3)
= g_{32}

g_{31} = negative of the conductance sum between nodes (3) and (1)
= g_{13}

g_{32} = negative of the conductance sum between nodes (3) and (2)
= g_{23}

Solving (1.8), (1.9) and (1.10), we would get the values of V_1, V_2 and V_3 and hence, currents in different branches can be available.

Example 1.2. *Analyse the circuit as shown in* Fig. 1.20.

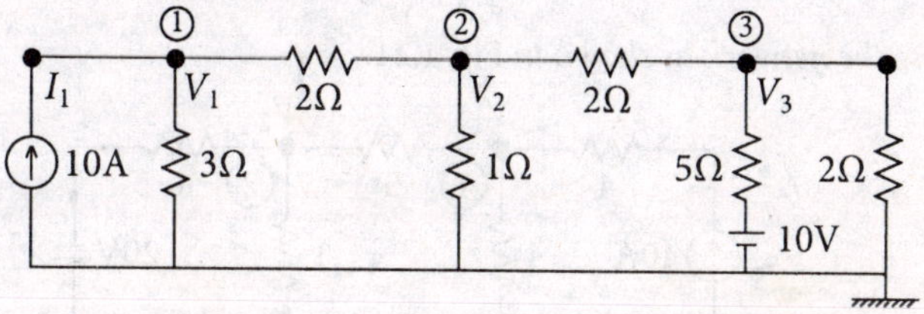

Fig. 1.20

Solution: The nodes 1,2,3 are identified in the figure. Following the scheme of writing (1.8), (1.9) and (1.10), we write the conductance matrix equation as $\begin{bmatrix} g_{11} & g_{12} & g_{13} \\ g_{21} & g_{22} & g_{23} \\ g_{31} & g_{32} & g_{33} \end{bmatrix}\begin{bmatrix} V_1 \\ V_2 \\ V_3 \end{bmatrix} = \begin{bmatrix} I_1 \\ I_2 \\ I_3 \end{bmatrix}$.

In compact notation, $[g][V]=[I]$.

Equivalently we can write $[V]=[g]^{-1}[I]$, where $[g]^{-1}$ is the inverse of the conductance matrix.

At node 1.

$$I_1 - \frac{V_1-0}{3} - \frac{V_1-V_2}{2} = 0 \Rightarrow V_1\left(\frac{1}{3}+\frac{1}{2}\right) - V_2\left(\frac{1}{2}\right) + V_3\times 0 = I_1 = 10$$

$$\Rightarrow V_1\left(\frac{5}{6}\right) + V_2\left(-\frac{1}{2}\right) + V_3\times 0 = 10. \tag{1.12}$$

At node 2.

$$\frac{V_1-V_2}{2} - \frac{V_2-0}{1} - \frac{V_2-V_3}{2} = 0 \Rightarrow V_1\left(-\frac{1}{2}\right) + V_2\left(\frac{1}{2}+1+\frac{1}{2}\right) + V_3\left(-\frac{1}{2}\right) = 0. \tag{1.13}$$

At node 3.

$$-\frac{V_2-V_3}{2} + \frac{V_3-10}{5} + \frac{V_3-0}{2} = 0 \Rightarrow V_1\times 0 + V_2\left(-\frac{1}{2}\right) + V_3\left(\frac{1}{2}+\frac{1}{5}+\frac{1}{2}\right) = 2$$

$$\Rightarrow V_1\times 0 + V_2\left(-\frac{1}{2}\right) + V_3\left(\frac{6}{5}\right) = 2. \tag{1.14}$$

$$\therefore V_1 = \frac{\Delta_1}{\Delta}, V_2 = \frac{\Delta_2}{\Delta}, V_3 = \frac{\Delta_3}{\Delta}.$$

So, the conductance matrix equation can be written as

$$\begin{bmatrix} g_{11} & g_{12} & g_{13} \\ g_{21} & g_{22} & g_{23} \\ g_{31} & g_{32} & g_{33} \end{bmatrix}\begin{bmatrix} V_1 \\ V_2 \\ V_3 \end{bmatrix} = \begin{bmatrix} I_1 \\ I_2 \\ I_3 \end{bmatrix} \Rightarrow \begin{bmatrix} 5/6 & -1/2 & 0 \\ -1/2 & 2 & -1/2 \\ 0 & -1/2 & 6/5 \end{bmatrix}\begin{bmatrix} V_1 \\ V_2 \\ V_3 \end{bmatrix} = \begin{bmatrix} 10 \\ 0 \\ 10/5 \end{bmatrix}$$

$$\Delta = \frac{5}{6}\left(\frac{12}{5}-\frac{1}{4}\right) + \frac{1}{2}\left(-\frac{3}{5}-0\right) = \frac{5}{6}\times\left(\frac{48-5}{20}\right) - \frac{3}{10} = \frac{5}{6}\times\frac{43}{20} - \frac{3}{10} = 1.49.$$

$$\Delta_1 = \begin{bmatrix} 10 & -1/2 & 0 \\ 0 & 2 & -1/2 \\ 2 & -1/2 & 6/5 \end{bmatrix} = 10\left(\frac{12}{5}-\frac{1}{4}\right) + \frac{1}{2}(1) = \frac{10(48-5)}{20} + \frac{1}{2} = \frac{43}{2} + \frac{1}{2} = 22.$$

$$\therefore\ V_1 = \frac{22}{1.49} = 14.75\text{V, etc.}$$

Example 1.3. *Analyse the network as shown in* Fig. 1.21.

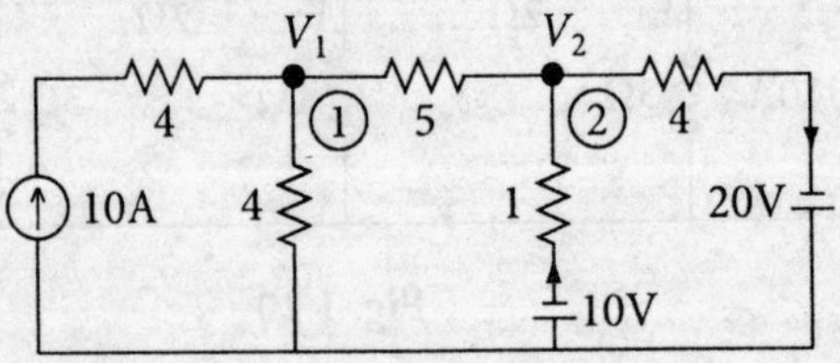

Fig. 1.21

The matrix equation for the node voltages can be developed as below.

At node 1.

$$10 - \frac{V_1 - 0}{4} - \frac{V_1 - V_2}{5} = 0 \Rightarrow V_1\left(\frac{1}{4}+\frac{1}{5}\right) + V_2\left(-\frac{1}{5}\right) - 10 \Rightarrow V_1\left(\frac{9}{20}\right) + V_2\left(-\frac{1}{5}\right) = 10. \qquad (1.15)$$

At node 2.

$$\frac{V_1 - V_2}{5} + \frac{V_2 + 10}{1} + \frac{V_2 + 20}{4} = 0 \Rightarrow V_1\left(-\frac{1}{5}\right) + V_2\left(\frac{1}{5} - 1 - \frac{1}{4}\right) = +15 \Rightarrow V_1\left(-\frac{1}{5}\right) + V_2\left(-\frac{21}{20}\right) = +15. \qquad (1.16)$$

$$V_1 = \frac{\Delta_1}{\Delta}, \quad V_2 = \frac{\Delta_2}{\Delta}$$

$$\Delta = \begin{vmatrix} 9/20 & -1/5 \\ -1/5 & -21/20 \end{vmatrix} = -\frac{9\times 21}{400} - \frac{1}{25} = -\frac{189-8}{400} = -\frac{197}{400};$$

$$\Delta_1 = \begin{vmatrix} 10 & -1/5 \\ +15 & -21/20 \end{vmatrix} = -\frac{21}{2} + 3 = -\frac{15}{2}.$$

$$\therefore\ V_1 = \frac{-(15/2)}{-(197/400)} = 15.22\text{V}.$$

Please remember the following rules.

When no voltage source is included in the circuit

Rule 1. Find the reference node.

Rule 2. Label the remaining non-reference nodes numbered as 1, 2, 3, etc.

Rule 3. Assign arbitrarily voltages V_1, V_2, V_3, etc. at the non-reference nodes.

Rule 4. Incoming currents to a node (income) are assumed to be positive and outgoing currents from a node are taken as negative (same as *credit* is reckoned as positive, *debit* as negative) and sum them at a node to zero according to KCL.

Rule 5. Find the matrix equation and hence, solve for V_1, V_2, V_3, etc.

1.5 Method of Supernode Analysis

If there is a voltage source between any two nodes in a network, the nodes are said to constitute supernodes. Such supernodes are reduced to one node and then equations are formed by applying KCL. See the circuit as shown in Fig. 1.22 in which there is existence of a supernode. The solution of such a network is through a technique called *supernode analysis* (cf. 2, 3 in Fig. 1.22).

Example 1.1.

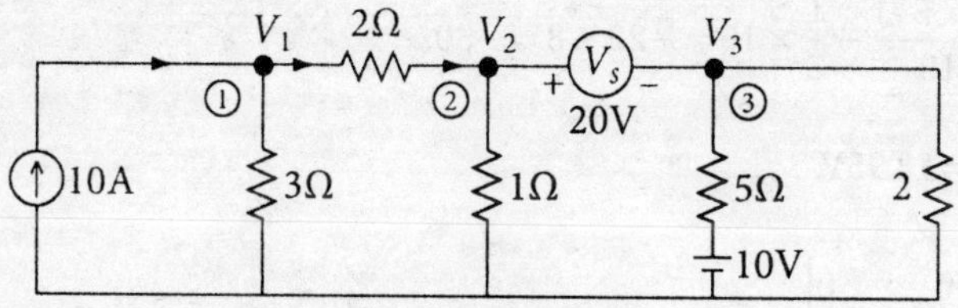

Fig. 1.22

There is a voltage source between nodes (2) and (3), hence they constitute a supernode.

At node 1:

$$\frac{V_1}{3}+\frac{V_1-V_2}{2}=10 \Rightarrow V_1\left(\frac{1}{3}+\frac{1}{2}\right)+V_2\left(-\frac{1}{2}\right)=10. \tag{1.17}$$

The nodes 2 and 3 include a voltage source V_s, hence they are supernodes. Applying KCL at these two nodes simultaneously, we get (considering 2, 3 nodes coalesce to a single node)

$$-\frac{V_1-V_2}{2}+\frac{V_2-0}{1}+\frac{V_3-10}{5}+\frac{V_3-0}{2}=0 \Rightarrow V_1\left(-\frac{1}{2}\right)+V_2\left(\frac{1}{2}+1\right)+V_3\left(\frac{1}{5}+\frac{1}{2}\right)=0$$

$$\Rightarrow V_1\left(-\frac{1}{2}\right)+V_2\left(\frac{3}{2}\right)+V_3\left(\frac{7}{10}\right)=2. \tag{1.18}$$

The third equation can be written as

$$V_2-V_3=20 \Rightarrow V_1\times 0+V_2\times 1-V_3=20. \tag{1.19}$$

Thus, we have

$$V_1\left(\frac{5}{6}\right)+V_2\left(-\frac{1}{2}\right)+V_3\times 0=10. \tag{1.20}$$

$$V_1\left(-\frac{1}{2}\right)+V_2\left(\frac{3}{2}\right)+V_3\left(\frac{7}{10}\right)=2. \tag{1.21}$$

$$V_1\times 0+V_2\times 1-V_3\times 1=20. \tag{1.22}$$

$$V_1=\frac{\Delta_1}{\Delta},\quad V_2=\frac{\Delta_2}{\Delta},\quad V_3=\frac{\Delta_3}{\Delta}$$

$$\Delta=\begin{vmatrix}5/6 & -1/2 & 0\\ -1/2 & 3/2 & 7/10\\ 0 & 1 & -1\end{vmatrix}=\frac{5}{6}\left(-\frac{3}{2}-\frac{7}{10}\right)+\frac{1}{2}\left(\frac{1}{2}\right)=-\frac{5}{6}\times\frac{22}{10}+\frac{1}{4}=-\frac{55}{30}+\frac{1}{4}$$

$$=-\frac{11}{6}+\frac{1}{4}=\frac{-44+6}{24}=-\frac{19}{12}=-1.58.$$

$$\Delta_1 = \begin{vmatrix} 10 & -1/2 & 0 \\ 2 & 3/2 & 7/10 \\ 20 & 1 & -1 \end{vmatrix} = 10\left(-\frac{3}{2}-\frac{7}{10}\right)+\frac{1}{2}\left(-2-\frac{7}{10}\times 20\right)$$

$$= -10\times\frac{(15+7)}{10}-\frac{1}{2}\times 16 = -22-8 = -30.$$

$$\therefore\ V_1 = \frac{-30}{-19/12} = 18.95\text{V}.$$

$$\Delta_2 = \begin{vmatrix} 5/6 & 10 & 0 \\ -1/2 & 2 & 7/10 \\ 0 & 20 & -1 \end{vmatrix} = \frac{5}{6}\left(-2-\frac{7}{10}\times 20\right)-10\left(\frac{1}{2}\right) = -\frac{5}{6}\times 16-5 = -\frac{40}{3}-5 = -\frac{55}{3}.$$

$$V_2 = \frac{-55/3}{-19/12} = \frac{55}{3}\times\frac{12}{19} = \frac{220}{19} = 11.58\text{V}.$$

$$V_2 - V_3 = 20 \Rightarrow V_3 = 11.58-20 = -8.42\text{V}.$$

1.6 Method of Mesh Analysis

Mesh analysis is another method for solving network problems. If a network contains a large number of voltage sources, mesh analysis method can be suitably used, but if it contains a large number of current sources, the node analysis method is recommended. The point to remember is that mesh analysis method is applicable only when the network is a *planar one,* which means it can be drawn on a surface without any crossover. A nonplanar circuit is one, which cannot be drawn on a plane surface without a crossover.

1.7 Essence of the Method

This method is more frequently applied to solve network problems subject to the applicability restriction as mentioned above.

Step 1. Select suitable closed circuits in the network.

Step 2. Apply KVL to each of the closed circuits in terms of assigned cyclic currents in the meshes.

Step 3. We will get a number of equations equal to the number of meshes identified.

Step 4. Solve the simultaneous equations formulated on the basis of Kirchhoff's voltage law by Cramer's rule to get mesh currents.

Points to store in memory

1. Sum of resistances or impedances belonging to a mesh is referred to as self-impedance.
2. The sum of resistances or impedances common to all pairs of meshes is referred to as coupling impedances.
3. If loop currents in a pair of meshes flow in the same direction in a coupling impedance, that impedance is taken as positive but if the currents are in opposite directions, it is taken as negative.
4. The voltages in the meshes will be taken as positive if loop current flows in the direction of the source, but negative if the direction of loop current is in opposition to the direction of the source voltage.

Example 1.1. *Apply the mesh analysis to the network as shown in* Fig. 1.23.

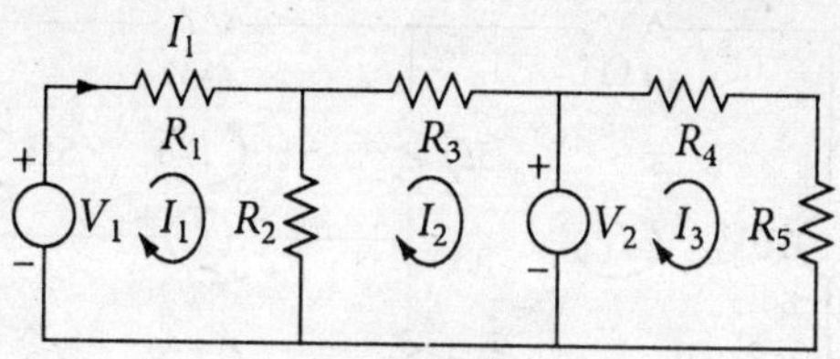

Fig. 1.23

Let the loop currents be denoted by I_1, I_2, I_3 in the three loops

Loop 1. $I_1 R_1 + (I_1 - I_2) R_2 = V_1$

$$\Rightarrow I_1(R_1 + R_2) - I_2 R_2 = V_1. \tag{1.23}$$

Loop 2. $(I_2 - I_1) R_2 + I_2 R_3 = -V_2$

$$\Rightarrow -I_1 R_2 + I_2(R_2 + R_3) = -V_2. \tag{1.24}$$

Loop 3. $I_3 R_4 + I_3 R_5 = V_2$

$$\Rightarrow I_3(R_4 + R_5) = V_2. \tag{1.25}$$

$(R_1 + R_2) = R_{11}$, self-resistance of Loop 1.

R_2 = coupling resistance between *Loop* 1 and *Loop* 2. Through this coupling resistance, the loop currents flow in opposite direction, hence R_2 is negative.

There is no common resistance between *Loop* 1 and *Loop* 3, hence, $R_{13} = 0 = R_{23} = R_{32} = R_{31}$.

$(R_2 + R_3) = R_{22}$, self-resistance of *Loop* 2

$(R_4 + R_5) = R_{33}$, self-resistance of *Loop* 3. So, in general, we can write the equations as

$$I_1 R_{11} \pm R_{12} I_2 \pm R_{13} I_3 = V_1 \tag{1.26}$$

$$\pm I_1 R_{21} + I_2 R_{22} \pm R_{23} I_3 = V_2 \tag{1.27}$$

$$\pm I_1 R_{31} \pm R_{32} I_2 + R_{33} I_3 = V_3 \tag{1.28}$$

$$\Rightarrow \begin{bmatrix} R_{11} & \pm R_{12} & \pm R_{13} \\ \pm R_{21} & R_{22} & \pm R_{23} \\ \pm R_{31} & \pm R_{32} & R_{33} \end{bmatrix} \begin{bmatrix} I_1 \\ I_2 \\ I_3 \end{bmatrix} = \begin{bmatrix} V_1 \\ V_2 \\ V_3 \end{bmatrix}. \tag{1.29}$$

R_{11} = sum of resistances through which loop current I_1 flows

R_{22} = sum of resistances through which I_2 flows

R_{33} = sum of resistances through which I_3 flows

$R_{12} = R_{21}$, sum of coupling resistances between *Loop* 1 and *Loop* 2 is considered positive if loop currents I_1 and I_2 flow in the same direction; negative sign is attached if the two currents flow in opposite directions.

Example 1.2. *Solve the network shown in* Fig. 1.24.

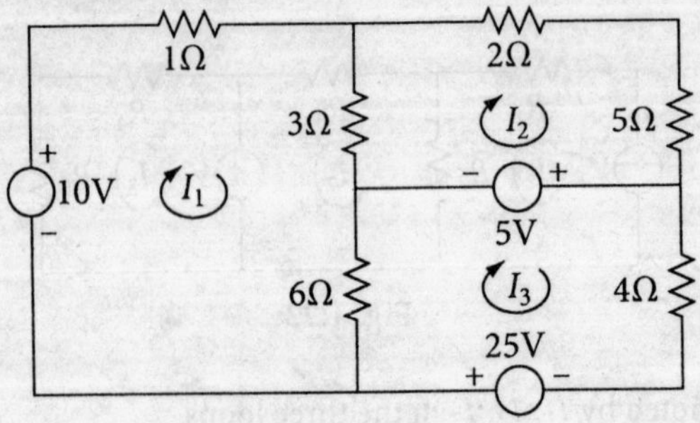

Fig. 1.24

Here, $R_{11} = 10\Omega$, $R_{12} = -3\Omega$, $R_{13} = -6\Omega$

$R_{21} = -3$, $R_{22} = 10\Omega$, $R_{23} = 0$

$R_{33} = 10$, $R_{31} = -6\Omega$, $R_{32} = 0$

So, we have $\begin{bmatrix} 10 & -3 & -6 \\ -3 & 10 & 0 \\ -6 & 0 & 10 \end{bmatrix} \begin{bmatrix} I_1 \\ I_2 \\ I_3 \end{bmatrix} = \begin{bmatrix} 10 \\ -5 \\ 25 \end{bmatrix}$.

By Cramer's rule, we get $I_1 = \frac{\Delta_1}{\Delta}$, $I_2 = \frac{\Delta_2}{\Delta}$, $I_3 = \frac{\Delta_3}{\Delta}$, where

$$\Delta = \begin{vmatrix} 10 & -3 & -6 \\ -3 & 10 & 0 \\ -6 & 0 & 10 \end{vmatrix} = 10 \times 100 + 3 \times (-30) - 6(60) = 1000 - 90 - 360 = 550.$$

$$\Delta_1 = \begin{vmatrix} 10 & -3 & -6 \\ -5 & 10 & 0 \\ 25 & 0 & 10 \end{vmatrix} = 10 \times 100 + 3 \times (-50) - 6 \times (-250) = 1000 - 150 + 1500 = 2350.$$

$$\therefore\ I_1 = \frac{2350}{550} = 4.27\text{A}.$$

$$\Delta_2 = \begin{vmatrix} 10 & 10 & -6 \\ -3 & -5 & 0 \\ -6 & 25 & 10 \end{vmatrix} = 10 \times (-50) - 10 \times (-30) - 6 \times (-75 - 30) = -500 + 300 + 630 = 430.$$

$$\therefore\ I_2 = \frac{430}{550} = 0.78\text{A}.$$

$$\Delta_3 = \begin{vmatrix} 10 & -3 & 10 \\ -3 & 10 & -5 \\ -6 & 0 & 25 \end{vmatrix} = 10 \times 250 + 3 \times (-75 - 30) + 10 \times 60 = 2500 - 315 + 600 = 2785.$$

$$\therefore\ I_3 = \frac{2785}{550} = 5.06\text{A}.$$

1.8 Supermesh Analysis

If any branch of the network contains a *current source* shared by two adjacent loops, the two loops are said to constitute a *supermesh*. To solve such a network, *supermesh analysis* is employed, which consists in writing KCL for branches other than that containing the source current and finally writing the source current as difference between the loop currents in proper sense.

Technique. Apply KCL to the loops leaving out the branch containing the current source. Consider the network shown in Fig. 1.25.

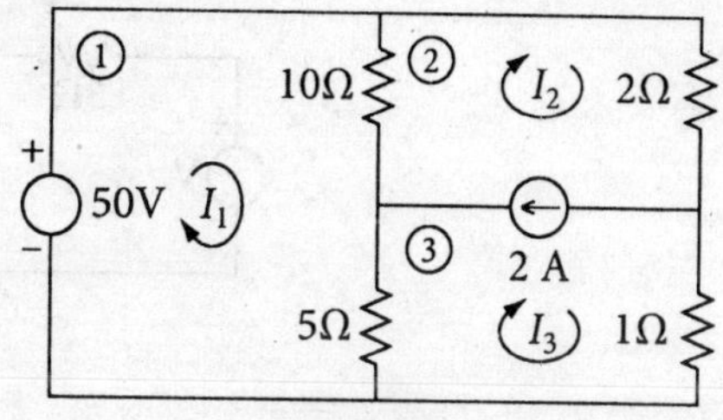

Fig. 1.25

In the network *Loop* 2 and *Loop* 3 form a *supermesh* having a current source of 2A being shared between them. So writing KCL for *Loop* 1:

$$10\times(I_1-I_2)+(I_1-I_3)\times 5=50 \Rightarrow 15I_1-10I_2-5I_3=50. \tag{1.30}$$

KCL applied to *Loop* 2 and *Loop* 3 leaving out the branch containing the current source 2A,

$$I_2\times 2+(I_2-I_1)\times 10+I_3\times 1+(I_3-I_1)\times 5=0 \Rightarrow -15I_1+12I_2+6I_3=0. \tag{1.31}$$

Finally, w.r.t. the branch containing the current source,

$$I_2-I_3=2. \tag{1.32}$$

Solving (1.30), (1.31) and (1.32) by Cramer's rule, we get $I_1=19.99$A, $I_2=17.33$A, $I_3=15.33$A.

1.9 Illustrative Examples

Example 1.1. *Analyse the given network to find the node voltages V_1, V_2, V_3 [Fig. 1.26].*

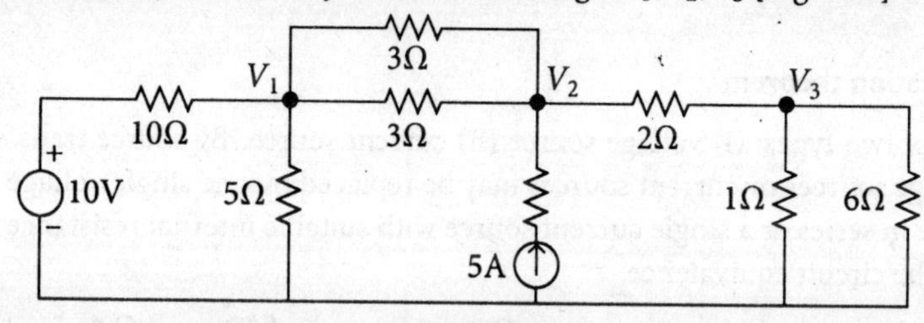

Fig. 1.26

Solution:

Node 1.

$$-\frac{V_1-10}{10}-\frac{V_1-V_2}{3}-\frac{V_1-0}{5}-\frac{V_1-V_2}{3}=0 \Rightarrow V_1\left(\frac{1}{10}+\frac{2}{3}+\frac{1}{5}\right)-V_2\left(\frac{2}{3}\right)=1.$$

Node 2.

$$-\frac{V_2-V_3}{2}+(+5)-\frac{V_2-V_1}{3}-\frac{V_2-V_1}{3}=0 \Rightarrow -\frac{2V_1}{3}+V_2\left(\frac{1}{2}+\frac{2}{3}\right)-\frac{V_3}{2}=5.$$

Node 3.

$$-\frac{V_3-V_2}{2}-\frac{V_3-0}{1}-\frac{V_3-0}{6}=0 \Rightarrow -V_2\left(\frac{1}{2}\right)+V_3\left(1+\frac{1}{2}+\frac{1}{6}\right)=0.$$

Solve these equations by Cramer's rule to get the values of V_1, V_2 and V_3.

Example 1.2. *For what value of V [Fig. 1.27] the current flowing through 10Ω resistor is zero.*

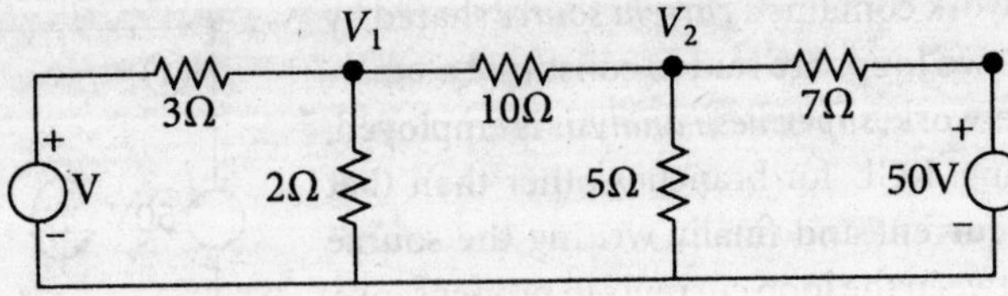

Fig. 1.27

Solution: Since no current flows thorough 10Ω resistor, it implies $V_1 = V_2$.

At *Node* 1.

$$-\frac{V_1 - V}{3} - \frac{V_1 - 0}{2} - \frac{V_1 - V_2}{10} = 0 \Rightarrow V_1\left(\frac{1}{3} + \frac{1}{2} + \frac{1}{10}\right) - \frac{V_2}{10} = \frac{V}{3}.$$

At *Node* 2.

$$-\frac{V_2 - V_1}{10} - \frac{V_2 - 0}{5} - \frac{V_2 - 50}{7} = 0 \Rightarrow -\frac{V_1}{10} + V_2\left(\frac{1}{10} + \frac{1}{5} + \frac{1}{7}\right) = \frac{50}{7}.$$

$$\because V_1 = V_2, -\frac{V_1}{10} + \frac{V_1}{10} + V_1 \times \frac{12}{35} = \frac{50}{7} \Rightarrow V_1 = \frac{50}{7} \times \frac{35}{12} = \frac{250}{12}.$$

$$\therefore \frac{250}{12} \times \frac{5}{6} = \frac{V}{3} \Rightarrow V = \frac{250}{12} \times \frac{5}{6} \times 3 = 52.08\text{V}.$$

Source transformation theorem

Energy source is of two types: (i) voltage source (ii) current source. By source transformation technique, a number of voltage sources or current sources may be replaced by one single voltage source with suitable internal resistance in series or a single current source with suitable internal resistance in parallel. Thus, we get [Fig. 1.28] as the circuit equivalence.

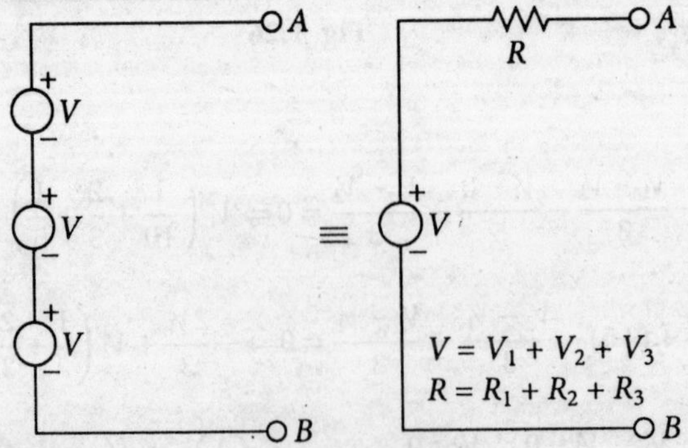

Fig. 1.28

Replacement of a number of voltage sources by a single source and internal resistance *R*.

Conversion of a voltage source into equivalent current source

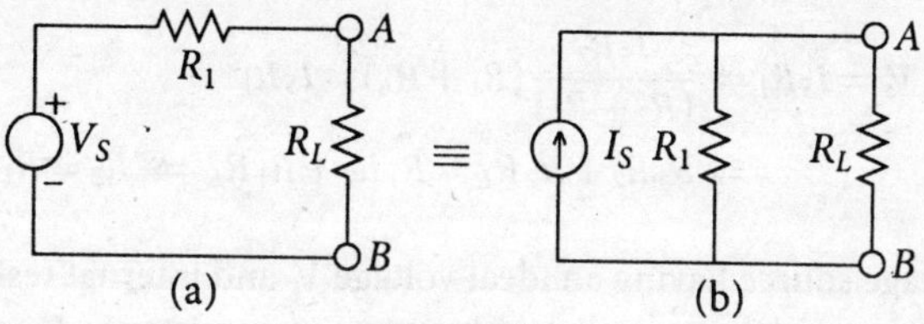

Fig. 1.29

A practical voltage source is represented by an ideal voltage source V_s with an internal resistance R_1 in series [Fig. 1.30(a) and (b)] shows corresponding current source I_s with internal resistance R_2 in parallel.

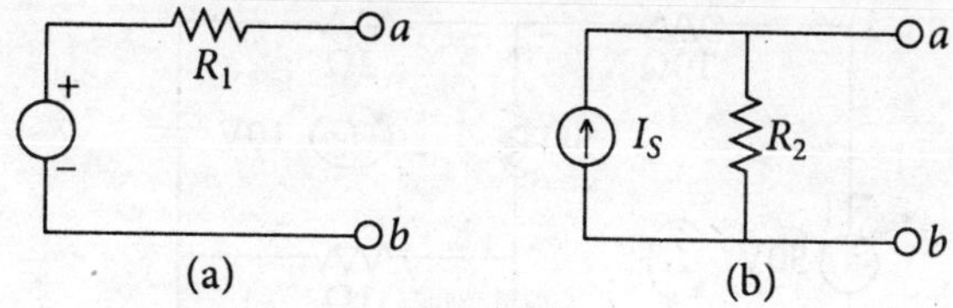

Fig. 1.30

If a load resistance R_L is connected across the terminals (a, b) of both the circuits and current flowing through R_L in both the circuits is same and also the voltage developed across the load resistance is also same, the voltage source representation and the current-source representation of a circuit will be said to be identical. Then, if R_L represents the load resistance bridging the gap between a, b, we would get the circuits as shown in Fig. 1.31(a) and (b).

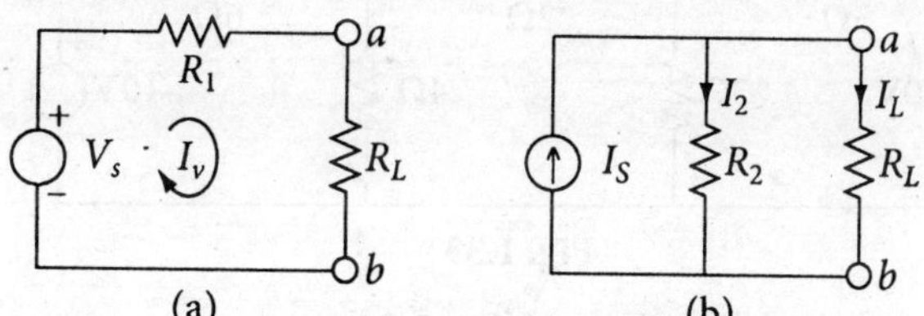

Fig. 1.31

These two representatives would be identical if the current through R_L and the voltage developed across R_L are identical. Thus, applying KVL to the circuit 1.31(a), we get

$$I_v R_1 + I_v R_L = V_s \Rightarrow I_v = \frac{V_s}{R_1 + R_L}.$$

Applying KCL to the circuit [Fig. 1.31(b)], we get

$$I_L = I_S \times \frac{R_2}{R_2 + R_L}.$$

$\therefore$ identity between the two circuits requires

$$I_v = I_L \Rightarrow \frac{V_s}{(R_1 + R_L)} = I_S \times \frac{R_2}{(R_2 + R_L)}.$$

Voltage developed across (a, b) in the current-source representation is given by

$$V_s = I_S R_1 \Rightarrow \frac{I_S R_2}{(R_2 + R_L)}(R_1 + R_L) = I_S R_1$$
$$\Rightarrow R_1 R_2 + R_2 R_L = R_1 R_2 + R_1 R_L \Rightarrow R_2 = R_1.$$

Thus, any practical voltage source having an ideal voltage V_s and internal resistance R_1 in series is equivalent to a current source $I_S = V_S/R_1$ in parallel with an internal resistance $R_2 = R_1$.

1.10 Numerical Problems

1. Find the current I_1 in the circuit [Fig. 1.32] by mesh analysis.

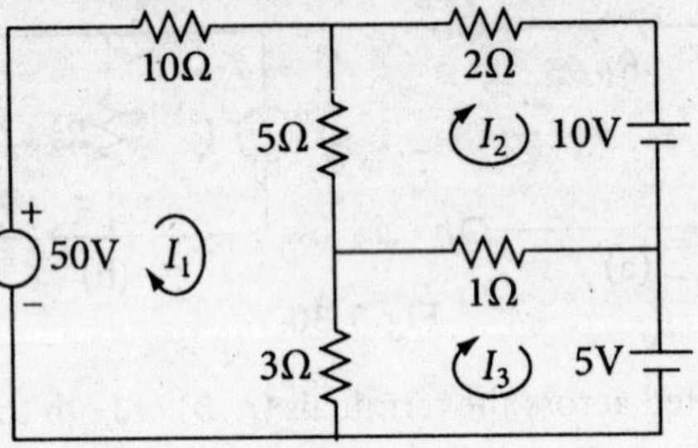

Fig. 1.32 [$I_1 = 3.3$A]

2. Using mesh analysis find the dissipation in 4Ω resistor in the circuit [Fig. 1.33].

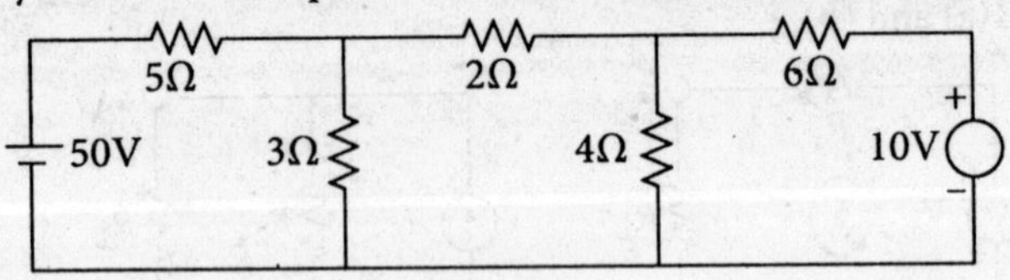

Fig. 1.33 [$P_4 = 23.23$W]

3. Find by mesh analysis method the voltage V [Fig. 1.34].

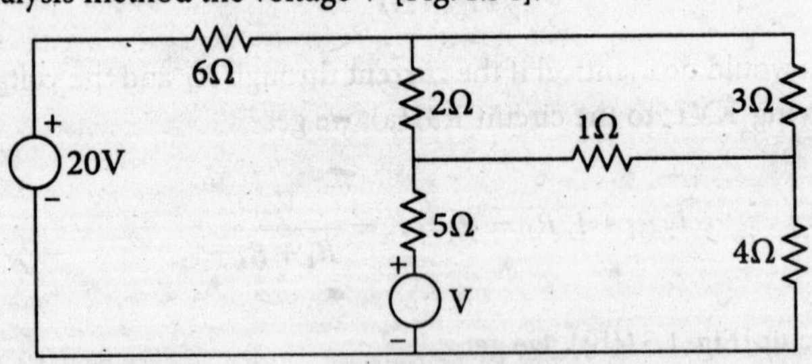

Fig. 1.34 [$V = 43.7$V]

4. Determine the power dissipated by 5Ω resistor in the circuit [Fig. 1.35].

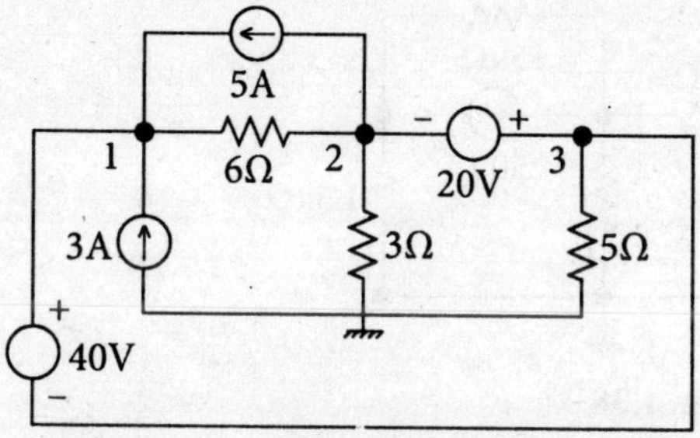

Fig. 1.35

[P_5 = 65.52W]

5. By using nodal analysis method, find the power delivered by the 5A current source [Fig. 1.36].

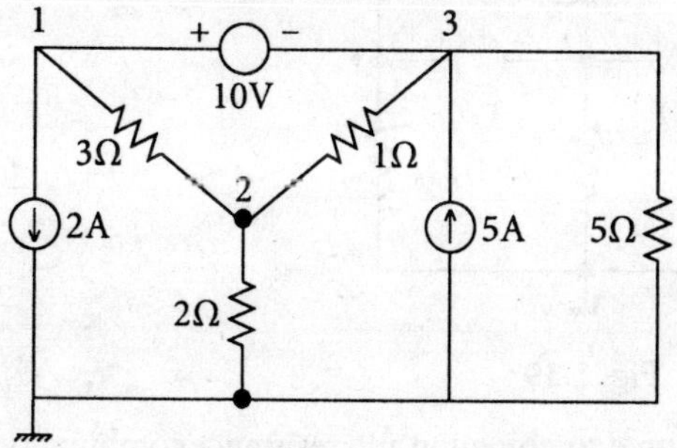

Fig. 1.36

[P = 18.6W]

6. By applying source transformation method, find the power delivered by 50V voltage source in the circuit [Fig. 1.37].

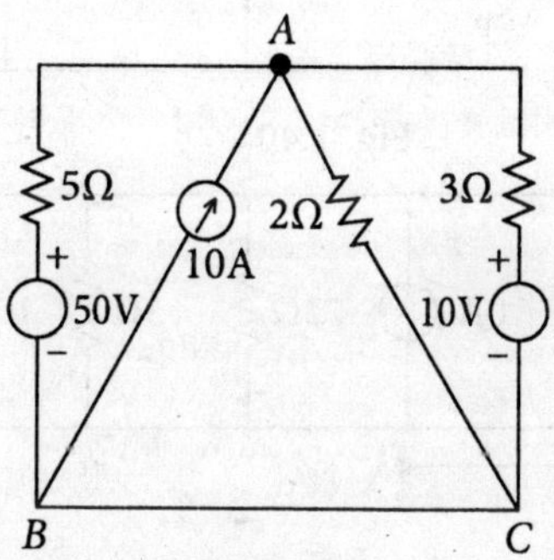

Fig. 1.37

[P = 273.5W]

7. Find the node voltages v_1, v_2, v_3 in the circuit [Fig. 1.38].

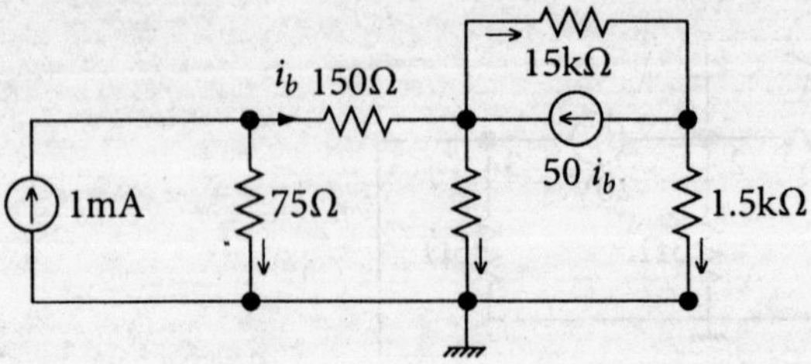

Fig. 1.38

[$V_1 = 0.071$V, $V_2 = 0.063$V, $V_3 = -3.69$V]

8. Find the node voltages v_1, v_2, v_3 [Fig. 1.39].

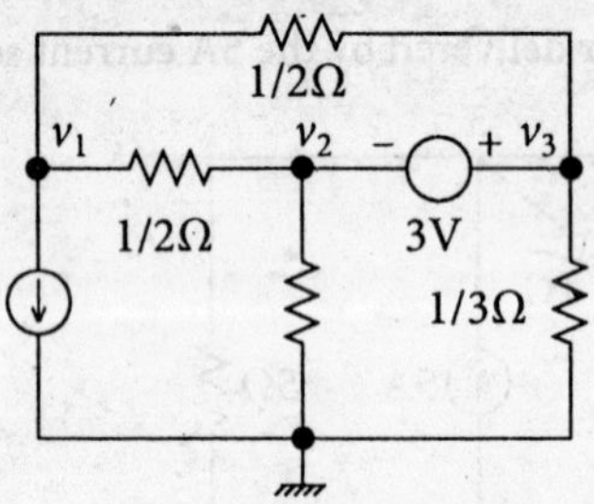

Fig. 1.39

[$v_1 = -\frac{4}{7}$V, $v_2 = -\frac{11}{7}$V, $v_3 = \frac{10}{7}$V]

9. Using source transformation, source combination and resistance combination, convert the circuit [Fig. 1.40].

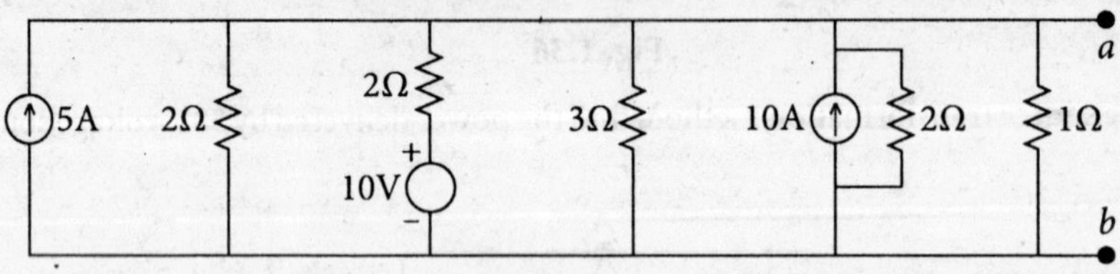

Fig. 1.40

Ans. ≡

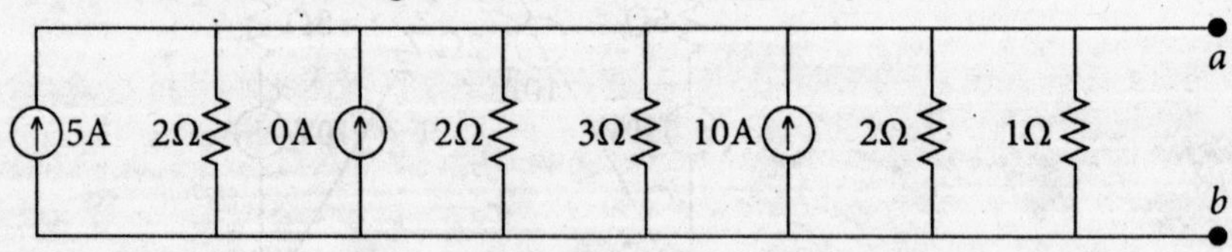

≡

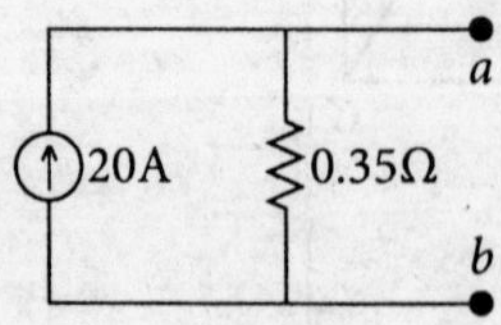

1.11 Solution of Mixed Numerical Problems

No voltage source: Only current sources (No super-node present)

Problem 1.

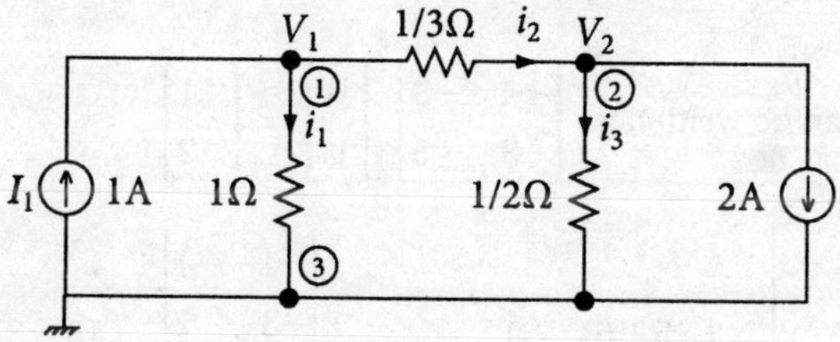

Fig. 1.41

Solution: At node 1:

$$+I_1 - i_1 - i_2 = 0 \Rightarrow i_1 + i_2 = I_1 \Rightarrow \frac{V_1}{1} + \frac{V_1 - V_2}{(1/3)} = 1 \Rightarrow 4V_1 - 3V_2 = 1. \tag{1.33}$$

At node 2:

$$+i_2 - i_3 - I_2 = 0 \Rightarrow i_2 - i_3 = I_2 \Rightarrow \frac{V_1 - V_2}{(1/3)} - \frac{V_2}{(1/2)} = I_2 \Rightarrow 3V_1 - 5V_2 = 2. \tag{1.34}$$

Conductance matrix equation takes the form as

$$\begin{bmatrix} g_{11} & g_{12} \\ g_{21} & g_{22} \end{bmatrix} \begin{bmatrix} V_1 \\ V_2 \end{bmatrix} = \begin{bmatrix} I_1 \\ I_2 \end{bmatrix} \Rightarrow \begin{bmatrix} 4 & -3 \\ +3 & -5 \end{bmatrix} \begin{bmatrix} V_1 \\ V_2 \end{bmatrix} = \begin{bmatrix} 1 \\ 2 \end{bmatrix}.$$

$$v_1 = \frac{\begin{vmatrix} 1 & -3 \\ 2 & -5 \end{vmatrix}}{\begin{vmatrix} 4 & -3 \\ 3 & -5 \end{vmatrix}} = \frac{-5+6}{-20+9} = \frac{1}{-11}\text{V}; \quad v_2 = \frac{\begin{vmatrix} 4 & 1 \\ 3 & 2 \end{vmatrix}}{\begin{vmatrix} 4 & -3 \\ 3 & -5 \end{vmatrix}} = \frac{5}{-11} = -\frac{5}{11}\text{V}.$$

Both of V_1 and V_2 are negative and $V_2 < V_1$; so i_1 will be reversed in direction, but not i_2. If we take this fact into account, then redrawing the circuit accordingly we would get the circuit [Fig. 1.42].

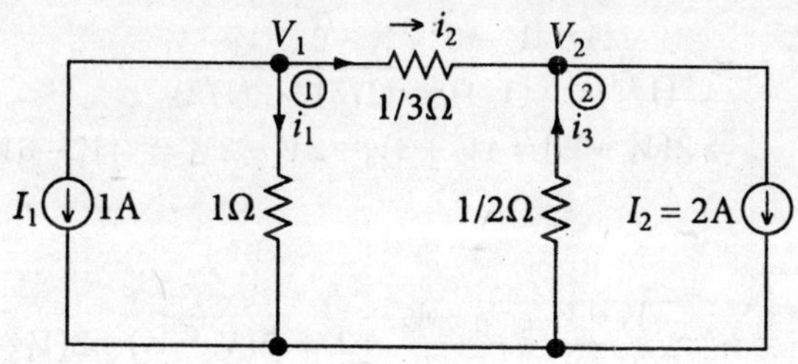

Fig. 1.42

At node 1: $i_1 = i_2 + I_1$

$$i_1 - i_2 = I_1 \Rightarrow \frac{0 - V_1}{1} - 3(V_1 - V_2) = 1 \Rightarrow -4V_1 + 3V_2 = 1 \Rightarrow V_1 - 3V_2 = -1.$$

At node 2: $i_2 + i_3 + I_2 = 0$

$$\frac{(V_1 - V_2)}{(1/3)} + \frac{0 - V_2}{(1/2)} + I_2 = 0 \Rightarrow 3V_1 - 5V_2 = -I_2 = -2 \Rightarrow -3V_1 + 5V_2 = 2.$$

So, the matrix equation can be written as $\begin{bmatrix} +4 & -3 \\ -3 & 5 \end{bmatrix}\begin{bmatrix} V_1 \\ V_2 \end{bmatrix} = \begin{bmatrix} -1 \\ 2 \end{bmatrix}$.

$$V_1 = \frac{\begin{vmatrix} -1 & -3 \\ 2 & 5 \end{vmatrix}}{\begin{vmatrix} 4 & -3 \\ -3 & 5 \end{vmatrix}} = \frac{(-5+6)}{(20-9)} = \frac{1}{11}\text{V}; \quad V_2 = \frac{\begin{vmatrix} 4 & -1 \\ -3 & 2 \end{vmatrix}}{\begin{vmatrix} 4 & -3 \\ -3 & 5 \end{vmatrix}} = \frac{(8-3)}{(20-9)} = \frac{5}{11}\text{V}.$$

If we look back to the matrix equation, called the *conductance matrix equation*, we see that $g_{21} = g_{12} =$ −ve of the sum of the reciprocals of the resistance common to both node 1 and node 2.

Problem 2. Solve the circuit [Fig. 1.43] by nodal analysis technique. Since there is voltage source between nodes 2 and 3, they are said to form a supernode pair.

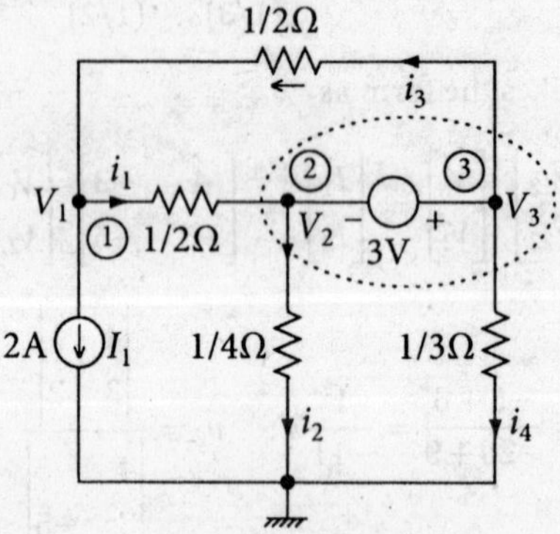

Fig. 1.43

Solution: Applying KCL to the supernodes 2 and 3 and treating them as a single node, we get

$$i_1 = i_2 + i_3 + i_4 \Rightarrow \frac{V_1 - V_2}{(1/2)} = \frac{V_2 - 0}{(1/4)} + \frac{V_3 - 0}{(1/3)} + \frac{V_3 - V_1}{(1/2)}$$

$$\Rightarrow 2(V_1 - V_2) = 4V_2 + 3V_3 + 2V_3 - 2V_1 \Rightarrow 4V_1 - 6V_2 - 5V_3 = 0. \tag{1.35}$$

At node 1: $i_3 = i_1 + I_1$

$$\Rightarrow i_3 - i_1 = I_1 \Rightarrow \frac{V_3 - V_1}{(1/2)} - \frac{V_1 - V_2}{(1/2)} = 2 \Rightarrow 2(V_3 - V_1) - 2(V_1 - V_2) = 2$$

$$\Rightarrow -4V_1 + 2V_2 + 2V_3 = 2 \Rightarrow 2V_1 - V_2 - V_3 = -1. \tag{1.36}$$

The third equation is written w.r.t. the voltage source as

$$V_2 + 3 = V_3 \Rightarrow -V_2 + V_3 = 3. \tag{1.37}$$

The conductance matrix equation from the above three equations can be written as

$$\begin{bmatrix} 4 & -6 & -5 \\ 2 & -1 & -1 \\ 0 & -1 & 1 \end{bmatrix} \begin{bmatrix} V_1 \\ V_2 \\ V_3 \end{bmatrix} = \begin{bmatrix} 0 \\ -1 \\ 3 \end{bmatrix}.$$

$$V_1 = \frac{\begin{vmatrix} 0 & -6 & -5 \\ -1 & -1 & -1 \\ 3 & -1 & 1 \end{vmatrix}}{\begin{vmatrix} 4 & -6 & -5 \\ 2 & -1 & -1 \\ 0 & -1 & -1 \end{vmatrix}} = \frac{6(-1+3)-5(1+3)}{4(1-1)+6(-2)-5(-2)} = \frac{12-20}{-12+10} = \frac{-8}{-2} = 4\text{V}.$$

Problem 3. Find the current in the 5Ω resistor [Fig. 1.44].

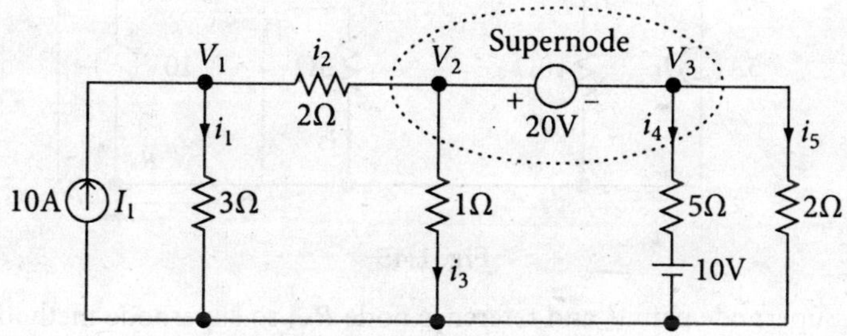

Fig. 1.44

Solution: Applying KCL to the supernodes (2) and (3) and treating them as one single node, we get

$$i_2 = i_3 + i_4 + i_5 \Rightarrow \frac{V_1 - V_2}{2} = \frac{V_2 - 0}{1} + \frac{V_3 - 10}{5} + \frac{V_3 - 0}{2}$$

$$\Rightarrow 5(V_1 - V_2) = 10V_2 + 2V_3 - 20 + 5V_3 \Rightarrow 5V_1 - 15V_2 - 7V_3 = -20$$

$$\Rightarrow 5V_1 - 15V_2 - 7V_3 = -20 \Rightarrow 0.5V_1 - 1.5V_2 - 0.7V_3 = -2. \quad (1.38)$$

At node 1: $i_1 + i_2 = I_1$

$$\Rightarrow \frac{V_1}{3} + \frac{V_1 - V_2}{2} = 10 \Rightarrow V_1\left(\frac{1}{3} + \frac{1}{2}\right) - 0.5V_2 = 10 \Rightarrow 0.83V_1 - 0.5V_2 = 10. \quad (1.39)$$

With respect to the voltage source, the third equation can be written as

$$V_2 - V_3 = 20. \quad (1.40)$$

$$\therefore V_3 = \frac{\begin{vmatrix} 0.5 & -1.5 & -2 \\ 0.83 & -0.5 & 10 \\ 0 & 1 & 20 \end{vmatrix}}{\begin{vmatrix} 0.5 & -1.5 & -0.7 \\ 0.83 & -0.5 & 0 \\ 0 & 1 & -1 \end{vmatrix}}$$

$$= \frac{0.5(-0.5\times 20-10)+1.5(0.83\times 20)-2(0.83)}{0.5(0.5)-1.5(0.83)-0.7(0.83)} = \frac{-10+24.9-1.66}{0.25-1.24-0.58} = \frac{13.24}{-1.57} = -8.43\text{V}.$$

$\therefore i_4 = \frac{V_3-10}{5} = \frac{-18.43}{5} = -3.68$A, the negative sign shows that the current through 5Ω resistor is opposite to the direction of i_4 [Fig. 1.44].

Problem 4. Write the node voltage equations and determine the currents in each branch of the network [Fig. 1.45].

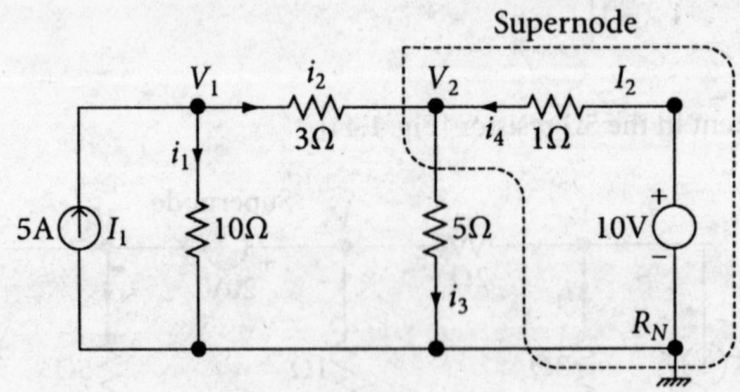

Fig. 1.45

Solution: There is a supernode pair V_2 and reference node R_N, so supernode method of solution may be adopted here. But this circuit problem may be solved in a simpler method as given below.

Here source currents I_1 and I_2 are given by $I_1 = 5$A and $I_2 = 10/1 = 10$A.

At node 1: $I_1 = i_1 + i_2$

$$\Rightarrow \frac{V_1}{10} + \frac{V_1 - V_2}{3} = I_1 = 5\text{A} \Rightarrow 0.43V_1 - 0.33V_2 = 5. \quad (1.41)$$

At node 2: $i_2 + I_2 = i_3$

$$\Rightarrow i_3 - i_2 = I_2 \Rightarrow \frac{V_2}{5} - \frac{(V_1 - V_2)}{3} = 10$$

$$\Rightarrow -(0.33)V_1 + 0.53V_2 = \frac{(10 - V_2)}{1} \Rightarrow -(0.33)V_1 + (1.53)V_2 = 10. \quad (1.42)$$

So, the conductance matrix equation is written as

$$\begin{bmatrix} 0.43 & -0.33 \\ -0.33 & 1.53 \end{bmatrix}\begin{bmatrix} V_1 \\ V_2 \end{bmatrix} = \begin{bmatrix} 5 \\ 10 \end{bmatrix} \equiv \begin{bmatrix} g_{11} & g_{12} \\ g_{21} & g_{22} \end{bmatrix}\begin{bmatrix} V_1 \\ V_2 \end{bmatrix} = \begin{bmatrix} I_1 \\ I_2 \end{bmatrix}.$$

$$g_{11} = \left(\frac{1}{10} + \frac{1}{3}\right) = 0.43\mho, \quad g_{12} = -\left(\frac{1}{3}\right) = -0.33, \quad I_1 = 5\text{A}$$

$$g_{21} = -\frac{1}{3} = -0.33, \quad g_{22} = \left(1 + \frac{1}{3} + \frac{1}{5}\right) = 1.53\mho, \quad I_2 = \frac{10}{1} = 10\text{A}.$$

Problem 5. In this circuit use the supernode technique for finding the voltages and currents [Fig. 1.46].

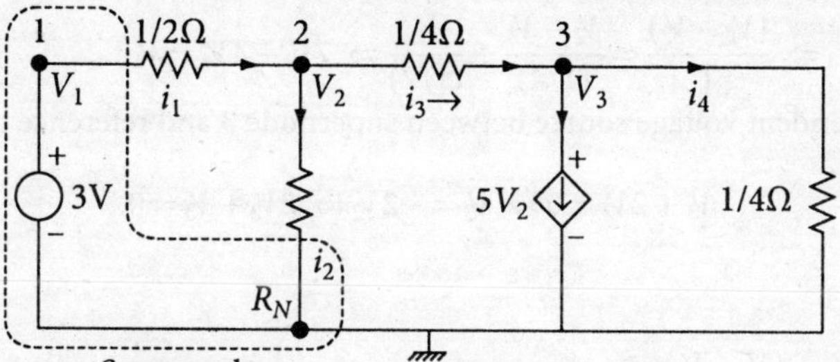

Fig. 1.46

Solution: There is a supernode between node 1 and the reference node R_N. Considering the supernode 1', we get $V_1 = 3\text{V}$ (First nodal equation).

At node 2: $i_1 = i_2 + i_3$

$$\rightarrow \frac{(3-V_2)}{(1/2)} = \frac{V_3}{(1/2)} + \frac{V_2 - V_3}{(1/4)} \Rightarrow 2(3-V_2) = 2V_2 + 4(V_2 - V_3) \Rightarrow 8V_2 - 4V_3 = 6 \Rightarrow 4V_2 - 2V_3 = 3. \quad (1.43)$$

At node 3: $i_3 = 0.5V_2 + \frac{V_3}{(1/4)}$

$$\Rightarrow \frac{(V_2 - V_3)}{(1/4)} = 0.5V_2 + 4V_3 \Rightarrow 4(V_2 - V_3) = 0.5V_2 + 4V_3 \Rightarrow 3.5V_2 - 8V_3 = 0. \quad (1.44)$$

So, the matrix equations can be written as

$$\begin{bmatrix} 4 & -2 \\ 3.5 & -8 \end{bmatrix} \begin{bmatrix} V_2 \\ V_3 \end{bmatrix} = \begin{bmatrix} 3 \\ 0 \end{bmatrix}.$$

$$\therefore V_2 = \frac{\begin{vmatrix} 3 & -2 \\ 0 & -8 \end{vmatrix}}{\begin{vmatrix} 4 & -2 \\ 3.5 & -8 \end{vmatrix}} = \frac{-24}{-32+7} = \frac{24}{25} = 0.96\text{V}; \qquad V_3 = \frac{\begin{vmatrix} 4 & 3 \\ 3.5 & 0 \end{vmatrix}}{\Delta} = \frac{-10.5}{-25} = 0.42\text{V}.$$

Problem 6. In the circuit [Fig. 1.47], there are two pairs of supernodes between 1 and reference node R_N (containing the independent voltage source v_1) and another between nodes 3 and R_N. Find the voltages and currents in the different branches.

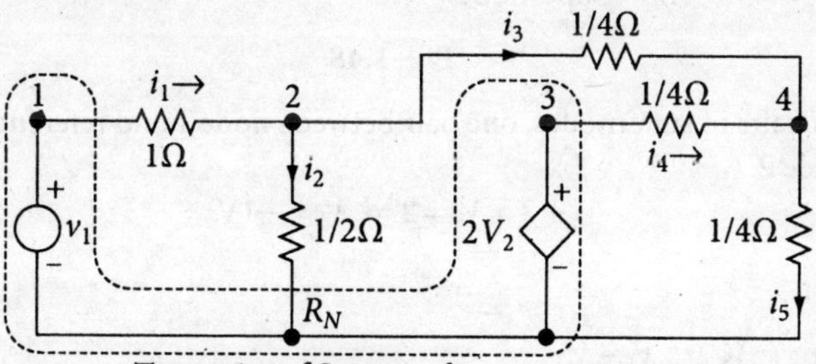

Fig. 1.47

For the first supernode 1 and R_N, $V_1 = v_1$.

At node 2: $i_1 = i_2 + i_3 \Rightarrow \dfrac{(V_1 - V_2)}{1} = \dfrac{V_2 - V_4}{(1/4)} + \dfrac{V_2}{(1/2)} \Rightarrow 7V_2 - 4V_4 = V_1$.

Considering the dependent voltage source between supernode 3 and reference node R_N, we get

$$V_3 + 2V_2 = 0 \Rightarrow V_3 = -2V_2 \Rightarrow 2V_2 + V_3 = 0. \quad (1.45)$$

At node 4: $i_3 + i_4 = i_5$

$$\Rightarrow \frac{V_2 - V_4}{(1/4)} + \frac{(V_3 - V_4)}{(1/4)} = \frac{V_4}{(1/4)} \Rightarrow (V_2 - V_4) + (V_3 - V_4) = V_4 \Rightarrow V_2 + V_3 - 3V_4 = 0$$

$$\Rightarrow V_2 + (-2V_2) - 3V_4 = 0 \Rightarrow V_2 + 3V_4 = 0. \quad (1.46)$$

So, the matrix equation can be written as

$$\begin{bmatrix} 7 & -4 \\ 1 & 3 \end{bmatrix} \begin{bmatrix} V_2 \\ V_4 \end{bmatrix} = \begin{bmatrix} V_1 \\ 0 \end{bmatrix}.$$

$$\therefore V_2 = \frac{\begin{vmatrix} v_1 & -4 \\ 0 & 3 \end{vmatrix}}{\begin{vmatrix} 7 & -4 \\ 1 & 3 \end{vmatrix}} = \frac{3v_1}{(21+4)} = \frac{3v_1}{25} = 0.12v_1;\ V_4 = \frac{\begin{vmatrix} 7 & v_1 \\ 1 & 0 \end{vmatrix}}{\Delta} = \frac{-v_1}{25} = -0.04\text{V};\ V_3 = -2V_2 = \frac{-6v_1}{25} = -0.24v_1.$$

Problem 7. Using supernode technique to solve the circuit [Fig. 1.48].

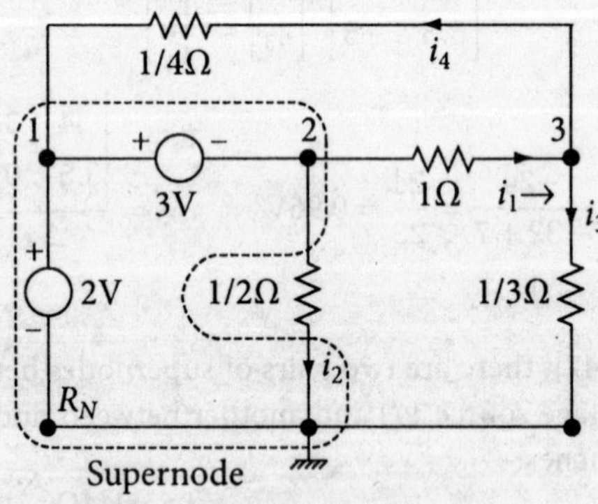

Fig. 1.48

Solution: There are two pairs of supernodes, one pair between node 1 and reference node R_N and another between node 1 and node 2.

$$V_2 + 3 = V_1 = 2 \Rightarrow V_2 = -1\text{V}.$$

At node 3: $i_1 = i_3 + i_4$

$$\Rightarrow \frac{(V_2 - V_3)}{1} = \frac{V_3}{(1/3)} + \frac{V_3 - V_1}{(1/4)} \Rightarrow (V_2 - V_3) = 3V_3 + 4(V_3 - V_1) \Rightarrow 4V_1 + V_2 - 8V_3 = 0$$

$$\Rightarrow 4 \times 2 - 1 - 8V_3 = 0 \Rightarrow 8V_3 = 7 \Rightarrow V_3 = \tfrac{7}{8}\text{V}.$$

Problem 8. Solve the circuit [Fig. 1.49].

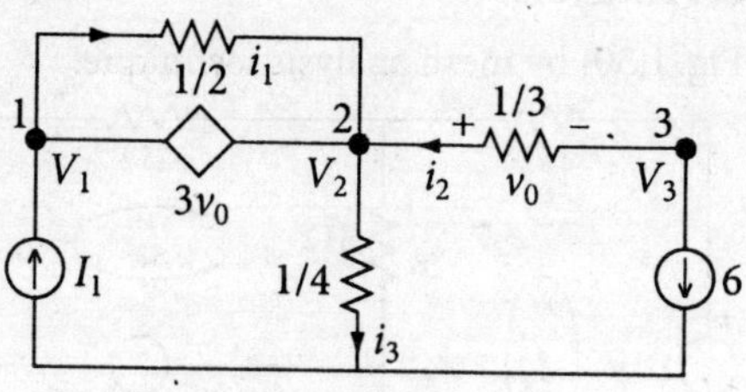

Fig. 1.49

Solution: The relations, which are evident from the circuit is

(i) $v_0 = V_2 - V_3$.

(ii) $I_1 = i_1 + 3v_0$.

(iii) At node 2: $i_1 + i_2 + 3v_0 = i_3$

$$\Rightarrow \frac{V_1 - V_2}{(1/2)} + \frac{V_3 - V_2}{(1/3)} + 3v_0 = \frac{V_2}{(1/4)} \Rightarrow -V_1 + 3V_2 = 0. \tag{1.47}$$

(iv) At node 1: $I_1 = i_1 + 3v_0$

$$\Rightarrow 7 = \frac{V_1 - V_2}{(1/2)} + 3v_0 \Rightarrow 2V_1 - 2V_2 + 3v_0 = 7 \Rightarrow 2V_1 - 2V_2 + 3(V_2 - V_3) = 7 \Rightarrow 2V_1 + V_2 - 3V_3 = 7. \tag{1.48}$$

At node 3: $i_2 + 6 = 0$

$$\Rightarrow 3(V_3 - V_2) = -6 \Rightarrow -3V_2 + 3V_3 = -6. \tag{1.49}$$

So, the matrix equation can be written as $\begin{bmatrix} 2 & 1 & -3 \\ -1 & 3 & 0 \\ 0 & -3 & 3 \end{bmatrix} \begin{bmatrix} V_1 \\ V_2 \\ V_3 \end{bmatrix} = \begin{bmatrix} 7 \\ 0 \\ -6 \end{bmatrix}$.

$$\therefore V_1 = \frac{\begin{vmatrix} 7 & 1 & -3 \\ 0 & 3 & 0 \\ -6 & -3 & 3 \end{vmatrix}}{\begin{vmatrix} 2 & 1 & -3 \\ -1 & 3 & 0 \\ 0 & -3 & 3 \end{vmatrix}} = \frac{7(9) - 1(0) - 3(18)}{2(9) - 1(-3) - 3(3)} = \frac{63 - 54}{18 + 3 - 9} = \frac{9}{12} = \frac{3}{4}\text{V}.$$

$$V_2 = \frac{\begin{vmatrix} 2 & 7 & -3 \\ -1 & 0 & 0 \\ 0 & -6 & 3 \end{vmatrix}}{\Delta} = \frac{2(0) - 7(-3) - 3(6)}{12} = \frac{3}{12} = \frac{1}{4}\text{V}.$$

$$V_3 = \frac{\begin{vmatrix} 2 & 1 & 7 \\ -1 & 3 & 0 \\ 0 & -3 & -6 \end{vmatrix}}{12} = \frac{-36 - 1 \times 6 + 7 \times 3}{12} = \frac{-21}{12} = -\frac{7}{4}\text{V}.$$

1.12 Problems on Mesh Analysis

Problem 9. Solve the circuit [Fig. 1.50] by mesh analysis technique.

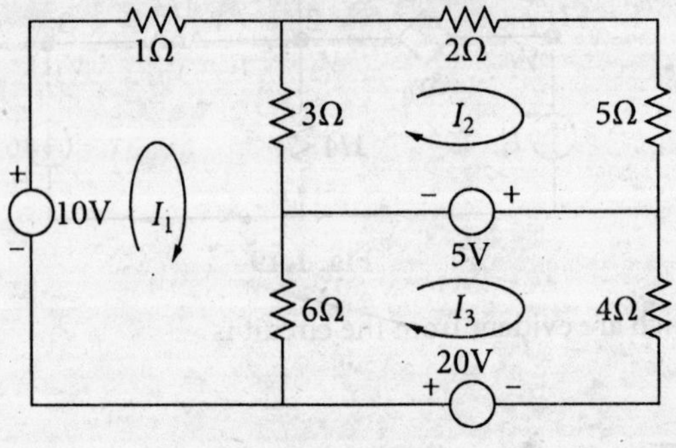

Fig. 1.50

Solution: There are three meshes and hence, three mesh currents are assigned as I_1, I_2 and I_3, all are supposed to flow in the clockwise direction. The mesh equations in the general form of matrix can be written as under:

$$\begin{bmatrix} R_{11} & \pm R_{12} & \pm R_{13} \\ \pm R_{21} & R_{22} & \pm R_{23} \\ \pm R_{31} & \pm R_{32} & R_{33} \end{bmatrix} \begin{bmatrix} I_1 \\ I_2 \\ I_3 \end{bmatrix} \begin{bmatrix} V_1 \\ V_2 \\ V_3 \end{bmatrix}.$$

R_{11} = self-resistance of the *Loop* 1 = (1+3+6)Ω = 10Ω

R_{12} = mutual resistance common to both of loops 1 and 2, ± sign is indicated depending on the direction of currents through it; in the same direction + sign is used and in opposite direction – sign be attached
= –3Ω (I_1 and I_2 flow in opposite direction)

R_{13} = –6Ω (I_1 and I_3 flow in opposite direction)

$R_{21} = -3\Omega$

$R_{22} = 3+2+5 = 10\Omega$

R_{23} = 0 (no common resistance between 2 and 3)

$R_{31} = -6\Omega$

$R_{32} = 0$

$R_{33} = 10\Omega$

V_1 = driving voltage in *Loop* 1 = 10V

V_2 = algebraic sum of the voltages in *Loop* 2 = –5V

V_3 = algebraic sum of the cottages in *Loop* 2 and *Loop* 3 = (5 + 20) = 25V.

∴ the matrix equation is written as $\begin{bmatrix} 10 & -3 & -6 \\ -3 & 10 & 0 \\ -6 & 0 & 10 \end{bmatrix} \begin{bmatrix} 10 \\ -5 \\ 25 \end{bmatrix} = \begin{bmatrix} I_1 \\ I_2 \\ I_3 \end{bmatrix}.$

$$\therefore I_1 = \frac{\begin{vmatrix} 10 & -3 & -6 \\ -5 & 10 & 0 \\ 25 & 0 & 10 \end{vmatrix}}{\begin{vmatrix} 10 & -3 & -6 \\ -3 & 10 & 0 \\ -6 & 0 & 10 \end{vmatrix}} = \frac{10(100)+3(-50)-6(-250)}{10(100)+3(-30)-6(60)} = \frac{1000-150+1500}{1000-90-360} = \frac{2350}{550} = 4.27\text{A}.$$

$$I_2 = \frac{\begin{vmatrix} 10 & 10 & -6 \\ -3 & -5 & 0 \\ -6 & 25 & 10 \end{vmatrix}}{\Delta} = \frac{10(-50)-10(-30)-6(-75+30)}{\Delta} = \frac{-500+300+270}{550} = \frac{70}{550} = 0.127\text{A}.$$

$$I_3 = \frac{\begin{vmatrix} 10 & -3 & 10 \\ -3 & 10 & -5 \\ -6 & 0 & 25 \end{vmatrix}}{\Delta} = \frac{10(250)+3(-75+30)+10(60)}{550}\ \frac{2965}{550} = 5.4\text{A}.$$

Problem 10. Apply supernode analysis technique to solve the circuit [Fig. 1.51].

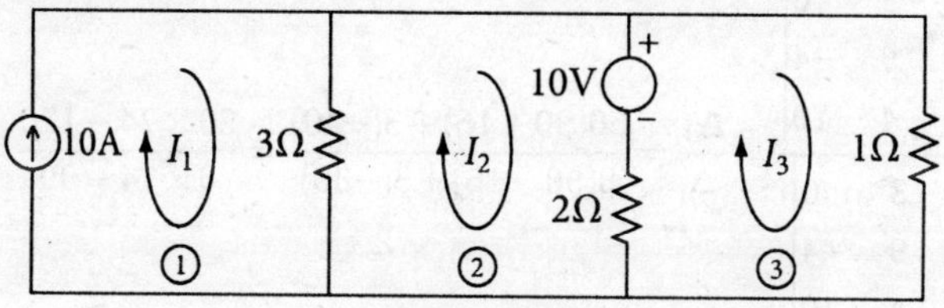

Fig. 1.51

Solution: Mesh 1 is a supermesh and current flow $I_1 = 10\text{A}$.

For mesh 2: $(I_2 - I_3)\times 2 + 3(I_2 - 10) = -10 \Rightarrow 5I_2 - 2I_3 = 20.$

For mesh 3: $I_3 \times 1 + (I_3 - I_2)\times 2 = 10 \Rightarrow 3I_3 - 2I_2 = 10 \Rightarrow -2I_2 + 3I_3 = 10.$

$$\therefore \begin{bmatrix} 5 & -2 \\ -2 & 3 \end{bmatrix}\begin{bmatrix} 20 \\ 10 \end{bmatrix} = \begin{bmatrix} I_2 \\ I_3 \end{bmatrix}.$$

$$I_2 = \frac{\begin{vmatrix} 20 & -2 \\ 10 & 3 \end{vmatrix}}{\begin{vmatrix} 5 & -2 \\ -2 & 3 \end{vmatrix}} = \frac{80}{\Delta} = \frac{80}{15-4} = \frac{80}{11} = 7.27\text{A}, \quad I_3 = \frac{\begin{bmatrix} 5 & 20 \\ -2 & 10 \end{bmatrix}}{\Delta} = \frac{90}{11} = 8.18\text{A}.$$

So, we have the solution as $I_1 = 10\text{A}$, $I_2 = 7.27\text{A}$, $I_3 = 8.18\text{A}$.

Problem 11. Find the power dissipated in the 4Ω resistor.

Solution: Three mesh currents I_1, I_2, I_3 are [Fig. 1.52].

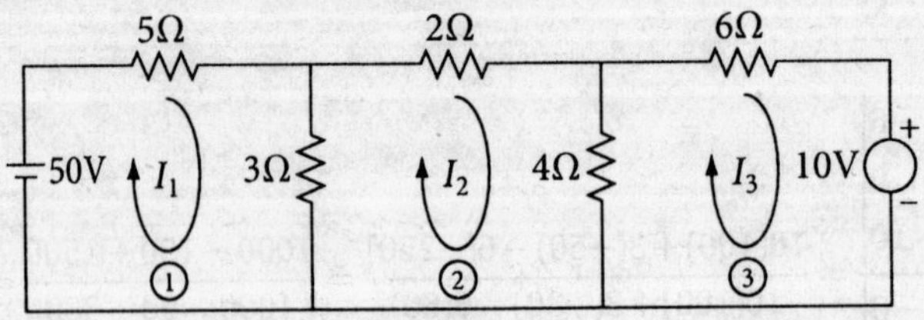

Fig. 1.52

Mesh 1. $5I_1 + (I_1 - I_2) \times 3 = 50 \Rightarrow 8I_1 - 3I_2 = 50.$

Mesh 2. $2I_2 + (I_2 - I_3) \times 4 + (I_2 - I_1) \times 3 = 0 \Rightarrow -3I_1 + 9I_2 - 4I_3 = 0.$

Mesh 2. $6I_3 + (I_3 - I_2) \times 4 = -10 \Rightarrow 10I_3 - 4I_2 = -10 \Rightarrow -4I_2 + 10I_3 = -10.$

So, the matrix equation can be written as

$$\begin{bmatrix} 8 & -3 & 0 \\ -3 & 9 & -4 \\ 0 & -4 & 10 \end{bmatrix} \begin{bmatrix} I_1 \\ I_2 \\ I_3 \end{bmatrix} = \begin{bmatrix} 50 \\ 0 \\ -10 \end{bmatrix}.$$

$$\therefore I_1 = \frac{\begin{vmatrix} 50 & -3 & 0 \\ 0 & 9 & -4 \\ -10 & -4 & 10 \end{vmatrix}}{\begin{vmatrix} 8 & -3 & 0 \\ -3 & 9 & -4 \\ 0 & -4 & 10 \end{vmatrix}} = \frac{\Delta_1}{\Delta} = \frac{50(90-16)+3(-40)}{8(90-16)+3(-30)} = \frac{50 \times 74 - 120}{8 \times 74 - 90} = \frac{3580}{502} = 7.13\text{A}.$$

$$I_2 = \frac{\begin{vmatrix} 8 & 50 & 0 \\ -3 & 0 & -4 \\ -0 & -10 & 10 \end{vmatrix}}{502} = \frac{8(-40) - 50(-3 \times 10)}{502} = \frac{-320 + 1500}{502} = 2.35\text{A}.$$

$$I_3 = \frac{\begin{vmatrix} 8 & -3 & 50 \\ -3 & 9 & 0 \\ 0 & -4 & 10 \end{vmatrix}}{\Delta} = \frac{8(-90) + 3(30) + 50(12)}{502} = -\frac{30}{502} = -0.06\text{A}.$$

Current passing through 4Ω resistor = $2.35 + 0.06 = 2.41$A.

Power dissipated in 4Ω resistor = $(2.41)^2 \times 4 = 23.23$W.

1.13 Numerical Problems

1. Use mesh analysis to solve the circuit [Fig. 1.53] for finding the power developed to 4Ω resistor. To what voltage should the 100V battery be charged so that no power develops in the 4Ω resistor?

[2580W; −32V.]

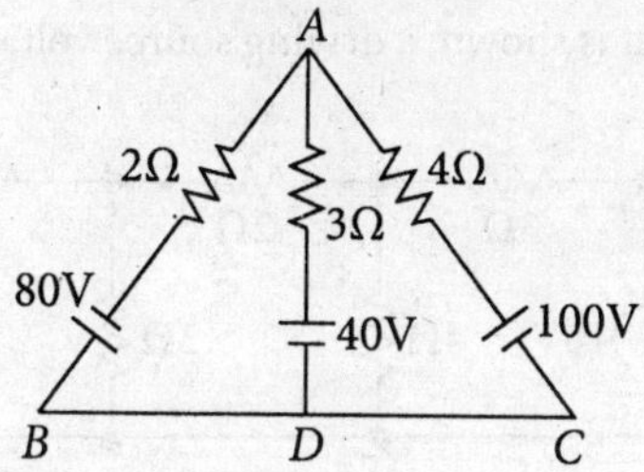

Fig. 1.53

2. Use the nodal analysis to find out the voltage across 4Ω resistor and the power supplied by the 5A source [Fig. 1.54]. [−60.9V; 195.7W]

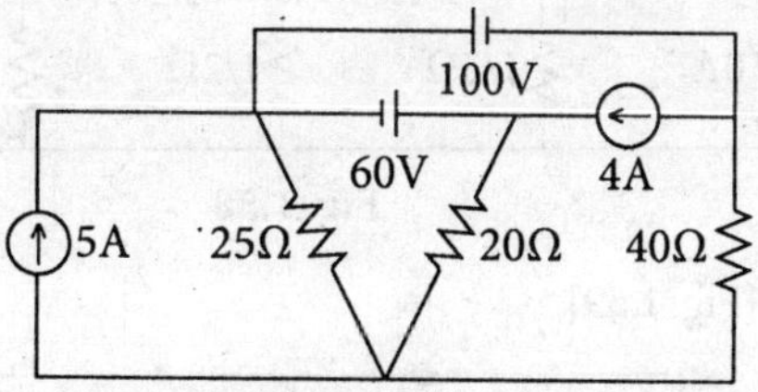

Fig. 1.54

3. Solve by applying mesh analysis method the circuit [Fig. 1.55] for finding the nodal voltages V_1, V_2, V_3 and also voltage V_4 across the $\frac{1}{2}\Omega$ resistor. [$V_1 = 1V, V_2 = -\frac{11}{7}V, V_3 = \frac{10}{7}V, V_4 = 2V$]

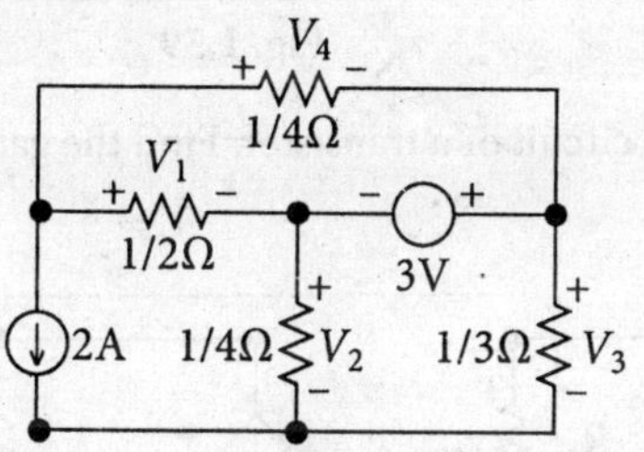

Fig. 1.55

4. By applying loop analysis, solve the circuit [Fig. 1.56] to find v, the voltage across the 3Ω resistor.

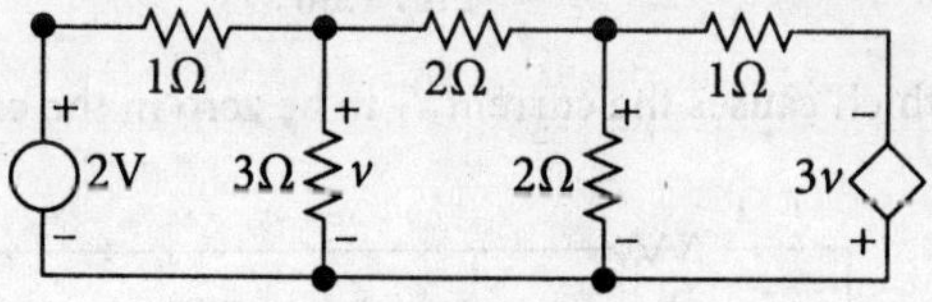

Fig. 1.56

5. Use nodal analysis method to solve the circuit [Fig. 1.56]. [$v = 0.36V$]

6. In [Fig. 1.57] a resistive circuit is shown; a driving source voltage 10V is included in the circuit. Find v and i. $[\frac{2}{3}\text{V}, 2\text{A}]$

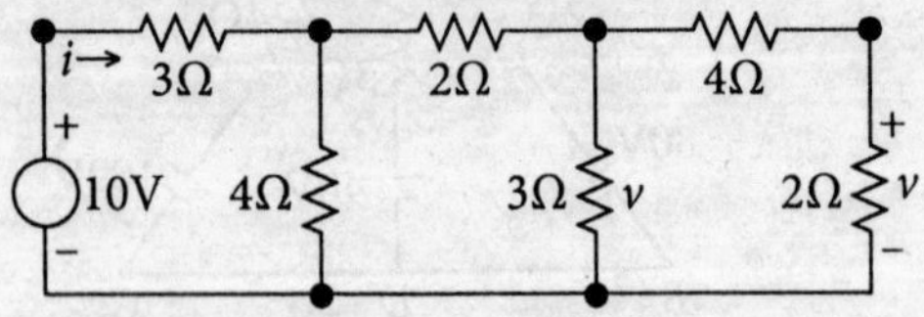

Fig. 1.57

7. Find v and i in the circuit [Fig. 1.58].

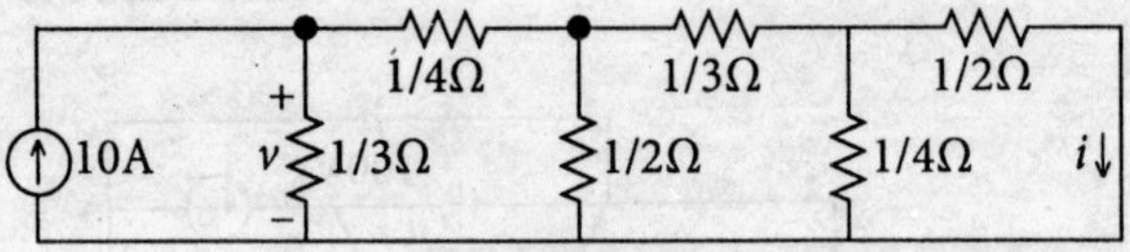

Fig. 1.58

8. Find v and also i_1 and i_2 [Fig. 1.59]. $[-\frac{5}{2}\text{V},\ i_1 = -\frac{5}{6}\text{A},\ i_2 = -0.5\text{A}]$

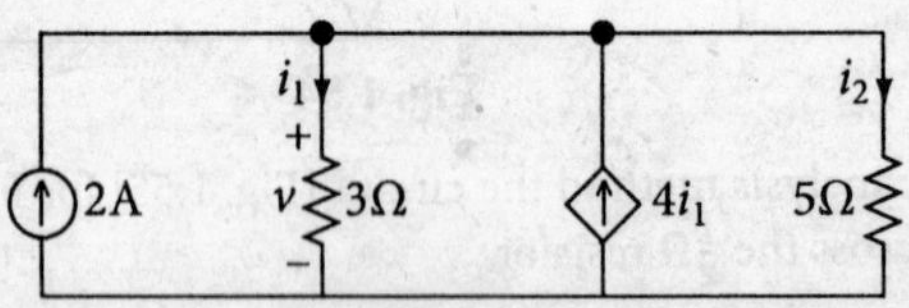

Fig. 1.59

9. Fig. 1.60 shows an amplified circuit of a transistor. Find the gain of such an amplifier.

$$\left[\frac{v_2}{v_1} = -\frac{g_m R_d R_g R_l}{(R_g + R_1)(R_d + R_l)}\right]$$

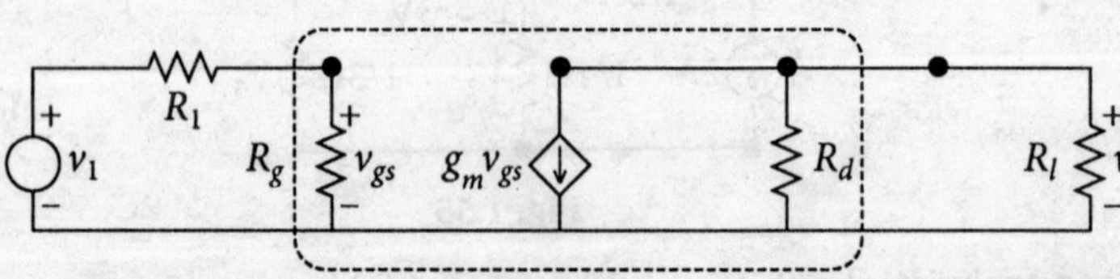

Fig. 1.60

10. Determine voltage V, which causes the current I_1 to be zero in the circuit [Fig. 1.61]. Use loop analysis. $[V = 43.7\text{V}]$

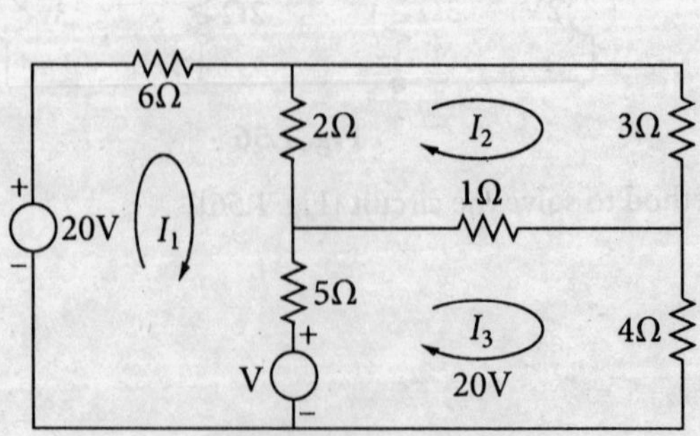

Fig. 1.61

11. Find the power delivered by the voltage source and the current in the 10Ω resistor in the circuit [Fig. 1.62]. [$I_1 = 9.76$A; $I_2 = 6.76$A; $I_3 = 10$A; $P = 488$W]

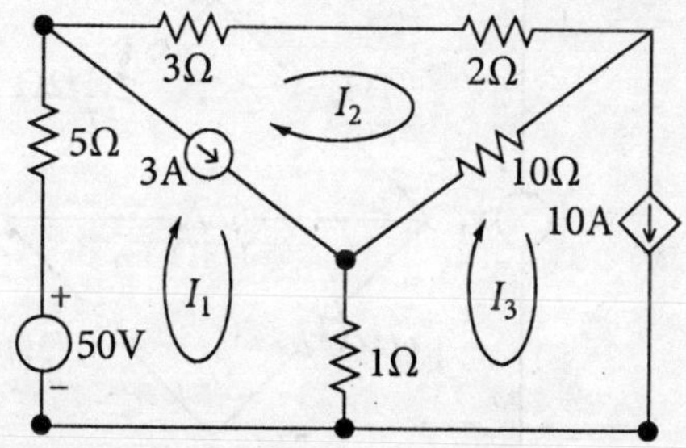

Fig. 1.62

12. Find the voltage V in the circuit [Fig. 1.63] for which no current flows through the 10Ω resistor. [$V = 52.06$V]

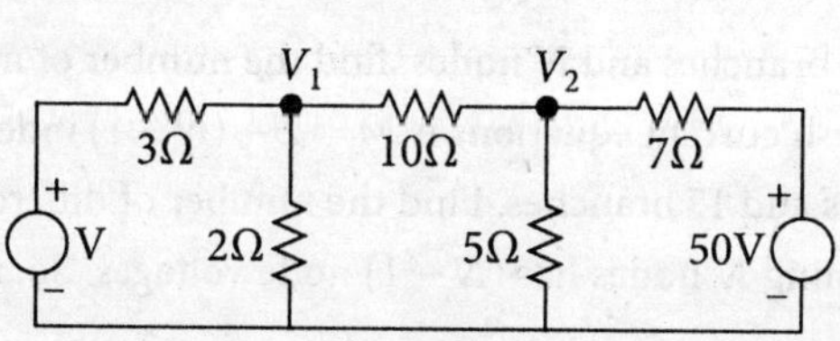

Fig. 1.63

13. The circuit [Fig. 1.64] is a BJT-amplifier with feedback. The portion in the dashed box is an approximate T model of a transistor in CE-node. Use the nodal analysis technique to find: (a) voltage gain $\left(\frac{v_2}{v_1}\right)$ (b) input resistance R_i. [(a) −47.76, (b) 253Ω]

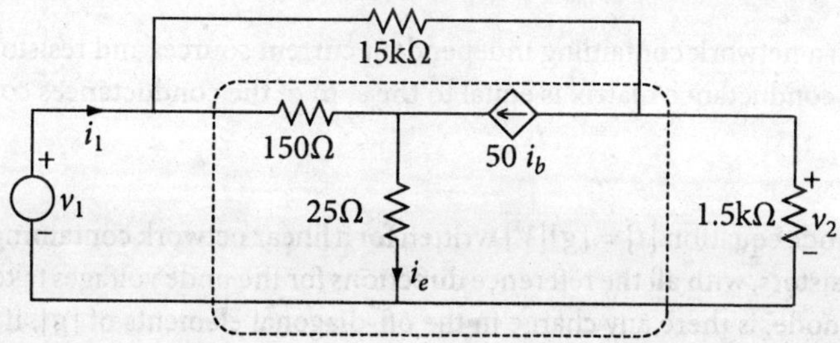

Fig. 1.64

14. The circuit [Fig. 1.65] is an h-parameter model of a transistor amplifier. Use loop analysis to find: (i) voltage gain $\left(\frac{v_2}{v_1}\right)$, (ii) Input resistance R_i.

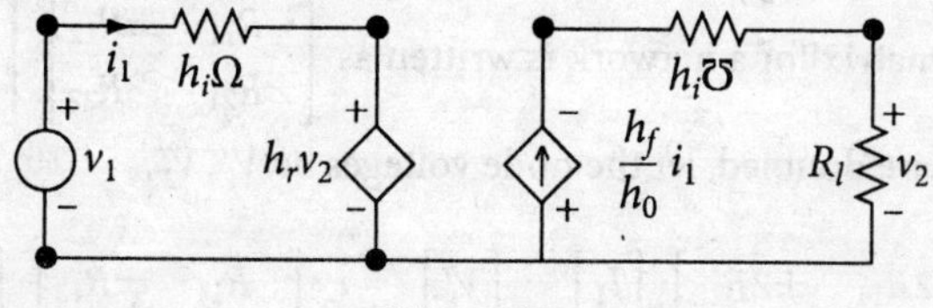

Fig. 1.65

15. Use loop analysis method to find voltage gain: (i) $\frac{v_2}{v_1}$, (ii) Input resistance $R_i = \frac{v_1}{i}$. $[R_i = \frac{v_1}{i} = \frac{15}{31}\Omega]$

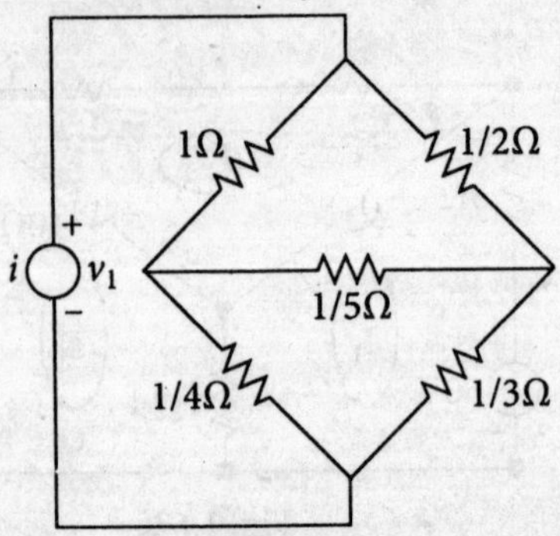

Fig. 1.66

1.14 Short Theory Questions with Answers

1. If a network contains B branches and N nodes, find the number of mesh current equations.

   Ans. The number of mesh current equations is $M = B - (N-1)$ independent equations.

2. A network has 10 nodes and 17 branches. Find the number of different node pair voltages.

   Ans. A network containing N nodes has $(N-1)$ node voltages. So, the number of node voltages is $(10-1) = 9$.

3. How do you define a practical voltage source w.r.t. an ideal voltage source?

   Ans. A practical voltage source is an ideal voltage with an internal impedance.

4. How do you define a practical current source w.r.t. an ideal current source?

   Ans. A practical current source may be defined as an ideal current source with an impedance in parallel.

5. Is it true that in a network containing independent current sources and resistors each main diagonal element of the conductance matrix is equal to the sum of the conductances connected to the related node?

   Ans. Yes.

6. In the matrix node equation $[i] = [g][V]$ written for a linear network containing independent current sources and resistors, with all the reference directions for the node voltages taken with a negative sign at the ground node, is there any charge in the off-diagonal elements of $[g]$, if the reference polarity for one of the node voltages is reversed?

   Ans. Yes.

7. In a network containing only independent current sources and resistors, the values of all resistors are doubled. What happens to the value of the node voltages?

   Ans. The resistance matrix for a network is written as $\begin{bmatrix} R_{11} & \pm R_{12} \\ \pm R_{21} & R_{22} \end{bmatrix} \begin{bmatrix} I_1 \\ I_2 \end{bmatrix} = \begin{bmatrix} V_1 \\ V_2 \end{bmatrix}$.

   If all the resistances are doubled, let the node voltages be V_1', V_2'.

$$\therefore \begin{bmatrix} 2R_{11} & \pm 2R_{12} \\ \pm 2R_{21} & 2R_{22} \end{bmatrix} \begin{bmatrix} I_1 \\ I_2 \end{bmatrix} = \begin{bmatrix} V_1' \\ V_2' \end{bmatrix} \Rightarrow 2 \begin{bmatrix} R_{11} & \pm R_{12} \\ \pm R_{21} & R_{22} \end{bmatrix} \begin{bmatrix} I_1 \\ I_2 \end{bmatrix} = \begin{bmatrix} V_1' \\ V_2' \end{bmatrix}.$$

$$\therefore\ \frac{1}{2}\begin{bmatrix} V_1' \\ V_2' \end{bmatrix} = \begin{bmatrix} V_1 \\ V_2 \end{bmatrix} \Rightarrow V_1' = 2V_1,\ V_2' = 2V_2.$$

8. If the same constant value of current be added to all the independent current sources in a network, what happens to the node voltages?

Ans. The resistance matrix is written as $\begin{bmatrix} R_{11} & \pm R_{12} \\ \pm R_{21} & R_{22} \end{bmatrix}\begin{bmatrix} I_1 \\ I_2 \end{bmatrix} = \begin{bmatrix} V_1 \\ V_2 \end{bmatrix}$.

If i_c be the constant current to each independent current values, we would get the above equation as

$$\begin{bmatrix} R_{11} & \pm R_{12} \\ \pm R_{21} & R_{22} \end{bmatrix}\begin{bmatrix} I_1 + i_c \\ I_2 + i_c \end{bmatrix} = \begin{bmatrix} V_1' \\ V_2' \end{bmatrix} \Rightarrow \begin{bmatrix} R_{11} & \pm R_{12} \\ \pm R_{21} & R_{22} \end{bmatrix}\begin{bmatrix} I_1 \\ I_2 \end{bmatrix} + \begin{bmatrix} R_{11} & \pm R_{12} \\ \pm R_{21} & R_{22} \end{bmatrix}\begin{bmatrix} i_c \\ i_c \end{bmatrix} = \begin{bmatrix} V_1' \\ V_2' \end{bmatrix}$$

$$\Rightarrow \begin{bmatrix} V_1 \\ V_2 \end{bmatrix} + \begin{bmatrix} R_{11} & \pm R_{12} \\ \pm R_{21} & R_{22} \end{bmatrix}\begin{bmatrix} i_c \\ i_c \end{bmatrix} = \begin{bmatrix} V_1' \\ V_2' \end{bmatrix}$$

$$\Rightarrow V_1' = V_1 + \text{constant value}$$

$$\Rightarrow V_2' = V_2 + \text{constant value.}$$

9. Of the two analysis techniques for a network being available, discuss, which one is more suitable for a particular network.

Ans. There are two methods available for analysing a network: (a) *Nodal analysis* (b) *Mesh* or *loop analysis.*

The suitability of analysis technique is decided by the number of voltage sources or current sources included in the network. For greater voltage sources in the network, the mesh technique is conveniently employed, whereas for greater number of current sources, nodal analysis technique is recommended. However, the mesh analysis is valid only for planar circuits and not applicable for nonplanar circuits.

A planar circuit is one, where the circuit can be drawn on a plane surface without crossovers; a nonplanar circuit cannot be drawn on a plane surface without crossovers.

10. Mesh analysis or nodal analysis — which is more general and why?

Ans. Nodal analysis technique is more general, because its application is not restricted to planar circuits only but the mesh analysis technique is restricted to planar circuits and not valid for nonplanar circuits.

11. A network consists of resistors and voltage sources. If the values of the voltage sources are doubled, what will happen to the mesh currents?

Ans. The resistance matrix can be written as $\begin{bmatrix} R_{11} & \pm R_{12} \\ \pm R_{21} & R_{22} \end{bmatrix}\begin{bmatrix} I_1 \\ I_2 \end{bmatrix} = \begin{bmatrix} V_1 \\ V_2 \end{bmatrix}$.

If the voltage values are doubled, then the above equation becomes

$$\begin{bmatrix} R_{11} & \pm R_{12} \\ \pm R_{21} & R_{22} \end{bmatrix}\begin{bmatrix} I_1 \\ I_2 \end{bmatrix} = \begin{bmatrix} 2V_1 \\ 2V_2 \end{bmatrix} = 2\begin{bmatrix} V_1 \\ V_2 \end{bmatrix} \Rightarrow \begin{bmatrix} R_{11} & \pm R_{12} \\ \pm R_{21} & R_{22} \end{bmatrix}\begin{bmatrix} \frac{1}{2}I_1 \\ \frac{1}{2}I_2 \end{bmatrix} = \begin{bmatrix} V_1 \\ V_2 \end{bmatrix}.$$

So, the mesh currents are halved.

12. We have shown below a number of circuits (a), (b), (c) and (d). Can mesh analysis be applied to all of them? If not, why?

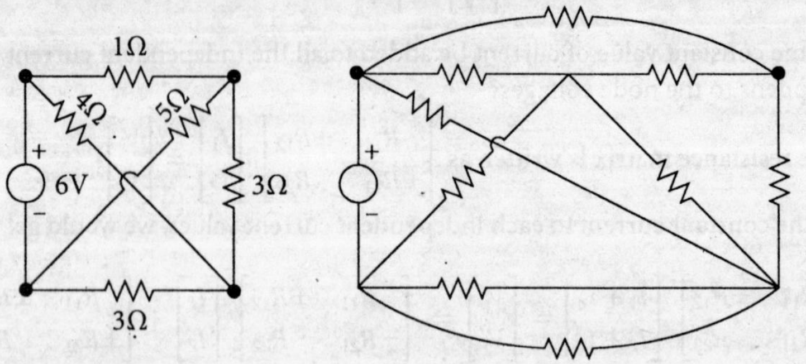

(a) (b)

Fig. 1.67

Ans. The circuit (a) shows a crossover but can be drawn in a plane without crossover, hence can be teated by mesh analysis. But the circuit (b) is a nonplanar circuit, because it cannot be drawn in a plane surface without crossover. Hence, there is no scope of mesh analysis technique to be applied.

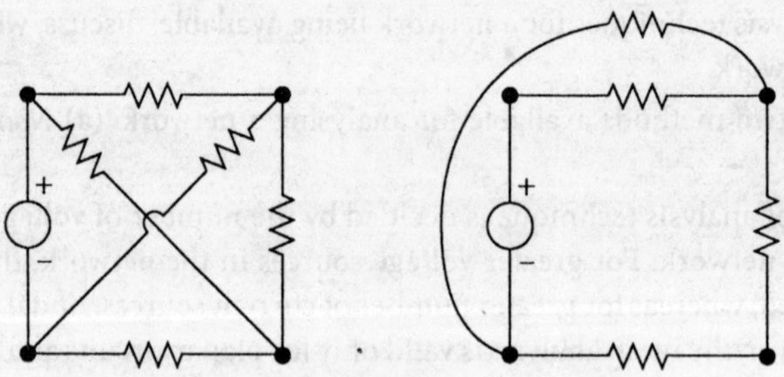

(c) (d)

Fig. 1.68

1.15 More Mixed Problems Solved (Taken from Univ. Exams.)

1. Using the superposition theorem for a network, find the current through 40Ω resistor in the circuit shown.

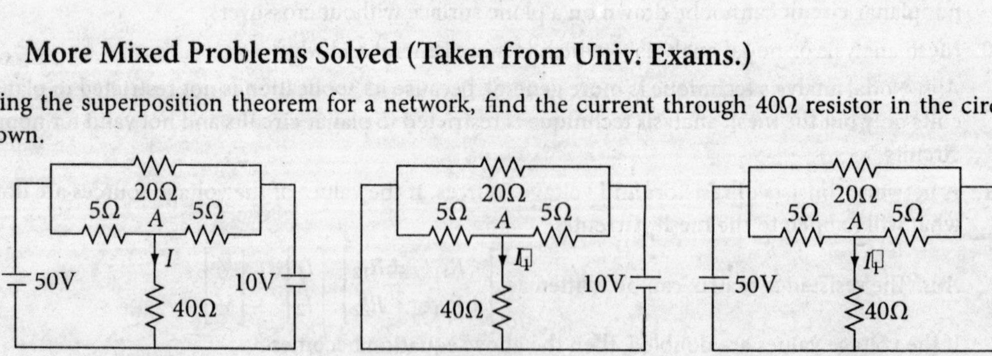

(a) (b) (c)

Fig. 1.69

Solution: According to [Fig. 1.69(a) and (b)], superposition theorem gives $I = I_1' + I_1''$.

To find I_1', the 50V source is replaced by its internal resistance and the equivalent circuit is drawn as shown in [Fig. 1.69(b)]. So, the total current drawn $I_1 = \frac{10}{R_t} = 1.56\text{A}$.

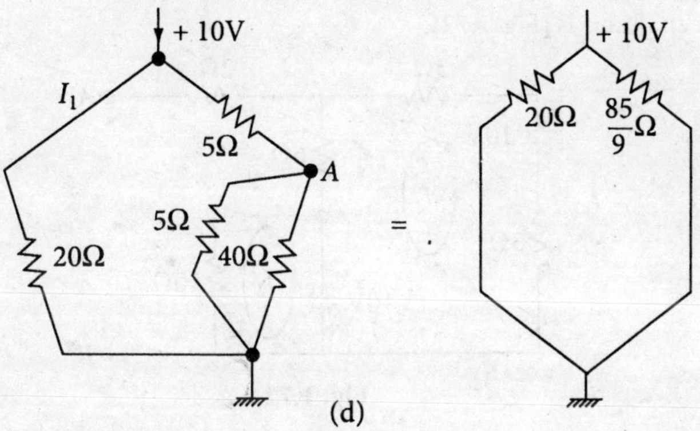

(d)

Fig. 1.70

The current through the right branch of the circuit

$$I_{1r} = I_1 \times \frac{20}{\left(5+\frac{40}{9}\right)+20} = 1.56 \times \frac{20\times 9}{265} = 1.06\text{A};\quad I_1' = 1.06 \times \frac{5}{45} = 0.12\text{A}.$$

To find I_1'' : The 10V source is replaced by its internal resistance (= 0) and the equivalent circuit is drawn [Fig. 1.69(c)]

$$R_e' = \frac{20\times\frac{85}{9}}{\left(20+\frac{85}{9}\right)} = \frac{20\times 85}{265} = 6.42\Omega.$$

$$I_t = \frac{50}{R_e'} = 7.79\text{A};\quad I_2' = 7.49 \times \frac{20}{\left(20+\frac{85}{9}\right)} = 5.09\text{A};\quad I_1'' = 5.09\times\frac{5}{45} = 0.57\text{A}.$$

$$\therefore\ I = I_1' + I_1'' = 0.12 + 0.57 = 0.69\text{A}.$$

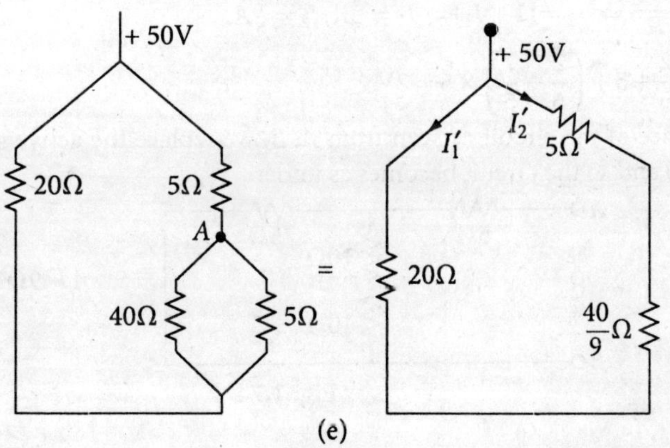

(e)

Fig. 1.71

2. Find the PD between A and B [Fig. 1.72].

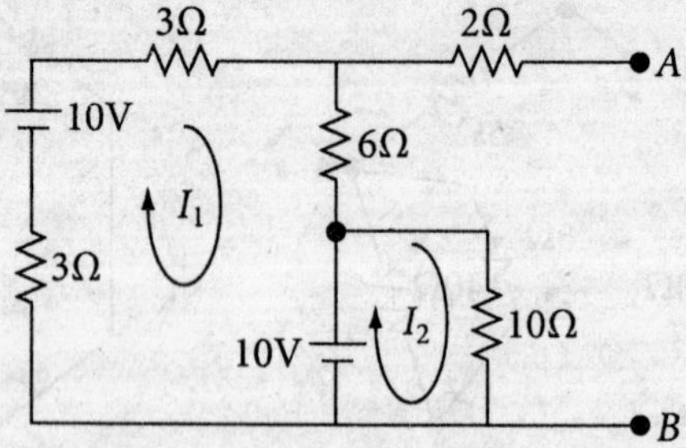

Fig. 1.72

Solution: Let I_1 and I_2 be the cyclic currents flow in the closed circuits [Fig. 1.72].
Then, $-3I_1 - 6I_1 - 3I_1 + (-10+10) = 0 \Rightarrow I_1 = 0$; $-10\times I_2 + 10 = 0$; $I_2 = 1$A.
$\therefore\ 10\times 1 + 6\times 0 = V_{AB} \Rightarrow V_{AB} = 10$V. The resistance of the network as seen from the terminals A, B replacing all the active sources of emf by their internal resistance $r_1 = 0$ is $2 + (6\|6)5\Omega = 3 + 2 = 5\Omega$.

3. Find V_{AB} [Fig. 1.73].

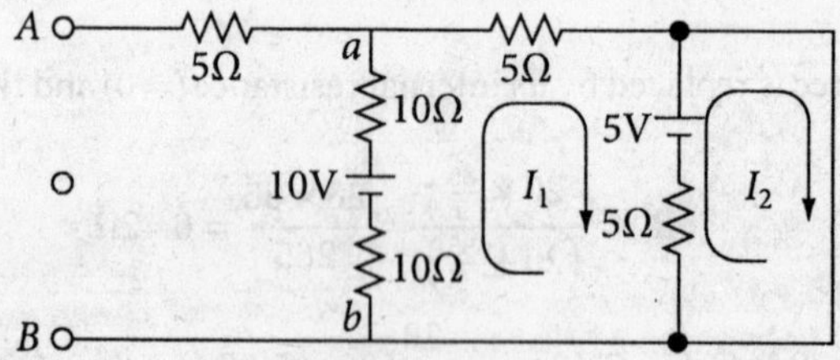

Fig. 1.73

Solution: $-10I_1 - 10I_1 - 5I_1 - 5(I_1 - I_2) + (10-5) = 0 \Rightarrow 25I_1 + 5(I_1 - I_2) = 5 \Rightarrow 6I_1 - I_2 = 1$.
Again, $(I_2 - I_1)\times 5 = 5 \Rightarrow I_2 = (1 + I_1) \Rightarrow I_1 = \frac{2}{5}A, I_2 = \frac{7}{5}$A.

So $V_{AB} = V_{ab} = 5\times\dfrac{2}{5} + 5 + \left(\dfrac{2}{5} - \dfrac{7}{5}\right)\times 5 = 2$V.

To find the resistance of the circuit as seen from A, B, we replace the active sources by their internal resistances (= zero) and so the circuit becomes as under:

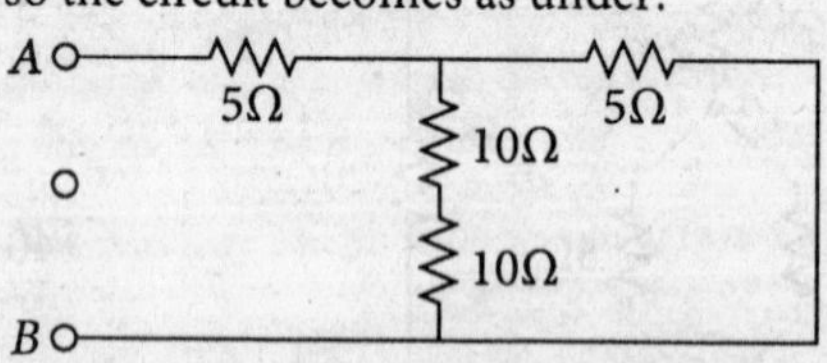

$\equiv 5 + 4 = 9\Omega$

Fig. 1.74

4. Find V_{AB} and the equivalent resistance of the network [Fig. 1.75].

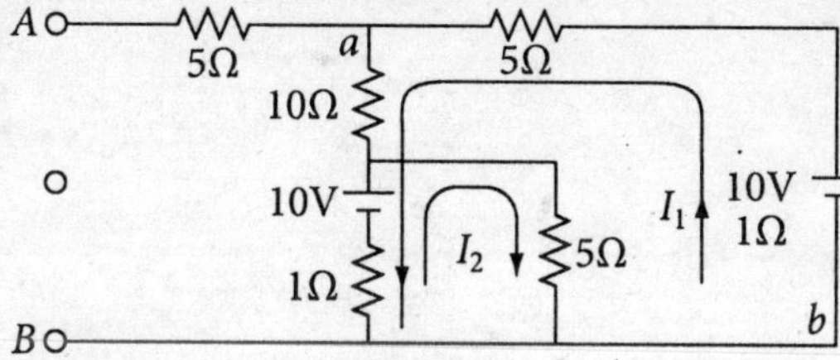

Fig. 1.75

Solution: Let I_1 and I_2 be the cyclic currents flowing in the closed meshes [Fig. 1.75].

Applying KVL, $-I_1\times 1 - I_1\times 5 - I_1\times 10 - (I_1 - I_2)\times 1 + (10-10) = 0 \Rightarrow 17I_1 = I_2.$

$$-I_a\times 5 - (I_1 - I_2)\times 1 + 10 = 0 \Rightarrow 4I_2 - I_1 = 10 \Rightarrow 68I_1 - I_1 = 10 \Rightarrow I_1 = \tfrac{10}{67} = 0.15.$$

$$V_{AB} = V_{ab} = 10 - 6\times 0.15 = 10 - 0.90 = 9.1\text{V}.$$

5. Applying Thévenin's theorem determine the current flowing in the branch *AB* of the circuit [Fig. 1.76].

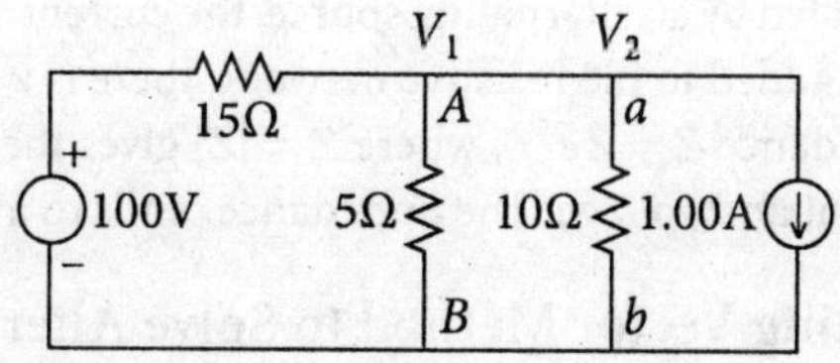

Fig. 1.76

Solution: Remove the resistor of 5Ω between *A*, *B* thus making them open-circuited. The resulting circuit is redrawn [Fig. 1.77].

$$i = \tfrac{100}{15+10} = 4\text{A} \Rightarrow i_1 = i - 1 = 3\text{A};\quad V_{AB} = V_{Th} = 30\text{V};\quad R_{Th} = 15\|10 = \tfrac{15\times 10}{25} = 6\Omega.$$

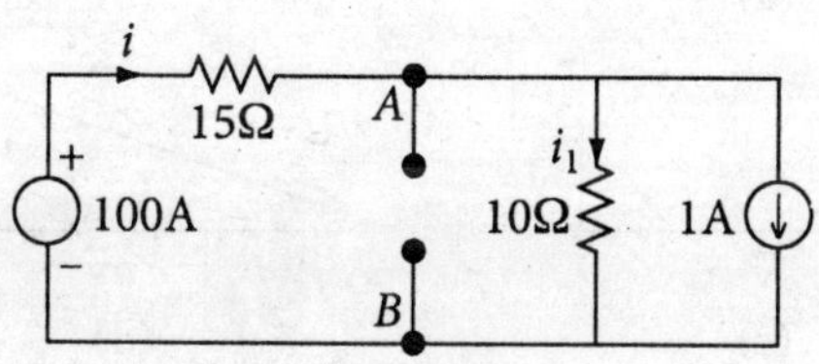

Fig. 1.77

So Thévenin's circuit is shown cf Fig. 1.78.

$$[\text{Fig. 1.78}] \rightarrow i_5 = \frac{V_{Th}}{6+5} = \frac{30}{11} = 2.72\text{A}.$$

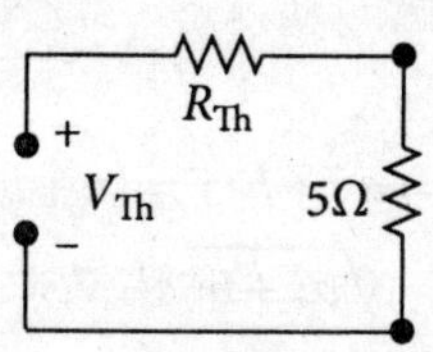

Fig. 1.78

Chapter 2

Complex Impedance

2.1 Introduction

Only in case of resistive network fed by an alternating source, the current and voltage are in phase. But if any of the other circuit elements be added to the resistive network, there is a phase difference between current and voltage. The complex impedance $Z = Ze^{j\phi}$, where $Z = |Z|$ gives the magnitude and ϕ represents the phase difference between current and voltage. The impedance tends to a maximum as $\phi \to 0$.

2.2 Essence of the Rotating Vector Method to Solve Alternating Current Problems

A unit vector $\vec{v}$ is shown in a specified direction [Fig. 2.1(a)].

A vector $a\vec{v}$ is shown, which is a vector a times as great in the same direction [Fig. 2.1(b)].

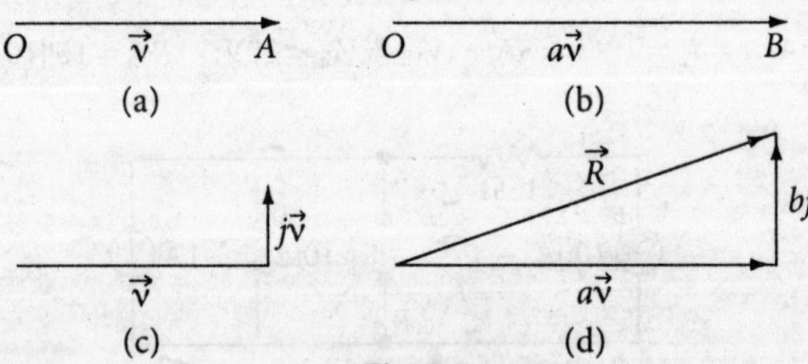

Fig. 2.1

We have shown a vector $j\vec{v}$ and let us look into its meaning [Fig. 2.1(c)].

$j^2(\vec{v}) = -\vec{v} = j\,(j\vec{v})$, so j^2 may be looked upon as an operator, which operating on a vector $\vec{v}$ reverses its direction. So j may be looked upon as an operator, which operating on the unit vector $\vec{v}$ turns it through an angle $(\pi/2)$ in the positive sense. Thus, $j\vec{v}$ is a unit vector perpendicular to the directions of $\vec{v}$.

By addition theorem,

$$\vec{R} = a\vec{v} + bj\vec{v} = (a + bj)\,\vec{v} \tag{2.1}$$

$$= \sqrt{a^2 + b^2}\, e^{j\phi}\vec{v}. \tag{2.2}$$

$a = c\cos\phi,\ b = c\sin\phi \Rightarrow (a + bj) = c(\cos\phi + j\sin\phi) = \sqrt{a^2 + b^2}\, e^{j\phi}$.

Thus, $\vec{R}$ is a vector of magnitude $c = \sqrt{a^2+b^2}$ and it makes an angle ϕ w.r.t. $\vec{v}$-direction such that $\tan\phi = \frac{b}{a}$.

Write $\vec{R} = \sqrt{a^2+b^2}\, e^{j\phi(t)}\vec{v} = \hat{R}\, e^{j\phi(t)}\vec{v}$.

It is a vector of magnitude $\sqrt{a^2+b^2} = \hat{R}$ and makes an angle ϕ at the time instant t and it is a function of time. So it is a rotating vector, because its inclination w.r.t. $\vec{v}$-direction varies with time.

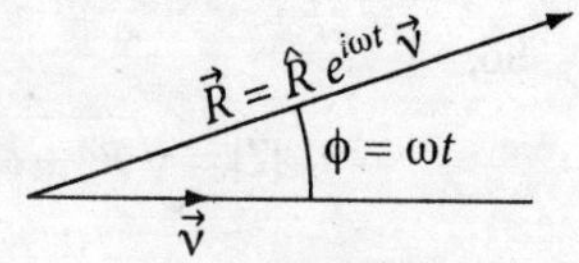

Fig. 2.2

If the vector rotates with constant angular speed ω, we have $\phi = \omega t$, $\vec{R} = \hat{R}\, e^{j\omega t}\vec{v}$ [Fig. 2.2].

Its resolution on the $\vec{v}$-axis denoted by $\vec{R}\cdot\vec{v}$ gives the instantaneous value of the rotating vector on the $\vec{v}$-axis.

$\therefore$ instantaneous value $= \vec{R}\cdot\vec{v} = \hat{R}\cos\omega t$; so in this formalism, any alternating voltage $v = \hat{V}\cos\omega t$ can be represented by a rotating vector of magnitude $\hat{V}$ rotating with angular speed ω and at some instant t, its projection on the direction of $\vec{v}$ [Fig. 2.3] gives its instantaneous value. If on the other hand, $\phi = \left(\frac{\pi}{2} - \omega t\right)$, then

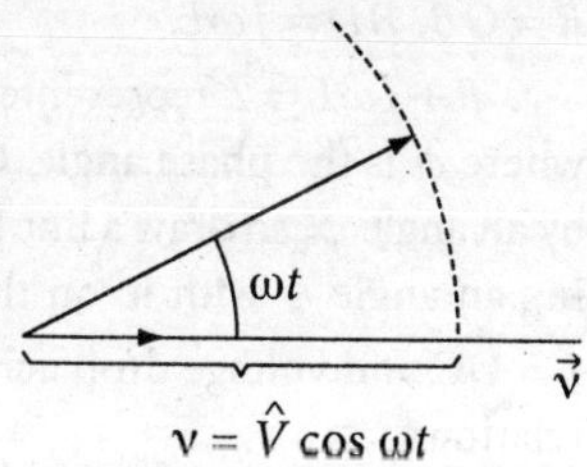

Fig. 2.3

$$v = \hat{V}\sin\omega t = \hat{V}\, e^{j\left(\frac{\pi}{2}-\omega t\right)}\vec{v}\cdot\vec{v}. \tag{2.3}$$

Differentiation of a rotating vector

Let the rotating vector be represented by $\vec{R} = \hat{R}\, e^{j\omega t}\vec{v}$.

$$\therefore\ \frac{d\vec{R}}{dt} = j\omega\hat{R}\, e^{j\omega t}\vec{v} + \hat{R}\, e^{j\omega t}\underset{\substack{\| \\ \odot}}{\frac{d\vec{v}}{dt}} \Rightarrow \frac{d\vec{R}}{dt} = j\omega\vec{R}. \tag{2.4}$$

Integration of a rotating vector

$$\vec{R} = \hat{R}\, e^{j\omega t}\vec{v} \Rightarrow \int \vec{R}\,dt = \hat{R}\vec{v}\int e^{j\omega t}\,dt = \left(\frac{\hat{R}}{j\omega}\right) e^{j\omega t}\vec{v} \Rightarrow \boxed{\int \vec{R}\,dt = \frac{1}{j\omega}\vec{R}}. \tag{2.5}$$

2.3 LR Circuit

The instantaneous circuit emf equation is written as

$$L\frac{di}{dt} + Ri = v = \hat{V}\cos\omega t$$

Fig. 2.4

Let the alternating voltage be $\vec{V} = \hat{V}\, e^{j\omega t}\vec{v}$. The current flowing in the circuit is also alternating in nature, so we can represent it by suitable rotating vector $\vec{I}$ such that the instantaneous value i is written as $i = \vec{I}\cdot\vec{v}$.

Transforming the instantaneous emf equation in terms of rotating vector language, we get

$$j\omega L\vec{I}\cdot\vec{v} + R\vec{I}\cdot\vec{v} = \hat{V}\, e^{j\omega t}\vec{v}\cdot\vec{v} - \vec{V}\cdot\vec{v} \Rightarrow (R + j\omega L)\vec{I}\cdot\vec{v} = \vec{V}\cdot\vec{v} \Rightarrow \vec{I} = \frac{\vec{V}}{(R + j\omega L)} = \frac{\vec{V}}{Z}. \tag{2.6}$$

$Z =$ complex impedance

$= (R + j\omega L) = (a + bj)$

$= \sqrt{R^2 + \omega^2 L^2}\, e^{j\phi}$, where $\tan\phi = \frac{\omega L}{R} \Rightarrow \phi = \tan^{-1}\left(\frac{\omega L}{R}\right)$.

So,

$$|Z| = \sqrt{R^2 + \omega^2 L^2} \angle\phi \Rightarrow \vec{I} = \left(\frac{\hat{V}}{\sqrt{R^2 + \omega^2 L^2}}\right) e^{j(\omega t - \phi)}\vec{v} \Rightarrow i = \hat{I}\cos(\omega t - \phi). \quad (2.7)$$

Phasor diagram

Let us take voltage directions as reference. Take R along the x-axis, it happens to be also the axis of voltage. $R = OB$, $BD = j\omega L$.

$\therefore$ $R + j\omega L = Z$ represented by OD [Fig. 2.5]; $\tan\phi = \frac{\omega L}{R}$, where ϕ is the phase angle. Current lags behind the voltage by an angle ϕ, so draw a line below the voltage axis and making an angle ϕ with it; on this line mark off current vector $\vec{I} = \overrightarrow{OC}$ and voltage drop across R, $\vec{V_R} = \overrightarrow{OG}$. Then from the relation,

$$\vec{V} = \vec{V_R} + \vec{V_L} = \vec{I}(R + j\omega L)$$

$$\Rightarrow \vec{V_L} = j\omega L\vec{I} = (0 + bj)\,\vec{I} = \sqrt{0^2 + \omega^2 L^2}\, e^{j\theta}\,\vec{I}$$

$$= \omega L e^{j\theta}\,\vec{I}, \text{ where } \tan\theta = \frac{b}{0} = \infty \Rightarrow \theta = \frac{\pi}{2}.$$

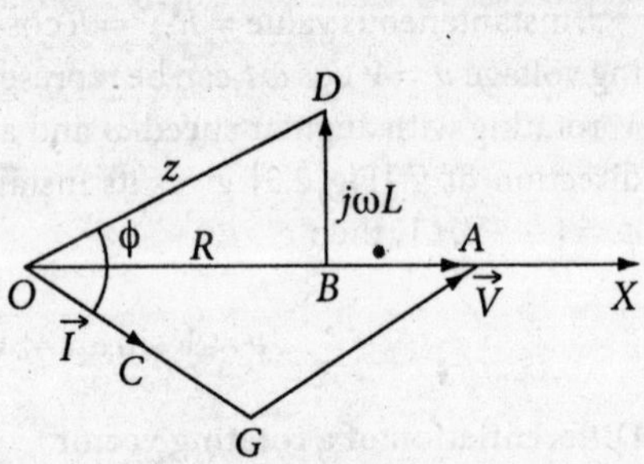

Fig. 2.5

Thus, the voltage drop across L can be represented by a suitable rotating vector $\vec{V_L}$ in quadrature with the current rotating vector $\vec{I}$. Thus, w.r.t. [Fig. 2.5], we have $V = v\angle O$, $Z = |Z| = \sqrt{R^2 + \omega^2 L^2}\,\angle + \phi$.

$$\therefore\ I = \frac{V}{Z} = \frac{\hat{V}\angle O}{\sqrt{R^2 + \omega^2 L^2}\angle\phi} = \frac{\hat{V}}{\sqrt{R^2 + \omega^2 L^2}}\angle - \phi \Rightarrow i = \frac{\hat{V}}{\sqrt{R^2 + \omega^2 L^2}}\cos(\omega t - \phi). \quad (2.8)$$

Example 2.1. *A circuit contains a resistor of* 1kΩ, *an inductor* 50mH. *A* 10V (rms), 10kHz *signal is applied to it. Find impedance* Z, *phase angle* ϕ, V_R, *and* V_L.

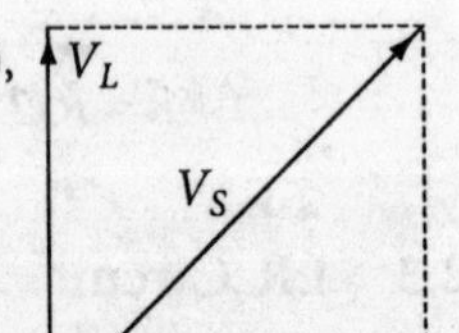

Fig. 2.6

Solution: $X_L = \omega L = 2\pi f L = 2\pi \times 10^4 \times 50 \times 10^{-3} = 3142\Omega$

$Z = 1000 + j3142$

$|Z| = \sqrt{10^6 + 3142^2} = 3297\Omega$

$I = \frac{V}{Z} = \frac{10}{3297} = 3.03\text{mA}.$

Phase angle $\phi = \tan^{-1}\left(\frac{V_L}{V_R}\right) = \tan^{-1}\left(\frac{\omega L}{R}\right) = \tan^{-1}\left(\frac{3142}{10^3}\right) = 72.34°$.

So, $Z = 3297\angle 72.34$

$$I = \frac{V}{Z} = \frac{10\angle 0°}{3297\angle 72.34} = 3.03 \times 10^{-3}\angle - 72.34.$$

Current lags behind the voltage by an angle equal to 72.34°.

$$V_L = X_L I = 3142 \times 3.03 \times 10^{-3}\angle - 72.34° = 9.52\angle - 72.34°$$

$$V_R = IR = 1 \times 10^{+3} \times 3.03 \times 10^{-3}\angle - 72.34° = 3.03\angle - 72.34°.$$

Example 2.2. *In the circuit* [Fig. 2.7], $V=20\angle 0$, 50Hz, $L=0.1$H. *A* 20Ω *resistor is in parallel to the reactance of the inductance* $X_L=j\omega L$; 10Ω *is in series with this parallel combination. Find* (i) Z (ii) I (iii) ϕ.

Solution: $Z=10+\dfrac{20\times j\omega L}{20+j\omega L}$.

$X_L=\omega L=2\pi f L=6.28\times 50\times 0.1=31.40\Omega$.

$$\Rightarrow Z=10+\frac{20\times j31.40}{20+j31.40}=10+\frac{628\angle 90°}{37.23\angle 57.5°}=10+16.87\angle 32.5°=10+Z'.$$

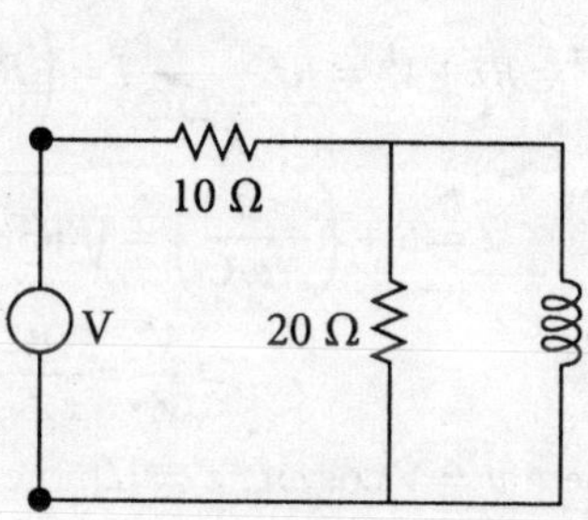

Fig. 2.7

$Z'=a+bj$; $\frac{b}{a}=\tan 32.5$.

$\therefore\ a^2+b^2=16.87^2$; $\frac{b}{a}=0.64\Rightarrow a^2+0.64^2a^2=16.87^2\Rightarrow a=14.21\Rightarrow b=0.64a=0.64\times 14.21=9.09$.

$Z=10+14.21+9.09j=24.21+j9.09=25.87\angle 20.5°$.

$$I=\frac{V}{Z}=\frac{20\angle 0}{25.87\angle 20.5}=0.77\angle -20.5°.$$

So, the current value is 0.77A lagging behind the voltage by an angle $\phi=20.5°$.

2.4 CR Circuit

The applied voltage is represented by a suitable rotating vector $\vec{V}$ such that the instantaneous value is given by $v=\hat{V}\cos\omega t=\hat{V}e^{j\omega t}\vec{v}\cdot\vec{v}=\vec{V}\cdot\vec{v}$.

The current can be represented by suitable rotating vector $\vec{I}$. The emf equation can be written as

$$Ri+\frac{q}{C}=v\Rightarrow R\vec{I}\cdot\vec{v}+\frac{1}{C}\int i\,dt=\vec{V}\cdot\vec{v}\Rightarrow R\vec{I}\cdot\vec{v}+\frac{1}{j\omega C}\vec{I}\cdot\vec{v}=\vec{V}\cdot\vec{v}$$

$$\Rightarrow\vec{I}\left(R+\frac{1}{j\omega C}\right)=\vec{V}\Rightarrow\vec{I}=\frac{\vec{V}}{R+\frac{1}{j\omega C}}=\frac{\vec{V}}{R-\frac{j}{\omega C}}.$$

$$Z=R-\frac{j}{\omega C}=a-bj=\sqrt{a^2+b^2}\,e^{-j\theta},$$

$$\tan\theta=\frac{b}{a}=\sqrt{R^2+\frac{1}{\omega^2C^2}}\,e^{-j\theta}=Z\angle -\theta,$$

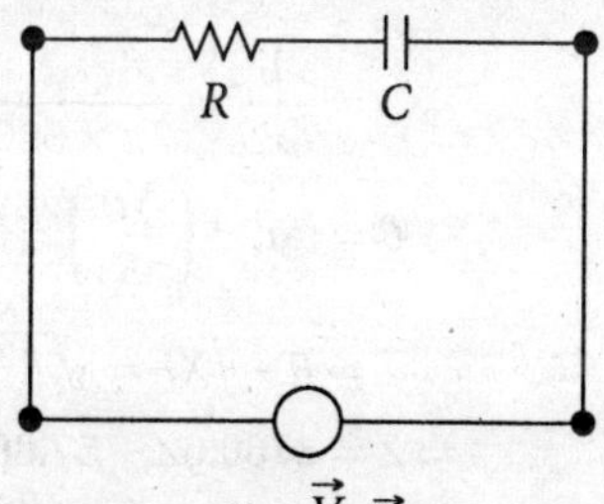

Fig. 2.8

where $Z=|Z|=\sqrt{R^2+\frac{1}{\omega^2C^2}}$.

$$\vec{I}=\frac{\vec{V}}{Z}=\frac{\hat{V}e^{j(\omega t)}\vec{v}}{\sqrt{R^2+\frac{1}{\omega^2C^2}}\,e^{-j\theta}}=\frac{\hat{V}}{\sqrt{R^2+\frac{1}{\omega^2C^2}}}e^{j(\omega t+\theta)}\vec{v}=\hat{I}e^{j(\omega t+\theta)}\vec{v}.$$

$$i=\hat{I}\cdot\vec{v}=\hat{I}\cos(\omega t+\theta) \quad (2.9)$$

$$=\hat{I}\angle\theta. \quad (2.10)$$

So, the current leads over the voltage by an angle $\theta=\tan^{-1}\left(\dfrac{1}{\omega CR}\right)$.

Phasor diagram

Let us represent the vector diagram w.r.t. $\vec{V}$-direction [Fig. 2.9]

$$\vec{V} = R\vec{I} + \vec{V}_C = R\vec{I} - \frac{1}{\omega C}\vec{I} = \left(R - \frac{j}{\omega C}\right)\vec{I} \Rightarrow \vec{V}_C = \left(-\frac{j}{\omega C}\right)\vec{I}.$$

$$Z = R + \left(-\frac{j}{\omega C}\right) = \sqrt{R^2 + \frac{1}{\omega^2 C^2}}\, e^{-j\theta} = |Z|\angle -\theta.$$

$$I = \frac{V}{Z} = \frac{v\angle 0}{z\angle -\theta} = \frac{v}{z}\angle\theta,$$

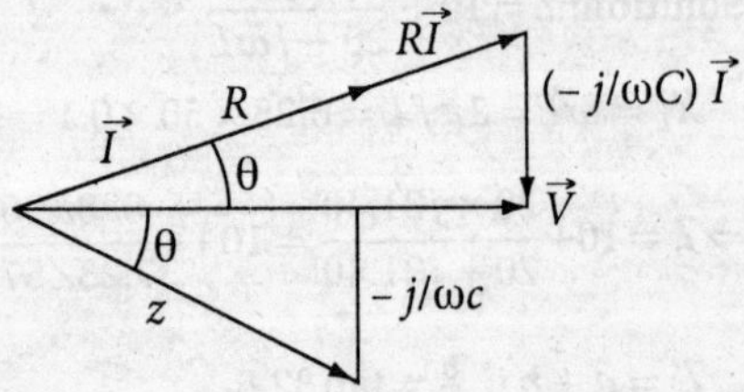

where $v = \hat{V}\cos\omega t$, $z = |Z|$, the current leads the voltage by an angle $\theta = \tan^{-1}\left(\frac{1}{\omega CR}\right)$.

Fig. 2.9

Example 2.1. *The circuit* [Fig. 2.10] is *fed by a sine wave generator of output* V = 10V, 50Hz *through a* 2kΩ *resistor and* 0.1μF *capacitor, all in series. Determine the total impedance* (i) *Z, current I, phase angle θ*, V_C *and* V_R.

Solution: $V = 10\angle 0°$.

$$Z = \sqrt{R^2 + \frac{1}{\omega^2 C^2}}\angle -\theta = |Z|\angle -\theta.$$

$$z = \sqrt{R^2 + \frac{1}{\omega^2 C^2}} = \sqrt{(2\times 10^3)^2 + \frac{1}{(2\pi\times 500\times 0.1\times 10^{-6})^2}}$$

$$= 10^3\sqrt{4 + \frac{10^6}{(2\pi\times 500\times 0.1)^2}} = 10^3\times 3.7606\Omega.$$

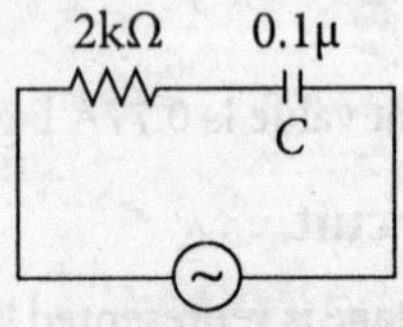

Fig. 2.10

$$X_C = \frac{1}{\omega C} = \frac{1}{2\pi\times 500\times 0.1\times 10^{-6}} = 3183.09\Omega$$

$$\theta = \tan^{-1}\left(\frac{X_C}{R}\right) = 57.86°$$

$$Z = R - jX_C = \sqrt{R^2 + X_C^2}\, e^{-j\theta} = \sqrt{(2\times 10^3)^2 + (3.183\times 10^3)^2}\, e^{-j\theta} = 3.7606\times 10^3 e^{-j\theta}$$

$$\Rightarrow Z = 3760.6\angle -57.86°$$

$$I = \frac{V\angle 0}{Z} = \frac{10\angle 0°}{3760.6\angle -57.86°} = 2.66\times 10^{-3}\angle 57.86°.$$

$$V_R = IR = 2.66\times 10^{-3}\times 2\times 10^3\angle 57.86° = 5.32\angle 57.86° \tag{2.11}$$

$$\vec{V}_C = \left(-\frac{j}{\omega C}\right) I = X_C e^{-j\pi/2}\vec{I} = 2.66\times 10^{-3}\times 3183.09\angle -90°$$

$$V_C = 8.47\angle -90°. \tag{2.12}$$

2.5 LCR Circuit

All the circuit elements are connected in series [Fig. 2.11]

$$v = \vec{V}\cdot\vec{v},\ i = \vec{I}\cdot\vec{v}$$

The emf equation at any instant is given by

$$L\frac{di}{dt}+\frac{q}{C}+Ri=v=\vec{V}\cdot\vec{v}$$

$$\Rightarrow j\omega L\vec{I}\cdot\vec{v}+\frac{1}{j\omega C}\vec{I}\cdot\vec{v}+R\vec{I}\cdot\vec{v}=\vec{V}\cdot\vec{v}$$

$$\Rightarrow\left[R+j\left(\omega L-\frac{1}{\omega C}\right)\right]\vec{I}\cdot\vec{v}=\vec{V}\cdot\vec{v}$$

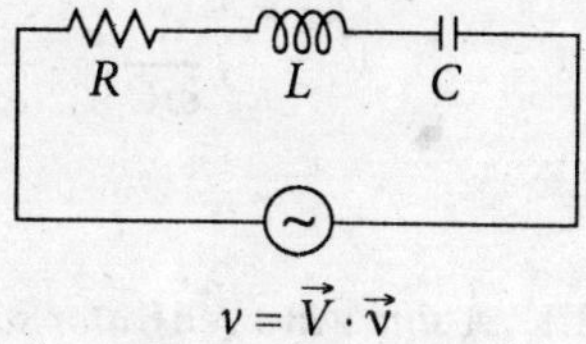

Fig. 2.11

$$\Rightarrow\vec{I}=\frac{\vec{V}}{R+j\left(\omega L-\frac{1}{\omega C}\right)}=\frac{\vec{V}}{Z}=\frac{\hat{V}e^{j\omega t}\vec{v}}{\sqrt{R^2+\left(\omega L-\frac{1}{\omega C}\right)^2}\,e^{j\theta}}=\frac{\hat{V}e^{j(\omega t-\theta)}\vec{v}}{\sqrt{R^2+\left(\omega L-\frac{1}{\omega C}\right)^2}}=\hat{I}\,e^{j(\omega t-\theta)}\vec{v}$$

$$i=\vec{I}\cdot\vec{v}=\hat{I}\cos(\omega t-\theta)=\frac{\hat{V}}{\sqrt{R^2+\left(\omega L-\frac{1}{\omega C}\right)^2}}\cos(\omega t-\theta). \tag{2.13}$$

$$Z=R+j\left(\omega L-\frac{1}{\omega C}\right)=R+\underbrace{j\omega L}_{\text{inductive reactance}}+\underbrace{(-j/\omega C)}_{\text{capacitive reactance}}.$$

$$\tan\theta=\frac{\omega L-\frac{1}{\omega C}}{R}$$

Vector or Phasor diagram

The direction of voltage rotating vector $\vec{V}$ is taken as reference and then the vector diagram representing the different physical quantities like Z, $\vec{I}$, θ, $\vec{V}_R$, $\vec{V}_L$, $\vec{V}_C$, etc. are in Fig. 2.12. When $X_L>X_C$, $\overrightarrow{OA}=\vec{V}$, $OP=R$

$$PQ=j\omega L=X_L, QS=-j/\omega C=X_c$$

$$\Rightarrow OS=Z=R+j\left(\omega L-\frac{1}{\omega C}\right)$$

$$\tan\theta=\frac{\omega L-\frac{1}{\omega C}}{R},$$

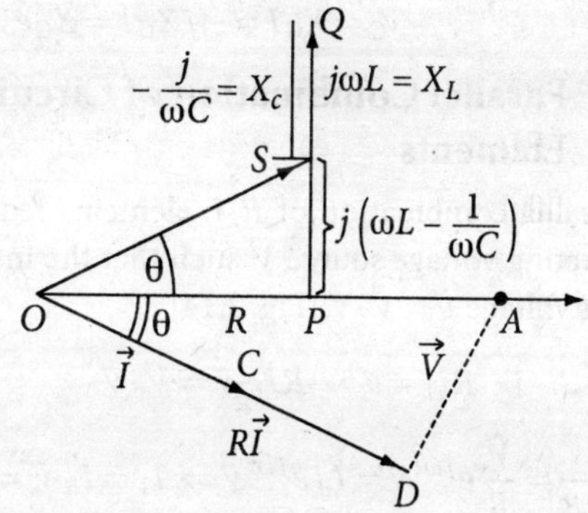

Fig. 2.12

θ is called the *phase angle.* Draw a line making an angle θ below the reference voltage line OA, $\overrightarrow{OC}=\vec{I}$, $\overrightarrow{OD}=R\vec{I}$

$$\overrightarrow{OA}=\vec{V}=\overrightarrow{OD}+\overrightarrow{DA}=R\vec{I}+\overrightarrow{DA}\Rightarrow R\vec{I}+j\left(\omega L-\frac{1}{\omega C}\right)\vec{I}=\vec{V}=R\vec{I}+\overrightarrow{DA}$$

$$\Rightarrow\overrightarrow{DA}=j\left(\omega L-\frac{1}{\omega C}\right)\vec{I}=\underbrace{j\omega L\vec{I}}_{\vec{V}_L}+\underbrace{(-j/\omega C)\vec{I}}_{\vec{V}_C}$$

$$\vec{V}_L=j\omega L\vec{I}=\omega Le^{j\pi/2}\vec{I};\quad V_L=\omega LI\angle 90^\circ. \tag{2.14}$$

$$\vec{V}_C = -j\frac{1}{\omega C}\vec{I} = \frac{1}{\omega C}e^{-j\pi/2}\vec{I}; \quad V_C = \frac{1}{\omega C}I\angle -90°; \quad V_S = V\angle 0°.$$

$$I = \frac{V\angle 0°}{Z\angle\theta} = \left(\frac{V}{Z}\right)\angle -\theta. \tag{2.15}$$

Example 2.1. *A sine wave generator supplies 50V, 50Hz signal to a series circuit* [Fig. 2.13]. *Determine the impedance Z, the line current and the phase angle θ in the given circuit and also the voltage across each element V_R, V_L, V_C.*

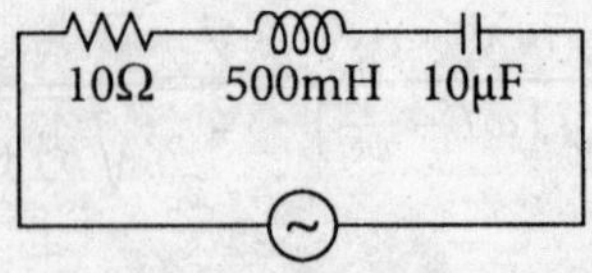

Fig. 2.13

Solution: $V = 50\angle 0°, 50\text{Hz}$. $X_L = 2\pi \times 50 \times 0.5 = 157\Omega$.

$$X_C = \frac{1}{\omega C} = \frac{1}{2\pi \times 50 \times 10 \times 10^{-6}} = 318.3\Omega.$$

$$Z = R + j\left(\omega L - \frac{1}{\omega C}\right) = 10 + j(157 - 318.3)$$

$$= 10 - j161.3 = 161.6\angle -86.45°.$$

$$I = \frac{V\angle 0°}{Z} = \frac{50}{161.6}\angle 86.45 = 0.3\angle 86.45°.$$

Current leads the voltage by an angle 86.45°

$$V_R = IR = 0.3 \times 10\ \angle 86.45° = 3\angle 86.45°.$$

$$V_L = jX_L I = I \times 157e^{j\pi/2} = I \times 157\angle 90° = 0.3 \times 157\angle 90° = 47.1\angle 90°.$$

$$V_C = X_C I = -jX_C I = X_C e^{-j\pi/2} I = 318.3 \times 0.3\angle -90° = 95.5\angle -90°.$$

2.6 Parallel Combination of Circuit Elements

A parallel combination of R, L elements fed by an alternating voltage source $\vec{V}$ such that the instantaneous voltage $v = \vec{V}\cdot\vec{v}$ [Fig. 2.14]

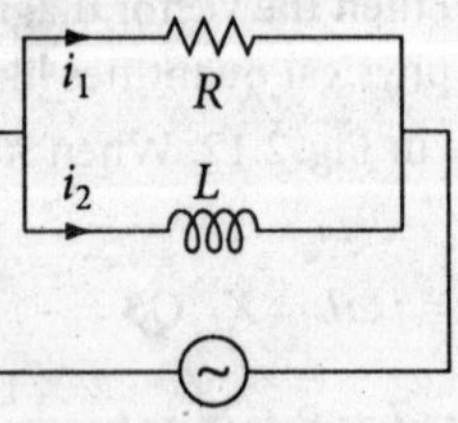

Fig. 2.14

$$Ri_1 = v \Rightarrow R\vec{I}_1\cdot\vec{v} = \vec{V}\cdot\vec{v}.$$

$$\vec{I}_R = \frac{\vec{V}}{R} = \frac{\hat{V}}{R}e^{j\omega t}\vec{v} = \hat{I}_R e^{j\omega t}\vec{v} \Rightarrow i_1 = \vec{I}_R\cdot\vec{v} = \hat{I}_R\cos\omega t = \hat{I}\angle 0°.$$

Similarly,

$$\vec{I}_L = \frac{\vec{V}}{j\omega L} = \frac{\hat{V}e^{j(\omega t)}\vec{v}}{\omega L e^{j\pi/2}} = \frac{\hat{V}}{\omega L}e^{j(\omega t - \frac{\pi}{2})}\vec{v} \Rightarrow i_2 = \vec{I}_2\cdot\vec{v} = \frac{\hat{V}}{\omega L}\cos\left(\omega t - \frac{\pi}{2}\right) = \frac{\hat{V}}{\omega L}\angle -\pi/2.$$

$$I = \vec{I}_R + \vec{I}_L = \frac{\vec{V}}{R} + \frac{\vec{V}}{j\omega L} = \vec{V}\left(\frac{R + j\omega L}{j\omega LR}\right) = \frac{\sqrt{R^2 + \omega^2 L^2}\,e^{-j\phi}}{\omega L R e^{j\pi/2}}\vec{V} = \frac{\sqrt{R^2 + \omega^2 L^2}}{\omega LR}\vec{V}e^{-j(\phi + \frac{\pi}{2})}.$$

$$i = \hat{I}\angle -(\pi/2 + \phi)°, \ \tan\phi = \frac{\omega L}{R}. \tag{2.16}$$

Example 2.1. *The circuit shows a parallel combination of R and L fed by an alternating source. Find the total impedance, line current* [Fig. 2.15].

Solution:

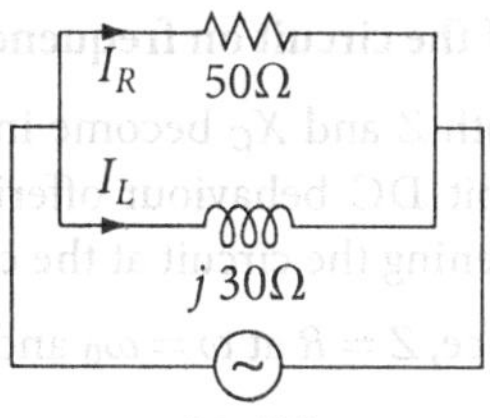

Fig. 2.15

$$I_R = \frac{V}{R} = \frac{20\angle 0°}{50} = 0.4\angle 0°$$

$$I_L = \frac{V\angle 0°}{j30} = \frac{20\angle 0°}{30e^{j\pi/2}} = 0.66\angle -\frac{\pi}{2}.$$

Total current

$$I_T = I_R - j0.66 = 0.41 - 0.66j = 0.77\angle -58\cdot 2°$$

$$Z = \frac{V\angle 0°}{0.77\angle -58.2} = \frac{20\angle 0°}{0.77\angle -58.2} = 25.99\angle 58.2°.$$

2.7 Series and Parallel Resonant Circuit

Series resonant circuit is the series combination of circuit elements R, L and C giving maximum current flow in the circuit.

Parallel resonant or anti-resonant circuit is a parallel combination of R, L and C such that the current flowing through the circuit is a minimum. We will now analyse the two circuits individually in details.

Series resonant circuit

A series combination is of R, L, C [Fig. 2.16] and we have done its rotating vector treatment earlier to get the expression for the current flow as

$$I = \frac{V}{\sqrt{R^2 + (X_L - X_C)^2}}\angle -\theta, \text{ for } X_L > X_C$$

$$Z = \sqrt{R^2 + (X_L - X_C)^2} = \frac{1}{Y},$$

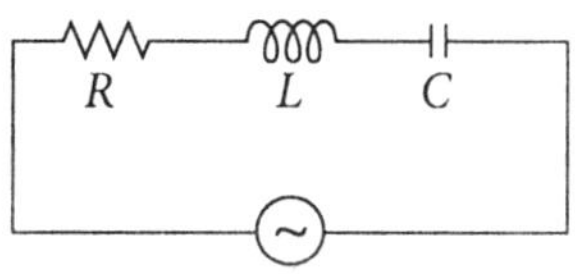

Fig. 2.16

where Y is called the *admittance of the circuit.* When Z is maximum, Y will be minimum and acceptance of current by the circuit becomes minimum.

Resonance is, in general, the condition when the current and voltage are in phase. Let us see what such a condition implies.

The phase difference between the voltage and current is given by

$$\tan\theta = \frac{\omega L - \frac{1}{\omega C}}{R}.$$

For zero phase difference $\theta = 0 \Rightarrow \omega L - \frac{1}{\omega C} = 0 \Rightarrow X_L = X_C$. Implications = ?

1. Z is minimum.
2. Y is maximum.
3. Current maximum is accepted by the circuit.
4. Resonant frequency of the series resonant circuit is given by

$$(X_L)_{\omega_0} = (X_C)_{\omega_0} \Rightarrow \omega_0 = \frac{1}{\sqrt{LC}} \Rightarrow f_0 = \frac{1}{2\pi\sqrt{LC}}.$$

Behaviour of the circuit on frequency

At $\omega = 0$, both Z and X_C become infinitely large and $X_L = 0$, because at this frequency, the voltage and current exhibit DC behaviour offering infinitely high capacitive reactance and zero inductive reactance thus as if, opening the circuit at the capacity point and shorting the circuit at the inductance point.

At resonance, $Z = R$ at $\omega = \omega_0$ and at lower values (of $\omega < \omega_0$), Z decreases along with X_C.

Z* vs. *ω

The impedance of the circuit is dependent on the frequency [Fig. 2.17], X_L vs. ω is a straight line having slope equal to L; X_C vs. ω is a rectangular hyperbola. At resonance $\omega = \omega_0$, $X_L = X_C$ shown as $PQ = PR$, Z becomes equal to R, which is the minimum value of Z.

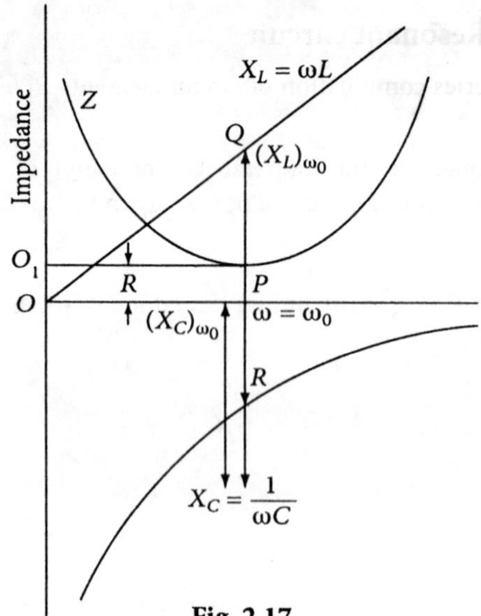

Fig. 2.17

The variation of phase angle θ with frequency [Fig. 2.18]. At frequencies lower than the resonant frequency, the phase angle θ decreases and becomes equal to zero at $\omega = \omega_0$.

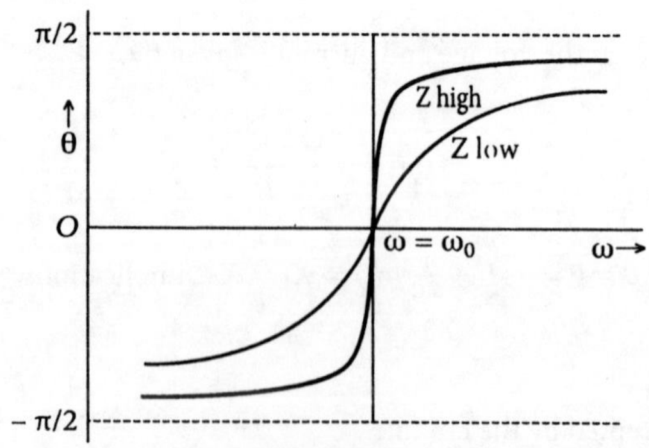

Fig. 2.18

In the frequency domain, $\omega > \omega_0$, the angle θ increases and attains a value +90°.

When $1/\omega C > \omega L$, θ is negative and current leads over the voltage for $\omega \rightarrow 0$, $\theta \rightarrow -\pi/2$.

$\tan\theta = \dfrac{\omega L - \frac{1}{\omega C}}{R}$, when $1/\omega C$ from greater value decreases down to ωL, θ becomes $-\pi/2$,

$$1/\omega C > \omega L \Rightarrow \omega < \omega_0.$$

For $\omega > \omega_0$, $\theta \to \pi/2$ for large value of ω [cf. Fig. 2.18].

Resonance curve (series resonance)

$I_r = \dfrac{V_r}{\sqrt{R^2 + \left(\omega L - \frac{1}{\omega C}\right)^2}}$, where I_r and V_r are the rms values of current and voltage.

At resonance, $\omega = \omega_0$,

$$(I_r)_{\max} = \frac{V_r}{R} \Rightarrow \frac{I_r^2}{(I_r)_{\max}^2} = \frac{R^2}{R^2 + \left(\omega L - \frac{1}{\omega C}\right)^2}.$$

For $I_r = \frac{1}{\sqrt{2}} \cdot (I_r)_{\max}$,

$$\frac{1}{2} = \frac{R^2}{R^2 + \left(\omega L - \frac{1}{\omega C}\right)^2} \Rightarrow \left(\omega L - \frac{1}{\omega C}\right)^2 = R^2 \Rightarrow \left|\left(\omega L - \frac{1}{\omega C}\right)\right| = R. \tag{2.17}$$

(2.17) is taken as the series resonance condition. It is equivalent to two equations:

$$\left(\omega L - \frac{1}{\omega C}\right) = R \tag{2.18}$$

$$\text{and } \left(\frac{1}{\omega C} - \omega L\right) = R. \tag{2.19}$$

(2.18) gives us $\omega^2 LC - 1 - \omega CR = 0$

$$\Rightarrow \omega = \frac{CR \pm \sqrt{C^2R^2 + 4LC}}{2LC} = \frac{R}{2L} \pm \sqrt{\frac{1}{LC} + \frac{R^2}{4L^2}}. \tag{2.20}$$

Since ω is to be a positive quantity, we get one solution for the resonance (2.17) as

$$\omega_2 = \frac{R}{2L} + \sqrt{\frac{1}{LC} + \left(\frac{R^2}{4L^2}\right)}. \tag{2.21}$$

(2.19) gives us $\omega^2 LC + \omega CR - 1 = 0$

$$\Rightarrow \omega = \frac{-CR \pm \sqrt{C^2R^2 + 4LC}}{2LC} \Rightarrow \omega = -\frac{R}{2L} \pm \sqrt{\frac{1}{LC} + \frac{R^2}{4L^2}}.$$

For the same reason the plausible solution of the resonance (2.19) is written as

$$\omega_1 = -\frac{R}{2L} + \sqrt{\left(\frac{R^2}{4L^2}\right) + \frac{1}{LC}}. \tag{2.22}$$

So $\omega_2 > \omega_1$.

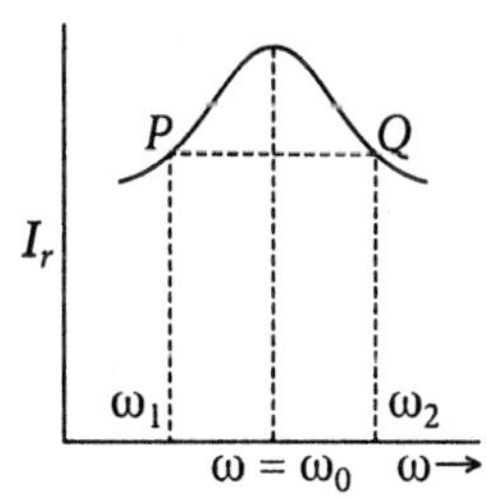

Fig. 2.19

The behaviour of variation of current vs. $\omega(-2\pi f)$ [Fig. 2.19], which shows that at two frequencies ω_1 and ω_2, the current in the SRC (series resonant circuit) drops to $\frac{1}{\sqrt{2}}$ times its resonant value. The bandwidth is defined by $\Delta\omega = \omega_2 - \omega_1 = R/L$.

2.8 Parallel Resonant Circuit or Anti-resonant Circuit

A series combination of resistance (may be the resistance of the coil) and inductance coil is shunted by a capacitor and an alternating voltage source is placed across the parallel combination [Fig. 2.20]. R_C is a measure of the imperfectness of the capacitor

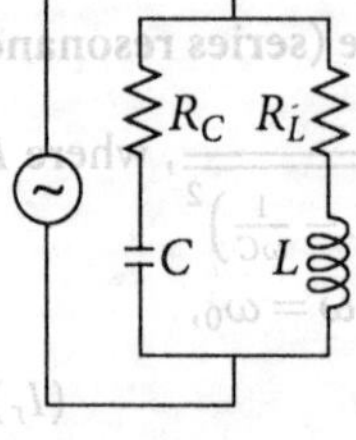

Fig. 2.20

$$\frac{1}{Z} = \frac{1}{Z_{RL}} + \frac{1}{Z_{RC}} \Rightarrow Y = \frac{1}{R_L + j\omega L} + \frac{1}{R_C - \frac{j}{\omega C}} = Y_{RL} + Y_{RC},$$

where Y is the vector admittance.

$$Y_{RL} = \frac{1}{R_L + j\omega L} = \frac{R_L - j\omega L}{\left(R_L^2 + \omega^2 L^2\right)}; \quad Y_{RC} = \frac{1}{R_C - \frac{j}{\omega C}} = \frac{R_C + \frac{j}{\omega C}}{\left(R_C^2 + \frac{1}{\omega^2 C^2}\right)}$$

$$\therefore Y = \frac{R_L - j\omega L}{\left(R_L^2 + \omega^2 L^2\right)} + \frac{R_C + \frac{j}{\omega C}}{\left(R_C^2 + \frac{1}{\omega^2 C^2}\right)} = \frac{R_L}{\left(R_L^2 + \omega^2 L^2\right)} + \frac{R_C}{\left(R_C^2 + \omega^2 L^2\right)} + j\underbrace{\left[\frac{\left(\frac{1}{\omega C}\right)}{R_C^2 + \frac{1}{\omega^2 C^2}} - \frac{\omega L}{R_L^2 + \omega^2 L^2}\right]}_{\text{susceptance } S_C}$$

For minimum admittance, the susceptance part of the admittance Y is to be zero and that occurs at a frequency ω_r, called the *resonant frequency* of the parallel resonant circuit.

So, we write,

$$\frac{\frac{1}{\omega_0 C}}{R_C^2 + \frac{1}{\omega_0^2 C^2}} - \frac{\omega_0 L}{R_L^2 + \omega_0^2 L^2} = 0.$$

Simplifying, we get

$$\omega_0^2 R_C^2 - \omega_0^2 \frac{L}{C} = \frac{R_L^2}{LC} - \frac{1}{C^2} \Rightarrow \omega_0 = \frac{1}{\sqrt{LC}}\left[\frac{R_L^2 - \frac{L}{C}}{R_C^2 - \frac{L}{C}}\right]^{1/2}. \tag{2.23}$$

(2.23) gives the resonant frequency in general but in case $R_C = 0$, we get the resonant frequency as

$$\omega_0 = \frac{1}{\sqrt{LC}}\left[\left(\frac{L}{C} - R_L^2\right)\Big/\left(\frac{L}{C}\right)\right]^{1/2} = \left[\frac{1}{\sqrt{LC}}\left(1 - \frac{CR_L^2}{L}\right)^{1/2}\right] = \left(\frac{1}{LC} - \frac{R_L^2}{L^2}\right)^{1/2}. \tag{2.24}$$

(2.24) is the resonant frequency for a Tank circuit. A Tank circuit is a series combination of L and R shunted by a capacitor without any leakage. The energy oscillates between magnetic field energy of the coil and capacitive energy of the capacitor, the frequency of oscillation being determined by (2.24).

2.9 *Q*-Factor and Its Effect on the Bandwidth $\Delta\omega$ of a Series Resonant Circuit

$$\Delta\omega = \frac{R}{L} \Rightarrow \frac{\omega_0}{\Delta\omega} = \frac{\omega_0 L}{R},$$

this defines the Q-factor of the coil. So this factor $\rightarrow$ large value if $\Delta\omega \rightarrow$ small quantity and L is large. Narrowness of the resonance curve is a measure of sharpness of resonance. Since power is proportional to I_r^2, we can say that if Q is greater, power or energy stored in the coil falls off more rapidly on either side of the resonant frequency.

Since $\frac{P_r}{P_{max}} = \frac{I_r^2}{I_{max}^2}$, the two frequencies ω_2 and ω_1 are said to define half-power frequencies.

Q-factor is, in general, defined by

$$Q = \frac{2\pi \times \text{ maximum energy stored in (inductor/capacitor)}}{\text{energy dissipated in the resistor/cycle}}$$

Maximum energy stored in the inductor is given by $\frac{1}{2}LI_0^2$.

Maximum energy stored in the capacitor per cycle $= \frac{1}{2}CV_c^2 = \frac{1}{2}C\left(\frac{I_0}{\omega C}\right)^2$

Energy dissipated in the resistor/cycle $= \frac{1}{2}I_0^2RT = \frac{1}{2}I_0^2R/f$.

$$\therefore\ Q = 2\pi \times \frac{1}{2}LI_0^2 \Big/ \left(\frac{1}{2}RI_0^2/f_0\right) = 2\pi f_0 \frac{L}{R} = \omega_0 \frac{L}{R} \text{(coil)} \tag{2.25}$$

$$Q = 2\pi \times \frac{\frac{1}{2} \times C\left(\frac{I_0}{\omega_0 C}\right)^2}{\frac{1}{2}\left(RI_0^2/f_0\right)} = 2\pi f_0 \cdot \frac{1}{\omega_0^2 CR} = \frac{1}{\omega_0 CR} \text{ (capacitor).} \tag{2.26}$$

Q-factor may be looked upon as a magnification factor

At resonance,

$$V_L = \omega_0 L I_0 = \omega_0 L \left(\frac{V}{R}\right) = QV. \tag{2.27}$$

(2.27) interprets Q as a magnification factor, which multiplies the applied voltage to obtain voltage across the inductor. Similarly,

$$V_C = IX_C = \left(\frac{V}{R}\right) \cdot \frac{1}{\omega_0 C} = QV,$$

the interpretation of Q is also same as above.

2.10 Parallel Resonance Circuit: Q-factor in such a Circuit

Refer to Fig. 2.21, where

$$\frac{1}{Z} = \frac{1}{R} + \frac{1}{j\omega L} + j\omega C \Rightarrow Y = \frac{1}{R} + j\left(\omega C - \frac{1}{\omega L}\right)$$

$$\Rightarrow Y = \left[\left(\frac{1}{R}\right)^2 + \left(\omega C - \frac{1}{\omega L}\right)^2\right]^{1/2}. \tag{2.28}$$

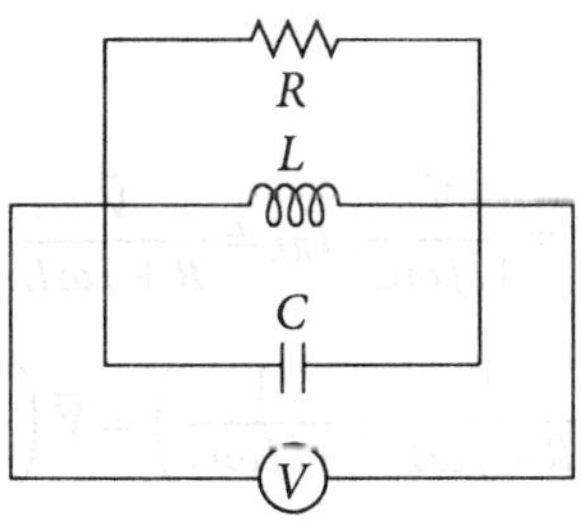

Fig. 2.21

From (2.28), the admittance will be minimum at resonance, when

$$\omega_0 C = \frac{1}{\omega_0 L} \Rightarrow \omega_0 = \frac{1}{\sqrt{LC}}.$$

Current at resonance

$$I_R = VY_0 = \frac{V}{R}.$$

Current at a frequency ω ($\neq \omega_0$)

$$I = V\left\{\left(\frac{1}{R}\right)^2 + \left(\omega C - \frac{1}{\omega L}\right)^2\right\}^{1/2}.$$

$$\therefore \frac{I_r^2}{I^2} = \frac{V/R}{\frac{V}{R}\left\{1+\left(\omega CR - \frac{R}{\omega L}\right)^2\right\}} = \frac{1}{2} \text{ (say)} \tag{2.29}$$

$$\Rightarrow \left\{1+\left(\omega CR - \frac{R}{\omega L}\right)^2\right\} = 2 \Rightarrow R^2\left(\omega C - \frac{1}{\omega L}\right)^2 = 1 \Rightarrow \omega C - \frac{1}{\omega L} = \pm\frac{1}{R}.$$

$$\therefore \omega C - \frac{1}{\omega L} = \frac{1}{R} \text{ giving maximum value of the current} \tag{2.30}$$

$$\text{and } \omega C - \frac{1}{\omega L} = -\frac{1}{R} \text{ giving minimum value of the current} \tag{2.31}$$

Solving (2.30), the plausible solution is

$$\omega_2 = \frac{1}{2CR} + \sqrt{\frac{1}{CL} + \frac{1}{4C^2R^2}}.$$

Solving (2.31), the plausible solution is

$$\omega_1 = \frac{1}{2CR} - \sqrt{\frac{1}{CL} + \frac{1}{4C^2R^2}}.$$

$$\Delta\omega = \text{bandwidth} = \omega_2 - \omega_1 = \frac{1}{CR}.$$

$$\text{Quality factor } Q = \frac{\omega_0}{\omega_2 - \omega_1} = \omega_0 CR. \tag{2.32}$$

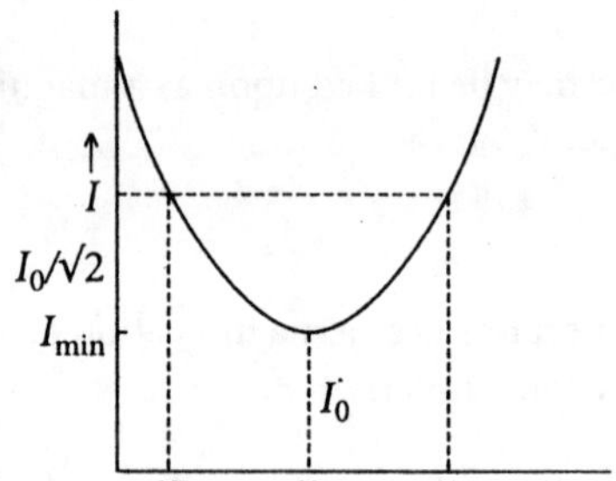

Fig. 2.22

The variation of current vs. ω in case of parallel resonance circuit [Fig. 2.22].

Suppose, the current vector be $\vec{I}$. This current vector is divided into two parts $\vec{I}_C$ and $\vec{I}_{RL}$.

$$\therefore \vec{I} = \vec{I}_C + \vec{I}_{RL}.$$

$$\text{Now, } \vec{I}_C = \frac{\vec{V}}{1/j\omega C}; \ \vec{I}_{RL} = \frac{\vec{V}}{R + j\omega L}.$$

$$\therefore \vec{I} = \vec{V}\left(\frac{1}{R+j\omega L} + \frac{1}{1/j\omega C}\right) = \vec{V}\left(\frac{1}{R+j\omega L} + j\omega C\right) = \vec{V}\left[\frac{R}{R^2+\omega^2L^2} + j\left(\omega C - \frac{\omega L}{R^2+\omega^2 L^2}\right)\right] = \frac{\vec{V}}{\mathsf{Z}}.$$

$$\therefore \frac{1}{\mathsf{Z}} = \frac{R}{R^2+\omega^2L^2} + j\left(\omega C - \frac{\omega L}{R^2+\omega^2L^2}\right).$$

The reciprocal of impedance operator is known as *admittance operator* Y.

$$\therefore \mathsf{Y} = \frac{R}{R^2+\omega^2L^2} + j\left(\omega C - \frac{\omega L}{R^2+\omega^2L^2}\right).$$

So, magnitude of admittance

$$|Y|=\sqrt{\left(\frac{R}{R^2+\omega^2L^2}\right)^2+\left(\omega C-\frac{\omega L}{R^2+\omega^2L^2}\right)^2}=\frac{1}{R^2+\omega^2L^2}\sqrt{R^2+(\omega CR^2+\omega^3CL^2-\omega L)^2}.$$

For minimum value of $|Y|$; we must have

$$\omega CR^2+\omega^3CL^2-\omega L=0 \text{ or, } \omega=\sqrt{\frac{1}{LC}-\frac{R^2}{L^2}}.$$

Hence, for this value of frequency, admittance of the circuit is minimum and therefore, impedance of the circuit maximum, hence, it is called *rejector circuit*. The circuit current is minimum.

Under this condition, the admittance becomes

$$|Y|=\frac{R}{\omega^2L^2+R^2}.$$

Impedance $=\frac{\omega^2L^2+R^2}{R}$, called *dynamic resistance*, where

$$\omega^2=\frac{1}{LC}-\frac{R^2}{L^2}.$$

$$\therefore \text{dynamic resistance}=\frac{R^2+L^2\left(\frac{1}{LC}-\frac{R^2}{L^2}\right)}{R}=\frac{L}{CR}.$$

The dimension of the right-hand quantity is that of resistance, because $\left[\frac{L}{R}\right]=[T],[CR]=[T]$, so, $\left[\frac{L}{CR}\right]=[R]$.

Therefore, at resonance, the peak current from the supply source is $\frac{\hat{V}}{L/CR}=\frac{\hat{V}CR}{L}$, called *make up current*.

The resonance curve in such a case [Fig. 2.22]. The minimum current occurs corresponding to frequency $\omega_0=\sqrt{\frac{1}{LC}-\frac{R^2}{L^2}}$. The variation of I in the neighbourhood of resonant frequency is depicted in the figure.

2.11 Power Factor — Wattless Component of Current

The power supplied to and dissipated by a circuit is at each instant equal to the product of instantaneous emf and current.

∴ the instantaneous power is given by

$$p=\left(\vec{V}\cdot\vec{v}\right)\left(\vec{I}\ \vec{v}\right)=vi-\hat{V}\cos\omega t\,\hat{I}\cos(\omega t\mp\phi).$$

Here, the current lags or leads the emf by an angle ϕ.

$$\therefore\ p=\frac{\hat{V}\hat{I}}{2}[2\cos\omega t\cos(\omega t\mp\phi)]=\frac{\hat{V}\hat{I}}{2}\cos\phi+\frac{\hat{V}\hat{I}}{2}\cos(2\omega t\mp\phi).$$

This shows that instantaneous power oscillates with angular frequency 2ω. The instantaneous power is positive when both the cosine terms are positive; when ωt and $(\omega t-\phi)$ may have any value θ such that

$0 < \theta < 2\pi$ the instantaneous power will at certain times be negative. In the case when p is positive, the generator supplies energy to the circuit, which is stored up partly as energy of the electromagnetic field, partly transformed into mechanical work and the rest dissipated as Joulean heat. In the second case, the generator receives back energy from the electromagnetic field, the store-house of energy.

The instantaneous power is not of any practical sense, the quantity of any sense is the real power, which is defined as the mean of the instantaneous power over the periodic time.

$$\therefore\ P_m = \frac{1}{T}\int_0^T p\,dt = \frac{\hat{V}\hat{I}}{2T}\cos\phi\int_0^T dt + \int_0^T \frac{\hat{V}\hat{I}}{2T}\cos(2\omega t \mp \phi)\,dt$$

$$= \frac{\hat{V}\hat{I}}{2}\cos\phi + 0 = \frac{\hat{V}}{\sqrt{2}}\cdot\frac{\hat{I}}{\sqrt{2}}\cos\phi = V_r\cdot I_r\cdot\cos\phi.$$

So, the real power in an ac circuit does not depend on the current and voltage only, the product of which gives the apparent power, but also depends on the phase difference between current and voltage. In order to get the real power, the apparent power is to be multiplied by $\cos\phi$ and hence, it is known as *power factor.*

$\therefore$ real power = apparent power × power factor.

In purely resistive circuit, since current and the emf are always in phase, the power factor is unity, hence, power absorption in resistive load is maximum. But in case $\phi = \pm\pi/2$ (e.g., capacitance or inductance), $\cos\phi = 0$ and hence, pure reactances cannot absorb power. The energy involved in such case of exchange does not leave the system and represents the conserved power of the circuit. Thus, in an inductive circuit, the circuit contains energy, which is stored as magnetic energy $\frac{1}{2}Li^2$ during the growth, but during the falling of voltage, this energy flows back to the generator from the magnetic field.

In the circuits, where $\cos\phi = 0$, the current is said to be *wattless current.* Even if $\cos\phi \neq 0$, the current vector can be resolved into two components — one in phase with emf contributing to the power; the other in quadrature with it, contributing nothing to the power. Hence, the latter is termed as wattless component.

Maximum power transfer. In many a case of electrical power transfer, it becomes necessary to adjust the load elements (comprising the load impedance) such that the power delivered to the load is maximum. This is called *matching of load impedance* to the source of power. Let us investigate the theoretical aspect of matching when the load is resistive, called the *resistive matching* and when the impedance is complex, called *impedance matching.*

(i) Load is resistive (resistive matching)

Any source of electrical power is equivalent to a simple voltage generator V in series with an internal impedance X_g.

When the impedance are purely resistive, $X_g = R_g$, $X = R$.

The load current $I = \frac{V}{R+R_g}$.

$\therefore$ power supplied to the load

$$P = I^2R = \frac{V^2R}{\left(R+R_g\right)^2};\quad \frac{dP}{dR} = V^2\frac{(R_o+R)-2R}{(R_o+R)^3}.$$

For stationary value, $\frac{dP}{dR} = 0$.

$$\therefore\ R = R_g.$$

Putting this value of R, $\frac{d^2P}{dR^2}$ is negative, which means, for this value power delivered is maximum.

$$\therefore\ P_{\max} = \frac{V^2 R_g}{4R_g^2} = \frac{V^2}{4R_g}.$$

Thus, when the external resistance in the circuit is made equal to the internal resistance of the generator, the power developed across R is maximum.

When the impedances are complex

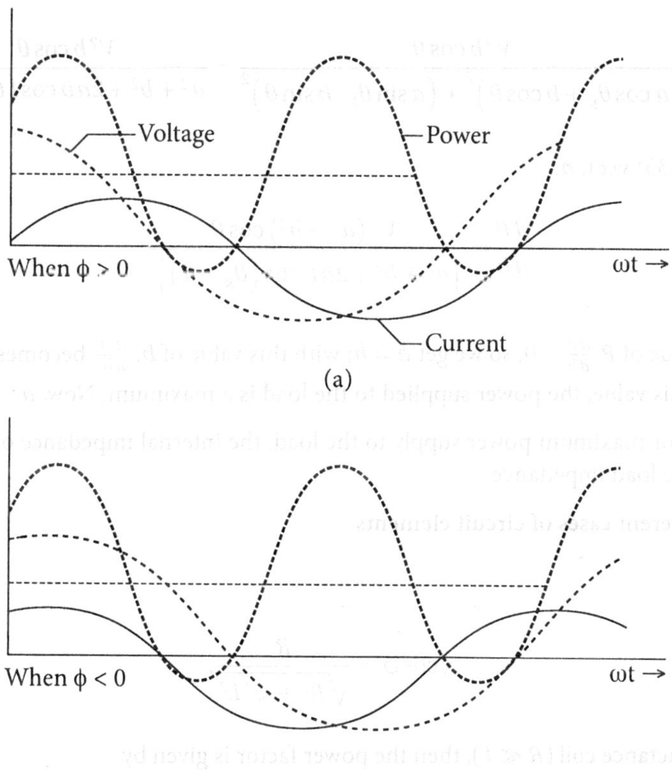

Fig. 2.23

Let the internal impedance of the generator [Fig. 2.23] $Z_g = R_g + jX_g$.
External impedance of the circuit $Z = R + jX$.

$$\therefore\ \vec{I} = \frac{\vec{V}}{X_g + Z} = \frac{\vec{V}}{(R+R_g)+j(X_g+X)} = \frac{\hat{V}\,e^{j(\omega t-\phi)}\vec{\gamma}}{\sqrt{(R+R_g)^2+(X_g+X)^2}},$$

$$\tan\phi = \frac{X_g+X}{R+R_g}.$$

Power across the load,

$$P = V_r I_r \cos\phi = I_r Z\cdot I_r\cdot\frac{(R+R_g)}{Z} = I_r^2\,(R+R_g),$$

which shows that the whole amount of power is developed across the resistive element of the impedance.

We are now to consider the condition so that power across the load becomes a maximum.

The power supplied to the load is

$$P = I^2 R = \frac{V^2 R}{Z'^2}.$$

Now, $Z_g = R_g + jX_g = a\left(\cos\theta_0 + j\sin\theta_0\right)$; $Z = (R + jX) = b(\cos\theta + j\sin\theta)$.

$$\therefore\ P = \frac{V^2 b\cos\theta}{\left(a\cos\theta_g + b\cos\theta\right)^2 + \left(a\sin\theta_g \cdot b\sin\theta\right)^2} = \frac{V^2 b\cos\theta}{a^2 + b^2 + 2ab\cos\left(\theta_g - \theta\right)}. \tag{2.33}$$

Differentiating (2.33) w.r.t. b,

$$\frac{dP}{db} = \frac{V^2\left(a^2 - b^2\right)\cos\theta}{\left[a^2 + b^2 + 2ab\cos\left(\theta_g - \theta\right)\right]^2}.$$

For stationary value of P, $\frac{dP}{db} = 0$, so we get $a = b$; with this value of b, $\frac{d^2P}{db^2}$ becomes a negative quantity, which means, for this value, the power supplied to the load is a maximum. Now, $a = \sqrt{R_g^2 + X_g^2}$ and $b = \sqrt{R^2 + X^2}$. Hence, for maximum power supply to the load, the internal impedance of the generator must be made equal to the load impedance.

Power factor in different cases of circuit elements

(i) R-L circuit

Here, $\tan\phi = \frac{\omega L}{R}$.

$$\therefore \cos\phi = \frac{R}{\sqrt{R^2 + \omega^2 L^2}}.$$

If we take an inductance coil ($R \ll L$), then the power factor is given by

$$\cos\phi = \frac{R}{\omega L} = \frac{1}{\frac{\omega L}{R}} = \frac{1}{Q},$$

where Q is usually referred to as *figure of metrit* of the inductor.

(ii) C-R circuit

Here, $\tan\phi = \frac{1}{\omega CR}$.

$$\therefore \cos\phi = \frac{\omega CR}{\sqrt{1 + \omega^2 C^2 R^2}} = \frac{R}{\sqrt{R^2 + \frac{1}{\omega^2 C^2}}}.$$

In case of a capacitor, $R \ll \frac{1}{\omega C}$

$$\therefore \cos\phi = \frac{R}{\frac{1}{\omega C}} = \omega CR.$$

2.12 Numerical Problems

1. A resistor of resistance 20Ω is fed from an ac supply of amplitude 340V and frequency 50c/s. Obtain an expression for the instantaneous current.

   Solution: $\omega = 2\pi f = 2\pi \times 50 = 100\pi$.

   If $v = \hat{V}\cos\omega t$, then

$$i = \frac{\hat{V}}{R}\cos\omega t = \frac{340}{20}\cos 100\pi t = 17\cos 100\pi \text{A}.$$

2. Two factory buildings are supplied from the same ac mains. The first takes a load of 250A at a power factor of 0.75. The two factories together have a load of 400A and a power factor of 0.62. Assuming that the current lags in each case, calculate the load and power factor for the second factory.

   Solution: Total load = I at power factor $\cos\phi$. The first factory takes a load I_1 at power factor $\cos\phi_1$; the second factory takes a load of I_2 at power factor $\cos\phi_2$.

$$I\cos\phi = I_1\cos\phi_1 + I_2\cos\phi_2$$
$$I\sin\phi = I_1\sin\phi_1 + I_2\sin\phi_2,$$

   where ϕ is the phase angle [Fig. 2.24].

   Now, $I = 400$A, $I_1 = 250$A, $\cos\phi_1 = 0.75$, $\cos\phi = 0.62$.

   $400 \times 0.62 = 250 \times 0.75 + I_2\cos\phi_2$.

   $400\sqrt{1-(0.62)^2} = 250 \times \sqrt{1-(0.75)^2} + I_2\sin\phi_2$.

   $\therefore\ I_2\cos\phi_2 = 60.5;\ I_2\sin\phi_2 = 148.5$.

   $\therefore\ I_2 = 160.4$A, $\cos\phi_2 = 0.37$ (lag).

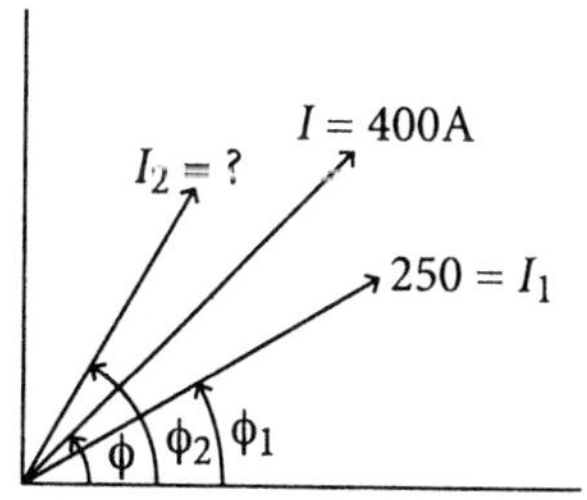

Fig. 2.24

3. An electric lamp, which runs at 100V dc and 10A current is connected to 200V ac mains of 50c/s frequency and the current is found to be 15A. Calculate the inductance of the lamp.

   Solution: W = V × A = 100 × 10 = 1000.

   Resistance of the lamp = $\frac{100}{10} = 10\Omega$.

   Now, when connected to 200V ac, the current is

$$I = \frac{V}{\sqrt{R^2+\omega^2L^2}} \text{ or, } 15 = \frac{200}{\sqrt{100+\omega^2L^2}} \text{ or, } 100+\omega^2L^2 = \frac{200\times200}{15\times15} = \frac{1600}{9} \text{ or, } \omega^2L^2 = \frac{700}{9}.$$

   Now, $\omega = 2\pi = 2 \times 3.14 \times 50 = 314$.

$$\therefore\ L^2 = \frac{700}{9}\cdot\frac{1}{314\times314}.$$

4. An alternating emf of rms value 100V and frequency 50c/s is applied to a circuit consisting of an inductance of 5H, resistance 1000Ω in series. What will be the rms value of the current flowing in the circuit and its phase lag w.r.t. the emf? What power will be consumed by the circuit?

   Solution: $I = \dfrac{V}{\sqrt{R^2+\omega^2L^2}} = \dfrac{100}{\sqrt{10^6+(314)^2\times25}} = \dfrac{100}{1861} = 0.054$A.

$$\tan\phi = \frac{\omega L}{R} = \frac{314\times5}{1000} = 1.570.$$

   $\therefore\ \phi = \tan^{-1}(1.570) = 57°30.03;\ P = 100 \times 0.054 \times \cos\phi = 2.90$W.

5. A 100V, 60W lamp is to be operated by 220V, 50 cycles per sec mains. What (a) pure resistance (b) pure inductance, placed in series with lamp will enable it to be correctly used?

Solution: W = V × A. ∴ current through the lamp $= \frac{60}{100} = 0.6$A.

Resistance of the lamp $R_L = \frac{1000}{6} = \frac{500}{3}\Omega$.

(a) $6 = \frac{220}{R+\frac{500}{3}}$ or, $0.6R + 100 = 200$ or, $R = \frac{120}{0.6} = 200\Omega$.

(b) $0.6 = \frac{220}{\sqrt{R_L^2+\omega^2 L^2}}$ or, $0.36 = \frac{220\times 220}{\frac{500\times 500}{3\times 3}+314\times 314\times L^2}$ or, $L = 1.043$H [Fig. 2.25].

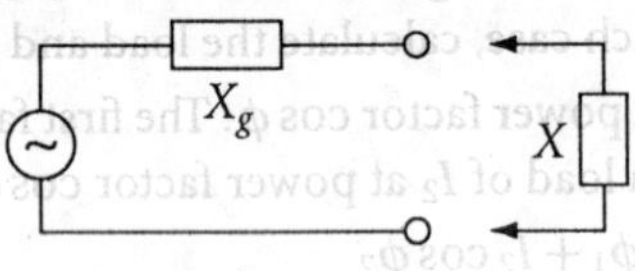

Fig. 2.25

6. A series circuit consisting of a coil having resistance of 400Ω and inductance of 2.5H and condenser having capacitance of 10μF is connected across a variable frequency source of emf kept constant at 250V. Calculate the frequency at which resonance takes place, and determine the value of the current at resonance.

Solution: $I_r = \frac{V_r}{\sqrt{R^2+\left(\omega L - \frac{1}{\omega C}\right)^2}}$.

At resonance, $\omega L - \frac{1}{\omega C} = 0$ or, $\omega = \frac{1}{\sqrt{LC}}$

$$\text{or, } f = \frac{1}{2\pi}\sqrt{\frac{1}{LC}} = \frac{1}{2\times 3.14} \times \frac{1}{\sqrt{2.5\times 10\times 10^{-6}}} = \frac{1}{6.28}\times\frac{10^3}{5} = 32\text{c/s}.$$

$$I_{res} = \frac{V_r}{R} = \frac{250}{400} = 0.625\text{A}.$$

7. A source of emf of 50V (rms) is connected across an iron-cored coil. When the supply frequency is 50c/s, the power consumed is found to be 100W, whereas when the frequency is increased to 100c/s, the power consumed becomes 50W. Find the self-inductance and the resistance of the coil [$\pi^2 = 10$].

Solution: $P = VI\cos\phi$.

$$100 = \frac{(50)^2 R}{R^2+\omega_1^2 L^2}. \tag{2.34}$$

Again,

$$50 = \frac{(50)^2 R}{R^2+\omega_2^2 L^2}. \tag{2.35}$$

Solving (2.34) and (2.35), we get $R = \sqrt{20}\times 10^2 L$ and $L = 0.0373$H, so, $R = 16.7\Omega$.

8. An inductance of 0.5H and variable resistance are in series with an ac supply of 250V, 50c/s. Calculate the value of resistance required for maximum power in the circuit and the phase of the current and power factor.

Solution: $P = \frac{V^2 R}{2(R^2+\omega^2 L^2)}$.

For maximum of P, $\omega L = R$.

$\therefore\ R = 2\pi f L = 2 \times 50 \times 3.14 \times 0.5 = 0.5 \times 314 = 157\Omega.$

$$\tan\phi = \frac{\omega L}{R} = 1 \Rightarrow \phi = \frac{\pi}{4}.$$

The current in such a case will lag the emf by $\frac{\pi}{4}$.
Power factor $\cos\phi = \cos\frac{\pi}{4} = \frac{1}{\sqrt{2}}$.

Power $= \frac{V^2}{4R} = \frac{250\times250}{4\times157} = 995\text{W}.$

9. An arc lamp (noninductive) takes 10A at 50V. Calculate impedance of a choke of 1Ω to be placed in series with it in order that it may be worked by a 200V, 50c/s supply. Calculate total power and power factor used.

Solution: Resistance of the lamp $R = \frac{50}{10} = 5\Omega$ [Fig. 2.26].

$V\cos\phi = RI,\ 200 \times \cos\phi = 6 \times 10 = 60.$

$\therefore\ \cos\phi = \frac{60}{200} = 0.3.\ \therefore\ \phi = 72.54°.$

$\therefore$ total power consumed – $VI\cos\phi$ – $200 \times 10 \times 0.3 = 600\text{W}$.

$\frac{\omega L}{6} = \tan\phi\ \tan 72.54° = 3.1793;\ \omega L = 19.0758.$

$\therefore\ Z = \sqrt{36 + (1908)^2} = 20\Omega.$

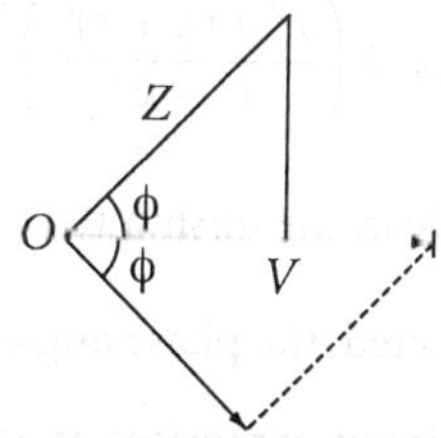

Fig. 2.26

10. A choking coil of resistance 5Ω and inductance 0.6H is in series with a capacitance 10μF. If a voltage 200V is applied and the frequency adjusted to resonance, find the current, voltages across inductance and capacitance.

Solution: $f_{\text{res}} = \frac{1}{2\pi}\sqrt{\frac{1}{LC}} = \frac{1}{2\pi}\sqrt{\frac{10^6}{6 \times 10}} = 65\text{c/s}.$

$I_0 = \text{resonance current} = \frac{V}{R} = \frac{200}{5} = 40\text{A}.$

Voltage across choke $= \left(\sqrt{R^2 + \omega^2 L^2}\right) I = 40\sqrt{25 + (2 \times 3.14 \times 65 \times 0.6)^2} = 9799\text{V}.$

Voltage across the capacitor $\frac{1}{\omega C} I = \frac{40 \times 10^6}{10 \times 2 \times 314 \times 65} = 9794\text{V}.$

11. A sinusoidal voltage of rms value 250V and frequency 50c/s is applied to the circuit shown. Calculate the current in the circuit and its phase w.r.t. that of the applied voltage [Fig. 2.27].

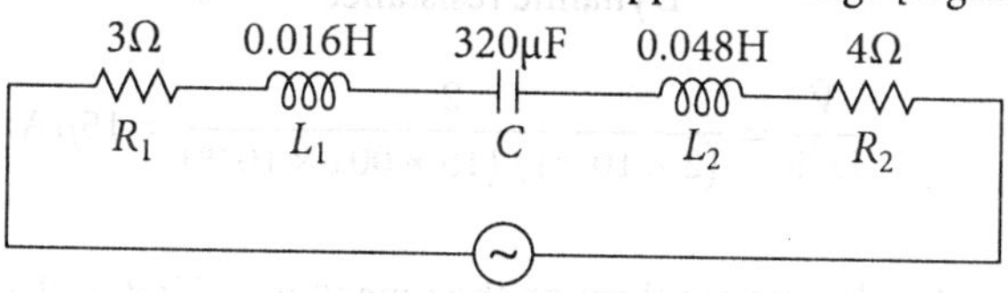

Fig. 2.27

Solution: The impedance operator of the circuit element is

$$Z = R_1 + R_2 + j\omega L_1 + j\omega L_2 - \frac{j}{\omega C} = 7 + 5j + 15j - 10j = 7 + 10j.$$

$$\therefore |Z| = \sqrt{7^2+10^2} = 12.2\Omega; \quad \therefore I = \frac{V}{Z} = \frac{250}{122} = 20.5\text{A}; \quad \tan\phi = \frac{10}{7}.$$

or, ϕ = phase difference $= \tan^{-1}\left(\frac{10}{7}\right) = 55°$ (voltage leads the current).

12. An alternating current passes through a resistance of kΩ and an inductance whose reactance at the working frequency also equals kΩ. If the magnitude of the current remains unaltered when a shunt resistance of r Ω is connected across the inductance, calculate r in terms of k. Also, find the phase angles between the voltage and current in the shunted and unshunted condition.

Solution: In the unshunted condition, the impedance operator is $z_1 = k + j\omega L = k + jk$. $\therefore |z_1| = \sqrt{2}k$.

When shunted $z_2 = k + \frac{rjk}{r+jk} = k\left(\frac{r+j(k+r)}{r+jk}\right)$.

$$\therefore |z_2| = k\left(\frac{r^2+(k+r)^2}{r^2+k^2}\right)^{1/2}. \quad \therefore \frac{V}{|z_1|} = \frac{V}{|z_2|} \text{ or, } \sqrt{2}k = k\left(\frac{r^2+(k+r)^2}{r^2+k^2}\right)^{1/2}. \quad \therefore r = \frac{k}{2}.$$

Phase angle in the unshunted condition: $\tan^{-1}(k/k) = 45°$.

When shunted, the phase angle is given by $\tan^{-1}\left(\frac{k+r}{r}\right) - \tan^{-1}\left(\frac{k}{r}\right) = \tan^{-1}3 - \tan^{-1}2 \simeq 8°$.

13. A series circuit comprises an inductance of 200μH, a capacitance 0.0005μF, and a resistance 10Ω. Find: (a) resonant frequency (b) voltage magnification when the circuit is tuned to resonance.

Solution: f_0 = resonance frequency $= \frac{1}{2\pi}\frac{1}{\sqrt{LC}} = \frac{1}{2\times 3.14} \times \frac{1}{\sqrt{200\times 10^{-6} \times 0.0005 \times 10^{-6}}} = 502\text{kc/s}.$

Voltage magnification at resonance

$$= \frac{2\pi f_0 L}{R} = \frac{2\pi \times 502 \times 10^3 \times 200 \times 10^{-6}}{10} = 63.2$$

14. A coil of inductance $L = 2\times 10^{-3}$H, $R = 15\Omega$ is connected in parallel with a capacitor of $C = 0.001\mu$F. Find: (a) the frequency for which current in the circuit is minimum (b) peak value of the make up current for peak supply emf 2V (c) peak oscillatory current under these conditions.

Solution: (a) $f_0 = \frac{1}{2\pi}\sqrt{\frac{1}{LC} - \frac{R^2}{L^2}} = \frac{10^6}{6.28\sqrt{2}} = 113\text{kc/s}.$

(b) Make up current at resonance $= \frac{\text{Peak supply emf}}{\text{Dynamic resistance}}$.

$$\frac{\hat{V}}{\text{LCR}} = \frac{2}{(2\times 10^{-3})/(15\times 001 \times 10^{-6})} = 15\mu\text{A}.$$

(v) Oscillatory current is the current through the capacitor $= \hat{V}\omega C = 2\times 2\pi f_0 \times 0.001 \times 10^{-6}$

$$= 2\times 2\times 3.14 \times 113 \times 10^{-3} \times 10^6$$

$$= 15\mu\text{A}.$$

15. The p.d. as measured with an ac voltmeter across a coil is 100V, while the current, as measured by an ac ammeter is 20A. The ohmic resistance of the coil is 2Ω and frequency of the supply 50c/s. Calculate:

(a) the impedance (b) admittance (c) self-inductance L. Also, determine the active components of current and also the power factor.

Solution: $V_{rms} = 100$, $I_{rms} = 20A$.

$$\therefore\ Z = \text{impedance} = \frac{100}{20} = 5\Omega;\quad Y = \frac{1}{Z} = \frac{1}{5} = 0.2℧.$$

Now, $R^2 + \omega^2 L^2 = 25$ or, $L = \sqrt{\frac{21}{4 \times 50 \times 50 \times 10}} = \frac{1}{100}\sqrt{\frac{21}{10}} = 14.5\text{mH}.$

Power factor $= \cos\phi = \frac{R}{Z} = \frac{2}{5}.$

$\therefore$ active component of curent $= I\cos\phi = 20 \times \frac{2}{5} = 8\text{A}.$

16. A sinusoidal emf of 100V (rms) at a frequency of 100 cycles per second is applied to a circuit containing a condenser of $C = 2\mu F$, and resistance R in series. The current is 0.125A. Calculate the value of R and the power factor of the circuit.

Solution: $I = \frac{V}{\sqrt{R^2 + \frac{1}{\omega^2 C^2}}}.$

$$0.125 = \frac{100}{\sqrt{R^2 + \frac{1}{4 \times 10 \times 10^4 \times 4 \times 10^{-12}}}} \text{ or, } \left(125 \times 10^{-3}\right)^2 = \frac{10^4}{R^2 + \frac{10^7}{16}}$$

$$\text{or, } R^2 + \frac{10^7}{16} = \frac{10^4}{\left(125 \times 10^{-3}\right)^2} = \frac{10^{10}}{125 \times 125}.$$

$$R^2 = \frac{10^7 \times 3}{125 \times 16}.\ \therefore\ R = 122.5\Omega.$$

17. An emf of 240V (rms) value at 50c/s is applied to a circuit containing capacitance of $5\mu F$ in series with a resistance of 1000Ω. Find the rate at which energy is dissipated in the circuit.

Solution: $Z = \sqrt{R^2 + \frac{1}{\omega^2 C^2}} = \sqrt{10^6 + \frac{1}{4 \times 10 \times 2500 \times 25 \times 10^{-12}}} = \sqrt{10^6 + \frac{10^7}{25}}$

$$= \sqrt{10^6 + 0.4 \times 10^6} = 10^3\sqrt{1.4} = 1183\Omega.$$

$$\therefore\ P = I^2 R = \frac{V^2}{Z^2}R = \frac{240 \times 240 \times 10^3}{10^6 \times 1.4} = 41.1.$$

18. An inductance $L = 6\text{mH}$ is in series with a resistance $R = 100\Omega$. Calculate power factor of the circuit at a frequency of 50c/s. What value of capacitance connected in parallel with the circuit increases the power factor to unity?

Solution: $\omega = 2\pi f = 2 \times 3.14 \times 50 = 314;\ L\omega = 6 \times 10^{-3} \times 314 = 1.9\Omega.$

$\therefore\ Z = \sqrt{R^2 + \omega^2 L^2} = 100.01\Omega.$

The power factor $= \frac{R}{Z} = \frac{100}{10001} = 0.9998$ with a capacitance C in parallel with the original circuit, the impedance is

$$\frac{1}{Z} = \frac{1}{R + j\omega L} + j\omega C = \frac{R}{R^2 + \omega^2 L^2} + J\left(\omega C - \frac{\omega L}{R^2 + \omega^2 L^2}\right).$$

For unit power factor,

$$\left(\omega C - \frac{\omega L}{R^2 + \omega^2 L^2}\right) = 0 \Rightarrow C = \frac{L}{R^2 + L\omega^2 l^2} = \frac{L}{z^2} \text{ or, } C = 0.6\mu\text{F}.$$

19. A purely inductive reactance $X_L = 5\Omega$, a purely capacitive reactance $X_C = 8\Omega$, and resistance 3Ω are connected in parallel and 120V (rms) is applied across it. Calculate the combined current and current in each branch.

Solution: $\vec{V} = (R + j\omega L)\vec{I}_{L_R} = \dfrac{1}{j\omega C}\vec{I}_C.$

$$\therefore \vec{I}_L = \frac{\vec{V}}{R + j\omega L} = \frac{\hat{V} e^{j\omega t}\vec{n}}{\sqrt{R^2 + \omega^2 L^2}\, e^{j\phi}} = \frac{\hat{V} e^{j(\omega t - \phi)}\vec{n}}{\sqrt{R^2 + \omega^2 L^2}}.$$

$$\vec{I}_C = j\omega C\vec{V} = \omega C e^{j\pi/2}\hat{V} e^{j(\omega t)}\vec{n} = \omega C\hat{V} e^{j(\pi/2 + \omega t)}\vec{n}.$$

$$\therefore i_L = \frac{\hat{V}}{\sqrt{R^2 + \omega^2 L^2}}\cos(\omega t - \phi) = \frac{\hat{V}}{\sqrt{R^2 + X_L^2}}\cos(\omega t - \phi); \quad i_C = -\omega C\hat{V}\sin\omega t.$$

$$I_L(\text{rms}) = \frac{V(\text{rms})}{\sqrt{R^2 + X_L^2}} = \frac{120}{\sqrt{9 + 25}} = \frac{120}{\sqrt{34}} = 20.58\text{A}$$

$$I_C(\text{rms}) = X_C V(\text{rms}) = 8 \times 120 = 960\text{A}.$$

20. A heating coil of 100Ω, a choke of $L = 0.5$H and a capacitance 8μF are connected in series across 100V, 50 cycles per second mains. Calculate the power factor and the amount of heating produced in 10s. Calculate what should have been the frequency of the supply as to produce the maximum heating.

Solution: $i = \dfrac{\hat{V}}{\sqrt{R^2 + \left(\omega L - \dfrac{1}{\omega C}\right)^2}}\cos(\omega t - \phi); \quad \tan\phi = \dfrac{\left(\omega L - \dfrac{1}{\omega C}\right)}{R}.$

$$\therefore i = \hat{I}\cos(\omega t - \phi); \quad \hat{I} = \frac{\hat{V}}{\sqrt{R^2 + \left(\omega L - \dfrac{1}{\omega C}\right)^2}}.$$

The power factor $= \cos\phi = \dfrac{R}{\sqrt{R^2 + \left(\omega L - \dfrac{1}{\omega C}\right)^2}}.$

Now, $\omega L = 2\pi f L = 2\pi \times 50 \times 0.5 = 137.08.$

$$\frac{1}{\omega C} = \frac{1}{2\pi f C} = \frac{10^6}{2\pi \times f \times C} = \frac{10^6}{2\pi \times 50 \times 8} = \frac{10^6}{2513.3} = 3.98 \times 10^2 = 398\Omega.$$

$$\therefore \cos\phi = \frac{100}{\left[10^4 + (15708 - 398)^2\right]^{1/2}} = 0.383.$$

The amount of heating in 10s $= 100 \times I^2_{\text{rms}} \times 10 = 100(769)^2 \times 10 = 59\dot{1}.4$J.

The heating will be maximum when I_{rms} is maximum. This becomes maximum when $\omega L - \frac{1}{\omega C} = 0$ or, when $\omega = \frac{1}{\sqrt{LC}}$.

21. A 240V single phase generator has a current carrying capacity of 100A. What is the maximum rate at which energy can be taken from this generator by a system with a power factor 0.64? How is this maximum affected if the power factor of the system is raised to a value 0.92?

Solution: The power developed $= V_r \times I_r \times \cos\phi$.

Here, $I_r = 100$A, $V_r = 240$V, $\cos\phi = 64$. $\therefore\ P = 100 \times 240 \times 0.64 = 15.4$kW.

In the second case, $P = 100 \times 240 \times 0.92 = 22.1$kW.

2.13 Exercises

1. Show that the power absorbed by a coil carrying an alternating current is $EI\cos\phi$, where E and I are the rms values of the emf and current and ϕ is the phase difference between them.
2. Discuss the distinction between *reactance* and *resistance* in an electrical circuit. Derive an expression for the impedance of an ac circuit consisting of an inductance and a resistance.
3. An emf of 100V (rms) at 50c/s is applied to a circuit containing an inductance of 80mH and a resistance of 20Ω in series. Calculate: (a) the magnitude and phase angle of the current and (b) rate at which energy is dissipated in the circuit. [(a) 3.12A, $\phi = 51°28$ (b) 194.4W]
4. Explain the terms *mean value of the current, root-mean-square current.* A resistance of 100Ω and capacitance 20μF are arranged in series and connected to a 240V ac supply whose frequency is 60c/s. Calculate the current, power and power factor, also p.d. across the resistor and condenser.
5. An alternating voltage $\hat{V}\cos\omega t$ is applied across a coil of inductance L and resistance R. Show that the amplitude of the current flowing through the coil is $\frac{\hat{V}}{\sqrt{R^2+\omega^2L^2}}$ and it lags behind the voltage by an angle $\tan^{-1}\frac{\omega L}{R}$.
6. Define the *form factor* of an alternating wave. Calculate the value of the same for a pure sinusoidal current. What is its significance?

   A sinusoidal voltage of 100V (rms) and a frequency of 100c/s is applied to a coil and a condenser in series. Calculate: (a) rms current (b) voltage across the condenser (c) the power factor.

   Given, $R = 40\Omega$, $L = 3$H, $C = 5\mu$F. [(a) 0.735A (b) 234V (c) 0.294]
7. Derive an equation for the instantaneous value of current under steady state conditions in a series circuit having 10Ω resistance 2H inductance and 5μF capacitance, when connected to a source of alternating emf whose instantaneous value can be represented by the equation.

   [$v = 300\cos 314t$, where t = time in sec.]
8. Find expression for the frequency at which resonance takes place in (a) a series circuit (b) a parallel circuit.

   A 5μF condenser and a 10Ω coil are connected in series to a 10V ac supply of frequency $\frac{100}{2\pi}$c/s. Calculate the power supplied. [0.25 mW]
9. An alternating voltage of 50V (rms) at a frequency 159kc/s is applied to a circuit consisting of an inductance $L = 0.9$mW, $C = 0.001\mu$F, $R = 10\Omega$ in series. Find power factor, real power, apparent power. [0.0995, 2.475W, 24.88W]
10. Define the following terms in connection with an ac circuit: Peak value of the emf, Pulsatance; Reactance; Complex Vector Impedance; Transient and Steady State Currents.

11. An electrical circuit containing a noninductive resistance of 10Ω, an inductance of 450μH and a capacitance of 0.0002μF all in series, is connected to an ac voltage supply whose frequency can be altered. Find the frequency for which maximum current can be obtained. Find also the voltage magnification at this maximum current. Why is it called an *acceptor circuit*?

 [*Hints*. See the series resonant circuit].

12. *In general an alternating emf produced by a generator is not a simple sinusoidal emf of a single frequency.* — Justify the statement mathematically or otherwise.

 [*Hints*. See the text].

13. What is a parallel resonant circuit? State its characteristics and uses. What is *Q*-factor?

14. Distinguish between mean value, peak value and root mean square of an alternating current. Justify the statement that 220V ac is more dangerous than 220V dc.

 Prove that in an ac circuit the average power is equal to the product of the virtual current, virtual emf and the cosine of the angle of lag or lead between the emf and the current.

Chapter 3

Inductively Coupled Circuits

3.1 Introduction

By means of inductively coupled circuits energy can be transferred from one circuit to the other due to mutual inductance of the two circuits. Energy transference takes place in such a circuit due to transformer action, i.e., ac flowing in one circuit produces change of magnetic flux linked up with the second thus setting up induced current, and a transfer of energy takes place from the first coil, called *primary* to the second coil, called the *secondary.* The importance of such coupled circuits is that it forms the basis of constructions of a transformer so common an equipment in electrical and electronic engineering.

3.2 Transformers

The purpose of the so-called device is to step-up or step-down the voltage applied to the primary to appear across the secondary. If V_2 and V_1 represent the rms values of the voltages developed across the secondary due to the impressed voltage across the primary, then if it so happens that $V_2/V_1 < 1$, the transformer is a step-down transformer while, if $V_2/V_1 > 1$, it acts as a step-up transformer.

Depending on the range of frequencies over which the transformers are used, they are classified mainly into the following:

(i) *Radio-frequency transformer* (*r-f transformer*): Used in the radio-frequency range from 40kc/s to some megacycles/s. Core material is usually dust iron consisting of minute particles of iron in the form of a powder in a matrix of bakelite. But here in the dust-iron core, the hysteresis and eddy current losses increase with frequency. So it is better to have no core material within the coupling coils, and as such, termed as air-core transformer.

(ii) *Audio frequency transformer* (*a-f transformer*): Used for transformation in the frequency range of 20c/s to 20kc/s. As for core material, ferromagnetic materials such as iron or permalloy are used. Eddy current losses are reduced by core material taken in the form of laminated sheets.

(iii) *Power transformer*: Used at frequency range of 50–60c/s, here also laminated sheets of iron form the core material.

Depending on the disposition of the core material, transformers are named as:

(a) *Core type*: It consists of one single magnetic circuit [Fig. 3.1(a)].

(b) *Shell type*: It consists of double magnetic circuits [Fig. 3.1(b)].

(c) *Berry type*: It consists of distributed magnetic circuits.

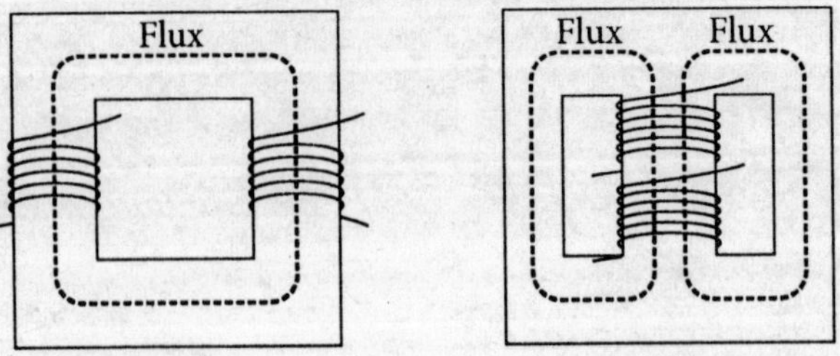

Fig. 3.1

Action of the transformer: The action of a transformer will be understood better if it is considered as a mutual inductor.

In a transformer, there are two circuits — one, primary and the other — secondary. The variation of current in the primary will cause an induced emf in the secondary, the magnitude of which depends on the extent of flux produced by the primary and linking up with the secondary. The extent of such linkage is determined by the number of turns, disposition and the medium separating the coils. In case of an ideal transformer, the magnetic leakage is negligible, so that the tubes of induction are closed paths within the core material. Whether a transformer is a step-up or step-down one, will be decided by the the turn ratio, i.e., $\frac{n_2}{n_1}$, where n_2 is the number of turns in the secondary, n_1 those of the primary.

3.3 Mathematical Theory of the Transformer

We will now develop the theory of an ideal transformer of the core type. Let n_1 = number of primary turns, n_2 = number of secondary turns, L_p = self-inductance of the primary coil, L_s = self-inductance of the secondary coil; R_p, R_s = resistances of the primary and secondary circuits respectively, M = mutual inductance between P and S [Fig. 3.2].

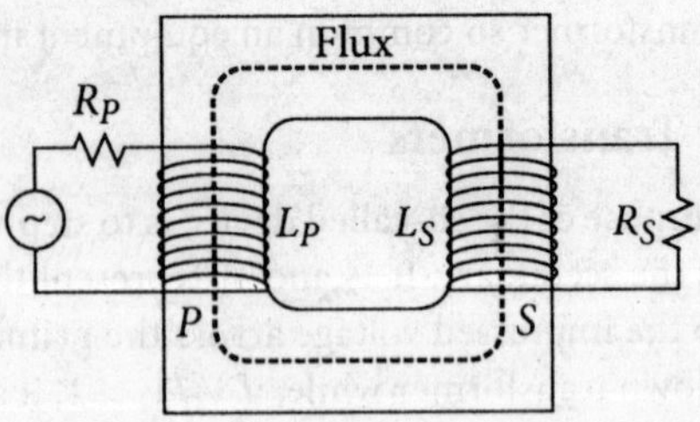

Fig. 3.2

Let the instantaneous emf applied to the primary be v_p given by $v_p = \hat{V}\cos\omega t = \hat{V}\,e^{j\omega t}\vec{n}\cdot\vec{n} = \vec{V}_p\cdot\vec{n}$.

i_p = instantaneous current in the primary circuit = $\vec{I}_p\cdot\vec{n}$, $\vec{I}_p$ is the primary current vector to be determined.

Then, the instantaneous emf equations in the primary and secondary circuits are

$$R_p i_p + L_p\frac{di_p}{dt} + M\frac{di_s}{dt} = v_p \tag{3.1}$$

$$\text{and } R_s i_s + L_s\frac{di_s}{dt} + M\frac{di_p}{dt} = 0. \tag{3.2}$$

Writting down in terms of rotating vector notations,

$$R_p\vec{I}_p\cdot\vec{n} + j\omega L_p\vec{I}_p\cdot\vec{n} + j\omega M\vec{I}_s\cdot\vec{n} = \vec{V}_p\cdot\vec{n}$$

$$\text{or, } \left(R_p + j\omega L_p\right)\vec{I}_p + j\omega M\vec{I}_s = \vec{V}_p \text{ or, } Z_pI_p + j\omega M\vec{I}_s = \vec{V}_p. \tag{3.3}$$

Similarly,

$$Z_s\vec{I}_s + j\omega M\vec{I}_p = 0. \tag{3.4}$$

Solving (3.3) and (3.4),

$$\vec{I}_p = \frac{\begin{vmatrix} \vec{V}_p & j\omega M \\ 0 & Z_s \end{vmatrix}}{\begin{vmatrix} Z_p & j\omega M \\ j\omega M & Z_s \end{vmatrix}} \text{ and } \vec{I}_s = \frac{\begin{vmatrix} Z_p & \vec{V}_p \\ j\omega M & 0 \end{vmatrix}}{\begin{vmatrix} Z_p & j\omega M \\ j\omega M & Z_s \end{vmatrix}}.$$

$$\therefore\ \vec{I}_p = \frac{\vec{V}_p Z_s}{Z_p Z_s + \omega^2 M^2} = \frac{\vec{V}_p}{Z_p + \frac{\omega^2 M^2}{Z_s}} \text{ and } \vec{I}_s = \frac{-j\omega M \vec{V}_p}{Z_p Z_s + \omega^2 M^2} = \frac{-j\omega M \vec{V}_p / Z_p}{Z_s + \omega^2 M^2 / Z_p}.$$

Thus, as a result of coupling between the two circuits, the impedance of the primary becomes $Z'_p = Z_p + \frac{\omega^2 M^2}{Z_s}$ and the secondary circuit behaves as if it were under an alternating voltage $\frac{-j\omega M \vec{V}_p}{Z_p}$ with impedance of the secondary given by $Z'_s = Z_s + \frac{\omega^2 M^2}{Z_p}$. The equivalent primary and secondary circuits [Fig. 3.3(a) and (b)] are shown below.

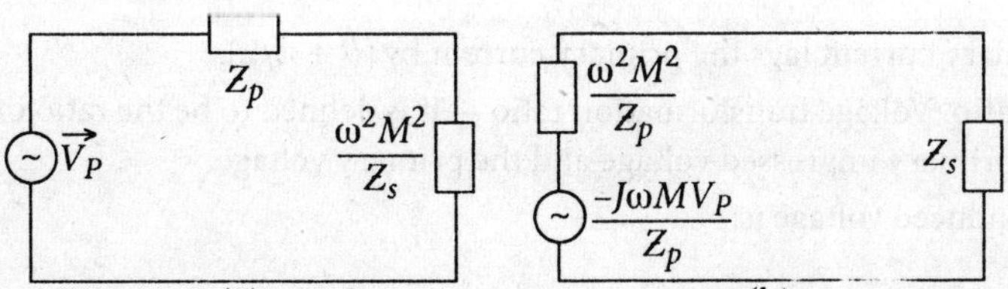

(a) (b)

Fig. 3.3

Z'_p and Z'_s are the effective primary and secondary impedances and $\frac{\omega^2 M^2}{Z}$ is known as *transferred* or *reflected impedance* from the secondary and $\frac{\omega^2 M^2}{Z_p}$ is the reflected impedance to the secondary from the primary.

Now

$$Z'_p = Z_p + \frac{\omega^2 M^2}{Z_s} = (R_p + j\omega L_p) + \frac{\omega^2 M^2}{R_s^2 + \omega^2 L_s^2}(R_s - j\omega L_s) = \left(R_p + \frac{\omega^2 M^2 R_s}{R_s^2 + \omega^2 L_s^2}\right) + j\omega\left(L_p - \frac{L_s \omega^2 M^2}{R_s^2 + \omega^2 L_s^2}\right).$$

$$\therefore\ R'_p = \text{effective primary resistance} = \text{primary inductance + reflected resistance} = R_p + \frac{\omega^2 M^2 R_s}{R_s^2 + \omega^2 L_s^2}.$$

$$L'_p = \text{effective primary inductance} = \text{primary inductance} - \text{reflected inductance from the secondary}$$

$$= L_p - \frac{\omega^2 M^2 L_s}{R_p^2 + \omega^2 L_s^2}.$$

So due to mutual coupling between the two circuits

$$\vec{I}_p = \frac{\vec{V}_p}{Z_p + \frac{\omega^2 M^2}{Z_s}} = \left[\frac{\vec{V}_p}{R_p + \frac{\omega^2 M^2 R_s}{R_s^2 + \omega^2 L_s^2} + j\omega\left(L_p - \frac{L_s \omega^2 M^2}{R_s^2 + \omega^2 L_s^2}\right)}\right].$$

$$\therefore\ i_p = \vec{I}_p \cdot \vec{n} = \frac{\hat{V}_p e^{j\omega t}\vec{n}\cdot\vec{n}}{\left[\left(R_p + \frac{\omega^2 M^2 R_s}{R_s^2+\omega^2 L_s^2}\right)^2 + \omega^2\left(L_p - \frac{L_s\omega^2 M^2}{R_s^2+\omega^2 L_s^2}\right)^2\right]^{1/2} e^{j\phi}} = \frac{\hat{V}_p}{|Z'_p|}\cos(\omega t - \phi),$$

where $\tan\phi = \omega L'_p / R'_p$.

Similarly,

$$\vec{I}_s = \frac{-j\omega M\vec{V}_p}{Z_s Z_p + \omega^2 M^2} = \frac{\frac{-j\omega M\vec{V}_p}{Z_p}}{Z_s + \frac{\omega^2 M^2}{Z_p}} = \frac{-j\omega M\vec{V}_p}{Z_p}\cdot\frac{\vec{I}_p}{\vec{V}_p} = \frac{-j\omega M\vec{I}_p}{Z_s}$$

$$= \vec{I}_p \frac{\omega M e^{\frac{-j\pi}{2}}}{\sqrt{R_s^2+\omega^2 L_s^2}}\cdot e^{-j\theta}\left(\because -j\omega M = \omega M e^{-j\pi/2}\right) \tag{3.5}$$

$$= \vec{I}_p \frac{\omega M}{\sqrt{R_s^2+\omega^2 L_s^2}}\cdot e^{-j(\theta+\pi/2)} = \frac{\omega M\hat{V}_p}{|Z_s|\,|Z'_p|}\cdot e^{j(\omega t-\phi-\theta-\pi/2)}\vec{n} = \frac{\omega M\hat{V}_p}{|Z_s|\,|Z'_p|}\cdot e^{[j\omega t-\phi-(\theta+\pi/2)]}\vec{n}. \tag{3.6}$$

Thus, the secondary current lags the primary current by $(\theta+\pi/2)$.

Transformation ratio. Voltage transformation ratio — it is defined to be the ratio of the secondary voltage developed due to primary impressed voltage and the primary voltage.

The secondary induced voltage is:

$$\vec{V}_s = \frac{j\omega M\vec{V}_p}{Z_p} = \frac{\hat{V}_p\omega M e^{j(\omega t-\phi-\pi/2)}}{|Z_p|}\vec{n}. \quad \therefore \quad \frac{V_s}{V_p}(\text{rms values}) = \frac{\omega M}{\sqrt{R_p^2+\omega^e L_p^2}}.$$

If $R_p \ll \omega L_p$, the above ratio becomes

$$T_v = \frac{\omega M}{\omega L_p} = \frac{M}{L_p}. \tag{3.7}$$

Now, if there is no leakage, coupling coefficient tends to unity. The coupling coefficient is defined by the relation $k = M/\sqrt{L_p L_s}$.

If the primary current be I, and there is no leakage of lines of force, we get

$$\oint \vec{H}\cdot d\vec{l} = Hl = N_1 I \Rightarrow H = \frac{N_1 I}{l},$$

N_1 = total number of turns in the primary.

H = field due to I.

The flux linked with secondary is given by $\phi_s = H\alpha N_2 = \dfrac{N_2\alpha N_1 I}{l}$.

$\therefore$ mutual inductance, which is the flux linked up with the secondary for unit current flow in the primary is $M = \alpha N_1 N_2 / l$.

Self-inductances of the primary and secondary are given by

$$L_p = \frac{\alpha N_1^2}{l}, \quad L_s = \frac{\alpha N_2^2}{l}. \tag{3.8}$$

$\therefore$ in case of ideal transformer having no leakage, the coefficient of coupling

$$k=\frac{M}{\sqrt{L_p L_s}}=1. \tag{3.9}$$

$\therefore$ from (3.7), $T_v=\dfrac{M}{L_p}=\sqrt{\dfrac{L_s}{L_p}}=\dfrac{N_2}{N_1}=T_n$, called *turn-ratio.*

Current transformation ratio (T_c): $\vec{I_s}=\dfrac{-j\omega M\vec{I_p}}{Z_s}$.

$$\therefore\ T_c=\frac{I_s}{I_p}\text{ rms values}=\frac{\omega M}{\sqrt{R_s^2+\omega^2 L_s^2}}.$$

If $R_s \ll L_s\omega$, then $T_c=\dfrac{\omega M}{\omega L_s}=\sqrt{\dfrac{L_p L_s}{L_s^2}}=\sqrt{\dfrac{L_p}{L_s}}=\dfrac{N_1}{N_2}=\dfrac{1}{T_n}$.

Thus, the current transformation ratio is the inverse of the voltage transformation ratio when the non-inductive load is of small value.

Impedance transformation. Now,

$$I_p=\frac{V_p}{|Z'_p|},\quad I_s=\frac{V_s}{|Z'_s|},\quad \frac{I_s}{I_p}=\frac{V_s}{V_p}\cdot\frac{|Z'_p|}{|Z'_s|}\ \text{or,}\ \frac{|Z'_p|}{|Z'_s|}=\frac{N_1}{N_2}\cdot\frac{N_1}{N_2}=\left(\frac{N_1}{N_2}\right)^2\ \text{or,}\ \frac{|Z'_s|}{|Z'_p|}=\left(\frac{N_2}{N_1}\right)^2=(T_n)^2. \tag{3.10}$$

Primary no-load current. It is the current in the primary when the secondary is open-circuited, i.e., $I_s=0$. In such a case, the primary current i_p lags the emf by an angle $\tan^{-1}\frac{\omega L_p}{R_p}$. Since, $R_p \ll \omega L_p$, the angle of lag tends to become $\pi/2$. When the secondary is closed through a noninductive load of small value, the angle of lag becomes

$$\tan\phi=\frac{\omega\left(L_p-\dfrac{\omega^2M^2L_s}{R_s^2+\omega^2L_s^2}\right)}{R_p+\dfrac{\omega^2M^2R_s}{R_s^2+\omega^2L_s^2}}.$$

Since, inductance of the primary decreases and the effective resistance increases, $\tan\phi$ decreases, which means, under load condition of the transformer, i.e., when the secondary is closed, the angle of lag of the current behind the emf decreases and hence, power factor increases, so the power taken from the ac source increases. If the secondary load is very small $R_s\to 0$, then

$$\tan\phi=\frac{\omega\left(L_p-\dfrac{\omega^2M^2L_s}{\omega^2L_s^2}\right)}{R_p}=\frac{\omega\left(L_p-L_p\right)}{R_p}=0.$$

$\therefore\ \phi=0$ and hence, power factor becomes unity and the power drawn from ac source becomes maximum.

Effective primary resistance and inductance in terms of turn-ratio:

$$R'_p=R_p+\frac{\omega^2M^2R_s}{R_s^2+\omega^2L_s^2},\quad L'_p=L_p-\frac{\omega^2M^2L_s}{R_s^2+\omega^2L_s^2}.$$

When $R_s \ll \omega L_s$ (R_s is very small)

$$R'_p = R_p + \left(\frac{L_p}{L_s}\right) R_s = R_p + \frac{R_s}{T_n^2}, \quad L'_p = L_p - L_p = 0.$$

This shows that when the secondary of high inductance is closed by a small resistance, the primary inductance is effectively reduced to zero.

Relation between effective impedance of the primary, the impedance of the secondary and the transformation ratio:

As a result of coupling between the primary and secondary of the transformer, both primary and the secondary circuit impedances now become $Z'_p = Z_s + \dfrac{\omega^2 M^2}{Z_s}$; $Z'_s = Z_s + \dfrac{\omega^2 M^2}{Z_p}$.

$$Z'_p = R_p + j\omega L_p + \frac{\omega^2 M^2 (R_s - j\omega L_s)}{R_s^2 + \omega^2 L_s^2} = \left[R_p + \frac{\omega^2 M^2 R_s}{(R_s^2 + \omega^2 L_s^2)}\right] + j\omega\left(L_p - \frac{\omega^2 M^2 L_s}{R_s^2 + \omega^2 L_s^2}\right).$$

$$\therefore \left(Z_e^p\right)^2 = \left(R_p + \frac{\omega^2 M^2 R_s}{(R_s^2 + \omega^2 L_s^2)}\right)^2 + \omega^2\left(L_p - \frac{\omega^2 M^2 L_s}{R_s^2 + \omega^2 L_s^2}\right)^2 = \left(R_p + T_c^2 R_s\right)^2 + \omega^2\left(L_p - T_c^2 L_s\right)^2.$$

Similarly,

$$Z'_s = (R_s + j\omega L_s) + \frac{\omega^2 M^2 (R_p - j\omega L_p)}{R_p^2 + \omega^2 L_p^2} = \left[R_s + \frac{\omega^2 M^2 R_p}{R_p^2 + \omega^2 L_p^2}\right] + j\omega\left(L_s - \frac{\omega^2 M^2 L_p}{R_p^2 + \omega^2 L_p^2}\right).$$

$$\therefore \left(Z_e^s\right)^2 = \left[R_s + \frac{\omega^2 M^2 R_p}{R_p^2 + \omega^2 L_p^2}\right]^2 + \omega\left[L_s - \frac{\omega^2 M^2 L_p}{R_p^2 + \omega^2 L_p^2}\right]^2 = \left(R_p + T_v^2 R_p\right)^2 + \omega^2\left(L_s - T_v^2 R_p\right)^2, \quad (3.11)$$

where T_v and T_c are the voltage and current transformation ratio.

3.4 Practical Transformer

What we have considered previously is an ideal transformer, which is far from practical realisation. A practical transformer differs from an ideal one in some of the following aspects (a) finite values of primary and secondary inductances (b) primary and secondary resistances (c) eddy current and hysteresis losses in the core material, known as *iron loss.* There is also loss of some useful energy in the resistance of the coils in the form of heat, known as the *copper loss* (d) shunt capacitances across the windings and (e) mutual capacitance between the windings.

To consider such non-ideal practical transformer, it is usual to consider inductances in each circuit to consist of two parts: (i) a leakage inductance having no coupling with the other circuit (ii) coupled inductances, which has $k = 1$. Since, $M = k\sqrt{L_p L_s} = \sqrt{kL_p \cdot kL_s}$ the coupled inductances are kL_p and kL_s and the corresponding leakage inductances are $(1-k)L_p$ and $(1-k)L_s$ respectively.

A practical transformer can be represented Fig. 3.4. Any loss due to eddy currents and hysteresis may be represented by a resistance in parallel with the primary coupled inductance. We represent by [Fig. 3.4]

R_p = resistance of the primary winding;
$(1-k)L_p$ = primary leakage inductance;
kL_p = primary coupled inductance;
r_e = resistance equivalent to eddy current loss, hysteresis loss or core loss;
R_s = load across the secondary;
r_2 = resistance of the secondary winding;
$(1-k)L_s$ = leakage inductance of the secondary;
kL_s = coupled secondary inductance;
C_1 = self-capacitance of the primary coils;
C_2 = self-capacitance of the secondary coils;
C = mutual capacitance between the primary and secondary;
M = mutual inductance.

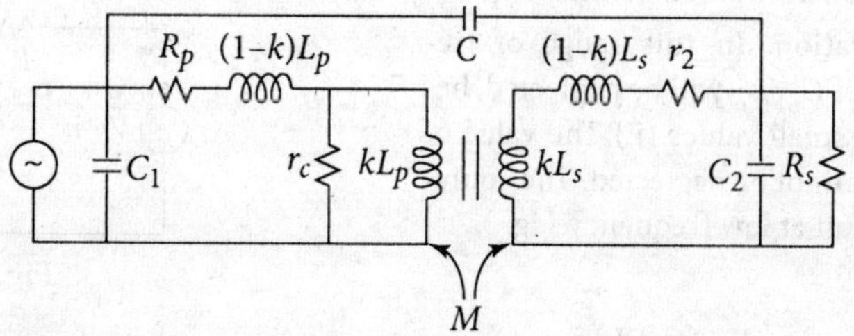

Fig 3.4

Neglecting the effect of C and transferring the circuit elements from the secondary side to the primary side we can draw the equivalent circuit of a practical transformer. In the process of drawing the equivalent circuit, where elements of the secondary have been shown to be referred to the primary side, we have taken help of the transformation ratio.

We have shown that in case of transformer, $\dfrac{V_2}{V_1} = T_n$, $\dfrac{I_2}{I_1} = \dfrac{1}{T_n}$, $\dfrac{Z_2}{Z_1} = (T_n)^2$.

So a voltage V_2 in the secondary is equivalent to V_2/T_n in the primary, a current I_2 in the secondary is equivalent to I_2T_n in the primary and an impedance Z_2, in the secondary is equivalent to Z_2/T_n^2 in the primary. So, the equivalent circuit is shown in Fig. 3.5.

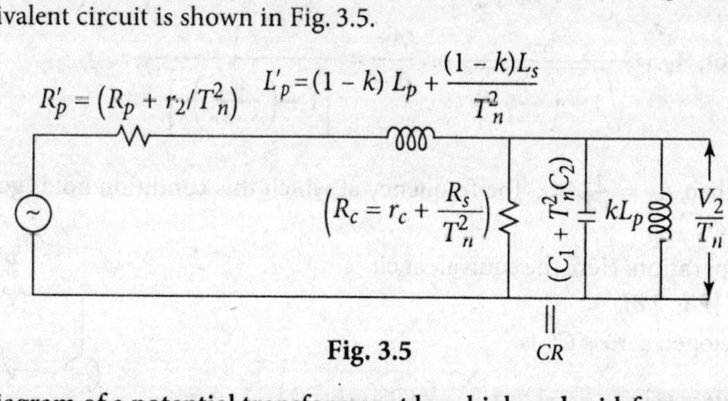

Fig. 3.5

Equivalent circuit diagram of a potential transformer at low, high and mid-frequencies.

The equivalent circuit of a potential transformer [Fig. 3.4] and the equivalent circuit of the same with the secondary circuit elements are referred to the primary side [Fig. 3.5]. The operation of the transformer will be determined by the frequency range.

Medium frequency. In equivalent circuit diagram [Fig. 3.5]:

(i) The stray capacitance C'_p is very small, and neglected at mid-frequency.

(ii) L'_p is of small value, hence the reactance at mid-frequency is also small, hence it does not affect the gain at mid-frequency.

(iii) kL_p is large, but since it is connected in parallel in the circuit, its reactance is small, hence neglected.

(iv) We will, in general, neglect the effect of R_c.

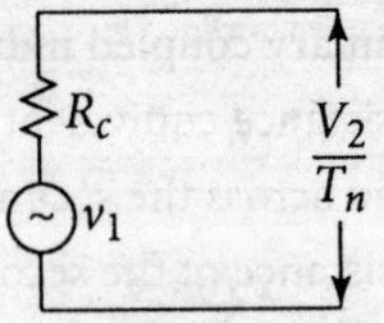

Fig. 3.6

The equivalent circuit diagram at mid-frequency is shown in Fig. 3.6.

∴ voltage gain or transformation at mid-frequency $A_{\text{mf}} = V_2/V_1 = T_n$.

Low-frequency operation. In this range of frequency: (i) the effect of C'_p, L'_p will be neglected, because all of them are small values (ii) The value of kL_p is large, hence, cannot be neglected. The equivalent appropriate circuit at low frequency Fig. 3.7.

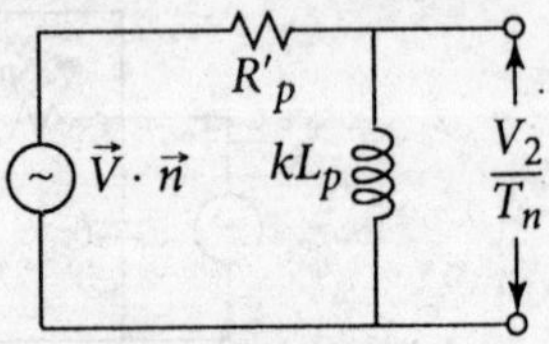

Fig. 3.7

$$\frac{\hat{V}\cos(\omega t-\phi)}{\left[R_p'^2+(k\omega L_p)^2\right]^{1/2}}\cdot k\omega L_p = \text{voltage across } kL_p.$$

The rms voltage developed is

$$\frac{\frac{\hat{V}}{\sqrt{2}}\cdot k\omega L_p}{\left[R_p'^2+(k\omega L_p)^2\right]^{1/2}} = \frac{V_2}{T_n} \text{ or, } \frac{V_1}{\left[R_p'^2+(k\omega L_p)^2\right]^{1/2}}\cdot k\omega L_p = \frac{V_2}{T_n}$$

$$\text{or, } \frac{V_1}{k\omega L_p\left[1+\left(\frac{R'_p}{k\omega L_p}\right)^2\right]^{1/2}}\cdot k\omega L_p = \frac{V_2}{T_n} \text{ or, } \frac{V_2}{V_1} = \frac{T_n}{\sqrt{1+\left(\frac{R'_p}{k\omega L_p}\right)^2}}$$

$$\text{or, } A_{\text{lf}} = \frac{A_{\text{mf}}}{\sqrt{1+\left(\frac{R'_p}{k\omega L_p}\right)^2}} \text{ or, } \frac{A_{\text{lf}}}{A_{\text{mf}}} = \frac{1}{\left[1+\left(\frac{R'_p}{k\omega L_p}\right)^2\right]^{1/2}}.$$

If $k\omega L_p = R'_p$, then $A_{\text{lf}} = \frac{1}{\sqrt{2}}A_{\text{mf}}$. The frequency at which this condition holds good is known as *lower half-power frequency*.

High-frequency operation: Here the equivalent circuit is represented [Fig. 3.8].

The voltage developed across C'_p is

$$V_c = \frac{\hat{V}\cos(\omega t-\phi)}{\left[R_p'^2+\left(\omega L'_p-\frac{1}{\omega C'_p}\right)^2\right]^{1/2}}\cdot\frac{1}{\omega C'_p} = \frac{\hat{V}\cos(\omega t-\phi)}{\left[R_p'^2+(X_L-X_C)^2\right]^{1/2}}\cdot X_c.$$

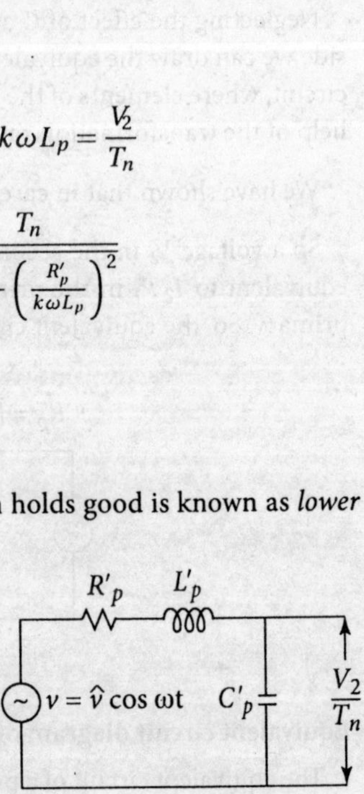

Fig. 3.8

$$\therefore \text{ rms value of the voltage} = \frac{\hat{V}/\sqrt{2}}{\left[R_p'^2 + (X_L - X_C)^2\right]^{1/2}} \cdot X_C = \frac{V_1}{\left[R_p'^2 + (X_L - X_C)^2\right]^{1/2}} X_C.$$

$$\therefore \frac{V_1}{\left[R_p'^2 + (X_L - X_C)^2\right]^{1/2}} \cdot X_C = \frac{V_2}{T_n} \text{ or, } A_{hf} = \frac{V_2}{V_1} = \frac{T_n X_C}{\left[R_p'^2 + (X_L - X_C)^2\right]^{1/2}}$$

$$\text{or, } A_{hf} = \frac{A_{mf} X_C}{\left[R_p'^2 + (X_L - X_C)^2\right]^{1/2}} \text{ or, } \frac{A_{hf}}{A_{mf}} = \frac{X_C}{\left[R_p'^2 + (X_L - X_C)^2\right]^{1/2}}.$$

When $X_L = X_C$, we get $\dfrac{A_{hf}}{A_{mf}} = \dfrac{X_C}{R_p'}$.

Again, when $R_p' = \omega L_p' - \dfrac{1}{\omega C_p'}$, we get $A_{hf} = \dfrac{X_C A_{mf}}{\sqrt{2}(X_L - X_C)}$.

The frequency, where this condition holds good, is known as *upper half-power frequency*.

Auto-transformer. It is a transformer with one winding only, part of it being common to both primary and secondary. Theory and action is similar to two-winding transformer. It can act as an auto step-up transformer [Fig. 3.9(a)], or as an auto step-down transformer [Fig. 3.9(b)].

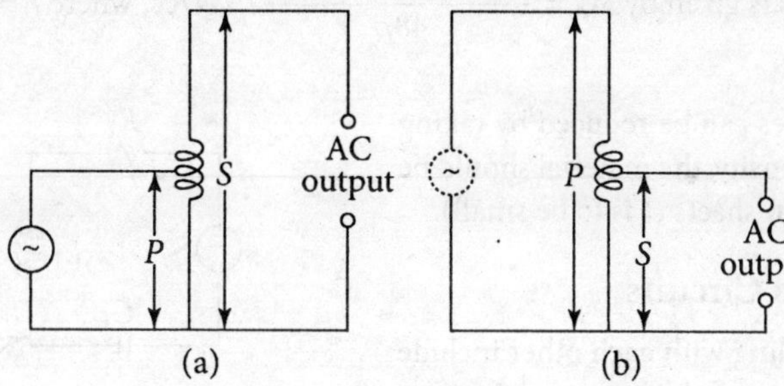

Fig. 3.9

Since this transformer uses one coil it uses less copper, hence it is cheaper. When it is used as a step-up transformer, the entire winding acts as the secondary and a part is used as the primary. In case of step-down transformer, the entire winding is the primary and only a part of it acts as the secondary. Thus, in both cases it is seen that one part of the winding is always common to both primary and secondary.

Such a transformer Is usually used:

(i) as auto starter transformer to give 50 to 60 % of full voltage to an induction motor during starting.
(ii) as furnace transformer.
(iii) to give a small boost to a distribution cable to correct for the voltage drop.
(iv) as line output transformer (LOT) in TV receiver producing a voltage of 18–20kV for picture tube.

3.5 Efficiency of a Transformer — All-day Efficiency

The efficiency of a transformer in commercial sense is defined as $\dfrac{\text{Output in W}}{\text{Intake in W}}$.

The all day efficiency is the ratio $\dfrac{\text{Output in kW/h}}{\text{Intake in kW/h}}$.

The all day efficiency is not equal to the commercial efficiency, it is less than the commercial efficiency, because the primary windings are connected to the line for 24 hours/day and so there is always core loss going on, whereas copper losses occur only under load condition.

3.6 Transformer Losses and Their Remedy

In course of working of a transformer, useful energy is lost in two forms,

(i) Iron loss or core loss due to eddy current, hysteresis, etc. of the core material.

(ii) Copper loss due to Joulean heating.

Hysteresis loss and eddy current loss. Hysteresis loss is given by Steinmetz formula: $W_\eta = \eta B_m^{1.6}$ ergs per cc per cycle,

where B_m = maximum value of magnetic induction;

η = constant, called *Steinmetz coefficient*;

$\eta = 0.002$ for good quality transformer iron;

$= 0.00076$ for alloyed steel containing 5 % silicon (called *Stalloy*).

Hence, for minimisation of hysteresis loss, stalloy is preferable and low frequency is desirable.

The eddy current loss is given by $W_e = \dfrac{(8.8)^2 B_m^2 f^2 \times 10^{-16}}{48\rho} t^2$ W/cc, where f = frequency, t = thickness, ρ = density.

So, eddy current losses can be reduced by taking core material of large density, the material should be taken in the form of thin sheets (t is to be small).

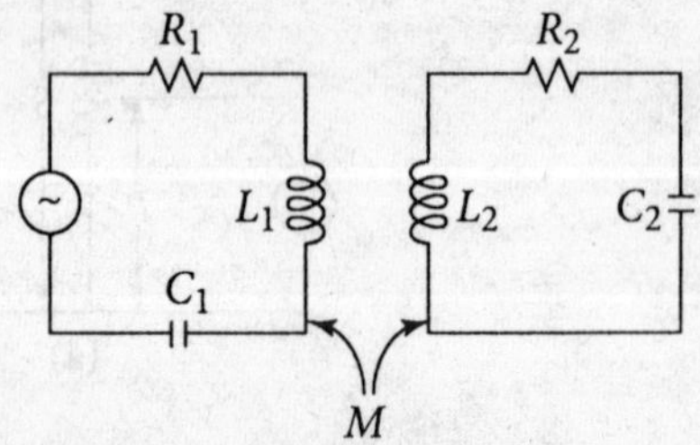

Fig. 3.10

3.7 LCR Coupled Circuits

If the two circuits coupling with each other include capacitances C_1 and C_2 in addition to resistances (R_1, R_2) and inductances (L_1, L_2) the circuits will be as shown in Fig. 3.10.

The emf equations for the circuits can be written as

$$R_1\vec{I}_1 \cdot \vec{n} + j\omega L_1 \vec{I}_1 \cdot \vec{n} + \frac{1}{j\omega C_1}\vec{I}_1 \cdot \vec{n} + j\omega M\vec{I}_2 \cdot \vec{n} = \vec{V} \cdot \vec{n}$$

$$\text{or, } \left[R_1 + j\left(\omega L_1 - \frac{1}{\omega C_1}\right)\right]\vec{I}_1 + j\omega M\vec{I}_2 = \vec{V} \text{ or, } Z_1\vec{I}_1 + j\omega M\vec{I}_2 = \vec{V}. \tag{3.12}$$

$\vec{I}_1$ and $\vec{I}_2$ are the primary and secondary current vectors to be determined.

$$R_2\vec{I}_2 \cdot \vec{n} + j\omega L_2 \vec{I}_2 \vec{n} + j\omega M\vec{I}_1 \cdot \vec{n} + \frac{1}{j\omega C_2}\vec{I}_2 \cdot \vec{n} = 0$$

$$\text{or, } \left[R_2 + j\left(\omega L_2 - \frac{1}{\omega C_2}\right)\right]\vec{I}_2 + j\omega M\vec{I}_1 = 0 \text{ or, } j\omega M\vec{I}_1 + Z_2\vec{I}_2 = 0. \tag{3.13}$$

Solving (3.11) and (3.12),

$$\vec{I}_1 = \frac{\begin{vmatrix} \vec{V} & j\omega M \\ 0 & Z_2 \end{vmatrix}}{\begin{vmatrix} Z_1 & j\omega M \\ j\omega M & Z_2 \end{vmatrix}} = \frac{Z_2\vec{V}}{Z_1Z_2+\omega^2M^2} \text{ and } \vec{I}_2 = \frac{-j\omega M\vec{I}_1}{Z_2} = \frac{-j\omega M\vec{V}}{Z_1Z_2+\omega^2M^2} = \frac{\frac{-j\omega M\vec{V}}{Z_1}}{Z_2+\frac{\omega^2M^2}{Z_1}}. \tag{3.14}$$

(3.14) shows that the current $\vec{I}_2$ in the secondary is due to a voltage generator of voltage $\frac{-j\omega M\vec{V}}{Z_1}$ with internal impedance $\frac{\omega^2M^2}{Z_1}$. So, the equivalent secondary circuit [Fig. 3.11].

Now, $\vec{I}_2 = \dfrac{-j\omega M\vec{V}}{Z_1Z_2+\omega^2M^2}$.

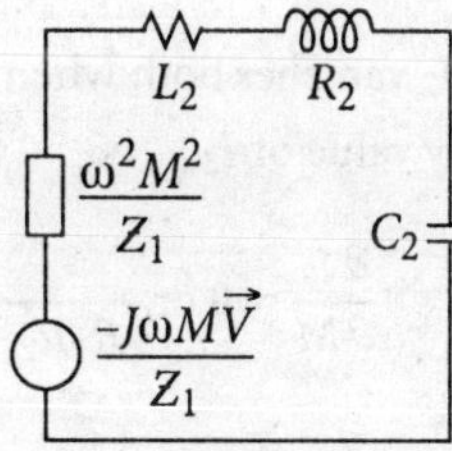

Fig. 3.11

In our case,

$$Z_1 = R_1 + j\left(\omega L_1 - \frac{1}{\omega C_1}\right) = R_1 + jX_1,$$

$$Z_2 = R_2 + jX_2.$$

$$\therefore\ \vec{I}_2 = \frac{-j\omega M\vec{V}}{(R_1+jX_1)(R_2+jX_2)+\omega^2M^2} = \frac{-j\omega M\vec{V}}{(R_1R_2+\omega^2M^2-X_1X_2)+j(R_1X_2+X_1R_2)}. \tag{3.15}$$

$$\vec{I}_1 = \frac{\vec{V}}{z_1+\frac{\omega^2M^2}{z_2}} = \frac{\vec{V}}{R_1+jX_1+\frac{\omega^2M^2(R_2-jX_2)}{R_2^2+X_2^2}}$$

$$= \frac{\vec{V}}{\left(R_1+\frac{\omega^2M^2R_2}{R_2^2+X_2^2}\right)+j\left(X_1-\frac{\omega^2M^2X_2}{R_2^2+X_2^2}\right)} = \frac{\hat{V}e^{j\omega t}\vec{n}}{R_1'+jX_1'} = \frac{\hat{V}e^{j(\omega t-\phi)}\vec{n}}{\sqrt{R_1'^2+X_1'^2}},$$

where

$$\tan\phi = \frac{X_1'}{R_1'} = \frac{X_1-\frac{\omega^2M^2X_2}{R_2^2+X_2^2}}{R_1+\frac{\omega^2M^2R_2}{R_2^2+X_2^2}}. \tag{3.16}$$

(3.16) shows that $\vec{I}_1$ lags $\vec{V}$ by an angle ϕ. In order that the current vector is in phase with $\vec{V}$, ϕ must be zero.

$\therefore$ from (3.16), we may have: (i) $X_1 = X_2 = 0$ (ii) $X_1 = \dfrac{\omega^2M^2X_2}{R_2^2+X_2^2}$.

Now, $X_1 = X_2 = 0$ implies

$$\omega L_1 - \frac{1}{\omega C_1} = \omega L_2 - \frac{1}{\omega C_2} = 0 \text{ or, } \omega = \omega_0 = \frac{1}{\sqrt{L_1C_1}} = \frac{1}{\sqrt{L_2C_2}},$$

it means the natural frequencies of both must be same. Under this condition, then $\vec{I}_1 = \dfrac{\hat{V}\, e^{j\omega t}\, \vec{n}}{R_1'}$.

$$\therefore\ I_1 = \frac{V}{R_1'} = \frac{VR_2}{R_1R_2 + \omega_0^2 M^2}\ \text{(rms values)};\qquad \vec{I}_2 = \frac{-j\omega_0 M\vec{V}}{R_1R_2 + \omega_0^2 M^2} = \frac{\omega_0 M\hat{V}\, e^{j(\omega t - \pi/2)}}{\left(R_1R_2 + \omega_0^2 M^2\right)}\vec{n}.$$

$$\therefore\ I_2 = \frac{\omega_0 MV}{R_1R_2 + \omega_0^2 M^2}. \tag{3.17}$$

From (3.17) I_2 vanishes both when $M = 0$ and $M = \infty$, so there is a value of M for which I_2 is maximum. For stationary value of I_2

$$\frac{\partial I_2}{\partial M} = 0 \text{ or, } \frac{1}{R_1R_2 + \omega_0^2 M^2} - \frac{2M^2\omega_0^2}{\left(R_1R_2 + \omega_0^2 M^2\right)} = 0 \text{ or, } \omega_0^2 M^2 = R_1R_2.$$

It can be shown that for this value of M, I_2 is a maximum. This is the condition of optimum coupling. Under this condition of optimum coupling, the coefficient of coupling is given by

$$k = \frac{M}{\sqrt{L_1L_2}} = \frac{\sqrt{R_1R_2}}{\omega_0\sqrt{L_1L_2}} = \sqrt{\frac{R_1}{\omega_0 L_1}\cdot\frac{R_2}{\omega_0 L_2}} = \frac{1}{\sqrt{Q_1Q_2}},$$

where Q_1 and Q_2 are the Q-factors of the primary and secondary. Then

$$I_1 = \frac{VR_2}{2R_1R_2} = \frac{V}{2R_1} \text{ and } I_2 = \frac{V}{2\sqrt{R_1R_2}}.$$

Again, when condition (ii) is satisfied, we get (cf. pp.77)

$$X_1 = \frac{\omega^2 M^2 X_2}{R_2^2 + X_2^2}. \tag{3.18}$$

Usually, as the case is, the resistances of the primary and secondary circuits are small compared to the reactances X_1 and X_2 and in radio-communication practice, the inductance and capacitance of both the circuits are made equal. This gives rise to coupled tuned circuit, $X_1 = X_2 = X$ and (3.18) becomes

$$X^2 = \omega^2 M^2 \text{ or, } \omega^2 M^2 = \left(\omega L - \frac{1}{\omega C}\right)^2 = L^2\omega^2\left(1 - \frac{1}{\omega^2 LC}\right)^2 \text{ or, } M^2 = L^2\left(1 - \frac{\omega_0^2}{\omega^2}\right)^2.$$

Now, $k^2 = \dfrac{M^2}{L^2}$.

$$\therefore\ k^2 = \left(1 - \frac{\omega_0^2}{\omega^2}\right)^2 \text{ or, } \omega_0^2 = (1 \pm k)\omega^2 \text{ or, } f_0^2 = f^2(1 \pm k).$$

$$\therefore\ f = \frac{f_0}{(1 \pm k)^{1/2}} \simeq \frac{f_0}{1 \pm \frac{k}{2}}. \tag{3.19}$$

(3.19) shows that there are two resonant frequencies — one slightly less, one slightly greater than f_0, the resonant frequency of either tuned circuit is considered separately.

Behaviour of the resonant circuits coupled together. From the foregoing analysis we understand that the behaviour of the two circuits depends on the value of k, the degree of coupling.

For very small value of k, which may be due to small M value or very large secondary impedance, the coupled impedance is small and so the primary current is very nearly the same as if there is no secondary circuit and the effect of the primary current on the secondary will be likewise small. When k is large (may be due to large M value or small secondary impedance), the coupled impedance is great and great is the effect on the current and voltage in the primary as well as in the secondary.

1. *Small coupling* ($k = 0.005$) : In this case, the coupled impedance is small and the primary current is not affected to a large extent, the curve of primary current vs. frequency exhibits the characteristics of series resonance curve of the primary, [Fig. 3.12(a)] for $k = 0.005$. The secondary current under this condition is small having much more weakness than the usual resonance curve for the secondary alone [Fig. 3.12(b) for $k = 0.005$].

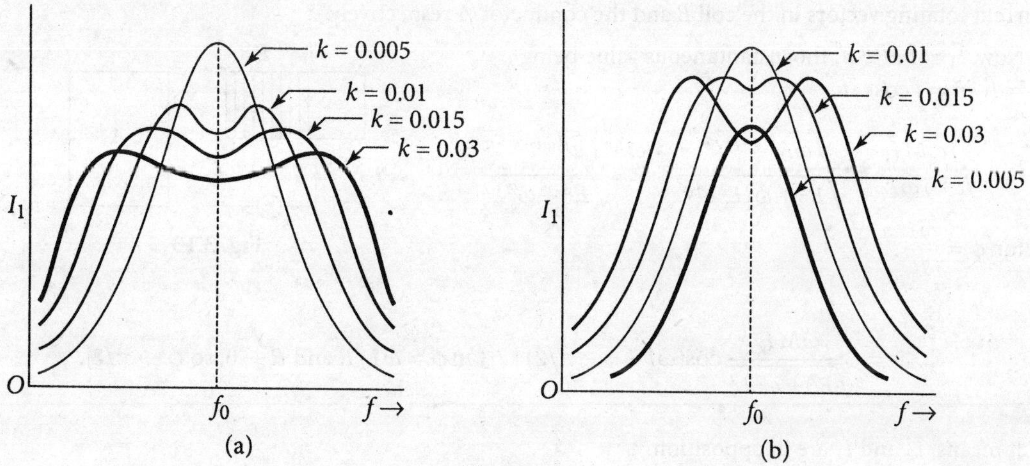

Fig. 3.12

2. *Phase coupling*: As the value of k increases, the coupled impedance increases, the primary current curve becomes broader; peak value is reduced. The secondary current at the same time increases and the curve becomes broader until the current becomes maximum for $k = 0.01$. The primary current at this stage shows double humps due to the reflected resistance from the secondary being equal to the resistance of the primary (case of critical coupling or optimum coupling).

As coupling goes beyond the critical values the primary and secondary currents decrease more at resonant frequency. The primary current double humps are more prominent and become separated. The secondary current also shows double humps. At resonance, the coupled resistance in the primary becomes large, the current becomes relatively small while on each side of the resonance, there is a frequency, where the coupled reactance exactly neutralises the primary reactance thus giving zero reactance in the primary impedance causing large currents corresponding to these frequencies. A double-hump in the primary current is thus explained. Since the secondary current is related to the primary current, the corresponding curve also shows double humps [Fig. 3.12(b) for $k = 0.015, k = 0.03$].

Width of bandpass. We have shown that the secondary current becomes maximum corresponding to two frequencies f_1 and f_2 on either side of resonant frequency f_0, where f_1 and f_2 are given by

$$f_1 = \frac{f_0}{1 + k/2}, \quad f_2 = \frac{f_0}{1 - k/2}.$$

$$\therefore\ f_2 - f_1 = \frac{f_0}{1-k/2} - \frac{f_0}{1+k/2} = f_0\frac{(1+k/2-1+k/2)}{1-k^2/4} \text{ or, } \Delta f \simeq f_0 k \text{ since } k^2/4 \ll 1.$$

The above relation decides the width of the band of frequencies between the double humps. Since for k small, the fall in secondary current between the frequencies is not much, we get band pass circuit of immense importance, used in connection with improving the fidelity of receivers of radio signals.

3.8 Force on a Conductor Due to ac flowing in a Neighbouring Circuit

Let an alternating current flow through the coil B and a conductor A of self-inductance L, and resistance R is in the neighbourhood, the mutual inductance between them being M. The conductor as a result will experience a force of repulsion.

The emf equation in the conductor A is $R\vec{I}_2\cdot\vec{n} + j\omega L\vec{I}_2\cdot\vec{n} + j\omega M\vec{I}_1\cdot\vec{n} = 0$, where $\vec{I}_2$ and $\vec{I}_1$ are the current rotating vectors in the coil B and the conductor A respectively.

Now, $\vec{I}_1 = \hat{I}e^{j\omega t}\vec{n}$, the instantaneous value being $i_1 = \vec{I}_1\cdot\vec{n} = \hat{I}\cos\omega t$.

$$\therefore\ \vec{I}_2 = \frac{-j\omega M\vec{I}_1}{R+j\omega L} = \frac{\hat{I}\omega M e^{-j\pi/2}}{\sqrt{R^2+\omega^2L^2}e^{j\phi}} = \frac{\omega M\hat{I}\,e^{j(\omega t-\phi-\pi/2)\vec{n}}}{\sqrt{R^2+\omega^2L^2}},$$

$$\tan\phi = \frac{\omega L}{R}.$$

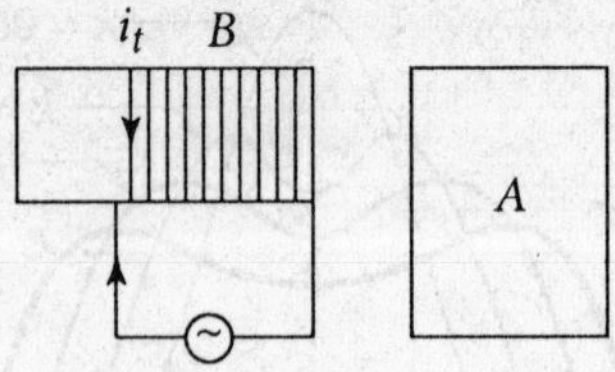

Fig. 3.13

$$\therefore\ i_2 = \frac{\omega M\hat{I}}{\sqrt{R^2+\omega^2L^2}}\cos(\omega t-\phi-\pi/2)\ [\because\ \tan\phi = \omega L/R \text{ and } R\to 0, \text{ so } \phi\to\pi/2].$$

It means, i_2 and i_1 are in opposition.

The force F between B and A is proportional to the average value of the products of i_1 and i_2 and so

$$F = \text{const.}\frac{1}{T}\int_0^T i_1 i_2\,dt = K\left[\frac{1}{T}\int_0^T \frac{\hat{I}\cos\omega t\cdot\omega M\hat{I}\cos(\omega t-\phi-\pi/2)}{\sqrt{R^2+\omega^2L^2}}dt\right]$$

$$= K\cdot\frac{1}{2}\cdot\frac{\omega M\hat{I}^2}{\sqrt{R^2+\omega^2L^2}}\left[\frac{1}{T}\int_0^T\cos(2\omega t-\phi-\pi/2)\,dt + \frac{1}{T}\int_0^T\cos(\phi+\pi/2)\,dt\right]$$

$$= K\cdot\frac{1}{2}\frac{\omega M\hat{I}^2}{\sqrt{R^2+\omega^2L^2}}\cos(\pi/2+\phi), \text{ since } \int_0^T\cos(2\omega t-\phi-\pi/2)\,dt = 0$$

$$= -K\cdot\frac{1}{2}\frac{\omega M\hat{I}^2}{\sqrt{R^2+\omega^2L^2}}\cdot\frac{\omega L}{\sqrt{R^2+\omega^2L^2}} = -\frac{K}{2}\cdot\frac{\omega^2LM\hat{I}^2}{(R^2+\omega^2L^2)}.$$

This shows that the force is a force of repulsion. The negative sign appears, because the current induced in the neighbouring conductor is in opposite direction to the current flowing in the coil. If $R\gg\omega L$, then $F = \frac{K}{2}\frac{\omega^2LM\hat{I}^2}{R^2}$ thus force of repulsion is greater, less is the resistance of the conductor and higher the frequency. Thus, if we place a ring on the end of an electromagnet, then when the alternating emf is switched on, the metal ring will be seen to shoot off, in conformity with our analysis.

3.9 Numerical Problems

1. The current in one circuit is $\hat{I}\cos\omega t$. A neighbouring circuit consists of a coil of self-inductance L and negligible resistance in series with a condenser C. If M be the mutual inductance between the circuits, show that the instantaneous charge on the condenser is $\frac{\omega MC\hat{I}}{\omega^2 LC-1}\sin\omega t$ in magnitude.

Solution: The secondary circuit equation is $\left(j\omega L+\frac{1}{j\omega C}\right)\vec{I_2}+j\omega M\vec{I_1}=0$. $\vec{I_1}$ = current in the primary.

$$\therefore\ \vec{I_2}=\frac{-j\omega M\vec{I_1}}{j\left(\omega L-\frac{1}{\omega C}\right)}=\frac{\omega Me^{j\pi/2}\vec{I_1}}{\left(\omega L-\frac{1}{\omega C}\right)e^{j\pi/2}}=\frac{\omega M\hat{I}\,e^{j(\omega t-\pi)}}{\left(\omega L-\frac{1}{\omega C}\right)}\vec{n}.$$

$$\therefore\ i_2=\frac{\omega M\hat{I}\cos(\omega t-\pi)}{\left(\omega L-\frac{1}{\omega C}\right)}=-\frac{\omega M\hat{I}\cos\omega t}{\left(\omega L-\frac{1}{\omega C}\right)}\ \text{or,}\ \frac{dp}{dt}=-\frac{\omega M\hat{I}\cos\omega t}{\left(\omega L-\frac{1}{\omega C}\right)}.$$

$$\therefore\ q=-\frac{\omega M\hat{I}}{\left(\omega L-\frac{1}{\omega C}\right)}\int\cos\omega t\,dt=\frac{-\omega M\hat{I}}{\left(\omega L-\frac{1}{\omega C}\right)}\cdot\frac{1}{\omega}\sin\omega t=\frac{-M\hat{I}\sin\omega t}{(\omega^2 CL-1)}\cdot\omega C=\frac{-\omega CM\hat{I}\sin\omega t}{(\omega^2 CL-1)}.$$

The instantaneous charge is of magnitude $\omega MC\hat{I}\sin\omega t$.

2. In an iron core transformer, the self-inductances of the primary and secondary are 10 and 4H respectively. Assuming that the windings have negligible resistances and that there is no magnetic leakage, calculate the primary current due to a 220V, 50 cycles per second voltage applied in the primary when the secondary is closed through a 100Ω resistance.

Solution: Under load condition, the primary current is given by

$$I_p(\text{rms})=\frac{V_p(\text{rms})}{z'_p}=\frac{V_p}{\sqrt{R_p+\frac{\omega^2M^2R_s}{R_s^2+\omega^2L_s^2}+\omega^2\left(L_p-\frac{\omega^2M^2L_s}{R_s^2+\omega^2L_s^2}\right)}}.$$

In this case, $k^2=\frac{M^2}{L_sL_p}=1, R_p\simeq 0, R_s=100\Omega$

$$\therefore\ I_p=\frac{V_p}{\sqrt{\frac{\omega^2L_pL_sR_s}{R_s^2+\omega^2L_s^2}+\omega^2\left(L_p-\frac{\omega^2L_pL_s^2}{R_s^2+\omega^2L_s^2}\right)^2}}$$

$$=\frac{220}{\sqrt{\frac{4\pi^2\times 50\times 50\times 10\times 4\times 10^2}{10^4+4\pi^2\times 50\times 50\times 4^2}+4\pi^2\times 50\times 50\left(10-\frac{4\pi^2\times 50\times 50\times 10\times 4\times 10^2}{10^4+4\pi^2\times 50\times 50\times 4^2}\right)}}$$

$=8.70$A.

3. In an ideal iron-core transformer, the secondary coil gives 400V, when the primary voltage is 50V. If the secondary current is 5A, calculate the primary current. If a resistance of 400Ω is now connected to the secondary, what will be its reflected value on the primary side.

Solution: From voltage transformation expression $T_v=\frac{V_s}{V_p}=\frac{N_2}{N_1}=T_n$ or, $\frac{N_2}{N_1}=\frac{400}{50}=8=T_n$.

From current transformation, $T_v = \frac{I_s}{I_p} = \frac{1}{T_n}$ or, $\frac{5}{I_p} = \frac{1}{T_n} = \frac{1}{8}$.

$\therefore\ I_p = 40\text{A}$.

From impedance transformation expression, $R_p = \left(\frac{L_p}{L_s}\right) R_s = \left(\frac{N_1}{N_2}\right)^2 R_s = \frac{R_s}{T_n^2} = \frac{40}{8\times 8} = 0.625\Omega$.

4. A single phase step-down transformer has a core of 100 sq cm cross section. The maximum flux density through the core is 10,000 lines/cm². If the supply is 220V, 50c/s and the turns ratio is 1/5, find the number of turns of each winding, assuming its resistance to be negligible.

Solution: Since the flux is alternating, it can be written as $\phi = \phi_{\max}\sin\omega t$.

$\therefore$ emf induced in the primary — its instantaneous value is $e_1' = -\frac{d\phi}{dt}\cdot N_1 A$ volt, where N_1 = number of primary turns.

$\therefore\ e_1' = -\phi_{\max}\omega N_1 A\cos\omega t = -2\pi f\phi_{\max} N_1 A\cos\omega t$.

The rms value of this $E_1' = -\frac{2\pi\phi_{\max}}{\sqrt{2}} f N_1 A = -4.44\phi_{\max} N_1 f A$.

Since, the resistance is small, applied primary voltage has only to oppose the induced primary voltage. Hence, $E_1 = -E_1' = 4.44\phi_{\max} N_1 f A$.

Similarly, E_2' = secondary induced voltage = $-4.44\phi_{\max} N_2 f A$, N_2 = number of secondary turns and this is the voltage appearing at the secondary terminals on no-load. So at no-load, $E_2 = -E_2'$.

$\therefore\ E_2/E_1 = N_2/N_1$.

Now, in the given problem $E_2/E_1 = N_2/N_1$ or, $E_2/220 = 1/5$.

$\therefore\ E_2 = 44\text{V}$.

Now $E_2 = 4.4\phi_{\max} N_2 f A$ or, $44 = 4.4\times 10\times 10^3\times 10^2\times N_2\times 50\times 10^{-4}$ or, $N_2 = \frac{44\times 10}{4.4\times 5} = 20$ turns.

$\therefore\ N_1 = 5N_2 = 100$ turns.

5. Two identical resonant circuits each consisting of a coil having $L = 100\mu\text{H}$, Q-factor of 100, are tuned by a 200F lossless condenser. An emf of 1V is applied to one circuit and the current is measured in the other. If $k = 0.005$, find the frequency for which the secondary current is maximum and values of primary and the secondary currents at this frequency.

Solution: The secondary current is given by $I_2 = \frac{\omega_0 M V}{R_1 R_2 + \omega_0^2 M^2}$ [from (3.16)].

For maximum value of $I_2:\ \omega_0^2 M^2 = R_1 R_2$.

$$\therefore\ (I_2)_{\max} = \frac{V}{2\sqrt{R_1 R_2}};\quad (I_1)_{\max} = \frac{V}{2R_1},\quad 0.005 = \frac{M}{100\times 10^{-6}}.$$

$$\text{In the present case, } Q = \frac{\sqrt{LC}}{R} = \sqrt{\frac{10^2\times 10^{-6}}{200\times 10^{-12}}}\cdot\frac{1}{R}.$$

$$\therefore\ 100 = \frac{1}{R}\sqrt{\frac{10^{-4}}{2\times 10^{-10}}} \text{ or, } R = \frac{1}{100}\sqrt{\frac{10^{-4}}{2\times 10^{-20}}} = \sqrt{\frac{10^2}{2}} = \frac{10}{\sqrt{2}}.$$

$$\therefore\ \omega_2^2 = \frac{R_1 R_2}{M^2} = \frac{R^2}{M^2} = \frac{100}{2}\cdot\frac{1}{(5\times 10^{-7})^2}.$$

$$\therefore\ \omega_0 = \frac{10}{\sqrt{2}}\frac{1}{5\times 10^{-7}} = \sqrt{2}\times 10^7 = 1.41\times 10^7.$$

$$\therefore f = \frac{1}{2\pi} \times 1.41 \times 10^7 = 2.25 \text{ Mc/s}.$$

$$(I_2)_{max} = \frac{V}{2R} = \frac{1}{2 \times \frac{10}{\sqrt{2}}} = \frac{1}{1.41 \times 10} = \frac{1}{14} = 71.4\text{mA}; \quad (I_1)_{max} = \frac{V}{2R} = 71.4\text{mA}.$$

6. Two identical circuits resonant at 1 Mc/s, having Q-factor 100 and inductance 200μH are coupled by a mutual inductance of 6μH. An emf of 1V from a generator with zero internal impedance is connected in series with the primary; calculate (a) the frequencies at which the secondary current shows peaks (b) the magnitude of peak currents (c) secondary current at 1 Mc/s.

Solution: (a) The frequencies are given by $f_1 = \dfrac{f_0}{\sqrt{1+k}}$, $f_2 = \dfrac{f_0}{\sqrt{1-k}}$.

Now, $k = \dfrac{M}{\sqrt{L_1 L_2}} = \dfrac{M}{L} = \dfrac{6 \times 10^{-6}}{200 \times 10^{-6}} = 0.03.$

$$f_0 = 1 \times 10^6 \text{ c/s}$$

$$\therefore f_1 = \frac{10^6}{\sqrt{1+0.03}} = 10^6 \left(1 - \frac{1}{2} \times 0.03\right) = 985 \text{ kc/s}$$

$$f_2 = 10^6 \left(1 + \frac{1}{2} \times 0.03\right) = 1015 \text{ kc/s}.$$

(b) $(I_1)_m = V/2R$.

Now, $Q = \dfrac{\omega_0 L}{R} = \dfrac{2\pi f_0 L}{R}$.

$$\therefore R = \frac{2\pi f_0 L}{Q}.$$

$$\therefore (I_1)_m = \frac{QV}{4\pi f_0 L} = \frac{100 \times 1}{4 \times 3.14 \times 10^6 \times 2 \times 10^{-4}} = 39.8\text{mA}.$$

(c) $I_2 = \dfrac{\omega_0 M V}{R^2 + \omega_2^2 M^2} = \dfrac{2n f_0 M V}{\left(4\pi^2 f_0^2 L^2\right) Q^2 + 4\pi^2 f_0^2 M^2}$.

Putting the values, we get $I_2 = \dfrac{12\pi}{160\pi^2} = \dfrac{3}{40\pi} = 23.8\text{mA}.$

7. Two coils whose resistances are R_1 and R_2, self-inductances L_1, L_2 respectively are connected in parallel and fed from a potential source of pure sine waveform. Prove that the currents in the coils are in phase with each other if the mutual inductance between them is $(R_1 L_2 - R_2 L_1)(R_1 - R_2)$ [Fig. 3.14].

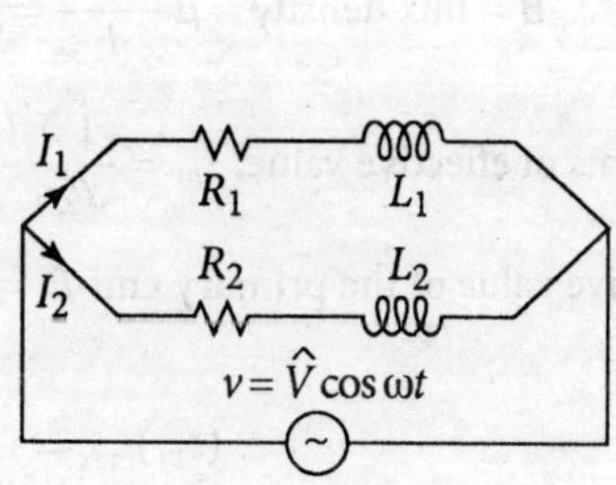

Fig. 3.14

Solution: The emf equations in the two coils can be written as

$$R_1\vec{I}_1\cdot\vec{n}+j\omega L_1\vec{I}_1\cdot\vec{n}+j\omega M\vec{I}_2\cdot\vec{n}=\vec{V} \qquad \therefore \quad (R_1+j\omega L_1)\,\vec{I}_1+j\omega M\vec{I}_2=\vec{V}$$

and $$R_2\vec{I}_2\cdot\vec{n}+j\omega L_2\vec{I}_2\cdot\vec{n}+j\omega M\vec{I}_1\cdot\vec{n}=\vec{V}. \qquad j\omega M\vec{I}_1+(R_2+j\omega L_2)\,\vec{I}_2=\vec{V}.$$

$$\therefore\ \vec{I}_1=\frac{\begin{vmatrix}\vec{V} & j\omega M\\ \vec{V} & z_2\end{vmatrix}}{z_1z_2+\omega^2M^2}=\frac{(z_1-j\omega M)\vec{V}}{z_1z_2+\omega^2M^2}.$$

Now, $(z_2-j\omega M))\,\vec{V}=[R_2+j(\omega L_2-\omega M)]\,\vec{V}=\sqrt{R_2^2+(\omega L_2-\omega M)^2}\,\hat{V}\,e^{j(\omega t-\phi_2)}\vec{n}.$

$$(z_1-j\omega M)\,\vec{V}=\sqrt{R_1^2+(\omega L_1-\omega M)^2}\,\hat{V}\,e^{j(\omega t-\phi_2)}\vec{n}.$$

$$\therefore\ \vec{I}_2=\frac{\begin{vmatrix}Z_1 & \vec{V}\\ j\omega M & \vec{V}\end{vmatrix}}{Z_1Z_2+\omega^2M^2}=\frac{(1-j\omega M)\vec{V}}{Z_1Z_2+\omega^2M^2}.$$

In order that $\vec{I}_1$ and $\vec{I}_2$ are in phase,

$$\phi_1=\phi_2 \text{ or, } \tan\phi_1=\tan\phi_2 \text{ or, } \frac{\omega L_2-\omega M}{R_2}=\frac{\omega L_1-\omega M}{R_1} \text{ or, } M=\frac{R_1L_2-R_2L_1}{R_1-R_2}.$$

8. A transformer working on a 2000V, 50c/s circuit has 300 primary turns. The core has a mean magnetic path of 100cm and cross section 10^3 cm^2, the iron having permeability of 1800. The iron loss is 400W. Calculate the primary no-load current.

Solution: When the secondary is open, the primary current I_p has two components, one in phase with V_p, called the *working component* producing necessary power to supply hysteresis and eddy current losses in the core; we denote this component by I_ω; the second, the magnetising component I_m lags $\pi/2$ behind V_p, its function is to produce flux. These components can be found as below:

The magnetising current I_m is given by $H=\frac{I_mN_1}{l}$, l = magnetic path length.

$$\therefore\ B=\text{flux density}=\mu\frac{I_mN_1}{l}=\mu_0\mu_r\frac{I_mN_1}{l};\quad \therefore\ B_{\max}=\frac{N_1\mu_0\mu_r(I_m)_{\max}}{l}\ \ [\mu=\mu-0\mu_r].$$

In terms of effective value, $I_m=\dfrac{1}{\sqrt{2}}\cdot\dfrac{l\,B_{\max}}{N_1\mu}.$

Effective value of the primary emf $E_1=\dfrac{2\pi}{\sqrt{2}}B_{\max}N_1f\times\alpha.$

$$\therefore\ (I_m)_{\max}=\frac{l}{\sqrt{2}N_1\mu}\cdot\frac{\sqrt{2}E_1}{2\pi\alpha N_1f}=\frac{E_1l}{2\pi N_1^2\mu f\alpha}=0.3\text{A}.$$

$$\text{Working component, } I_\omega=\frac{\omega_h+\omega_e}{E_1}=\frac{\text{hysteresis loss + eddy current loss}}{\text{primary voltage}}=\frac{\text{iron loss}}{\text{primary voltage}}$$

$$=\frac{400}{2000}=0.2\text{A}.$$

$$\therefore\ \text{no-load current}=\sqrt{(I_m)_{\max}^2+(I_\omega)^2}=\sqrt{(0.3)^2+(0.2)^2}=0.36\text{A}.$$

3.10 Exercises

1. Explain the principle of action of an ac transformer and obtain a relation between its primary and secondary voltages.
2. Work out the simple theory of a transformer and show how the effective primary resistance and inductance are affected by the secondary circuit when closed.
3. Give the theory of a transformer and establish a relationship between effective impedance of the primary, the impedance of the secondary and the transformation ratio.
   [*Hints.* For the second part, see (3.11).]
4. Given two identical series circuits with resistance 10Ω, $L = 20$mH and variable capacitance. They are coupled magnetically with coefficient of coupling being equal to 1. If a sinusoidal emf of frequency 10^4c/s is injected into one of the circuits, find how the capacitances must be adjusted to give a maximum current in the other circuit. [$C_1 = C_2 = 0.0141\mu$F or, $C_1 = C_2 = 0.015\mu$F]
5. An emf $E_0 \cos pt$ is applied to the primary of an air-cored transformer; the external load of the secondary circuit is purely resistive. Derive expressions for the primary and secondary currents and show that the effect of the presence of the secondary is equivalent to that of certain change in the primary resistance and inductance.
6. Two tuned circuits each consisting of L, R and C in series are coupled inductively. An emf of the form $v = \hat{V} \cos \omega t$ is set up in one of them. Obtain expression for the potential differences across the condensers at any instant and discuss the manner of their variation with that of ω, the coupling being assumed to be small.
7. Give the theory of a transformer and establish a relationship between the effective impedance of the primary, the impedance of the secondary and the transformation ratio.
   Hence, draw the equivalent circuits of an imperfect potential transformer at (a) low frequency (b) high frequency.
8. Explain the principle of action of an ac transformer and obtain a relation between its primary and secondary voltages. Is transformer a device that violates conservation of energy?
9. Explain the terms: copper losses, core losses, magnetic leakage and percentage efficiency of a transformer. Discuss how these losses can be minimised. What is the meaning of the reflected inductances from the secondary?

Chapter 4

Rotating Magnetic Field

4.1 Introduction

Because of many practical applications, in general, and in the construction of electro-motors, in particular, the present topic finds a place for inclusion in our discourse.

The rotating magnetic field of utility may be conveniently produced with the help of a polyphase alternating current. When a single coil rotates with a constant angular velocity in a magnetic field, we get a single phase alternating current and emf.

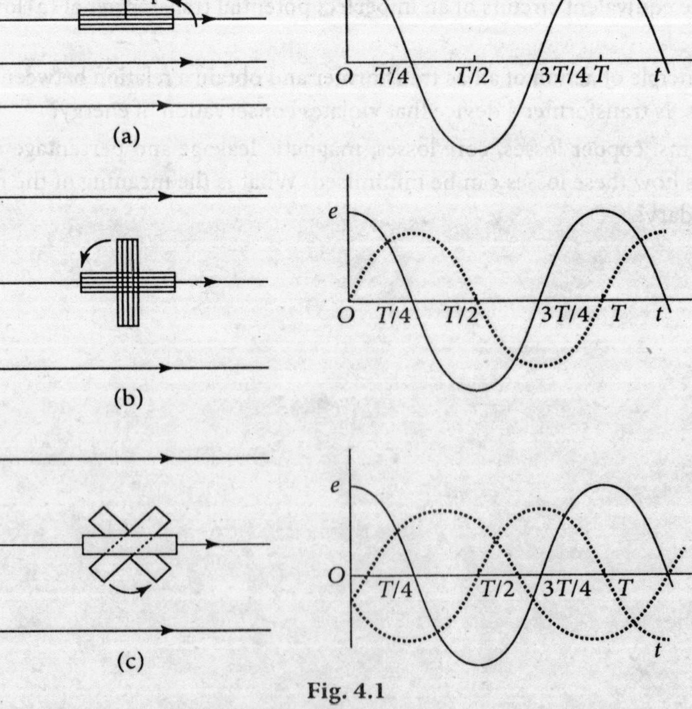

Fig. 4.1

A single coil rotates in a magnetic field, $\hat{n}_1$ denotes unit vector normal to its plane; the corresponding $e \sim t$ curve is shown to the right [Fig. 4.1(a)]. Two coils are at right angles with each other [Fig. 4.1(b)] such that when emf in one is maximum, the other is zero, this is a two-phase ac generator. In Fig. 4.1(c) the coils are relatively displaced by $\frac{2\pi}{3}$ w.r.t. each other. The emfs in the coils differ in phase by $\frac{2\pi}{3}$ w.r.t. each other. So, the emfs can be written as below:

Fig. 4.1(a), $v_1 = \vec{V}\cos\omega t = \vec{V}e^{j\omega t}\vec{n}\cdot\vec{n}$

Fig. 4.1(b), $v_1 = \vec{V}\cos\omega t = \vec{V}e^{j\omega t}\vec{n}\cdot\vec{n}$ $\qquad v_2 = \vec{V}\cos\left(\omega t - \frac{\pi}{2}\right) = \hat{V}e^{j\left(\omega t - \frac{\pi}{2}\right)}\vec{n}\cdot\vec{n}$

Fig. 4.1(c), $v_1 = \hat{V}\cos\omega t = \hat{V}e^{j\omega t}\vec{n}\cdot\vec{n}$ $\qquad v_2 = \hat{V}\cos\left(\omega t - \frac{2\pi}{3}\right) = \hat{V}e^{j\left(\omega t - \frac{2\pi}{3}\right)}\vec{n}\cdot\vec{n}$

$v_3 = \hat{V}\cos\left(\omega t - \frac{4\pi}{3}\right) = \hat{V}e^{j\left(\omega t - \frac{4\pi}{3}\right)}\vec{n}\cdot\vec{n}$

Suppose that there are two coils fed from a two-phase ac generator such that the currents in them differ in phase by ϕ. The coils are placed [Fig. 4.2], their planes being at right angles to each other. These currents may be represented by $i_1 = \hat{I}_1\cos\omega t$; $i_2 = \hat{I}_2\cos(\omega t - \phi)$.

In terms of rotating vector notation, $i_1 = \hat{I}_1\cdot\hat{n} = \hat{I}_1 e^{j\omega t}\vec{n}\cdot\vec{n}$; $i_2 = \hat{I}_2 e^{j(\omega t - \phi)}\vec{n}\cdot\vec{n} = \hat{I}_2\cdot\vec{n}$.

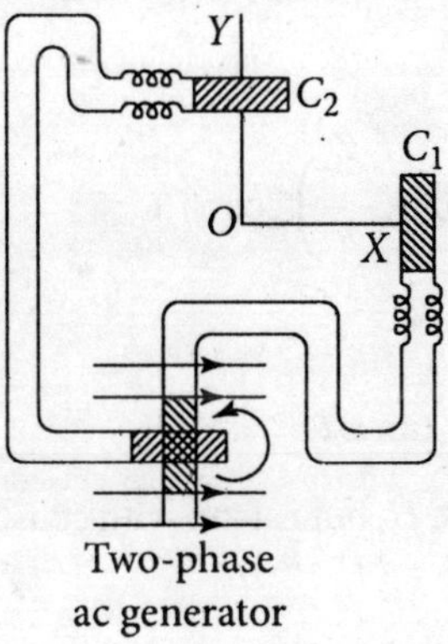

Fig. 4.2

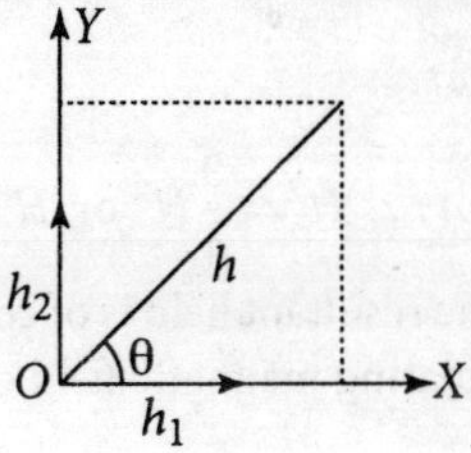

Fig. 4.3

The magnetic fields at O, the point of intersection of the axes of the coils will be in phase with the current and hence, they may be written as

$$\vec{H}_1 = \hat{H}_1 e^{j\omega t}\vec{n}$$

$$h_1 = \hat{H}_1 e^{j\omega t}\vec{n}\cdot\vec{n} = \hat{H}_1\cdot\vec{n}$$

$$\text{and } h_2 = \hat{H}_2 e^{(j\omega t - \phi)}\vec{n}\cdot\vec{n} = \hat{H}_2\cdot\vec{n}.$$

$\therefore$ resultant field at O is $\vec{H} = \hat{H}_1 + \hat{H}_2$.

So, the resultant magnetic field is of magnitude given as $h^2 = h_1^2 + h_2^2$ and its inclination relative to OX-direction is given by $\tan\theta = h_2/h_1$.

If $\hat{H}_1 = \hat{H}_2 = \hat{H}$ and $\phi = \frac{\pi}{2}$, actually this is the most important case, then the resultant field is given by $h^2 = \hat{H}^2\cos^2\omega t + \hat{H}^2\sin^2\omega t$ or, $h = \hat{H}$ and

$$\tan\theta = \frac{\hat{H}\sin\omega t}{\hat{H}\cos\omega t} = \tan\omega t. \qquad \therefore \theta = \omega t.$$

Hence, the resultant field is of constant magnitude $\hat{H}$ and rotates with constant angular velocity ω. As t increases, θ increases, the rotation, therefore, is anticlockwise. So, the resultant rotating magnetic field can be written as $\vec{H} = \hat{H} e^{j\omega t} \vec{n}$.

4.2 Rotating Magnetic Field by a Three-Phase ac Source

Here, the resulting magnetic field due to three-phase ac can be written as

$$h_1 = \hat{H}\cos\omega t;$$
$$h_2 = \hat{H}\cos\left(\omega t - \frac{2\pi}{3}\right);$$
$$h_3 = \hat{H}\cos\left(\omega t - \frac{4\pi}{3}\right).$$

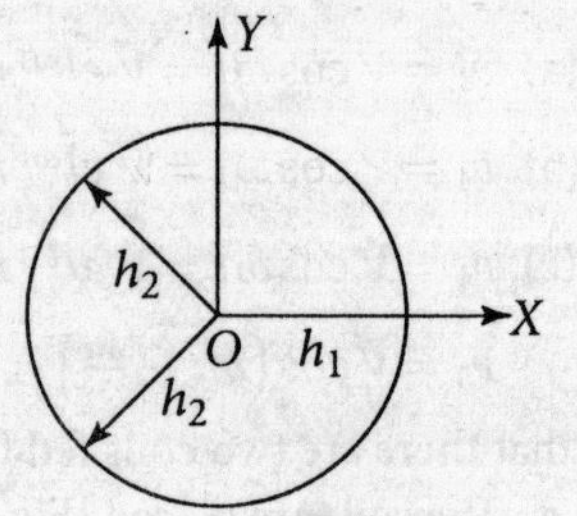

Fig. 4.4

Let the resultant magnetic field be H and it makes an angle θ with OX. Then resolving along OX and OY, we get

$$H\cos\theta = h_1 - h_2\cos\frac{\pi}{3} - h_3\cos\frac{\pi}{3}$$
$$= \hat{H}\cos\omega t - \hat{H}\cos\left(\omega t - \frac{2\pi}{3}\right)\cos\pi/3 - \hat{H}\cos\left(\omega t - \frac{4\pi}{3}\right)\cos\pi/3$$
$$= \frac{3}{2}\hat{H}\cos\omega t.$$

$$H\sin\theta = \frac{3}{2}\hat{H}\sin\omega t, \therefore\ H^2 = \frac{9}{4}H^2 \text{ or, } H = \frac{3}{2}\hat{H} \text{ and } \tan\theta = \tan\omega t.$$

So, in this case, the resultant field is of constant magnitude $\frac{3}{2}\hat{H}$, but rotates with constant angular velocity ω. Hence, such a rotating magnetic field can be written as $\hat{H} = \frac{3}{2}\hat{H}\, e^{j\omega t}\vec{n}$.

4.3 The Coupling Acting on a Small Coil Placed in a Rotating Magnetic Field

Let a small coil of N turns, effective area A is placed at O, where the rotating magnetic field is of magnitude H (produced as previously discussed). The coil rotates with angular velocity ω_c while the magnetic field rotates with angular velocity ω_m. At any instant of time t, the relative positions of the field and the plane of the coil are shown in Fig. 4.5, $\hat{n}$ is the unit vector perpendicular to the plane of the coil.

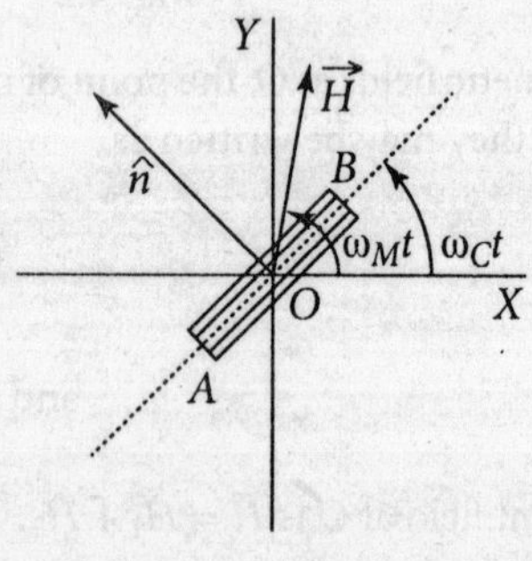

Fig. 4.5

Then, the number of linkages associated with the coil is given by

$$\phi = N\int \hat{H}\cdot\vec{ds} = NHA\cos\left(\frac{\pi}{2} - \overline{\omega_m - \omega_c\, t}\right) = NHA\sin\left(\overline{\omega_m - \omega_c\, t}\right).$$

If $\omega_m - \omega_c = \omega$ = relative angular velocity of the magnetic field w.r.t. the coil, the induced emf at that instant,

$$v = -\frac{d\phi}{dt} = -NHA\omega\cos\omega t = NH\omega A\cos(\omega t - \pi) = \hat{V}\cos(\omega t - \pi) = \hat{V}\, e^{j(\omega t - \pi)}\vec{n}\cdot\vec{n}.$$

If R and L be the resistance and inductance in the coil, the current through the coil at the instant is given by

$$R\hat{I}\cdot\hat{n}+j\omega L\hat{I}\cdot\hat{n}=\hat{V}e^{j(\omega t-\pi)}\vec{n}\cdot\vec{n} \text{ or, } \vec{I}=\frac{\vec{V}}{R+j\omega L}=\frac{\vec{V}e^{-j\theta}}{\sqrt{R^2+\omega^2L^2}}=\frac{\hat{V}e^{j(\omega t-\pi-\theta)}\vec{n}}{\sqrt{R^2+\omega^2L^2}}.$$

$$\therefore\ i=\vec{I}\cdot\vec{n}=\frac{\hat{V}\cos(\omega t-\pi-\theta)}{\sqrt{R^2+\omega^2L^2}}=\frac{NAH\omega\cos(\omega t-\pi-\theta)}{\sqrt{R^2+\omega^2L^2}},$$

where $\tan\theta=\omega L/R$.

Now, the current carrying coil is equivalent to a magnetic shell of moment $\vec{M}=\hat{n}NiA$. So, the couple acting on the coil is given by

$$\vec{\Gamma}=\vec{M}\times\vec{H}=NiA\hat{n}\times\vec{H}=\hat{k}NiAH\sin\left(\frac{\pi}{2}-\omega t\right),\ \hat{k} \text{ is a unit vector perpendicular to } \hat{n} \text{ and } \vec{H}.$$

$$=\hat{k}\frac{N^2A^2H^2\omega\cos(\omega t-\theta)\cos\omega t}{\sqrt{R^2+\omega^2L^2}}.$$

From the diagram it is clear that the couple is anti-clockwise.

∴ the magnitude of the average anti-clockwise couple

$$\left|\vec{\Gamma}_{\text{av}}\right|=\frac{N^2A^2H^2\omega}{\sqrt{R^2+\omega^2L^2}}\frac{1}{T}\int_0^T\cos(\omega t-\theta)\cos\omega t\,dt$$

$$=\frac{N^2A^2H^2\omega}{\sqrt{R^2+\omega^2L^2}}\frac{1}{T}\left[\int_0^T\cos^2\omega t\cos\theta+\sin\omega t\cos\omega t\sin\theta)\,dt\right]$$

$$=\frac{N^2A^2H^2\omega}{\sqrt{R^2+\omega^2L^2}}\cdot\frac{1}{2}\cos\theta=\frac{N^2A^2H^2\omega}{2\sqrt{R^2+\omega^2L^2}}\cos\theta$$

$$=\frac{N^2A^2H^2\omega}{2\sqrt{R^2+\omega^2L^2}}\cdot\frac{R}{\sqrt{R^2+\omega^2L^2}}=\frac{N^2A^2H^2\omega R}{2(R^2+\omega^2L^2)}. \tag{4.1}$$

So, when a coil is placed in a rotating magnetic field, as in the figure, the coil will rotate in the anti-clockwise fashion in the direction of the field. This is a must, because this is the demand of Lenz's law, which says that the forces due to the magnetic effects between the field due to induced currents and the original field are such that they tend to prevent any relative motion of the field and the conductor. The system will ultimately move with a constant relative velocity ω attained when the average rate of work done by the couple equals the rate of frictional and other external losses.

From (4.1) it is evident that the average couple is zero both when $\omega=0$ and $\omega=\infty$, so the average couple will have maximum value for a certain value of ω such that

$$\frac{d}{d\omega}\left(\left|\vec{\Gamma}_{\text{av}}\right|\right)=0.$$

Now,

$$\frac{d}{d\omega}\left(\frac{N^2A^2H^2\omega R}{2(R^2+\omega^2L^2)}\right)=\frac{N^2A^2H^2R}{2(R^2+\omega^2L^2)}\left(1-\frac{2L^2\omega^2}{L^2\omega^2+r^2}\right).$$

$\therefore$ for $\dfrac{d}{d\omega}\left(\left|\vec{\Gamma}_{av}\right|\right)=0$, we have $\omega=\dfrac{R}{L}$.

Under this condition, the average couple is given by $\left|\vec{\Gamma}_{\text{av}}\right|=\dfrac{N^2A^2H^2}{4L}$ and $\theta=\dfrac{\pi}{4}$.

4.4 Application of the Rotating Magnetic Field

There are many applications of this couple experienced by a closed coil or conductor in a rotating magnetic field, the most important being possibly the constructions of *induction motors*. The polyphase induction motors are used more extensively than any other form of ac motor. The name of the motor as such is derived from the fact that in such a case the current in the rotor is not drawn from the supply but is induced by relative motion of the rotor conductors and the magnetic field produced by stator currents. The stator is fed by a poly-phase ac and a rotating magnetic field is produced thereby. The lines of the stator field cut the rotor conductors and induce currents in them; hence, the rotor follows the stator field. From the expression of the torque on the conductor in a rotating field derived, we know that its value depends on the relative velocity of the stator and rotor field, hence even on no-load the rotor speed is less than the speed of the stator field, because of friction and iron losses. If now some load is applied to the motor, the torque must increase, which can be possible only at the cost of decrease of the rotor speed, because the stator field speed is fixed and hence, for increase in the relative angular velocity, the rotor speed must decrease.

If ω_1 = angular velocity of stator field, ω_2 = angular velocity of the rotor, then $(\omega_1 - \omega_2)$ = absolute *slip*. The ratio $\frac{\omega_1-\omega_2}{\omega_1}$ = fractional slip and $\frac{\omega_1-\omega_2}{\omega_1} \times 100$ is the percentage slip.

$\sigma = \frac{\omega_1-\omega_2}{\omega_1}$ becomes equal to unity when the rotor is stationary.

Now, average torque τ is:

$$\tau = \frac{N^2A^2H^2\omega R}{R^2+\omega^2L^2} = \frac{N^2A^2H^2\sigma\omega_1 R}{R^2+\sigma^2\omega_1^2L^2}.$$

So, at the moment of start, $\sigma = 1$, and the starting torque $\propto \frac{R}{R^2+\omega_1^2L^2}$. This is a maximum for $R = \omega_1 L$. So to obtain the maximum starting torque, a sufficient resistance is to be connected in series with the rotor so as to make the total resistance equal to the reactance of the rotor. A variation of the starting torque vs. rotor resistance is shown in Fig. 4.6.

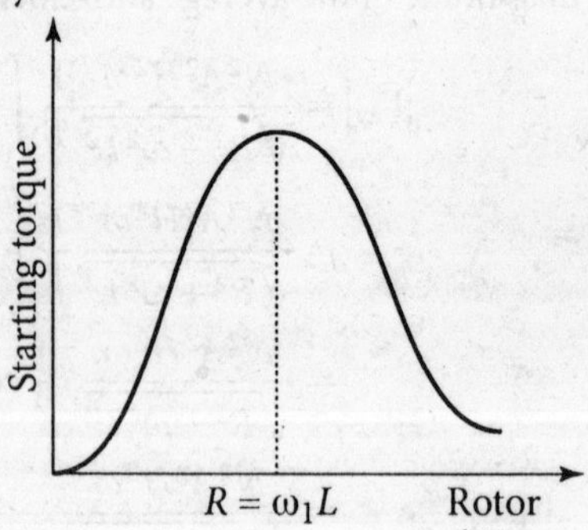

Fig. 4.6

A *squirrel-cage* or short-circuited rotor has a small starting torque, under running condition of the motor

$$\tau \propto \frac{\sigma R}{R^2+\sigma^2\omega_1^2L^2}.$$

This is a maximum, when $R = \sigma\omega_1 L$.

$$\therefore \tau_{max} \propto \frac{\frac{R^2}{\omega_1 L}}{R^2+\frac{R^2}{\omega_1^2L^2}\omega_1^2L^2} \propto \frac{R^2}{2R^2\omega_1 L} \propto \frac{1}{2\omega_1 L} \propto \frac{1}{\omega_1 L}.$$

Since $\omega_1 L$ is a constant, τ_{max} is also a constant independent of rotor resistance.

4.5 Construction of an Induction Motor

The conductor R is the rotor of the induction motors, mounted on an axle in the region, where a rotating magnetic field is produced by electromagnets, called *stators*, S_1 and S_2; S_3 and S_4.

The pairs of electromagnets are fed from the same voltage source but the currents feeding S_1 and S_2 pass through a condenser and that feeding S_3 and S_4 pass through the inductance coil L_1 and therefore, these two currents must be in quadrature and hence, the resulting magnetic field will be a rotating field of constant magnitude in the region occupied by the rotor. Hence, it will experience a couple, which will rotate it in the direction of the stator field.

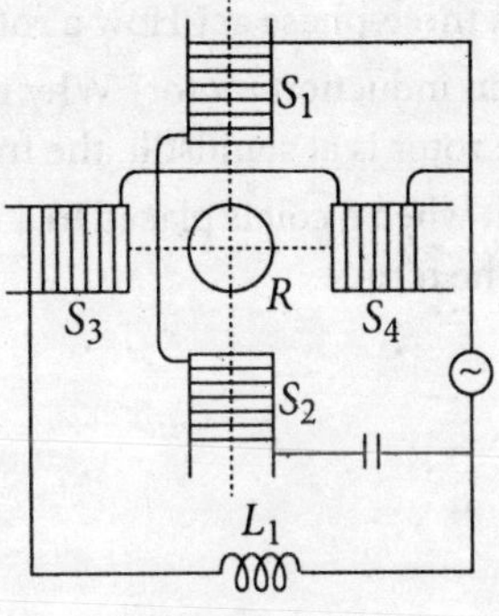

Fig. 4.7

4.6 Numerical Problems

1. A squirrel-cage motor has a ratio of rotor standstill reactance to resistance per phase of 4 to 1 and the maximum torque is $2\frac{1}{4}$ times normal full-load torque. Calculate: (a) the full-load slip (b) ratio of starting torque to normal torque.

Solution: (a) With normal voltage on the stator, the torque on the rotor, $\tau = k\dfrac{\sigma R}{R^2+\sigma^2 X_2^2}$, σ = fractional slip, X = reactance. Now, $\tau = \tau_{\max}$, when $R = \sigma X_2$.

$$\therefore\ \tau_{\max} = \frac{k}{2X_2},\ \text{if } \tau_{\max} = 100\,\%,\ k = 200X_2.$$

$$\frac{\tau_{\max}}{\tau_{\text{normal}}} = 2\frac{1}{4}\ \text{or, } \tau_{\text{normal}} = \frac{4}{9}\tau_{\max} = \frac{4}{9}\cdot\frac{k}{2X_2} = \frac{4}{9}\times 100 = 44.4,\ \therefore\ 44.4 = 800R\cdot\frac{\sigma R}{R^2+\sigma^2X_2^2}.$$

Here $X_2 = 4R$.

$$\therefore\ 44.4 = 800R\cdot\frac{\sigma R}{R^2+16R^2\sigma^2}\ \text{ or, } \sigma = 5.7 \text{ or } 107.$$

This shows that the motor will run normally with a full-load slip of 5.7 %. In some region, the slip also becomes 107 % (it is the counter current breaking region).

(b) At starting, $\sigma = 1$.

$$\therefore \tau_{\text{st}} = 200X_2\cdot\frac{R}{R^2+X_2^2} = \frac{800R^2}{R^2+16R^2};\quad X_2 = 4R = 47.$$

At the full stator voltage, τ_{st} is given by the above value of 47 % of the maximum possible torque.

$$\therefore\ \tau_{\text{st}} = \frac{47}{44.4} = 1.06\tau_{\text{normal}}.$$

4.7 Exercises

1. Explain how a rotating magnetic field can be produced and show how this principle is utilised in the construction of an induction motor. What do you understand by the slip of the motor?
2. Show that two alternating magnetic fields at right angles to each other is equivalent to a rotating magnetic field. Explain how this equivalence is utilised in the construction of an ac motor.

3. What is a three-phase ac? How a rotating magnetic field is produced by a three-phase ac?
4. What is an induction motor? Why is it so-called? What is the meaning of *slip* of the motor? Show that when the rotor is at standstill, the fractional slip is unity.
5. Show that when a coil is placed in a rotating magnetic field, it experiences a couple. Find the maximum value of the torque.

Chapter 5

AC Measurements

5.1 Introduction

AC measurements involve the determination of self-inductance, capacitance, resistance, mutual inductance and also the frequency of ac source. These quantities are seriously affected by variation of frequency. For example, a condenser offering infinite resistance to a direct current yields to an ac of high frequency and bypasses it. The resistance offered by an inductance is very great for a high frequency ac. The resistance also changes with frequency of the alternating current, giving rise to the so-called *skin effect.* In this chapter, we will have a discourse on how these quantities can be determined experimentally laying out all the necessary theoretical background.

5.2 Kirchhoff's Laws in ac Network

Kirchhoff's two laws applied to network fed by dc source with slight modification can be extended to circuits associated with ac source. This modified form can be applied to frequencies not too high, in the range of 1000c/s or in its neighbourhood but not in the region of *high frequency,* where the phenomenon of *skin effect* creeps in.

1st law. The vector sum of alternating currents of sinusoidal waveform, each of pulsatance ω meeting at a point is zero.

If $\vec{I}_1$, $\vec{I}_2$ and $\vec{I}_3$, etc. be the alternating currents in conductor meeting at a point, then $i_1 = \vec{I}_1 \cdot \vec{n}$, $i_2 = \vec{I}_2 \cdot \vec{n}$, etc.

$$\therefore\ i_1 + i_2 + i_3 + \cdots = \vec{I}_1 \cdot \vec{n} + \vec{I}_2 \cdot \vec{n} + \vec{I}_3 \cdot \vec{n}. \tag{5.1}$$

This expression (5.1) is true at all instants of time and therefore,

$$\vec{I}_1 + \vec{I}_2 + \vec{I}_3 + \cdots = 0 \text{ or, } \sum_i \vec{I}_j = 0.$$

2nd law. In any closed network of conductors carrying ac of sinusoidal form of pulsatance ω, the vector sum of the p.d. vectors equals the vector sum of the emfs encountered in traversing that loop.

Let $\vec{I}_j$ be the current rotating vector that flows through a network of conductors having impedances Z_1, Z_2, etc. and emf included be $\vec{V}$, then

$$\vec{I}_1 Z_1 + \vec{I}_2 Z_2 + \vec{I}_3 Z_3 + \cdots = \vec{V} \text{ or, } \sum \vec{I}_j Z_j = \vec{V}.$$

5.3 Laws of Series and Parallel Impedances

Let z_1, z_2, z_3 be the impedance operators in series, then $\vec{V} = \vec{I}Z_1 + \vec{I}Z_2 + \vec{I}Z_3$.

If Z be the equivalent impedance operator, then $\vec{I}Z = \vec{V}$.

$\therefore\ Z = Z_1 + Z_2 + Z_3$ or, $Z = \Sigma Z_i$.

If Z_1, Z_2, Z_3 be connected in parallel, then $\vec{I}_1 = \vec{V}/Z_1$, $\vec{I}_2 = \vec{V}/Z_2$, $\vec{I}_3 = \vec{V}/Z_3$.

Now,

$$\vec{I} = \vec{I}_1 + \vec{I}_2 + \vec{I}_3 = \vec{V}\left(\frac{1}{Z_1} + \frac{1}{Z_2} + \frac{1}{Z_3}\right).$$

If equivalent impedance operator be Z, $\vec{I} = \vec{V}/Z$.

$$\therefore\ \frac{1}{Z} = \frac{1}{Z_1} + \frac{1}{Z_2} + \frac{1}{Z_3} \text{ or, } \frac{1}{Z} = \sum \frac{1}{Z_i}.$$

5.4 Alternating Current Bridges

Wheatstone's bridge with ac source. The Wheatstone's bridge for comparison of resistance, we have already talked about in our previous discourse on direct current theory. But now we will discuss the working of Wheatstone's bridge when the emf applied is alternating. With the help of such a bridge, almost all types of possible impedances can be compared. Here, in place of a galvanometer as with dc measurements, suitable detector (e.g., vibration galvanometer, Duddell oscillograph) is to be used.

Let Z_1, Z_2, Z_3, Z_4 be the four impedances forming the arms of the Wheatstone's bridge, Z_5 and Z_6 are the impedances [Fig. 5.1]. The source of emf is an alternating source $v = \vec{V} \cdot \vec{n}$.

Let the currents in the different branches be [Fig. 5.1].

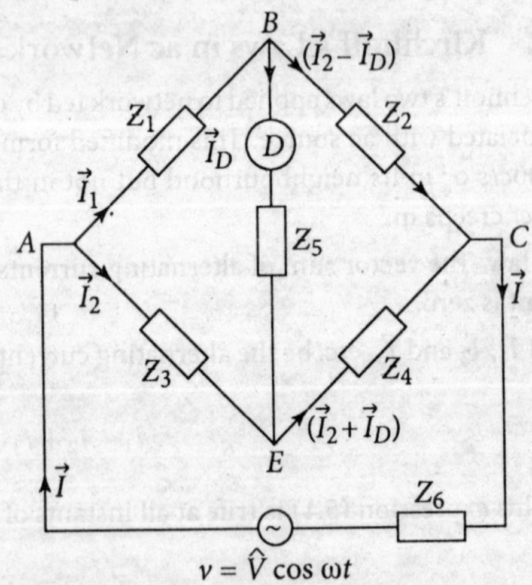

Fig. 5.1

In mesh ABE,

$$\vec{I}_1 Z_1 + \vec{I}_d Z_5 - \vec{I}_2 Z_3 = 0. \quad (5.2)$$

In mesh BCE,

$$\left(\vec{I}_1 - \vec{I}_d\right) Z_2 - \vec{I}_d Z_5 - \left(\vec{I}_2 + \vec{I}_d\right) Z_4 = 0. \quad (5.3)$$

In mesh $ABCA$,

$$\vec{I}_1 Z_1 + \left(\vec{I}_1 - \vec{I}_d\right) Z_2 + \left(\vec{I}_1 + \vec{I}_2\right) Z_6 = \vec{V}. \quad (5.4)$$

Arranging the terms, we get

$$\vec{I}_d Z_5 + \vec{I}_1 Z_1 - \vec{I}_2 Z_3 = 0$$
$$-(Z_2 + Z_5 + Z_4)\vec{I}_d + \vec{I}_1 Z_2 - \vec{I}_2 Z_4 = 0$$
$$-I_d Z_2 + \vec{I}_1 (Z_1 + Z_6 + Z_2) + \vec{I}_2 Z_6 = \vec{V}.$$

$$\vec{I}_d = \frac{\begin{vmatrix} 0 & Z_1 & -Z_3 \\ 0 & Z_2 & -Z_4 \\ \vec{V} & (Z_1+Z_2+Z_6) & Z_6 \end{vmatrix}}{\begin{vmatrix} Z_5 & Z_1 & -Z_3 \\ -(Z_2+Z_5+Z_4) & Z_2 & -Z_4 \\ -Z_2 & (Z_1+Z_2+Z_6) & Z_6 \end{vmatrix}} = \frac{\vec{V}(+Z_3Z_2-Z_1Z_4)}{\Delta}.$$

Δ is the determinant in the denominator of the expression for $\vec{I}_d$.

For balance of the bridge, $\vec{I}_d = 0$. So, we get $Z_1Z_4 = Z_2Z_3$, $Z_1 = Z_1 e^{j\phi_1}$, $Z_2 = Z_3 e^{j\phi_2}$ and etc.

$\therefore\ Z_1Z_4 e^{j(\phi_1+\phi_4)} = Z_2Z_3 e^{j(\phi_2+\phi_3)}$.

So for balance, the magnitudes of the impedances must satisfy $Z_1/Z_2 = Z_3/Z_4$ and also $\phi_1+\phi_4 = \phi_2+\phi_3$. When these conditions are satisfied, there is zero current in the detector.

All the ac bridges work on this principle and we will now investigate the different bridges meant for determining different circuit elements.

But before that let us discuss the sensitivity of the Wheatstone's bridge arrangement, because if we know the measures of sensitivity and the factors controlling the sensitivity, the bridges designed to determine the different quantities may be suitably adjusted.

Sensitivity of the Wheatstone arrangement

If for a slight departure from the balance condition, a large current flows through the detector, the bridge is said to be sensitive.

The best condition for the sensitivity is when all the six impedances (Z_1, Z_2, Z_3, Z_4, Z_5 and Z_6) are equal.

We will discuss the most simple case by taking Z_5 and Z_6 very small. The bridge is balanced by varying Z_4, which is a resistance arm. Then

$$\vec{I}_d = \frac{(Z_2Z_3 - Z_1Z_4)\vec{V}}{\Delta}.$$

Hence,

$$\frac{d\vec{I}_d}{dZ_4} = \frac{Z_1\vec{V}}{\Delta} - \frac{(Z_2Z_3 - Z_1Z_4)\vec{V}}{\Delta^2}\frac{d\Delta}{dZ_4}.$$

$\therefore$ at balance $\dfrac{d\vec{I}_d}{dZ_4} = \dfrac{Z_1\vec{V}}{\Delta_1}$, where Δ_1 is the value of the determinant by putting $Z_1Z_4 = Z_2Z_3$.

If Z_5 and Z_6 are negligible, then

$$\Delta_1 = \begin{vmatrix} 0 & Z_1 & -Z_3 \\ -(Z_2+Z_4) & Z_2 & -Z_4 \\ -Z_2 & (Z_1+Z_2) & 0 \end{vmatrix} = Z_4(Z_1+Z_3)(Z_2+Z_1).$$

$$\therefore\ \frac{d\vec{I}_d}{dZ_4} = \frac{Z_1\vec{V}}{\Delta_1} = \frac{Z_1\vec{V}}{Z_4(Z_1+Z_3)(Z_2+Z_1)} = \frac{Z_2Z_3}{Z_4}\cdot\frac{1}{Z_4}\frac{\vec{V}}{(Z_1+Z_3)(Z_2+Z_1)}$$

$$= \frac{1}{Z_4}\cdot\frac{\vec{V}}{(1+Z_4/Z_2)(Z_2+Z_1)} = \frac{1}{Z_4}\cdot\frac{\vec{V}}{(Z_2+Z_1+Z_4+Z_3)} = \frac{\vec{V}}{Z_4(Z_1+Z_2+Z_3+Z_4)}$$

$$\text{or, } \Delta\vec{I}_d = \frac{\delta Z_4}{Z_4}\cdot\frac{\vec{V}}{(Z_1+Z_2+Z_3+Z_4)}.$$

Thus, the factor $1/(Z_1+Z_2+Z_3+Z_4)$ is a measure of the sensitivity of the bridge. So, the sensitivity of the bridge increases when the sum of the impedances decreases. Let the ratio of the capacities of the two arms (as in de Sauty's bridge) containing Z_1 and Z_2 be n. Then

$$Z_1 = \frac{1}{j\omega C_1}, \quad Z_2 = \frac{1}{j\omega C_2}.$$

$\therefore\ C_1 = nC_2.$

So, $Z_2 = nZ_1$, and at balance, $Z_4 = nZ_3$, since Z_3 is the resistance R_3 and Z_4 is the resistance R_4.

$\therefore\ R_4 = nR_3.$

$$\therefore\ S = \frac{1}{\frac{1}{j\omega C_1} + \frac{1}{j\omega C_2} + R_3 + R_4} = \frac{1}{R_3(1+n) - \frac{j}{\omega C_1} - \frac{j}{\omega C_2}} = \frac{1}{(1+n)R_3 - \frac{j}{\omega C_1} - \frac{jn}{\omega C_1}}$$

$$= \frac{1}{(1+n)R_3 - \frac{1}{\omega C_1}(1+n)} = \frac{1}{(1+n)\left(R_3 - \frac{j}{\omega C_1}\right)}$$

$$= \frac{1}{(1+n)\sqrt{R_3^2 + \frac{1}{\omega^2 C_1^2}}\, e^{-j\theta}} = \frac{e^{j\theta}}{(1+n)A} \text{ where } A = \sqrt{R_2^2 + \frac{1}{\omega^2 C_1^2}} \ \tan\theta = \frac{1}{\omega C_1 R_3}.$$

$$\therefore\ \delta \vec{I}_d = \frac{\delta Z_4}{Z_4} \cdot \frac{\vec{V} e^{j\theta}}{(1+n)A} = \frac{\delta R_4}{R_4} \cdot \frac{\vec{V} e^{j\theta}}{(1+n)A}.$$

The current amplitude through the detector is

$$\delta i_0 = \frac{\delta R_4}{R_4} \cdot \frac{\hat{V}}{(1+n)\sqrt{R_3^2 + \frac{1}{\omega^2 C_1^2}}} = \frac{\delta R_3}{R_3} \cdot \frac{\hat{V}}{(1+n)\sqrt{R_3^2 + \frac{1}{\omega^2 C_1^2}}}.$$

So, the sensitivity increases as R_3 decreases until R_3^2 becomes small compared to $1/\omega^2 C_1^2$, when it tends to a constant value $(1/\omega C_1 \simeq 3200\Omega)$. A further increase in sensitivity can be made by making n small, which means Z_2 and Z_1 to be made small. Therefore, we infer the general tendency that for the sensitivity increase, all the impedances should be equal. In this case, then it means the impedances should be small.

5.5 Different Bridges for Determination of Different Circuit Elements

For the measurement of different quantities like: Self-inductance, Capacitance C, Mutual inductance M, at different frequencies different bridges are designed.

For the measurement of	*The bridge recommended*
1. (a) Self-inductance L (b) Comparison of self-inductance	(i) Maxwell LC bridge (ii) Owen's bridge (iii) Anderson's bridge (iv) Maxwell's bridge (v) Schering's bridge.
2. (a) Capacitance C (b) Comparison of self-inductances	(i) de Sauty's bridge (ii) Wien's bridge (iii) de Sauty's bridge with imperfect condensor
3. Mutual inductance	Carey Foster's bridge modified by Campbell.
4. Frequency of ac source	(i) Arno's method (ii) A simple method.

1. (a) *Measuring self-inductance by*

(i) *Maxwell LC bridge.* This bridge is designed by Maxwell to determine inductance L in terms of a capacitance and two standard resistances. Such a bridge is shown in Fig. 5.2. The important point worthy of note is that in all such bridges aimed at the determination of L, the resistances must have residual inductances as low as possible.

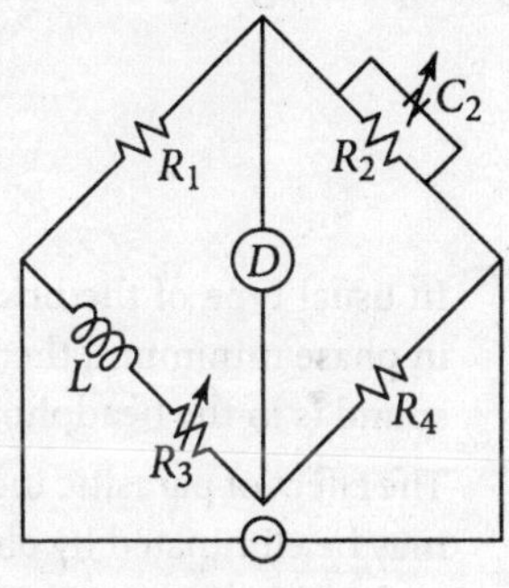

Fig. 5.2

Theory of bridge. C_2 = variable condenser shunting the resistance R_2. R_3 = variable resistance, it also includes resistance of the coil.

Now, $1/Z_2 = (1/R_2) + j\omega C_2$.

At balance, we know $\frac{Z_1}{Z_2} = \frac{Z_3}{Z_4}$ or, $R_1\left(\frac{1}{R_2} + j\omega C_2\right) = \frac{R_3 + j\omega L}{R_4}$.

Separating the real and imaginary parts, we get the condition of balance

$$\frac{R_1}{R_2} = \frac{R_3}{R_4} \text{ and } R_1 \omega C_2 = \frac{\omega L}{R_4} \text{ or, } R_1 R_4 = \frac{L}{C_2}.$$

$\therefore L = R_1 R_4 C_2$.

Condition of balance:

$$R_1/R_2 = R_3/R_4. \tag{5.5}$$

Equation for measuring ***L***:

$$L = R_1 R_4 C_2. \tag{5.6}$$

The bridge is balanced if (5.5) and (5.6) are satisfied. There are two conditions, because the current components in phase with ac supply must be balanced out separately from the components in quadrature with it (idle component). One of these balances is usually the balance of the resistive components.

(5.5) is the balance of the resistive components. In practice, R_1/R_2 is set to a simple ratio, and R_3 is varied to obtain an in-phase minimum and then C_2 is varied to obtain a quadrature minimum. Perfect silence is not obtained, because of coupling between the various branches of the circuit. The balance in this bridge is independent of frequency.

(ii) *Owen's bridge.* This bridge is also designed for the purpose of determining L in terms of standard capacitance. The actual circuit is shown in Fig. 5.3.

For balance of the bridge,

$$\frac{R_1 + (1/j\omega C_1)}{(1/j\omega C_2)} = \frac{R_3 + j\omega L}{R_4}$$

$$\text{or, } R_1 j\omega C_2 + \frac{C_2}{C_1} = \frac{R_3}{R_2} + j\frac{\omega L}{R_4}.$$

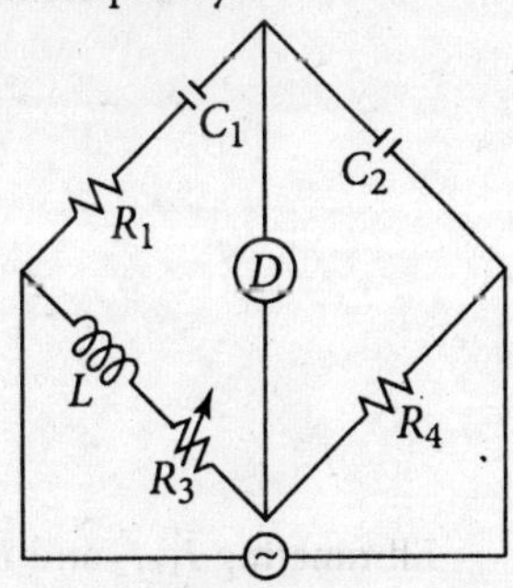

Fig. 5.3

Separating the real and imaginary parts,

$$C_2/C_1 = R_3/R_4 \tag{5.7}$$

$$\text{and } L = R_1 R_4 C_2. \tag{5.8}$$

In usual type of the bridge, $C_1 = C_2 = \frac{1}{3}\mu$F; the resistance R_3 is adjusted so as to obtain an in phase minimum, then R_1 is varied to obtain a quadrature minimum so that barely audible sound is in the headphone. Perfect silence is not realisable, because of mutual coupling.

The effect of parasitic inductance in the resistor coils and the dielectric losses in the capacitors may be eliminated by obtaining a balance first with the inductance and them without it. If R_1' and R_1'' be the two values in the cases, then $L = \left(R_1' - R_1''\right) R_4 C_2$.

This feature of the bridge is of importance in the accurate determination of $L \simeq 100\mu$H. Here, the effect of the earth's capacitance is also small, because C_1 and C_2 are large.

(iii) *Anderson's bridge.* This bridge is most commonly employed in the laboratory for the measurement of inductance in terms of a standard capacitance. Though it is a modification of Maxwell LC bridge, its advantage is that the double balance can be obtained by adjusting resistances only.

The actual circuit of the bridge is shown in Fig. 5.4.

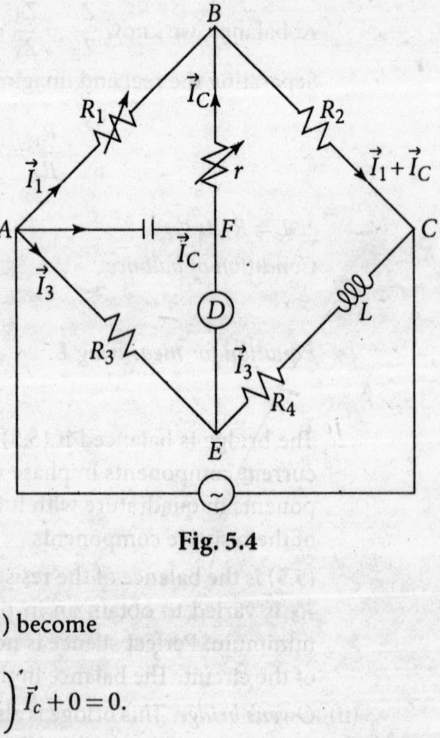

Fig. 5.4

Applying Kirchhoff's second law to the closed meshes $\overrightarrow{ABFA}$, $\overrightarrow{AFEA}$ and $\overrightarrow{EBCE}$, we get

$$\vec{I}_1 R_1 - \frac{\vec{I}_c}{j\omega C} - r\vec{I}_c = 0. \tag{5.9}$$

$$\frac{1}{j\omega C}\vec{I}_c - R_3\vec{I}_3 = 0. \tag{5.10}$$

$$\left(\vec{I}_1 + \vec{I}_c\right) R_3 + r\vec{I}_c - (R_4 + j\omega L)\vec{I}_3 = 0. \tag{5.11}$$

Rearranging the terms, the (5.9), (5.10) and (5.11) become

$$\vec{I}_1 R_1 - \left(r + \frac{1}{j\omega C}\right)\vec{I}_c + 0 = 0.$$

$$0 + \frac{1}{j\omega C}\vec{I}_c - R_3\vec{I}_3 = 0.$$

$$\vec{I}_1 R_2 + (r + R_2)\vec{I}_c - (R_4 + j\omega L)\vec{I}_3 = 0.$$

Eliminating $\vec{I}_1$, $\vec{I}_2$ and $\vec{I}_3$, we get
$$\begin{vmatrix} R_1 & -\left(r + \frac{1}{j\omega C}\right) & 0 \\ 0 & \frac{1}{j\omega C} & -R_3 \\ R_2 & (r + R_2) & -(R_4 + j\omega L) \end{vmatrix} = 0.$$

Expanding the determinant,

$$R_1\left[-\frac{(R_4+j\omega L)}{j\omega C}+R_3(r+R_2)\right]+\left(r+\frac{1}{j\omega C}\right)R_3R_2=0$$

$$\text{or, } -R_1(R_4+j\omega L)+R_1R_3(r+R_2)\,j\omega C+rR_3R_2j\omega C+R_3R_2=0.$$

Separating the real and imaginary parts, we get the double balance condition of the bridge.

Condition 1.

$$\frac{R_1}{R_3}=\frac{R_2}{R_4}=\frac{R_1+R_2}{R_3+R_4}. \tag{5.12}$$

Condition 2.

$$L=C\left[R_3R_2+\frac{R_3}{R_1}(R_1+R_2)\,r\right]=C\left[R_3R_2+(R_3+R_4)\,r\right]. \tag{5.13}$$

Thus, the first condition is satisfied by adjusting the resistance R_1 to obtain in phase balance; then r is adjusted to get quadrature balance so as to get barely audible sound in the headphone. From (5.13), we get the value of the inductance L.

Note. It is the best method for finding the self-inductance.

1. The ac balance condition is independent of ω.
2. The two balance conditions are mutually independent.
3. For maximum sensitivity to changes in r, we should have $LC=2r^2$ and $R_1=R_2=\frac{1}{2}R_4=\frac{1}{2}R_3$ (Butterworth's criterion).

(b) *Comparison of self-inductance: Maxwell.* This bridge compares an unknown inductance with a standard inductance. The circuit diagram is represented in Fig. 5.5.

For the balance of the bridge,

$$\frac{R_1+j\omega L_1}{R_2+j\omega L_2}=\frac{R_3}{R_4}\ \text{or,}\ \frac{R_1+j\omega L_1}{R_3}=\frac{R_2+j\omega L_2}{R_4}$$

$$\text{or, } R_1=\frac{R_3}{R_4}\times R_2 \tag{5.14}$$

and

$$L_1=\frac{R_3}{R_4}L_2. \tag{5.15}$$

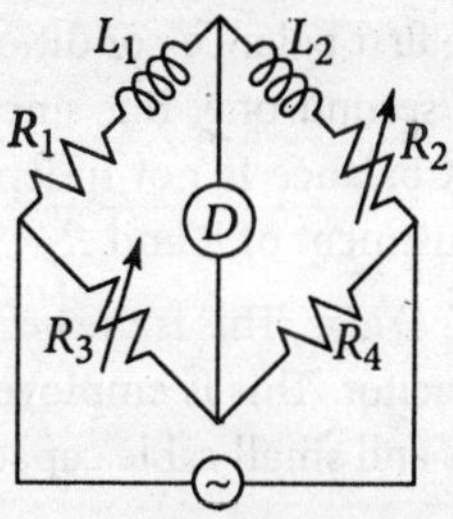

Fig. 5.5

It appears from (5.14) and (5.15) that the two balance conditions are not independent of each other, because the ratio $\frac{R_3}{R_4}$ occurs both in (5.14) and (5.15). So this is done by alternate adjustment of R_3 and R_2. If it were easy to make a variable inductance over a wide range, this bridge then would become of practical value. So bridges, which give expression of the inductance in terms of capacitance and resistance were designed to suit the purpose (e.g., Owen's bridge, Anderson's bridge).

2. (a) *Comparison of capacitance.*

(i) *de Sauty's bridge.* This bridge is used for the comparison of capacitances [Fig. 5.6].

For balance,

$$\frac{R_1}{R_2}=\frac{1/j\omega C_1}{1/j\omega C_2}=\frac{C_2}{C_1}.$$

There is, therefore, only one condition of balance, which can be obtained by adjusting R_1 and R_2.

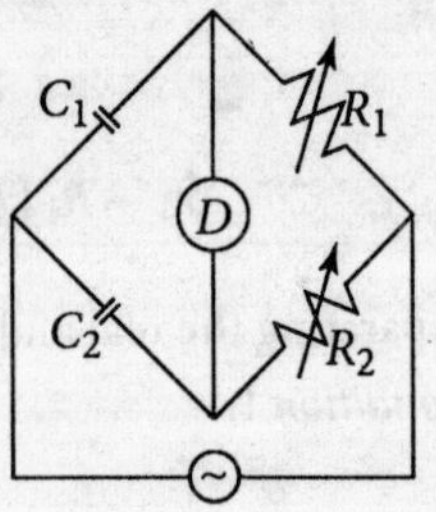

Fig. 5.6

(ii) *Wien's capacitance bridge.* This bridge is derived from de Sauty's bridge, with resistances in series with the condensers C_1 and C_2.

For balance,

$$\frac{R_1-(j/\omega C_1)}{R_2}=\frac{R-(j/\omega C_2)}{R_3}.$$

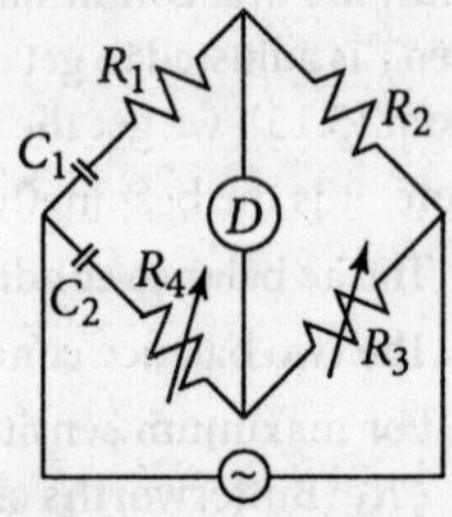

Fig. 5.7

Separating real and imaginary parts, we get the double conditions of balance.

Condition 1. $R_1/R=R_2/R_3$.

Condition 2. $C_1=(R_3/R_2)C_2$.

The first balance condition is obtained by varying R, which affects the first equation but not the second one. But since the second balance condition also includes the ratio R_3/R_2, the first balance is not independent of the second. Hence, the balance is adjusted by alternate adjustment of R and R_2.

(b) *Schering's bridge.* This is the best suited bridge for the determination of capacitance in terms of standard capacitor. This is employed to find the relative permittivities or dielectric constants specially of liquids and small cable capacitances. The circuit is shown in Fig. 5.8.

For balance of the bridge,

$$\frac{r+(1/j\omega C_1)}{(1/j\omega C_3)}=\frac{R_2}{R_4/(1+jC_4R_4)},$$

because the fourth arm has the impedance consisting of resistances R_4 and $\not{C}_4$ in parallel, hence

$$1/Z_4=(1/R_4)+j\omega\not{C}_4\,.$$

From the above, we get

$$j\omega C_3 r+\frac{C_3}{\not{C}_4}=\frac{R_2}{R_4}\left(1+j\omega\not{C}_4\,R_4\right).$$

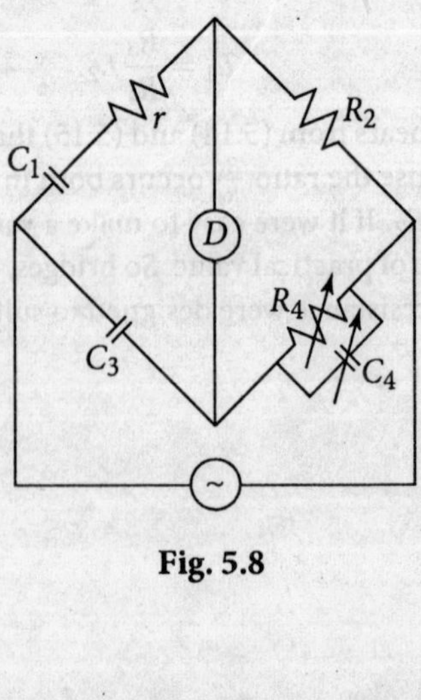

Fig. 5.8

So, separating the real and imaginary parts,
Condition 1. $C_3/C_1 = R_2/R_4$.
Condition 2. $rC_3 = R_2 \not{C}_4$.
These two balance conditions are independent, the first adjustment for balance being down by R_4 and the second condition being attained by adjustment of $\not{C}_4$, a variable air condenser. C_3 is also a standard air condenser. So, when the bridge is balanced, we can find the unknown capacitance C_1.

3. *Mutual inductance* (*Campbell's bridge*). This bridge originally due to Carey Foster but modified by Campbell is designed to determine the mutual inductance between two coils. The circuit be shown in Fig. 5.9.

A variable resistance is in series with the condenser C and this is the modification made in Carey Foster's original design meant for determining the mutual inductance using a dc source. The condition of balance is that there will be no flow of current through the detector.

Let the currents in the different branches of the bridge be as shown in Fig. 5.9.

Applying Kirchhoff's law in the mesh,

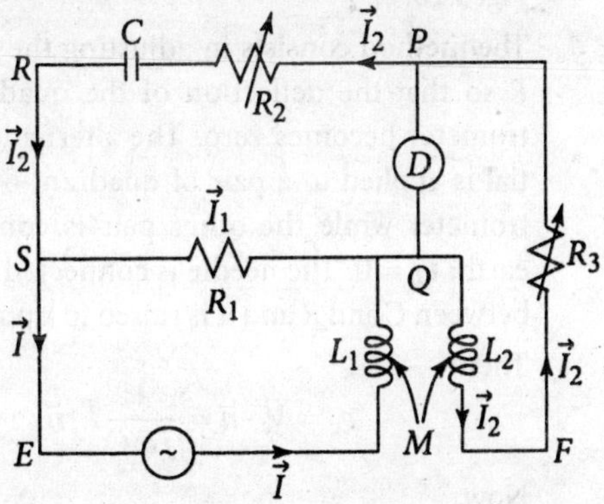

Fig. 5.9

1. $PQFP$: $j\omega L_2\vec{I}_2 + R_3\vec{I}_2 - j\omega M\vec{I} = 0$.

2. $QPRSQ$: $-\left(R_2 + \frac{1}{j\omega C}\right)\vec{I}_2 + R_1\vec{I}_1 = 0$.

From the first equation, we get

$$(j\omega L_2 + R_3 - j\omega M)\,\vec{I}_2 - j\omega M\vec{I}_1 = 0; \quad -\left(R_2 + \frac{1}{j\omega C}\right)\vec{I}_2 + R_1\vec{I}_1 = 0.$$

Eliminating $\vec{I}_1$ and $\vec{I}_2$,

$$\begin{vmatrix} R_3 + j(\omega L_2 - \omega M) & -j\omega M \\ -(R_2 - j/\omega C) & R_1 \end{vmatrix} = 0 \text{ or, } R_1R_3 + jR_1(\omega L_2 - \omega M) - j\omega M\left(R_2 - \frac{j}{\omega C}\right) = 0.$$

Equating real and imaginary parts,

$$R_1R_3 - \frac{M}{C} = 0 \text{ or, } R_1R_3 = \frac{M}{C} \text{ and } R_1(\omega L_2 - \omega M) = \omega MR_2.$$

$$R_1L_2 - R_1M = MR_2 \text{ or, } M = \left(\frac{R_1}{R_1 + R_2}\right)L_2 \text{ or, } L_2 = \left(\frac{R_1 + R_2}{R_1}\right)M.$$

Thus, for balance,
Condition 1. $R_1R_3 = M/C$.
Condition 2. $L_2 = \left(\frac{R_1+R_2}{R_1}\right)M$.
The first condition is obtained by varying R_3, which does not have any relation to the condition 2 and the second condition of balance is obtained varying R_2 thus the two conditions of balance are mutually exclusive.

Note. M must be inserted in the correct way and of correct value so as to obtain a balance at all condition. If the first adjustments indicate impossibility of getting a balance, this is to be re-examined by interchanging the connections to one of the coils L_1 or L_2. The second condition requires that $L_2 > M$. So we must know the approximate values of L_2 and M before doing the experiment.

4. *Frequency measurement.*

(a) *Arno's method.* This method can be used to measure a capacitance or frequency of the supply source. The circuit of this method is shown in Fig. 5.10.

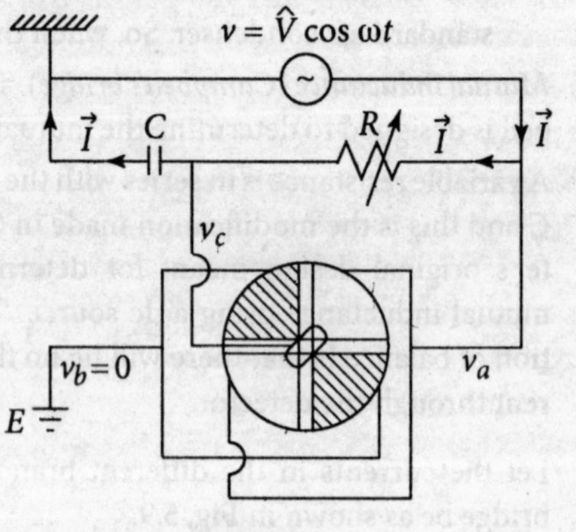

Fig. 5.10

The method consists in adjusting the resistance R so that the deflection of the quadrant electrometer becomes zero. The alternating potential is applied to a pair of quadrant of the electrometer while the other pair is connected to earth, $v_b = 0$. The needle is connected to a point between C and R and it is raised to a potential v_c. Then

$$v_c = \vec{V}_c \cdot \vec{n} = \frac{1}{j\omega C}\vec{I} \cdot \vec{n}.$$

Now,

$$\vec{I} = \frac{\vec{V}}{\left(R - \frac{1}{\omega C}\right)}.$$

$$\therefore\ v_c = \frac{1}{j\omega C}\frac{\vec{V} \cdot \vec{n}}{\left(R - \frac{1}{\omega C}\right)} = \frac{-\vec{V} \cdot \vec{n}}{1 + j\omega CR} = \frac{\hat{V}e^{j\omega t}\vec{n} \cdot \vec{n}}{\sqrt{1+\omega^2C^2R^2}\,e^{j\phi}}, \quad \tan\phi = \omega CR$$

$$= \frac{\hat{V}e^{j(\omega t-\phi)}\vec{n} \cdot \vec{n}}{\sqrt{1+\omega^2C^2R^2}} = \frac{\hat{V}\cos(\omega t - \phi)}{\sqrt{1+\omega^2C^2R^2}}. \tag{5.16}$$

(5.16) gives the instantaneous p.d. applied to the needle. The p.d. applied to one pair of quadrants is $v_a = \hat{V}\cos\omega t$ and the other pair is earthed, $v_b = 0$. Hence, the instantaneous couple on the needle is

$$\tau \propto (v_a - v_b)\left[v_c - \frac{v_a + v_b}{2}\right],$$

from the theory of quadrant electrometer.

$$\therefore\ \tau \propto \hat{V}\cos\omega t\left[\frac{\hat{V}\cos(\omega t-\phi)}{\sqrt{1+\omega^2C^2R^2}} - \frac{\hat{V}\cos\omega t}{2}\right] \propto \left[\frac{\hat{V}^2\cos\omega t\cos(\omega t-\phi)}{\sqrt{1+\omega^2C^2R^2}} - \frac{\hat{V}^2\cos^2\omega t}{2}\right].$$

So, the average couple is

$$\tau_{\text{av}} \propto \left[\frac{\hat{V}^2\cos\phi}{2\sqrt{1+\omega^2C^2R^2}} - \frac{\hat{V}^2}{4}\right].$$

This couple vanishes so that there is no deflection of the needle, when

$$\frac{\cos\phi}{\sqrt{1+\omega^2C^2R^2}}=\frac{1}{2}$$

$$\text{or, }\frac{1}{2}=\frac{1}{\sqrt{1+\omega^2C^2R^2}}\cdot\frac{1}{\sqrt{1+\omega^2C^2R^2}}\left[\because \cos\phi=\frac{1}{\sqrt{1+\omega^2C^2R^2}}\right]$$

$$=\frac{1}{\omega^2C^2R^1+1}$$

$$\text{or, } \omega^2C^2R^2=1 \text{ or, } \omega CR=1. \qquad (5.17)$$

(5.17) is the balance condition of the bridge. This is obtained by varying R. So, if C is a standard condenser of known value, the frequency could be determined. If ω be known, the capacitance could be found out.

(b) *A Simple method.* The circuit is shown in Fig. 5.11.

For balance, $\frac{R_1}{R_2}=\frac{R_3}{R_4+j\left(\omega L-\frac{1}{\omega C}\right)}$.

Condition 1. $R_1/R_2=R_3/R_4$.

Condition 2. $\omega L=1/\omega C$.

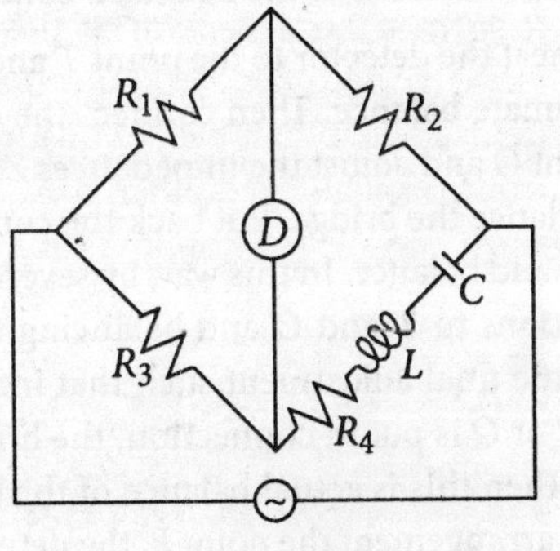

Fig. 5.11

First condition is obtained by adjusting the resistive components and the second condition realised by adjustment of either L and C.

5.6 Wagner's Earthing Device for Accurate Measurement with ac Bridge

In all the ac bridges, we have described, we have neglected one effect, which needs to be considered for measurement of small reactance at high frequency. It is the effect of *earth capacitance.* Because of this effect, it is very difficult to obtain the balance and the balance condition as it stands in the different bridges described does not hold. In order to remedy this stray capacitances, one very simple and by far the best method is that of Wagner's device, which is nothing but an auxiliary balancing branch of vector impedances Z_5 and Z_6 [Fig. 5.12(a) and 5.12(b)]; Fig. 5.12(a) being general network with Wagner's arrangement and 5.12(b) is a particular bridge, called *Schering's bridge,* with *Wagner's device.* The *stray capacitances* may be for the sake of analysis represented by small capacitances shown as C_a, C_b, C_c, C_d, C_e, etc. [Fig. 5.12(a)].

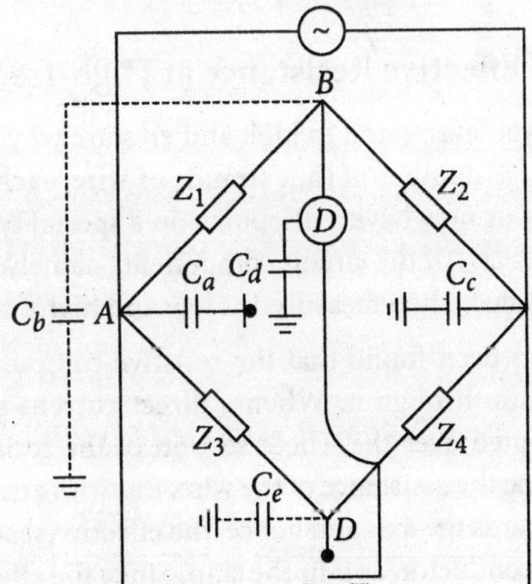

Fig. 5.12(a)

Because of the capacitances between the terminals of the bridge and the earth, there is always leakage of current. The leakage of current at A and C, entry-point and exit-point of current from the bridge network does affect the balance condition. But if there is leakage from the terminals like B and its opposite terminal T, then the amount of currents in AB and BC (or AT and TC) will not be same and hence, the balance condition will not be obtained as previously. To remove these stray capacitances, B, T and D are to be brought to the same potential, that of the potential of the earth. If we connect the point T directly to earth, which we may think of as the solution of the present problem, then also the currents in the branches AT and TC will not be equal and the balance condition will not hold. Such an impasse could be solved by an auxiliary impedance branch $APQC$, containing impedances Z_5 and Z_6, which are duplicates of Z_3 and Z_4, the midpoint being connected to earth. This is known as *Wagner's earthing device*, which brings the points B, D and T to earth in the balanced condition. The balancing is done as follows:

Connect the detector to the point T and adjust for approximate balance. Then connect the detector to the point O and adjust the impedances Z_5 and Z_6 so as to balance the bridge. Put back the connection to T again and balance. In this way, by several alternate connections to T and O and balancing the bridge, we get the final adjustment such that in both cases when T or O is put in connection, the bridge is balanced. Then this is actual balance of the bridge and by such arrangement the point B, the detector D and the point T are all brought to zero potential without actually connecting the point T to direct earth.

The modified Schering's bridge with Wagner's device has been shown [Fig. 5.12(b)].

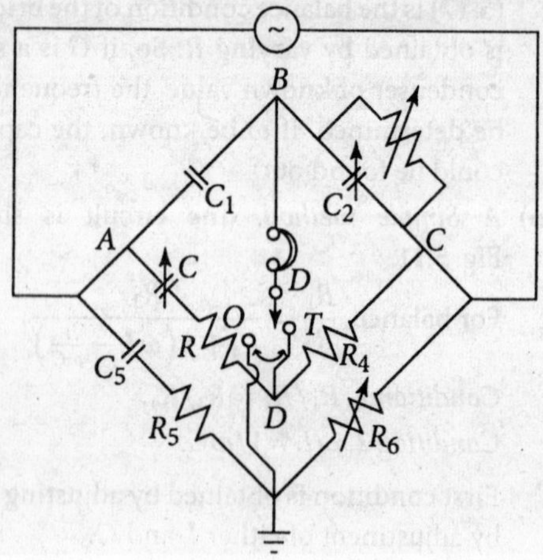

Fig. 5.12(b)

5.7 Effective Resistance at High-frequency Skin Effect

Have you ever cared to look and enquire why the cables used in high frequency work consist of a bundle of a large number of thin strands of wire, each separately insulated and joined at the ends o the cables. If not, let us now have a discourse on a special type of effect usually found in high frequency works. It is the skin effect. All the circuit components namely, the inductance, capacitance and resistance do not have the same value when measured at low and high frequencies.

It has been found that the resistive component of a circuit is greatly changed when a high frequency ac passes through it. When a direct current passes through a resistor, the current density is uniformly distributed over the whole section of the resistive wire and the whole section of the wire is effective in deciding the resistance of the wire. But with an ac, the current density near the surface is maximum and falls off towards the axis and hence, the effective section of the wire containing the current flow is the outer shell of the conductor, as if in the skin. Since the effective area of cross section decreases with an ac through the conducting wire, its resistance increases and this is known as *skin effect.* The effective depth of distribution of current density, called *skin depth* can be estimated from a simpler consideration or more rigorously from the electromagnetic wave propagation through a continuous solid. We will consider here the simpler method, the more rigorous one being referred to a chapter (cf. chapter on *electromagnetic theory*, in *Theory of Electricity and Magnetism*, Scitech Publication, by the author).

Physical Interpretation of the Effect

The fact that at high frequency the resistance varies can be given a simple explanation. Consider the current-flow through an isolated cylindrical wire. We know that current flowing through a wire produces magnetic flux, which are concentric circles with centre on the axis of the wire. These magnetic field lines are produced not only in the surrounding medium but also within the conductor. Just think of an electron along the axis of the conductor. It will be surrounded by greater number of lines of force than the one near the surface of the conductor and hence, when the current changes, the induced back emf due to change of flux the current changes, the induced back emf due to change of flux and proportional to the rate of change of flux will be of greater magnitude near the axis than that towards the surface. The inductive reactance, therefore, is greater near the the axis and hence, most of the current flows through a thin outer shell, so to say, the skin of the conductor, and since the effective area decreases, the resistance becomes higher, higher the frequency of the ac. It has been found that above 10Mc/s, skin effect becomes of any real significance.

Effective Resistance of a Conductor Carrying an ac

Let us try to calculate the effective resistance of a conductor of circular section of certain radius a, through which an alternating current passes.

l = length of the conducting wire. Let r = radius of the wire, ρ = specific resistance, μ = permeability of the material of the wire.

If a direct current or a low frequency ac passes along a wire, current density is uniform over the cross section, but when the frequency increases, the current-density distribution alters, being greater near the surface and less near the axis. So it may be said that the current density is a function of the radial distance from the axis. So we can put, $J = f(r)$.

$$I = \hat{I}\cos\omega t.$$

∴ the current density, if uniform, would be given by

$$J = \frac{I}{\pi r^2} = \frac{\hat{I}\cos\omega t}{\pi r^2} = J_p \cos\omega t.$$

Consider a point at a distance a from the axis. The magnetic field H is given by Ampere's circuital relation,

$$\oint \vec{B}\cdot\vec{dl} = \mu i \text{ or, } 2\pi a B = \mu J \times \pi a^2 \text{ or, } B = \frac{\mu J a}{2}. \tag{5.18}$$

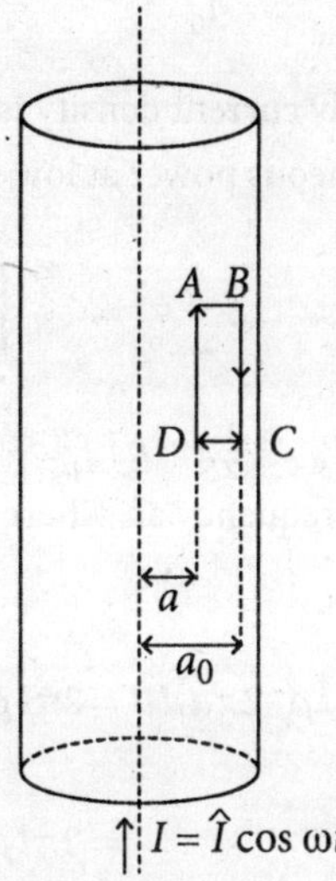

Fig. 5.13

The flux through the rectangle $ABCD$ [Fig. 5.13] of unit height is

$$\phi = \int_a^{a_0} \mu \mathrm{H}\, da = \mu \int_a^{a_0} 2\pi a J\, da = \pi\mu\left(a_0^2 - a^2\right) J_p \cos\omega t,$$

J_p = peak current-density.

$$\therefore\ e = -\frac{d\phi}{dt} = -\mu\frac{d}{dt}\left[\pi\left(a_0^2 - a^2\right) J_p \cos\omega t\right] = \mu\omega\pi\left(a_0^2 - a^2\right) J_p \sin\omega t.$$

$$\therefore \phi = \int_a^{a_0} B\,da = \int_a^{a_0} \frac{\mu J a}{2}\,da = \frac{\mu J}{4}\left(a_0^2 - a^2\right) = \frac{\mu}{4}\left(a_0^2 - a^2\right) J_p \cos\omega t = \frac{\mu J_p}{4}\left(a_0^2 - a^2\right)\cos\omega t.$$

$$\therefore e = -\frac{d\phi}{dt} = \frac{\mu\omega}{4} J_p \left(a_0^2 - a^2\right)\sin\omega t.$$

This induced emf will produce an eddy current of density J_e (eddy current is produced whenever a conductor is in a varying magnetic field) given by $e = \rho J_e$.

This current density will be in addition to current density due to external supply emf across the conductor. This eddy current necessitates an additional expenditure of energy in the form of heat but this does not cause any additional current to flow in and out of the conductor, i.e., it does not add to the current but simply alters the distribution. The lines of eddy current-flow will be closed-loops, where within a distance a_0 the resultant flow is in one direction; outside a_0, it is in the opposite direction, the net effect being thus equal to zero.

$$\therefore \int_0^r J_e 2\pi a\,da = 0 \text{ or, } 2\pi \int_0^r \frac{\mu\omega\left(a_0^2 - a^2\right) J_p \sin\omega t}{4\rho}\,a\,da = 0$$

$$\text{or, } \frac{2\pi\mu\omega J_p}{4\rho}\sin\omega t \int_0^r \left(a_0^2 - a^2\right) a\,da = 0 \text{ or, } \int_0^r \left(a_0^2 - a^2\right) a\,da = 0 \text{ or, } \frac{a_0^2 r^2}{2} - \frac{r^4}{4} = 0. \therefore a_0 = \frac{r}{\sqrt{2}}.$$

Thus, the eddy current density is zero at a distance equal to $\frac{r}{\sqrt{2}}$ from the axis of the wire.

The instantaneous power at low frequency ac, where eddy currents are negligible, is

$$\int_0^r J^2 l\rho \cdot 2\pi a\,da = \pi r^2 \rho l J_p^2 \cos^2\omega t.$$

$\therefore$ average $< P_{dc} \geq \frac{1}{2}\pi r^2 \rho l J_p^2$.

With a high frequency ac, where eddy current is of considerable effect, the instantaneous power dissipation is given by

$$\int_0^r (J + J_e)^2 l\rho \cdot 2\pi a\,da = 2\pi l\rho \int_0^r \left[J_p \cos\omega t + \frac{\mu\omega}{4\rho}\left(\frac{r^2}{2} - a^2\right) J_p \sin\omega t\right]^2 a\,da$$

$$= 2\pi l\rho \int_0^r \left[J_p^2 \cos^2\omega t + \frac{\mu^2\omega^2}{16\rho^2}\left(\frac{r^2}{2} - a^2\right)^2 J_p^2 \sin^2\omega t\right.$$

$$\left. + 2J_p^2 \cos\omega t \sin\omega t \frac{\mu\omega}{4\rho}\left(\frac{r^2}{2} - a^2\right)\right] a\,da.$$

$$(P_{ae})_{\text{mean}} = 2\pi l\rho J_p^2 \int_0^p \left[\frac{1}{2} + \frac{\mu^2\omega^2}{16\rho^2}\left(\frac{r^2}{2} - a^2\right)^2\right] a\,da = 2\pi l\rho J_p^2 \left[\frac{r^2}{4} + \frac{\mu^2\omega^2}{16\rho^2} \cdot \frac{r^6}{24}\right]$$

$$= \pi l\rho J_p^2 \left[\frac{r^2}{2} + \frac{\mu^2\omega^2}{8\rho^2} \cdot \frac{r^6}{24}\right].$$

Now since $< P_{ac} > \alpha R_{ac}, < P_{dc} > \alpha R_{dc}$,

$$\therefore \frac{P_{ac}}{P_{dc}} = \frac{R_{ac}}{R_{dc}} = 2\pi l\rho \frac{\left(\frac{r^2}{2} + \frac{\mu^2\omega^2}{8\rho^2} \cdot \frac{r^6}{24}\right)}{\pi l\rho r^2} = \left(1 + \frac{\mu^2\omega^2 r^4}{96\rho^2}\right) = 1 + \frac{\mu^2\pi^2 f^2 r^4}{24\rho^2}.$$

It shows, therefore, that resistance of a conductor increases when it carries an ac than when it carries a dc. It is also evident that this effect is small if f is small. This is why in high frequency works, the transmission cable is in the form of a bundle of thin strands of wire, called *Litz Wire.*

5.8 Exercises

1. Derive the balance condition for Anderson's bridge by direct application of Kirchhoff's laws to the four current loops.
2. One arm of a Wheatstone's bridge contains a resistance $R_1(1+\delta)$, while the other three arms all contain resistances equal to R. Show that the current through the detector of resistance r is approximately $\frac{\delta}{4(R+r)} \cdot V$, when a voltage V is applied to the bridge from a source of negligible impedance.
3. Explain how self-inductance of a coil can be measured in terms of a known capacitance by means of an ac bridge. Describe such a bridge and give the theory of the method.
4. Obtain the condition for balance in the generalised Wheatstone's bridge in which each arm has resistance, inductance and capacitance. Apply it for the solutions of the Maxwell's bridge giving the circuit diagram.
5. Derive the condition of balance for the general form of the ac bridge.
6. When to design an ac bridge determine a physical parameter of a network. What requisite characteristic will you have to consider?

   [*Hints.* ac bridges are designed such that two conditions of balance must be mutually exclusive and the conditions must not have any dependence on frequency of the source.]

   Describe an ac bridge method for the determination of (a) self-inductance of a coil (b) capacitance of a condenser.

It follows, therefore, that resistance of a conductor increases when it carries ac rather than when it carries dc. It is also evident that this effect is small if r is small. This is why in high frequency works, the transmission cable is in the form of a bundle of thin strands of wire, called Litz Wire.

5.8 Exercises

1. Derive the balance condition for Anderson's bridge by the application of Kirchhoff's laws to the four critical loops.
2. One arm of a Wheatstone's bridge contains a resistance $R(1 + \delta)$, while the other three arms all contain resistances equal to R. Show that the current through the detector of resistance r is approximately $\frac{\delta}{4(R+r)} V$ when a voltage V is applied to the bridge from a source of negligible impedance.
3. Explain how self inductance of a coil can be measured in terms of a known capacitance by means of an ac bridge. Describe such a bridge and give the theory of the method.
4. Obtain the condition for balance in the generalized Wheatstone's bridge in which each arm has resistance, inductance and capacitance. Apply it for the solutions of the Maxwell's bridge giving the circuit diagram.
5. Derive the condition of balance for the general form of the ac bridge.
6. When to design an ac bridge to determine a physical parameter of a network. What requisite characteristics will you have to consider?
7. How ac bridges are designed such that two conditions of balance must be mutually exclusive and the conditions must not carry any dependence on frequency of the source?
8. Describe an ac bridge method for the determination of (a) self-inductance of a coil (b) capacitance of a condenser.

Part II

Semiconductor Electronics
[Analog and Digital]

Chapter 1

Electron Ballistic

1.1 What is Electronics?

The history of the development of this branch of science is of fifty to sixty years, and within this brief span this branch has established its claim so much so that this age is rightly described as the age of electronics. From the Radio, the TV, the video decks to the manoeuvring of spacecrafts by the remote control devices electronics works its miracles. It is the study of the motion of electrons through vacuum, gas, or semiconductors and their control to put them to different useful purposes that makes this branch so useful and fascinating. The first appearance of this discipline was in the form of valve electronics, which is superseded at present by semiconductors. Semiconductor electronics is of two types: discrete and integrated. The latter because of its multidimensional advantage has been gaining grounds and has given birth to electronic calculator, electronic watches, various control devices, etc.

In the present few chapters we try to expose as far as possible the branch of electronics in the sequence of its developments, but remembering the vastness of the fields it covers, our attempt will be only to give a bird's eye view of the subject.

We start our discussion with the ballistics of electrons, i.e., the motion of electrons under different field configurations.

1.2 Electron Ballistics

(A) Motion of Electrons in Electrostatic Field (along positive Y-direction)

Let us consider the motion of an electron in an electrostatic field of intensity $\vec{X}$, the equation can be written as

$$m\frac{d^2\vec{r}}{dt^2} = -e\vec{X} \text{ or, } \frac{d^2\vec{r}}{dt^2} = -\frac{e\vec{X}}{m}. \tag{1.1}$$

In terms of components, we can write:

$$\frac{d^2x}{dt^2} = 0. \tag{1.2}$$

$$\frac{d^2y}{dt^2} = -\frac{e}{m}X \text{ (negative } y\text{-direction)} \tag{1.3}$$

$$\frac{d^2z}{dt^2} = 0 \text{ (the field is along } y\text{-axis).} \tag{1.4}$$

Let v_0 be the velocity along x-axis at the entry to the field [Fig. 1.1]. (1.2) when integrated twice w.r.t. time gives

$$x = a_2 t + a_2. \tag{1.5}$$

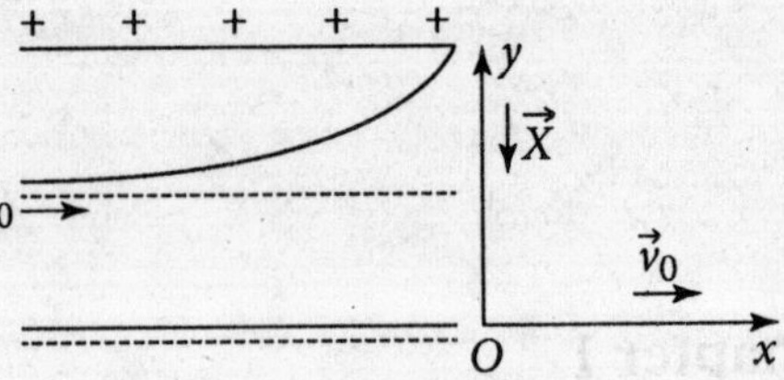

Fig. 1.1

Now, $x = 0$ at $t = 0$, $\dot{x} = v_0$ at $t = 0$. $\therefore$ $a_2 = 0$, $a_1 = v_0$.

$$x = v_0 t. \tag{1.6}$$

If $t = t_2$ for $x = l$, we get $l = v_0 t_2$.

$$t_2 = 1/v_0. \tag{1.7}$$

(1.4) when solved gives $z = b_1 t + b_2$.

At $t = 0$, $z = 0$ and also $\dot{z} = 0$. $\therefore$ $b_1 = 0 = b_2$.

So, the motion of the electron beam is confined to the xy-plane. (1.3) gives solution when integrated twice w.r.t. time as

$$y = -\frac{1}{2} \cdot \frac{e}{m} X t^2 + c_1 t + c_2. \tag{1.8}$$

At $t = 0$, $y = 0$, $\dot{y} = 0$. $\therefore$ $c_2 = 0 = c_1$.

$$\therefore y = \frac{1}{2} \cdot \frac{e}{m} X t^2 = \frac{1}{2} \cdot \frac{eX}{m} \cdot \frac{x^2}{v_0^2} \tag{1.9}$$

Let at the exit the deflection of the electron beam be y_0, then

$$y_0 = \frac{1}{2} \cdot \frac{eX}{m} \cdot \frac{l^2}{v_0^2}, \tag{1.10}$$

where l = length of the field. At the exit-point the electrons move along the tangent to the trajectory given by (1.9).

$$\therefore \tan\theta = \left.\frac{dy}{dx}\right|_{x=l} = \frac{eXl}{m v_0^2}. \tag{1.11}$$

(B) Motion of Electrons in a Magnetic Field

Let the electron beam move along x-axis with velocity $\vec{V}$ from left to right and a magnetic field of magnitude $\vec{B}$ be applied perpendicular to the direction of motion and along z-axis. Then the equation of motion is given by

$$m \frac{d^2 \vec{r}}{dt^2} = -e\left(\vec{V} \times \vec{B}\right) \text{ or, } \frac{d^2 \vec{r}}{dt^2} = -\frac{e}{m}\left(\vec{V} \times \vec{B}\right). \tag{1.12}$$

$$\therefore i \frac{d^2 x}{dt^2} + j \frac{d^2 y}{dt^2} + k \frac{d^2 z}{dt^2} = -\frac{e}{m} \begin{vmatrix} i & j & k \\ \dot{x} & \dot{y} & \dot{z} \\ 0 & 0 & B \end{vmatrix}.$$

In terms of components, we can write,

$$\frac{d^2x}{dt^2} = -\frac{Be}{m}\dot{y}. \tag{1.13}$$

$$\frac{d^2y}{dt^2} = \frac{Be}{m}\dot{x}. \tag{1.14}$$

$$\frac{d^2z}{dt^2} = \frac{Be}{m} \times 0 = 0. \tag{1.15}$$

So, the motion is confined in the xy-plane. Considering (1.13), we get

$$\frac{d}{dt}\left(\frac{dx}{dt}\right) = -\frac{Be}{m}\frac{dy}{dt}.$$

Integrating w.r.t. t, we get

$$\frac{dx}{dt} = -\frac{Be}{m}y + a_1. \tag{1.16}$$

Similarly, from (1.14) we can get

$$\frac{dy}{dt} = \frac{Be}{m}x + a_2. \tag{1.17}$$

Now, suppose, at $t = 0$, $x = 0$, $y = 0$, $z = 0$ and $\dot{x} = v_0$, $\dot{y} = 0$, $\dot{z} = 0$. Using these conditions, we get $a_1 = v_0$ and $a_2 = 0$.

$$\therefore \ \frac{dx}{dt} = -\frac{Be}{m}y + v_0 = \left(v_0 - \frac{Be}{m}y\right).$$

$$\frac{dy}{dt} = \frac{Be}{m}x.$$

Now

$$\frac{dx}{dt} = \frac{dx}{dy} \cdot \frac{dy}{dt}.$$

$$\therefore \ \frac{dx}{dy} \cdot \frac{dy}{dt} = \left(v_0 - \frac{Be}{m}y\right) \text{ or, } \frac{dx}{dy} \cdot \frac{Be}{m}x = \left(v_0 - \frac{Be}{m}y\right) \text{ or, } x\frac{dx}{dy} = \frac{mv_0}{Be} - y \text{ or, } x\,dx = \frac{mv_0}{Be}dy - y\,dy.$$

Integrating, we get

$$\frac{x^2}{2} = \frac{mv_0}{Be}y - \frac{y^2}{2} + a_3.$$

If the electron has deflection from the original track equal to s when it comes out of the magnetic field region extending over a length equal to l, then $x = l$, $y = s$.

$$\therefore \ \frac{l^2}{2} = \frac{mv_0}{Be}s - \frac{s^2}{2} + a_3 \text{ or, } a_3 = \frac{l^2 + s^2}{2} - \frac{mv_0}{Be}s.$$

$$\therefore \ \frac{x^2}{2} = \frac{mv_0}{Be}y - \frac{y^2}{2} + \frac{l^2 + s^2}{2} - \frac{mv_0 s}{Be} \text{ or, } x^2 + y^2 - \frac{2mv_0}{Be}y = \left(l^2 + s^2 - \frac{mv_0 s}{Be}\right)$$

$$\text{or, } x^2 + \left(y - \frac{mv_0}{Be}\right)^2 = l^2 + s^2 - \frac{mv_0 s}{Be} + \frac{m^2 v_0^2}{B^2 e^2}.$$

So, the trajectory of the electron is a circle whose centre is at the point $\left(0, \frac{mv_0}{Be}\right)$ and radius is given by

$$R = \sqrt{l^2 + s^2 - \frac{mv_0}{Be}\left(s - \frac{mv_0}{Be}\right)}.$$

Combined Electric and Magnetic Field (Cross Electric and Magnetic Field)

Here also we will suppose that the electron moves in xy-plane, electric field is along x-axis, the magnetic field along z-axis. The equation of motion of the electron is given by

$$m\frac{d^2\vec{r}}{dt^2}=-e\vec{E}-\left(\vec{V}\times\vec{B}\right) \text{ or, } (i\ddot{x}+j\ddot{y}+k\ddot{z})=-\frac{e}{m}\vec{E}-\frac{e}{m}\left(\vec{V}\times\vec{B}\right)=-i\frac{e}{m}E-\left(\frac{e}{m}\right)\begin{vmatrix} i & j & k \\ \dot{x} & \dot{y} & \dot{z} \\ 0 & 0 & B \end{vmatrix}.$$

$$\therefore \ddot{x}=-\frac{e}{m}E-\frac{e}{m}B\dot{y} \tag{1.18}$$

$$\ddot{y}=\frac{e}{m}B\dot{x} \tag{1.19}$$

$$\ddot{z}=0. \tag{1.20}$$

Taking (1.20) and integrating it twice w.r.t. time, we get $z=\frac{1}{2}a_1t^2+b_1$, now since at $t=0, z=0$ and also $\dot{z}=0$, we have $a_1=b_1=0$.

Thus, the motion is confined to the xy-plane. So (1.18) and (1.19) define the motion of the electron.

Put $X=x+jy, \therefore \ddot{X}=\ddot{x}+j\ddot{y}$.

Multiplying (1.19) by j and adding to (1.18), we get

$$\ddot{X}-j\frac{Be}{m}\dot{X}=-\frac{Ee}{m}.$$

Set $\dot{X}=\eta$, then

$$\frac{d\eta}{dt}-j\frac{Be}{m}\eta=-\frac{Ee}{m}.$$

Multiplying throughout by the integrating factor $\exp\left(-j\frac{Be}{m}t\right)$ and integrating w.r.t. time, we write

$$\eta\exp\left(-j\frac{Be}{m}t\right)=-\frac{E}{jB}\exp\left(-j\frac{Be}{m}t\right)+A \text{ or, } \eta=-\frac{E}{jB}+A\exp\left(J\frac{Be}{m}t\right).$$

Now, at $t=0, x=y=0, \dot{x}=\dot{y}=0.$ $\therefore \eta=0$ at $t=0.$ $\therefore A=E/jB.$

$$\eta=-\frac{E}{jB}+\frac{E}{jB}\exp\left(j\frac{Be}{m}t\right)=-\frac{E}{jB}\left[1-\exp\left(j\frac{Be}{m}t\right)\right]$$

$$\therefore X=-\frac{E}{jB}\left[t-\frac{m}{jBe}\exp\left(J\frac{Be}{m}t\right)\right]+A_1.$$

Now, at $t=0, x=0, y=0. \therefore X=0.$

$$\therefore A_1=-\frac{mE}{B^2e}.$$

$$\therefore X=-\frac{E}{jB}\left[t-\frac{m}{jBe}\exp\left(j\frac{Be}{m}t\right)\right]+\frac{mE}{B^2e}$$

$$=j\frac{E}{B}t-\frac{mE}{B^2e}\exp\left(j\frac{Be}{m}t\right)+\frac{mE}{B^2e}=j\frac{E}{B}t+\frac{mE}{B^2e}\left[1-\exp\left(j\frac{Be}{m}t\right)\right].$$

$$\therefore\ x=\frac{mE}{B^2e}\left(1-\cos\frac{Be}{m}t\right)=\frac{b}{a^2}(1-\cos at). \tag{1.21}$$

$$\therefore\ y=\frac{E}{B}t-\frac{mE}{B^2e}\sin\frac{Be}{m}t=\frac{E}{\frac{B^2e}{m}}\left[B\frac{e}{m}t-\sin\frac{Be}{m}t\right]=\frac{b}{a^2}(at-\sin at), \tag{1.22}$$

where $a=Be/m$, $b=eE/m$.

Thus, the trajectory of the electron represented by (1.21) and (1.22) is cycloid.

For further details, see *Theory of Electricity and Magnetism* by the author, Scitech Publication, Chennai.

Energy in Electron Volts

The energy gained by an electron in falling through a potential difference of 1V is known as *one electron volt.*

$$1\text{eV}=1.6\times10^{-19}\times1=1.6\times10^{-19}\text{J}.$$

1.3 Analog Electronics vs. Digital Electronics

Analog electronics refers to process or system accomplished by valves (in the past) or transistor (at present) in a continuous fashion. Digital electronics, on the other hand, refers to a process or system accomplished using discrete units or it refers to a system of counting using discrete units.

Most of the modern electronic gadgets starting from camera to computer are based on digital electronics. The basic concept of digital circuits is related to the work of George Boole who proposed his idea contained in what is known as *Boolean Algebra.* It is an algebra of logic containing two variables OFF and ON or YES and NO. So it is a two-state logic system for electronic device like transistors. The importance of this logic system remained unrealised for about a century after which it was Claude Shannon who realised its application in switching circuits. In analog electronics, the basic hardware is made of semiconductor elements like diodes, transistors, etc. biased in such a way that it operates in the active region so as to avail an output corresponding to an input at every instant. What it means really is that the output is a continuous function of the input. In digital electronics, however, the output of the system is a discrete function of the input and so the semiconductor elements are biased between *cut-off* and *saturation* states only having two configurations of OFF and ON as a switch. In analog electronics, the counting system is based on decimal number system, where we can recognise ten states or ten numerals from 0 to 9, whereas in digital counting system, the basis is the Boolean Algebra using only two states 0 (*Off*) and I (*On*); so the output of a digital device consists of a stream of *zeros* and *ones,* called *binary number system,* which, of course, can be converted to decimal number system, so familiar and convenient to us.

In the arena of signal transmission, we have analog signal processing (ASP) and digital signal processing (DSP). Analog signal is a signal that exists at every instant of time, and amplitude may or may not be constant w.r.t. time. The digital signal, on the other hand, is one that is discrete w.r.t. time as well as amplitude. Since the analog signal is a continuous function of time, greater amount of energy is involved for its creation and transmission. But digital signal being a discrete function of time, lesser power is expended for its creation and transmission.

1.4 Numerical Problems

1. An electron is projected with an initial velocity $v_0=10^7$m/s into a uniform field between two parallel plates. The direction of the field is vertically downwards and the field is zero except in the space between the plates. The electron enters the field at a point midway between the plates. If the electron just misses the upper plate as it emerges from the field, find the electric intensity. The length of the field region is 2cm and the separation of the plates 1cm, $e/m=5.27\times10^{17}$esu/gm.

Solution: $v_0 t = 2$, $\frac{1}{2}\cdot\frac{Xe}{m}t^2 = 0.5$. Solving, we get $X = 142.3$V/cm.

2. Electrons initially at rest are accelerated by a p.d. of 200V. Then they pass through an electric field and a magnetic field at right angles to each other and also at right angles to the electron beam so arranged that the electrons emerge out of the field straight. The electric field was found to be (160/1.4)V/cm and the magnetic field was 13.6 Oe. Calculate their specific charge.

Solution: $\frac{1}{2}mv^2 = eV$ or, $\dfrac{e}{m} = \dfrac{v^2}{2V} = \dfrac{v^2}{2\times\frac{200}{300}} = \dfrac{3}{4}v^2$.

Again, $v = X/H$.

Now,

$$X = \frac{160}{1.4}\text{V/cm} = \frac{160}{1.4}\times 10^8\text{emu/cm}.$$

$$H = 13.6 \text{ Oe (magnetic field in emu)}.$$

$$\therefore\ v = \frac{160\times 10^8}{1.4\times 13.6} = 8.40\times 10^8\text{cm/s}.$$

$$\therefore\ \frac{e}{m} = \frac{3}{4}\times(8.40)^2\times 10^{16} = 5.29\times 10^7\text{esu/gm}.$$

3. An electron is accelerated through a p.d. of 400V before it enters a magnetic field of flux density 10^{-2}Wb/m^2 at an angle of 30° to the field. Explain why the path of the electron in the magnetic field will be a helix? Deduce an expression for the pitch and diameter of the helix, $e = 1.6\times 10^{-19}$C, $m = 9.1\times 10^{-32}$kg.

Solution: The electron's motion makes an angle of $\theta = 30°$ with the field direction. Let the velocity at the entry to the field be v_0. This velocity can be resolved into two components parallel and perpendicular to the field.

$$u_{\parallel} = v_0\cos\theta \text{ (parallel)}.$$

$$u_{\perp} = v_0\sin\theta \text{ (perpendicular)}.$$

For motion along the field, the electron experiences no force, hence $v_0\cos\theta$ remains constant. For velocity component perpendicular to the field, the electron experiences a force $= Bev_0\sin\theta$. So to an observer looking along X-direction, the electron appears to be in a circular path of radius R, given by

$$\frac{mv_0^2\sin^2\theta}{R} = Bev_0\sin\theta \text{ or, } R = \frac{mv_0\sin\theta}{Be} \text{ or, } T = \frac{2\pi m}{Be}\text{ sec.}$$

The forward distance moved during this time is the pitch of the helical path.

$\therefore\ p = T\times v_0\cos\theta = \frac{2\pi m}{Be}\times v_0\cos\theta \simeq \frac{2\pi m}{Be}v_0$, when θ is small.

Putting the values in the expression for R, we get

$$R = 0.337\times 10^{-2}\text{m}.$$

$$\therefore\ D = 0.674\times 10^{-2}\text{m}.$$

$$p = 6.455\times 10^{-2}\text{m}.$$

4. The spacing between the parallel plates of a capacitor in vacuum is 1cm. There is a steady p.d. of 1.76kV between the plates and a uniform magnetic field of 10^{-2}Wb/m^2 directed parallel to the plates. An electron is released from the negative plate with zero velocity. Find: (a) the transit time of the electron (b) the

velocity of impact on the positive plates (c) the angle of incidence on impact (d) position of the electron at $t = 0.5 \times 10^{-9}$ sec.

Solution: Here, $x = 1\text{cm} = 10^{-2}\text{m}$.

$$E = \frac{1.76 \times 10^2}{10^{-2}} = 1.76 \times 10^5 \text{V/m}.$$

$$\frac{e}{m} = 1.76 \times 10^{11} \text{C/kg}, \; B = 10^{-2} \text{Wb/m}^2.$$

$$\therefore\ a = \frac{Be}{m} = 1.76 \times 10^{11} \times 10^{-2} = 1.76 \times 10^9$$

$$b = \frac{Ee}{m} = 1.76 \times 10^{11} \times 1.76 \times 10^5 = 1.76^2 \times 10^{16}.$$

(a) $x = \frac{b}{a^2}(1 - \cos at), y = \frac{b}{a^2}(at - \sin at)$.

$$\therefore\ 10^{-2} = 10^{-2}(1 - \cos at) \text{ or, } \cos at = 0 \text{ or, } at = \frac{\pi}{2}.$$

$$\therefore\ t = \frac{\pi}{2} \times \frac{1}{1.76} \times 10^{-9} = 0.892 \times 10^{-9} \text{s}.$$

(b) $u_x = \frac{dx}{dt} = \frac{b}{a}\sin at = \frac{dy}{dt} = \frac{b}{a^2}(a - a\cos at) = \frac{b}{a}(1 - \cos at)$.

So, when $at = \frac{\pi}{2}, u_x = \frac{b}{a} = 1.76 \times 10^7, \; u_y = \frac{b}{a} = 1.76 \times 10^7$.

$\therefore$ velocity of impact $= u = \sqrt{u_x^2 + u_y^2} = 2.489 \times 10^7 \text{m/s}$.

(c) $\tan\alpha = \frac{u_y}{u_x} = 1 = \tan 45^\circ . \therefore\ \alpha = 45^\circ$.

(d) $x_1 = \frac{b}{a^2}(1 - \cos 1.76 \times 0.5) = 10^{-2} \times 1.179 \times 10^{-4} = 1.179 \times 10^{-3}\text{m}$ and

$y_1 \frac{b}{a^2}(1.76 \times 0.5 - \sin 1.76 \times 0.5) = 0.865 \times 10^{-2}\text{m}$.

1.5 Exercises

1. The anode and cathode of a diode are coaxial cylinders of radii R and r respectively with $R > r$. A p.d. V_1 is maintained between them. An electron leaves the cathode radially with negligible velocity.
   (a) Derive an expression for the subsequent velocity of the electron as a function of the radius x from the axis of the cylinders.
   (b) Show that the velocity reaches half of its final value when $x = (Rr^3)^{1/4}$.
2. An electron having an initial energy of 1000eV is projected horizontally into a uniform electrostatic field of intensity 10^5V/m acting vertically downwards. Determine the position of the electron after 10^{-9}s. [$x = 1.88$cm, $y = 0.88$cm]
3. A sinusoidal voltage with $f = 1$Mc/s and a maximum value of 10V is applied to the plates of a parallel-plate capacitor, which are 2cm apart. If an electron is released from one plate with a velocity of 10^8cm/s along a line of force at the instant when the voltage is zero, determine the position of the electron at any subsequent time. [$x = 1.5 \times 10^7 t - 2.23\sin(6.28 \times 10^6 t)$m]
4. A beam of electron enters the field between two parallel plates A, B at an angle 30° with B. Find: (a) electrostatic field (b) magnetic flux density so as to cause the electron beam to leave the field at a distance 6cm from the point of entry. The incoming electron beam possesses an energy of 70eV. [2.02×10^{-3}V/m, 47mWb/m^2]

5. An electron moving with an energy of 400eV moves in a magnetic field of uniform flux density 10^{-1}Wb/m^2, the field and velocity being mutually perpendicular. Show that the path is a circle and calculate its radius. [6.74cm]
6. A magnetic field of uniform flux density 10^{-2}Wb/m^2 is applied between and parallel to two plates spaced 2cm apart. An electron enters the magnetic field through a hole in one plate with a velocity normal to the plates. Calculate the energy in eV that the electron must acquire prior to entering the field if its path is to be tangential at the opposite plate, $e/m = 1.76 \times 10^{11}\text{C/kg}$. [3.52keV]

Chapter 2

Semiconductor Electronics

2.1 Introduction

In its early form we had to study valve electronics. But nowadays valve electronics is superseded by semiconductor electronics, which is again split up into: (i) discrete electronics (ii) integrated electronics. In discrete semiconductor electronics we utilise the semiconductor elements like diode, transistor, etc. to serve different functions as discrete units in an electronic circuit. During 1960s techniques were developed for the construction of integrated circuits (ICs) in which whole amplifier or electronic system is fabricated on one small piece or chip of silicon. Such fabrication has brought forth tremendous saving of space, reliability and extension of operational range. A digital computer using vacuum tubes in 1955 occupied the space of several rooms, but using discrete transistors, one room would suffice; nowadays with ICs the same computer fits on a desk top. Thus, miniaturisation attempt led to the development of microelectronics technique.

We will consider first the discrete circuitry using semiconductors (diodes, transistors, etc.) and then introduce the ICs.

2.2 Semiconductors and Insulators

There is a class of substances whose conductivity lies between conductors on one side and insulators on the other. This class is known as *semiconductor.* There are two classes of semiconductors: (a) intrinsic semiconductors (b) extrinsic semiconductors. Intrinsic semiconductors are germanium, silicon, etc. at room temperature. At zero temperature, these substances behave as insulators, because the four valence electrons of germanium or silicon form covalent bonds with the electrons of the neighbouring atoms in the crystal lattice thus leaving no free electrons to give rise to conduction. But when the temperature is raised such that the thermal energy exceeds the ionisation potential, some of the covalent bonds are broken, because of the thermal vibration of the crystal lattice. It thus produces a *free electron,* which migrates from the site, and a *hole* is thus created, the number of such thermally generated *electron-hole* pair is very small and they are referred to as *minority carriers.*

In the light of band theory of solids, the above fact is explained in a slightly different way.

When atoms are widely separated each atom has its own valence electrons with discrete energy levels available to each electron. But as the atoms come closer, the energy levels spread out into bands. The band-structure characteristic of metals, semiconductors and insulators are as shown in Fig. 2.1(a) and 2.1(b), the difference between the band structure of a metal and semiconductor or insulator is that in case of metal the valence band is partially filled or there is a region of overlap between conduction band and valence band, whereas there is a gap, called *forbidden gap,* between the conduction band and valence band in case of semiconductor or insulator band structure. In insulator the forbidden gap E_g (also called *ionisation potential*) is of the order of 6eV, whereas in case of semiconductor the band-gap is of the order of 0.72 to 1.22eV.

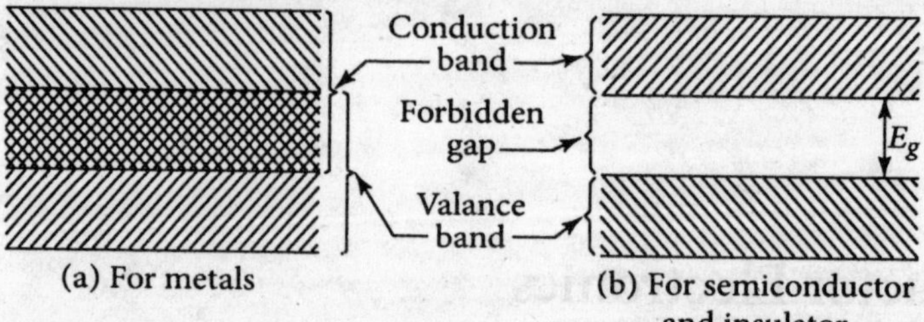

Fig. 2.1

As stated above, the germanium crystal at zero deg. Kelvin behaves as an insulator, because all the four valence electrons are bound by covalent bond. This implies in band theory that there is no possibility of an electron to jump across the forbidden gap to cross over to the conduction band, where an electron behaves as *free*. But with the rise of temperature the electrons in the valence band may acquire energy sufficient to cross the gap and pass over to the conduction band and thus leave a *hole* in the valence band [Fig. 2.2]. Thus, an electron-hole pair is created and some conductivity is possessed by the crystal. Such a semiconductor is known as *intrinsic semiconductor* whose conductivity is due to small number of electrons and holes generated as a result of thermal energy.

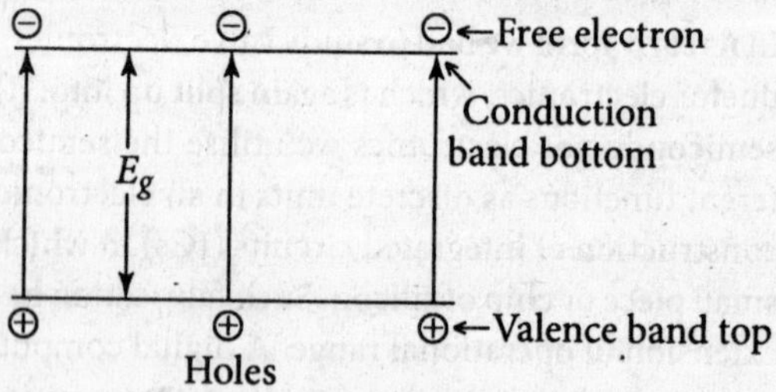

Fig. 2.2

2.3 Extrinsic or Doped Semiconductor

The extrinsic or doped semiconductor is that in which the carriers of one kind either holes or electrons predominate. The predominant carriers are called *majority carriers*. The semiconductor, where the majority carriers are the electrons is known as the *n-type semiconductor* and the semiconductor, where the majority carriers are *holes* is known as the *p-type semiconductor*. The doped chemical imperfection plays a very important role in the electrical properties of the material. Since the impurity doped is minute, the basic crystal structure is not appreciably altered.

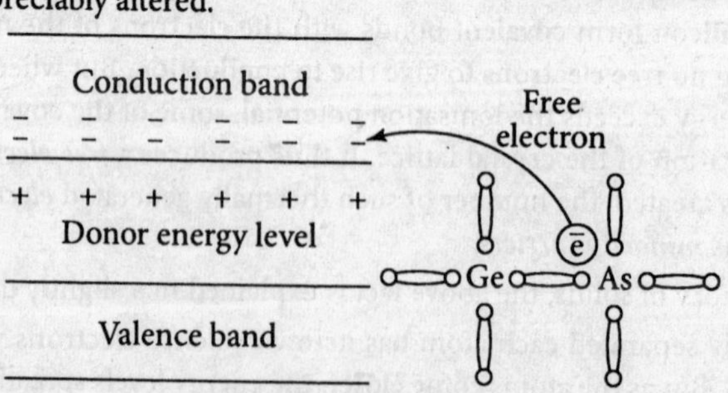

Fig. 2.3

If the doped element is pentavalent (e.g., As, Sb or phosphorus), the four electrons of the foreign element complete covalent bonds and the fifth electron is becoming free. The band model and bond model in this case is [Fig. 2.3]. Such a semiconductor is known as *n-type semiconductor.* The impurity atom, which has donated an electron is ionised.

If the doping element is trivalent (Al, B, Ca, Indium, etc.) the impurity atom lacks an electron in its outermost shell and is in a position to accept an electron and thus a covalent bond vacancy or a hole is created. In this case, the band model and bond model is shown in Fig. 2.4. Here the majority carriers are the holes, which behave as positively charged particles. Such semiconductor is known as the *p-type semiconductor.*

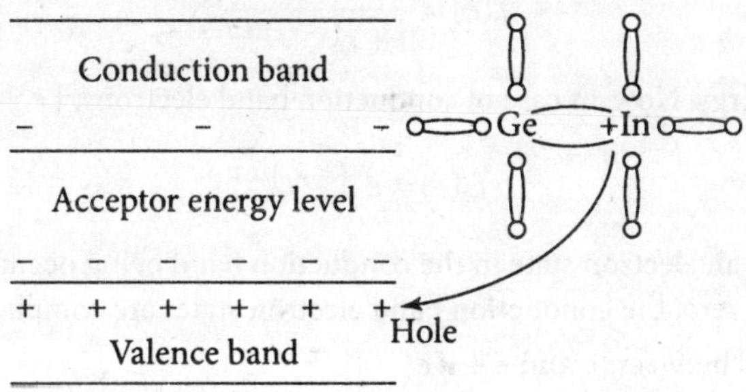

Fig. 2.4

It is to be noted that minute addition of impurity adds discrete energy levels, because in case the atoms are close, the energy levels spread out into bands but for small concentration, the atoms are widely apart such that wave function overlap is negligible. If the addition of the energy levels is in the forbidden gap, then they may be close to the conduction or valence band. In the former case electrons may occupy the impurity level close to the conduction band and with rise of temperature these electrons are excited into the conduction band and become free electrons. The impurity levels from which the electrons come are known as *donor levels.* In case the impurity levels lie close to the valence band, electrons are excited from the valence band into these levels with the rise of temperature and so these levels are now the acceptor levels. This process leaves holes in the valence band and the material is known as the *p-type semiconductor.*

2.4 Electron Distribution and Fermi level in Semiconductors

We have already stated that electrons in a metal cannot be treated as forming an electron gas whose distribution can be represented by classical statistics. Same is true in case of electrons in a semiconductor. So we apply Fermi-Dirac statistics, which is applicable in such a case. In FD statistics one important feature comes in and it is the concept of Fermi energy. Even at absolute zero temperature, according to this statistics, there are some electrons possessing energy, but there is a maximum limit of this energy, called *Fermi energy.* This is a new concept. In an intrinsic semiconductor, the Fermi level lies midway between the bottom of the conduction band and top of the valence band. In the extrinsic semiconductor the Fermi level at absolute zero is exactly midway between the bottom of the energy levels donated by the donor impurity atoms and the bottom of the conduction band (in case of n-type semiconductor) or midway between the top of the valence band and top of the energy levels donated by the acceptor impurity atoms (in case of p-type semiconductor). But as T increases, the Fermi level in n-type semiconductor descends and even goes below the bottom of the donor energy levels and tends to occupy position midway between the conduction band and valence band and then an n-type semiconductor becomes an intrinsic semiconductor. This is why solid-state elements like diode or transistor are restricted by temperature. Similarly, in a p-type semiconductor the Fermi level ascends with temperature and at a certain temperature it occupies a position midway between the conduction band and valence band.

Fermi level in intrinsic semiconductor

The number of electrons per unit volume between energy intervals ϵ and $\epsilon + d\epsilon$ is given by

$$dn = f(\epsilon)\, g(\epsilon)\, d\epsilon, \tag{2.1}$$

where $f(\epsilon)$ is the Fermi-Dirac distribution and $g(\epsilon)$ is the energy-density of states. The Fermi-Dirac distribution is given as

$$f(\epsilon) = \frac{1}{1 + e^{(\epsilon - \epsilon_f)/kT}},$$

where ϵ_f is the Fermi energy. Now, in case of conduction band electrons, $(\epsilon - \epsilon_f) \gg kT$, so

$$f_c(\epsilon) = e^{-(\epsilon - \epsilon_f)/kT} \tag{2.2}$$

$f_c(\epsilon)$ is the probability of an electron state in the conduction band being occupied. So at $T = 0$, $f_c(\epsilon) \to 0$, which means, at absolute zero, the conduction band electron states are completely empty.

Density of electron states between ϵ and $\epsilon + d\epsilon$

The particles having energy between ϵ and $\epsilon + d\epsilon$ have momentum between p and $p + dp$, where $\epsilon = p^2/2m$. So, the number of particles between energy ϵ and $\epsilon + d\epsilon$ will be proportional to the volume of the spherical shell between p and $p + dp$ in momentum space [Fig. 2.5], the volume of the spherical shell is $4\pi p^2 dp$. The volume containing an electron is given by h^3 in view of the uncertainty principle. Hence, the density of states between ϵ and $\epsilon + d\epsilon$ is given by $g(\epsilon) = 2 \times \frac{4\pi p^2 dp}{h^3}$, the factor 2 accounts for the two directions of spin.

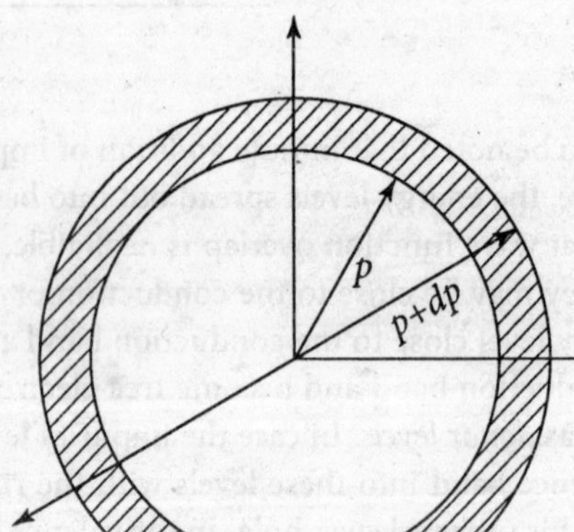

Fig. 2.5

Since, $\epsilon = p^2/2m$, we get

$$d\epsilon = \frac{p\,dp}{m} \Rightarrow p\,dp = m\,d\epsilon.$$

$$\therefore\ g(\epsilon) = 2 \times \frac{4\pi(2m\epsilon)^{1/2}\, m\, d\epsilon}{h^3} = \frac{1}{2\pi^2}\left(\frac{2m}{\hbar^2}\right)^{3/2} \epsilon^{1/2}\, d\epsilon, \text{ where } \hbar = h/2\pi.$$

So, the density of state function for conduction band electron is written as

$$g_c(\epsilon) = \frac{1}{2\pi^2}\left(\frac{2m}{\hbar^2}\right)^{3/2} (\epsilon - \epsilon_c)^{1/2}\, d\epsilon,$$

where ϵ_c is the energy corresponding to the bottom of the conduction band.

$$\therefore\ dn_c = f_c(\epsilon)\, g_0(\epsilon)\, d\epsilon = \frac{1}{2\pi^2}\left(\frac{2m}{\hbar^2}\right)^{3/2} e^{-(\epsilon-\epsilon_f)/kT}(\epsilon - \epsilon_c)^{1/2}\, d\epsilon$$

$$\Rightarrow n_c = \frac{1}{2\pi^2}\left(\frac{2m}{\hbar^2}\right)^{3/2} \int_{\epsilon_c}^{\infty} (\epsilon - \epsilon_c)^{1/2} e^{-(\epsilon-\epsilon_f)/kT}\, d\epsilon.$$

Set, $(\epsilon-\epsilon_c)/kT=x \Rightarrow d\epsilon = kT\,dx$.

$$\therefore \int_{\epsilon_c}^{\infty}(\epsilon-\epsilon_c)^{1/2}e^{-(\epsilon-\epsilon_f)/kT}\,d\epsilon=\int_0^{\infty}x^{1/2}(kT)^{1/2}e^{-(\epsilon_c-\epsilon_f)/kT}\,e^{-x}(kT)\,dx$$

$$=(kT)^{3/2}\,e^{-(\epsilon_c-\epsilon_f)/kT}\int_0^{\infty}x^{1/2}\,e^{-x}\,dx$$

$$=(kT)^{3/2}\,e^{-(\epsilon_c-\epsilon_f)/kT}\Gamma\left(\frac{3}{2}\right)=(kT)^{3/2}\,e^{-(\epsilon_c-\epsilon_f)/kT}\cdot\frac{1}{2}\sqrt{\pi}.$$

$$\therefore\ n_c=\frac{4\pi}{h^3}(2m)^{3/2}\cdot(kT)^{3/2}e^{-(\epsilon_c-\epsilon_f)/kT}\cdot\frac{1}{2}\sqrt{\pi}=2\left(\frac{2\pi mkT}{h^2}\right)^{3/2}e^{-(\epsilon_c-\epsilon_f)/kT}.$$

Now, the electron is moving in a periodic potential field and the motion in such a field may be considered as the motion of a free electron if the mass of the electron is taken as m_e^*, called the *effective mass of the electron*. So, the electron density in the conduction band can be written as

$$n_c=2\left(\frac{2\pi m_e^*kT}{h^2}\right)^{3/2}e^{-(\epsilon_c-\epsilon_f)/kT}=N_c e^{-(\epsilon_c-\epsilon_f)/kT},\tag{2.3}$$

where $N_c=2\left(\frac{2\pi m_e^*kT}{h^2}\right)^{3/2}$.

Conclusion. For the intrinsic semiconductor, greater the difference $(\epsilon_c-\epsilon_f)$, less will be the number of conduction electrons.

Density of holes in the valence band

The valence electron has energy such that $(\epsilon_f-\epsilon)\gg kT$, so that we have $e^{(\epsilon-\epsilon_f)/kT}=e^{-(\epsilon_f-\epsilon)/kT}$, a small quantity.

$$\therefore\ \frac{1}{1+e^{(\epsilon-\epsilon_f)/kT}}\simeq 1-e^{(\epsilon-\epsilon_f)/kT}.$$

$$\therefore\ f_v(\epsilon)=1-e^{(\epsilon-\epsilon_f)/kT}\Rightarrow 1-f_v(\epsilon)=e^{(\epsilon-\epsilon_f)/kT}.$$

$[1-f_v(\epsilon)]$ represents the probability of no occupancy of the energy state in the valence band by the electron or it can be taken as the probability of energy state being occupied by a hole. The density of state function for the valence band is given as

$$g_v(\epsilon)=\frac{1}{2\pi^2}\left(\frac{2m_h^*}{\hbar^2}\right)^{3/2}(\epsilon_v-\epsilon)^{1/2}\,d\epsilon.$$

$$\therefore\ dn_h=[1-f_v(\epsilon)]\ g_v(\epsilon)\,d\epsilon=e^{(\epsilon-\epsilon_f)/kT}\cdot\frac{1}{2\pi^2}\left(\frac{2m_h^*}{\hbar^2}\right)^{3/2}(\epsilon_v-\epsilon)^{1/2}\,d\epsilon$$

$$=4\pi\left(\frac{2m_h^*}{h^2}\right)^{3/2}(\epsilon_v-\epsilon)^{1/2}e^{-(\epsilon_f-\epsilon)/kT}\,d\epsilon.$$

$$\therefore\ n_h=4\pi\left(\frac{2m_h^*}{h^2}\right)^{3/2}\int_{-\infty}^{\epsilon_v}(\epsilon_v-\epsilon)^{1/2}e^{-(\epsilon_f-\epsilon)/kT}\,d\epsilon.$$

$$\frac{\epsilon_v-\epsilon}{kT}=x\Rightarrow d\epsilon=-kT\,dx.$$

$$\epsilon_f - \epsilon = \left(\epsilon_f + xkT - \epsilon_v\right) \Rightarrow \frac{\epsilon_f - \epsilon}{kT} = \frac{\epsilon_f - \epsilon_v}{kT} + x.$$

$$\therefore \int_{-\infty}^{\epsilon_v} (\epsilon_v - \epsilon)^{1/2} e^{-(\epsilon_f - \epsilon)/kT}\, d\epsilon = \int_{\infty}^{0} x^{1/2} (kT)^{1/2} e^{-x} \cdot e^{-(\epsilon_f - \epsilon_v)/kT} (-kT)\, dx$$

$$= (kT)^{3/2} e^{-(\epsilon_f - \epsilon_v)/kT} \int_0^{\infty} x^{1/2} e^{-x}\, dx = (kT)^{3/2} e^{-(\epsilon_f - \epsilon_v)/kT} \cdot \frac{1}{2}\sqrt{\pi}.$$

$$\therefore n_h = 4\pi \left(\frac{2m_h^*}{h^2}\right)^{3/2} \cdot (kT)^{3/2} e^{-(\epsilon_f - \epsilon_v)/kT} \cdot \frac{1}{2}\sqrt{\pi} = 2\left(\frac{2\pi m_h^* kT}{h^2}\right)^{3/2} e^{-(\epsilon_f - \epsilon_v)/kT}$$

$$= N_v e^{-(\epsilon_f - \epsilon_v)/kT} \qquad (2.4)$$

In case of intrinsic semiconductor, $n_h = n_c$, which fact can be used to locate the Fermi level in such a semiconductor

$$\therefore N_c e^{-(\epsilon_c - \epsilon_f)/kT} = N_v e^{-(\epsilon_f - \epsilon_v)/kT} \Rightarrow \left(\frac{N_v}{N_c}\right) = e^{-(\epsilon_c - \epsilon_f)/kT} \cdot e^{(\epsilon_f - \epsilon_v)/kT} \Rightarrow \left(\frac{m_h^*}{m_e^*}\right)^{3/2} = e^{(2\epsilon_f - \overline{\epsilon_c + \epsilon_v})/kT}.$$

Taking log on both sides,

$$2\epsilon_f - \overline{\epsilon_c + \epsilon_v} = \frac{3}{2} kT \ln \frac{m_h^*}{m_e^*} \Rightarrow \epsilon_f = \frac{\epsilon_c + \epsilon_v}{2} + \frac{3}{4} kT \ln \frac{m_h^*}{m_e^*}. \qquad (2.5)$$

(2.5) says that as in intrinsic semiconductor $m_h^* \simeq m_e^*$,

$$\epsilon_f = \frac{\epsilon_c + \epsilon_v}{2}.$$

But actually, however, $m_h^* > m_e^*$, hence with rise of temperature, there is a slight rise in the position of the Fermi level.

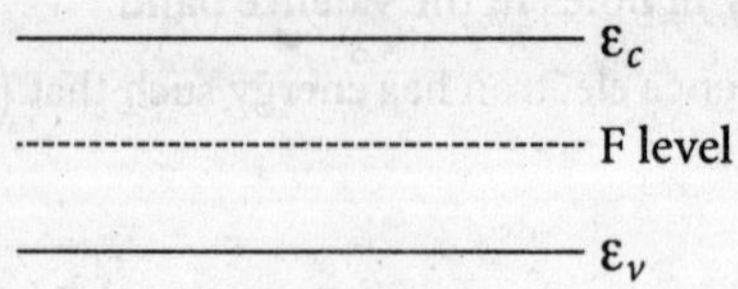

Fig. 2.6

With this value of ϵ_f putting into (2.3), we would get $n_c = N_c e^{-(\epsilon_c - \epsilon_f)/kT}$

$$\Rightarrow n_c = N_c e^{-(\epsilon_c - \epsilon_v)/2kT} e^{+\ln\left(\frac{m_h^*}{m_e^*}\right)^{3/4}} = N_c \epsilon^{-(\epsilon_c - \epsilon_f)/2kT} \cdot \left(\frac{m_h^*}{m_e^*}\right)^{3/4}$$

$$= 2\left(\frac{2\pi kT}{h^2}\right)^{3/2} \left(m_e^* m_h^*\right)^{3/4} e^{-(\epsilon_c - \epsilon_v)/2} = 2\left(\frac{2\pi kT}{h^2}\right)^{3/2} \left(m_e^* m_h^*\right)^{3/4} e^{-E_g/2kT}, \qquad (2.6)$$

where $(\epsilon_c - \epsilon_v) = E_g$ = forbidden gap energy.

Also,

$$n_c \cdot n_h = N_c \cdot N_v e^{-(\epsilon_c - \epsilon_f)/kT} e^{-(\epsilon_f - \epsilon_v)/kT} = 4\left(\frac{2\pi kT}{h^2}\right)^3 \left(m_e^* m_h^*\right)^{3/4} e^{-(\epsilon_c - \epsilon_v)/kT}. \qquad (2.7)$$

(2.7) shows that the product of the two densities is independent of the Fermi level.

Extrinsic semiconductor

The extrinsic semiconductor is prepared by adding impurities either a pentavalent element (like As) or trivalent element (like Boron or Al). These impurities add energy levels in the forbidden gap. In the first case, we would get an *n*-type semiconductor and in the second case, we get a *p*-type semiconductor.

***N* type semiconductor**

If ϵ_d represents the energy of the bottom of the energy levels donated by the impurity atoms, then the density of the conduction band electrons is given by (2.3) as $n_c = N_c e^{-(\epsilon_c - \epsilon_f)/kT}$.

If n_d represents the density of states of the donor levels, then the number of the ionised donors is $n_d\,[1 - f_d(\epsilon_d)] \simeq n_d e^{-(\epsilon_f - \epsilon_d)/kT}$.

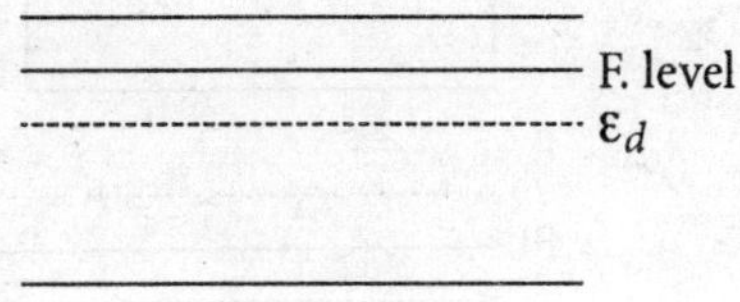

Fig. 2.7

The number of conduction band electrons per cc is equal to the number of ionised donors/cc.

$$\therefore\ n_d e^{-(\epsilon_f - \epsilon_d)/kT} = N_c e^{-(\epsilon_c - \epsilon_f)/kT} \Rightarrow \left(\frac{n_d}{N_c}\right) = e^{-(\epsilon_c - \epsilon_f)/kT} \,.\, e^{(\epsilon_f - \epsilon_d)/kT} = e^{+(2\epsilon_f - \epsilon_c + \epsilon_d)/kT}$$

$$\Rightarrow kT \ln\left(\frac{n_d}{N_c}\right) = +2\epsilon_f - (\epsilon_c + \epsilon_d)$$

$$\Rightarrow \epsilon_f = \frac{\epsilon_c + \epsilon_d}{2} + \frac{1}{2} kT \ln\left(\frac{n_d}{N_c}\right). \tag{2.8}$$

Now, at $T = 0$, $\epsilon_f = \frac{\epsilon_c + \epsilon_d}{2}$, which shows that at absolute zero, the Fermi level lies at the midway between the bottom of the conduction band and the bottom of the donor level [Fig. 2.7]. Since n_d is less than N_c, with rise of temperature, (2.8) suggests that the Fermi level descends; goes below the donor levels and tends to occupy the position midway between the conduction band and the valence band. Then it behaves like an intrinsic semiconductor.

Putting this value of ϵ_f in n_c, we get

$$n_c = (2n_d)^{1/2} \left(\frac{2nm_e^* kT}{h^2}\right)^{3/4} e^{-(\epsilon_c - \epsilon_d)/kT}$$

$$= (2n_d)^{1/2} \left(\frac{2\pi m_e^* kT}{h^2}\right)^{3/4} e^{-\Delta E/2kT}. \tag{2.9}$$

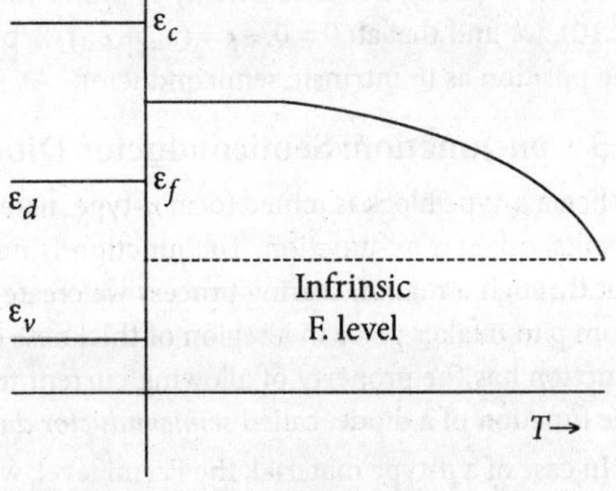

Fig. 2.8

It shows that density of electrons in the conduction band depends on the donor concentration. The variation of the Fermi level position with temperature [Fig. 2.8], where starting from the midway position between ϵ_d and ϵ_c, the Fermi level falls with temperature and is found to occupy the position of intrinsic Fermi level positions.

***P* type semiconductor**

When a trivalent impurity is added, we get some addition of energy levels near the valence band, called the *acceptor levels*.

If n_a represents the density of energy states in the acceptor levels and ϵ_a be the energy of the top of the acceptor levels, then exactly in the same way as we have treated in *n*-type material, we can write $n_h = N_v\, e^{-(\epsilon_f - \epsilon_v)/kT}$ and the number of ionised atoms in the acceptor levels is

$$n_a\,[1 - f_a(\epsilon_a)] = n_a e^{-(\epsilon_a - \epsilon_f)/kT}.$$

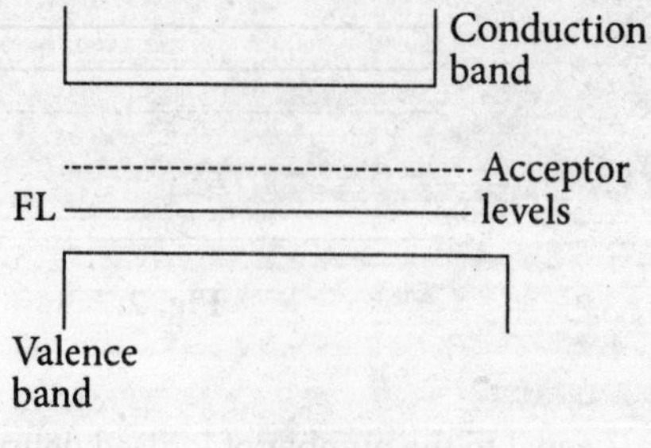

Fig. 2.9

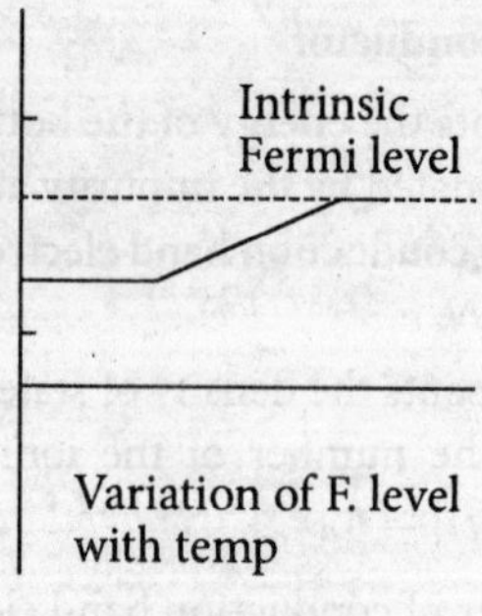

Fig. 2.10

$$\therefore\ n_a e^{-(\epsilon_a-\epsilon_f)/kT} = N_v e^{-(\epsilon_f-\epsilon_v)/kT} \Rightarrow \left(\frac{n_a}{N_v}\right) = e^{(\epsilon_a+\epsilon_v-2\epsilon_f)/kT}$$

$$\Rightarrow kT\ln\left(\frac{n_a}{N_v}\right) = \epsilon_a+\epsilon_v-2\epsilon_f$$

$$\Rightarrow \epsilon_f = \frac{\epsilon_a+\epsilon_v}{2} - \frac{1}{2}kT\ln\left(\frac{n_a}{N_v}\right). \tag{2.10}$$

Using this value of ϵ_f, we can easily get

$$n_h = (2n_a)^{1/2}\left(\frac{2\pi m_h^* kT}{h^2}\right)^{3/4} e^{-(\epsilon_a-\epsilon_v)/2kT}. \tag{2.11}$$

So from (2.11), the hole density is found to depend on the acceptor concentration levels. Also, from (2.10), we find that at $T=0$, $\epsilon_f=(\epsilon_a+\epsilon_v)/2$, but since $n_a < N_v$ as T increases, the Fermi level ascends to the position as in intrinsic semiconductor.

2.5 *pn*-Junction: Semiconductor Diode

When a *p*-type block is joined to an *n*-type material, a fundamental component of many electronic devices results, called a *pn*-junction. The junction is not formed simply by placing the two materials in contact, but through a manufacturing process we create transition from *p* to *n* within a single crystal. The change from *p* to *n* takes place in a region of thickness of the order of 10^{-3}mm known as *depletion region*. Such a junction has the property of allowing current to flow through it in one direction only and thus performs the function of a diode, called *semiconductor diode.*

In case of a *p*-type material, the Fermi level, we have seen, lies between the valence band top and the top of the acceptor level and with rise of temperature Fermi level rises. The Fermi energy level is defined in a *p*-type extrinsic semiconductor by an equation as

$$\epsilon_f = \frac{\epsilon_a+\epsilon_v}{2} - \frac{1}{2}kT\ln\left(\frac{n_a}{N_v}\right) = \frac{\epsilon_a+\epsilon_v}{2} + \frac{1}{2}kT(\ln N_e - \ln n_a).$$

So at a particular temperature, if n_a decreases due to hole-movement, Fermi level rises. The band structure of a *p*-type material with its characteristic Fermi level position at temperature T is shown in Fig. 2.11(a).

In case of *n*-type material the Fermi level is defined by an equation

$$\epsilon_f = \frac{\epsilon_c+\epsilon_d}{2} + \frac{1}{2}kT\ln\left(\frac{n_d}{N_c}\right) = \frac{\epsilon_c+\epsilon_d}{2} - \frac{1}{2}kT(\ln N_c - \ln n_d).$$

Usually $\ln N_c > \ln n_d$, so when n_d decreases due to electron-movement, the Fermi level descends [Fig. 2.11(b)].

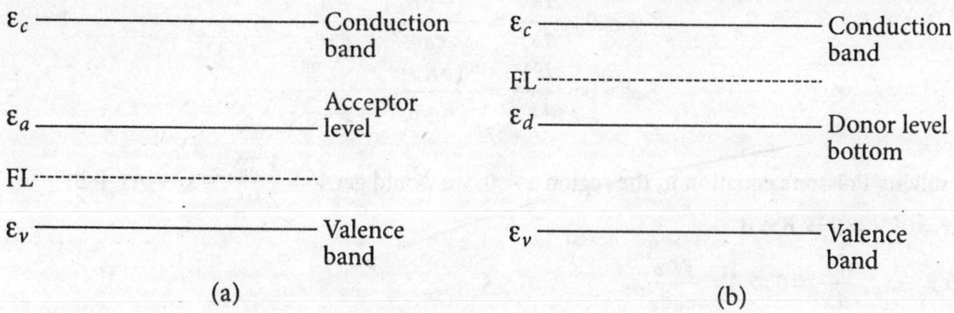

Fig. 2.11

Now, consider the formation of a *pn*-junction [Fig. 2.12] with the circuit symbol to the right. Across the junction layer, holes will diffuse from *p*- to *n*-side and electrons would diffuse from *n* to *p* side thus there will be elevation of Fermi level in *p* and depression of the level in *n* and ultimately both sides equalise when diffusion would stop. This state of affair is represented in Fig. 2.13.

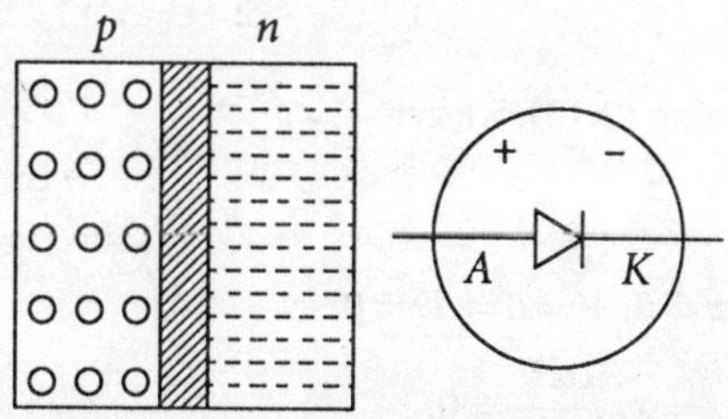

Fig. 2.12

When Fermi level is equalised in both materials a steady p.d. v_0 is established, called *barrier potential.*

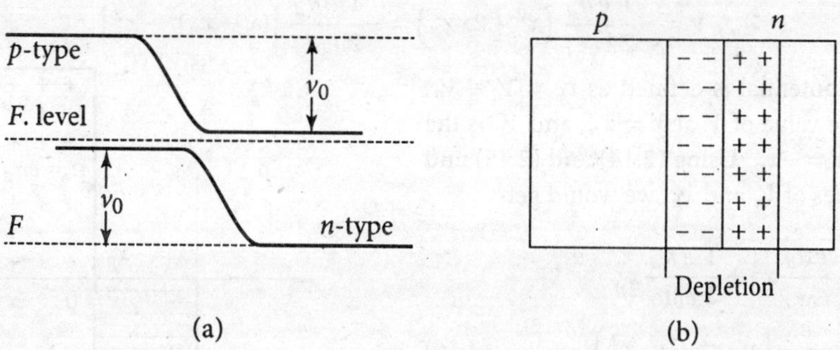

Fig. 2.13

In the depletion region, there is development of negative charge of the *p*-side and positive charge on the *n*-side, setting up an electric field directed from *n* to *p*, called the *potential barrier* or *junction potential,* which prevents diffusion of electrons and holes [Fig. 2.13(a)].

Let the impurity concentration in *p*-material be n_a and that in *n*-material be n_d; so removal of holes in the barrier layer on the *p*-type side leaves a negative space charge of $-en_a$ per unit volume, and removal of electrons on the *n*-type side leaves a positive space charge of en_d per unit volume. Let us consider the transition from *p* to *n* to take place discontinuously at the place $x = 0$ and the barrier thickness on the *p*-side be x_p and on the *n*-side be x_n. Then, the boundary conditions are $V = 0$ at $x = 0$ and $dV/dx = 0$ at $x = -x_p$ and $x = x_n$.

So applying Poisson's equations in the two regions

$$x<0: \frac{d^2V}{dx^2} = -\frac{(-en_a)}{\epsilon_0\epsilon_r} = \frac{en_a}{\epsilon_0\epsilon_r}. \tag{2.12}$$

$$x>0: \frac{d^2V}{dx^2} = -\frac{(en_d)}{\epsilon_r\epsilon_0}. \tag{2.13}$$

So, solving Poisson's equation in the region $x<0$, we would get $V = \frac{1}{2}\frac{en_a}{\epsilon_0\epsilon_r} - x^2 - Ax + B$.

At $x=0$, $V=0 \Rightarrow B=0$.

At $x=-x_p$, $\frac{dV}{dx} = 0 \Rightarrow A = \frac{en_a}{\epsilon_0\epsilon_r}x_p$.

$$\therefore\ V = \frac{1}{2}\cdot\frac{en_a}{\epsilon_0\epsilon_r}\left(x^2+2xx_p\right) = \frac{1}{2}\cdot\frac{en_a}{\epsilon_0\epsilon_r}\left\{\left(x+x_p\right)^2 - x_p^2\right\}. \tag{2.14}$$

Solving (2.13), we get

$$V = -\frac{1}{2}\frac{en_d}{\epsilon_0\epsilon_r}x^2 + Cx + D.$$

At $x=0$, $V=0 \Rightarrow D=0$.

At $x=x_n$, $\frac{dV}{dx} = 0$.

So, we get $C = \frac{1}{2}\cdot\frac{en_d}{\epsilon_0\epsilon_r}x_n$.

$$\therefore\ V = -\frac{1}{2}\frac{en_d}{\epsilon_0\epsilon_r}\left(x^2-2xx_n\right) = -\frac{1}{2}\frac{en_d}{\epsilon_0\epsilon_r}\left\{(x-x_n)^2 - x_n^2\right\}. \tag{2.15}$$

The barrier potential is defined as $v_0 = V_n - V_p$, where V_n is the value of V at $x = x_n$, and V_p is the value of V at $x = -x_p$. Using (2.14) and (2.15) and giving the values of V_n and V_p, we would get

$$v_0 = \frac{1}{2}\frac{en_d}{\epsilon_0\epsilon_r}x_n^2 + \frac{1}{2}\frac{en_a}{\epsilon_0\epsilon_r}x_p^2$$
$$= \frac{1}{2}\frac{e}{\epsilon_0\epsilon_r}\left(n_d x_n^2 + n_a x_p^2\right). \tag{2.16}$$

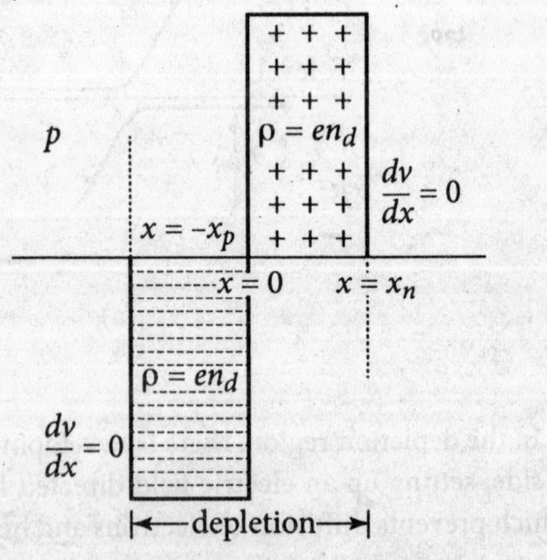

Fig. 2.14

The above mathematical consideration is represented in Fig. 2.14. The thickness of the depletion region is given by $\left(x_n + x_p\right)$. So, the barrier potential depends on the depletion thickness as well as the donor and acceptor level concentration.

Forward biasing and reverse biasing of *pn*-junction

As the *pn*-junction is formed, a potential barrier is created of value v_0 directed from *n* side to *p* side. Now if an external emf is so connected that the positive is connected to the *p*-end and the negative to the *n*-end such as to act in opposition to the barrier height, the junction is forward-biased [Fig. 2.15(a)]. But when the barrier potential and the applied p.d. are in the same direction thus increasing the barrier potential, the junction is reverse-biased. What happens actually when the junction is forward-biased is that the holes are repelled from the positive terminal of the battery and drift towards the junction; on the other side, the electrons are repelled from the negative terminal of the battery and also drift towards the junction. Thus, under the influence of the battery voltage electrons and holes penetrate the junction and combine with each other. For each combination an electron from the negative terminal enters the n-type semiconductor, at the same time an electron from an electron-pair bond in the crystal, near the positive terminal of the battery, breaks its bond and enters the positive terminal. For each electron that breaks its bond, a hole is created, which drifts toward the junction. Thus, there is a continuous flow of electron current in the external circuit. The current in the p-type consists of holes, the current in the n-type consists of electrons.

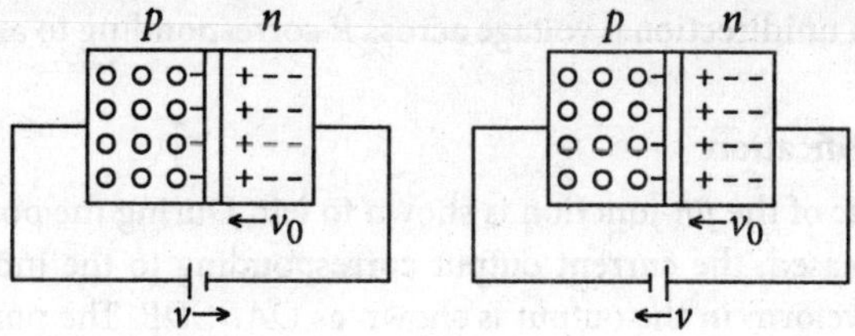

(a) (b)

Fig. 2.15

In the reverse biasing the holes are attracted by the negative terminal and not repelled towards the junction; the electrons from the n-type germanium are attracted by the positive terminal of the battery and not repelled towards the junction. Thus, there is no current flow through the germanium.

The diode current

When the junction is forward-biased with an external bias v, the potential barrier at the junction is also reduced from v_0 to $(v_0 - v)$, and in the reverse biasing, the potential barrier is increased from v_0 to $(v_0 + v)$. The voltage and current are related by

$$i_D = I_0\left(\exp\frac{ev_d}{mkT} - 1\right), \tag{2.17}$$

where i_D = diode current, v_d = voltage across the diode, I_0 = reverse saturation current, m = empirical constant varying between 1 and 2.

Let

$$V_r = \frac{kT}{e} = \frac{1.38\times10^{-16}\times300}{1.6\times10^{-19}} \simeq 25\text{mV at 300k}.$$

$$\therefore\ I = I_0\left(\exp\frac{v_d}{V_r} - 1\right),\ m = 1. \tag{2.18}$$

If v_d is positive and large compared to V_r, then (2.14) gives $I = I_0 \exp v_d/V_r$ and this represents the current in the forward biasing. But in the reverse biasing, v_d is negative, and then I reduces to $-I_0$. The typical $(v - i)$ characteristic of a diode is shown in Fig. 2.16.

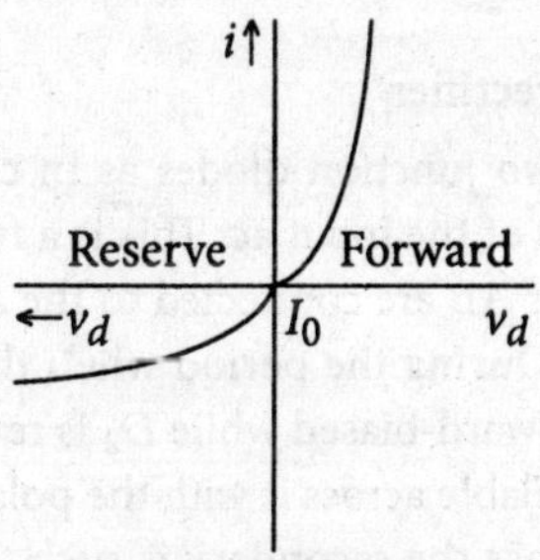

Fig. 2.16

2.6 Rectification Action of a Junction Diode

The *pn*-junction is a unilateral device; when forward-biased, the junction will conduct and it will permit current to flow. But when reverse-biased, it will not permit current to flow. These are the features of a diode. So, when an alternating source v_s is applied across the *pn*-junction, then when *p*-end is positive, the junction is forward-biased and current flows and p.d. appears across the load *R* with the polarities indicated. But with the reversal of the direction of emf of the source, when the *p*-end becomes negative, the junction is reverse-biased and no current flows. So no voltage is available during this half of the input. Again, when the *p*-end becomes positive, the junction will permit current making voltage once more available across *R*, keeping the positive and negative polarities unchanged. Thus, we get a unidirectional voltage across *R* corresponding to an ac input.

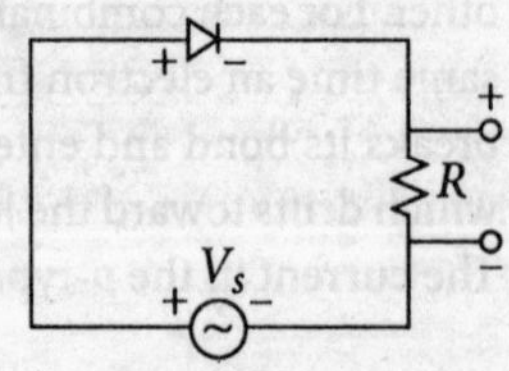

Fig. 2.17

Graphical analysis of rectification

The dynamic characteristic of the *pn*-junction is shown to left. During the positive half of the input when the junction is forward-biased, the current output corresponding to the increasing biasing is shown to the right. The current waveform in the output is shown as *OABCDE*. The portion *BC* corresponds to the no-conduction period when the junction is reverse-biased. The above mentioned circuit using a junction diode giving unidirectional voltage across the load during one half of the input ac is known as *half-wave rectifier* [Fig. 2.18].

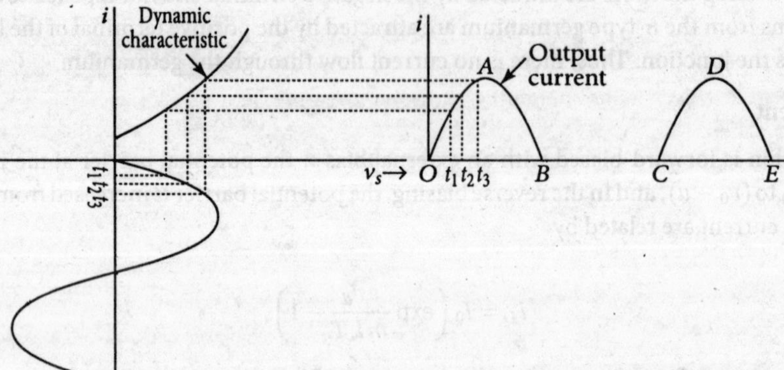

Fig. 2.18

Full-wave rectifier

By using two junction diodes as in circuit [2.18(a)], we can have unidirectional voltage output during both halves of the input ac. This is a full-wave rectifier. The two ends of the secondary of a centre-tapped transformer *AB* are connected to the anodes (*p*-type) and the centre tap point *C* is connected to the load as shown. During the period when the secondary induced voltage is such that *A* is positive w.r.t. *B*, D_1 diode is forward-biased while D_2 is reverse-biased. So D_1 conducts and current flows through *R* making a voltage available across it with the polarity indicated. During the other half of the input when the induced voltage across the secondary is such that *B* is positive w.r.t. *A*, the diode D_2 is forward-biased and D_1 is reverse-biased. So D_2 allows current to flow through *R* making voltage available across *R* with the polarity remaining unchanged. Thus, we are getting unidirectional voltage across *R* during both halves of the input. The input voltage across the secondary v_s vs. *t* and output voltage V_R vs. *t* are shown in Fig. 2.18(a).

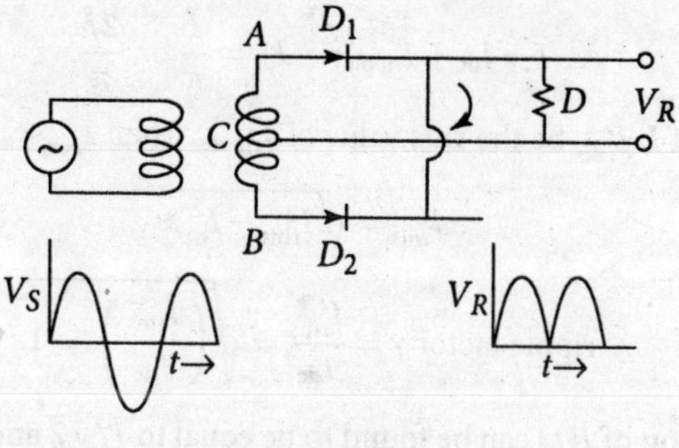

Fig. 2.18(a)

Ripple factor in half-wave and full-wave rectifier

The purpose of a rectifier is to convert ac into dc. The ripple factor is a measure of how successful a circuit is in achieving this purpose. The ripple factor is defined as

$$\gamma = \frac{\text{ripple voltage}}{\text{DC voltage}} = \frac{\text{rms value of ac components}}{\text{DC voltage}} = \frac{V_{\text{rms}}}{V_{\text{DC}}} \text{ or, } \frac{I_{\text{rms}}}{I_{\text{dc}}}.$$

In case of half-wave rectifier, the output function is mathematically defined as

$$I(t) = \begin{cases} \hat{I}\sin\omega t, & 0 \le \omega t \le \pi \\ 0, & \pi \le \omega t \le 2\pi. \end{cases}$$

In case of full-wave rectifier, the mathematical representation is

$$I = \begin{cases} \hat{I}\sin\omega t, & 0 \le \omega t \le \pi \\ -\hat{I}\sin\omega t, & \pi \le \omega t \le 2\pi. \end{cases}$$

Now, the output of a rectifier, whether a half-wave or a full-wave, is a periodic function and therefore, according to Fourier's theorem, it can be expressed as

$$I(t) = a_0 + a_1\sin\omega t + a_2\sin 2\omega t + \cdots + b_1\cos\omega t + b_2\cos 2\omega t + b_3\cos 3\omega t + \ldots. \tag{2.19}$$

The evaluation of the coefficients have been shown elsewhere[1]. We can write:
For half-wave rectifier,

$$I = \hat{I}\left(\frac{1}{\pi} + \frac{1}{2}\sin\omega t - \frac{2}{3\pi}\cos 2\omega t - \frac{2}{15\pi}\cos 4\omega t \ldots\right) \tag{2.20}$$

$$= \hat{I}\left(\frac{1}{\pi} + \frac{1}{2}\sin\omega t - \frac{2}{\pi}\sum_{2,4,6}\frac{\cos n\omega t}{(n+1)(n-1)}\right)$$

and for full-wave rectifier,

$$I = \hat{I}\left(\frac{2}{\pi} - \frac{4}{3\pi}\cos 2\omega t - \frac{4}{15\pi}\cos 4\omega t \ldots\right) = \hat{I}\left(\frac{2}{\pi} - \frac{4}{\pi}\sum_{2,4,6}\frac{\cos n\omega t}{(n+1)(n-1)}\right). \tag{2.21}$$

[1]A Textbook of Waves and Acoustics by Chakrabarti and Chowdhury, Central Book Agency (P) Ltd, Kolkata

$$\therefore\ I = I_{dc} + I_{ripple}; \quad I_{dc} = \frac{\hat{I}}{\pi} \text{ or, } \frac{2\hat{I}}{\pi}.$$

Let I_{rms} be the rms value of I, I'_{rms} be the rms value of I_{ripple}. Then $I^2_{rms} = I^2_{dc} + I'^2_{rms}$.

$$\therefore\ I'_{rms} = \sqrt{I^2_{rms} - I^2_{dc}}.$$

$$\therefore \text{ ripple factor } \gamma = \frac{I'_{rms}}{I_{dc}} = \sqrt{\left(\frac{I_{rms}}{I_{dc}}\right)^2 - 1}.$$

Now, I_{rms} from the definition of $I(t)$ can be found to be equal to $\hat{I}/\sqrt{2}$ and I_{dc} from (2.20) is $\hat{I}/\pi$. In case of full-wave rectifier the corresponding quantities are $\hat{I}/\sqrt{2}$ and $2\hat{I}/\pi$.

$$\therefore\ \gamma\,(\text{half-wave}) = \sqrt{\left(\frac{\hat{I}/2}{\hat{I}/\pi}\right)^2 - 1} = \sqrt{(1.57)^2 - 1} = 1.21.$$

$$\gamma\,(\text{full-wave}) = \sqrt{\left(\frac{\hat{I}/\sqrt{2}}{2\hat{I}/\pi}\right)^2 - 1} = \sqrt{(1.11)^2 - 1} = 0.48.$$

Thus, in the full-wave rectifier, the ripple factor is lower.

2.7 Peak Inverse Voltage (PIV)

This is a very important parameter of a diode. A diode is repeatedly thrown into the reverse bias region when it is used in a rectifier. The PIV required for a diode used in half-wave and full-wave rectification is different.

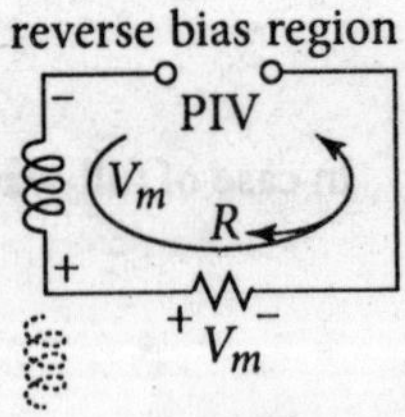

Fig. 2.19

Looking at Fig. 2.17 for half-wave rectification, the diode is subjected to the full input voltage in the reverse bias, and so PIV rating $\geq V_m$ for half-wave rectification.

Considering Fig. 2.18 redrawn as Fig. 2.19, it is evident that $+IR + V_m - \text{PIV} = 0$ or, $V_m + V_m = \text{PIV}$ or, $\text{PIV} = 2V_m$.

So, in this full-wave rectifier circuit, the diode used must have $\text{PIV} \geq 2V_m$.

2.8 Bridge Rectifier

It is also a popular full-wave rectifier circuit differing from the conventional full-wave rectification in many important aspects. In conventional circuit the secondary of the transformer is to be a centre-tapped one, whereas in bridge rectification, no centre tapping is needed. In bridge rectification current output is less but voltage output greater than those in conventional full-wave rectification.

In solid state TV or a stereo amplifier, the power supply unit consists of bridge rectifier arrangement.

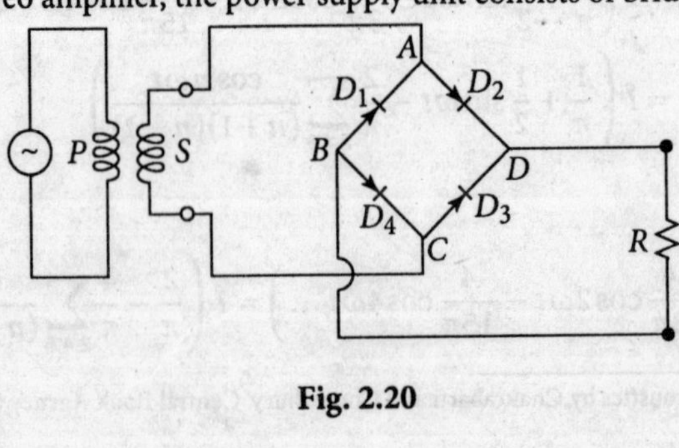

Fig. 2.20

We have shown a bridge arrangement. The four diodes D_1, D_2, D_3 and D_4 forming the arms of a Wheatstone's bridge make the arrangement as a full-wave rectifier [Fig. 2.20].

The functioning of the device can be understood w.r.t. [Figs. 2.21, 2.22 and 2.23].

When A is positive w.r.t. C, D_2 conducts and also D_4, because both of them are forward-biased, so current now flows in the direction of arrow [Fig. 2.21], while D_1 and D_3 are off. So current through the load resistor R flows downwards producing a voltage across R [Fig. 2.21].

During the next half of the input when A is negative and C is positive, the diodes D_1 and D_3 are forward-biased, hence they are in conduction but D_2 and D_4 are reverse-biased, hence they are off. The current flows in the direction of arrow, hence again flows down the load resistance R. Thus, we are getting unidirectional voltage across the load resistor.

The average value or dc value is double of that in half-wave rectification.

$\therefore$ average value = $0.636 V_m$.

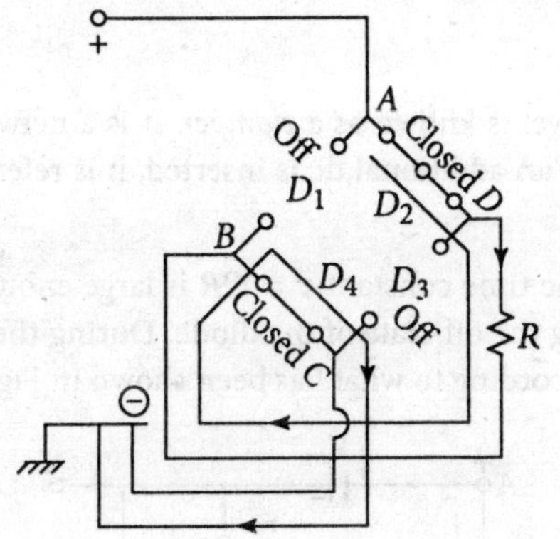

Fig. 2.21

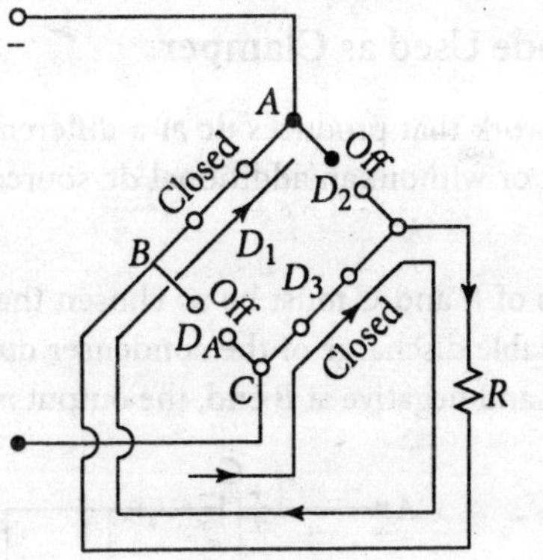

Fig. 2.22

PIV for the diodes

To find the PIV for the diodes to be used in the bridge circuit, consider the situation when A is positive and C is negative. So w.r.t. Fig. 2.23, we get IR–PIV = 0 or, PIV = V_m.

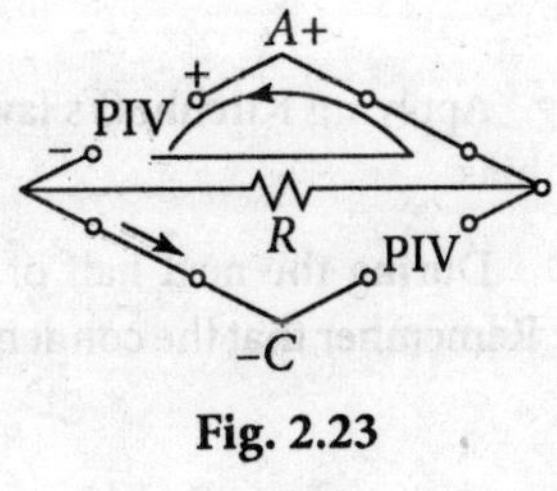

Fig. 2.23

This is another important point of difference between the conventional full-wave rectifier circuit and the bridge rectifier circuit. The PIV for the diodes to be used in conventional arrangement is twice the value as required in case of bridge arrangement.

2.9 Diodc as Clipper

Series and parallel connection of diodes

The series or parallel connection of a diode and resistor always clips one half of the input alternating voltage of a variety of forms. This functioning is referred to as *clipper*. A half-wave rectifier circuit is the *simplest clipper* circuit.

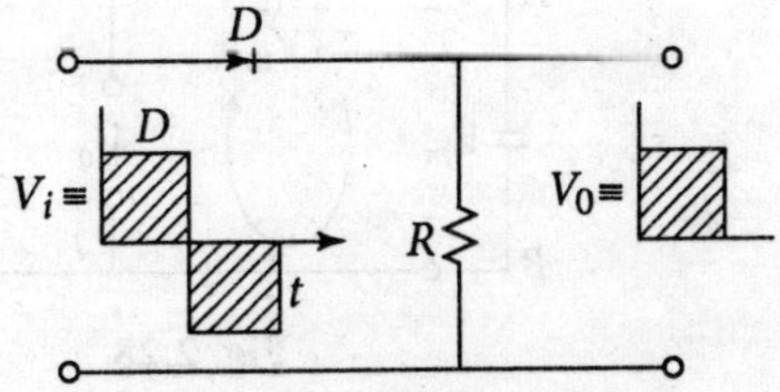

Fig. 2.24

Series clipper

A diode is in series with the load resistor R forming a series clipper, which clips the negative half of the input square-wave voltage. During positive half, the diode is in forward bias and hence, output is available across R, but during negative half, the diode is in reverse bias and $V_0 = 0$.

Parallel clipper

The parallel clipper [Fig. 2.25] clips the positive half of the input voltage. During the positive half of the input, the diode remains in the on-state, hence current flows and voltage drop occurs across R, producing zero voltage output; but during the negative half of the input, the diode remains in the off-state and voltage appears in the output.

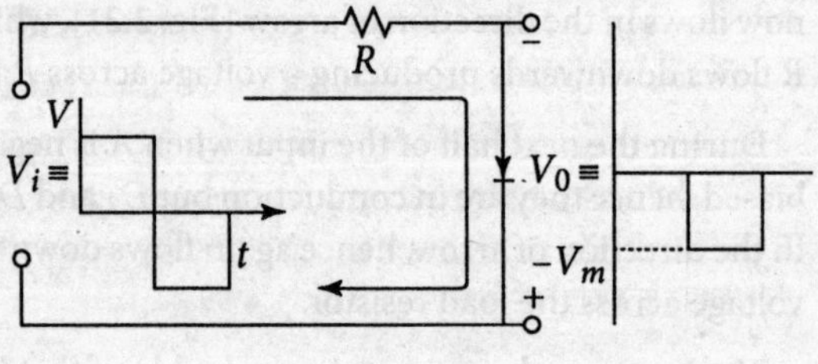

Fig. 2.25

2.10 Diode Used as Clamper

A diode network that produces dc at a different level is known as a *clamper.* It is a network of C, R and a diode with or without an additional dc source. If an additional dc is inserted, it is referred to as *biased clamper.*

The values of R and C must be so chosen that the time constant $\tau = CR$ is large enough so that there is no appreciable discharge of the condenser during the off-state of the diode. During the half when V_i is positive at A and negative at B end, the output is according to what has been shown in Fig. 2.27.

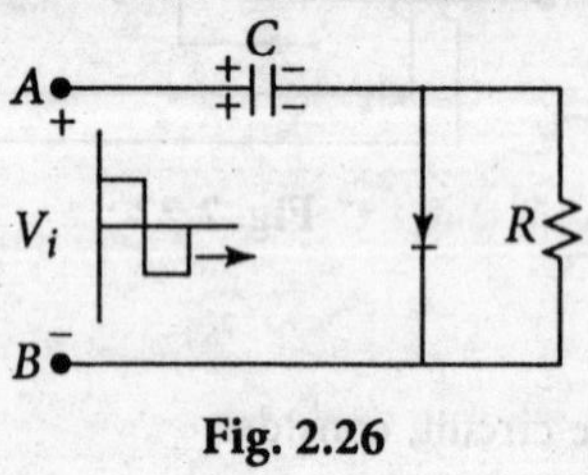

Fig. 2.26

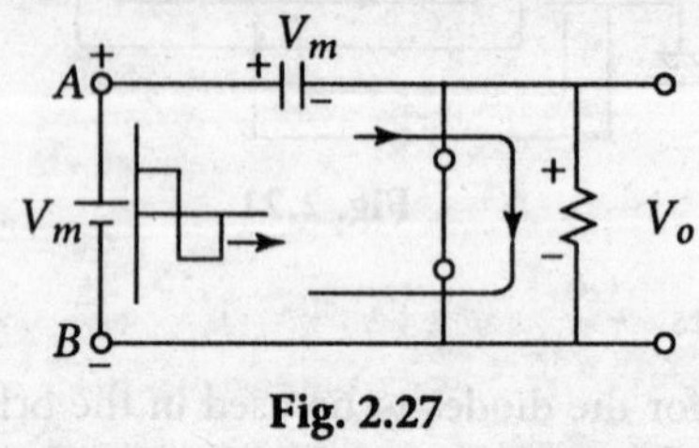

Fig. 2.27

Applying Kirchhoff's law, $V_0 + V_m - V_m = 0$ or, $V_0 = 0$. This is obvious, because the diode is in forward bias.

During the next half of the input, A reverses in polarity, hence the situation is as shown in Fig. 2.28. Remember that the condenser retains its charged state. Kirchhoff's law gives $V_0 + V_m + V_m = 0$ or, $V_0 = -2V_m$.

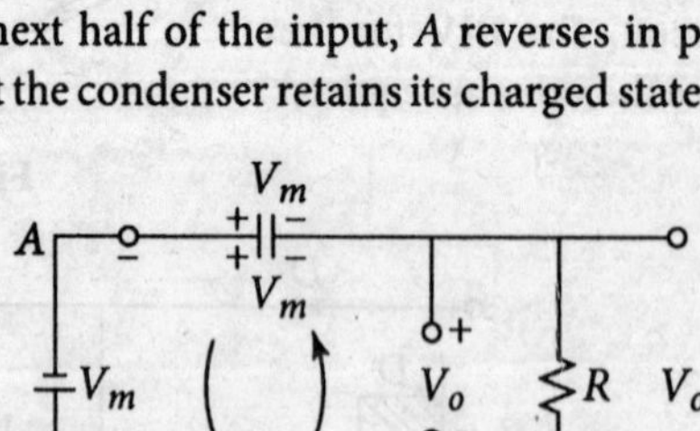

Fig. 2.28

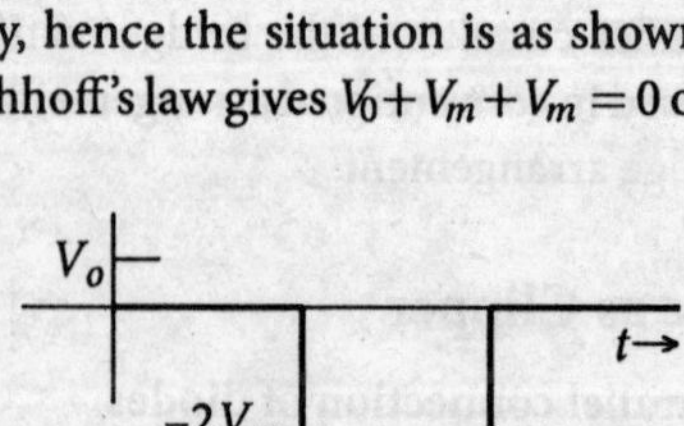

Fig. 2.29

So, the output waveform is as shown in Fig. 2.29, the output is twice the peak value of the input.

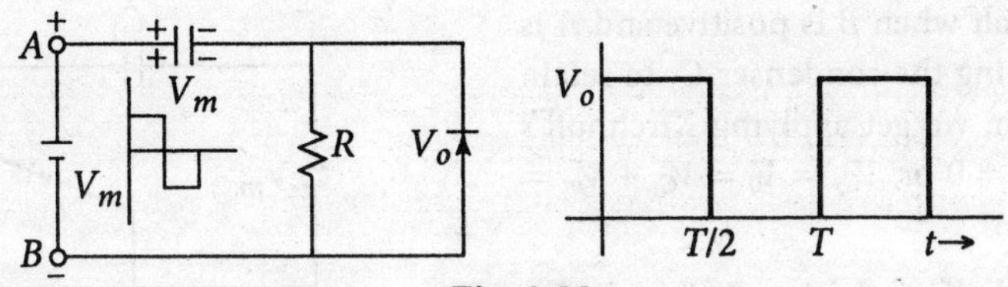

Fig. 2.30

In order to have positive half clamped, we consider the circuit as shown in Fig. 2.30.

During the negative half of the input, the condenser is charged immediately to V_m and the diode is off, so the situation is as shown in Fig. 2.31. Applying Kirchhoff's law, we get $V_0 - V_m - V_m = 0$ or, $V_0 = 2V_m$.

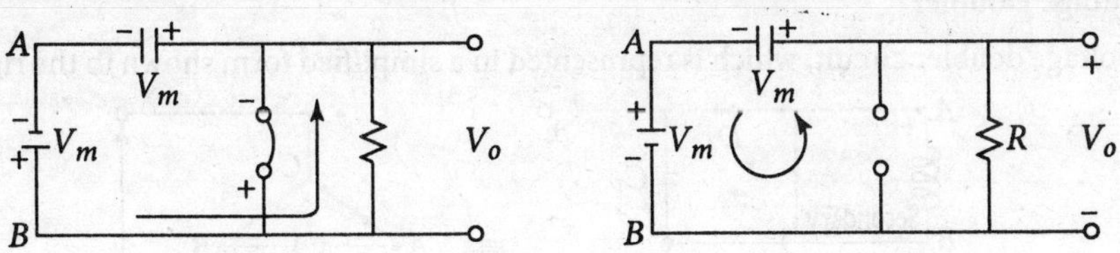

Fig. 2.31 **Fig. 2.32**

Now, during the next half of the input [Fig. 2.32] application of Kirchhoff's law gives $+V_0 - V_m + V_m = 0$ or, $V_0 = 0$.

Thus, the clamper circuit changes the dc level output.

2.11 Diode Used as Voltage Multiplier

Diodes and condensers producing dc output as double, triple or quadruple of the input ac peak.

Voltage Doubler

This is in effect similar to diode clamper [Fig. 2.32]; the ac input between A, B is of peak value of voltage equal to V_m.

During the positive half of the input when A is positive and B is negative, the diode D_1 shorts and D_2 is off and there is no output between C, D but C_1 is charged.

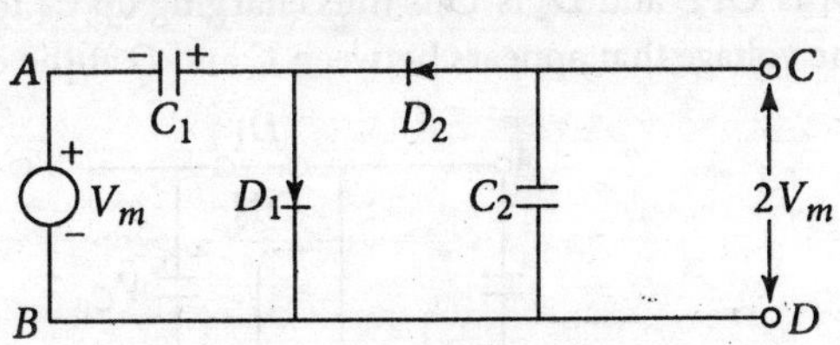

Fig. 2.33

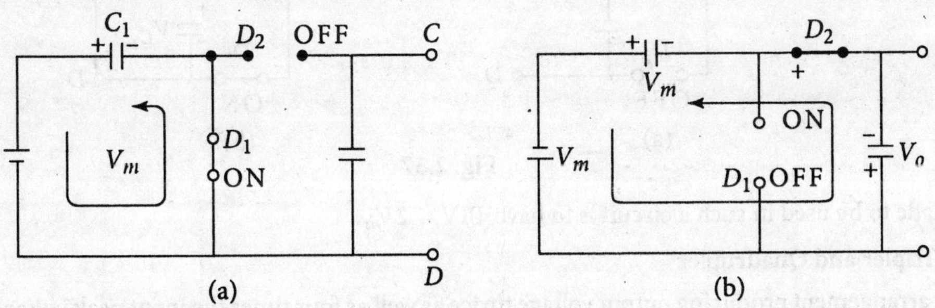

Fig. 2.34

During the next half when B is positive and A is negative, then assuming the condenser C_1 to retain its charged condition, we get applying Kirchhoff's law $-V_{C_2} + V_{C_1} + V_m = 0$ or, $V_{C_2} = V_0 = V_{C_1} + V_m = V_m + V_m = 2V_m$.

Since during one half of the input, there is no voltage output, this can be referred to as half-wave voltage doubler, which can be conventionally represented by the arrangement [Fig. 2.35].

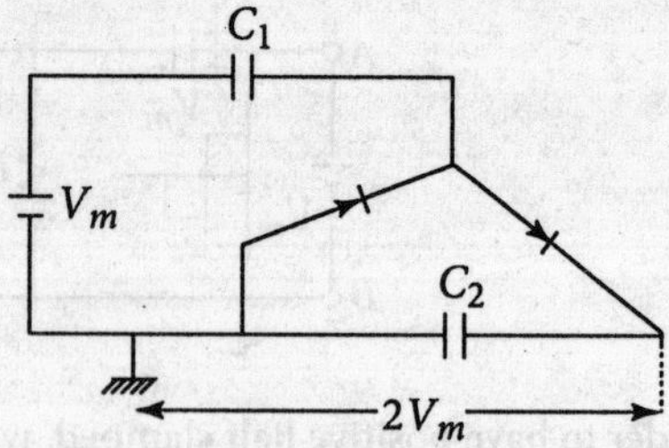

Fig. 2.35

Full-wave Voltage Doubler

A full-wave voltage doubler circuit, which is represented in a simplified form shown to the right [Fig. 2.36].

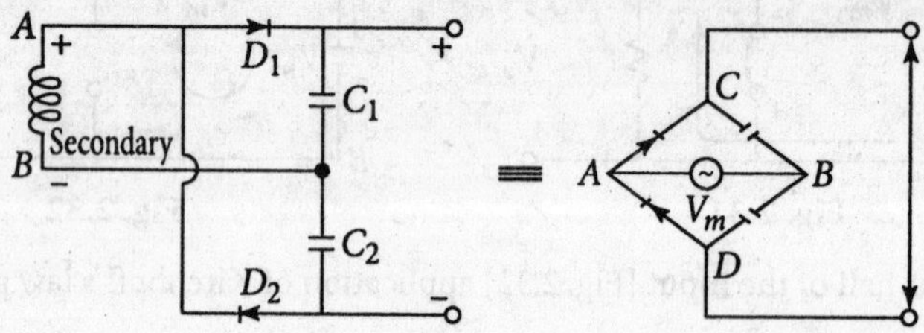

Fig. 2.36

When the induced voltage in the secondary is such that A is positive, B is negative, the situation will be as shown in Fig. 2.37(a). The diode D_1 is ON, charging C_1 to the peak input voltage V_m. During the next half, the situation changes to that of Fig. 2.37(b), when A is negative and B is positive; this time the diode D_1 is OFF and D_2 is ON thus charging up C_2 to the peak value V_m. The condenser C_1 retaining its state, the voltage that appears between C and D at the end of a cycle of the input is of value, $V_0 = 2V_m$.

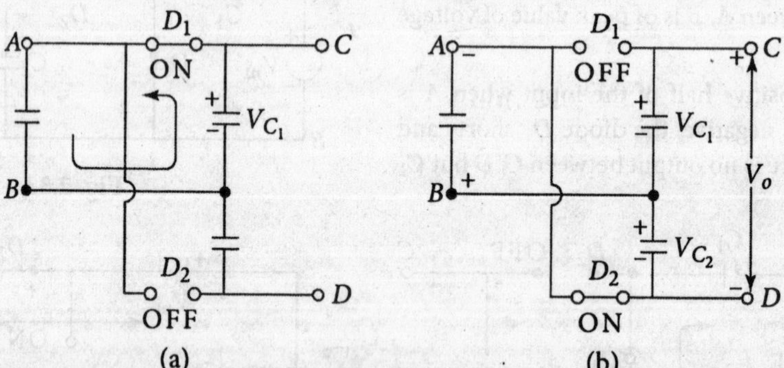

Fig. 2.37

The diode to be used in such a circuit is to have PIV $= 2V_m$.

Voltage Tripler and Quadrupler

A circuit arrangement producing output voltage thrice as well as four times the input peak voltage is shown in Fig. 2.38. The voltage tripler and quadrupler circuits can be represented in a simpler circuit design [Fig. 2.39]. The odd multiplies of input peak voltage will be available measured from the top of the secondary windings and the even multiplies of the input will be available measured w.r.t. the bottom of the secondary winding [Fig. 2.38].

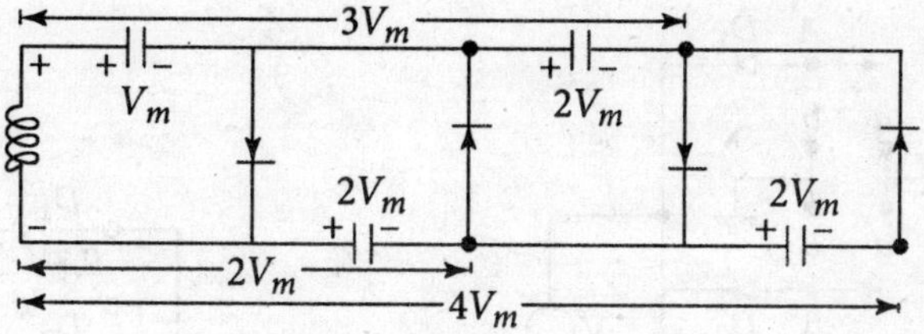

Fig. 2.38

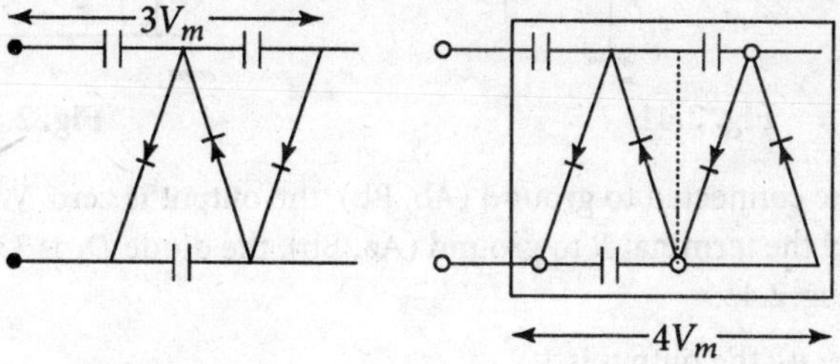

Voltage tripler Voltage quadrupler

Fig. 2.39

2.12 Diode Used as Damper

This purpose served by a diode is a very important one in view of the fact that here it acts as a life-saver of electronic or electromechanical devices.

In case of rapid interruption of current in an inductive load as in case of relay, transformer or electric motor, the inductor generates substantial back emf, which is detrimental to the relay coil, etc. To save life of the associated electronic components, a diode is used as a damper. Though a single diode is adequate for most of the practical cases, two diodes form a full protection network.

We have shown a relay operated by a transistor as electronic switch [Fig. 2.40]. The relay coil generates back emf of hundreds of volts when the coil current is suddenly broken, this back emf can damage the electronic coil-driver circuitry. The two diodes D_1 and D_2, called *damping diodes,* connected as shown in Fig. 2.40, act as a life-saver of the associated circuitry. The diode D_1 keeps the junction of the relay and the transistor within the safe limit of 0.6V above the positive supply and the diode D_2 keeps the junction within the safe limit of 0.6V below zero-volt value.

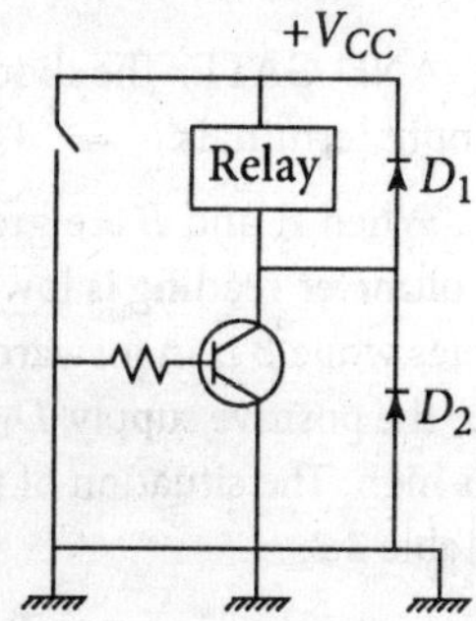

Fig. 2.40

2.13 Diode Used as GATE — OR and AND GATES

A gate is an electronic device, which produces a required output for a predetermined combination of inputs. Two of the fundamental gates are: (i) OR-gate and (ii) AND-gate.

OR GATE. The two diodes D_1 and D_2 are arranged [Fig. 2.41] with their cathodes joined together and connected to ground through a resistor R. The voltage across R read by the voltmeter V is the output. When the input terminals A and B are connected to the source, the inputs are ON or in state 1, while if they are connected to ground, they are in OFF-state or in state 0; similarly, when the output voltage is high, output is said to be in state 1, when low or zero, it is in state zero.

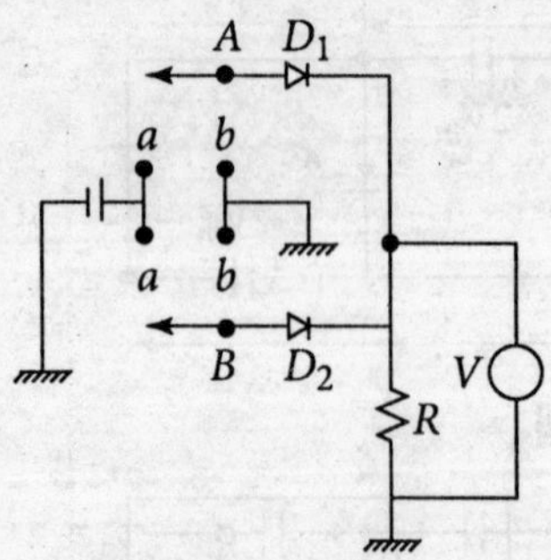

Fig. 2.41

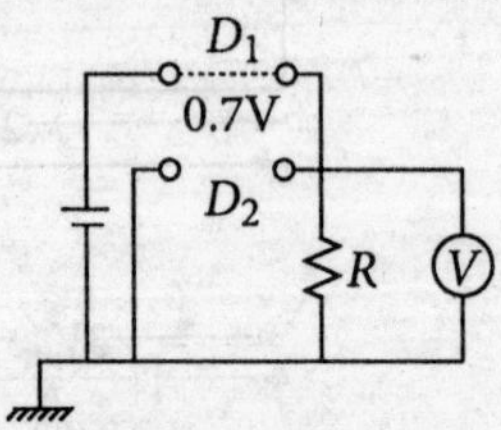

Fig. 2.42

When both A and B are connected to ground (Ab, Bb), the output is zero. When A is connected to the positive of the supply and the terminal B to ground (Aa, Bb), the diode D_1 is ON, D_2 remains off. So, the situation is as shown in Fig. 2.42.

So $v_i - 0.7 = v_0$ or, $v_i \simeq v_0$, the output is 1.

When both inputs are in state 1, the output is also 1. Thus, the behaviour of this gate is represented by in Table 2.1.

Table 2.1

A	B	Y
0	0	0
1	0	1
0	1	1
1	1	1

AND GATE. The diodes D_1 and D_2 are joined anode to anode, the free cathode terminals A, B are the input terminals.

When A and B are grounded, both are in the forward bias and hence, in the conduction mode, so the voltmeter reading is low (0.7V). If A is connected to positive supply while B is grounded, D_1 is in reverse bias while B is in forward bias, so again the voltmeter reading is low. But when both A and B are connected to the positive supply, D_1 and D_2 are off and the voltmeter reads practically the supply voltage, so the output is high. The situation of the output corresponding to possible combination of the inputs is as shown in the Table 2.2.

Table 2.2

A	B	Y
0	0	0
1	0	0
0	1	0
1	1	1

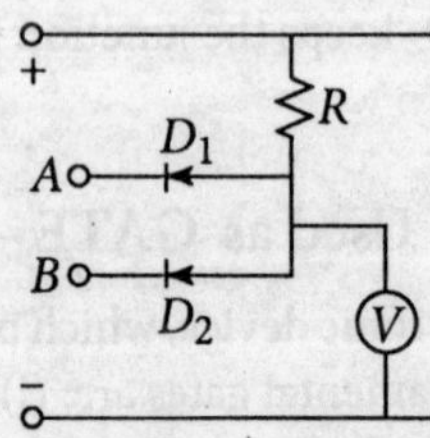

Fig. 2.42(a)

These two gates are very fundamental and using these two gates we can implement a large variety of logic functions. The details of the gates and implementation of the different logic functions have been discussed at length in later chapter.

2.14 Zener Diode

This is a special type of diode used in the reverse biasing arrangement for electronic voltage stabilisation (as in automatic stabiliser). The $(v\text{-}i)$ characteristic is shown in Fig. 2.43. The forward bias characteristic is similar to the usual *pn*-junction characteristic. But in the reverse bias connection the characteristic exhibits some region, where the terminal voltage is almost independent of the diode current. At this voltage the $(v\text{-}i)$ characteristic shows a very steep rise. This voltage is known as *Zener voltage* or *breakdown voltage*, which may be made sharp and distinct by suitably controlling the amount of impurity and making the *pn*-junction free from surface imperfection. Some useful specifications of a Zener are V_Z, the breakdown voltage; I_{ZT}, the corresponding current, known as the *test current.*

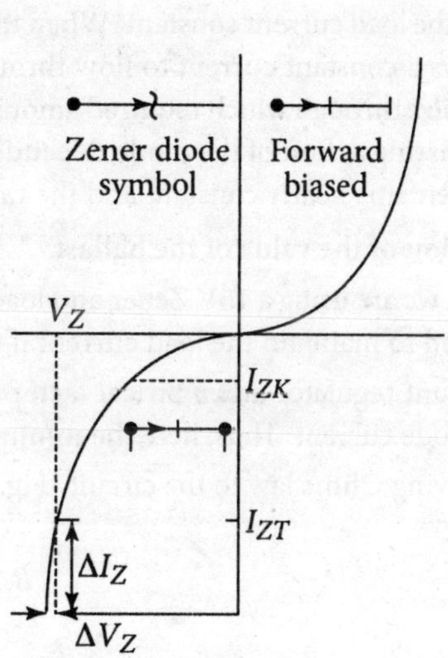

Fig. 2.43

Breakdown or Zener voltage V_Z

It is the reverse voltage corresponding to the linear portion of the reverse characteristic. This voltage is controlled by the amount of doping applied in the manufacturing process. Zeners of V_Z values ranging from about 2V to 200V with power capacity up to 100 watts are available.

Dynamic impedance

This specification of the Zener describes the steepness of the reverse characteristic. For a perfectly vertical breakdown curve, the dynamic resistance is zero. It is defined as

$$Z_Z = \frac{\Delta V_Z}{\Delta I_Z}, \tag{2.22}$$

i.e., the reciprocal of the slope of the reverse characteristic.

Voltage regulation by Zener

Fig. 2.44 shows a simple voltage regulating arrangement using Zener diode Z_1. The functioning as a voltage regulator can be explained qualitatively with reference to the reverse $v \sim i$ characteristic of the Zener [Fig. 2.44]. The voltage across the load R_L remains sensibly constant at the Zener voltage even though the input voltage V_{DC} and R_L vary over a wide range.

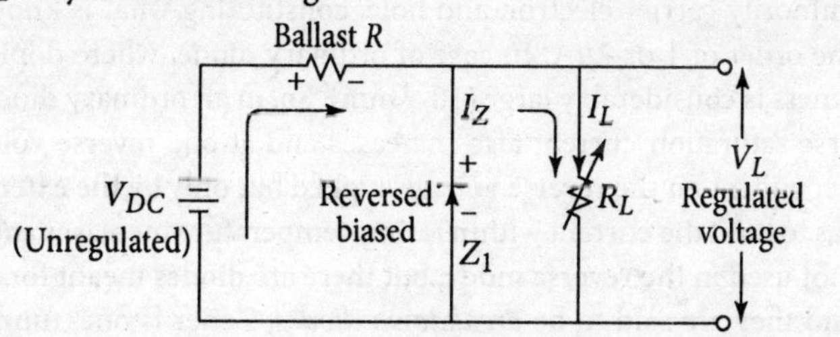

Fig. 2.44

Suppose, the input voltage increases, the diode tends to maintain the voltage V_L constant, so the voltage drop across the ballast R must increase and the resultant increase in I_i must flow through the diode thus keeping the load current constant. When the input voltage decreases, the diode agrees to accept less current but allows a constant current to flow through the load. In this regard the Zener may be looked upon as a *bypass valve* through which required amount of current is allowed such as to keep the load current constant. The regulating action of the diode depends on its use beyond the knee of the (v-i) curve, where the diode voltage remains nearly constant and the range of regulation depends on the value of the ballast resistor R.

Calculation of the value of the ballast

Suppose, we are using a 15V Zener and load current varies from 10mA to 100mA. We are to find the ballast R required to maintain the load current if the supply voltage V_{DC} = 20V.

For shunt regulator an empirical factor of 10 % of the maximum load current is used as the minimum Zener diode current. Thus, here the minimum value of I_Z = 10mA.

∴ applying Ohm's law to the circuit [Fig. 2.44], we get

$$R = \frac{V_{DC} - V_L}{I_i} = \frac{V_{DC} - V_L}{I_Z + I_L}.$$

The minimum Zener current will occur when I_L is maximum.

$$\therefore\ R = \frac{(V_{DC} - V_L)}{[(I_Z)_{min} + (I_L)_{max}]} = \frac{20 - 15}{(10 + 100) \times 10^{-3}} = 45.5\Omega.$$

$$\text{Resistor power rating} = \frac{V^2}{R} = \frac{5^2}{45.5} = 0.55\text{W}.$$

If accidentally the load resistor becomes open-circuited, $I_L = 0$, I_Z = 110mA, the Zener must be capable of dissipating power equal to $15 \times 110 \times 10^{-3}$ = 1.65W.

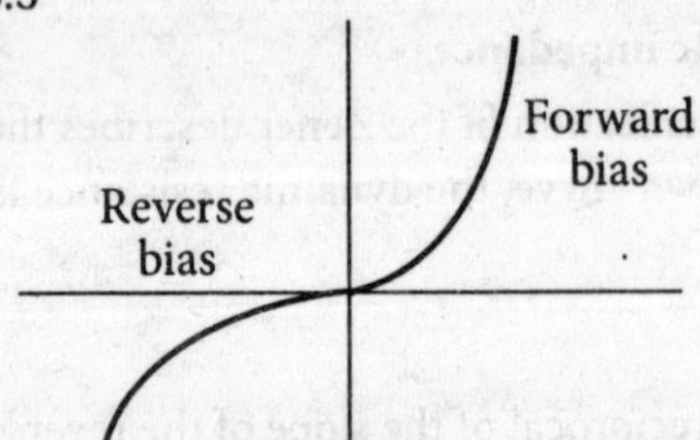

Fig. 2.45

Mechanism of breakdown

A diode on reverse bias breaks at a certain value. In case of an ordinary *pn*-junction diode containing 10^{15}–10^{16} as particle density of the impurity, the forward current vs. voltage curve and reverse voltage vs. current curve are shown in Fig. 2.45.

The forward current starts to rise at V_{app} > 0.7V but in the reverse bias, the current is very feeble due to the movement of minority carrier electron and hole, constituting what is known as *reverse saturation current,* which is of the order of 1 or 2μA. In case of ordinary diode, where doping is not very high, the depletion region thickness is considerably large (10^{-3} mm). So, in an ordinary diode, as the reverse voltage is increased, the reverse saturation current also increases and at one reverse voltage it breaks when the current is no longer dependent on the reverse voltage applied but only by the external circuit. If the circuit resistance is not such as to limit the current within rating, temperature increases and the diode is destroyed. So ordinary diode is not used in the reverse mode, but there are diodes meant for use in the reverse mode for special purpose and they are said to be *breakdown diodes*, Zener is one, tunnel diode is the other of this type. *Back diode* is also a breakdown diode with breakdown voltage at −0.1V. *Tunnel diode* is having the breakdown voltage at 0V and Zener diode is a breakdown diode with breakdown voltage >2V. The difference in the breakdown voltages in the case of breakdown diodes (Zener, back diode, tunnel diode) lies in the doping concentration; in ordinary diode, the doping concentration is of the order 10^{15}–10^{16}/cc, while in Zener diodes it is 10^{17}–10^{18} cc. In back or tunnel diodes the concentration is 10^{18}–10^{19}/cc. The breakdown diodes work on three mechanisms — (i) Punch through (ii) Zener breakdown (iii) Avalanche breakdown.

Of these three mechanisms, the first is related to diode, where one side at least is narrow and has low doping; the second mechanism is operative when both sides of the diode are highly doped and the third mechanism is operative when the doping is less than that in case of Zener diode.

Punch through

This type of breakdown occurs in a diode, where at least on one side there is low doping and small thickness. Depletion region in such a junction semiconductor diode extends mostly into the low doped side. If the low doped side is made narrow, a voltage is soon reached when the depletion reaches the Ohmic contact at the far end. Further increase of voltage makes the diode to act as a short-circuit and the current now is limited by the external circuit.

Zener breakdown

This is an important mechanism of breakdown operative when the breakdown voltage is less than 5V and both *n*- and *p*-sides of the diode are heavily doped. The depletion thickness is very much reduced, and the maximum junction field (at $x = 0$) may grow as high as MV/cm. At this narrowness of the region and high field even for a small reverse voltage of as small as 5V, there is abrupt increase in the current and the diode is said to break. The breakdown mechanism is the Zener mechanism, which can be explained classically and quantum mechanically. If a reverse bias is applied equal to $-V_R$, then total potential $V_T = V_0 - (-V_R) = (V_0 + V_R)$ and the depletion thickness increases. But for small reverse field, the increase in thickness is also small. So two different types of mechanism would be operative for breakdown voltages below 5V and for voltages above 5V. Zener mechanism is applicable for lower voltages and Avalanche mechanism for higher ones.

For high doping concentration on the *n*-type side and *p*-type side, the depletion region is very narrow $\simeq$ nm and as a result, a small reverse bias produces a tremendously high value of the electric field (MV/cm). This field in the classical or bond picture tears off electron from the covalent bond and creates a large number of electron-hole pair increasing the reverse saturation current and the diode is now said to break. Heavy doping of the *p*- and *n*-type sides of a *pn*-junction, the resulting narrowness of the depletion region, combined with a high electric field of the order of mV/cm prepare a broad background for tunnelling leading to the breakdown of the junction.

Avalanche breakdown

This is the most common type of mechanism of breakdown of a diode or a transistor.

When the reverse voltage is high, there is slight increase in the depletion thickness than when it is in equilibrium without bias. As a result of the high field of $\simeq$200kV/cm and widening of the depletion region, some of the minority carriers taking part in the normal leakage current flow, are accelerated to such an extent that they knock off valence electrons from the impurity atom on collision; the electron-hole pair thus created are also accelerated in opposite directions by the field and they also produce fresh electron-hole pairs. The effect is cumulative and in a very short time, a large number of electron-hole pairs are created. This is known as *avalanche*.

2.15 A Few Other Types of Diodes

A Photo diode

This type of diode depends for its operation on the light intensity. It is also a *pn*-junction diode so constructed that it can be exposed to light. Used in the reverse bias, the resistance of a photo diode decreases with intensity of light. In a conventional diode the reverse leakage current is due to the thermally generated minority carriers swept through the depletion region by the barrier voltage. So, in a photo diode as its resistance decreases, the reverse leakage current increases, the additional minority carriers are generated due to light intensity. Greater the intensity of light, greater is the reverse leakage current, hence less is the effective resistance. So it can be used as light detector.

Tunnel diode

This is also a special type of diode having characteristic feature of conduction for any forward bias. In this type of diode, the impurity concentration is of the order of 10^{18}-10^{19} per cc in both *p*- and *n*-regions. The barrier width is very small. The (*v*-*i*) characteristic of such a diode is shown in the figure. It shows that conduction occurs for almost 0V forward bias.

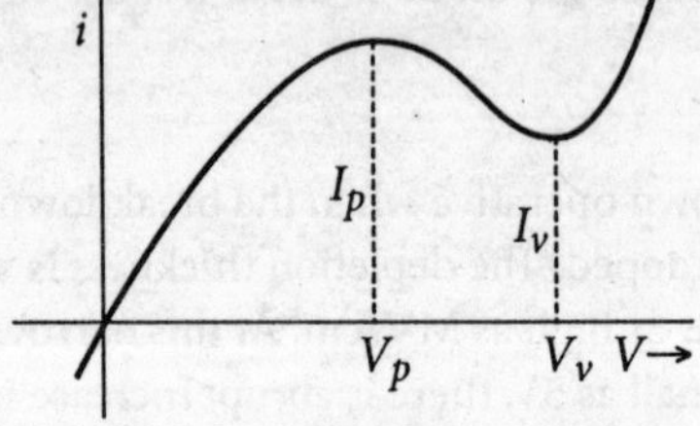

Fig. 2.46

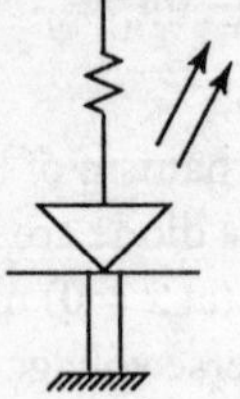

Fig. 2.47

From the characteristic curve, we get two important parameters of the diode — (i) peak current I_p and its corresponding voltage V_p, (ii) the valley current I_v and its corresponding voltage V_v. The region between the peak and the valley points is called a *negative resistance region* (recall this negative resistance in case of tetrode valve). It means, in this region an increase in voltage is associated with decrease in current. This phenomenon of negative resistance can be explained by a quantum mechanical process called *tunnelling* and as such this diode is called *tunnel diode*. Tunnel diodes (also called *Esaki diodes*) are used in making high frequency oscillators or high speed switching devices.

Light Emitting Diode (LED)

It is an opto-electronic device, where we get radiation through an electronic process.

In a forward-biased diode, free electrons cross across the *pn*-junction and fall into holes. As these electrons fall from higher energy level to lower, they radiate energy. So, in an ordinary diode, the energy that is radiated is in the form of heat. In a light emitting diode the energy is radiated in the form of light.

The circuit symbol of such an LED is shown in the Fig. 2.47. It is to be forward-biased and its colour depends on the material taken for the *pn*-junction. Ordinary diodes are made of silicon, which is an opaque material that blocks the passage of light. But with germanium, arsenic and phosphorus, we get LED's that have radiation red, green, yellow, blue, orange or, infrared radiations. Visible LED's are used in instruments, calculators, etc. while the infrared LED's are used in burglar alarm, etc. Typical voltage drop in an LED is between 1.5V and 2.5V for currents between 10mA and 50mA. Brightness of an LED depends on current.

Schottky diode

This diode is a metal-semiconductor junction used very frequently in digital computers. Its construction and circuit symbol is shown in Fig. 2.48. A junction is made by bonding Al to a suitably doped *n*-type silicon. Like *pn*-junction this *ms*-junction (metal-semiconductor) also exhibits high resistance when reverse-biased and low resistance when forward-biased. At depletion region a barrier potential also is created in this junction almost in the same way as in a *pn*-junction. The difference is that in this case carrier diffusion consists of electrons only from the semiconductor side.

In the switching circuits when diodes are employed, they must respond very quickly from high resistance state to low resistance state. Ordinary diodes have finite recovery time in going from on-state to off-state and vice versa, but metal semiconductor junction diodes are able to respond more rapidly most probably, because only majority carriers (electrons in n-type) are involved in the process.

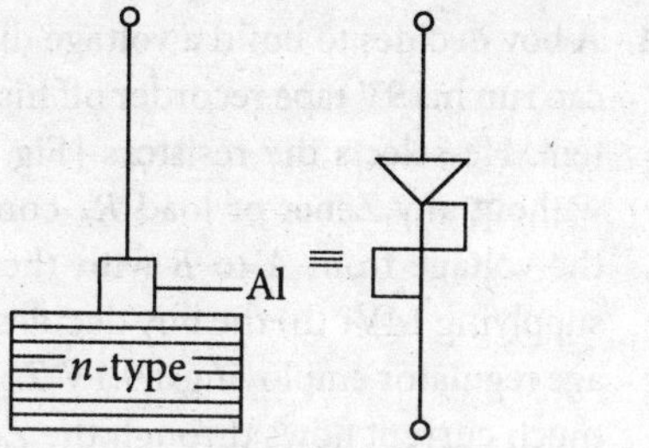

Fig. 2.48

2.16 Numerical Problems

1. **A power supply has a no-load voltage of 12V. What will be its full-load voltage if its voltage regulation is 10 %?**

Solution: Voltage regulation $= \dfrac{V_{\text{no-load}} - V_{\text{full-load}}}{V_{\text{full-load}}}$

$$\therefore \frac{1}{10} = \frac{12 - V_{\text{full-load}}}{V_{\text{full-load}}} = \frac{12}{V_{\text{full-load}}} - 1 \text{ or, } V_{\text{full-load}} = \frac{120}{11} = 10.9\text{V}.$$

2. **What is the necessary ac input power from the transformer secondary used in a half-wave rectifier to deliver 500W of dc power to the load?**

Solution: Rectification efficiency in half-wave rectifier is given by

$$\frac{P_{\text{dc}}}{P_{\text{ac}}} = \frac{4}{\pi^2} \cdot \frac{R_l}{R_d + R_l} \simeq \frac{4}{\pi^2}, \text{ if } R_d \ll R_l,$$

where R_d = diode resistance.

$$\therefore 0.406 = \frac{P_{dc}}{P_{\text{ac}}} \text{ or, } P_{\text{ac}} = \frac{500}{0.406} = 1232\text{W}.$$

3. **Design a voltage regulator circuit to provide 12V across a load whose current varies from 5mA to 35mA. An unregulated 18V dc source is to be used. Assume a Zener diode has been chosen with a Zener voltage of 12V at $I_{ZT} = 20$mA. Calculate:**

(a) The voltage dropping resistor and its power rating using an average value for I_L when $I_Z = I_{ZT}$.

(b) The minimum power rating of the Zener.

(c) Maximum increase in supply voltage that can be tolerated if the diode is rated at 1W.

Solution: $R = \frac{V_{\text{in}} - V_{\text{out}}}{I_i}$. The average load current = 20mA.

$\therefore$ total current I_i = 20 + 20 = 40mA. $\therefore$ $R = \frac{18-12}{40} \times 10^3 = 150\text{W}$.

Power rating $= \frac{V^2}{R} = \frac{6^2}{R} = \frac{36}{150} = 0.24\text{W}$.

Minimum current that flows through the diode = (40−35) = 5mA.

$\therefore$ minimum rating $= I_Z V_Z = 5 \times 10^{-8} \times 12 = 60\text{mW}$.

Maximum value of $I_Z = \frac{1}{12}$ = 83mA. Minimum load current = 5mA.

Thus, the maximum permissible value of $I_i = (83+5) = 88$mA.

Then the maximum voltage drop across $R = 88 \times 10^{-3} \times 150 = 13.20$V.

$\therefore$ maximum input voltage = 13.20 + 12 = 25.20V.

4. A boy decides to build a voltage divider so that he can run his 9V tape recorder off his car voltage system. He selects the resistors [Fig. 2.49]. Then (a) without any Zener or load R_L connected, what is the voltage from A to B with the car's generator supplying 14V? (b) the boy decides to make a voltage regulator employing a 9.1V Zener diode. How much current flows through the Zener when connected from A to B and what should be its power rating (assume no-load connected yet). (c) what is the maximum load R_l that can be applied without losing effective voltage regulation if the current through the Zener must be at least 25mA to maintain a 9.1V drop across it.

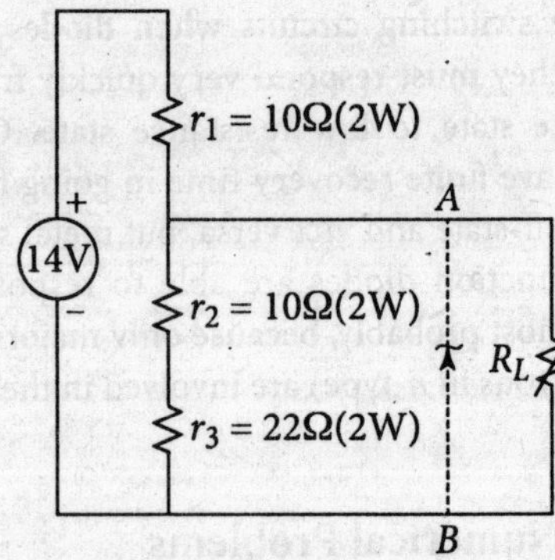

Fig. 2.49

Solution: (a) Total resistance of the potential divider network = 42Ω.

$\therefore$ p.d. between A and $B = \frac{14}{42} \times 32 = \frac{32}{3}$V.

(b) 25×10^{-3} A flowing through the diode maintains a p.d. = 9.1V.

$\therefore$ impedance of the diode $= \frac{9.1\times10^3}{25} = 364\Omega$.

$\therefore$ current flowing through the diode when connected across A, $B = \frac{32}{3}/364 = 29$mA.

(c) $4 \times 10^{-8} \times R_l = 9.1$. $\therefore$ $R_l = \frac{9.1}{4} \times 10^3 = 2.275$kΩ.

5. A dc power supply is known to have a ripple factor of 10 %. If the dc output voltage is 10V, what is rms value of the ripple voltage in the output?

Solution: Ripple factor $\gamma = \sqrt{\left(\frac{I_{\rm rms}}{I_{\rm dc}}\right)^2 - 1}$ or, $\frac{1}{10} = \sqrt{\frac{I_{\rm rms}^2}{100} - 1}$.

$\therefore$ $I_{\rm rms} = 10.05$V.

6. What is the necessary ac input power from the transformer secondary used in a half-wave rectifier to deliver 500W of dc power to the load?

Solution: Rectification efficiency in half-wave rectifier is given by

$$\frac{P_{\rm dc}}{P_{\rm ac}} = \frac{4}{\pi^2} \cdot \frac{R}{R_d + R_l} \simeq \frac{4}{\pi^2}, \text{ if } R_d \ll R_l,$$

where R_d = diode resistance.

$$\therefore\ 0.406 = \frac{P_{\rm dc}}{P_{\rm ac}} \text{ or, } P_{\rm ac} = \frac{500}{0.406} = 1232\text{W}.$$

7. Design a voltage regulator circuit to provide 12V across a load whose current varies from 5mA to 35mA. An unregulated 18V dc source is to be used. Assume a Zener diode has been chosen with a Zener voltage of 12V at I_{ZT} =20mA. Calculate:

(a) The voltage dropping resistor and its power rating using an average value of I_L, when $I_Z = I_{ZT}$.

(b) The maximum power rating of the Zener.

(c) Maximum increase in supply voltage that can be tolerated if the diode is rated at 1W.

Solution: $R = \frac{V_{\rm in} - V_{\rm out}}{I_l}$. The average load current = 20mA.

$\therefore$ total current $I_i = 20 + 20 = 40$mA. $\therefore$ $R = \frac{18-12}{40} \times 10^3 = 150\Omega$.

Power rating $= \frac{V^2}{R} = \frac{6^2}{R} = \frac{36}{150} = 0.24$W.

Minimum current that flows through the diode = (40−35) = 5mA.

$\therefore$ minimum rating $= I_Z V_Z = 5 \times 10^{-3} \times 12 = 60$mW. The rating of the given Zener = 1W.

Maximum value of $I_Z = \frac{1}{12} = 83$mA. Minimum load current = 5mA.

Thus, the maximum permissible value of $I_i = (83 + 5) = 88$mA.

Then the maximum voltage drop across $R = 88 \times 10^{-3} \times 150 = 13.20$V.

$\therefore$ maximum input voltage = 13.20 + 12 = 25.20V.

2.17 Exercises

1. What do you mean by semiconductor? How many types of semiconductors are there? Explain the formation of an extrinsic semiconductor from both bond and band models.
2. Show theoretically that in case of intrinsic semiconductor, the Fermi level lies at the middle of the conduction and valence bands but in case of extrinsic semiconductor, the position depends on the concentration of the impurities.
3. When a p-type semiconductor is put in contact with an n-type semiconductor, a potential barrier is created. Find an expression for the height of the potential barrier and explain forward biasing and reverse biasing.
4. Explain the functioning of a junction diode as a rectifier.
5. Explain the function of a bridge rectifier. What is its advantage over the usual full-wave rectifier?
6. What is a Zener diode? What purpose is served by such a diode? Design a voltage regulator with input voltage 20V, output voltage required is 15V, over a load current ranging from 10mA to 20mA. Assume a 15V Zener with $1_{ZT} = 6$mA.
7. What is the polarity of voltage across a Zener diode that is being used in a voltage regulating circuit?
8. The Zener diode has a fixed voltage drop if 18V across it as long as I_Z is maintained between 200mA and 2A. Find: (a) R so that V remains at 17V while V_{DC} is free to vary from 22V to 28V (b) Maximum power rating of the diode.

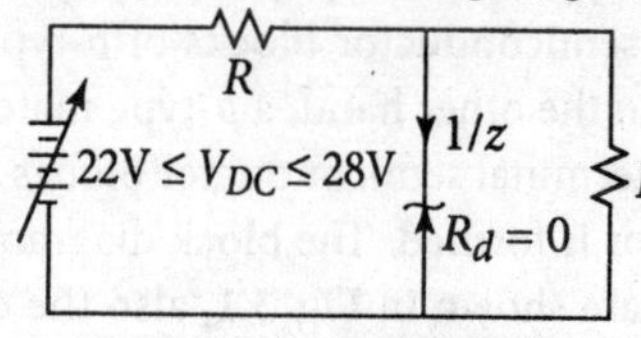

Fig. 2.50

9. An unregulated supply varies between 20V and 25V and has a 10Ω internal impedance. A 10V Zener is to regulate this voltage for use in a tape recorder. The recorder draws 30mA while recording and 50mA when playing back. The Zener diode has a resistance of 10Ω when the Zener current is 30mA. The knee of the Zener characteristic occurs at 10mA. In addition, the Zener diode can dissipate a maximum power of 0.8W. Find: (a) ballast resistor R, so that the diode regulates continuously (b) the maximum peak-to-peak output ripple.

Chapter 3

Transistors — Basic Characteristics and as Amplifying Device

3.1 What is a Transistor?

A transistor (transfer resistor) has the basic property of impedance transfer and hence, the name. It is the device that initiated a veritable revolution in the field of electronics since early 1950s. Starting with bipolar junction transistor (BJT) and passing through several stages of development producing new types of transistors in the name of field effect transistor (FET) and ultimately metaloxide semiconductor transistor or MOSFET, the transistor electronics has become a technically well-developed branch of electronic science. We will start our discussion with the bipolar junction transistor and then end with FETs and MOSFETs.

3.2 Elements of a Transistor

A transistor is made of two *pn*-junction diodes back to back or front to front. Bipolar junction transistors may be of two types: *npn* or *pnp* type. When thin layer of *n*-type semiconductor is sandwiched between two-terminal semiconductor blocks of *p*-type, a *pnp* transistor is formed; if on the other hand, a *p*-type material is sandwiched between two-terminal semiconductor blocks of *n*-type material, a *npn* transistor is formed. The block diagrams of the two types of transistors are shown in Fig. 3.1, also the circuit symbols for *pnp* and *npn* transistors are shown in the same figure. The central layer is termed as *base*, which compares with the control grid of a triode valve, one of the terminal blocks is the *emitter* comparable to the cathode, the other terminal block, called the *collector*, which compares the plate of the triode valve. In some transistor the outer layers are symmetrical and can be used either as the emitter or collector depending on the biasing voltages. In usual practice, however, the junctions are asymmetrical and there are indications given for emitter, base and collector.

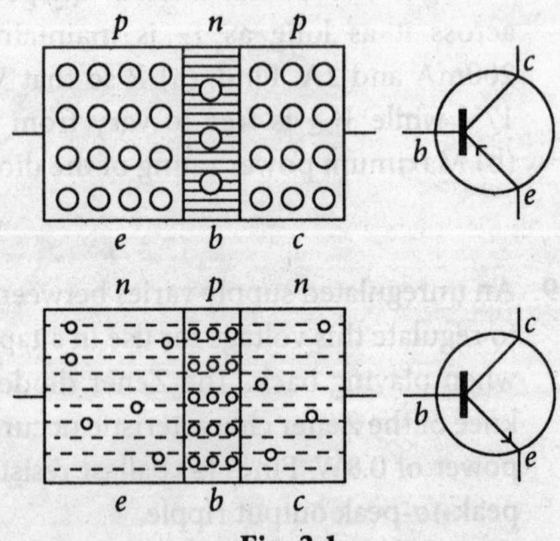

Fig. 3.1

3.3 Transistor Fabrication

Transistors are fabricated by introducing impurities of known amount and type at required points on the *p*- and *n*-types of semiconductor (viz., Silicon, Germanium) substrates and interconnecting various points by thin metallic layers or wires.

The construction of a *npn* transistor is represented [Fig. 3.2]. In the figure, P_1 is the substrate of *p*-type semiconductor, N_1 is a large *n*-type area, which is diffused into the substrate and becomes the collector (*C*). Next another *p*-type area P_2 is diffused into the collector area N_1 and becomes the base (*B*). Lastly, a small *n*-type area N_2 is diffused into the base P_2 and becomes the emitter (*E*). Then an oxide layer (SiO_2) is applied and the terminals are etched through it contacting the three areas — collector, base and emitter thus producing an *npn* transistor. The *pnp* transistor can be fabricated in the same way by taking the substrate of *n*-type semiconductor, then a layer of *p*-type diffused into the substrate becomes the collector and so on. Diffusion is the technique of introducing minute amounts of impurities on the substrate by thermal process. A slice of silicon is heated in a furnace to a high temperature (1000°C– 1200°C) and a vapour containing the impurity is passed through the furnace over the slice. Some of the impurity atoms will then come into contact with the surface of silicon and as the process is continued, will very slowly move or be diffused into the silicon.

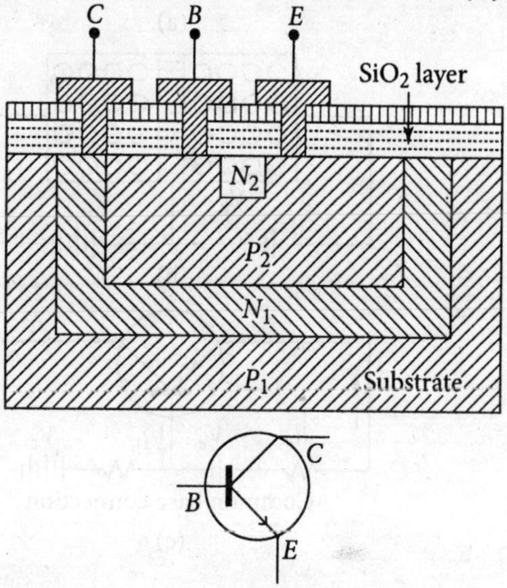

Fig. 3.2

3.4 Transistor Biasing and Mechanism of Current Flow

***pnp* transistor**

A *pnp* transistor may be looked upon as a combination of two junction diodes as shown. The emitter-base junction is forward-biased, the base-collector junction is reverse-biased (see the battery connection [Fig. 3.3]. The circuit symbol also is shown, the direction of the arrow in the emitter shows the conventional current direction. The forward bias to the emitter-base junction injects holes into the base region (most of the holes crossover the base region), and come to the right-hand negatively biased *p*-region, the collector. The number of electron-hole union in the emitter-base junction region is less than 5%. But, however, this union produces a small emitter-base current and the collector current is less than the emitter current by this amount. The emitter current is the flow of charge per unit time across the emitter-base junction, the collector current is measured by the flow of charge per unit time across the base-collector junction. Holes reaching the negative battery terminal at the collector combine with electrons from the battery. At the same time holes are created at the emitter by the electrons breaking their electron bonds and entering the positive of the battery terminal.

***npn* transistor**

The operation of an *npn* transistor is similar to that of a *pnp* transistor except the battery connection. Here negative terminal of the battery V_e is connected to the emitter (*n*-material) and the positive terminal of the battery V_c is connected to the collector (*n*-material) thus the emitter-base is forward-biased while the base-collector is reverse-biased [Fig. 3.4].

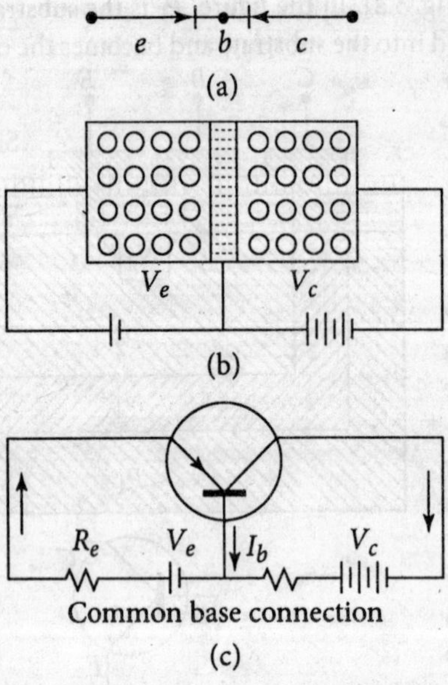

Fig. 3.3

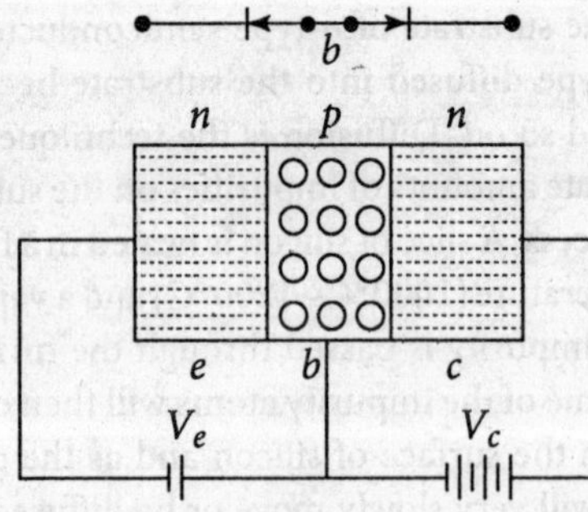

Fig. 3.4

3.5 Different Current Components in a Transistor

Usual biasing arrangement of *pnp* transistor is shown in the figure, the base-emitter junction (J_1) is forward-biased and the base-collector junction (J_2) is reverse-biased. The current components in a transistor are: (i) The emitter current (ii) The base current (iii) The collector current (iv) The reverse saturation current.

(i) *The Emitter Current* I_e. The emitter current consists of two components, one $i_e(n)$ due to electrons injected from base into the emitter, the other $i_e(p)$ due to the holes injected from the emitter to the base. Thus, $I_e = i_e(p) + i_e(n)$. Since in a commercial transistor, $i_e(n) \ll i_e(p)$, so, we get $I_e \simeq i_e(p)$.

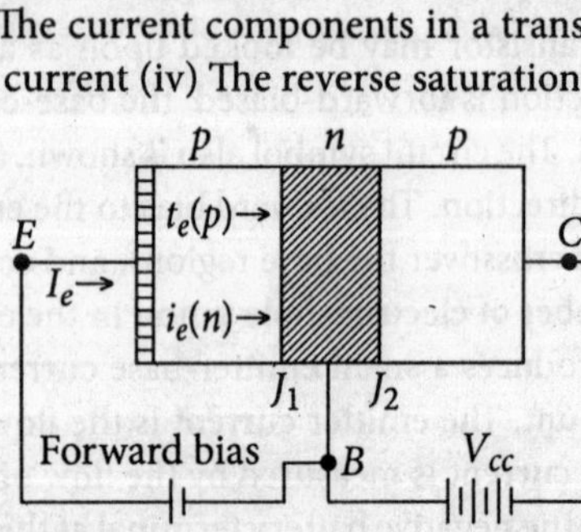

Fig. 3.5

(ii) *Base Current*. Of the holes injected into the base, recombination takes place between a few holes and electrons in the base giving rise to recombination base current of magnitude $I_b = i_e(p) - i_c(p)$, where $i_c(p)$ is the collector current, I_b is usually very small.

(iii) *Collector current*. It also consists of two components — one due to the holes that pass through the base and the other component is the reverse saturation current. This component needs some explanation.

Consider the emitter-base junction as open, i.e., $I_e = 0$ and the collector-base junction J_2 as reverse-biased. In this condition holes present as minority carriers in the base would move across the junction towards the collector and electrons as minority carriers in p-type collector would move towards the base. So this collector current denoted by I_{CEO} and termed as the reverse saturation current can be written as $I_{\text{CEO}} = I_{\text{CO}}(n) + I_{\text{CO}}(p)$.

Taking this component into account, the collector current I_c under working condition can be written as $I_C = I_c(p) + I_{CEO}$.

3.6 Different Transistor Configuration and Gains of a Transistor

Though the transistor is a current device, it does not mean that voltage amplification is not possible with a transistor. There are several gains associated with a transistor.

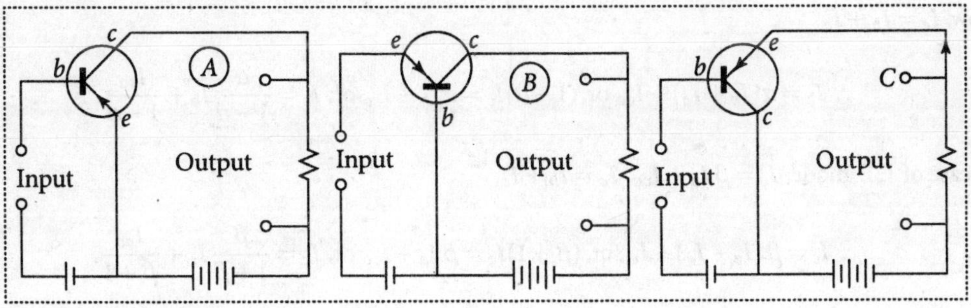

Fig. 3.6

Since a transistor has three terminals and input and output circuit must have four-terminal, in all transistor circuitry, either of emitter, base or collector is made common to both the input and output. Thus, there are three configurations (a) common emitter (b) common base (c) common collector shown respectively [Fig. 3.6].

Current Gains of a Transistor

(a) *Common Emitter Configuration:* dc *Current Gain.* When the transistor is used [Fig. 3.6(a)], the dc current gain is defined by the ratio I_c/I_B.

$$\therefore \beta_{dc} = h_{FE} = I_c/I_B.$$

ac *Current Gain.* The ac current gain in CE mode is given by the ratio

$$\beta_{ac} = h_{fe} = \left(\frac{\delta I_C}{\delta I_B}\right)_{V_{CE}}.$$

The dc current gain may vary between 10 to 500, as also the ac current gain.

(b) *Common Base Configuration:* dc *Current Gain.* The dc current gain in this configuration is defined as $\alpha_{dc} = h_{FB} = I_c/I_E$.

ac *Current Gain.* It is defined as the ratio

$$\alpha_{ac} = h_{fb} = \left(\frac{\delta I_c}{\delta I_E}\right)_{V_{CB}}.$$

Expression for the Collector Current in CE and CB Modes:

CE *Mode.* We have $I_c = \beta I_B + I_{CEO}$.

CB *Mode.* $I_C = \alpha I_E + I_{CBO}$.

More Exact Expression for the Collector Current in terms of α

The relation $I_c = \alpha I_e$ is an approximate relation. A more exact relationship may be derived with the introduction of the contribution by the reverse saturation current. Thus, in CB mode,

$$I_c = \alpha I_e + I_{\text{CEO}} = \alpha I_e + I_{\text{CEO}}.$$

Now, $I_e = I_b + I_c$

$$\therefore\ I_c = \alpha(I_b + I_c) + I_{co} \text{ or, } (1-\alpha)I_c = \alpha I_b + I_{co} \text{ or, } I_c = \frac{\alpha}{1-\alpha}I_b + \frac{I_{co}}{1-\alpha}. \tag{3.1}$$

In case of CE mode, $I_c = \beta I_b + I_{co}$, $I_c = I_b + I_c$.

$$\therefore\ I_c = \beta(I_e - I_c) + I_{co} \text{ or, } (\beta+1)I_c = \beta I_e + I_{co} \text{ or, } I_c = \frac{\beta}{1+\beta}I_e + \frac{I_{co}}{\beta+1}. \tag{3.2}$$

This reverse saturation current is a very important transistor parameter. In case of a silicon transistor, its value is of the order of 10^{-9} A, whereas for germanium transistor it is 10^{-6} A. The I_{co} is very susceptible to temperature changes. As temperature increases, I_{co} increases resulting in the increase of I_c, so the collector junction increases in temperature. This increase in temperature again increases the I_{co}, which again increases I_c, hence rise of temperature of the junction. The effect is cumulative and it might so happen that the junction temperature exceeds the rated value. The junction is then said to be destroyed. This phenomenon is referred to as *thermal runaway* of the transistor. With regard to this effect a silicon transistor is preferred to a germanium transistor.

Relation between α and β

Considering the common base mode, $I_E = I_C + I_B$.

Now $\alpha = I_C/I_E$, $\beta = I_C/I_B$.

$$\therefore\ \beta = \frac{I_C}{I_B} = \frac{I_C}{I_E - I_C} \text{ or, } \frac{I_E - I_C}{I_C} = \frac{1}{\beta} \text{ or, } \frac{1}{\alpha} - 1 = \frac{1}{\beta}.$$

$\therefore\ \alpha = \frac{\beta}{\beta+1}$ thus knowing β accurately, we can find out the value of α.

3.7 Transistor Characteristics

For either type of transistor, there are three characteristics — output characteristic (I_c vs. V_c with I_B constant), transfer characteristic (I_c vs. I_B, with V_c constant), input characteristic (I_B vs. V_B, V_c constant). Of the different configurations, the common emitter characteristics of the transistor are very useful, because in most of the amplifier application, this mode is used.

The characteristics can be drawn w.r.t. Fig. 3.7, where we have used a *pnp* transistor.

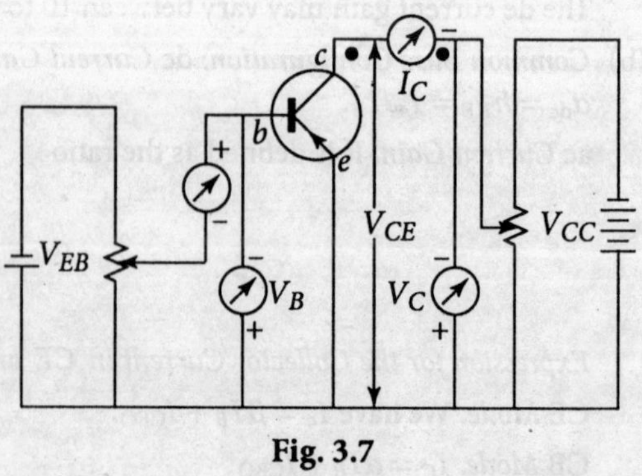

Fig. 3.7

Output characteristics

With reference to Fig. 3.7 we draw the output characteristics showing the variation of I_C vs. V_{CE} at constant I_B. Maintain the base current values $I_B = 50\mu A$, $60\mu A$, $80\mu A$, etc. and for each setting, note down I_C vs. V_C. Plot your observations to get output characteristics shown in Fig. 3.8(a).

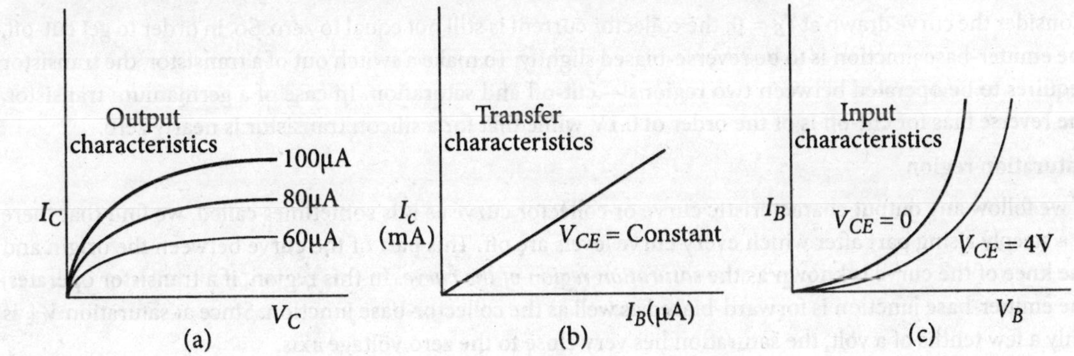

Fig. 3.8

Transfer characteristics

Keeping V_c constant and record I_C vs. I_B to get the transfer characteristics. It is a straight line graph [Fig. 3.8(b)].

Input characteristics

It shows the variation of I_B vs. B_B at different constant collector voltages. The input characteristics are shown in Fig. 3.8(c).

3.8 Distinct Regions of the Characteristic Curves of a Transistor

Common emitter output characteristics

If we draw the output characteristics in the CE-mode, i.e., collector current vs. collector-emitter voltage taking base current as parameter, we get curves as shown in Fig. 3.9.

Four distinct regions of the characteristic curves are to be identified.

They are: (i) Active region (ii) Saturation region (iii) Cut-off region (iv) Breakdown region. All of these regions are of important considerations. The discussion is given below:

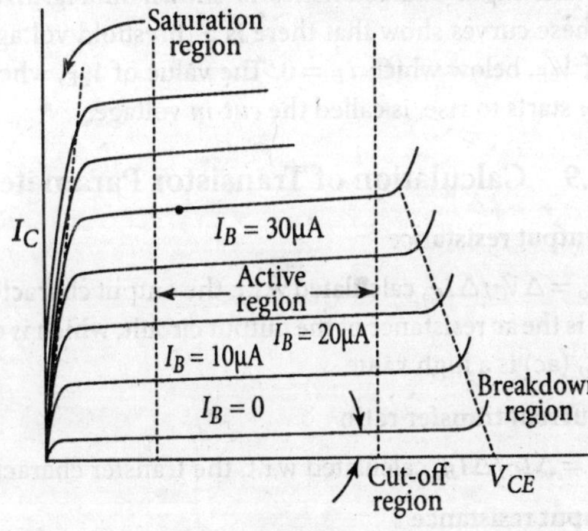

Fig. 3.9

Active region

This region refers to the actual operating region of a transistor when it is used as an amplifier. When i_B is low, the characteristic curve levels are off but with increase in the value of i_B, the characteristic curves appear to possess finite slope. It means, with greater i_B, i_C value for the same V_{CE} is greater. This is due to what is known as *Early effect* due to the internal feedback from collector diode to the base diode. Under this condition, the emitter-base junction remains forward-biased and the base-collector junction is reverse-biased. So for a *pnp* transistor to operate in the active region, the emitter is at a positive potential w.r.t. the base and the collector is at a negative potential w.r.t. it.

Cut-off region

Consider the curve drawn at $i_B = 0$, the collector current is still not equal to zero. So, in order to get cut-off, the emitter-base junction is to be reverse-biased slightly. To make a switch out of a transistor, the transistor requires to be operated between two regions — cut-off and saturation. In case of a germanium transistor, the reverse bias for cut-off is of the order of 0.1V while that for a silicon transistor is nearly zero.

Saturation region

If we follow any output characteristic curve or collector curve as it is sometimes called, we find that there is a steeply rising part after which every curve levels are off. This part of the curve between the origin and the knee of the curve is known as the *saturation region of the curve.* In this region, if a transistor operates, the emitter-base junction is forward-biased as well as the collector-base junction. Since at saturation V_{CE} is only a few tenths of a volt, the saturation lies very close to the zero voltage axis.

Breakdown region

This region refers to the sharply rising part beyond the limit of the active region. This region is to be avoided at all costs.

Input characteristics

A curve drawn with i_B against V_{BE} taking V_{CE} as parameter is said to constitute the input characteristic curve of the transistor.

The input characteristics is shown in Fig. 3.10. These curves show that there is a threshold voltage of V_{BE}, below which $i_B = 0$. The value of V_{BE}, when i_B starts to rise, is called the *cut-in* voltage.

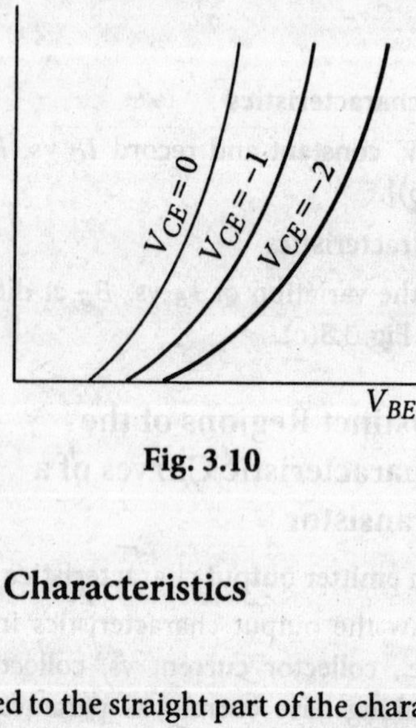

Fig. 3.10

3.9 Calculation of Transistor Parameters from the Characteristics

Output resistance

$R_o = \Delta V_C / \Delta I_C$, calculated w.r.t. the output characteristics, referred to the straight part of the characteristic. It is the ac resistance of the output circuit, which is different from the dc resistance defined as R_o (dc) V_C / I_C. R_o (ac) is a high value.

Current transfer ratio

$\beta = \Delta I_C / \Delta I_B$, calculated w.r.t. the transfer characteristic, which shows linear variation of I_C with I_B.

Input resistance

$R_i = \Delta V_B / \Delta I_B$, calculated w.r.t. the input characteristics showing variation of I_B vs. V_B. It is a non-linear curve and R_i varies.

Transistor used as an amplifier

We have seen previously that a transistor can be used in three configurations depending on the common terminal used. Whatever may be the configuration there should be two voltage sources supplying bias — one for the emitter-base, the other for collector-base.

3.10 Transistor Biasing and Thermal Stability of the Biasing Arrangement

A transistor has other functions than amplifications. Depending on the functioning for which the transistor is used, the transistor is made to operate in different regions of its characteristic curve. When it is used as an amplifier, biasing arrangement should be such as to forward bias the emitter-base junction and reverse bias the collector-base junction. But when it is used as a switch, it is to be operated between cut-off and saturation regions, so it requires to be properly biased.

There are different biasing arrangements but all of them are not guaranteed equally for thermal stability. We have discussed earlier the effect known as *thermal runaway of the transistor.* So, the biasing arrangement will be so designed and chosen that it becomes thermally stable.

Fixed-bias circuit

The positive voltage line provides both voltages, for the base as well as for the collector. The voltage supply for collector is denoted by V_{CC}. The collector to emitter (grounded) voltage is denoted by V_{CE} and p.d. between *A* and ground is denoted by V_{BB}, which is here equal to the barrier p.d. between emitter and base denoted by V_{BE}. The value of $V_{BE} = 0.7$V for a silicon transistor, whereas for germanium $V_{BE} = 0.4$V.

In Fig. 3.11 an *npn* transistor is biased through resistors R_c and R_b from the same supply voltage. If I_b be the base current and I_c be the collector current, then we can formulate two equations related to the input and output circuits as

$$I_b R_b + V_{BE} = V_{cc} \text{ input circuit equation} \quad (3.3)$$

$$\text{and } I_c R_c + V_{CE} = V_{cc} \text{ output circuit equation.} \quad (3.4)$$

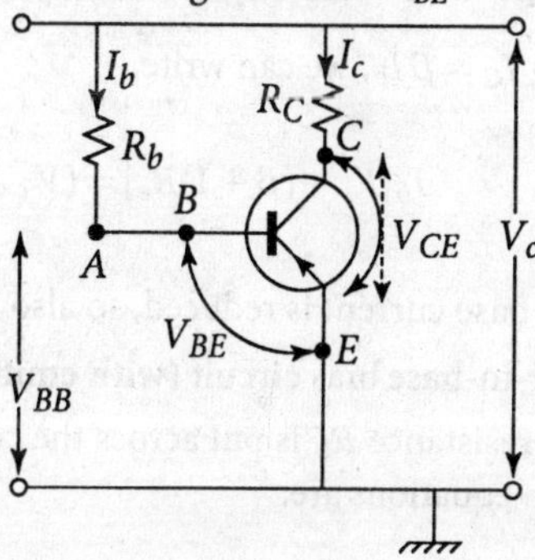

Fig. 3.11

The base current from (3.3) is given as

$$I_b = \frac{V_{cc} - V_{BE}}{R_b} \simeq \frac{V_{cc}}{R_b}. \quad (3.5)$$

If the transistor operates in the active region, $I_c = \beta I_b$ thus the collector current is not dependent on the collector resistance R_c. It is to remember that $I_c = \beta I_b$ relation applies when the transistor operates in the active region. So before application of these formulae we must be sure that the transistor is not in saturation region.

To check whether operation is in saturation region

For operation in saturation region, $V_{CE} = 0.1$V.

$$\therefore I_C = \frac{V_{cc} - V_{CE}}{R_C} \text{ or, } V_{CE} = V_{cc} - I_C R_C.$$

If $I_c < V_{cc}/R_C$, $V_{CE} > 0$ and if it is > 0.1V, the transistor is said to be operated in the active region. For $I_C = V_{cc}/R_C$, we have $V_{CE} = 0$, I_C has its maximum value. So check up whether $I_C < V_{cc}/R_C$ or, $I_C = V_{cc}/R_C$, if it is less, then (3.3) and (3.4) apply.

Thermal stability of the arrangement

We introduce a factor *S* to denote stability factor, which is a measure of change of collector current due to change of I_{CO}. If this factor is higher in some arrangement, the transistor is becoming more prone to undergo thermal runaway.

Mathematically, the stability factor S is defined as $S=\frac{dI_C}{dI_{CO}}$.

So using (3.1), we get

$$I_C=\frac{\alpha}{(1-\alpha)}I_b+\frac{I_{CO}}{(1-\alpha)} \text{ or, } S=\frac{dI_C}{dI_{CO}}=\frac{1}{1-\alpha}.$$

If we take $\alpha = 0.98$, $S = 50$, so this biasing arrangement is not suitable vis-à-vis the thermal stability. It is also to note that in this biasing R_C has no effect on the collector current I_C. What happens when we connect a resistor R_e is the emitter circuit [Fig. 3.12]. In this biasing arrangement, we get

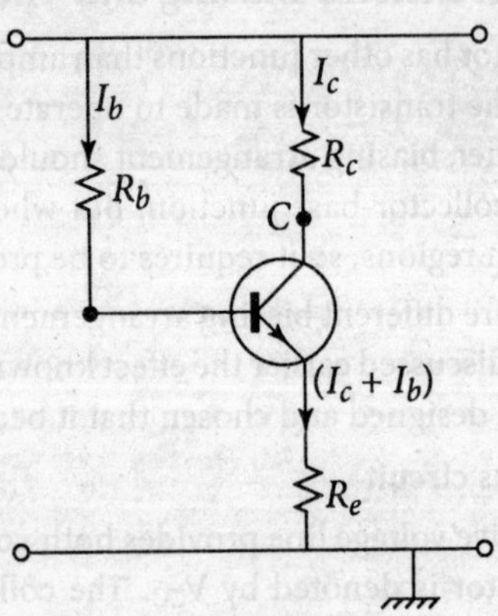

Fig. 3.12

$$I_bR_b+V_{BE}+(I_C+I_b)R_e=V_{cc}$$
$$\text{and } I_CR_C+V_{CE}+(I_C+I_b)R_e=V_{cc}.$$

Putting $I_C=\beta I_b$, we can write,

$$I_b[R_b+(\beta+1)R_e]=(V_{cc}-V_{BE}) \text{ or, } I_b=\frac{V_{cc}-V_{BE}}{R_b+(\beta+1)R_e}\simeq\frac{V_{cc}}{R_b+(\beta+1)R_e}. \quad (3.6)$$

So, the base current is reduced, so also is the collector current.

Collector-to-base bias circuit (with emitter bias)

Here the resistance R_b is put across the collector terminal and base and not across the V_{cc} supply and the base. Two equations are:

$$I_bR_b+(I_b+I_c)R_c+V_{BE}+(I_b+I_c)R_e=V_{cc}$$
$$(I_b+I_c)R_c+V_{CE}+(I_b+I_c)R_e=V_{cc}$$
$$V_{cc}=(I_b+I_c)R_c+(I_b+I_c)R_e+I_bR_b+V_{BE}=I_b(R_c+R_b+R_e)+I_c.$$

Since V_{BE} = a small quantity, we get from above

$$V_{cc}=I_b(R_c+R_e+R_b)+I_c(R_c+R_e)$$

$$\text{or, } I_b=\frac{V_{cc}-I_c(R_c+R_e)}{R_b+R_c+R_e}. \quad (3.7)$$

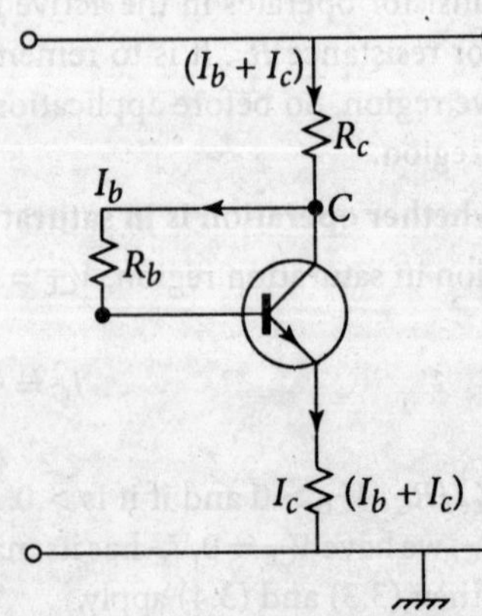

Fig. 3.13

$$I_c=\frac{\alpha}{1-\alpha}I_b+\frac{1}{1-\alpha}I_{co}$$

$$\text{or, } I_c=\frac{\alpha}{1-\alpha}\left[\frac{V_{cc}-I_c(R_c+R_e)}{R_b+R_c+R_e}\right]+\frac{I_{CO}}{(1-\alpha)}$$

$$\text{or, } I_c\left[(1-\alpha)+\frac{\alpha(R_c+R_e)}{(R_b+R_c+R_e)}\right]=I_{CO}+\frac{\alpha V_{cc}}{(R_b+R_c+R_e)}.$$

Differentiating w.r.t. I_{CO}, we get

$$S=\frac{1}{\left[(1-\alpha)+\frac{\alpha(R_c+R_e)}{R_c+R_b+R_e}\right]}. \quad (3.8)$$

$R_e = 0$, then the stability factor S is given by

$$S = \frac{1}{(1-\alpha) + \dfrac{\alpha R_c}{R_b + R_c}}. \tag{3.9}$$

It shows that the thermal stability is improved

$$(1-\alpha)I_c = \alpha I_b + I_{CO} \Rightarrow (1-\alpha)I_c = \alpha \frac{(V_{cc} - V_{be})}{(R_c + R_e + R_b)} - \alpha I_c \frac{(R_c + R_e)}{(R_c + R_b + R_e)} + I_{CO}$$

$$\Rightarrow \left[(1-\alpha) + \frac{\alpha(R_c + R_e)}{(R_c + R_b + R_e)}\right] I_c = \frac{\alpha(V_{cc} - V_{be})}{(R_c + R_b + R_e)} + I_{CO}$$

$$\Rightarrow I_c = \frac{\alpha(V_{cc} - V_{be})/(R_c + R_b + R_e)}{\left[(1-\alpha) + \dfrac{\alpha(R_c + R_e)}{R_c + R_b + R_e}\right]} + \frac{I_{CO}}{\left[(1-\alpha) + \dfrac{\alpha(R_c + R_e)}{(R_c + R_b + R_e)}\right]}. \tag{3.10}$$

$$\therefore S = \frac{dI_c}{dI_{CO}} = \frac{1}{\left[(1-\alpha) + \dfrac{\alpha(R_c + R_e)}{(R_c + R_b + R_e)}\right]}. \tag{3.11}$$

(3.11) shows that if $R_e \ll R_c$, it becomes equal to S given by (3.9).

Potential divider biasing arrangement

This biasing arrangement is often employed in amplifier circuit design we can make a simple analysis of the biasing without any consideration to the feedback or a more rigorous analysis with due consideration to the feedback involving the application of Thévenin's theorem. Whatever may be the mode of analysis this biasing arrangement is the best suited to make the operating point independent of current gain.

Fig. 3.14 shows the schematic diagram of the biasing arrangement. The resistors R_1 and R_2 connected across the supply from the potential divider arrangement; the p.d. developed across the resistor R_2 supplied the necessary bias voltage to the base.

This arrangement is widely used for the purpose of stabilisation, because in this case the operating point is almost independent of β-value of the transistor.

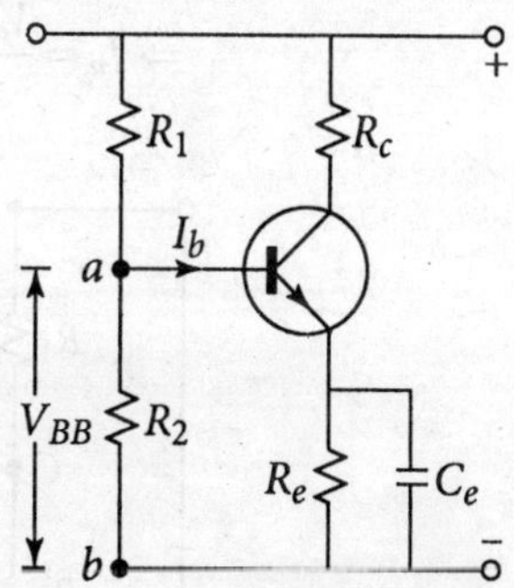

Fig. 3.14

Approximate analysis

If I_b is assumed to be small, (yes, it is in practice), then

$$V_{BB} = \frac{V_{cc} R_2}{(R_1 + R_2)} = V_{be} + (I_b + I_c) R_e \Rightarrow \frac{1}{R_e}\left[\frac{V_{cc} R_2}{(R_1 + R_2)} - V_{be}\right] = I_b + I_c. \tag{3.12}$$

$$I_b = \frac{V_{cc} R_2}{(R_1 + R_2) R_e} - \frac{V_{be}}{R_e} - I_c.$$

$$\therefore\ I_c(1-\alpha)=\alpha I_b+I_{CO} \Rightarrow I_c(1-\alpha)=\alpha\left[\frac{V_{CC}R_2}{(R_1+R_2)R_e}-\frac{V_{be}}{R_e}-I_c\right]+I_{CO}$$

$$\Rightarrow I_c(1-\alpha)+\alpha I_c=\alpha\left[\frac{V_{CC}R_2}{(R_1+R_2)R_e}-\frac{V_{be}}{R_e}\right]+I_{CO}$$

$$\Rightarrow I_c=\alpha\left[\frac{V_{CC}R_2}{(R_1+R_2)R_e}-\frac{V_{be}}{R_e}\right]+I_{CO}.$$

$$\therefore\ S=\frac{dI_c}{dI_{CO}}=1. \tag{3.13}$$

Exact analysis

The current I_b can be found by applying Thévenin's theorem. The Thévenin's voltage V_{TH} is the open voltage between a, b when the path of I_b is snapped.

$$V_{TH}=\frac{V_{CC}R_2}{(R_1+R_2)}.$$

To find the Thévenin's resistance we proceed as below.

Assuming the source to have zero internal resistance, we have to look into the network from the open-circuited terminals a, b replacing the active source by the internal resistance.

We have

$$V_{TH}=I_bR_{TH}+(I_b+I_c)R_e,\ \text{neglecting } V_{be}$$

$$\Rightarrow \frac{V_{CC}R_2}{R_1+R_2}=I_b(R_{TH}+R_e)+I_cR_e \Rightarrow I_b(R_{TH}+R_e)=\frac{V_{CC}R_2}{R_1+R_2}-I_cR_e$$

$$\Rightarrow I_b=\frac{V_{CC}R_2/(R_1+R_2)}{(R_{TH}+R_e)}-\frac{I_cR_e}{R_{TH}+R_e}.$$

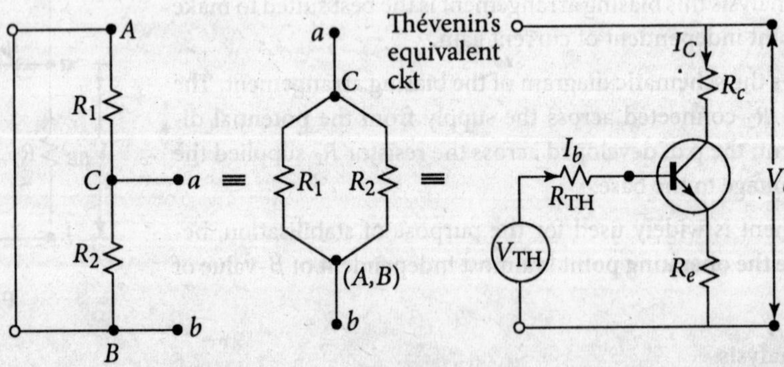

Fig. 3.15

$$\therefore (1-\alpha)I_c=\alpha I_b+I_{CO}=\alpha\left[\frac{V_{CC}R_2/(R_1+R_2)}{R_{TH}+R_e}-\frac{I_cR_e}{R_{TH}+R_e}\right]+I_{CO}$$

$$\Rightarrow \left[(1-\alpha)+\frac{\alpha R_e}{R_{TH}+R_e}\right]I_c=\frac{\alpha V_{CC}R_2/(R_1+R_2)}{R_{TH}+R_e}+I_{CO}.$$

$$\therefore S = \frac{dI_c}{dI_{CO}} = \frac{1}{(1-\alpha)+\dfrac{\alpha R_e}{(R_{TH}+R_e)}}.$$

Now

$$\frac{1-\alpha}{\alpha} = \frac{1}{\beta} \Rightarrow \frac{1}{\alpha} = \frac{1+\beta}{\beta} \Rightarrow \alpha = \frac{\beta}{(\beta+1)}.$$

$$\therefore (1-\alpha) = \frac{1}{\beta}\cdot\alpha = \frac{1}{(\beta+1)}.$$

$$\therefore S = \frac{1}{\dfrac{1}{\beta+1}+\dfrac{\beta R_e}{(\beta+1)(R_{TH}+R_e)}} = \frac{(\beta+1)}{1+\dfrac{\beta R_e}{(R_{TH}+R_e)}}.$$

3.11 Numerical Problems

1. **Determine the bias resistor R_b for fixed biasing arrangement and collector to base bias arrangement and compare the stability factor, given $V_{cc}=12\text{V}$, $R_c=330\Omega$, $I_b=0.3\text{mA}$, $\beta=100$, $V_{cc}=6\text{V}$.**

Solution: From (3.5), we get

$$I_b = \frac{V_{CC}-V_{BE}}{R_b} \Rightarrow R_b = \frac{V_{CC}-V_{BE}}{I_b} \simeq \frac{V_{CC}}{I_b} = \frac{12}{0.3\times10^{-3}} = 40\text{k}\Omega.$$

Also, we get

$$V_{cc} = I_bR_b + V_{BE} \Rightarrow R_b = \frac{V_{ce}-V_{BE}}{I_b} \simeq \frac{V_{cc}}{I_b} = \frac{6}{0.3\times10^{-3}} = 20\text{k}\Omega.$$

In the first case,

$$S = \frac{1}{(1-\alpha)} = \beta+1 = 100+1 = 101.$$

In the second case,

$$S = \frac{1}{(1-\alpha)+\dfrac{\alpha R_c}{R_c+R_b}} = \frac{1}{\dfrac{1}{(\beta+1)}+\dfrac{\beta R_c}{(\beta+1)(R_c+R_b)}} = \frac{\beta+1}{1+\dfrac{\beta R_c}{R_c+R_b}} = \frac{101}{1+\dfrac{100\times0.330}{20.3}} = 38.2.$$

2. **A base resistor biasing is shown in Fig. 3.16. Determine: (a) the collector current I_c (b) collector to emitter voltage V_{cc}, $\beta=60$, $V_{BE}\simeq0$ (c) If R_b is changed to 200kΩ and the transistor is considered to be silicon *npn*.**

Solution: (a) $I_b\times300\times10^{-3}+V_{BE}=6\text{V} \Rightarrow I_b = \dfrac{6}{300\times10^3} = 20\mu\text{A}.$

$\therefore I_c = \beta I_b = 20\times10^{-6}\times60 = 1.2\text{mA}.$

(b) $I_c\times R_c + V_{cc} = 12$

$\Rightarrow V_{cc} = 12 - I_cR_c = 12-1.2\times10^{-3}\times3\times10^3 = 8.4\text{V}.$

(c) For silicon $V_{BE}=0.7\text{V}.$

$$I_b = \frac{V_{BB}-V_{BE}}{200\times10^3} = \frac{6-0.7}{200\times10^3} = \frac{5.3}{200}\times10^{-3} = 26.5\mu\text{A}.$$

$$I_c = \beta\times I_b = 60\times26.5 = 1.59\text{mA}.$$

$V_{cc} = 12-1.59\times10^{-3}\times3\times10^3 = 12-4.77 = 7.23\text{V}.$

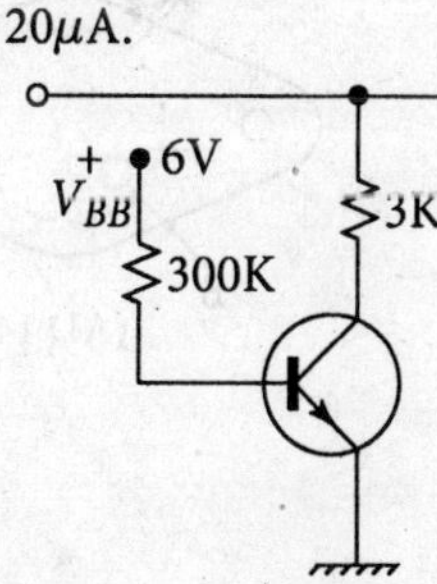

Fig. 3.16

3. Refer to circuit [Fig. 3.17] and draw the load-line (dc), what will be its stability factor?

Solution: The output circuit equation is given by

$$V_{CC} = I_C \times R_C + V_{CC} \Rightarrow \frac{V_{ce}}{V_{CC}} + \frac{I_C}{(V_{CC}/R_c)} = 1.$$

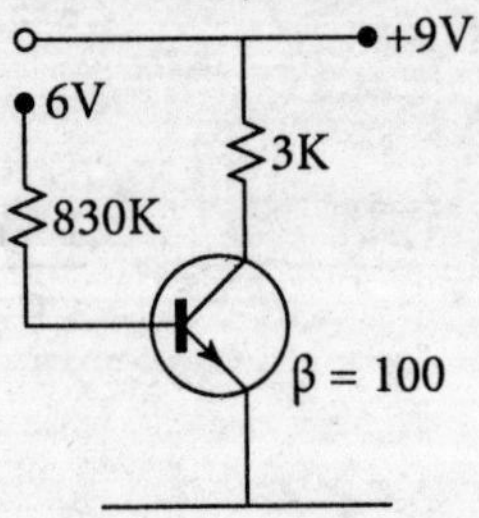

Fig. 3.17

This is the load-line equation, which is a straight line having intercept on the x-axis equal to V_{CC} and on the y-axis equal to $V_{CC}/R_c = 9/3 \times 10^3 = 3\text{mA}$ [Fig. 3.18].

Now,

$$I_b R_b = V_{CC} - V_{BE} \Rightarrow I_b = \frac{9-0.7}{830 \times 10^3} = 10\mu A$$

$$\Rightarrow I_c = \beta I_b = 100 \times 10 = 1\text{mA}.$$

$$V_{CC} = 9 - I_c R_c = 9 - 1 \times 10^{-3} \times 3 \times 10^3 = 6\text{V}.$$

$$S = (\beta + 1) = 101.$$

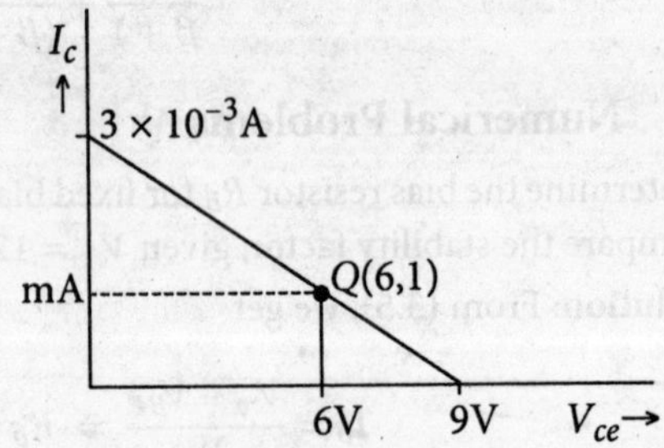

Fig. 3.18

Basing identification

To identify the leads an index or colour dot is placed next to the collector terminal. Some of the basing diagrams for the identification of leads [Fig. 3.19].

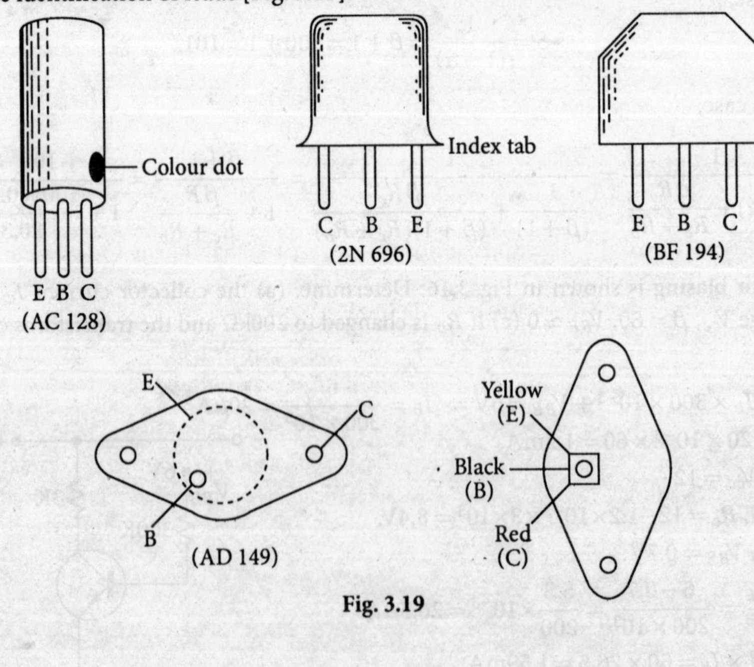

Fig. 3.19

3.12 Different Configurational Amplifier Circuits

The common base amplifier

We have shown a common base amplifier with two different sources of bias voltages [Fig. 3.20]. In practical circuit, however, the suitable biasing arrangement provides forward biasing between emitter and base and reverse biasing between collector and base.

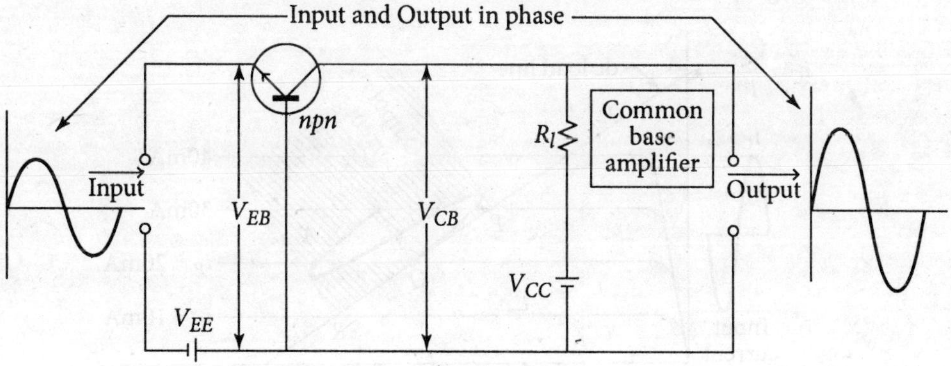

Fig. 3.20

Explanation of the working: Graphical analysis

This type of amplifier circuit, where base is common to both input and output circuit is commonly used, because of its relatively high voltage gains. This configuration has a very low input impedance from 30Ω to 150Ω and a very high output impedance of 300kΩ to 1MΩ. But disadvantage of this circuit is in the matching of the input and output impedances. However, this amplifier is suited to the amplification of the output of a low-impedance magnetic phono-pick-up.

In absence of any input signal a certain amount of collector current flows, because of the biases applied thus producing a voltage drop across the load resistor R_l. This voltage drop opposes the collector battery voltage and places the collector bias lower than the battery voltage. A signal is now injected between the emitter and base. During the positive half of the input, there will be opposition to the negative bias to the emitter thus reducing the forward bias and resulting in the reduction of the emitter current and hence, also the collector current. Since the collector current is reduced, the voltage drop across the load R_l also decreases. So, the collector becomes more positive. Thus, we see that during the positive half of the input, the output at the collector also passes through the positive half. In the same way it will be seen that as the input passes through the negative half, the output developed at the collector also passes through its negative half. This is the important feature of common base amplifier. Thus, the input and output voltages are in phase, which [Fig. 3.21] by similar wave shape of input and output.

Analysis w.r.t. the load-line

Let us hold the following discussion w.r.t. the output characteristics. When there is no signal, the transistor being properly biased, there is certain amount of collector current flowing through the load causing a p.d. across the load thus the equation involving the variables in the collector circuit is given by

$$V_{cc} - i_c R_l = V_{CB}. \tag{3.14}$$

Now, when $i_c = 0$, we get $V_{CB} = V_{cc}$.

Again, when $V_{CB} = 0$, we get $i_c = V_{cc}/R_l$.

Thus, two points of the dc load-line are fixed and when they are joined, we get a straight line known as the dc load-line (AB). This line cuts a number of characteristic curves, but the operating point or quiescent point is the point of intersection of the load-line with the characteristic drawn at i_E at the specified value, which is determined by V_{CE} and given by the equation $V_{EE} = i_E R_E + V_{EB}$, where R_E is the resistance of the emitter-base circuit. Suppose, $i_E = 20$mA, Q is the quiescent point. The biasing is so done that Q falls at the middle of the straight portion of the characteristic curves.

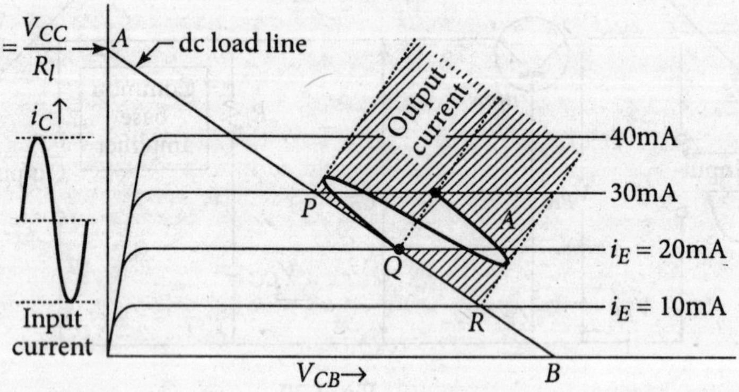

Fig. 3.21

Suppose, the input current is such that it makes the emitter current vary between 30mA and 10mA about the quiescent point Q. When the input grows, emitter current increases and it goes up to P along the load-line, then as the input decreases the variation of the output current takes place along the load-line from P to Q; during the negative half of the input, the variation of the output takes place along the load-line from Q to R and back again to Q. Thus, one complete cyclic change of the input produces cyclic variation of the output which is always in phase with the input. The output wave shape is represented to the right top [Fig. 3.21]. It is evident from the graph that peak-to-peak output current variation is a magnified version of the input.

Analytical solution of the amplifier circuit

The amplifier circuit may be analytically solved by drawing the equivalent circuit drawn w.r.t. constant current source or constant voltage source representation of a transistor. This equivalent representation is shown in Fig. 3.23(a) and (b) respectively with reference to common base configuration.

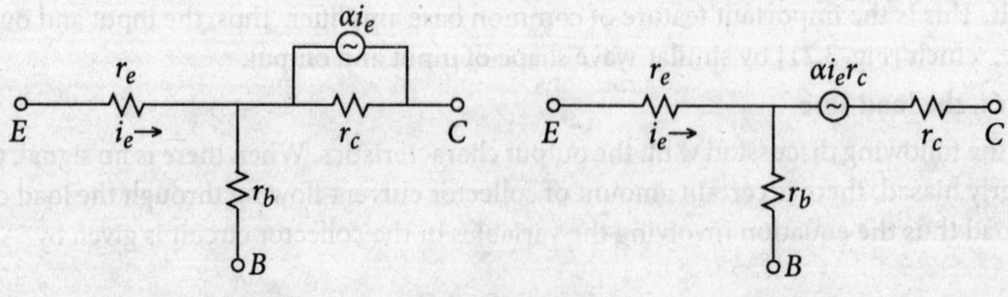

Fig. 3.22

Let us take once again the common base configurational amplifier [Fig. 3.23].

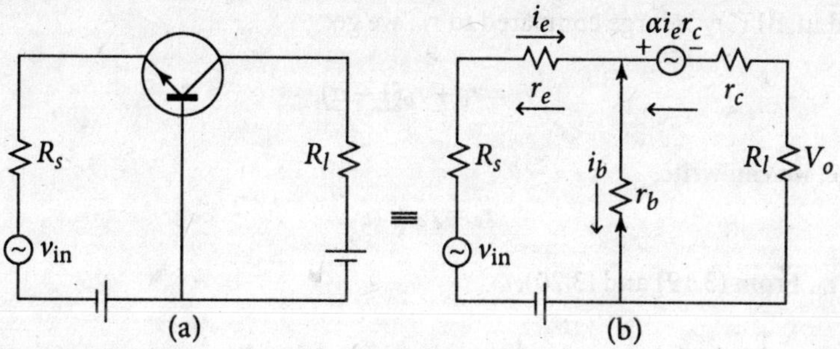

Fig. 3.23

Take the input side of the equivalent circuit and apply Kirchhoff's law:

$$v_{in} = i_e(R_s + r_e) + i_b r_b = i_e(R_s + r_e) + (i_e - i_c)\, r_b, \quad [\because i_e = i_c + i_b]$$
$$= i_e(R_s + r_e + r_b) - i_c r_b. \tag{3.15}$$

Considering the output side, we get

$$0 = i_c(R_l + r_c) - i_b r_b - \alpha i_e r_c = i_c(R_l + r_c) - (i_e - i_c)\, r_b - \alpha i_e r_c = i_c(r_b + R_l + r_c) - i_e r_b - \alpha i_e r_c.$$

$$\therefore \alpha i_e r_c = i_c(r_b + R_l + r_c) - i_e r_b. \tag{3.16}$$

Solving (3.15) and (3.16), we get

$$v_{in} = i_c(R_s + r_e + r_b) - r_b i_c = h_1 i_c - r_b i_c$$
$$h_1 = (R_s + r_c + r_b) \tag{3.17}$$
$$0 = i_c(r_b + R_l + r_c) - i_c(r_b + \alpha r_c) = i_c(r_b + \alpha r_c) + h_2 i_c$$
$$h_2 = (r_b + R_l + r_c). \tag{3.18}$$

$$\therefore i_c = \frac{\begin{vmatrix} v_{in} & -r_b \\ 0 & h_2 \end{vmatrix}}{\begin{vmatrix} h_1 & -r_b \\ -(r_b + \alpha r_e) & h_2 \end{vmatrix}} = \frac{h_2 v_{in}}{h_1 h_2 - r_b(r_b + \alpha r_c)}$$

$$= \frac{v_{in}}{h_1 - \dfrac{r_b}{h_2}(r_b + \alpha r_c)} = \frac{v_{in}}{(R_s + r_c + r_b) - r_b\left(\dfrac{r_b + \alpha r_c}{r_b + R_1 + r_s}\right)}.$$

Let r_i = input resistance.

$$\therefore i_e = \frac{v_{in}}{R_s + r_i}.$$

$$\therefore R_s + r_i = R_s + r_c + r_b - \frac{r_b(r_b + \alpha r_c)}{(r_b + R_l + r_c)}$$

$$\text{or, } r_i = r_c + r_b \cdot \frac{(r_b + R_l + r_c - r_b - \alpha r_c)}{(r_b + R_l + r_c)} = r_c + r_b \cdot \frac{R_l + (1 - \alpha)\, r_c}{r_b + R_l + r_c}.$$

If R_l is small and in BJT r_c is large compared to r_b, we get

$$r_i = r_c + r_b(1-\alpha). \tag{3.19}$$

If R_l is very large, we can write,

$$r_e = r_c + r_b. \tag{3.20}$$

Output Resistance r_0. From (3.19) and (3.20),

$$\therefore\ i_c = \frac{\begin{vmatrix} h_1 & v_{\text{in}} \\ -(r_b+\alpha r_c) & 0 \end{vmatrix}}{\begin{vmatrix} h_1 & -r_b \\ -(r_b+\alpha r_c) & h_2 \end{vmatrix}} = \frac{v_{\text{in}}(r_b+\alpha r_c)}{h_1 h_2 - r_b(r_b+\alpha r_c)} = \frac{v_{\text{in}}(r_b+\alpha r_c)/h_1}{h_2 - r_b\left(\frac{r_b+\alpha r_c}{h_1}\right)}.$$

$$\therefore\ V = v_{\text{in}} \times (r_b+\alpha r_c)/h_1.$$

If r_0 be the output resistance, then

$$i_c = \frac{V}{r_0 + R_l}.$$

$$\therefore\ r_0 + R_l = h_2 - \frac{r_b(r_b+\alpha r_c)}{h_1} \quad \text{or, } r_0 = r_b + r_c - \frac{r_b(r_b+\alpha r_c)}{(R_s+r_0+r_b)}$$

$$\text{or, } r_0 = r_b + r_c - \frac{r_b(r_b+\alpha r_c)}{(R_s+r_c+r_b)} = r_c + r_b\frac{R_s+r_e-\alpha r_c}{(R_s+r_c+r_b)}. \tag{3.21}$$

If $R_s = 0$, then

$$r_0 = r_c + r_b\frac{(r_e-\alpha r_c)}{(r_e-r_b)} = r_c\frac{\left[1+\frac{r_b}{r_c}\cdot r_e - r_b\alpha\right]}{(r_e+r_b)}.$$

When $r_b r_e/r_c$ is neglected, it being small, we get

$$r_0 = r_c\left[1+\frac{-r_b\alpha}{r_e+r_b}\right] = r_0\left[\frac{r_e+r_b(1-\alpha)}{(r_e+r_b)}\right].$$

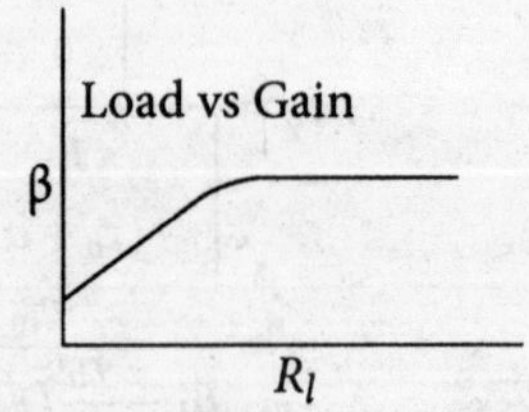

Fig. 3.24

In case of large R_s,

$$r_0 = r_b + r_c. \tag{3.22}$$

Voltage gain.

$$A_v = \frac{v_{\text{out}}}{v_{\text{in}}} = \frac{i_e R_l}{V_{\text{in}}} = \frac{R_l(r_b+\alpha r_c)/h_1}{h_2 - \frac{r_b(r_b+\alpha r_c)}{h_1}} = \frac{R_l(r_b+\alpha r_c)}{h_1 h_2 - r_b(r_b+\alpha r_c)}.$$

When β is plotted against R_l, we get a curve [Fig. 3.24]. r_c is large in practice, so

$$\frac{V_{\text{out}}}{V_{\text{in}}} = \frac{\alpha R_l r_c}{r_c(R_s+r_c+r_b) - \alpha r_b r_c} = \frac{\alpha R_l}{R_s+r_c+r_b-\alpha r_b}.$$

3.13 Solved Numerical Problems

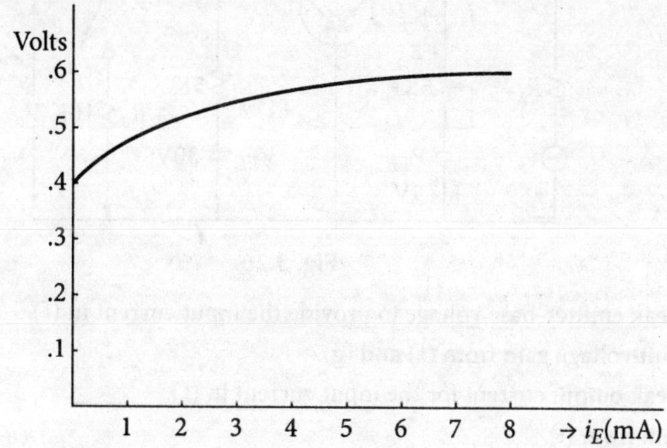

Fig. 3.25(a)

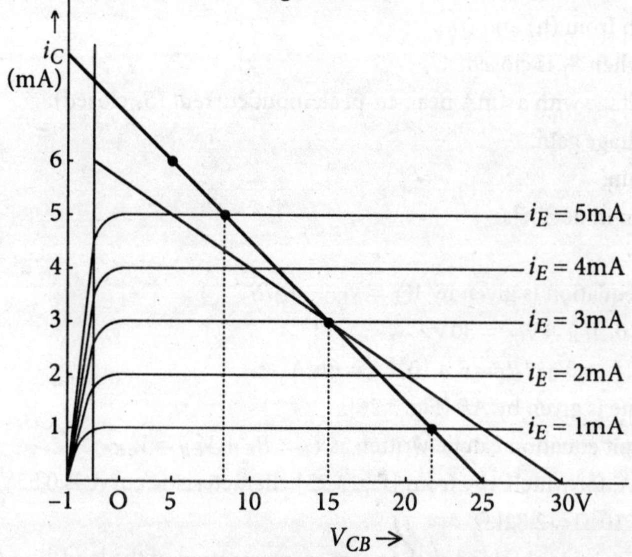

Fig. 3.25(b)

1. **A CB amplifier using 2N 3903 with its characteristic [Fig. 3.25(b)]. With reference to the said circuit, determine the following:**
   (a) **The dc load-line.**
   (b) **The value of R_E for which emitter bias current is 3mA.**
   (c) **The value of V_{CB} and i_C at the operating point in (b).**
   (d) **The emitter-base bias voltage for the operating point in (b).**
   (e) **dc input resistance to the transistor.**
   (f) **The open-circuit output voltage ($S-1$ open) when a 4mA peak-to-peak emitter current is applied.**

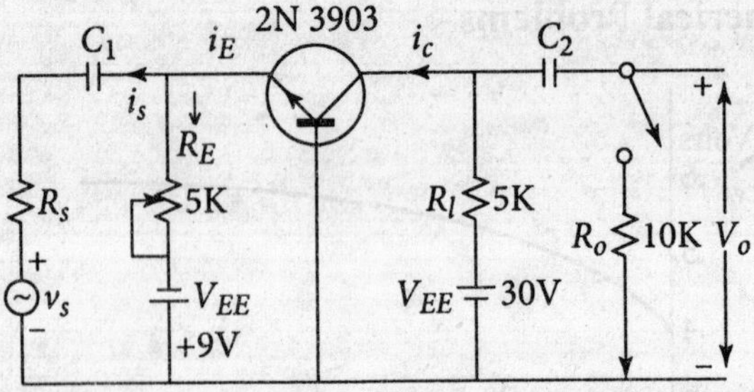

Fig. 3.26

(g) The peak-to-peak emitter-base voltage to provide the input current in (f).

(h) The open-circuit voltage gain from (f) and (g).

(i) The peak-to-peak output current for the input current in (f).

(j) The current gain from (i).

(k) The ac input resistance to the transistor from (g).

(l) The power gain from (h) and (j).

(m) ac load-line when S_1 is closed.

(n) The output voltage with a 4mA peak-to-peak input current (S_1 closed).

(o) The loaded voltage gain.

(p) The current gain.

(q) The power gain in decibels.

Solution:

(a) The load-line equation is given by $V_{CB} = V_{CC} - i_C R_l$.

So, when $i_C = 0$, $V_{CB} = V_{CC} = 30V$.

When $V_{CB} = 0, i_C = V_{CC}/R_l = 6 \times 10^{-3} A = 6mA$.

So, the load-line is given by AB [Fig. 3.26].

(b) The input circuit equation can be written as $i_E \times R_E + V_{EB} = V_{EE}$.

Here $i_E = 3mA$; for which V_{EB} from [Fig. 3.25] characteristic curve is 0.55V.

$\therefore\ R_E = 2.82 \times 10^3 \Omega = 2.82k\Omega$.

(c) The operating point is obtained as the point of intersection of the load-line AB and $i_C \sim V_C$ characteristic curve corresponding to $i_E = 3mA$. From the graph $V_{CB} = 15V$ and $i_C = 3mA$.

(d) $V_{EB} = 9 - 2.82 \times 10^3 \times 3 \times 10^{-3} = 0.54V$.

(e) $V_{EB} = i_E \times R_i, \therefore\ R_i = \frac{0.54}{3\times10^{-3}} = \frac{540}{3} = 180\Omega$

(f) A variation of applied emitter current of 4mA peak-to-peak means bias current to vary from 5mA to 1mA.

The output circuit equation is given by $v_c = i_c \times R_l + V_{cc}$.

When $i_c = 5mA$, $v_c = 5 \times 10^{-3} \times 5 \times 10^3 + 30 = 55V$.

When $i_c = 1mA$, $r_c = 1 \times 10^{-5} \times 5 \times 10^3 + 30 = 35V$.

$\therefore$ open-circuit output voltage = 20V pp.

(g) The emitter-base bias voltage is 0.55V for 3mA current. So for variation of emitter-base current by 4mA peak-to-peak, emitter-base current varies, between 1mA and 5mA corresponding to which the V_{EB} is 0.5V and 0.5625V. So peak-to-peak voltage is 0.0625V (from V_{EB} vs. I_E graph).

(h) Voltage gain $= \frac{v_0\,(\text{peak-to-peak})}{v_i\,(\text{peak-to-peak})} = \frac{20}{0.0625} = 320.$

(i) Peak to peak output current = 3.95mA from the i_C vs. V_{CB} characteristic.

(j) Current gain = 3.95/4 = 0.988.

(k) ac input resistance $= \frac{0.0622}{4} \times 10^3 = 15.6\Omega.$

(l) Power gain = Voltage gain × Current gain = 320 × 0.988 = 316.

(m) With respect to the ac behaviour, the effective resistance in the output consists of R_o and R_l in parallel.

$$\therefore\ R_e = \frac{R_o R_l}{R_o + R_l} = \frac{10 \times 5}{15} = \frac{50}{15} = 3.33\text{k}\Omega.$$

The output circuit equation is $v_c = i_c R_l + V_{cc}$ or, $I_c = \frac{v_0}{R_l} - V_{cc}$.
∴ the slope of the ac load-line

$$= \frac{1}{R_l}\text{A/V} = \frac{1}{3.33 \times 10^3}\text{A/V} = 0.300 \times 10^{-3}\ \text{A/V}.$$

(n) The coordinate of the point Q is (15V, 3mA). The slope of the ac load-line is 3mA per 10V. So moving 10V to the left, we vertically move up (3+3), i.e., 6mA to locate the point A on the load-line. Then join AQ and get the ac load-line. For variation of input current 4mA peak-to-peak, the input current varies from 1mA to 5mA. The ac load-line cuts the 5mA line and 1mA line at C and D. The voltage corresponding to C is 8.25V and 21.5V. So, the output voltage variation is 13.25V, peak-to-peak.

(o) Voltage gain $= \frac{13.25}{0.0625} = 212.$

(p) 3.95/4 = 0.988.

(q) Power gain $A_P = A_v A_i = 212 \times 0.988 = 209.46.$

Power gain in dB $= 10\log 312 = 29.2$dB.

2. An input signal of 1mV is applied to a transistor amplifier provided with a load resistance of 500kΩ. If the input resistance is 50Ω, calculate the voltage gain (assume $\alpha = 1$).

Solution: Current input $i_e = \frac{V_{in}}{R_i} = \frac{1\times10^{-3}}{50} = 20\mu\text{A}.$

$$i_c = \alpha \times i_e = 1 \times 20 = 20\mu\text{A}.$$

$$\therefore\ v_0 = 20 \times 10^{-6} \times 0.5 \times 10^6 = 10\text{V}.$$

$$\therefore\ \text{voltage gain} = \frac{v_0}{v_i} = \frac{10}{1 \times 10^{-3}} = 10^4.$$

3. A transistor used in the common base configuration has $\alpha = 0.988$, emitter resistance $r_e = 30\Omega$, base resistance $r_b = 300\Omega$. If the collector load $R_l = 220\text{k}\Omega$, calculate the input resistance and voltage gain for input current of $i_e = 20\mu\text{A}$.

Solution: $r_i = r_e + r_b(1-\alpha) = 30 + 300(1-0.988) = 30 + 300 \times 0.012 = 33.6\Omega$

$V_{in} = i_e r_i = 20 \times 10^{-6} \times 33.6 = 672 \times 10^{-6}\text{V}$

$i_c = \alpha i_e = 0.988 \times 20 \times 10^{-6} = 19.76 \times 10^{-6}$

$V_{out} = i_c \times R_l = 19.76 \times 10^{-6} \times 220 \times 10^3 = 4.35\text{V}$

$\therefore\ \text{gain} = \frac{4.35}{672} \times 10^6 = 6.47 \times 10^3.$

3.14 Exercises on CB Amplifier

1. What causes the input resistance to the overall amplifier to be lower than the input resistance to the transistor itself? Give a mathematical justification of the above mentioned statement.
2. What is the output resistance of the overall CB amplifier usually determined by? Establish the relation $r_0 = r_b + r_c$, where the symbols have usual significance.
3. If the input and output voltages from a CB amplifier are used to provide a Lissajous figure on an oscilloscope, what figure should result and what phase relationship does this indicate?
4. Name the three basic configurations in which transistors can be connected. A very low input impedance and a very high output impedance are characteristic of what type of amplifier configuration? What is the advantage or disadvantage of such configurational amplifier?
5. How does a low value of emitter bias supply voltage affect the current gain in a CB amplifier?
6. Draw a common base amplifier circuit using a *pnp* transistor. Derive an expression for the voltage gain of such an amplifier.
7. What do you mean by load-line? How can you draw the dc load-line in case of a transistor used in CB mode?
8. A common base amplifier is being used as a voltage amplifier. Its collector load resistance is 15k providing an input resistance of 50Ω. To increase the voltage gain, the load resistance is increased to 1MΩ, as a result there was an increase of distortion in the output. Explain why did the distortion increase.
9. (a) Explain why a loaded voltage gain is smaller than the open-circuit voltage gain of CB amplifier.
   (b) How can you draw an ac load-line with a slope of 2mA/V?
10. A *pnp* transistor used in the CB mode has the following circuit parameters: $R_s = 10\Omega$, $R_E = 60\Omega$, $V_{EE} = 1.5\text{V}$, $R_l = 0.4\text{k}$, $V_{cc} = 16\text{V}$ and output closed through $R_o = 1\text{k}$. Draw the circuit and determine the following:
    (a) The dc load-line.
    (b) The value of V_{EB} for an emitter bias current of 3mA.
    (c) dc input resistance to the transistor.
    (d) The open-circuit output voltage when a 4mA peak-to-peak emitter current is applied.
    (e) The operating point is (10V, 3mA). First calculate the slope of ac load-line. Next explain how to draw the ac load-line.

3.15 Common Emitter Amplifier

This type of transistor configuration is widely used in actual or practical circuits. The difference between this CE mode and CB mode is that in this mode, the output is 180° out of phase w.r.t. the input, whereas in CB mode, the input and output are in phase.

In this type of amplifier the emitter is common to both input and output circuits, the base-emitter junction is as usual forward-biased and the emitter-collector junction reverse-biased (see the connection of the bias batteries); with *pnp* type of transistor, the biasing connection is just reversed, the current in the collector circuit is about 95 % of the emitter current, the balance 5 % flowing in the base circuit.

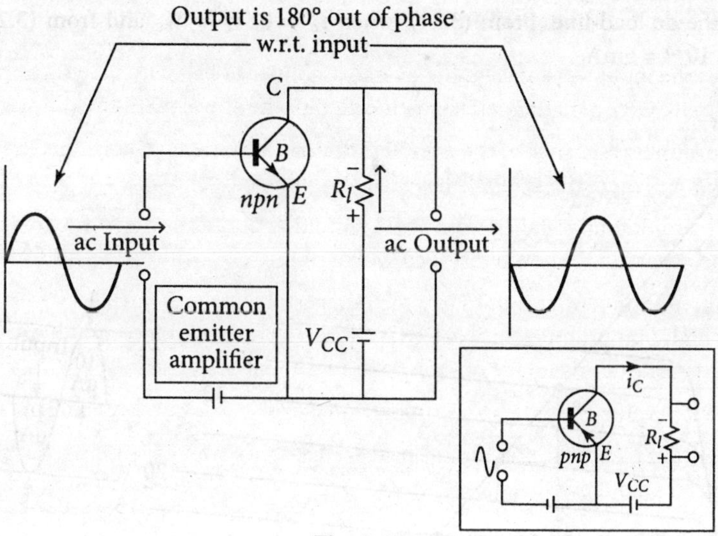

Fig. 3.27

Mathematical analysis

The emitter-collector voltage v_c can be written as (using *npn*)

$$i_c R_l = V_{cc} - v_c \text{ or, } v_c = V_{cc} - i_c R_l. \tag{3.23}$$

Using *pnp*,

$$v_c = V_{cc} + i_c R_l. \tag{3.24}$$

These equations may be utilised to analyse the action of this amplifier. With the signal injected between the base and emitter (similar to that between grid and cathode of a triode valve), when the signal grows in the positive half-cycle, the forward bias voltage increases (in *npn*) resulting in the increase of collector. So, the drop across the load also increases, hence v_c decreases or becomes less positive. During the other half of the input, the input signal opposes the forward bias, reducing the emitter and collector currents. The drop across the load decreases and hence, the collector voltage becomes more positive. Thus, we see that in CE mode, the input and output voltages are 180° phase out.

The input and output impedances of the common emitter circuit are considerably less severe than those of CB amplifier, and impedance matching is much simpler. The input resistance of a CE mode varies between 500Ω to 1500Ω while the output resistance is in the range of 50kΩ. Since the resistance gain of the amplifier in CE mode is very much less compared to that of CB mode, it might appear that the voltage gain would be much less. But the current gain of this type of amplifier is so high that even though the resistance gain is low, the voltage gain can be quite respectable. Since power gain is a function of the current squared, very high power gains ~ 40dB are very common in this type of common emitter configuration.

Analysis w.r.t. characteristic curves

The functioning of the amplifier can be explained with reference to characteristic curves of the transistor, one such characteristic curve may be taken as the output characteristic (i_c, V_c) or transfer characteristic (i_c, I_b).

Let us suppose in our circuit involving a *pnp* transistor, the collector supply V_{cc} =30V, load resistor R_l = 5k, peak-to-peak input current = 20μA, i_b = 25μA.

First construct the dc load-line. From (3.24), when $i_c = 0$, $v_c = V_{cc}$ and from (3.23), when $v_c = 0$, $i_c = V_{cc}/R_l = \frac{30}{5} \times 10^{-3} = 6\text{mA}$.

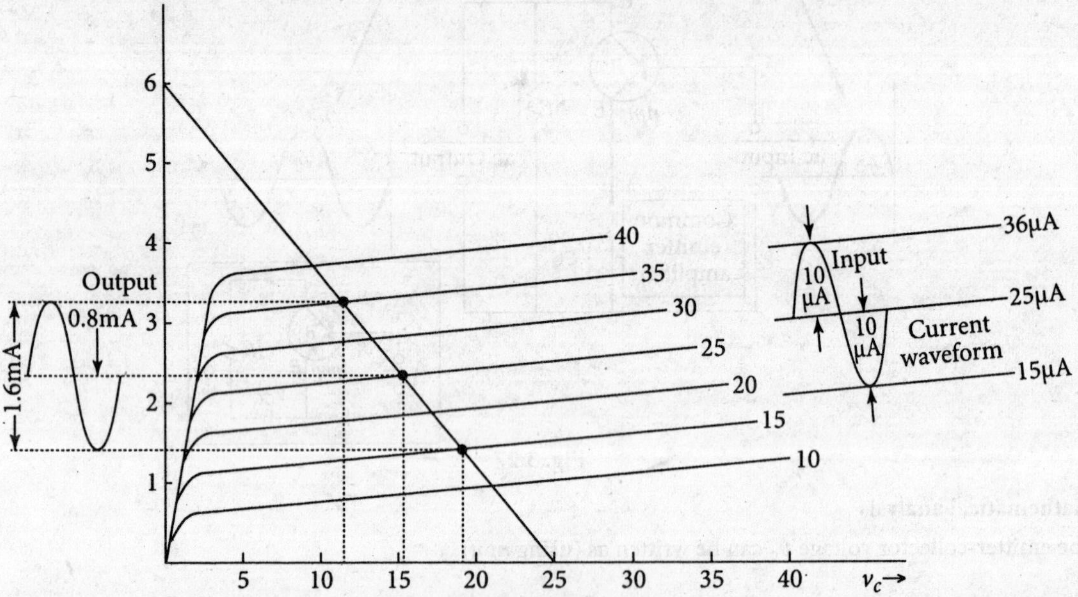

Fig. 3.28

So joining the points (6mA, 0) and (0, 30V), we get the dc load-line *AB*. This line cuts the 25μA base current characteristic at *Q*, where v_c = 17V. As the input current variation peak-to-peak is 20μA, the base current varies between 15μA below the *Q*-line and 35μA above the *Q*-line (which is 25μA). The input current waveform is shown to the right while the output current waveform is shown to the left. Thus, from the output characteristic it is evident that an input peak-to-peak variation of current equal to 20μA about 25μA base bias current results in peak-to-peak collector current, variation being equal to 1.6μA obtained by projecting on the collector current axis. Thus, current gain of $\frac{1.6}{0.02} = 80$ is obtained.

Voltage gain

20μA base current variation results in $20 \times 10^{-6} \times 500$, i.e., 0.01V, where the input resistance in 500Ω. Projecting on the v_e line we see that collector voltage variation is nearly 7V. So, the voltage gain = 700.

In terms of transfer dynamic characteristic of the transistor, equivalent to dynamic mutual characteristic of a triode valve, the transistor operates linearly if the operating point is suitably chosen and the base current variation is within the straight portion of the dynamic characteristic curve [Fig. 3.29].

Practical design of a common emitter amplifier

The common emitter amplifier is very widely used in practical design, because of easy provision of biasing. Such a practical circuit is shown in Fig. 3.30.

Here self-bias plus bias stabilisation is achieved through $R_b = 33$k and $R_e - C_e$ combination. Same voltage source supplies collector voltage and base bias. The input is injected through capacitor $C = 0.01\mu$F, the strength of the signal can be varied by the control potentiometer $R_h = 5$k. The amplified output appears across the load $R_l = 2.2$k. This amplifier output can be coupled to the next state of amplification.

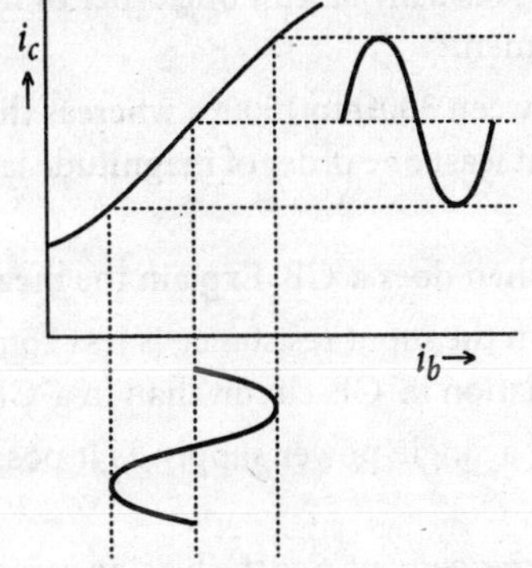

Fig. 3.29

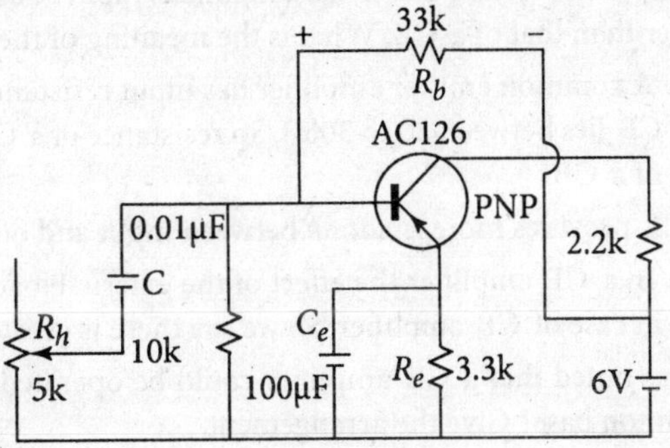

Fig. 3.30

Comparison between Common Base and Common Emitter Amplifier

Common Emitter	*Common Base*
1. $r_i \simeq 300$ to 1500Ω.	1. $r_i \simeq 30$ to 300Ω.
2. r_0 = Output resistance is not very high, varies from 20 to 100kΩ.	2. Output resistance is very high varies from 100kΩ to 1MΩ.
3. Thermal stability is poor.	3. Thermal stability is comparatively much better.
4. Because of high value of r_i, impedance matching becomes easy.	4. Because of low r_i and high r_0, the impedance matching becomes a problem.
5. Current gain is high ≃ 50.	5. Current gain is nearly equal to unity.
6. Because of high current gain, though resistance gain is low, the voltage gain becomes respectable and it increases with R_l.	6. Voltage gain increases with load resistance R_l.
7. Power gain is of the order of 40dB.	7. Power gain is of the order of 20–30dB.

3.16 Short Theory Questions with Answers

1. What is the main advantage of CE mode over CB mode?

Ans. The same battery may be used to supply collector bias as well as the base bias by means of a suitable resistance R_b. Since the input resistance is big, the impedance matching becomes easier in CE than in CB mode amplifier.

2. What is an approximate expression for the base bias current in a CE amplifier?

Ans. At the input circuit, we can write, $V_{BE} = i_b R_b + v_b$, where V_{BE} is the emitter-base bias battery voltage and v_b is the base bias.

$$\therefore\ I_b = \frac{V_{BE} - v_b}{R_b}.$$

For silicon transistor, $v_b = 0.6$V (approx.).

$$\therefore\ I_b = \frac{V_{BE} - 0.6}{R_b}.$$

3. It is commonly said that a CE amplifier's input resistance is usually at least one order of magnitude larger than that of a CB. What is the meaning of the statement?

   Ans. A common emitter amplifier has input resistance between 300Ω to 1500Ω, whereas that in case of a CB lies between 30 to 300Ω. So resistance of a CE is at least one order of magnitude larger than that of a CB.

4. A CE provides more *isolation* between input and output than does a CB. Explain the meaning.

   Ans. In a CE amplifier the effect of the external load R_l on the input resistance is less compared to that in case of CB amplifier. So we say there is greater isolation in CE circuit than in a CB.

5. It was noted that a CE amplifier could be operated from a single power supply. Is it possible by a common base? Give the arrangement.

   Ans. Yes, by voltage divider arrangement or by fixed bias arrangement. See the biasing arrangements discussed earlier.

3.17 Numerical Problems

1. Characteristics of a junction transistor are given below in the table.

Collector Voltages (volts)	*Collector Current I_c (mA)*		
(V_c)	$i_b = 0$	$i_b = 40\mu A$	$i_b = 80\mu A$
1.0	0.20	1.90	3.7
4.0	0.30	2.05	4.0
7.0	0.40	2.20	4.3

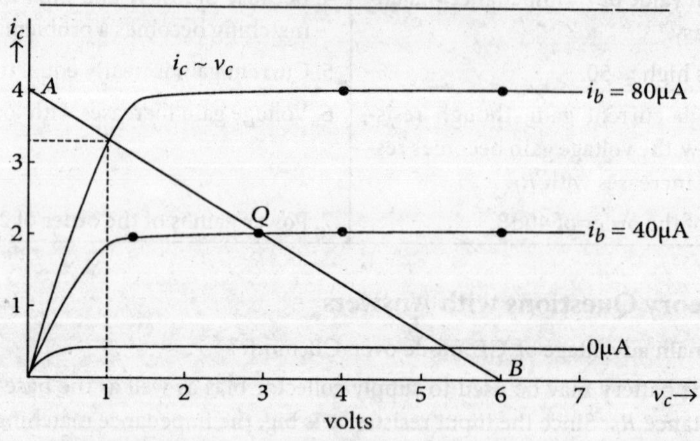

Fig. 3.31

The transistor is connected in CE mode. The load is $R_l = 1500\Omega$, a supply voltage of 6V and dc bias of $40\mu A$. Plot the characteristics and draw appropriate load-line. What will be the total voltage swing at the collector for an ac input signal current of $40\mu A$ peak (i.e., $80\mu A$ peak-to-peak) in the base?

Solution: The $i_c \sim v_c$ characteristic is plotted [Fig. 3.31]. The dc load-line is drawn w.r.t. the load-line equation for *npn* transistor (say) $v_c = V_{cc} - i_c R_l$, when $i_c = 0$, $v_c = V_{cc} = 6V$; when $v_c = 0$, $i_c = V_{cc}/R_l = 6/1500 = 4mA$.

Join the points (6, 0), (0, 4) and get the load-line. The Q-point is $v_c = 3V$, $i_c = 2mA$. The voltage swing in the collector is from 0.9V to 5.4V.

2. The behaviour of a transistor is represented [Fig. 3.32]. $V_{cc} = 10\text{V}$, $R_b = 10\text{k}\Omega$, $R_l = 1\text{k}\Omega$. The transistor has $\beta = 100$, $V_{BE} = 0.7\text{V}$ and saturation voltage $V_{CE,sat} = 0.1\text{V}$. Find the operating conditions when (a) $V_{BB} = 1.5\text{V}$ (b) 10.7V.

Solution: (a) $R_b I_B + V_{BE} = V_{BB}$.

$$\therefore\ I_B = \frac{V_{BB} - V_{BE}}{R_b} = \frac{1.5 - 0.7}{10 \times 10^3} = 0.08 \times 10^{-3} = 80\mu\text{A}.$$

$I_c = \beta I_B = 100 \times 80 = 8\text{mA}.$

$I_c \simeq I_E = 8\text{mA}.$

Again, in the output circuit, we get $I_c R_l + V_{CE} = V_{cc}$.

$\therefore\ V_{CE} = V_{cc} - I_c R_l = 10 - 8 \times 10^{-3} \times 10 = 2\text{V}.$

Thus, $V_{CE} > V_{CE,sat}$, hence, the transistor operates in the active region.

(b) $I_B = \frac{10.7 - 0.7}{10 \times 10^3} = 1\,\text{mA}.$

If $I_c = \beta I_B$ holds, then $I_c = 100\text{mA}$ and $V_{CE} = 10 - 100 \times 10^{-3} \times 10^3 = -90\text{V}$, which is an impossible fact. So, the transistor does not operate in the active region but it operates in the saturation region.

So $V_{CE} = V_{CE,sat} = 0.1\text{V}$.

$$I_c = \frac{9.9}{1 \times 10^3} = 9.9\text{mA}; \quad I_B = I_c + I_B = 10.9\text{mA}.$$

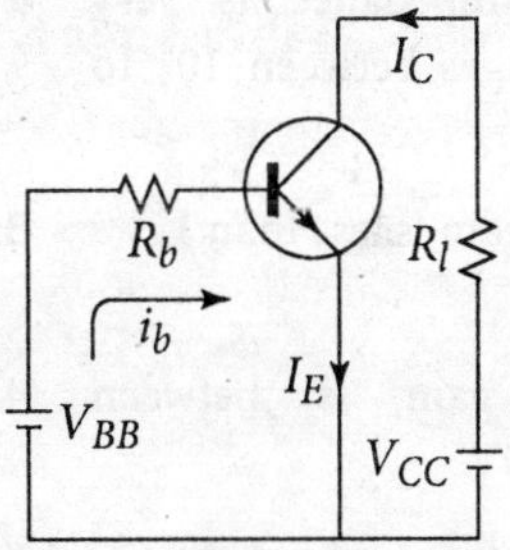

Fig. 3.32

3.18 Exercises

1. Why is common emitter amplifier commonly used? Give the usual amplifier circuit with a *pnp* transistor.
2. Discuss the merits and demerits of common base and common emitter amplifier.
3. Write a short note on a common emitter amplifier.
4. Deduce an expression for the voltage gain of a common emitter amplifier.

3.19 Common Collector Amplifier

In this type of amplifier the collector is common to both input and output circuits. The load R_l is in the emitter circuit similar to cathode follower in vacuum tube. In the circuit [Fig. 3.33] using an *npn* transistor, it shows that as the signal goes through its positive half, the forward bias of the input is increased. This increases the collector current, which flowing through R_l will make the emitter more positive when the signal is negative going, the input forward bias is reduced thus lowering the emitter and collector currents. This makes the emitter less positive. Thus, we see that input and output voltages are in phase.

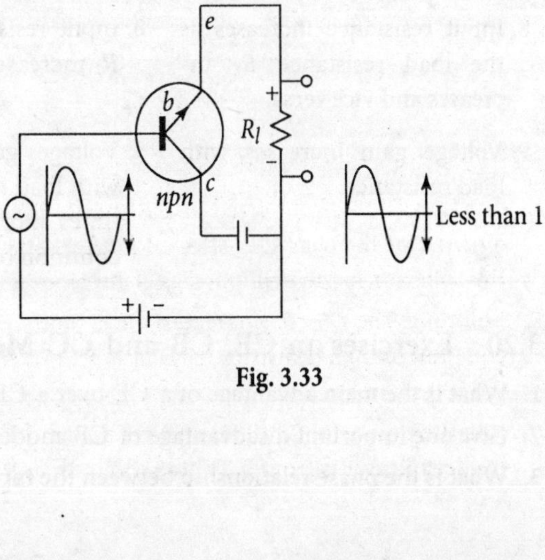

Fig. 3.33

The input resistance of this configuration is very high ranging from 100 to 500k, while the output resistance is low. The high value of the input resistance is due to the high resistance between base-collector circuit and low output resistance is due to low resistance in the emitter-collector circuit. So this circuit makes a very good isolation network. Though current gain is of the same order as that in case of common emitter amplifier, the voltage gain is usually less than unity, because of the degenerative effect of the load in the emitter circuit.

Comparative Study of Three Types of Transistor Amplifier

Common Base	*Common Emitter*	*Common Collector*
1. Input impedance is low, 30–150Ω	1. Input impedance is high, varies between 300–1500Ω	1. Input impedance is very high, varies from 100 to 500kΩ
2. Output impedance is very high, ranges between 10^5 to 10^6Ω.	2. Output impedance is low, ranges between 20 to 100k.	2. Output impedance is very low, ranges from 100 to 1000Ω.
3. Current gain is less than 1.	3. Current gain is of the order of 35.	3. It is also of the order of 35.
4. Voltage gain is between 300–1000V.	4. Voltage gain is less, between 250–300V.	4. Voltage gain is less than unity.
5. Power gain is in between 20–30dB.	5. Power gain is about 40dB.	5. Power gain is in between 15–30dB.
6. Thermal stability is good and hence, protective circuits are not required.	6. Thermal stability is poor, hence, protective circuits are needed.	6. None.
7. Interstage matching network becomes difficult, because of low input impedance and high output impedance.	7. Interstage matching network becomes easy.	7. Interstage matching also becomes much easy.
8. Input resistance increases as the load resistance R_l increases and vice versa.	8. Input resistance decreases as R_l increases.	8. Input resistance increases with R_l.
9. Voltage gain increases with load resistance.	9. Voltage gain also increases with load resistance, about 19 times more than that of the common base amplifier.	9. None.

3.20 Exercises on CE, CB and CC Modes

1. What is the main advantage of a CE over a CB as far as biasing is concerned?
2. Give one important disadvantage of CB mode.
3. What is the phase relationship between the input and output voltages in the CE amplifier.

4. What do you mean by self-bias? Draw a CE amplifier showing the necessary biasing arrangement.
5. Explain why the common emitter configuration is preferred for a transistor amplifier.
6. Discuss the relative merits and demerits of the three classes of amplifiers clearly indicating preference w.r.t. voltage gain, current gain and power gain.
7. Find the small signal equivalent circuit of a single stage grounded emitter transistor amplifier. Derive expressions for input impedance, output impedance, voltage gain and current gain.
8. Find the Q-point of the amplifier of the [Fig. 3.34] for (a) R_c = 1kΩ and (b) R_b = 10kΩ. Given V_{BE} = 0.7V, $\alpha = 0.99$, $I_{CBO} = 0$.

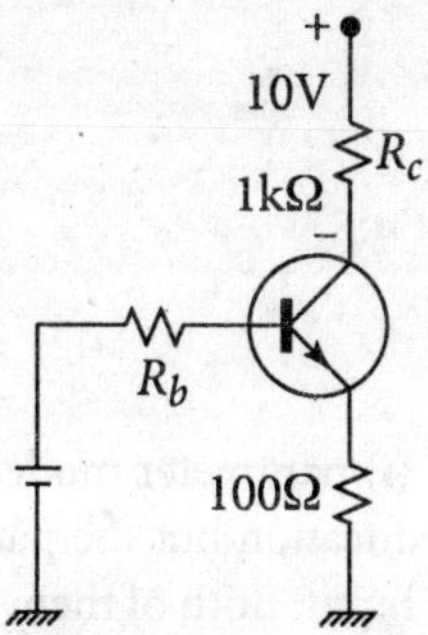

Fig. 3.34

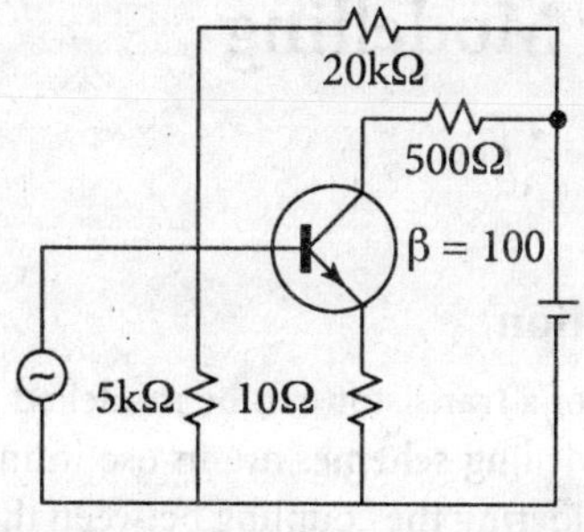

Fig. 3.35

9. The circuit [Fig. 3.35] is an amplifier in some mode. Name the transistor configuration and also discuss the biasing arrangement. Find: (a) the Q-point (b) another transistor of the same type is to be plugged into the same circuit. What is the minimum β that the new transistor may have if the quiescent collector current is not to change by more than 10 %?

Chapter 4

Transistor Modelling

4.1 Introduction

The ac operation of a transistor can be modelled in two ways: (i) parameter modelling (ii) r_e-modelling. Both of these modelling schemes are in use in industry and education but the parameter modelling has advantage of considering the coupling between the output and input. Both of them refer to small signal ac response of the transistor.

4.2 Essential Requirements of Parameter Modelling

1. The network must be a four-terminal network, two-input ports, two output ports.
2. The network must be linear, i.e., it must contain elements, which do not vary with voltage and current.

With regard to the above two requirements, a transistor operating in the active (linear) region satisfies condition (2). Condition (1) is also satisfied, because a transistor is used in the circuit taking one of the terminals common to both the input as well as the output thus providing us with two-input ports [Fig. 4.1] and two output ports.

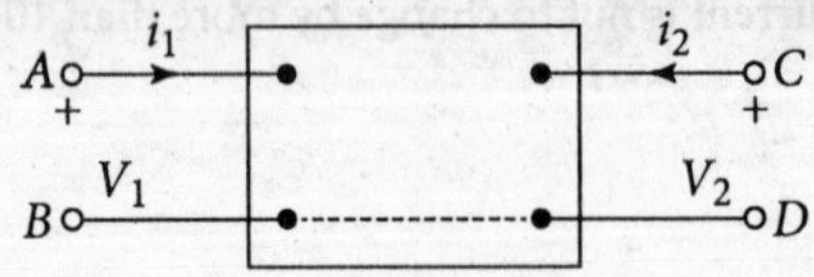

Fig. 4.1

The two-input ports are A and B; the output ones are C and D. The transistor converts the parameters related to the the input into parameters related to the output. Since it is said to operate in the linear region, the transformation is linear, hence the functioning of the transistor can be represented by suitable *transformation matrix* in the parameter modelling.

4.3 What are the Different Parametric Modelling?

Two quantities namely the voltage and current related to the input port characterise the input parameters, whereas the voltage and current related to the output port characterise the output parameters.

The currents in the input and output ports are assumed to flow into the network and V_1 and V_2 are the voltages at the input terminals and at the output terminals respectively. Depending on which two parameters out of four are to be transformed into the other two, the parameter modelling is designated differently.

Model 1: Impedance Parameter Modelling

In this modelling, V_1 and V_2 are dependent variables and i_1 and i_2 are independent variables and the transformation $(i_1, i_2) \rightarrow (V_1, V_2)$ is carried through the transformation matrix $[T]$, called *impedance parameter* matrix $[Z]$ represented by

$$\begin{bmatrix} V_1 \\ V_2 \end{bmatrix} = \begin{bmatrix} Z_{11} & Z_{12} \\ Z_{21} & Z_{22} \end{bmatrix} \begin{bmatrix} i_1 \\ i_2 \end{bmatrix} \Rightarrow \begin{array}{l} V_1 = Z_{11} i_1 + Z_{12} i_2 \\ V_2 = Z_{21} i_1 + Z_{22} i_2. \end{array} \tag{4.1}$$

To define Z_{11}, we set $i_2 = 0$, the output open-circuited, then

$$\begin{bmatrix} V_1 \\ V_2 \end{bmatrix} = \begin{bmatrix} Z_{11} & Z_{12} \\ Z_{21} & Z_{22} \end{bmatrix} \begin{bmatrix} i_1 \\ 0 \end{bmatrix} \Rightarrow \begin{array}{l} V_1 = Z_{11} i_1 \Rightarrow Z_{11} = \left.\dfrac{V_1}{i_1}\right|_{i_2=0} \\ V_2 = Z_{21} i_1. \end{array}$$

Also,

$$Z_{21} = V_2/i_1\Big|_{i_2=0}. \tag{4.2}$$

The quantities V_1 and i_1 both are of the input circuit, hence Z_{11} is said to be the input impedance; $Z_{21} = V_2/i_1\Big|_{i_2=0}$ is defined as *forward transfer impedance* (FTI).

If the input port is open-circuited, $i_1 = 0$, then

$$\begin{bmatrix} V_1 \\ V_2 \end{bmatrix} = \begin{bmatrix} Z_{11} & Z_{12} \\ Z_{21} & Z_{22} \end{bmatrix} \begin{bmatrix} 0 \\ i_2 \end{bmatrix} \Rightarrow \begin{array}{l} V_1 = Z_{12} i_2 \\ V_2 = Z_{22} i_2. \end{array}$$

$$\therefore\ Z_{12} = \left.\frac{V_1}{i_2}\right|_{i_1=0}, \tag{4.3}$$

it is called *reverse transfer impedance* (RTI) with input open.

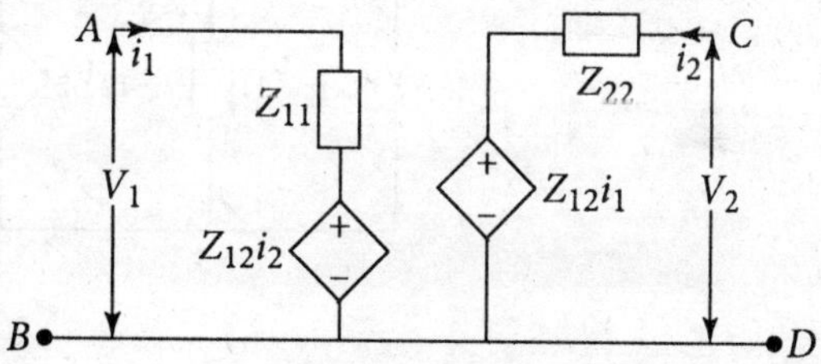

Fig. 4.2

$Z_{22} = \left.\frac{V_2}{i_2}\right|_{i_1=0}$ is called *open-circuit output impedance,* so the equivalent circuit of the two-port network in the impedance parameter model is shown in Fig. 4.2.

Model 2: Admittance Parameter Model: *Y*-Parameter Modelling

In this modelling, the two-port currents are taken as the dependent parameters, whereas the voltages are independent parameters. So, the modelling scheme is represented by $\begin{bmatrix} i_1 \\ i_2 \end{bmatrix} = \begin{bmatrix} Y_{11} & Y_{12} \\ Y_{21} & Y_{22} \end{bmatrix} \begin{bmatrix} V_1 \\ V_2 \end{bmatrix}$.

Input admittance Y_{11}. If the output port is short-circuited, $V_2 = 0$.

$$\therefore\ \begin{bmatrix} i_1 \\ i_2 \end{bmatrix} = \begin{bmatrix} Y_{11} & Y_{12} \\ Y_{21} & Y_{22} \end{bmatrix} \begin{bmatrix} V_1 \\ 0 \end{bmatrix} \Rightarrow \left.\begin{array}{l} i_1 = Y_{11} V_1 \\ i_2 = Y_{21} V_1 \end{array}\right| V_2 = 0.$$

$$\therefore\ Y_{11} = \left.\frac{i_1}{V_1}\right|_{V_2=0} \tag{4.4}$$

is called the *short-circuit input admittance.*

$$Y_{21} = \left.\frac{i_2}{V_1}\right|_{V_2=0} \tag{4.5}$$

is called the *short-circuit forward transfer admittance.*

If $V_1 = 0$, then we get

$$\begin{bmatrix} i_1 \\ i_2 \end{bmatrix} = \begin{bmatrix} Y_{11} & Y_{12} \\ Y_{21} & Y_{22} \end{bmatrix} \begin{bmatrix} 0 \\ V_2 \end{bmatrix} \Rightarrow \left. \begin{matrix} i_1 = Y_{12} V_2 \\ i_2 = Y_{22} V_2 \end{matrix} \right| V_1 = 0$$

$$\therefore\ Y_{12} = \left.\frac{i_1}{V_2}\right|_{V_1=0} \tag{4.6}$$

is called the *short-circuit reverse transfer admittance.*

$$Y_{22} = \left.\frac{i_2}{V_2}\right|_{V_1=0}$$

is called the *short-circuit output admittance.* The equivalent of the two-port network in this modelling is [Fig. 4.3].

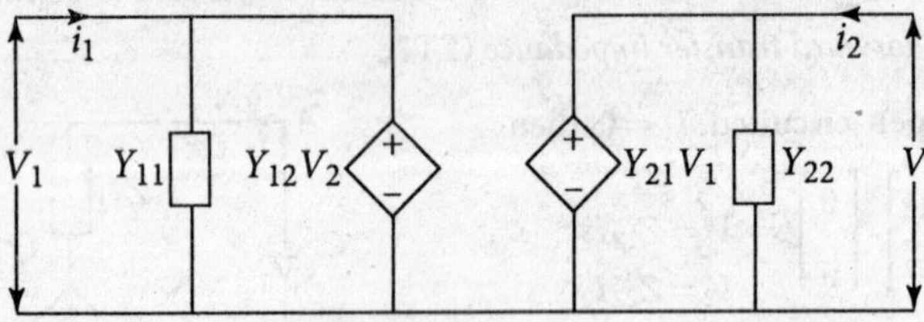

Fig. 4.3

Model 3: Hybrid (h) Parameter Model

In this modelling of two-port network, two dissimilar parameters such as V_1 and i_2 are taken as dependent variables, one related to the input port, the other related to the output port. The other dissimilar set is (i_1, V_2) as the independent parameters. This modelling scheme is represented by

$$\begin{bmatrix} V_1 \\ i_2 \end{bmatrix} = \begin{bmatrix} h_{11} & h_{12} \\ h_{21} & h_{22} \end{bmatrix} \begin{bmatrix} i_1 \\ V_2 \end{bmatrix}. \tag{4.7}$$

If $V_2 = 0$, then we get

$$\begin{bmatrix} V_1 \\ i_2 \end{bmatrix} = \begin{bmatrix} h_{11} & h_{12} \\ h_{21} & h_{22} \end{bmatrix} \begin{bmatrix} i_1 \\ 0 \end{bmatrix} \Rightarrow \left. \begin{matrix} V_1 = h_{11} i_1 \\ i_2 = h_{21} i_1 \end{matrix} \right| V_2 = 0. \tag{4.8}$$

$V_2 = 0$ implies that the output is short-circuited, then

$$h_{11} = \left.\frac{V_1}{i_1}\right|_{V_2=0} \tag{4.9}$$

called the *short-circuited input impedance.*

$$h_{21} = \left.\frac{i_2}{i_1}\right|_{V_2=0} \tag{4.10}$$

called the *forward current transfer ratio* (or *current gain*).

If $i_1 = 0$, which implies that the input is open-circuited, then we get

$$\begin{bmatrix} V_1 \\ i_2 \end{bmatrix} = \begin{bmatrix} h_{11} & h_{12} \\ h_{21} & h_{22} \end{bmatrix} \begin{bmatrix} 0 \\ V_2 \end{bmatrix} \Rightarrow \left. \begin{array}{l} V_1 = h_{12} V_2 \\ i_2 = h_{21} V_2 \end{array} \right| i_1 = 0. \tag{4.11}$$

$$\therefore h_{12} = \left. \frac{V_1}{V_2} \right|_{i_1=0} \tag{4.12}$$

is called the *open-circuit reverse voltage transformation ratio* (or *voltage gain*).

$$h_{22} = \left. \frac{i_2}{V_2} \right|_{i_1=0} \tag{4.13}$$

is called the *open-circuit output admittance* (Y_{22}). Thus, we see that all hybrids are not of similar physical quantities; h_{11} has the dimension of impedance, h_{22} has the dimension of admittance; h_{12} is the voltage gain and h_{21} is the current gain. So, they are mixtures of different quantities, hence this parameter model is called the *hybrid parameter.*

The equivalent two-port network in this model is shown in Fig. 4.4.

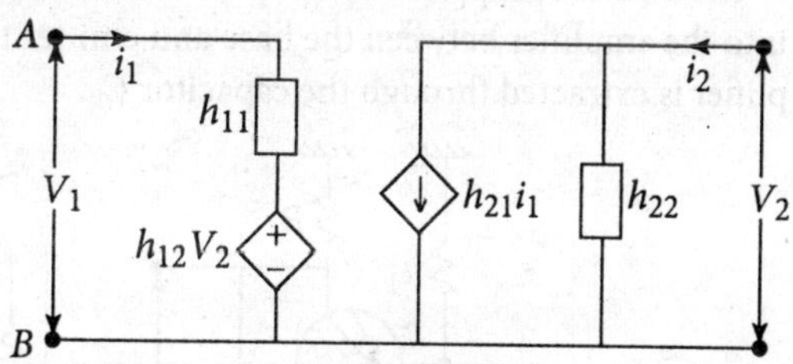

Fig. 4.4

4.4 Hybrid Parameter Model Details and its Application

The hybrid parameters for different configuration of the transistor are denoted differently.

In CE *mode.* In this case of using the transistor the different parameters are denoted by $h_{11} = h_{ie}$, the first suffix indicates the nature of the parameter, while the second denotes the terminal common to both the input and the output. The h_{ie} denotes the input impedance in CE mode

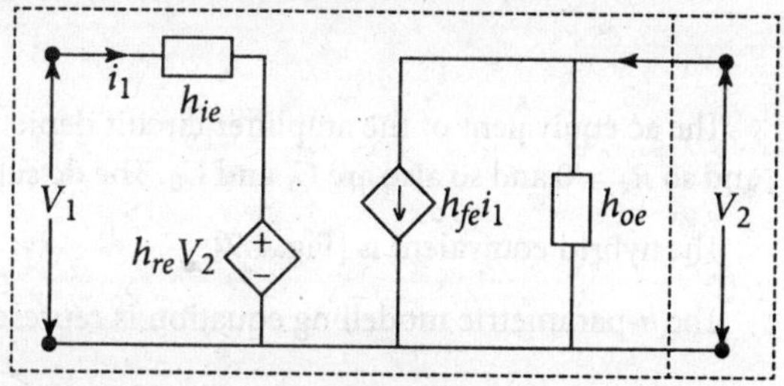

Fig. 4.5

$h_{11} = h_{ie}$, input impedance

$h_{12} = h_{re}$, reverse voltage gain

$h_{21} = h_{fe}$, forward current gain

$h_{22} = h_{oe}$, output admittance

— all in case of common emitter mode.

In CB *mode.* In this mode of use, the base is common to both input as well as output, so $h_{11} = h_{ib}$, $h_{12} = h_{rb}$, $h_{21} = h_{fb}$ and $h_{22} = h_{ob}$.

In CC *mode.* $h_{11} = h_{ic}$, $h_{12} = h_{re}$, $h_{21} = h_{fe}$, $h_{22} = h_{oc}$. So, the equivalent circuit [Fig. 4.4] is redrawn for CE mode [Fig. 4.5].

4.5 Analysis of the Transistor Amplifier in CE Mode

A circuit of an amplifier in CE mode is used in [Fig. 4.6], dc biasing being the conventional potential divider arrangement. The amplifier circuit is to be analysed w.r.t. finding the voltage gain, current gain, input impedance, output admittance, etc. using the hybrid parameter model.

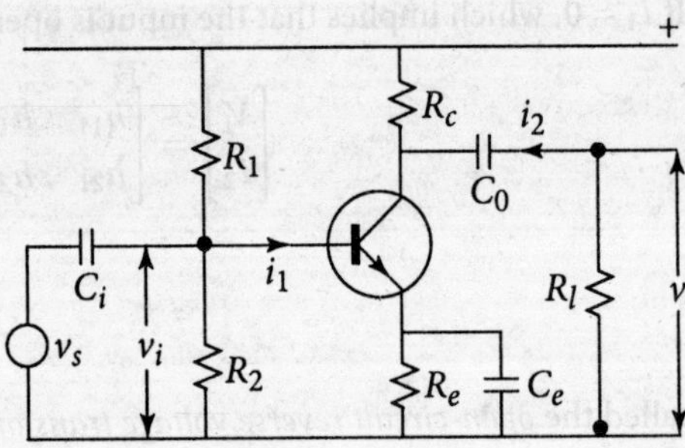

Fig. 4.6

The dc bias is supplied by the potential divider arrangement with emitter bias. The signal V_s is injected into the amplifier between the base and emitter through the input capacitor C_i and the output of the amplifier is extracted through the capacitor C_0.

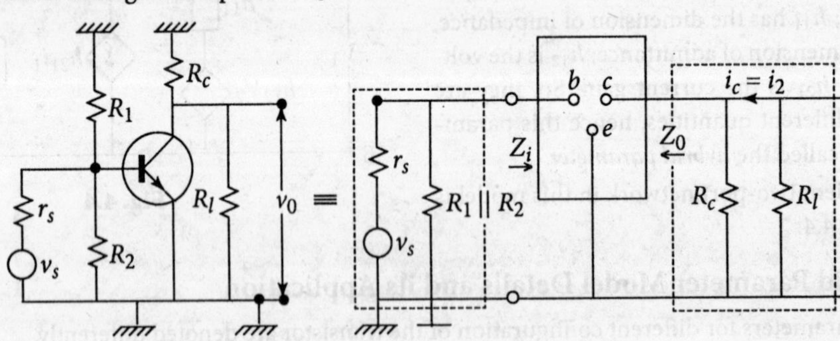

Fig. 4.7

The ac equivalent of the amplifier circuit depicted [Fig. 4.6] is redrawn using the fact that C_e is ac short and so $R_e = 0$ and so also are C_i and C_0. The dc supplies are replaced by zero potential equivalent [Fig. 4.7].

The hybrid equivalent is [Fig. 4.7].

The h-parametric modelling equation is represented by

$$\begin{bmatrix} V_1 \\ i_2 \end{bmatrix} = \begin{bmatrix} h_{ie} & h_{re} \\ h_{fe} & h_{oe} \end{bmatrix} \begin{bmatrix} i_1 \\ V_2 \end{bmatrix}, \tag{4.14}$$

where $V_1 = V_{be}$, $i_1 = i_b$, $i_2 = i_c = -i_L$, $V_2 = v_{ce} = i_L r_L$ [Fig. 4.8]

$$\Rightarrow \begin{bmatrix} v_{be} \\ i_c \end{bmatrix} = \begin{bmatrix} h_{ie} & h_{re} \\ h_{fe} & h_{oe} \end{bmatrix} \begin{bmatrix} i_b \\ v_{ce} \end{bmatrix}$$

$$\Rightarrow v_{be} = h_{ie} i_b + h_{re} v_{ce} = h_{ie} i_b - i_c r_L h_{re} \tag{4.15}$$

$$\text{and } i_c = h_{fe} i_b + h_{ae} v_{ce} = h_{fe} i_b - i_c r_L h_{0e}$$

$$\Rightarrow 0 = h_{fe} i_b - (1 + h_{oe} r_L)\, i_c. \tag{4.16}$$

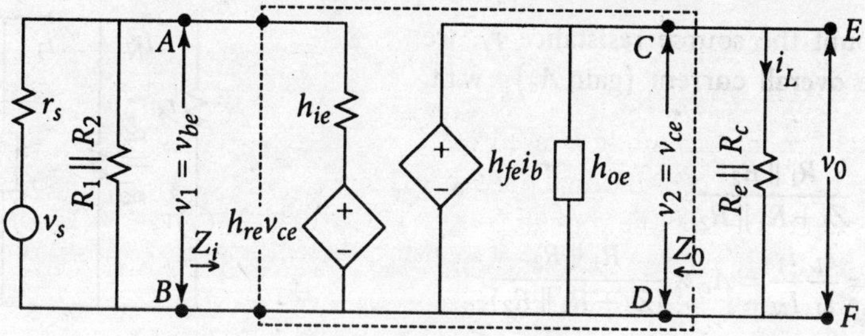

Fig. 4.8

$$\therefore\ i_b = \frac{\Delta_1}{\Delta} = \frac{\begin{vmatrix} v_{be} & -r_L h_{re} \\ 0 & -(h_{oe} r_L + 1) \end{vmatrix}}{\begin{vmatrix} h_{ie} & -r_L h_{re} \\ h_{fe} & -(1+h_{oe} r_L) \end{vmatrix}}; \qquad i_c = \frac{\begin{vmatrix} h_{ie} & v_{be} \\ h_{fe} & 0 \end{vmatrix}}{\begin{vmatrix} h_{ie} & -r_L h_{re} \\ h_{fe} & -(1+h_{oe} r_L) \end{vmatrix}},\ r_2 = R_c \| R_l.$$

$$\therefore\ \frac{i_c}{i_b} = \frac{\begin{vmatrix} h_{ie} & v_{be} \\ h_{fe} & 0 \end{vmatrix}}{\begin{vmatrix} v_{be} & -r_L h_{re} \\ 0 & -(1+h_{oe} r_L) \end{vmatrix}} = \frac{-v_{be} h_{fe}}{-v_{be}(1+h_{oe} r_L)} = \frac{f_{fe}}{1+h_{oe} r_L}. \tag{4.17}$$

Current Gain, A_c :

$$A_c = \frac{i_L}{i_1} = \frac{-i_c}{i_b} = -\frac{h_{fe}}{1+h_{oe} r_L}. \tag{4.18}$$

The negative sign indicates the direction of i_L w.r.t. i_c.

Input Impedance, Z_i :

$$Z_i = \frac{V_1}{i_1} = \frac{v_{be}}{i_b} = \frac{h_{ie} i_1 + h_{re} v_{ce}}{i_b} = \frac{h_{ie} i_b + h_{re} v_{ce}}{i_b} = h_{ie} + h_{re} \cdot \left(\frac{v_{ce}}{i_b}\right) = h_{ie} + h_{re} \times \left(-\frac{i_c r_L}{i_b}\right)$$

$$= h_{ie} - h_{re} r_L \left(\frac{i_c}{i_b}\right) = h_{ie} - \frac{h_{fe} h_{re} r_L}{(1+h_{oe} r_L)} = h_{ie} + A_c h_{re} r_L. \tag{4.19}$$

Voltage Gain, A_v :

$$A_v = \frac{v_{ce}}{v_{be}} = \frac{-i_c r_L}{v_{be}} = -\frac{(i_c/i_b)\, r_L}{(v_{be}/i_b)} = -\frac{h_{fe} r_L}{(1+h_{oe} r_L) Z_i}. \tag{4.20}$$

The negative sign indicates the phase inversion effect in the output w.r.t. the input.

Output Admittance

$$Y_0 = \frac{i_2}{v_2} = \frac{i_c}{v_{ce}} = \frac{h_{fe} i_b + h_{oe} v_{ce}}{v_{ce}} = h_{oe} + h_{fe} \frac{i_b}{v_{ce}} = h_{oe} + h_{fe} \frac{i_b}{-i_c r_L} = h_{oe} - \frac{i_b}{i_c} \cdot \frac{h_{fe}}{r_L} = h_{oe} + \frac{h_{fe}}{A_c r_L}. \tag{4.21}$$

Taking into account the source resistance r_s, we can calculate the overall current $(\text{gain } A_c)_g$ w.r.t. [Fig. 4.9].

$$i_1 = I_N \times \frac{R_1 \parallel R_2}{Z_i + R_1 \parallel R_2}.$$

$$(A_c)_g = \frac{i_L}{I_N} = \frac{i_L}{i_1}\frac{i_1}{I_N} = A_c \times \frac{R_1 \parallel R_2}{(Z_i + R_1 \parallel R_2)}.$$

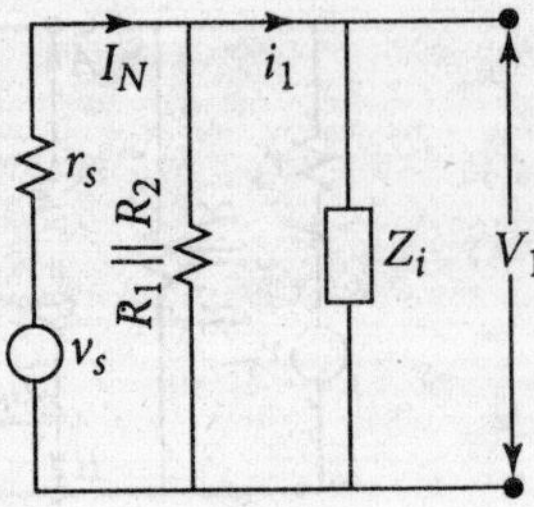

Fig. 4.9

Overall Voltage Gain $(A_v)_g$

$$v_{be} = v_1 = \frac{v_s z_e}{r_s + z_e}.$$

$$\therefore (A_v)_g = \frac{v_0}{v_s} = \frac{v_0}{v_{be}} \cdot \frac{v_{be}}{v_s} = A_v \cdot \frac{Z_e}{(r_s + z_e)}.$$

4.6 Numerical Problems

1. An amplifier circuit [Fig. 4.10]. Find: (a) ac emitter current (b) ac voltage at emitter, base and collector (c) voltage gain. Given $h_{ie} \simeq Z_{\text{in}} = 250\Omega$.

   v_i = signal voltage of peak value 5 mv.

   Solution: [Fig. 4.10].

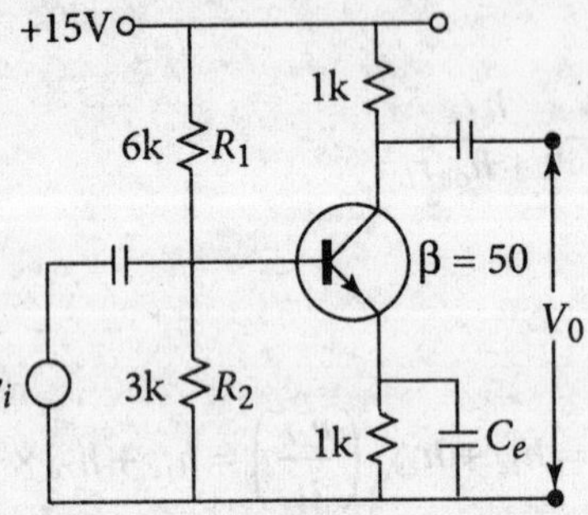

Fig. 4.10

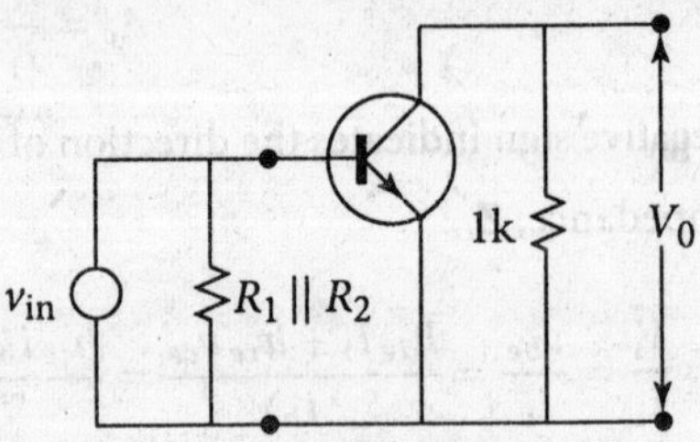

Fig. 4.11

The ac equivalent [Fig. 4.11].

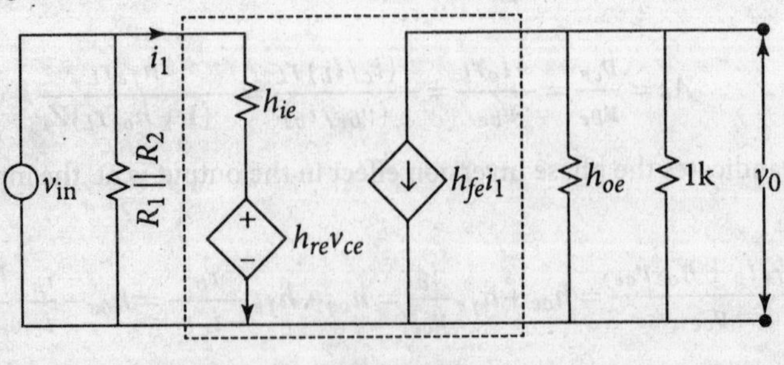

Fig. 4.12

Hybrid equivalent [Fig. 4.12].

(a) ac emitter voltage = 0, because C_e is ac short.

$$R_1 \parallel R_2 = \frac{6\times 3}{9} = 2\text{k}$$

$$i_b = \frac{v_{\text{in}}}{r_{\text{in}}} = \frac{5\times 10^{-3}}{250} \text{ (peak value)}$$

$$i_c = \beta\, i_b = \frac{50\times 5\times 10^{-3}}{250} = 1\text{mA (peak value)}$$

$$v_{be} = 5 \text{ mV (peak)}$$

$$v_{ce} = i_c R_L = 1\times 10^{-3}\times 1\times 10^{3} = 1\text{V}.$$

$$\text{voltage gain} = \frac{v_{ce}}{v_{be}} = \frac{1}{5\times 10^{-3}} = 0.2\times 10^{3} = 200.$$

2. A single stage amplifier is shown in Fig. 4.13. Draw its ac equivalent and calculate the voltage gain with or without R_L. Given $h_{fe} = 150, h_{ie} = 1.5\text{k} \simeq r_{\text{in}}, i_b = 10\mu\text{A}$.

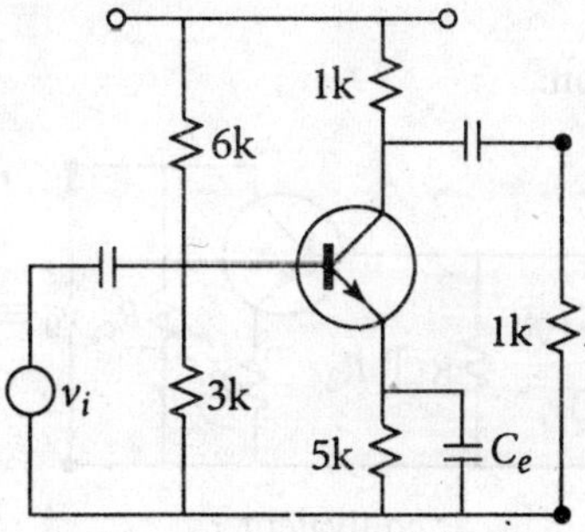

Fig. 4.13

Solution: Given:

$$i_b = 10\mu\text{A}$$

$$v_i = 10\times 10^{-6}\times r_{\text{in}} = 10\times 10^{-b}\times 1.5\times 10^{3} = 15\text{mV}$$

$$i_c = \beta\, i_b = 150\times 10\times 10^{-6} = 1.5\text{mA}$$

$$v_{ce} = i_c R_c = 1.5\times 10^{-3}\times 1\times 10^{3} = 1.5\text{V}$$

$$\therefore\ A_v = \frac{1.5\text{V}}{15\times 10^{-3}} = 100.$$

When load R_L is considered,

$$r_L = \frac{R_c R_L}{R_c + R_L} = \frac{1\times 1}{2} = 0.5\text{k}$$

$$\therefore\ v_{ce} = 1.5\times 10^{-3}\times 0{,}5\times 10^{3} = 0.75\text{V}$$

$$A_v = \frac{0.75}{15\times 10^{-3}} = 50.$$

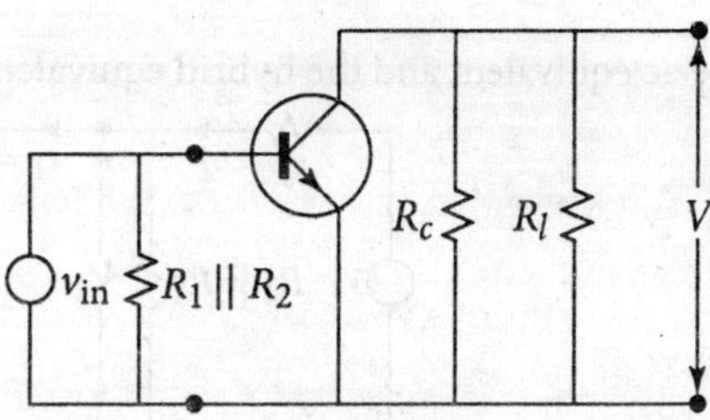

Fig. 4.14

3. For CE amplifier, the device has the following parameters: $h_{ie} = 1100\Omega$, $h_{re} = 2.5\times 10^{-4}$, $h_{fe} = 50$, $h_{oe} = 25\times 10^{-6}$A/V. Determine the maximum value of R_L for which R_i differs by not more than 10% of its value at $R_L = 0$.

Solution: $Z_i = R_i = h_{ie} - \frac{h_{fe}h_{re}r_L}{1+h_{oe}r_L}$.

When $r_L = 0$, $R_i = h_{ie} = 1100$, variation of R_i by 10% leads to $R_i = 0.9h_{ie}$.

$$\therefore 0.9h_{ie} = h_{ie} - \frac{h_{fe}h_{re}r_L}{1+h_{oe}r_L}$$

$$\Rightarrow 0.1h_{ie} = \frac{h_{fe}\times h_{re}\times r_L}{1+h_{oe}r_L}$$

$$\Rightarrow 0.1\times 1100 = \frac{50\times 2.5\times 10^{-4}\times r_L}{1+25\times 10^{-6}\times r_L}$$

$$\Rightarrow 110+25\times 110\times 10^{-6}\times r_L = 50\times 2.5\times 10^{-4} r_L$$

$$\Rightarrow (50\times 2.5-25\times 1.1)\times 10^{-4} r_L = 110$$

$$\Rightarrow r_L = \frac{110\times 10^4}{50\times 2.5-25\times 1.1} = 11.3\times 10^3\Omega = 11.3\text{k}\Omega.$$

4. A single stage amplifier with a *npn* transistor [Fig. 4.15]. The h-parameters are $h_{fe} = 50$, $h_{ie} = 1.1\text{k}\Omega$, $h_{re} = 2.5\times 10^{-4}$, $h_{oe} = 25\times 10^{-6}\mu\text{A/V}$. Find the values of current gain, voltage gain and input resistance.

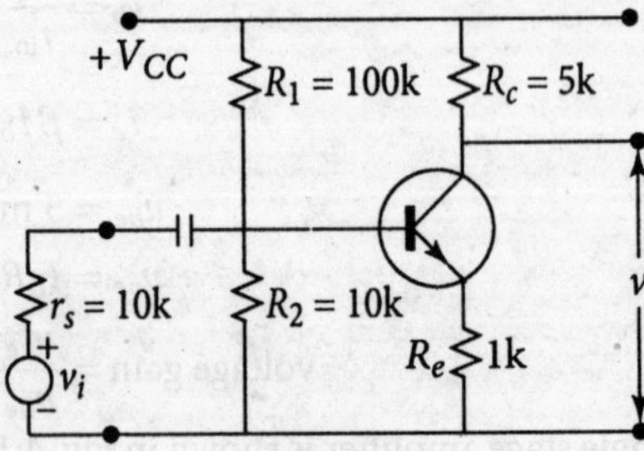

Fig. 4.15

Solution:

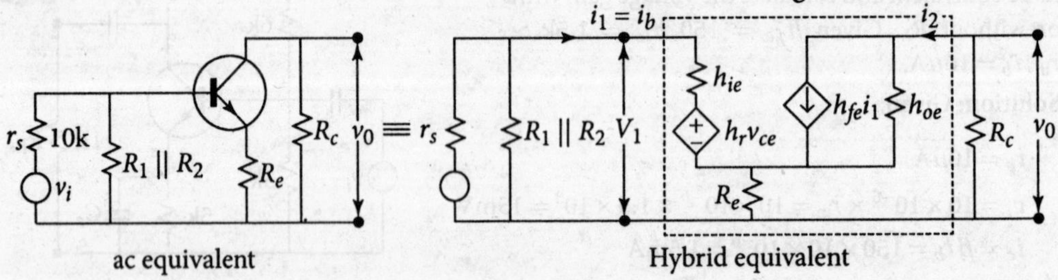

Fig. 4.16

The ac equivalent and the hybrid equivalent [Figs. 4.16 and 4.17].

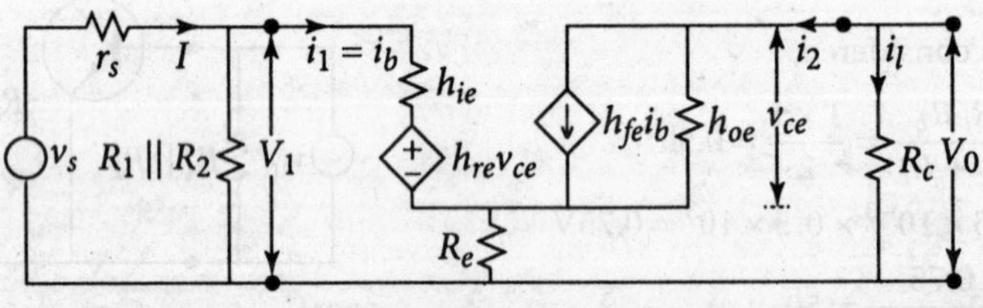

Fig. 4.17

With reference to Fig. 4.17, we would derive the input resistance, current gain, voltage gain, etc.

$$i_c = -i_l = h_{fe} i_b + h_{oe} v_{ce}$$

$$v_O = i_l R_c = v_{ce} + (i_b + i_c) R_e = v_{ce} + (i_b - i_l) R_e$$

$$\Rightarrow v_{ce} = i_l R_c - (i_b - i_l) R_e$$

$$\therefore -i_l = h_{fe} i_b + h_{oe}\left[i_l R_c - (i_b - i_l) R_c\right]$$

$$= h_{fe} i_b + h_{oe}(R_c + R_e)\, i_l - h_{oe} R_e i_b$$

$$\Rightarrow i_l\left[1 + h_{oe}(R_c + R_e)\right] = \left(h_{oe} R_e - h_{fe}\right) i_b$$

$$\Rightarrow \frac{i_l}{i_b} = \frac{h_{oe} R_e - h_{fe}}{1 + h_{oe}(R_c + R_e)} = -\frac{h_{fe} - h_{oe} R_e}{1 + h_{oe}(R_c + R_e)}$$

$$\Rightarrow A_c = -\frac{h_{fe} - h_{oe} R_e}{1 + h_{oe}(R_c + R_e)} = -\frac{50 - 25\times 10^{-6}\times 10^3}{1 + 25\times 10^{-6}\times 6\times 10^3} = -43.5. \tag{4.22}$$

Hybrid parameter analysis of transistor amplifiers

We first give the hybrid parameter model of the different amplifier configuration and then we will take up the details of the analysis. When applied to the specific configuration, the hybrid parameter will have two suffixes, the first suffix refers to the nature of the parameter, the second refers to the mode. Thus, h_{ie} (in place of h_{11}) refers to input resistance in the common emitter mode, h_{ib} for common base, h_{ie} for common collector mode respectively.

The Fig. 4.18 shows the transistor in different mode and the corresponding hybrid model.

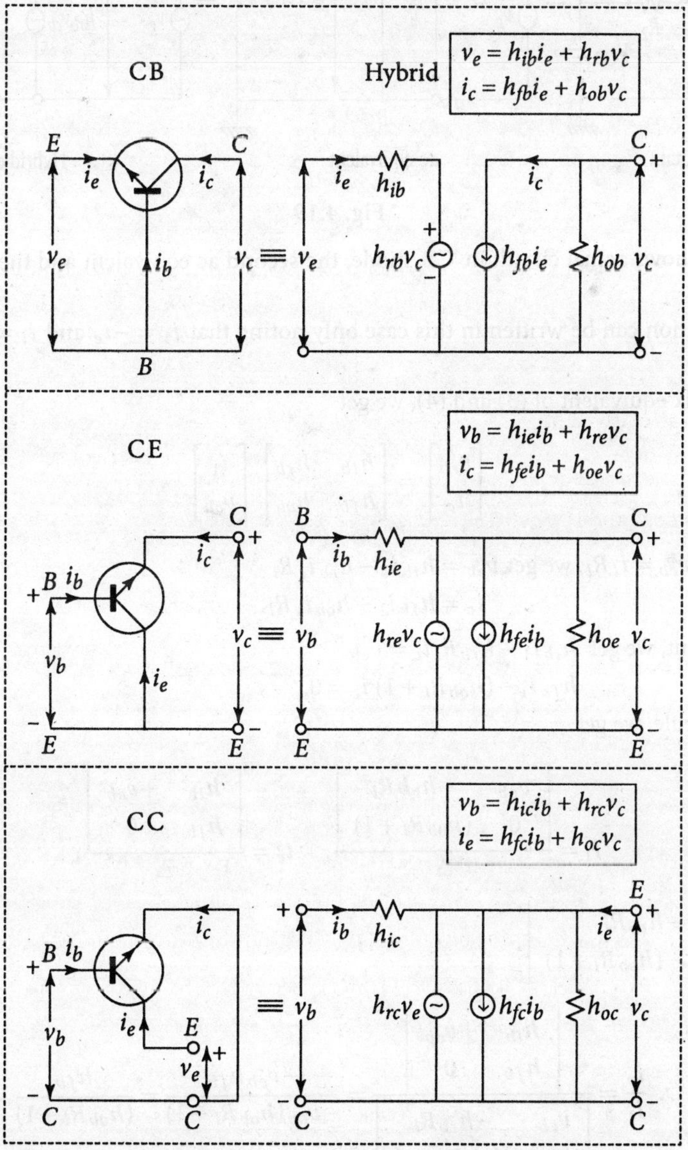

Fig. 4.18

Common base configuration

For complete and exact analysis we will take the signal source voltage v_g and resistance r_g [Fig. 4.19] for common base configuration.

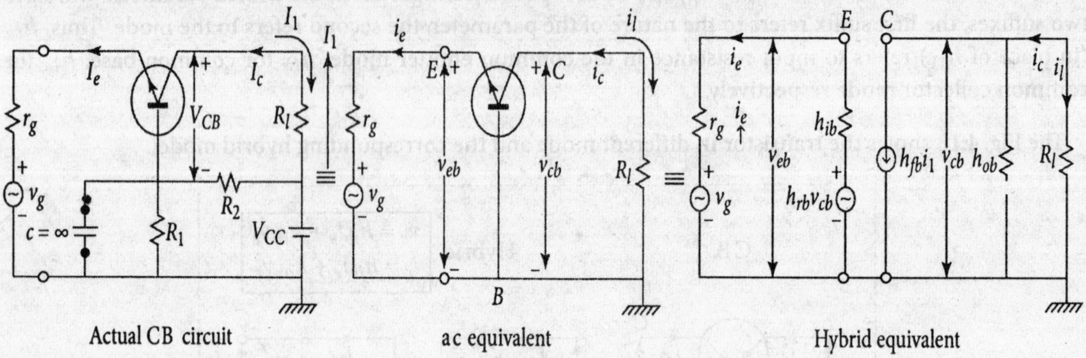

Fig. 4.19

The first figure shows actual circuit in CB mode, the second ac equivalent and the last diagram shows its hybrid model.

The hybrid equation can be written in this case only noting that $i_1 = -i_e$ and $i_l = -i_c$ [Fig. 4.19] CB mode.

Taking the matrix equivalent of (3) and (4), we get

$$\begin{bmatrix} v_{eb} \\ i_c \end{bmatrix} = \begin{bmatrix} h_{ib} & h_{rb} \\ h_{fb} & h_{ob} \end{bmatrix} \begin{bmatrix} i_1 \\ v_{cb} \end{bmatrix} \tag{4.23}$$

and using the fact $v_{cb} = i_l R_l$, we get $v_{eb} = h_{ib} i_1 - h_{rb} i_c R_l$

$$i_c = h_{fb} i_1 - h_{ob} i_c R_l.$$

On rearrangement, we get $h_{ib} i_1 - h_{rb} R_l i_c = v_{eb}$

$$h_{fb} i_1 - (h_{ob} R_l + 1)\, i_c = 0.$$

Using Cramer's rule, we get

$$i_1 = \frac{\begin{vmatrix} v_{eb} & -h_r b R_l \\ 0 & (h_{Ob} R_l + 1) \end{vmatrix}}{\Delta}, \quad i_c = \frac{\begin{vmatrix} h_{ib} & +v_{eb} \\ h_{fb} & 0 \end{vmatrix}}{\Delta},$$

where $\Delta = \begin{vmatrix} h_{ib} & -h_{rb} R_l \\ h_{fb} & -(h_{ob} R_l + 1) \end{vmatrix}$.

$$\therefore \frac{i_c}{i_1} = \frac{\begin{vmatrix} h_{ib} & +v_{eb} \\ h_{fb} & 0 \end{vmatrix}}{\begin{vmatrix} v_{eb} & -h_{rb} R_l \\ 0 & (h_{ob} R_l + 1) \end{vmatrix}} = \frac{-v_{eb} h_{fb}}{-v_{eb}(h_{ob} R_l + 1)} = \frac{h_{fb}}{(h_{ob} R_l + 1)}.$$

Current gain, A_c. Now, current gain is defined as

$$A_c = \frac{i_l}{i_1} = -\frac{i_c}{i_1} = -\frac{h_{fb}}{(h_{ob}R_l+1)}. \tag{4.24}$$

When the source resistance is taken into consideration, the current gain can be obtained w.r.t. Fig. 4.20.

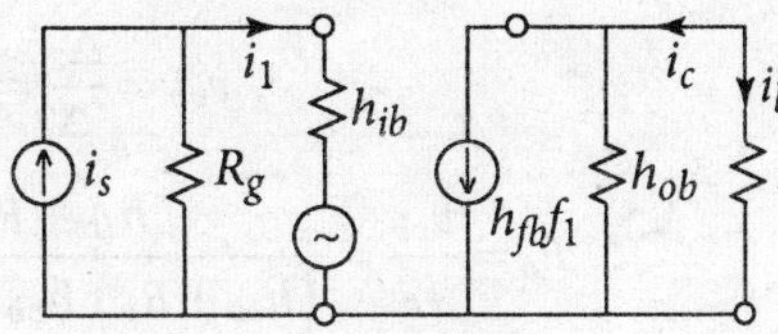

Fig. 4.20

Here, $i_1 = i_g \times \frac{R_g}{R_g+h_{ib}}$.

$$\therefore \text{ current gain } A_{cg} = \frac{i_c}{i_g} = \frac{i_c}{i_1} \times \frac{i_1}{i_g} = \frac{A_c R_g}{R_g + h_{ib}}. \tag{4.25}$$

Input resistance, R_i.

$$R_i = \frac{v_{eb}}{i_1} = h_{ib} - h_{rb}R_l \cdot \frac{h_{fb}}{(h_{ob}R_l+1)} = h_{ib} + A_c h_{rb} R_l. \tag{4.26}$$

Voltage gain.

$$A_v = \frac{v_c}{v_{eb}} = \frac{i_l R_l}{v_{eb}} = -\frac{i_c R_l}{v_{eb}} = -\frac{i_c}{i_1} \times \frac{i_1 R_l}{v_{eb}} = -\frac{h_{fb}}{(h_{ob}R_l+1)} \cdot \frac{R_l}{R_i} = \frac{A_c R_l}{R_i}. \tag{4.27}$$

Including the source resistance, the voltage gain becomes

$$A_{vg} = \frac{v_0}{v_g} = \frac{v_0}{v_{eb}} \times \frac{v_{eb}}{v_g} = A_v \frac{v_{eb}}{v_g}.$$

Now, $v_{eb} = v_g - R_g i_1$.

$$\therefore \left(1 + \frac{R_g i_1}{v_{eb}}\right) = \frac{v_g}{v_{eb}} \text{ or, } \left(1 + \frac{R_g}{R_i}\right) = \frac{v_g}{v_{eb}}.$$

$$\therefore A_{vg} = A_v \times \frac{1}{1 + R_g/R_i} = A_v \times \frac{R_i}{R_i + R_g}. \tag{4.28}$$

Output resistance. The output resistance R_o is defined w.r.t. the relation: $R_o = \left.\frac{v_{cb}}{i_c}\right|_{v_{eb}=0}$.

Now, we have

$$\begin{bmatrix} v_{eb} \\ i_c \end{bmatrix} = \begin{bmatrix} (h_{ib}+R_g) & h_{rb} \\ h_{fb} & h_{ob} \end{bmatrix} \begin{bmatrix} i_1 \\ v_{cb} \end{bmatrix}.$$

So, the output resistance will be obtained from the relation:

$$\begin{bmatrix} 0 \\ i_c \end{bmatrix} = \begin{bmatrix} (h_{ib}+R_g) & h_{rb} \\ h_{fb} & h_{ob} \end{bmatrix} \begin{bmatrix} i_1 \\ v_{cb} \end{bmatrix}.$$

Now

$$\Delta = \begin{vmatrix} h_{ib}+R_g & h_{rb} \\ h_{fb} & h_{ob} \end{vmatrix}, \quad \Delta_1 = \begin{vmatrix} 0 & h_{rb} \\ i_c & h_{ob} \end{vmatrix}, \quad \Delta_2 = \begin{vmatrix} h_{ib}+R_g & 0 \\ h_{fb} & i_c \end{vmatrix}.$$

$$\therefore\ v_{cb} = \frac{\Delta_2}{\Delta} = \frac{\left(h_{ib}+R_g\right) i_c}{\left(h_{ib}+R_g\right) h_{ob} - h_{rb}h_{fb}}.$$

$$\therefore\ R_o = \frac{v_{cb}}{i_c} = \frac{h_{ib}+R_g}{\left(h_{ib}+R_g\right) h_{ob} - h_{rb}h_{fb}} \text{ or, } G_0 = \frac{1}{R_o} = h_{ob} - \frac{h_{rb}h_{fb}}{\left(h_{ib}+R_g\right)}.$$

Note.

1. A few points to remember here are that in case of CE and CC modes, all the quantities found above will be valid, only the second suffix will denote the element common to the mode thus h_{ie} will denote the short-circuit input impedance for CE, and h_{ic} for CC mode, etc.
2. h_f is the forward current transfer gain of a transistor while A_c is the overall current gain. In the same way R_i, R_o, etc. are the overall input and output resistance respectively.
3. The negative sign with A_c in (4.18) is to be interpreted with reference to the direction of i_l and i_1. In case of CB mode, using *npn*, i_c flows into the base in the same direction as the current i_1 while i_c is in opposition to i_l.

Power gain. The power gain is the product of voltage gain and the current gain, i.e., $A_p = A_v \times A_i$, since A_v and A_i may be both negative or both positive, the power gain is always a positive quantity. Gain in decible is given by $10\log A_p$.

Example 4.1. *Consider* Fig. 4.31 *with a pnp transistor whose h-parameters are given as below:* $h_{ib} = 25\Omega$, $h_{rb} = 3\times10^{-4}$, $h_{fb} = -0.98$, $h_{ob} = 0.5\times10^{-6}$, $R_c = 600\Omega$, $v_g = 0.15$V, $R_l = 1.5$k.

Find: (a) A_c, (b) R_i, (c) A_v (d) R_o, (e) A_p, (f) A_{vs} *and* A_{cg}.

Solution: From (4.24), we have

$$A_c = -\frac{h_{fb}}{(h_{ob}R_l+1)} = \frac{0.98}{(0.5\times10^{-6}\times1.5\times10^{8}+1)} = 0.979 \simeq 0.98.$$

From (4.26), we have

$$R_i = h_{ib} + A_c h_{rb} R_l = 25 + 0.98\times3\times10^{-4}\times1.5\times10^{3} = 25.44\Omega.$$

From (4.27), we get

$$A_v = \frac{A_c R_l}{R_i} = \frac{0.98\times1.5\times10^{3}}{25.44} = 57.78.$$

$$G_o = \text{output conductance} = h_{ob} - \frac{h_{rb}h_{fb}}{\left(h_{ib}+R_g\right)} = 0.5\times10^{-6} - \frac{3\times10^{-4}\times(-0.98)}{(25+600)} = 0.97\times10^{-6}\mho.$$

$$\therefore\ R_o = 1.03\times10^{6}\Omega.$$

The power gain is given by $A_P = A_v \times A_c = 57.78\times0.98 = 56.62$.

$\therefore$ gain in dB $= 10\log A_P = 17.5$.

The voltage gain taking the source resistance

$$A_{vg} = A_v \times \frac{R_i}{\left(R_i+R_g\right)} = 57.78\times\frac{25.44}{(25.44+600)} = 2.35.$$

$$A_{cg} = A_c \times \frac{R_g}{R_g+h_{ib}} = 0.98\times\frac{600}{600+25} = 0.94.$$

4.7 Short Theory Questions with Answers

1. **What do you mean by a two-port network? What conditions must be met to represent a four-terminal complex circuit by a linear equivalent circuit? How these conditions are made to satisfy in a transistor amplifier circuit?**

   Ans. Any network having a single pair of accessible terminals is termed as a one-port network. So a network having two accessible terminals as the input and two for the output will be termed as a two-port network (It may be that one terminal becomes common to both input and output.)

   In order to model a linear equivalent circuit out of a two-port network, the conditions to be met are: (a) There must be a common connection between the input and output (b) The two-port network must contain linear elements.

   The first condition is satisfied, because depending on the mode, either the base or the emitter or the collector is made common. The second condition can be satisfied if the transistor is made to operate under small signal strength, so that the transistor operates in the active region, neither going into the saturation nor into the cut-off.

2. **Which parameters support the fact that a CB has a low input resistance and a CE has a high value of the input resistance?**

   Ans. The input resistance (overall) is given by $R_i = h_i + A_c h_r R_l$.

   Now, typical values of h_i for CB 29.4Ω and for CE $\simeq$ 4kΩ. So it is the parameter h_i which is the input resistance in CB $\ll$ resistance in CE.

3. **How can you show that $R_i < h_{ie}$ in CE and $R_i > h_{ib}$ in a CB?**

   Ans. $R_i = h_i + A_c h_r R_l$, A_c is negative in CE using *npn* whereas for the same transistor in CB, A_c is positive. ∴ $R_{ie} < h_{ie}$ but $R_{ib} > h_{ib}$.

4. **How can you show that in CE there is a phase inversion, while no phase change in CB?**

   Ans. $A_v = \frac{A_c R_l}{R_i}$. In CE, A_c is negative, which means, the output is phase inverse w.r.t. the input. But in case of CB, A_c is positive and hence, no phase change.

5. **What conditions would make an amplifier's overall voltage gain substantially the same as the transistor's voltage gain? Which amplifier connection provides this?**

   Ans. $A_{vg} = A_v \frac{R_i}{R_i + R_g} = \frac{A_v}{1 + R_g/R_i}$, so as $R_i \to \infty$, $A_{vg} \to A_v$. This is realised in CE mode, because $h_{ie} \simeq$ kΩ.

6. **Do the quantities A_c, A_v, R_i and R_o vary with R_l and R_g? Show by suitable graphs these variations in CB, CE and CC modes and from these graphs how can you conclude about the amplifier connection giving greatest isolation between the input and output?**

   Ans. Yes, all these quantities vary with R_l and R_g as evident from the theoretical expressions. These variations are represented in Fig. 4.21. Greater isolation means less variation of R_i with R_l and R_o with R_g. An inspection of the variation-curves shows that it is the CE mode that provides this isolation best.

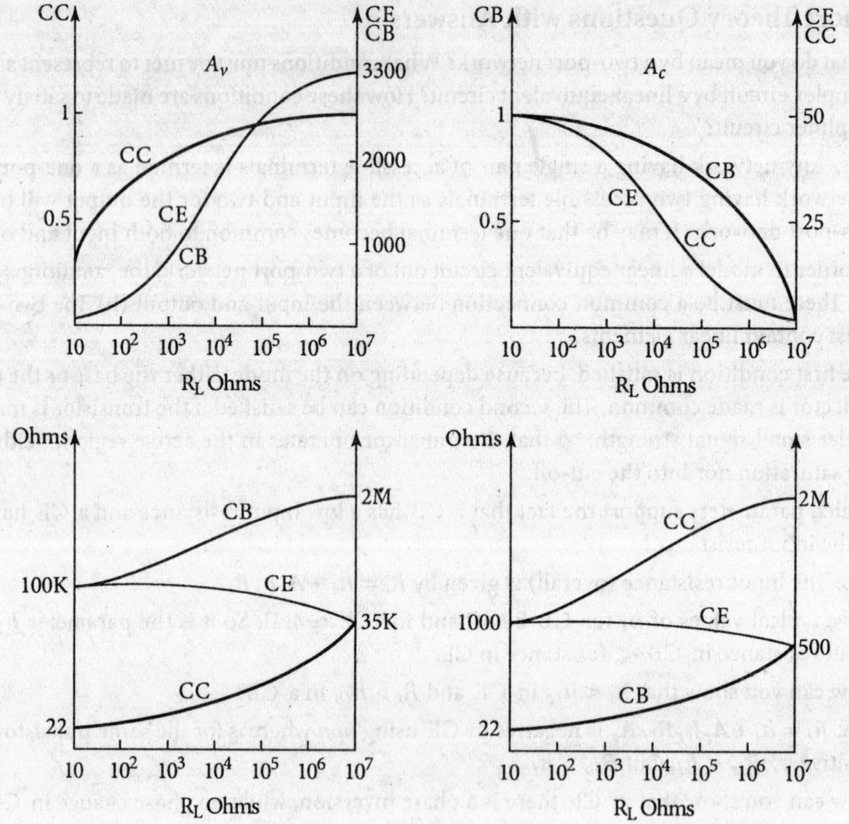

Fig. 4.21

4.8 Numerical Problems

1. A 2N3391 transistor is used in a CE amplifier at a quiescent collector current of 0.1mA. If the load resistance is 56kΩ, the source resistance $R_g = 600\Omega$ calculate A_c, R_i, A_p, R_o. Given $h_{ie} = 42\text{k}\Omega$, $h_{re} = 1.5\times10^{-4}$, $h_{oe} = 6\times10^{-6}\text{S}$ (S is the unit known as *Siemens*, SI unit of conductance or mho), $h_{fe} = 240$.

Solution: $A_c = -\dfrac{h_{fe}}{(h_{oe}R_l + 1)} = -\dfrac{240}{6\times10^{-6}\times56\times10^3+1} = -179.6$

$$R_i = h_{ie} + A_c h_{re} R_l = 42\times10^3 - 179.6\times1.5\times10^{-4}\times56\times10^3 = 40.5\text{k}\Omega$$

$$A_v = \frac{A_c R_l}{R_i} = -\frac{179.6\times56}{40.5} = -248$$

$$A_p = A_v\times A_c = 179.6\times248 = 44,540$$

$$G_o = \frac{1}{R_o} = h_{os} - \frac{h_{re}\times h_{fe}}{\left(h_{ie}+R_g\right)} = 6\times10^{-6} - \frac{1.5\times10^{-4}\times240}{(42\times10^3+600)} = 5.16\times10^{-6}.$$

$\therefore\ R_o = 193.7\text{k}\Omega.$

2. The hybrid parameters of a *pnp*-junction transistor used as an amplifier in the common emitter configuration are: $h_{ie} = 800\Omega$, $h_{fe} = 46$, $h_{oe} = 8\times10^{-5}/\Omega$, $h_{re} = 5.4\times10^{-4}$. If the load resistance $R_l = 5\text{k}\Omega$ and the effective source resistance 500Ω, calculate the voltage and current gains and the output resistance.

Solution: Current gain (transistor) $= A_c = \dfrac{-h_{fe}}{h_{oe}R_l + 1} = -\dfrac{46}{8\times10^{-5}\times5\times10^{3}+1} = -\dfrac{46}{1.4} = -33.$

Voltage gain, $A_v = \frac{A_c R_l}{R_i}$

$$R_i = h_{ie} + A_c h_{re} R_l = 800 - 33\times5.4\times10^{-4}\times5\times10^{3} = 710.9.$$

$$\therefore\ A_v = -\frac{33\times5\times10^{3}}{710.9} = -232.$$

Overall current gain,

$$A_{cg} = A_c \times \frac{R_g}{R_g + h_{ie}} = -33\times\frac{500}{500+800} = -12.7.$$

$$A_{vg} = A_v \times \frac{R_i}{R_i + R_g} = -232\times\frac{710.9}{710.9+500} = -136.2.$$

Output resistance is given by

$$\frac{1}{R_o} = h_{ve} - \frac{h_{re}h_{fe}}{\left(h_{ie}+R_g\right)} = 8\times10^{-5} - \frac{5.4\times10^{-4}\times46}{(800+500)} = 6.09\times10^{-5}.$$

$\therefore\ R_o = 16.4\text{k}\Omega.$

Taking load resistance R_l into account, the effective output resistance R_0' is given by

$$R_0' = \frac{R_o\times R_l}{R_o + R_l} = \frac{16.4\times5}{21.4} = 3.83\text{k}\Omega.$$

4.9 Exercises

1. Derive the expressions for A_c, A_v, R_i and R_o in terms of the hybrid parameters taking a common emitter configuration. How are these expressions modified if the source resistance and the load resistance are taken into consideration?
2. Are the A_c, A_v, R_i and R_o constant? Do they vary with source resistance and load resistance? Show graphically how do these quantities vary.
3. How do you relate the transistor current or voltage gain with the overall current or voltage gain of a given circuit?
4. Which amplifier configuration has a low input resistance and high input resistance? Which one serves better isolation network?
5. What do you mean by hybrid parameters? Give the schematic hybrid model of a common emitter amplifier. What is the circuit characteristic of this amplifier mode?

   A transistor used in CB mode has the following h-parameters: $h_{ib} = 28\Omega$, $h_{rb} = 5\times10^{-4}$, $h_{ob} = 0.34$ μmhos and $h_{fe} = 105$.

   Calculate A_c, R_i and A_v for a common collector amplifier, if $R_l = 1.2\text{k}\Omega$.

[*Hints.* Use the conversion table as given below:

Parameters	CE	CB	Parameters	CE	CB
h_{ie}	—	$\dfrac{h_{ib}}{1+h_{fb}}$	h_{fb}	$\dfrac{-h_{fc}}{1+h_{fe}}$	—
h_{re}	—	$\dfrac{h_{ib}h_{ob}}{1+h_{fb}}-h_{rb}$	h_{ob}	$\dfrac{h_{oe}}{1+h_{fe}}$	—
h_{fe}	—	$\dfrac{-h_{ib}}{1+h_{fb}}$	h_{ic}	h_{ie}	$\dfrac{h_{ib}}{1+h_{fb}}$
h_{oe}	—	$\dfrac{h_{ob}}{1+h_{fb}}$	h_{re}	$1-h_{re\simeq 1}$	1
h_{ib}	$\dfrac{h_{ie}}{1+h_{fe}}$	—	h_{fe}	$-(1+h_{fe})$	$-\dfrac{1}{1+h_{fb}}$
h_{rb}	$\dfrac{h_{ie}h_{oe}}{1+h_{fe}}-h_{re}$	—	h_{oc}	h_{oe}	$\dfrac{h_{ob}}{1+h_{fb}}$

Here, $h_{fe}=105, \therefore\ h_{fe}=-\left(1+h_{fe}\right)=-106$, $h_{fb}=-\dfrac{h_{fe}}{1+h_{fe}}=-\dfrac{105}{106}=-0.99$, etc.]

Chapter 5

Multistage Amplifier

5.1 Introduction

Amplifiers, which we discussed earlier may be divided into two types: (a) current amplifiers (b) voltage amplifiers. If an amplifier works as a voltage amplifier, it is necessary that the voltage of the signal source be not affected by the amplifier nor the output voltage of the amplifier be affected by the variation in the value of the load resistance. So it requires that the input resistance of the amplifier is high and output resistance low. In this respect a common collector amplifier having high input impedance is used as the voltage amplifier. On the other hand, the current amplifier requires that the current in the signal source be not affected by the amplifier, also the current output be independent of the load resistance. So for current amplifier the input impedance is to be low, while the output resistance is to be high. Amplifiers used in the CE mode or CB mode are thus current amplifiers.

When the output of a single stage amplifier is not enough, we cascade two or more stages, the interstage coupling may be made in a number of ways as we shall presently discuss.

5.2 RC Coupled Audio Frequency Amplifier

The most popular interstage coupling in transistor is done through resistance and capacitance (RC Coupling) or through a transformer (transformer-coupling). We have shown a typical coupled common emitter amplifier. [Fig. 5.1].

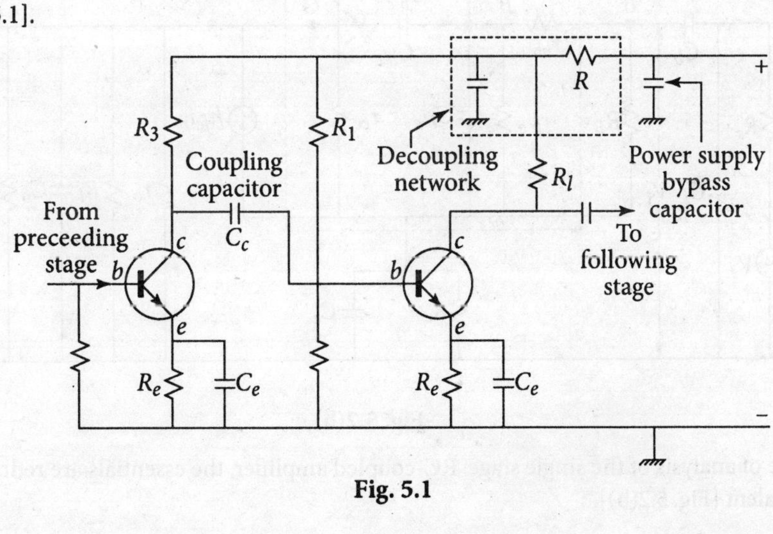

Fig. 5.1

In case of amplifications of weak signals the RC-coupling [Fig. 5.1] is preferred compared to transformer-coupling, which is more susceptible to hum pickup. Another advantage is that it takes small space also for connection. In the circuit shown, we have used the potential divider method for supplying the bias voltages and emitter stabilisation circuit. The coupling capacitor must be able to pass the lowest frequencies and the input and output resistances are relatively low, so the coupling capacitor must be fairly large (1 to 10μF). These capacitors are usually electrolytic types.

To prevent feedback, a decoupling network is often used. This is done by placing a resistor in series with the base resistor and then bypassing this resistor. Since the value of R is to be low (about 200Ω) so as not to lower the battery voltage to the preceding stages the bypassing condenser or decoupling condensers must be of large capacitance value (about 50–100μF). To prevent feedback, because of the voltage drop across the internal resistance of the battery, a bypass capacitor is placed across it.

Analytical explanation. Just as in case of RC-coupling version of the amplifier, the frequency response of a transistor stage is divided into three sections or bands: (i) lower frequency band (ii) middle frequency band (iii) higher or upper frequency band.

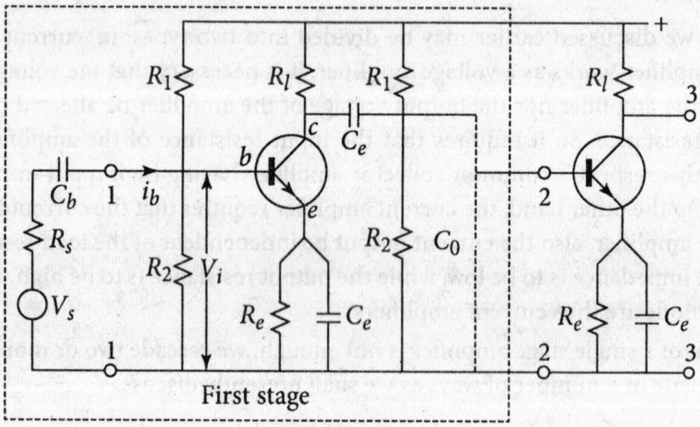

Fig. 5.2(a)

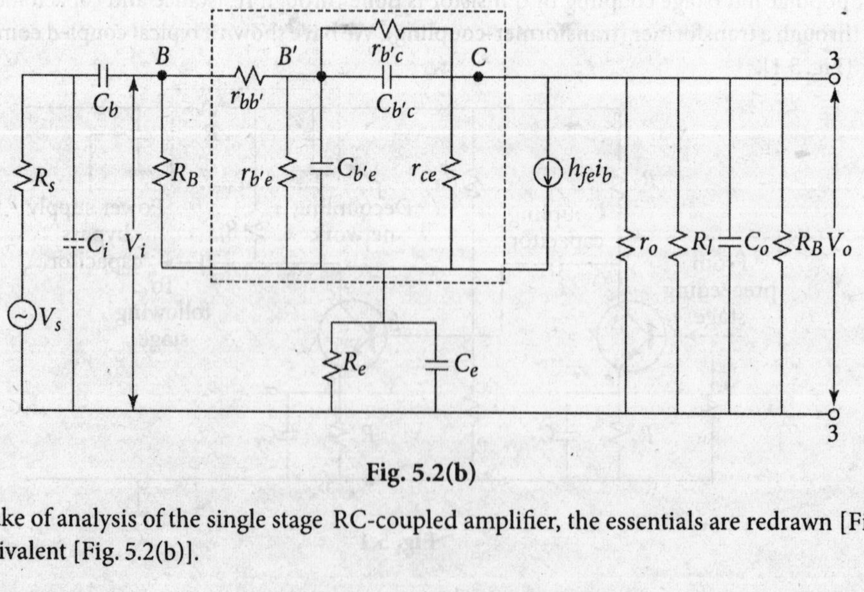

Fig. 5.2(b)

For the sake of analysis of the single stage RC-coupled amplifier, the essentials are redrawn [Fig. 5.2(a)] and the equivalent [Fig. 5.2(b)].

From the equivalent circuit represented [Fig. 5.2(b)], further reduction and simplification can be made depending on the effects of the different elements on the frequency response.

The elements within the dashed box refer to those inside the transistor and effect of these junction capacitances gains in prominence as the frequency approaches high frequency region.

Here, these elements refer to inside of the transistor:

- $r_{bb'}$ = base spreading resistance, which represents bulk of base resistance,
- $r_{b'e}$ = base-emitter resistance,
- $r_{b'c}$ = large value of resistance ($\simeq 4\,\text{M}\Omega$) between B', called *virtual base and collector*,
- r_{ce} = transistor output resistance,
- $C_{b'e}$ = base-emitter junction capacitance,
- $C_{b'c}$ = collector-base junction capacitance.

R_B = equivalent of parallel combination of R_1 and R_2 ($R_1 \parallel R_2$)

R_i = input resistance.

C_o, C_i, etc. are output and input capacitances of the device. Of these resistances, $r_{b'c}$ that shunts $C_{b'c}$ is very large, so that it can normally be omitted. The resistance $r_{ce} > R_l$, so we can usually omit r_{ce} when a load R_l is connected. At low or mid-frequencies the capacitative effects are negligible.

Mid-frequency (50 Hz–20 kHz) *response.* Here the reactance of the coupling condenser C_c is small and those of C_o and C_i are large so that C_e may be considered as short-circuit while C_o and C_i as open-circuited. So, the equivalent circuit from Fig. 5.2(b) in this case reduces to that shown in Fig. 5.2(c).

$$i_b = \frac{V_s}{R_s + R_i}, \text{ where } R_i = h_{ie}.$$

$$\therefore V_0 = -h_{fe} i_b \times R, \text{ where } R = (r_o \parallel R_l \parallel R_B) = -\frac{h_{fe} V_s R}{R_s + R_i}.$$

$\therefore$ voltage gain at mid-frequency

$$A_{mf} = -\frac{h_{fe} R}{R_s + R_i} = -\frac{h_{fe} R}{R_s + h_{ie}}.$$

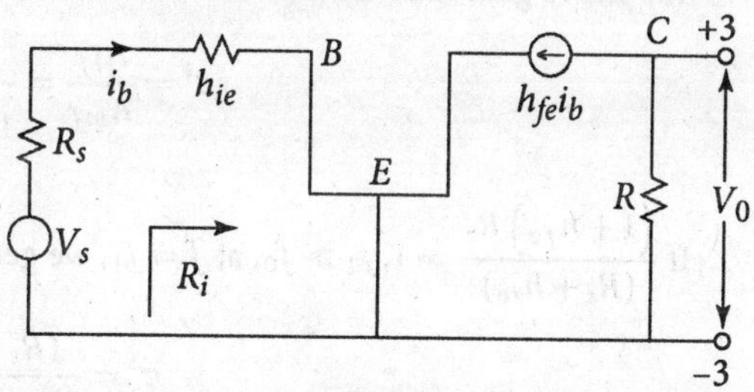

Fig. 5.2(c)

Low-frequency (<50 Hz) *response.* At low frequencies, the junction capacitance being small, their effects can be neglected, so the input resistance from base to emitter with the output shorted, i.e.,

$$h_{ie} = r_{bb'} + r_{b'e}. \tag{5.1}$$

•Assuming C_b is large so as to have negligible effect on the response and the value of C_c large enough that it may be considered as short-circuited. Then the equivalent circuit reduces to Fig. 5.2(d).

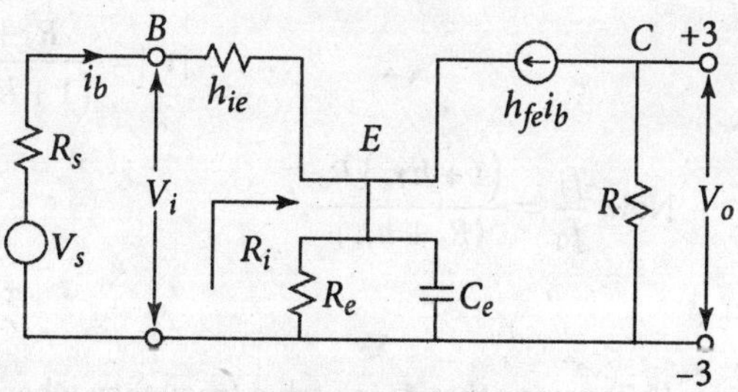

Fig. 5.2(d)

Here, $i_b = \frac{V_s}{R_s + R_i}$, where $R_i = h_{ie} + (1 + h_{fe}) Z_e$ and $Z_e = (R_e \parallel C_e)$.

$$\therefore \frac{1}{Z_e} = \frac{1}{R_e} + j\omega C_e \text{ or, } Z_e = \frac{R_e}{1 + j\omega C_e R_e}.$$

From [Fig. 5.2(d)], we get

$$V_0 = -h_{fe} i_b R, \text{ where } R = (r_o \parallel R_l \parallel R_B)$$
$$= -\frac{h_{fe} V_s R}{R_s + h_{ie} + (1+h_{fe}) Z_e}. \tag{5.2}$$

$$\therefore A_{lf} = \frac{V_o}{V_s} = -\frac{h_{fe} R}{R_s + h_{ie} + \frac{(1+h_{fe})R_e}{1+j\omega C_e R_e}}.$$

In terms of voltage gain at mid-frequency,

$$A_{lf}/A_{mf} = \frac{R_s + h_{ie}}{(R_s + h_{ie}) + \frac{(1+h_{fe})R_e}{1+j\omega C_e R_e}} = \frac{(R_s + h_{ie})(1+j\omega C_e R_e)}{(R_s + h_{ie})(1+j\omega C_e R_e) + (1+h_{fe}) R_e}$$
$$= \frac{(1+j\omega C_e R_e)}{(1+j\omega C_e R_e) + \frac{(1+h_{fe})R_e}{(R_s+h_{ie})}} = \frac{1}{1+\frac{(1+h_{fe})R_e}{R_s+h_{ie}}} \cdot \frac{1+j\omega C_e R_e}{\left[1+\frac{j\omega C_e R_e}{\frac{(1+h_{fe})R_e}{R_s+h_{ie}}}\right]}. \tag{5.3}$$

Let us now introduce two frequencies f_0 and f_1 defined respectively as

$$f_0 = \frac{1}{2\pi} \cdot \frac{1}{C_e R_e}, f_1 = \frac{1}{2\pi} \cdot \frac{1+\frac{(1+h_{fe})R_e}{R_s+h_{ie}}}{C_e R_e}. \tag{5.4}$$

The above gain ratio then can be written as

$$\tau^* = \frac{A_{lf}}{A_{mf}} = \frac{1}{1+\frac{(1+h_{fe})R_e}{(R_s+h_{ie})}} \cdot \frac{1+jf/f_0}{1+jf/f_1}. \tag{5.5}$$

If $\frac{(1+h_{fe})R_e}{(R_s+h_{ie})} \gg 1, f_1 \gg f_0$, at $f = f_1$, we get

$$\tau^* = \frac{(R_s+h_{ie})}{(1+h_{fe})R_e} \cdot \frac{1+jf_1/f_0}{1+j1}. \tag{5.6}$$

$$\therefore \tau = |\tau^*| = \frac{R_s+h_{ie}}{(1+h_{fe})R_e} \cdot \frac{f_1/f_0}{\sqrt{2}} \quad (\because f_1 \gg f_0).$$

Now, $\frac{f_1}{f_0} = \frac{(1+h_{fe})R_e}{(R_s+h_{ie})}$.

$$\therefore \tau = \frac{1}{\sqrt{2}} = \frac{A_{lf}}{A_{mf}}. \tag{5.7}$$

(5.7) means that f_1 is such a frequency where the voltage gain at low frequency is $\frac{1}{\sqrt{2}}$ times the voltage gain at mid-frequency, i.e.,

$$A_{lf} = \frac{1}{\sqrt{2}} A_{mf}.$$

In terms of decibel we can say that at f_1 the power gain has dropped 3dB from the gain at the mid-frequency. This frequency is, therefore, referred to as lower 3dB frequency.

Low frequency response when C_e is large but the effect of C_e is to be considered:

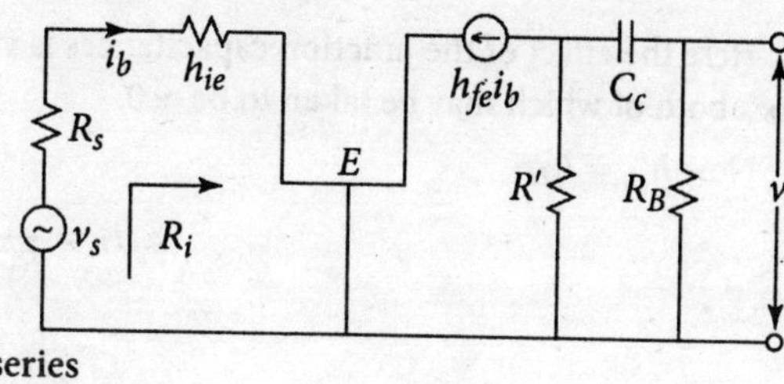

Fig. 5.2(e)

$$i_b = \frac{v_s}{R_s + R_i}, \text{ where } R_i = h_{ie}.$$

$$v_0 = -h_{fe} i_b \times \frac{Z_1}{Z_1 + Z_2} \times R_B, \text{where } Z_1 = R'(= r_o \parallel R_i),$$

$$Z_2 = R_B \text{ and } C_c \text{ in series}$$

$$= -\frac{h_{fe} v_s}{R_s + h_{ie}} \cdot \frac{R' R_B}{R' + R_B + \frac{1}{j\omega C_c}}.$$

$$\therefore A_{lf} = \frac{v_o}{v_s} = -\frac{h_{fe}}{R_s + h_{ie}} \cdot \frac{R' R_B}{(R' + R_B)} \cdot \frac{1}{\left[1 + \frac{1}{j\omega C_c} \cdot \frac{1}{(R' + R_B)}\right]}.$$

Now, $A_{mf} = -\dfrac{h_{fe} R}{R_s + h_{ie}}$.

$$\frac{1}{R} = \frac{1}{r_o} + \frac{1}{R_l} + \frac{1}{R_B}.$$

Again, $\frac{1}{R'} = \frac{1}{r_o} + \frac{1}{R_i}$.

$$\therefore \frac{1}{R} = \frac{1}{R'} + \frac{1}{R_B} \text{ or, } R = \frac{R' R_B}{R' + R_B}.$$

$$\therefore A_{lf} = A_{mf} \frac{1}{\left[1 + \frac{1}{j\omega C_c} \cdot \frac{1}{(R' + R_B)}\right]} = \frac{A_{mf}}{1 - j\frac{\omega_1}{\omega}},$$

where $\omega_1 = \dfrac{1}{C_c (R' + R_B)}$.

$$\therefore \frac{A_{lf}}{A_{mf}} = \frac{1}{\left[1 + \frac{f_1^2}{f^2}\right]^{1/2}}. \tag{5.8}$$

High-frequency (>20 kHz) *response.* This frequency range is such that the C_e can be taken as short-circuited and that the capacitances C_o, C_i only need be considered. Under this restriction the equivalent circuit from [Fig. 5.2(b)] can be drawn and it is shown in Fig. 5.2(f).

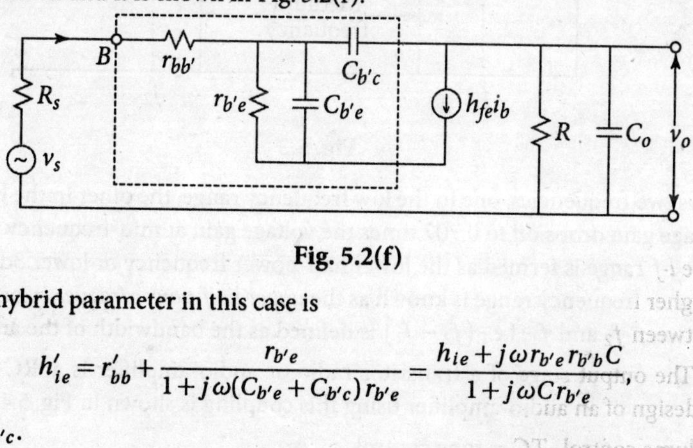

Fig. 5.2(f)

The short-circuit hybrid parameter in this case is

$$h'_{ie} = r'_{bb} + \frac{r_{b'e}}{1 + j\omega(C_{b'e} + C_{b'c}) r_{b'e}} = \frac{h_{ie} + j\omega r_{b'e} r_{b'b} C}{1 + j\omega C r_{b'e}}$$

where $C = C_{b'e} + C_{b'c}$.

Here the effect of the junction capacitances is very large compared to the spreading resistances $r_{b'b}$ and $r_{b'e}$ both of which may be taken to be $\simeq 0$.

Now $h'_{ie} = h_{ie}$,

$$\therefore\ i_b = \frac{v_s}{R_s + R_i} = \frac{v_s}{R_s + h_{ie}}. \tag{5.9}$$

$$v_o = -\frac{h_{fe} i_b Z_2}{Z_1 + Z_2}, \text{ here } Z_1 = R,\ Z_2 = \frac{1}{j\omega C_o}$$

$$= -\frac{h_{fe} R v_s}{(R_s + h_{ie})} \cdot \frac{1}{1 + \frac{Z_1}{Z_2}}.$$

$$\therefore\ \frac{A_{hf}}{A_{mf}} = \frac{1}{1 + j\omega C_o R} = \frac{1}{1 + j\frac{f}{f_2}}, \text{ where } 2\pi f_2 = \frac{1}{C_o R},$$

when $f = f_2$,

$$\left|\frac{A_{hf}}{A_{mf}}\right| = \frac{1}{\sqrt{2}}. \tag{5.10}$$

(5.10) means f_2 is such a frequency where voltage gain at high frequency range is $1/\sqrt{2}$ times the voltage gain at mid-range frequency. At this frequency the power gain has dropped 3dB from the gain at the mid-frequency. This frequency is, therefore, referred to as upper 3dB frequency.

Frequency response curve of a RC amplifier. The voltage gain vs. frequency in case of RC-coupled amplifier [Fig. 5.3], which clearly shows that over the mid-frequency range the response curve shows uniform gain and this gain decreases both in lower and higher frequency regions.

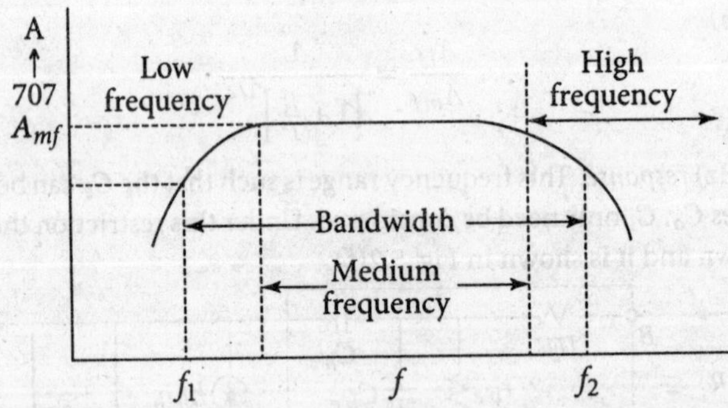

Fig. 5.3

The curve shows two frequencies, one in the low frequency range, the other in the high frequency range, such that the voltage gain drops off to 0.707 times the voltage gain at mid-frequency range, The frequency f_1 that falls on the *l-f* range is termed as the lower half-power frequency or lower 3dB frequency while f_2 that falls in the higher frequency range is known as the *upper half-power frequency* or *upper* 3dB *frequency.* The difference between f_2 and f_1, i.e., $(f_2 - f_1)$ is defined as the bandwidth of the amplifier.

Practical circuit. The output stage of a transistor radio or audio-amplifier is a RC-coupled amplifier. A typical practical design of an audio-amplifier using this coupling is shown in Fig. 5.4.

Here VC = volume control, TC = tone control.

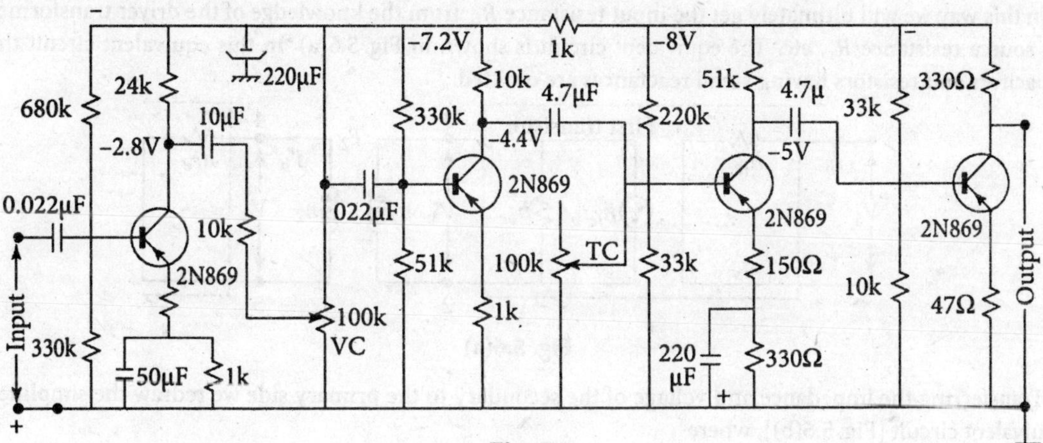

Fig. 5.4

5.3 Transformer Coupled Amplifier

Another method of interstage coupling, which is widely used is through a transformer and an amplifier so coupled is known as *transformer-coupled amplifier,* particularly useful in the low frequency audio-range. The collector impedance of a transistor being high compared to the base input impedance, the transformer-coupling provides excellent impedance matching and thus producing maximum power gain. Because of this, the power stage of a transistor receiver is in general transformer-coupled. Although a step-down transformer is used, it does not involve signal loss, because a transistor is a current device and so voltage step-down would entail current gain by the transformer.

A typical transformer-coupled amplifier is shown in Fig. 5.5. Here, R_1 and R_3 are bias resistors providing correct base biasing while R_2, C_2 and R_4, C_4 are emitter stabilising networks. Starting from the given load R_{l_1}, R_{l_2} can be obtained from the turn ratio of the transformer T_3. This is the load of the second transistor and for maximum power transfer, this R_{l_2} should be equal to the output resistance of the second transistor. R_{i_2} is the input resistance to the second transistor and will determine R_{l_1} from a knowledge of transformer T_2.

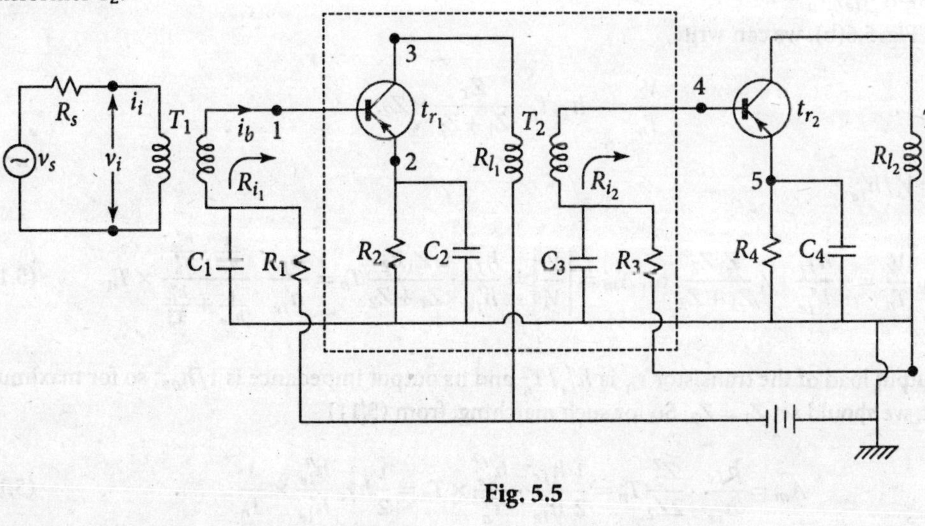

Fig. 5.5

In this way we will ultimately get the input resistance R_{i_1} from the knowledge of the driver transformer T_1, source resistance R_s, etc. The equivalent circuit is shown in Fig. 5.6(a). In this equivalent circuit the capacitors and resistors having small reactances are omitted.

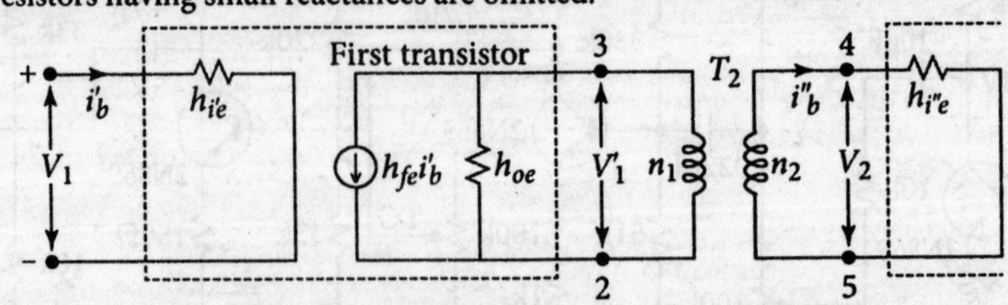

Fig. 5.6(a)

Transferring the impedance and voltage of the secondary to the primary side we redraw the simplified equivalent circuit [Fig. 5.6(b)], where

$$T_n = \left(\frac{n_2}{n_1}\right), \text{ called } \textit{turn ratio of the transformer.}$$

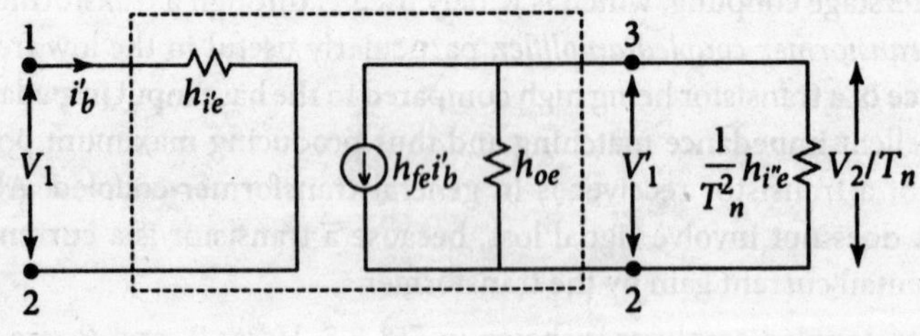

Fig. 5.6(b)

Mid-frequency gain. With reference to Fig. 5.6(b) we will develop an expression for the voltage gain at mid-frequency. Here the effect of inductance will not come into consideration, because its reactance is negligible in this frequency range. The voltage gain of the amplifier is given by $A_m = V_2/V_1$.

Let $1/h_{oe} = Z_1$, $h''_{ie}/T_n^2 = Z_2$.

Then, from Fig. 5.6(b), we can write,

$$\frac{V_2}{T_n} = -h_{fe}\, i'_b \cdot \frac{Z_1}{Z_1+Z_2} \times Z_2.$$

Again, $i'_b = V_1/h'_{ie}$.

$$\therefore \frac{V_2}{T_n} = -\frac{h_{fe}}{h'_{ie}} \cdot V_1 \frac{Z_1 Z_2}{Z_1+Z_2} \text{ or, } A_m = \left|\frac{V_2}{V_1}\right| = \frac{h_{fe}}{h'_{ie}} \cdot \frac{Z_1 Z_2}{Z_1+Z_2} T_n = \frac{h_{fe}}{h'_{ie}} \cdot \frac{\frac{1}{h_{oe}} \cdot \frac{h''_{ie}}{T_n^2}}{\frac{1}{h_{oe}} + \frac{h''_{ie}}{T_n^2}} \times T_n. \tag{5.11}$$

Effective output load of the transistor t_{r_1} is h''_{ie}/T_n^2 and its output impedance is $1/h_{oe}$, so for maximum power output, we should set $Z_1 = Z_2$. So for such matching, from (5.11)

$$A_m = \frac{h_{fe}}{h'_{ie}} \cdot \frac{Z_2^2}{2Z_2} T_n = \frac{1}{2}\frac{h_{fe}}{h'_{ie}} \cdot \frac{h''_{ie}}{T_n^2} \times T_n = \frac{1}{2} h_{fe} \frac{h''_{ie}}{h'_{ie}} \times \frac{1}{T_n}. \tag{5.12}$$

If identical transistors are used so that $h''_{ie} = h'_{ie}$, (5.12) gives us $A_m = \frac{1}{2} h_{fe}/T_n$. Suppose, in some case $h_{fe} = 50\Omega$, $h''_{ie} = h'_{ie} = 1\text{k}\Omega$, $h_{oe} = 25 \times 10^{-6}$ mho.

For maximum power transfer, $\frac{1}{h_{ce}} = \frac{h''_{ie}}{T_n^2}$.

$$\therefore \frac{1}{25 \times 10^{-6}} = \frac{1 \times 10^3}{T_n^2} \text{ or, } T_n^2 = 25 \times 10^{-6} \times 1 \times 10^3.$$

$$\therefore T_n = \sqrt{25 \times 10^{-3}} \text{ or, } \frac{n_1}{n_2} = \frac{1}{T_n} = 6.3.$$

So a step-down transformer with primary : secondary turn ratio 6.3 : 1 gives maximum power output.

Low frequency gain. At low frequency the reactance of the inductance cannot be neglected. So, the equivalent circuit in this case will be as shown in Fig. 5.6(c).

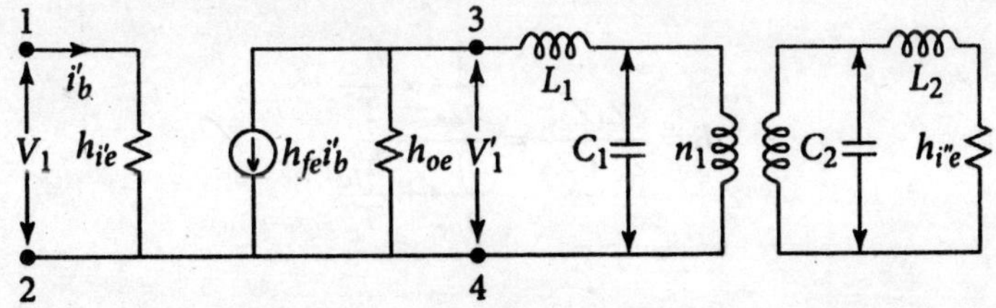

Fig. 5.6(c)

$$\frac{1}{h_{oe}} = Z_1, Z_2 = \frac{h''_{ie}}{T_n^2}, Z_3 = \omega L_p.$$

$i_1 Z_1 = i_3 Z_3 = i_2 Z_2$ and $i_1 + i_2 + i_3 = -h_{fe} i'_b$.

$$\therefore i_1 = \frac{i_2 Z_2}{Z_1}, i_3 = i_2 \frac{Z_2}{Z_3}.$$

$$\therefore i_2 \left(\frac{Z_2}{Z_1} + 1 + \frac{Z_2}{Z_3} \right) = -h_{fe} i'_b \text{ or, } i_2 = -\frac{h_{fe} i'_b \cdot Z_1 Z_3}{Z_2 Z_3 + Z_1 Z_3 + Z_1 Z_2}.$$

$$\therefore \frac{V_2}{T_n} = i_2 Z_2 = -\frac{h_{fe} i'_b Z_1 Z_2 Z_3}{(Z_2 Z_3 + Z_1 Z_3 + Z_1 Z_2)}.$$

Now, $i'_b = V_1 / h'_{ie}$.

$$\therefore \frac{V_2}{T_n} = -V_1 \times \frac{h_{fe}}{h'_{ie}} \cdot \frac{Z_1 Z_2}{(Z_1 Z_2 + Z_2 Z_3 + Z_3 Z_1)}$$

$$\text{or } \frac{V_2}{V_1} = -\frac{h_{fe}}{h'_{ie}} \cdot \frac{Z_1 Z_2 Z_3}{(Z_1 Z_2 + Z_2 Z_3 + Z_3 Z_1)} T_n = -\frac{h_{fe}}{h'_{ie}} \cdot \frac{Z_1 Z_2}{(Z_1 + Z_2)} T_n \cdot \frac{Z_3}{\left(\frac{Z_1 Z_2}{Z_1 + Z_2} + Z_3 \right)}.$$

$$\therefore |A_{lf}| = A_{mf} \cdot \frac{Z_3}{\left(\frac{Z_1 Z_2}{Z_1 + Z_2} + Z_3 \right)}. \qquad (5.13)$$

$\frac{Z_1 Z_2}{Z_1 + Z_2}$ represents the equivalent impedance of $\frac{1}{h_{oe}} \Big\| \frac{h''_{ie}}{T_n^2}$, which may be set equal to, say, R.

Then

$$A_{lf} = A_{mf}\frac{Z_3}{R+Z_3} = \frac{A_{mf}}{1+R/j\omega L_P}.$$

$$\therefore\ |A_{lf}| = \frac{|A_{mf}|}{\sqrt{1+R^2/\omega^2 L_P}}. \tag{5.14}$$

Now, when $R=\omega L_P$, we get $A_{lf}=\frac{1}{\sqrt{2}}A_{mf}$ and this frequency where low frequency gain is $0.707A_{mf}$, is known as *lower half-power frequency* or *lower* 3dB *frequency*. So, the lower 3dB frequency is given by

$$2\pi f_1 L_P = R = \frac{Z_1 Z_2}{Z_1+Z_2} \text{ or, } f_1 = \frac{R}{2\pi L_P}. \tag{5.15}$$

If $h_{oe}=25\times10^{-6}$ mho, $h'_{ie}=h''_{ie}=1\text{k}\Omega$, then

$$R = \frac{Z_1 Z_2}{Z_1+Z_2} = \frac{\dfrac{1}{h_{oe}}\times\dfrac{h''_{ie}}{T_n^2}}{\dfrac{1}{h_{oe}}+\dfrac{h''_{ie}}{T_n^2}}.$$

Here

$$\frac{1}{h_{oe}} = \frac{10^6}{25} = 40\times10^3; \qquad \frac{h''_{ie}}{T_n^2} = \frac{1\times10^3}{25\times10^{-3}} = \frac{10^6}{25} = 40\times10^3.$$

$\therefore\ R = 20\text{k}\Omega.$

So to obtain lower 3dB frequency at 200 cps, the primary inductance L_P is given by

$$L_P = \frac{R}{2\pi f_1} = \frac{20\times10^3}{2\pi\times200} = 15.9\text{H}.$$

High frequency gain. In the high frequency region the shunting effect of the primary inductance L_P is negligible but the leakage inductances of the primary and secondary and the distributed capacitances between the turns of the primary and secondary coils need to be considered. The equivalent circuit in this range is as shown in Fig. 5.6(d).

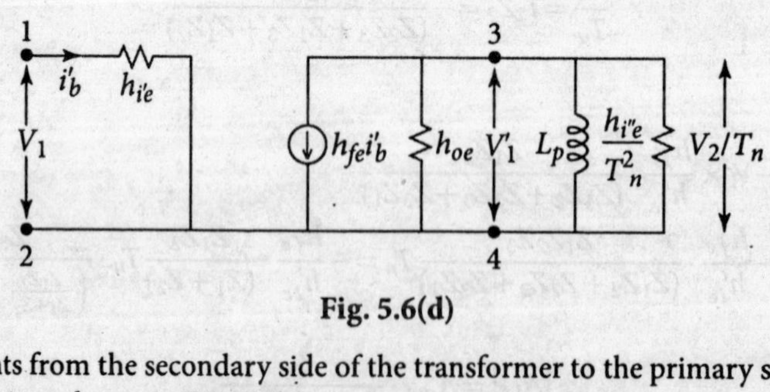

Fig. 5.6(d)

Transferring elements from the secondary side of the transformer to the primary side, we have the simplified equivalent circuit as shown in Fig. 5.6(e).

Let us put

$$\frac{1}{h_{oe}} = Z_1,\ \frac{h''_{ie}}{T_n^2} = Z_2,\ j\omega L'_1 = Z_3,\ \frac{1}{j\omega C'_1} = Z_4.$$

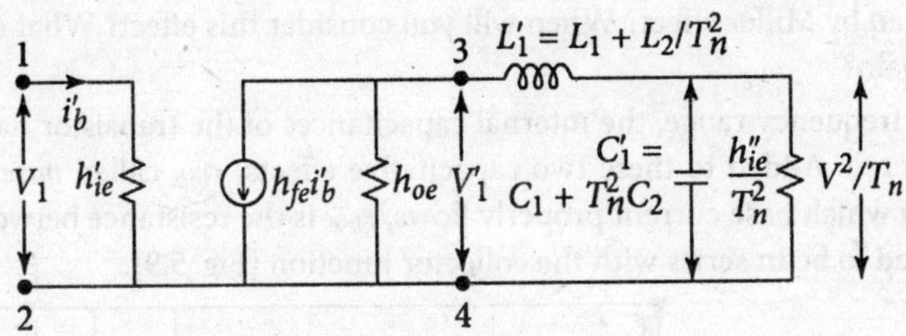

Fig. 5.6(e)

Resonance will take place when $X_L = X_C$, i.e., when

$$\omega L_1' = \frac{1}{\omega C_1'} \text{ or, } 2\pi f_0 = \frac{1}{\sqrt{L_2' C_1'}} \text{ or, } f_0 = \frac{1}{2\pi} \cdot \frac{1}{\sqrt{L_1' C_1'}}. \tag{5.16}$$

The upper 3dB frequency is somewhat greater than the resonant frequency f_0 as given by (5.16). The voltage gain of a transformer-coupled amplifier is shown in Fig. 5.7.

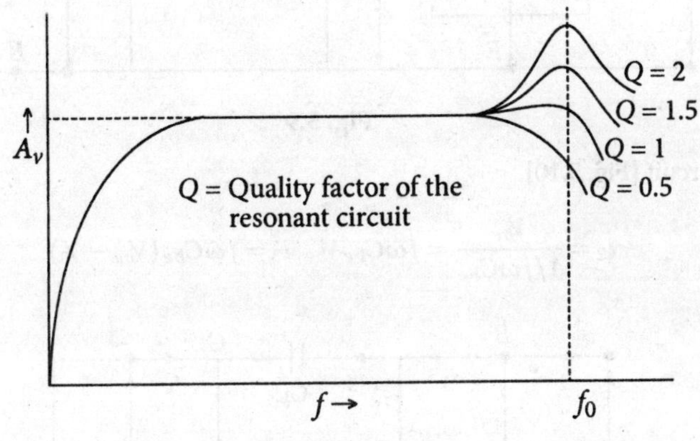

Fig. 5.7

5.4 Short Theory Questions with Answers

1. What do you mean by *virtual base*?

   Ans. At high frequency operation, capacitative effects of the *pn*-junction are to be taken into account. The inside of the transistor can be represented by [Fig. 5.8]. *E*, *B*, *C* are the external emitter, base and collector terminals. The capacitative effect of the forward-biased *pn*-junction is represented by C_{be}'. The capacitative effect of the reverse-biased *pn*-junction is represented by C_{bc}' between collector and base. The two capacitances have common termination at B', called *virtual base*.

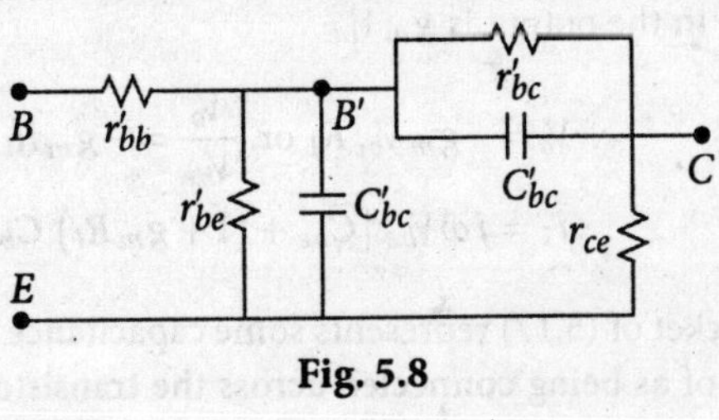

Fig. 5.8

2. **What do you mean by Miller effect? When will you consider this effect? What effect does it have on high frequency gain?**

Ans. In the high frequency range, the internal capacitances of the transistor namely $C_{b'e}$, $C_{b'c}$, etc. are to be considered. Added to these two capacitative effects, $r_{b'b}$ called *base-spreading resistance* ($\doteq 50\Omega$), through which base current properly flows, $r_{b'e}$ is the resistance between base-emitter and may be considered to be in series with the collector junction [Fig. 5.9].

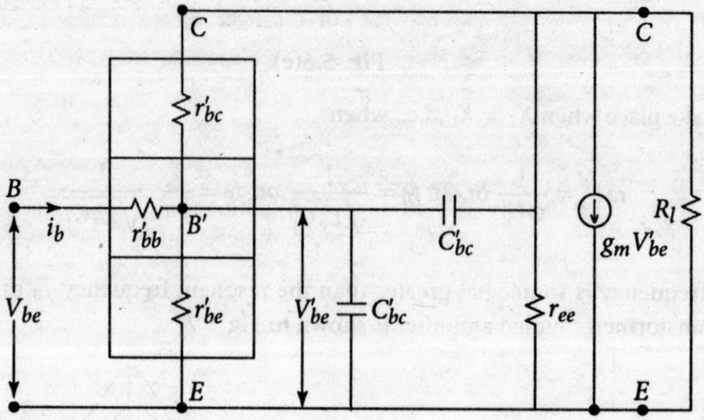

Fig. 5.9

The equivalent circuit [Fig. 5.10].

$$i_2 = \frac{V_{be}}{1/j\omega C_{be}} = j\omega C_{be} V_{be},\ i_3 = j\omega C_{be}(V_{be} - V_0).$$

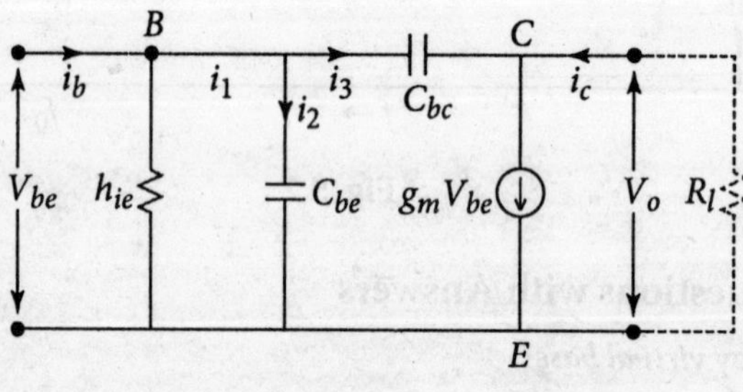

Fig. 5.10

Now,

$$i_1 = i_2 + i_3 = j\omega V_{be}\left(C_{be} + C_{bc} - \frac{V_o}{V_{be}} C_{bc}\right) = j\omega V_{be}\left[C_{be} + \left(1 - \frac{V_o}{V_{be}}\right) C_{bc}\right].$$

Now, the current source in the output is $g_m V_{be}$.

$$\therefore\ V_0 = -g_m V_{be} R_l \text{ or, } \frac{V_o}{V_{be}} = -g_m R_l.$$

$$\therefore\ i_1 = j\omega V_{be}\left[C_{be} + (1 + g_m R_l)\, C_{bc}\right]. \tag{5.17}$$

The term within the bracket of (5.17) represents some capacitance C given as $C = C_{be} + (1 + g_m R_l)$ C_{be}, it may be thought of as being connected across the transistor input. The active nature of the circuit has increased the value of C_{be} by multiplying it by the factor $(1+g_m R_l)$. This increase in capacitance is known as *Miller effect*.

3. What is the expression for lower 3dB frequency? Under what conditions is this expression valid?

Ans. The lower 3dB frequency is that frequency where the power gain has dropped 3dB from the gain at the mid-frequency. This frequency is given by $f_1 = \frac{(1+h_{fe})}{2\pi C_e(R_s+h_{ie})}$. This is valid only when f_2 defined approximately by $\frac{(1+h_{fe})R_e}{(R_s+h_{ic})\times 2\pi C_e R_e}$ is large compared to $f_0 = \frac{1}{2\pi C_e R_e}$.

4. What is the principal advantage of using transformers to cascade transistor amplifier stage? How would an interstage transformer differ from an output transformer?

Ans. Transformer-coupling is primarily advantageous, because it provides excellent impedance matching. The output impedance of one stage is matched with the input impedance of the next stage very effectively through this coupling.

Suppose, two transistors Q_1 and Q_2 are cascaded through transformer-coupling, the complete circuit consists of three transformers, driver (T_1), interstage and output transformer. The output transformer contains the load R_l. This will be such that the reflected impedance on the primary side R_{l_2} will be equal to the output impedance of the transistor Q_2. So, the output transformer will be an impedance matching transformer. The interstage transformer will be a step-down voltage transformer, so that current gain is achieved through it.

5. Why is current gain larger in transformer-coupled stages? Assume that transformers are chosen for maximum power transfer conditions, what is the expression for a stage's current gain and voltage gain if equal transistors are used in each case?

Ans. The current gain in transformer-coupling is usually larger than RC-coupling, because here the collector current magnified by the transformer turn ratio is the total base current input to the next stage, but in the latter mode of coupling, the collector current is divided between load resistance, bias resistors and base input. Under matching condition giving maximum power transfer, the current stage gain and voltage stage gain are given by the same expression and written as

$$A_i = \frac{h_{fe}}{2T_n} = A_v, \text{ where } T_n = \text{turn ratio} = N_2/N_1.$$

6. What is the high frequency response of a transformer-coupled CE amplifier usually determined by? What causes the possible rise in a transformer-coupled amplifier's frequency response curve at high frequencies?

Ans. The high frequency response of TC common emitter amplifier is normally determined by the characteristics of the transformer and not by the transistor's characteristics. The rise in the high frequency response curve is possibly, because of the Q-factor of the transformer. Greater the Q-factor, greater is the hump.

7. What do you mean by gain bandwidth product? How can you explain the fact that a transistor's bandwidth increases only at the expense of gain?

Ans. The bandwidth of the amplifier stage is defined as $(f_2 - f_1) \simeq f_2 = \Delta f$, where f_2 is the higher 3dB frequency and f_1 is the lower 3dB frequency. The Gain-Bandwidth-Product (GBW) denoted by f_T is defined as $f_T = |A_m f_2|$, where A_m is the gain at mid-frequency. Now, the mid-frequency gain A_m with source resistance $R_s = 0$ is given by

$$|A_m| = \frac{h_{fe} R_l}{R_s + h_{ie}} = \frac{h_{fe}}{h_{ie}} R_l.$$

$$f_T = \frac{h_{fe}}{h_{ie}} R_l \times \frac{1}{2\pi R_l C_{sh}} = \frac{h_{fe}}{h_{ie}} \cdot \frac{1}{2\pi C_{sh}} = \frac{g_m}{2\pi C_{sh}}. \tag{5.18}$$

Thus, f_T is a constant, which shows that the bandwidth increases at the cost of gain.

Table for the mid-frequency gains and lower 3dB frequency under various conditions:

When	$A_v = V_o/V_s$	$A_v = V_o/V_i$	f_1 Hz for V_o/V_s
$R_1 \parallel R_2 \gg R_i$ $R_o = 0$	$\frac{h_{fe}R_l}{R_s + h_{ie}}$	$\frac{h_{fe}R_l}{h_{ie}}$	$\frac{1+h_{fe}}{2\pi C_e(R_s+h_{ie})}$
$R_1 \parallel R_2 \gg R_i$ $R_o \neq 0$	$\frac{h_{fe}R_l}{R_s h_{ie} + (1+h_{fe})R_c}$	$\frac{h_{fe}R_l}{h_{ie} + (1+h_{fe})R_c}$	$\frac{1+h_{fe}}{2\pi C_e\left[R_s+h_{ie}+(1+h_{fe})R_c\right]}$
$R_1 \parallel R_2 \ngtr R_i$ $R_o = 0$	$\frac{h_{fe}R_l}{R_s + h_{ie} + \frac{h_{ie}R_s}{R_1 \parallel R_2}}$	$\frac{h_{fe}R_l}{h_{ie}}$	$\frac{1+h_{fe}}{2\pi C_e\left[R_s+h_{ie}+(1+h_{fe})R_c\right]}$
$R_1 \parallel R_2 \ngtr R_i$ $R_o \neq 0$	$\frac{h_{fe}R_l}{R_s + R'}$	$\frac{h_{fe}R_l}{h_{ie} + (1+h_{fe})R_c}$	$\frac{1+h_{fe}}{2\pi C_e(R_s+R')}$

where $R' = h_{ie} + h_{ie}R_s/(R_1 \parallel R_2) + (1+h_{fe})R_c$.

5.5 Numerical Problems

1. Find the value of C_e to provide a lower 3dB frequency of 200Hz, given $R_e = 0.5\text{k}\Omega$, $h_{fe} = 50$, $h_{ie} = 1\text{k}\Omega$ and $R_s = 200\Omega$.

Solution:

$$f_1 = \frac{1+h_{fe}}{2\pi C_e(R_s + h - ie)} \text{ or, } C_s = \frac{1+h_{fe}}{2\pi f_1(R_s + h_{ie})} = \frac{1+50}{2\pi \times 200(200+1000)} = 33.8\mu\text{F}.$$

The lower 3dB frequency is given by the above equation if $f_v/f_o \gg 1$.

Now,

$$\frac{f_v}{f_o} = \frac{(1+h_{fe})R_e}{(R_s+h_{ie})} = \frac{51 \times 500}{(200+1000)} \simeq 21.$$

So, the condition is fulfilled.

2. Consider the circuit diagram of a common emitter amplifier [Fig. 5.11] with following specifications.

$$V_{cc} = 12\,\text{V}, R_l = 4.7\text{k}\Omega$$

$$R_e = 1\text{k}\Omega, R_1 = 15\text{k}\Omega$$

$$R_2 = 2.2\text{k}\Omega, R_s = 600\Omega.$$

The transistor has h-parameters given by $h_{fe} = 400$, $h_{ie} = 10\text{k}\Omega$, $C_e = 50\mu\text{F}$ with no series resistance. Calculate

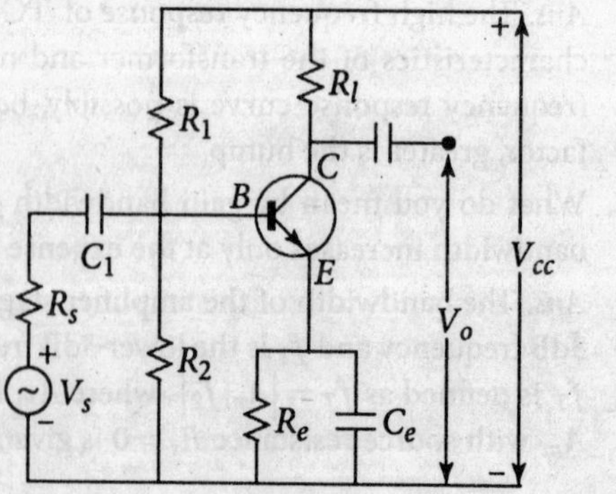

Fig. 5.11

(a) the mid-frequency voltage gain $A_{mf} = V_0/V_s$
(b) the mid-frequency voltage gain V_o/V_i
(c) $\frac{(1+h_{fe})R_e}{R_s+h_{ie}}$, is it much larger than 1? What does this condition imply?
(d) f_1.
Here, $R_i = h_{ie} = 10\text{k}, R_1 \parallel R_2 = 1.92\text{k}\Omega$.
$\therefore R_1 \parallel R_2 \not\gg R_i$ and $R_o = 0$.

Solution:

(a) $|A_{mf}| = \left|\frac{V_o}{V_s}\right| = \dfrac{h_{fe}R_l}{R_s + h_{ie} + \frac{h_{ie}R_s}{R_1\|R_2}} = 137.$

(b) $|A_{mf}|_{R_s=0} = \dfrac{h_{fe}\cdot R_l}{h_{ie}} = \dfrac{400\times 4.7\times 10^3}{10^4} = 188.$

(c) $\dfrac{(1+h_{fe})R_e}{(R_s+h_{ie})} = \dfrac{401\times 10^3}{(600+10^4)} = 37.8$, it is large compared to unity. In this case, the lower 3dB frequency f_1 is given by $f_1 = \dfrac{1+h_{fe}}{2\pi C_e(R_s+h_{ie})}$.

(d) $f_1 = \dfrac{1+h_{fe}}{2\pi C_e(R_s+h_{ie})} = \dfrac{401}{2\pi\times 50\times 10^{-6}(600+10^4)} = 120\text{Hz}.$

3. (a) Calculate the size of bypass capacitor C_e for the amplifier [Fig. 5.11] to have a lower 3dB frequency $f_1 = 50\text{Hz}$ for V_0/V_i (Assume $R_e = 0$).
(b) Mid-frequency voltage gain.
(c) What is the mid-frequency voltage gain V_o/V_i, if the capacitor C_e has a series resistance $R_c = 20\Omega$.
(d) What will be the value of f_1 when $R_c = 20\Omega$.

Solution:

(a) $f_1 = \dfrac{1+h_{fe}}{2\pi(R_e+h_{ie})C_e} = \dfrac{(1+h_{fe})}{2\pi h_{ie}C_e}$, because we want to have the value for V_0/V_i.

$$\therefore\ C_e = \frac{1+h_{fe}}{2\pi h_{ie}f_1} = \frac{401}{2\pi\times 10^4\times 50} = 127\mu\text{F}.$$

(b) $|A_{mf}| = \dfrac{h_{fe}R_l}{(R_e+h_{ie})} = \dfrac{400\times 4.7\times 10^3}{10^4}$ for $R_s = 0 = 188.$

(c) When $R_c \neq 0$, then the mid-frequency voltage gain is given by taking into account the effect of R_c, R_1 and R_2.

$$|A_{mf}| = \frac{h_{fe}R_l}{R_s + h_{ie} + (1+h_{fe})R_c + \frac{h_{ie}R_s}{R_1\|R_2}} = \frac{h_{fe}R_l}{h_{ie} + (1+h_{fe})R_c}, R_s = 0 = \frac{400\times 4.7\times 10^3}{10^4 + (401)\times 20} = 104.$$

(d) $f_1 = \dfrac{(1+h_{fe})}{2\pi C_e\{h_{ie} + (1+h_{fe})R_c\}}$, taking $R_s = 0$

$$= \frac{401}{2\pi\times 127\times 10^{-6}\{10^4 + 401\times 20\}} = 27.9\text{Hz}.$$

4. Assume the amplifier [Fig. 5.11] with the modification so that $R_e = 10\Omega$, and R_1 and R_2 are made large compared with h_{ie}, $C_e = 50\mu\text{F}$, $h_{ie} = 10\text{k}\Omega$, $h_{fe} = 400$, $R_s = 600\Omega$. Calculate

(a) $\frac{(1+h_{fe})R_e}{R_s+h_{ie}}$. Is this less than 0.414? What does this mean?

(b) f_0 and f_p.

Solution:

(a) $\frac{(1+h_{fe})R_e}{R_s+h_{ie}} = \frac{401\times 10}{600+10^4} = 0.378$. It is less than 0.414. It means that there is no 3dB point.

(b) $f_o = \frac{1}{2\pi C_e R_e} = \frac{1}{2\pi\times 50\times 10^{-6}\times 10} = 318\text{Hz}$

$$f_p = \frac{1^* + \frac{(1+h_{fe})R_e}{R_s+h_{ie}}}{2\pi C_e R_e} = (1+0.378)318 = 438\text{Hz}$$

$\therefore f_p \not> f_o$

5. An RC-coupled amplifier has mid-frequency gain of 50 with lower 3dB frequency f_1 = 50Hz and upper 3dB frequency 100kHz. Find the values of frequencies at which the gain reduces to 40 on either side of the mf region.

Solution:

$$\frac{A_{lf}}{A_{mf}} = \frac{1}{\left[1+(f/f_1)^2\right]^{1/2}} = \frac{40}{50}. \tag{5.19}$$

Also

$$\frac{A_{hf}}{A_{mf}} = \frac{1}{\left[1+(f/f_1)^2\right]^{1/2}} = \frac{40}{50} \tag{5.20}$$

Solving (5.19), we get $f = 66.7$Hz on the lower side.

Solving (5.20), we get f = 75kHz on the higher side.

6. Consider a transformer-coupled amplifier [Fig. 5.5], t_{r_1} and t_{r_2} are identical with $h_{ie} = 4\text{k}\Omega$, $h_{fe} = 100$, $h_{oe} = 50\times 10^{-6}$℧. The transformers are to be selected for maximum power transfer. Then,

(a) If $R_e = 1\text{k}\Omega$ and $R_l = 4\Omega$, specify the primary and secondary impedance of transformers T_1, T_2, T_3. Assume approximately $R_i = h_{ie}$.

(b) Determine the turn ratio of T_1, hence obtain current gain I_{b_1}/I_i for the input stage.

(c) Current gain I_{b_2}/I_{b_1} for the first stage.

(d) Current gain I_l/I_i, overall current gain.

(e) From the turns ratio in (b) obtain the voltage gain V_{b_1}/V_{io}.

(f) Voltage gain V_{b_2}/V_{b_1}.

(g) Voltage gain V_l/V_{b_2}.

(h) Overall voltage gain V_l/V_{in}.

(i) Determine the overall power gain in decibels.

Solution:

(a) The reflected impedance from the secondary to the primary of $T_3 = R_l/T_n^2$.

$\therefore$ primary impedance $R_{l_2} = R_l/T_n^2$.

For maximum power transfer $R_l/T_n^2 = 1/h_{oe} = \frac{1}{50}\times 10^6$.

$\therefore$ T_n for $T_3 = 14:1000$.

Hence, T_3 is a step-down transformer.

Now, R_{i_1} is the input resistance of t_{r_1} equal to h_{ie}, the reflected impedance on the primary = $R_{i_1}/R_n^2 = h_{ie}/T_n^2$. For maximum power transfer,

$$R_s = \frac{h_{ie}}{T_n^2} \text{ or, } T_n^2 = \frac{h_{ie}}{R_s} = \frac{4\times 10^3}{1\times 10^3} = 4.$$

$\therefore$ $T_n = 2$, hence T_1 is a step-up transformer. The transformer T_2 is also a step-down transformer. The load reflected to the primary of $T_2 = h_{ie}/T_n^2$.

$$\therefore \frac{h_{ie}}{T_n^2} = \frac{1}{h_{oe}} \text{ or, } T_n = \sqrt{h_{oe}\times h_{ie}} = \sqrt{50\times 10^{-6}\times 4\times 10^3} = 0.45.$$

(b) $I_{b_1}/I_i = 1/T_n$. $\therefore$ current gain = 1/2.

(c) $I_{b_2}/I_{b_1} = A_i = \dfrac{h_{fe}}{2T_n} = \dfrac{100}{2\times 0.45} = 111.$

(d) Overall current gain

$$A = \frac{I_l}{I_i} = \frac{I_l}{I_{b_2}} \times \frac{I_{b_2}}{I_{b_1}} \times \frac{I_{b_1}}{I_i} = \frac{h_{fe}}{2T_{n_2}} \times \frac{h_{fe}}{2T_{n_2}} \times \frac{1}{T_{n_1}} = \frac{100}{2\times\frac{1.41}{100}} \times \frac{100}{2\times 0.45} \times \frac{1}{2} = 19.7005.$$

(e) The transformer T_1 is a step-up transformer.

$$\therefore \frac{V_{b_1}}{V_i} = T_{n_1} = 2.$$

(f) $\dfrac{V_{b_2}}{V_{b_1}} = A_v = \dfrac{h_{fe}}{2T_{n_2}} = \dfrac{100}{2\times 0.45}.$

(g) $\dfrac{V_l}{V_{b_2}} = A_v = \dfrac{100}{2T_{n_2}} = \dfrac{100}{2\times\frac{1.41}{100}}.$

(h) $\therefore$ the overall voltage gain

$$V = \frac{V_l}{V_i} = \frac{V_l}{V_{b_2}} \times \frac{V_{b_2}}{V_{b_1}} \times \frac{V_{b_1}}{V_i} = \frac{100\times 100}{2\times 1.4} \times \frac{100}{2\times 0.45} \times 2 = 793650.$$

(i) $\therefore$ P = power gain = $A\times V = 1.56\times 10^{11}$.
Power gain in decibel = $10\log P$ = 112dB.

5.6 Exercises

1. An RC-coupled amplifier with a single stage has $f_1 = 70$Hz and $f_2 = 14$kHz. At what frequencies is the voltage gain only 5% less than its mid-range value?
2. What is the basic merit of a transformer-coupled amplifier? Derive an expression for the low frequency voltage gain in terms of the mid-frequency voltage gain?
3. Is the overall value of f_1 for a cascaded amplifier higher or lower than the value of f_1 for any of the individual stages? Is the bandwidth of the overall amplifier wider or narrower than that of one of the stages?
4. What is the high frequency response of a transformer-coupled CE amplifier usually determined by? What causes the possible rise in a transformer-coupled amplifier's frequency response curve at high frequencies? Under what condition will this not happen?

5. What do you mean by Gain-Bandwidth product? What does it indicate of a transistor? How that bandwidth increases at the expense of gain.
6. A given amplifier has an f_2 value of first stage = 25kHz, that of second stage = 15kHz. Find f_2 for the overall amplifier.

   (Use the formula $f_2(\text{overall}) = f_2 \times \sqrt{\frac{1}{2n} - 1}$, here $n = 2$).
7. Three identical cascaded stages have an overall upper 3dB frequency of 20Hz. What are f_2 and f_1 of each stage?
8. Sketch two RC-coupled CE transistor stages. Show the low frequency model for one stage. What is the expression for f_1?
9. A three-stage RC-coupled amplifier uses identical stages with cut-off frequencies of $f_1 = 100\text{Hz}$, $f_2 = 500\text{kHz}$. Find the overall cut-off frequency values assuming that the stages are non-interacting.
10. An amplifier of 45dB power gain has an input circuit resistance of 600Ω and load resistance of 5Ω. If the amplifier input voltage is 1V sinusoidal, find the output voltage and current in the load circuit.

5.7 Amplifiers Revisited

In the last chapter we have discussed the basic principle of the three principal amplifies devices — CE, CB and CC types. They are discussed at length giving the relative merits and demerits of the three modes of use of a transistor as amplifiers; of the three modes, common emitter mode amplifier is most frequently used.

Amplifiers classified: An amplifier produces an output as a magnified version of the input in phase with it or in opposite phase to it. They are classified in a number of ways as:

1. *Based on frequency.* (a) Audio frequency (AF): Such an amplifier is used to amplify signals having frequency in the range 20 c/s to 20 kc/s (b) Radio frequency (RF) amplifier: It is meant for amplification of signals having frequency ranging from 40 kc/s to some megacycles (c) Video frequency (VF) for frequency beyond RF.
2. *Based on the configuration* in which the transistor is used (a) CB mode (b) CE mode (c) CC mode CB mode amplifier: In this case of amplification, base is common to both the input as well as output. CE mode amplifier: Here, emitter is made common to both input as well as output.

   CC mode amplifier. In this case, collector is made common to both input as well as to the output.
3. *Based on the strength of the input signal* level (a) low signal amplifier (b) large signal amplifier.
4. *Based on the purpose.* (a) voltage amplifier (b) current amplifier (c) power amplifier.
5. *Based on the operating point of the amplifier.* In this classification we designate amplifiers as:
   (a) Class A. When the amplified output is available during the whole cycle of the input; distortion-less amplifier as in case of an audio amplification.
   (b) Class B. Amplified output is available during the angle θ such that $0 \le \theta \le 180°$.
   (c) Class C. Here, the output is available for $0 \le \theta \le \pi/2$.

   Amplifies may belong to a combination of any two of the above so that we may have amplifiers designated as AB, BC, CA.
6. *Based on interstage coupling.* If stages of a number of amplifies are hooked up for increase in gain, amplifiers are designated depending in the mode of coupling. Thus, we have (a) Direct coupled amplifier (b) Resistance capacitance (RC) amplifier (c) Transformer-coupled (TC) amplifier.

7. ***Based on the output load.*** **Amplifiers are classified on this basis as (a) tuned amplifier (b) untuned amplifiers.**

Of all these classifications we most frequently use the classification based on frequency and that based on configuration in which the transistor is used.

How to draw equivalent circuit of an amplifier?

To draw ac equivalent circuit, steps followed are the following:

Step 1. All dc sources are to be considered as absent, because they are supposed to be ac short to ground.

Step 2. All capacitors are also to be considered as being short.

To draw dc equivalent:

Step 1. All ac sources are to be taken as absent as if they are *grounded.*

Step 2. All capacitors arc considered to be *open*. The total current in any branch of the circuit is the sum of dc and ac currents; similarly, total voltage across any branch is the sum of the dc and ac voltages across that branch.

A few examples below will make the analysis more clear.

Example 1. We have shown an amplifier circuit, a CE amplifier having potential divider with emitter bias as the biasing arrangement. Draw the ac and dc equivalent circuit of the amplifier. Find the operating point of the amplifier.

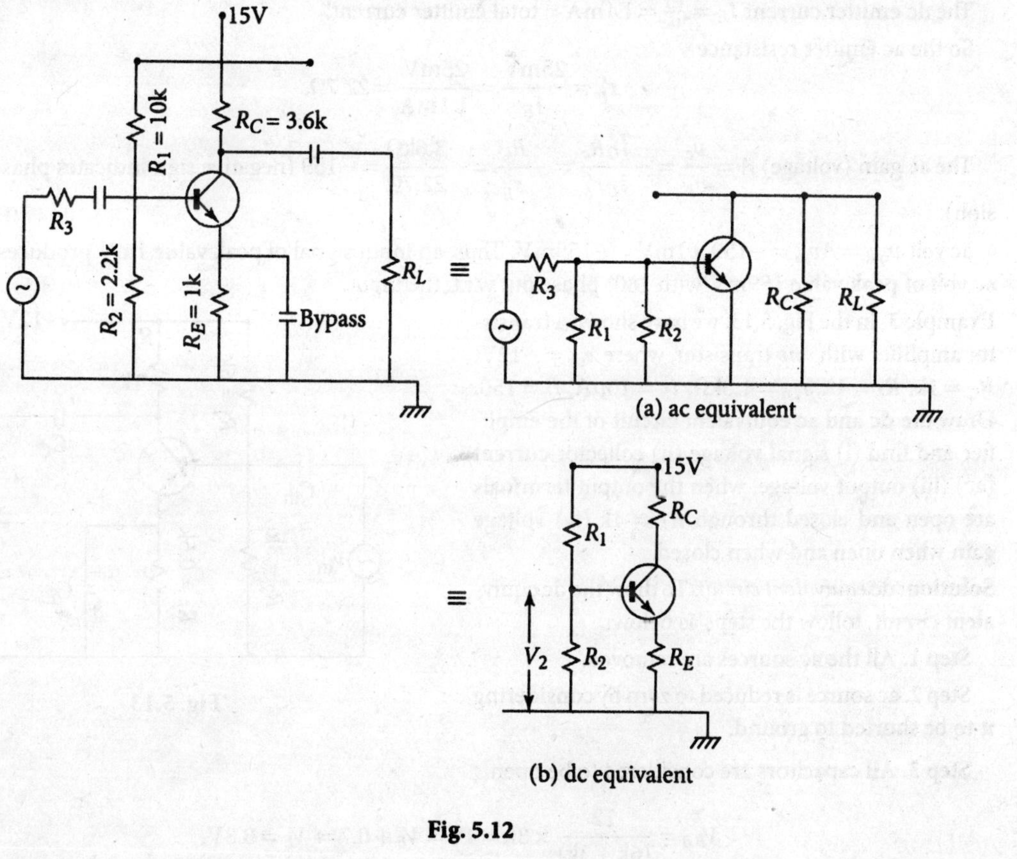

Fig. 5.12

To draw ac equivalent:

Step 1. Reduce 15V dc to zero by shorting to ground.

Step 2. All the capacitors are shorted.

With respect to dc equivalent

$$V_2 = \frac{15R_2}{R_1+R_2} = \frac{15\times 2.2}{(10+2.2)} = 2.705\text{V}.$$

Example 2. If in previous problem $V_{cc} = 10$V and the ac signal is a small sine wave, what are the voltages at the base, emitter and collector? Assume peak ac voltage of the input signal = 1mV (peak).

Solution: Assuming $R_3 = 0$, the whole of ac signal voltage appears at the base and this ac voltage is negligibly small = 1mV (peak).

So, the total voltage at the base is dc bias voltage + ac signal voltage.

$$V_B = \frac{R_2\times 10}{(R_1+R_2)} = \frac{2.2\times 10}{(10+2.2)} = 1.8\,\text{volt}.$$

So the total voltage = $V_B + v_B = 1.8$V.

The emitter is at a voltage 0.7V lower than that of base, hence, the total voltage at the emitter = $V_E + v_E = (1.8-0.7) \simeq 1.1$V, because no signal voltage appears at the emitter due to the bypass capacitor.

The dc emitter current $I_E = \frac{1.1}{1\text{k}} = 1.1$mA = total emitter current.

So the ac emitter resistance

$$r'_E = \frac{25\text{mV}}{I_E} = \frac{25\text{mV}}{1.1\text{mA}} = 22.7\Omega.$$

The ac gain (voltage) $A = \frac{v_c}{v_E} = -\frac{i_C R_c}{i_C r'_E} = -\frac{R_C}{r'_E} = -\frac{3.6\text{k}\Omega}{22.7\Omega} = -159$ (negative sign indicates phase inversion).

ac volt $v_{\text{out}} = Av_{\text{in}} = -159\times 1\text{mV} = -159$mV. Thus, an input signal of peak value 1mV produces output ac volt of peak value 159mV with 180° phase out w.r.t. the input.

Example 3. In the Fig. 5.13, we have shown a transistor amplifier with *pnp* transistor, where $V_{cc} = -12$V, $R_C = 1$k, $R_L = 1$k, $r_{in} = 1.5$kΩ, $i_b = 10\mu$A, $\beta = 150$. Draw the dc and ac equivalent circuit of the amplifier and find (i) signal voltage (ii) collector current (ac) (iii) output voltage, when the output terminals are open and closed through R_L = 1k (iv) voltage gain when open and when closed.

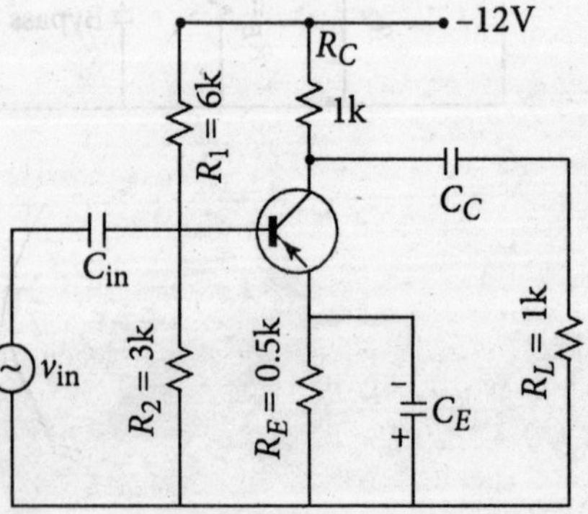

Fig. 5.13

Solution: dc *equivalent circuit.* To draw the dc equivalent circuit, follow the steps as below:

Step 1. All the ac sources are removed.

Step 2. ac source is reduced to zero by considering it to be shorted to ground.

Step 3. All capacitors are considered to be open.

$$V_{BB} = \frac{12}{(6\text{k}+3\text{k})}\times 3\text{k} = 4\text{V} = V_E + 0.7 \Rightarrow V_E = 3.3\text{V}.$$

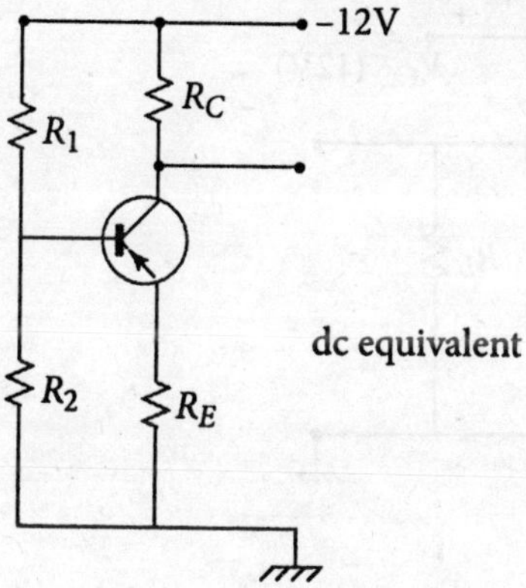

Fig. 5.14

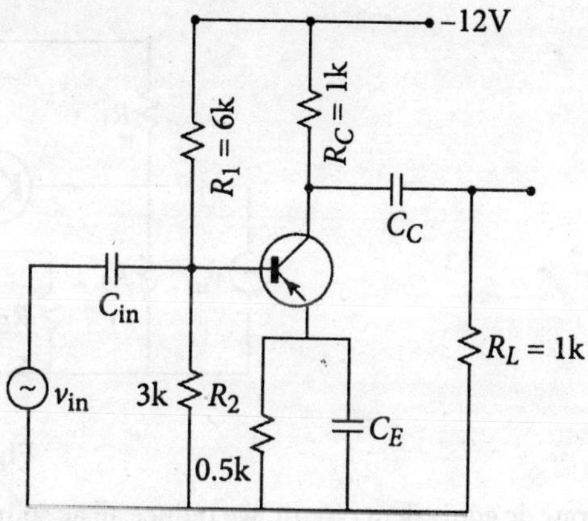

Fig. 5.15

To draw the ac equivalent of the circuit we follow the steps:

Step 1. All the dc sources are reduced to zero values by considering them to be shorted to ground.

Step 2. All the capacitors are supposed to be ac shorted.

Step 3. CE is ac shorted to ground, so no ac voltage in emitter.

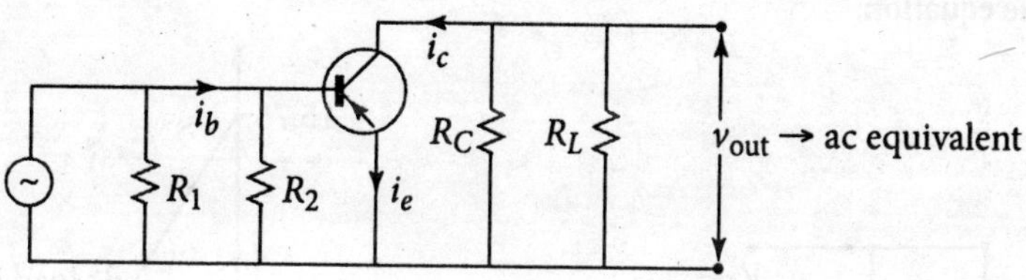

Fig. 5.16

The ac base current due to signal $i_b = 10\mu A$. Input resistance $r_{in} = 1.5k$.

(i) The signal voltage $v_{in} = 1.5 \times 10^3 \times 10 \times 10^{-6} = 15mV$.

(ii) ac collector current $i_c = \beta\, i_b = 150 \times 10^{-6} \times 10 = 1.5mA$.

(iii) ac $v_{out} = i_c \times R_{out}$.

Now, R_{out} when open ended $= R_C = 1 \times 10^3$ and

R_{out} when closed ended $= \dfrac{R_L \times R_C}{R_L + R_C} = \dfrac{1 \times 10^3 \times 1X^3}{2 \times 10^3} = 0.5k\Omega$.

Open ended voltage output $v_{out} = 1.5 \times 10^{-3} \times 1 \times 10^3 = 1.5V$

⇒ Open ended voltage gain $= \dfrac{1.5}{1.5 \times 10^{-3}} = 100$.

Closed ended voltage output $v_{out} = 1.5 \times 10^{-3} \times 0.5 \times 10^3 = 0.75V$

⇒ Closed ended voltage gain $= \dfrac{0.75}{15 \times 10^{-3}} = 50$.

Example 4. The circuit shows a transistor amplifier having different circuit components as $R_1 = 8k\Omega$, $R_2 = 4k\Omega$, $R_c = 1k\Omega$, $R_E = 2k\Omega$ and $R_L = 1k\Omega$.

If $V_{CC} = 12V$ and $V_{BE} = 0.7V$, then (i) draw dc load-line (ii) ac load-line (iii) determine the operating point.

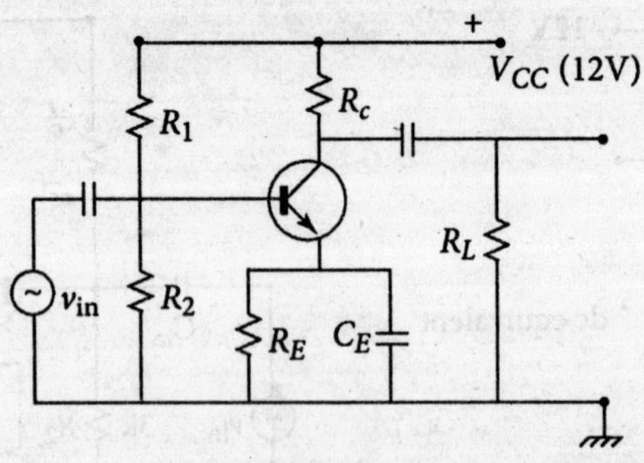

Fig. 5.17

To draw dc equivalent circuit, we reduce all ac sources to zero value by considering them to be shorted to ground; all capacitors are supposed to be open. So the dc equivalent of the actual circuit is as shown below:

The output loop of the dc equivalent of the amplifier circuit gives us

$$I_C R_C + V_{CE} + I_E R_E = V_{CC} \Rightarrow I_C(R_C + R_E) + V_{CE} = V_{CC} \text{ } (I_B \text{ is very small})$$

$$\Rightarrow \frac{I_C}{V_{CC}/(R_C+R_E)} + \frac{V_{CE}}{V_{CC}} = 1,$$

called dc load-line equation.

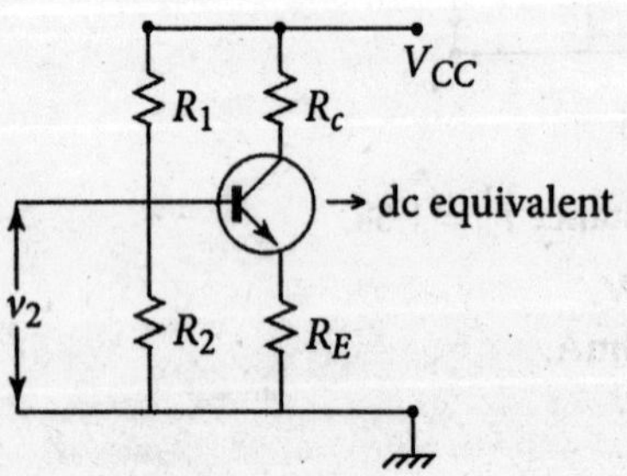

Fig. 5.18

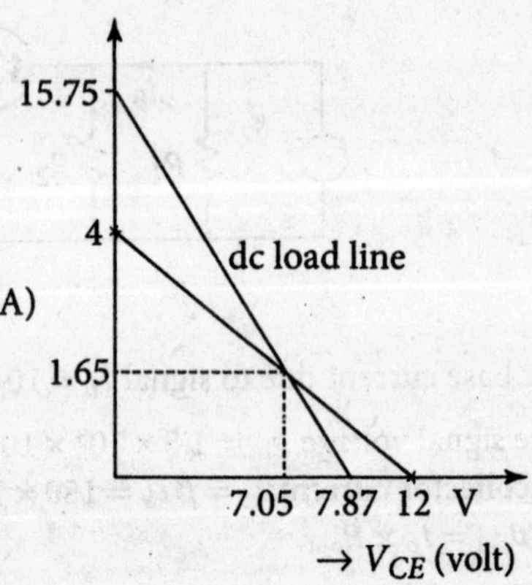

Fig. 5.19

So,

$$(I_C)_{max} = \text{intercept on the } I_C\text{-axis} = \frac{V_{CC}}{(R_C+R_E)} = \frac{I_2}{(1+2)\times 10^3} = 4\text{mA}$$

$$(V_{CE})_{max} = \text{intercept on the } V_{CE}\text{-axis} = V_{CC} = 12\text{V}.$$

From dc output loop equation, we get $V_{CE} = V_{CC} - I_C(R_C + R_E)$.

In absence of the signal, the collector current ($\simeq$ emitter current) can be calculated w.r.t. the input loop as indicated below:

$$V_2 = \frac{V_{CC}R_2}{(R_1+R_2)} = \frac{12\times 4}{(8+4)} = 4\text{V} = I_E R_E + V_{BE} = I_C \times 2\times 10^3 + 0.7$$

$$\Rightarrow I_C = \frac{(4-0.7)}{2} \times 10^{-3} = \frac{3.3}{2} \times 10^{-3} = \frac{3.3}{2} \times 10^{-3} = 1.65\text{mA}.$$

So, $V_{CE} = V_{CC} - I_C(R_C + R_E) = 12 - 1.65 \times 10^{-3} \times 3 \times 10^{-3} = 12 - 4.95 = 7.05\text{V}.$

The quiescent point Q of the amplifier lies on this dc load-line with coordinates as (1.65mA, 7.05V).

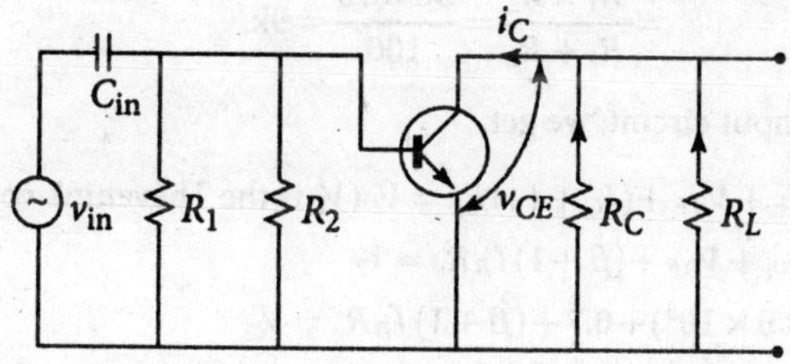

Fig. 5.20

ac load-line equation. We now draw the ac equivalent of the above amplifier circuit. In order to draw it, all the dc sources are reduced to zero value; all capacitors are taken to be shorted. The remaining circuit is the ac equivalent circuit as shown.

Effective resistance in the output $R_{ac} = R_c \| R_L = \frac{R_C \times R_L}{(R_C + R_L)} = \frac{1 \times 3}{2} \times 10^3 = 0.5\text{k}\Omega.$

$$i_c = 1.65\text{mA} + \frac{V_{CE}}{R_{ac}} = 1.65 + \frac{7.05}{0.5} = 1.65 + 14.10 = 15.75.$$

$$v_{CE} = V_{CE} + i_c \times \frac{R_C R_L}{R_C + R_L} = 7.05 + \frac{1 \times 1}{2} \times 10^{-3} = 7.87\text{V}.$$

Example 5. An amplifier having potential divider with emitter bias is shown in Fig. 5.21. Find the Q-point, drawing the dc and ac load-lines. $V_{cc} = 22.5\text{V}$, $R_1 = 90\text{k}$, $R_2 = 10\text{k}$, $R_C = 5.6\text{k}$, $R_L = 2\text{k}$, $\beta = 55$.

Solution: Drawing of dc load-lines.

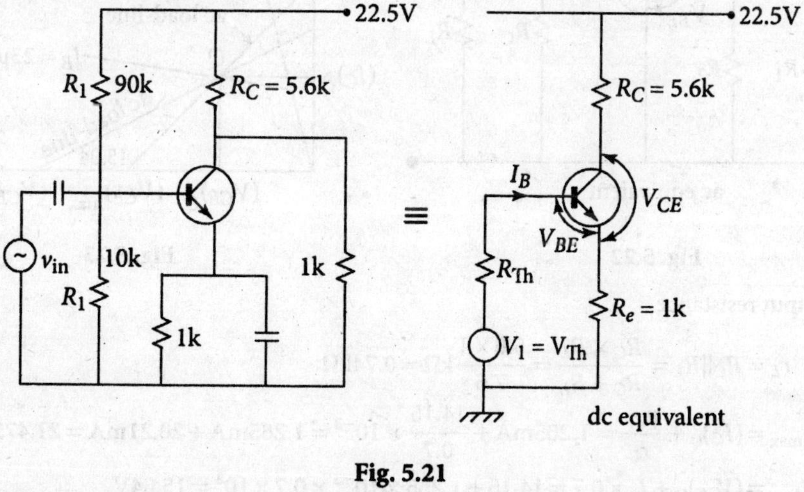

Fig. 5.21

$$V_2 = \frac{V_{ce}}{R_1 + R_2}, R_2 = \frac{22.5 \times 10}{100} = 2.25\text{V}$$

$$R_{Th} = R_1 \| R_2 \text{ called Thévenin's resistance}$$

$$= \frac{R_1 \times R_2}{R_1 + R_2} = \frac{90 \times 10}{100} = 9\text{k}.$$

So, applying KVL to the input circuit, we get

$$I_B R_{Th} + V_{BE} + (I_C + I_B) R_e = V_2 \text{ (V_2 is the Thévenin's voltage)}$$

$$\Rightarrow I_B R_{Th} + V_{BE} + (\beta + 1) I_B R_e = V_2$$

$$\Rightarrow (I_B \times 9 \times 10^3) + 0.7 + (\beta + 1) I_B R_e = V_2$$

$$\Rightarrow I_B (9 \times 10^3 + 56 \times 1 \times 10^3) + 0.7 = 2.25$$

$$\Rightarrow I_B \times 65 \times 10^3 = 1.55$$

$$\Rightarrow I_B = 23\mu\text{A} \ (\ll I_C).$$

$\therefore\ I_C = 55 \times 23 = 1.265\text{mA}.$

Applying KVL to the outputs, we get $I_C \times R_C + V_{CE} + (I_B + I_C) R_C = V_{CC} \Rightarrow V_{CE} = V_{CC} - I_C(R_c + R_e)$, remembering that $I_B R_e \ll I_c R_e$.

$\therefore\ V_{CE} = V_{ce}$, the maximum value of the V_{CE}, when $I_C = 0$.

But $V_{CE} = V_{cc} - I_c(R_c + R_e) = 22.5 - 1.265 \times 6.6 \times 10^{-3} \times 10^3 = 14.15\text{V}.$

So, the coordinates of the operating point Q are (1.2654mA, 14.15V) on the output characteristic at $I_B =$ 23μA.

ac load-line. In this case, there is the input signal, which varies the base current about its quiescent value of 23μA. The circuit in this case is shown below.

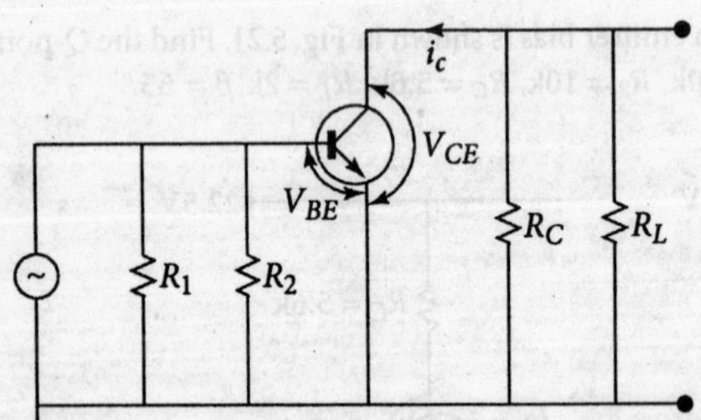

Fig. 5.22

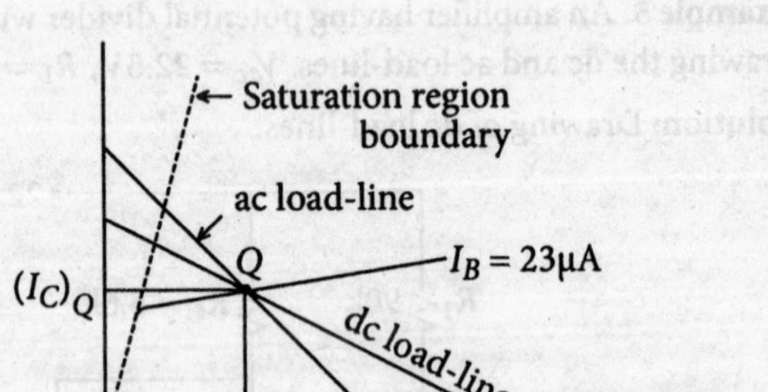

Fig. 5.23

Effective output resistance

$$r_L = R_C \| R_L = \frac{R_C \times R_L}{R_C + R_L} = \frac{5.6 \times 1}{7.6} \text{k}\Omega = 0.74\text{k}\Omega$$

$$(I_C)_{max} = (I_C)_Q + \frac{v_{CE}}{r_L} = 1.265\text{mA} + \frac{14.15}{0.7} \times 10^{-3} = 1.265\text{mA} + 20.21\text{mA} = 21.475\text{mA}$$

$$(V_{CE})_{max} = (V_{CE})_Q + I_c \times 0.7 = 14.15 + 1.265 \times 10^{-3} \times 0.7 \times 10^3 = 15.04\text{V}.$$

More about Quiescent point (Q-point)

There are two types of load-lines:

1. ***Static or dc load-line.*** **It represents the relationship between collector current I_c and collector to emitter voltage V_{CE} corresponding to a certain value of I_B, when no signal is applied. It is a straight line having a certain slope determined by the resistance in the collector-emitter loop. Refer to example 5, where we have shown**

$$V_{CE}-I_C(R_c+R_e)+V_{cc}\Rightarrow I_c(R_c+R_e)+V_{CE}=V_{CC}\Rightarrow I_c=-\frac{V_{CE}}{(R_c+R_e)}+\frac{V_{cc}}{(R_c+R_e)}.$$

It shows that when $I_c=0$, $V_{CE}=V_{CC}=$ maximum value of V_{CE} and $(I_c)_{max}=V_{cc}/(R_c+R_e)$. Thus, we get two points of the dc load-line, $\{(I_c)_{max}$ and $V_{CE}=0\}$ and $[0, V_{CE}(max)]$. Joining these two points, we would get a straight line $y=m_1x+c_1$, where

$$m_1=-\frac{1}{(R_c+R_e)},\quad c_1=\frac{V_{ce}}{(R_c+R_e)}.$$

I_c is plotted along y-axis and V_{CE} along x-axis. This straight line is called dc load-line and it intersects the output characteristic drawn at a certain value of I_B. The intersection is the Q-point.

2. ***ac load-line.*** **Refer the ac load-line drawn w.r.t. the example 5. Effective output resistance w.r.t. the operating point Q, we can write,**

$$(I_C)_{max}=(I_C)_Q+\frac{v_{CE}}{r_L},$$

where

$$r_L=R_C\|R_L=\frac{R_cR_L}{(R_C+R_L)},\quad (V_{CE})_{max}=(V_{CE})_Q+(I_C)_Q r_L.$$

These two equations give the end-points of the ac load-line.

Example 6. Find the operating point of the Si-transistor shown in Fig. 5.24. Draw the dc and ac load-lines.

Solution:

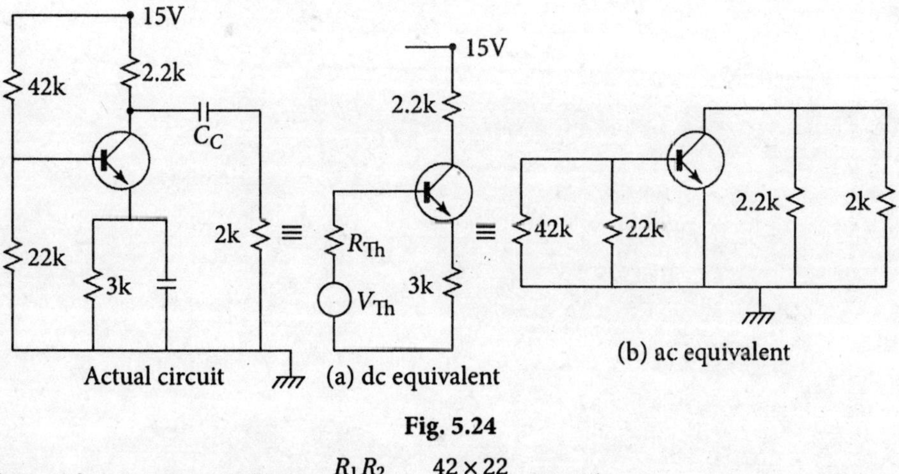

Fig. 5.24

$$R_{Th}=\text{Thévenin's resistance}=R_1\|R_2=\frac{R_1R_2}{R_1+R_2}=\frac{42\times 22}{64}=14.43\text{k}.$$

$$V_{Th}=\frac{V_{CC}}{R_1+R_2}\times R_2=\frac{15}{64}\times 22=6.11\text{V}.$$

Applying KVL to the base-emitter input loop, we would get

$$-I_B R_{Th} - V_{BE} - I_e R_e + V_{Th} = 0$$

$$\Rightarrow V_{Th} - V_{BE} - (I_C + I_B) R_e = I_B R_{Th}$$

$$\Rightarrow V_{Th} - B_{BE} - I_c R_e = I_B (R_e + R_{Th})$$

$$\Rightarrow \boxed{I_B = \frac{V_{Th} - V_{BE} - I_c R_e}{R_{Th} + R_e}}.$$

This equation gives the value of I_B.
Applying KVL again to collector-emitter output loop, we get

$$I_C R_C + V_{CE} + R_e I_e = V_{cc} \Rightarrow I_c R_c + V_{CE} + (I_C + I_B) R_e = V_{ce} \Rightarrow I_e R_c + V_{CE} + I_c R_e = V_{CC},$$

noting that $I_C \gg I_B$.

$$\boxed{I_C = -\frac{V_{CE}}{(R_c + R_e)} + V_{cc}}.$$

Chapter 6

Oscillators

6.1 Transistor Oscillators: Feedback in Amplifiers

A transistor oscillator consists of two principal sections — a frequency determining section, called *tank circuit* network and amplifying section. The amplifying section is of sufficient gain so as to compensate for the losses in frequency determining section. So a transistor oscillator is nothing but an amplifier, a portion of its output is fed in proper phase to the input. A transistor used as an amplifier amplifies the input, a portion of the output energy is fedback to the input and this feeding supplies transistor's own input signal and it oscillates at a frequency determined by the value of the feedback components. This basic action of the transistor as an oscillator is diagrammatically represented in [Fig. 6.1].

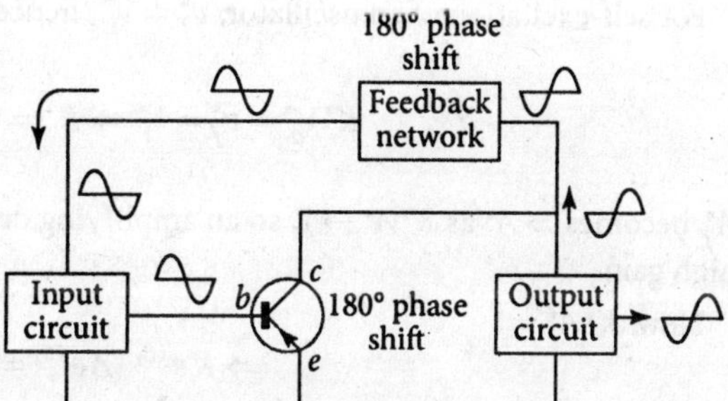

Fig. 6.1

6.2 Basic Components of an Oscillator

The basic components of an oscillator consists of: (i) tank circuit (ii) feedback network (iii) amplifier, which is, in this case, a transistor. The diagrammatic representation of these basic essentials is shown in Fig. 6.2. Since the transistor is operated in the active region, both input voltage V_i and output voltage V_o are sinusoidal, they can be represented by complex quantities; the voltage gain A_f of the amplifier is also complex given by $A^* = V_o^*/(v_i)_f^*$ with feedback, the input voltage $(v_i)_f^*$

$$(v_i)_f^* = V_i^* \pm v_f^* = V_i^* \pm K^* V_o^*,$$

where V_o^* is the output with feedback. With K^* positive $(v_i)_f^*$ is greater than V_i^*, hence it is a case of positive feedbacking while for K^* negative, $(v_i)_f^*$ is less than V_i^*, it is a case of negative feedbacking

$$\therefore A^* = \frac{V_o^*}{(v_i)_f^*} \text{ and } A_f^* = \frac{V_o^*}{V_i^*}$$

$$\Rightarrow \frac{1}{A^*} = \frac{(v_i)_f^*}{V_o^*} = \frac{V_i^* \pm K^* V_o^*}{V_o^*} = \frac{1}{A_f^*} \pm K^*.$$

$$\therefore \frac{1}{A_f^*} = \frac{1}{A^*} \mp K^* = \frac{1 \mp A^* K^*}{A^*} \Rightarrow A_f^* = \frac{A^*}{1 \mp A^* K^*}. \tag{6.1}$$

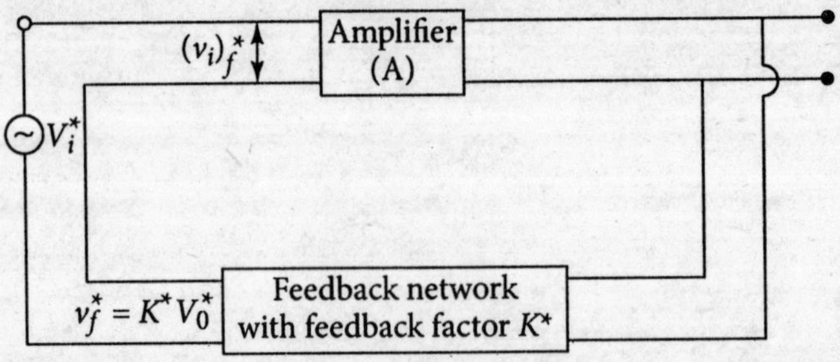

Fig. 6.2

From (6.1), it is evident that for K^* positive, the denominator is less and so the gain with feedback is greater than that without it, but for K^* negative, the denominator becomes greater, hence $A_f^* < A^*$. An electronic circuit behaves as a stable amplifier if its gain is less, hence there should be negative feedbacking. It behaves as an oscillator if its gain is very large, hence there should be provision for positive feedbacking.

For self-excitation of an oscillator, $v_f^* = V_i^*$, hence condition for self-excited oscillator is

$$K^* V_o^* = v_f^* = V_i^* \Rightarrow K^* = \frac{V_i^*}{V_o^*} \Rightarrow K^* = \frac{1}{A^*} \Rightarrow K^* A^* = 1. \tag{6.2}$$

A_f^* becomes $\gg A^*$ as $K^* A^* \to 1$, so an amplifying device tends to become an oscillator if it is of sufficiently high gain.

Now, $K^* A^* = 1$

$$\Rightarrow K e^{j\theta_1} \cdot A e^{j\theta_2} = 1 \Rightarrow A K e^{j(\theta_1+\theta_2)} = 1. \tag{6.3}$$

(6.3) embodies a very famous condition known as *Barkhausen criterion* for a self-excited oscillator.

Barkhausen criteria are: (i) $kA = 1$, called *amplitude condition*, (ii) $\theta_1 + \theta_2 = 0$, called the *phase condition*, which implies that the output of the amplifier and the feedback voltage must be in opposite phase so that positive feedbacking is achieved.

With positive feedback and with increase of gain, a critical stage is reached when the gain is infinitely large. In that case, condition for oscillation is $kA = 1$. Actually, however, if $kA > 1$, the gain of the whole system will adjust itself until kA becomes equal to unity. The condition of $kA = 1$ is known as *Barkhausen criterion.*

As in the case of valve oscillators, we have also equivalent transistorised oscillators, which can be grouped as under:

1. Sine wave oscillators: (a) Tuned-collector oscillators, (b) Tuned-base oscillators, (c) Tuned-base tuned-collector oscillators, (d) Hartley oscillators, (e) Colpitts oscillators, (f) Phase shift oscillators, (g) BF oscillators, (h) Crystal oscillators.
2. Relaxation oscillators: (a) Multivibrators (b) Blocking oscillators.

6.3 Commonly Used Oscillator Circuits

In Fig. 6.3, we have shown a Hartley oscillator using transistor as an amplifying unit. In the transistor oscillator circuit using common emitter configuration, positive feedback is done by placing the tank common to both input base and output collector circuits.

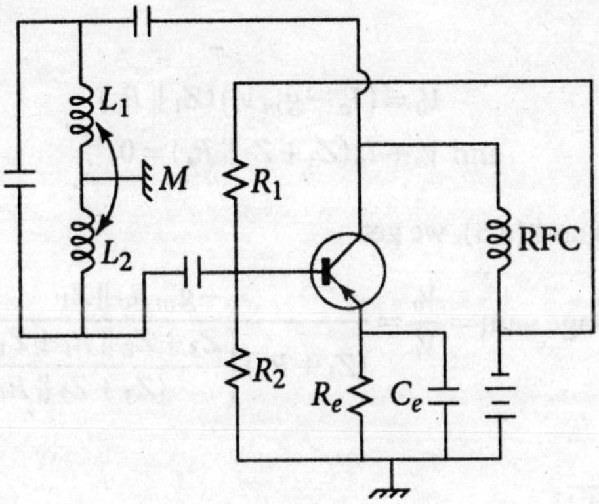

Fig. 6.3

Analysis of the oscillator. The LC tuned oscillator circuit analysis for any of the practical designs namely Colpitts or Hartley can be done with reference to the generalised LC tuned oscillator circuit [Fig. 6.4].

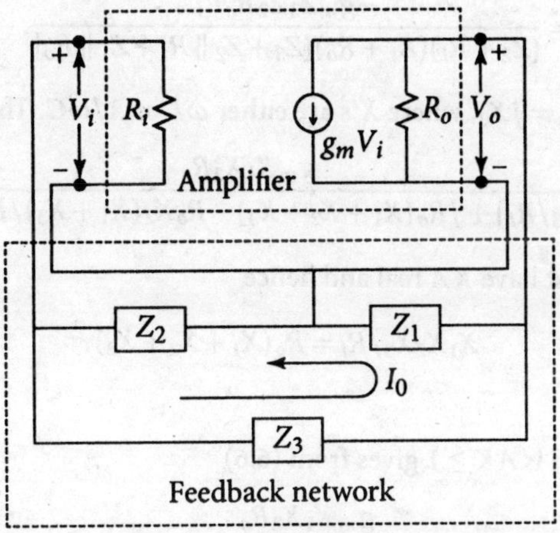

Fig. 6.4

The transistor used as amplifier is represented in transconductance model with $g_m = h_{fe}/h_{ie}$, R_o stands for output resistance of the transistor. Z_2, Z_3, Z_1 represent the input, feedback and output circuit impedances respectively. $g_m V_i$ is the current generator controlled by V_i. The simplified equivalent circuit [Fig. 6.4] is shown [Fig. 6.5].

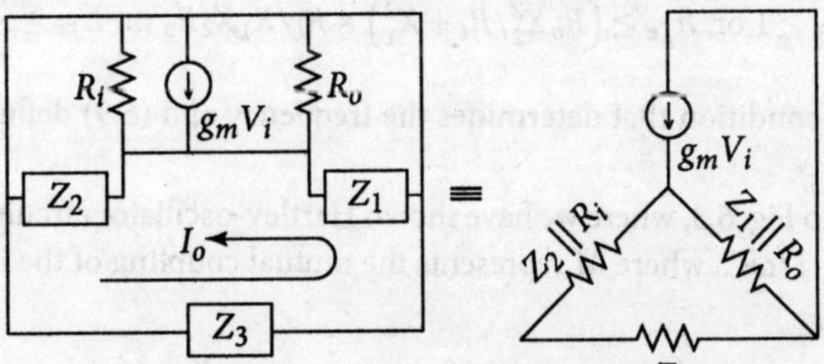

Fig. 6.5

So we can write,

$$V_0 = (I_o - g_m V_i)(Z_1 \parallel R_o) \tag{6.4}$$

$$\text{and } V_o + I_o(Z_3 + Z_2 \parallel R_i) = 0. \tag{6.5}$$

Eliminating I_o between (6.4) and (6.5), we get

$$A = \text{voltage gain} = \frac{V_o}{V_i} = \frac{-g_m R_o \parallel Z_1}{(Z_1 + R_o)\left[\dfrac{Z_3 + Z_2 \parallel R_i + Z_1 \parallel R_o}{(Z_3 + Z_2 \parallel R_i)}\right]}.$$

Again, $K = \dfrac{V_i}{V_o} = \dfrac{Z_2 \parallel R_i}{Z_3 + (Z_2 \parallel R_i)}$.

$$\therefore AK = \frac{-g_m R_o Z_1}{(Z_1 + R_o)\left[\dfrac{Z_3 + Z_2 \parallel R_i + Z_1 \parallel R_o}{Z_3 + Z_2 \parallel R_i}\right]} \times \frac{Z_2 \parallel R_i}{(Z_3 + Z_2 \parallel R_i)}$$

$$= \frac{-g_m Z_1 Z_2 R_o R_i}{(Z_2 + R_i)(Z_1 + R_o)[Z_3 + Z_2 \parallel R_i + Z_1 \parallel R_o]}.$$

Now, $Z_1 = jX_1$, $Z_2 = jX_2$, $Z_3 = jX_3$, where X's are either ωL or, $1/\omega C$. Then

$$AK = \frac{g_m X_1 X_2 R_o}{(-jX_1 X_2 X_3/R_i) + jR_o(X_1 + X_2 + X_3) - R_o X_2(X_1 + X_3)/R_i - X_1(X_2 + X_3)}. \tag{6.6}$$

For zero phase shift, we must have KA real and hence

$$X_1 X_2 X_3/R_i = R_o(X_1 + X_2 + X_3). \tag{6.7}$$

(6.7) determines frequency.

The condition for oscillation is $AK \geq 1$ gives from (6.6)

$$\frac{g_m X_1 X_2 R_o}{-R_o X_2(X_1 + X_3)/R_i - X_1(X_2 + X_3)} \geq 1.$$

For BJT circuits, $R_i = h_{ie}$, $h_{fe} = g_m h_{ie} = g_m R_i$ and for large R_i, (6.7) gives

$$X_1 + X_2 + X_3 = 0 \tag{6.8}$$

and

$$\frac{g_m X_1 X_2 R_o}{R_o X_2^2/R_i + X_1^2} \geq 1 \text{ or, } h_{fe} \geq \left(R_o X_2^2/R_i + X_1^2\right) \times R_i/X_1 X_2 R_o \text{ or, } h_{fe} \geq \frac{X_2}{X_1} + \frac{X_1}{R_o X_2} h_{ie}. \tag{6.9}$$

(6.8) is the simplified condition that determines the frequency and (6.9) defines the condition of oscillation.

Hartley oscillator. Refer to Fig. 6.3, where we have shown Hartley-oscillator circuit, where $X_1 = \omega(L_1 + M)$, $X_2 = \omega(L_2 + M)$, $X_3 = -1/\omega C$, where M represents the mutual coupling of the inductances L_1 and L_2.

$\therefore$ from (6.8), we get

$$\omega_0(L_1 + M) + \omega_0(L_2 + M) - \frac{1}{\omega_0 C} = 0 \text{ or, } \omega_0(L_1 + L_2 + 2M) - \frac{1}{\omega_0 C} = 0 \text{ or, } \omega_0^2 = \frac{1}{(L_1 + L_2 + 2M)C}.$$

$$\therefore\ \omega_0 = 2\pi f_0 = \frac{1}{[(L_1+L_2+2M)C]^{1/2}}. \tag{6.10}$$

From (6.9), condition of oscillation is given by

$$h_{fe} \geq \frac{(L_2+M)}{(L_1+M)} + \frac{L_1+M}{L_2+M}\cdot\frac{h_{ie}}{R_o}. \tag{6.11}$$

Colpitts oscillator. In Colpitts oscillator the reactances X_1 and X_2 are capacitative and X_3 is the reactance of an inductor (L) [Fig. 6.6]. Hence, we have

$$X_1 = -\frac{1}{\omega C_1},\ X_2 = -\frac{1}{\omega C_2},\ X_3 = \omega L.$$

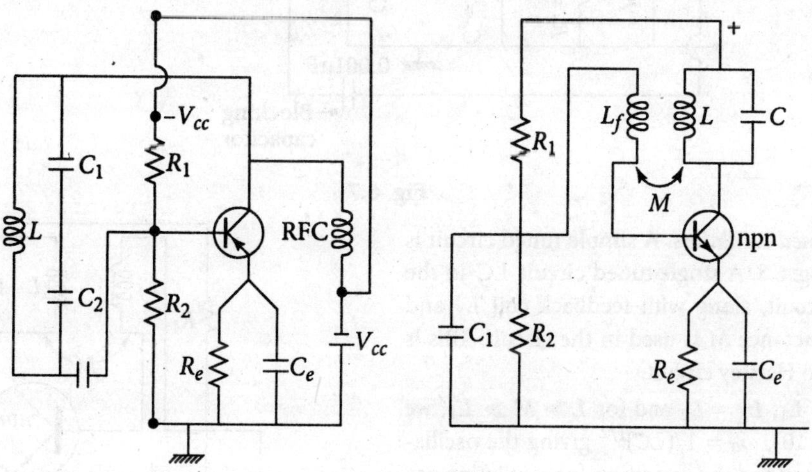

Fig. 6.6

$\therefore$ frequency of oscillation from (6.8) is given by

$$-\frac{1}{\omega_0 C_1} - \frac{1}{\omega_0 C_2} + \omega_0 L = 0 \text{ or, } \omega_0 L = \frac{(C_1+C_2)}{\omega_0 C_1 C_2} \text{ or, } \omega_0^2 L C_1 C_2 = C_1 + C_2.$$

$$\therefore\ \omega_0^2 = \frac{C_1+C_2}{LC_1C_2} \text{ or, } \omega_0 = \sqrt{\frac{C_1+C_2}{LC_1C_2}} = \sqrt{\frac{1}{L}\cdot\frac{1}{C_1C_2/(C_1+C_2)}} = \frac{1}{[L(C_1C_2)/(C_1+C_2)]^{1/2}}. \tag{6.12}$$

Condition for oscillation is

$$h_{fe} \geq \frac{C_1}{C_2} + \frac{C_2}{C_1}\cdot\frac{h_{fe}}{R_o}. \tag{6.13}$$

A practical local oscillator in a superheat receiver. We have drawn a practical local oscillator frequently used in a receiver [Fig. 6.7]. This is a Hartley type of oscillator, where AE is the oscillator coil. This generates unmodulated signal against which the modulated incoming signal is heterodyned in the mixer stage. The frequency determining network consists of coil L and the oscillator tuning and trimmer capacitors. The oscillator coil consists of the portion from A to C, the point C is at ground potential so far as ac potential is concerned, because of low reactance of C_1. The capacitor C_1 also prevents shorting of dc potential. The portion of the coil from B to C forms the output circuit since it is included between emitter and collector while the portion of the coil between C and E forms the input circuit. The point D is used as the take-off point of feeding the output of the oscillator to the mixer stage through the 0.05μF capacitor.

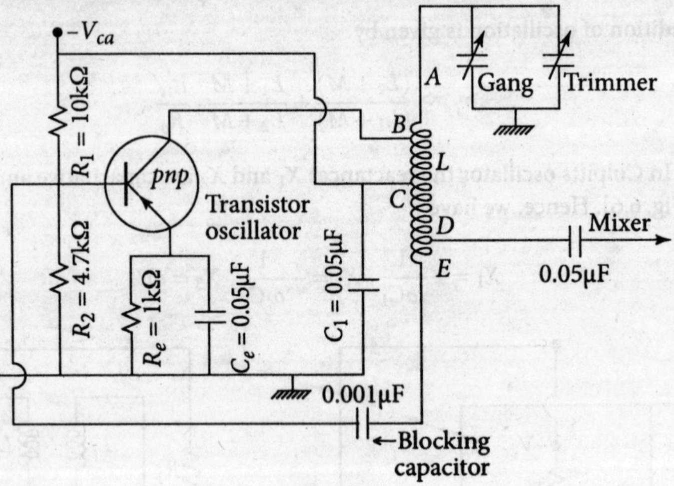

Fig. 6.7

Collector tuned oscillators. A simple tuned circuit is shown in Fig. 6.8. A single tuned circuit LC in the collector circuit, along with feedback coil L_f and mutual inductance M is used in the circuit. This is analogous to Hartley circuit.

Here $L = L_1$, $L_f = L_2$ and for $L \gg M \gg L_f$, we get from (6.10), $\omega_0 = 1/(LC)^{1/2}$ giving the oscillation frequency and the condition for oscillation as

$$h_{fe} \geq \frac{L}{M} \cdot \frac{h_{ie}}{R_o}. \qquad (6.14)$$

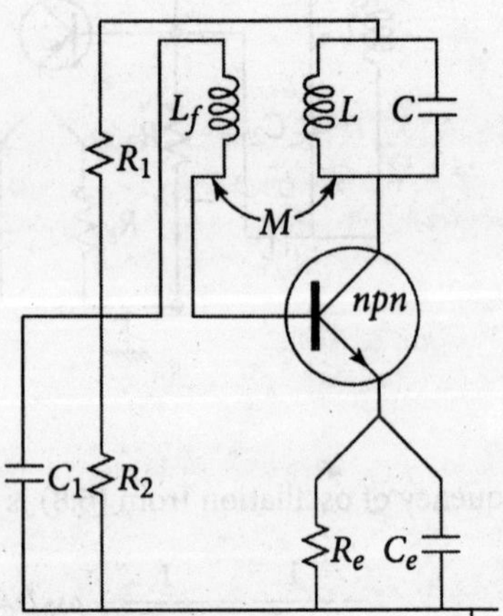

Fig. 6.8

Wien's Bridge oscillator. It is a variable frequency oscillator that uses two-stage CE amplifier providing an approximate 360° or 0° phase shift thus the frequency determining network has not to produce any additional phase shift. Such an oscillator is represented in Fig. 6.9.

The frequency determining network is shown enclosed in the dashed box. Let the equivalent impedance of $R \parallel C$ be denoted by Z_1, then the total impedance Z can be written as

$$Z = \left(Z_1 + R + \frac{1}{j\omega C} \right).$$

$$\therefore V_2 = \frac{V_1}{Z} \times Z_1 \text{ or, } \frac{V_2}{V_1} = \frac{Z_1}{Z} = \frac{R \parallel C}{R \parallel C + R + \frac{1}{j\omega C}}.$$

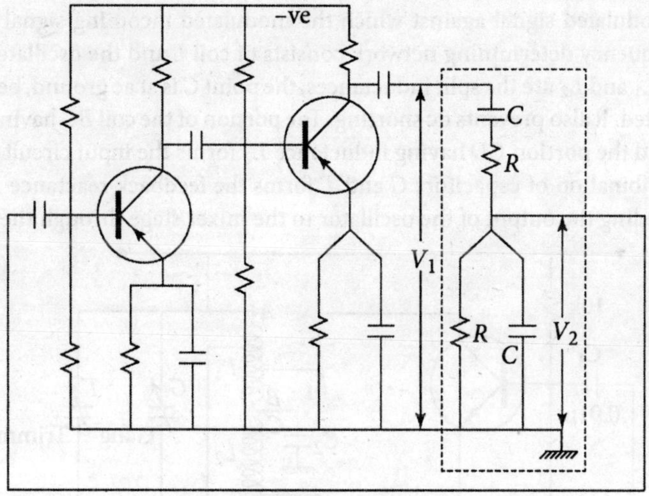

Fig. 6.9

Now,

$$R \parallel C = \frac{R \times \frac{1}{j\omega C}}{R + \frac{1}{j\omega C}} = \frac{R}{j\omega C} \times \frac{j\omega C}{(1 + j\omega CR)} = \frac{R}{(1 + j\omega CR)}.$$

$$\therefore \frac{V_2}{V_1} = \frac{\frac{R}{1+j\omega CR}}{\frac{R}{1+j\omega CR} + R + \frac{1}{j\omega C}} = \frac{R}{(1 + j\omega CR)} \times \frac{1}{\left(\frac{R}{1+j\omega CR} + R + \frac{1}{j\omega C}\right)} = \frac{R}{R + R + j\omega CR^2 + \frac{1}{j\omega C} + R}$$

$$= \frac{1}{3 + j\left(\omega CR - \frac{1}{\omega CR}\right)} = \frac{1}{3 + j\left(\omega CR - \frac{1}{\omega CR}\right)}. \tag{6.15}$$

Hence, for a phase shift of 0° or 360°, we must have $\omega_0 CR = 1/\omega_0 CR$.

$$\therefore \omega_0^2 = \frac{1}{C^2 R^2} \text{ or, } \omega_0 = \frac{1}{CR}.$$

At this frequency, $V_2/V_1 = 1/3 = K$.
(6.15) can also be put if $X_c = 1/\omega C$ as

$$\frac{V_2}{V_1} = \frac{RX_c}{3RX_c + j\left(R^2 - X_c^2\right)}. \tag{6.16}$$

When $R = X_c$,

$$\frac{V_2}{V_1} = K = \frac{1}{3}. \tag{6.17}$$

Hence, $\omega_0 = 1/CR = 2\pi f_0$.

$$\therefore f_0 = \frac{1}{2\pi CR}. \tag{6.18}$$

A practical local oscillator in a supernet radio-receiver. In Fig. 6.10 we have shown a practical local oscillator frequently used in a radio-receiver. This is a Hartley type of oscillator, where AE is the oscillator coil.

This generates unmodulated signal against which the modulated incoming signal is heterodyned in the mixer stage. The frequency determining network consists of coil L and the oscillator tuning and trimmer capacitors G and T. L_1 and L_2 are the split inductances, the point C is at ac ground, because of low reactance of C_2 w.r.t. ac generated. It also prevents dc shorting. The portion of the coil BC having inductance L_1 forms the output circuit and the portion CD having inductance L_2 forms the input circuit with reactance X_i; the reactance of the combination of capacitors G and T forms the feedback reactance X_f. The point D is the take-off point for feeding the output of the oscillator to the mixer stage through the 0.05μF capacitor.

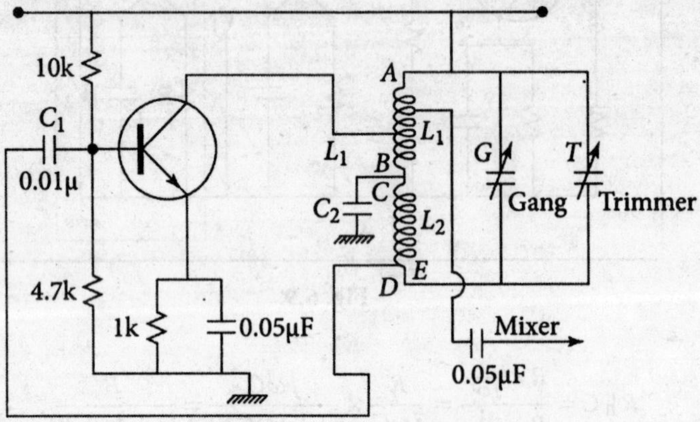

Fig. 6.10

Another form of practical Hartley oscillator. In [Fig. 6.11], we have shown a practical form of Hartley oscillator producing oscillation frequency of 15kHz. In this case, the tuning circuit of split inductance is included in the base-emitter circuit.

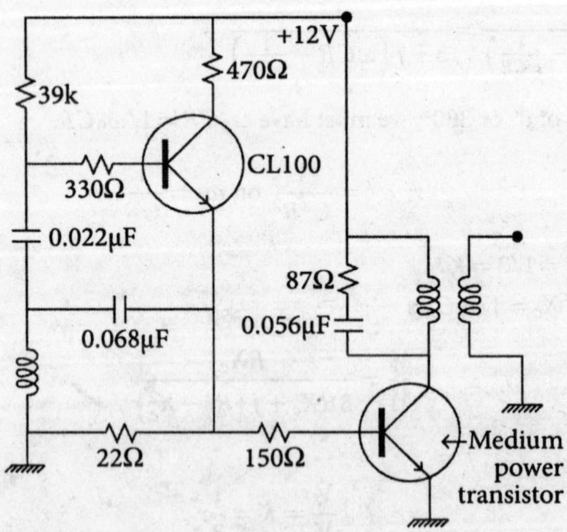

Fig. 6.11

Wien's Bridge oscillator (WBO). This oscillator is also a sinusoidal oscillator producing oscillation frequency ranging from 5Hz to 1MHz. It does not contain any LC tuning circuit or tank circuit. It is an RC phase shift oscillator and almost universally used in commercial audio generators and is preferred for other low frequency applications.

The WBO has two versions: (i) using transistors (T-version) (ii) using operational amplifier (Op-Amp version). Here we will give the description of transistor version of the WBO, the Op-Amp version will be described in a later chapter.

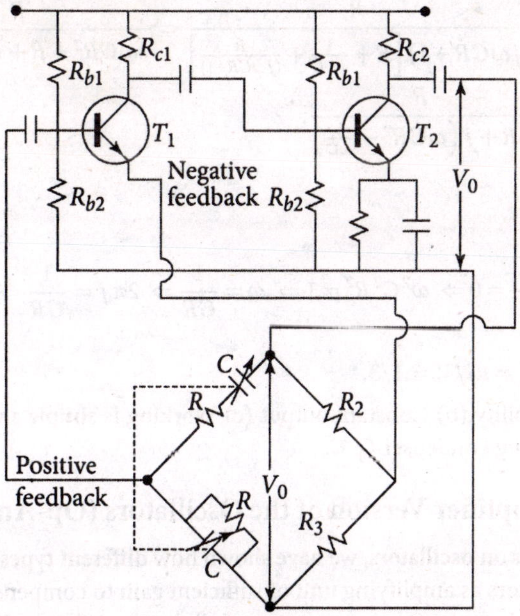

Fig. 6.12

Circuit analysis. The WBO [Fig. 6.12] is a two-stage amplifier using transistors T_1 acting as a CE amplifier and T_2 as an inverter. The two stages together produce a total phase difference of 360°, i.e., the output voltage is in phase with the input. The two capacitors C and C, one connected in series with R and the other connected in parallel with R are ganged so that the variation of their capacitances is same.

R, C in series and R, C in parallel form a Wheatstone's bridge with the two resistors R_2, R_3. The two arms of the bridge containing R, C in series and R, C in parallel form what is known as *lead-lag network.*

The functions of the lead-lag network. At low frequency, the capacitor C in series with R becomes open, so the feedback voltage $v_f = v_i$ becomes zero, the output V_o is also zero. At high frequencies C in series becomes short as well as the capacitor C in parallel. There is also no output voltage. Between these two extremes, there is a frequency f_r, called *resonance frequency,* when the output voltage reaches its maximum.

There are two types of feedback in this case: (i) positive feedback, which helps the oscillations to build up as the source is switched on; (ii) negative feedback to stabilise the output. The positive feedback is made through the lead-lag network while the negative feedback is provided through the potential divider network formed by R_2 and R_3. The resistor R_3 is usually made variable.

Theory. The impedance of R, C in series is $Z_S = R + \frac{1}{j\omega C}$.

The impedance of R, C in parallel is

$$Z_P = \frac{R \times \frac{1}{j\omega C}}{\left(R + \frac{1}{j\omega C}\right)}. \qquad \therefore\ v_f = v_i = \frac{V_o \cdot Z_P}{Z_S + Z_P}.$$

$$\Rightarrow K = \frac{v_f}{V_o} = \frac{Z_P}{Z_S + Z_P} = \frac{R \times \frac{1}{j\omega C}}{\left(R + \frac{1}{j\omega C}\right)} \times \frac{1}{\left[R + \frac{1}{j\omega C} + \frac{R}{j\omega C}\Big/\left(R + \frac{1}{j\omega C}\right)\right]}$$

$$= \frac{R}{(j\omega CR + 1)\left[R + \frac{1}{j\omega C} + \frac{R}{(j\omega CR+1)}\right]} = \frac{R}{j\omega CR^2 + R + R - \frac{j}{\omega C} + R}$$

$$= \frac{R}{3R + j\left(\omega CR^2 - \frac{1}{\omega C}\right)}.$$

For zero phase difference,

$$\omega CR^2 - \frac{1}{\omega C} = 0 \Rightarrow \omega^2 C^2 R^2 = 1 \Rightarrow \omega = \frac{1}{CR} \Rightarrow 2\pi f = \frac{1}{CR} \Rightarrow f = \frac{1}{2\pi} \cdot \frac{1}{CR}.$$

Under this condition, $K = v_f / V_o = 1/3$.

Advantages. (a) Better stability (b) Constant output (c) Working is simple and easy (d) Frequency can be changed by varying the gang condenser C, C.

6.4 Operational Amplifier Version of the Oscillators (Op-Amp Version)

In the previous discussions on oscillators, we have shown how different types of oscillators (tuned or non-tuned) work using transistors as amplifying unit of sufficient gain to compensate for the energy loss in the tank circuit. Here we will show the Op-Amp version of all these oscillators discussed earlier.

Plate tuned oscillators. In Fig. 6.13, we have shown a plate tuned oscillator using IC 741 Op-Amp as an amplifying unit of high gain. The tank circuit consisting of LC as the tuning elements is included in the inverting terminal of the Op-Amp. Once oscillation sets in having frequency determined by the values of L and C, the energy of oscillation is partly transferred to the secondary coil L_2 of the transformer so as to get feedback voltage.

The frequency of oscillation is given by

$$f = \frac{1}{2\pi} \cdot \frac{1}{\sqrt{LC}}.$$

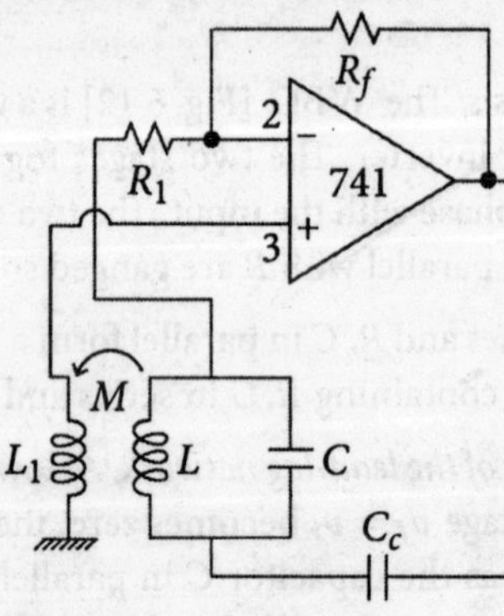

Fig. 6.13

Hartley oscillator. The frequency of oscillation is

$$f = \frac{1}{2\pi} \cdot \frac{1}{\sqrt{(L_1 + L_2)\,C}} \quad \text{[cf. Fig. 6.14]}.$$

Colpitts oscillator. The oscillation frequency of this oscillator is given by

$$f = \frac{1}{2\pi} \cdot \frac{1}{\sqrt{L[C_1 C_2/(C_1 + C_2)]}}.$$

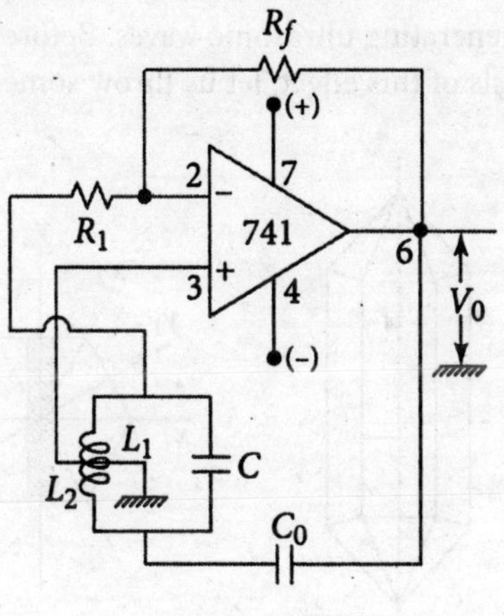

Fig. 6.14

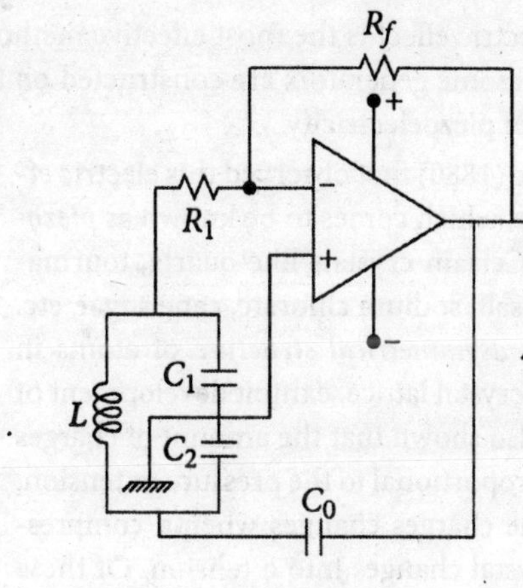

Fig. 6.15

Wien's Bridge oscillator. The Wien's Bridge oscillator using an operational amplifier is shown in Fig. 6.16. The elements R_3, R_4; R and C in series and R and C in parallel form the four-bridge arms as shown separately.

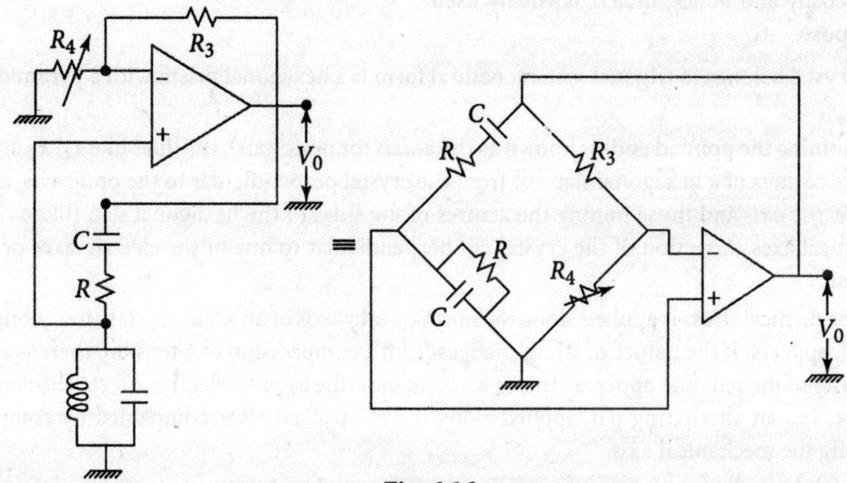

Fig. 6.16

Crystal oscillator. We have studied previously the tuned oscillators containing LC tuned circuit or RC tuned oscillator like Wien's Bridge oscillator. In either of them difficulty is that the oscillation frequency does not remain constant but it varies, because of changes in the value of R or L due to temperature change. So, in radio broadcasting, where frequency tolerance is required to be small, such oscillators cannot be recommended. Also, in some cases the frequency requirement cannot be met by resistances, capacitances and inductances. So such a problem is solved by using a piezoelectric crystal made of quartz in place of RL or RC.

What is piezoelectricity and what is the Geometry of a piezoelectric crystal?

The early methods of making high frequency oscillators are replaced by more improved and efficient techniques, one of which is the piezoelectric oscillator.

The piezoelectric effect is the most effective method of generating ultrasonic waves. Before we describe how these ultrasonic generators are constructed on the basis of this effect, let us throw some light on the phenomenon of piezoelectricity.

J and P Curie (1880) first observed this electric effect of pressure, which comes to be known as *piezoelectric effect.* Certain crystals like quartz, tourmaline, Rochelle-salt, sodium chlorate, cane sugar, etc. which possess *asymmetrical structure* of atoms in the respective crystal lattice, exhibit development of charges. It is also shown that the amount of charges developed is proportional to the pressure or tension. The sign of the charges changes when a compression of the crystal changes into a tension. Of these substances, it is true that Rochelle salt exhibits the largest piezoelectric effect, but this is not of very much use, as it is fragile and hygroscopic. It is mainly used in gramophone crystal pickups. Tourmaline is somewhat costly and hence, quartz is widely used for the purpose.

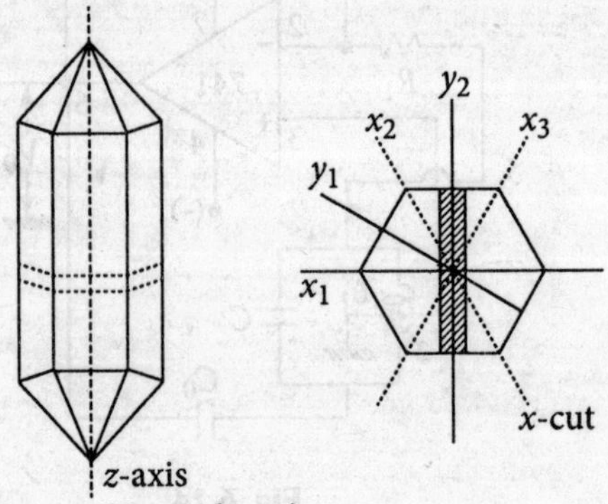

(a) (b)

Fig. 6.17

Quartz crystal belongs to trigonal system; natural form is a hexagonal prism with a pyramid at each end [Fig. 6.17].

The line joining the pointed ends is known as the *z-axis* (or optic axis); the lines like x_1, x_2 and x_3 joining the opposite corners of a hexagonal slab cut from the crystal perpendicular to the optic axis, are called the *x-axis* or *electric axis* and those joining the centres of the sides of the hexagonal slab (like y_1, y_2, etc.) are the mechanical axes. A section of the crystal cut perpendicular to one of the electric-axes or x-axes is an x-cut crystal.

Now, if mechanical stress is applied along the mechanical y-axis of an x-cut crystal, then along the electric axis x, a p.d. appears. If the nature of stress changes from a compression to a tension, there is a change also of the polarity of the p.d. that appears. This is what we describe as piezoelectric effect. The converse is also equally true, i.e., an alternating p.d. applied along the electric axis is accompanied by compression and tension along the mechanical axis.

This crystal depending on its geometry has a natural period and if the applied alternating p.d. has also this period, mechanical resonance takes place with a maximum of amplitude of vibration.

If such a piezoelectric quartz plate be placed in an alternating electric field with the direction of the field in the same sense as that of the electric axis, the quartz will undergo compression and expansion in each cycle.

The crystal thus will be set into elastic oscillation of the same period as the field and amplitude of oscillation becomes a maximum when the mechanical frequency (determined by the geometry of the crystal) coincides with the electric frequency. The vibration of the plate may take place in two possible varieties (1) Longitudinal elastic vibrations in the X-direction, known as *thickness vibration* (2) Longitudinal elastic vibrations in the Y-direction known as *length vibrations.*

In case of *length vibration*, the natural elastic frequency of the fundamental may be calculated, which is obtained from the consideration of the vibration of a bar. The frequencies of the possible vibrations are given by $f_n = \frac{n}{2l}\sqrt{\frac{E}{\rho}}$.

∴ for fundamental,

$$f_1 = \frac{1}{2l}\sqrt{\frac{E}{\rho}}.$$

Now, $E = 7.722 \times 10^{11}$ dynes/cm^2, $\rho = 2.65$ (for quartz).

$$\therefore\ f_1 = \frac{1}{2l}\sqrt{\frac{E}{\rho}} = \frac{1}{2l}\sqrt{\frac{7.722 \times 10^{11}}{2.65}} = \frac{272500}{l}\ \text{cycles/sec.}$$

In case of thickness vibration, the fundamental frequencies of vibration may be developed according to Voigt. It is given by

$$f_1' = \frac{1}{2t}\sqrt{\frac{C_{11}}{\rho}},$$

where t = thickness of the plate, C_{11} = characteristic modulus of elasticity for this type of vibration and direction = 8.546×10^{11} dynes/cm^2 (for quartz according to Voigt).

$$\therefore\ f_1' = \frac{1}{2t}\sqrt{\frac{8.546 \times 10^{11}}{2.65}} = \frac{285500}{t}\ \text{cycles/sec.}$$

Role of the crystal. The action of the crystal can be represented by its equivalent electrical characteristics as having R (resistance due to crystal structures internal friction); L and C represent electrical equivalents of its mass and compliance [Fig. 6.18]. The capacitor C_M shunting the series leg R, L, C represents the capacitive effect of the mounting of the crystal. Since R is very small, the quality factor Q of the circuit is very high. The equivalent representation of the crystal [Fig. 6.18] shows that there are two resonant frequencies f_1 and f_2 corresponding to the minimum and maximum values of the impedance of the RLC leg. The minimum impedance corresponds to series resonance when the reactances of L and C are equal and opposite leaving us with $Z_{min} = R$. The maximum impedance [Fig. 6.19] occurs at higher frequency f_2 when the reactance of the series leg happens to be equal to that of C_M. This is referred to as parallel or anti-resonant condition of the crystal.

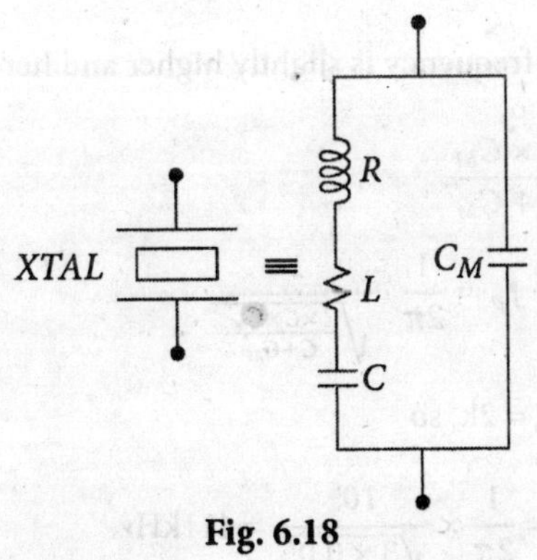

Fig. 6.18

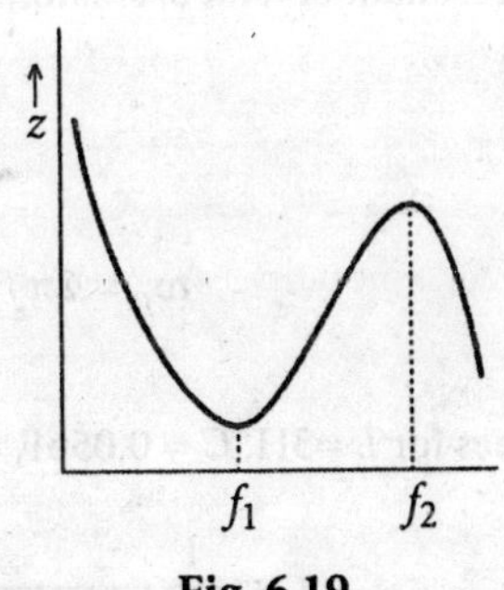

Fig. 6.19

Series resonant crystal oscillator circuit. [Fig. 6.20] shows a series resonant crystal oscillator, the crystal is connected as a series element in the feedback path. The electrical equivalent of the crystal, so far as its functioning is concerned, is represented by a series leg of LCR shown in the dashed box in the figure. RFC isolates the dc power line from the ac influence. The capacitor C_e acts as the ac bypass, so that dc bias of the transistor provided by R_1, R_2, and R_e is stabilised. The feedback from the collector to the base is a maximum when the impedance of the crystal XTAL is minimum (series resonant mode). The capacitor C_e acts as the dc block between the collector and base.

The frequency of the oscillator is set by the series resonant frequency of the crystal. The frequency remains constant in spite of variation of supply voltage, transistor parameters, etc.

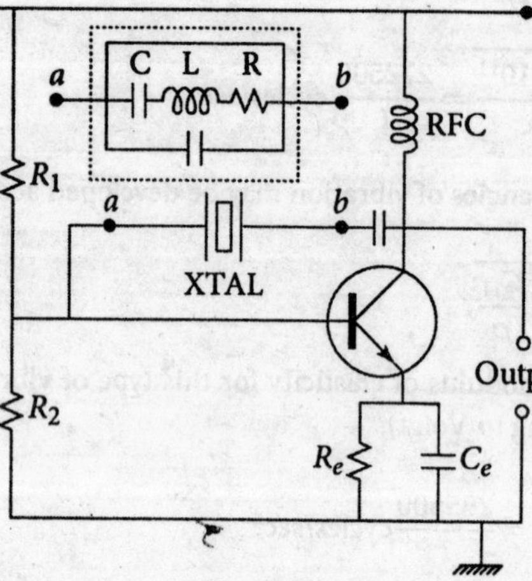

Fig. 6.20

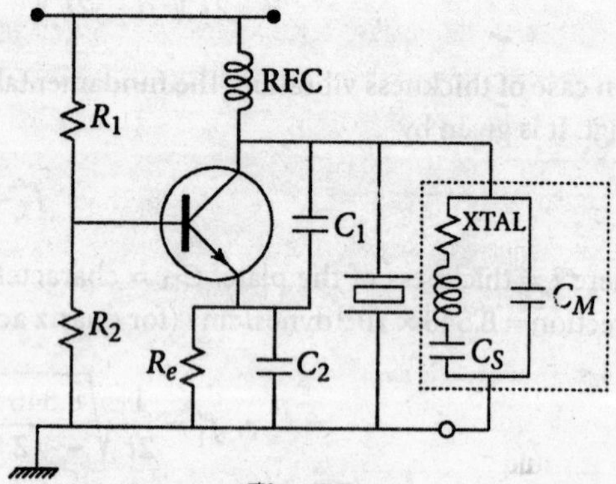

Fig. 6.21

Parallel-resonant circuit oscillator. This oscillator circuit is shown in Fig. 6.21. In parallel resonant circuit, the impedance is maximum, so the crystal behaves as an element of large L value. So it is in reality a Colpitts type of oscillator with C_1 and C_2 as the split capacitance and the voltage across C_2 is feedback to the emitter.

Calculation of frequency of the oscillator. In series resonant oscillator, the impedance of the crystal is a minimum and so the reactance of L and C of the crystal will cancel each other, leaving us with the resistance R (small) as the impedance.

$$\therefore\ \omega L = \frac{1}{\omega C} \Rightarrow \omega^2 = \frac{1}{LC} \Rightarrow f_s = \frac{1}{2\pi} \cdot \frac{1}{\sqrt{LC}}.$$

In parallel-resonant crystal oscillator, the resonant frequency is slightly higher and here, the equivalent capacitance is

$$C_e = \frac{C \times C_M}{C + C_M}.$$

$$\omega_p = 2\pi f_p = \frac{1}{\sqrt{LC_e}} \Rightarrow f_p = \frac{1}{2\pi} \cdot \frac{1}{\sqrt{\frac{L \times C \times C_M}{C + C_M}}}.$$

Typical values for $L = 3$H, $C = 0.05$pF, $C_M = 10$pF, $R = 2$k, so

$$f_s = \frac{1}{2\pi} \cdot \frac{1}{\sqrt{3 \times 0.05 \times 10^{-12}}} = \frac{1}{2\pi} \times \frac{10^6}{\sqrt{3 \times 0.05}} = 411\text{kHz}.$$

$$f_p = 412\text{kHz}.$$

6.5 Multivibrator (Transistor Version)

What is a Multivibrator?

An electronic device to generate non-sinusoidal oscillations of waveforms such as square, rectangular or sawtooth is called a *multivibrator.* Any non-sinusoidal but periodic oscillation like a square-wave, rectangular or sawtooth contains in it a large number of sinusoidal vibration of different frequencies, hence the device is named as multivibrator.

What is the Essence of a Multivibrator?

In essence a multivibrator is an electronic switch switching on one stage of amplifier or the other. Every multivibrator consists of two transistors, which are thrown alternately into ON-state or OFF-state.

In block diagram the two-transistor amplifier stages are shown in Fig. 6.22. It is a case of 100 % feedback, the output can be taken from across the first stage or the second stage. Let us take the output from the second. At any instant one transistor remains ON, the other is in the off-state. Suppose, Q_2 is on, collector current $I_{C(sat)}$ flows for a certain length of time represented by ab [Fig. 6.23] called the *on-time* after which it is thrown into off-state. The off-time is indicated by the length ce, during which the transistor Q_1 for the first time is thrown into conduction. Again, the transistor Q_2 is turned on and the sequence is repeated. So we would get a square-wave shown as $abcef$....

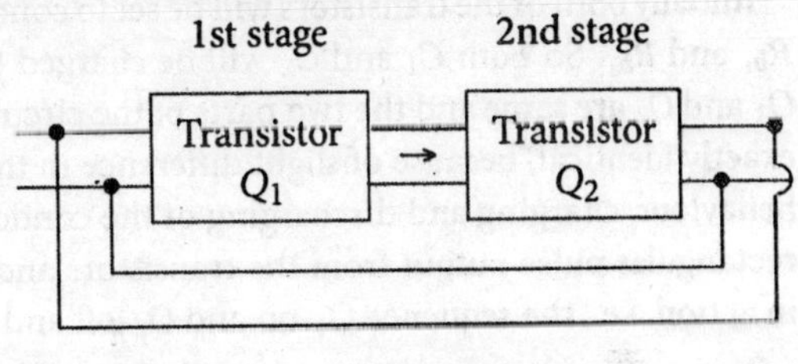

Fig. 6.22

Thus, we see that a multivibrator is a two-stage amplifier, the output of one is fedback to the input of the other offering two possible states of the transistors Q_1 and Q_2

$$\begin{matrix} Q_1 \\ Q_2 \end{matrix} \begin{Bmatrix} \text{off} \\ \text{on} \end{Bmatrix} \begin{Bmatrix} \text{on} \\ \text{off} \end{Bmatrix}$$

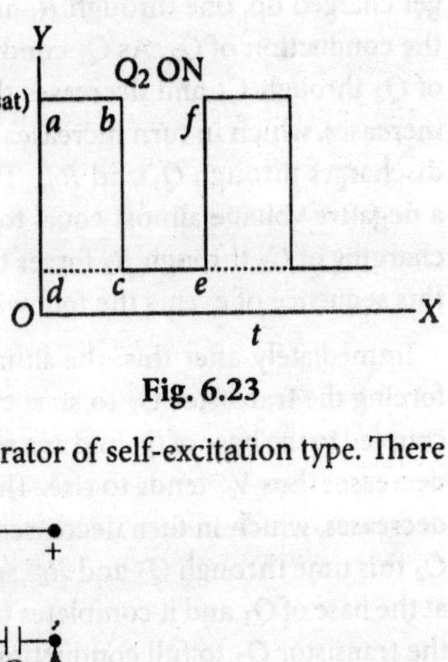

Fig. 6.23

Multivibrators are broadly classified as: (a) Astable or free running (b) Monostable or one-shot (c) Bistable or flip-flop.

(a) *Astable multivibrator.* In this case, the vibrator is a square-wave generator of self-excitation type. There is no external triggering needed for its operation.

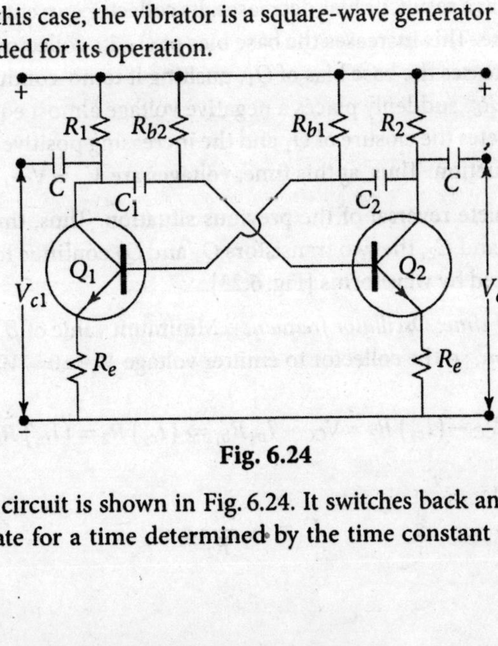

Fig. 6.24

The astable multivibrator circuit is shown in Fig. 6.24. It switches back and forth from one state to the other, remaining in each state for a time determined by the time constant of the discharging capacitive circuit. The circuit really consists of two RC-coupled amplifier stages in CE-mode; the feedbacking to any transistor is positive feedbacking.

Explanation of the circuit. The two amplifying units are identical, but the two transistors Q_1 and Q_2 though identical will differ in their characteristics. So, as the dc supply is switched on, one of them will conduct more than the other and this leads to a series of events.

Suppose, the transistor Q_1 conducts more than the other transistor Q_2. The result is that the collector voltage of Q_1, viz., V_{c_1} drops more rapidly than that of Q_2 and the collector voltage of Q_1 is reducing. This action decreases the forward bias to the base of the transistor Q_2 which will be pushed gradually to off-state, the output of Q_2 tends to attain a voltage equal to that of the supply V_{CC}.

Initially both of the transistors will be set to conduct, because of the base bias supplied by the bias resistors R_{b_1} and R_{b_2}. So both C_1 and C_2 will be charged through R_1 and R_2. But even though the two transistors Q_1 and Q_2 are same and the two parts of the circuit are identical, the behaviour of the two parts will not be exactly identical, because of slight difference in the characteristics of the transistors. This difference in the behaviour, charging and discharging of the condensers lead to a series of events to take place resulting in rectangular pulse output from the transistors and automatic changeover from one transistor to the other in action, i.e., the sequence Q_1-on and Q_2-off and Q_2-on and Q_1-off continues to be repeated.

Suppose, the transistor Q_1 conducts more than the other transistor Q_2. So both capacitors C_1 and C_2 get charged up, one through R_1 and the other through R_2. The capacitor C_1 charges more rapidly during the conduction of Q_2. As Q_1 conducts more, V_{c_1} falls more rapidly and this decrease is coupled to the base of Q_2 through C_1 and decreases the base bias of Q_2. Hence, the collector current of Q_2 decreases and V_{c_2} increases, which in turn increases the base bias of Q_1 causing more conduction of Q_1. In the mean time C_1 discharges through Q_1 and R_{b_1}. The initial pulse of discharge current from C_1 through R_{b_1} and Q_1 places a negative voltage almost equal to the supply voltage at the base of Q_2 forcing it to be completely off and charging of C_2 through R_2 forces the transistor Q_1 to pass into full conduction. Thus, we have at the end of this sequence of events the following voltage situations $V_{c_1} \simeq 0$, $V_{c_2} = V_{CC}$, $V_{b_1} \simeq 0$, $V_{b_2} = -V_{CC}$.

Immediately after this, the situation begins to change. The capacitor C_1 quickly recharges through R_1 forcing the transistor Q_2 to start conducting. As Q_2 turns on, V_{c_2} goes on decreasing, and this decrease is coupled to the base of Q_1 and as a result, its bias decreases. Its collector current (which was previously $I_{C(sat)}$) decreases thus V_{c_1} tends to rise. This increases the base bias of Q_2, its collector current increases, hence V_{c_2} decreases, which in turn decreases the base bias of Q_1, pushing it to no-conduction state. The discharge of C_2 this time through Q_2 and R_{b_2} suddenly places a negative voltage almost equal to the supply voltage V_{CC} at the base of Q_1 and it completes the closure of Q_1 and the increasing positive bias to the base of Q_2 pushes the transistor Q_2 to full conduction. Thus, at this time, voltages are $V_{c_1} \simeq V_{CC}$, $V_{c_2} \simeq 0$, $V_{b_1} = -V_{CC}$, $V_{b_2} \simeq 0$.

So, the situation is a complete reversal of the previous situation. Thus, through the charging and discharging of the capacitors C_1 and C_2, the two transistors Q_1 and Q_2 continue to be in the on- and off-states. These situations are represented by waveforms [Fig. 6.25].

Calculation of on-time and off-time: Oscillator frequency. Minimum value of β for which oscillation can set in: Suppose, Q_2 is in saturation, so the collector to emitter voltage $V_{c_2} \simeq 0 = V_{b_2}$.

$$\therefore\ V_{CC} - (I_{c_2})R_2 = V_{CC} - I_{b_2}R_{b_1} \Rightarrow (I_{c_2})R_2 = (I_{b_2})R_{b_1}.$$

$$I_{c_2} = I_{b_2}\frac{R_{b_1}}{R_2}.$$

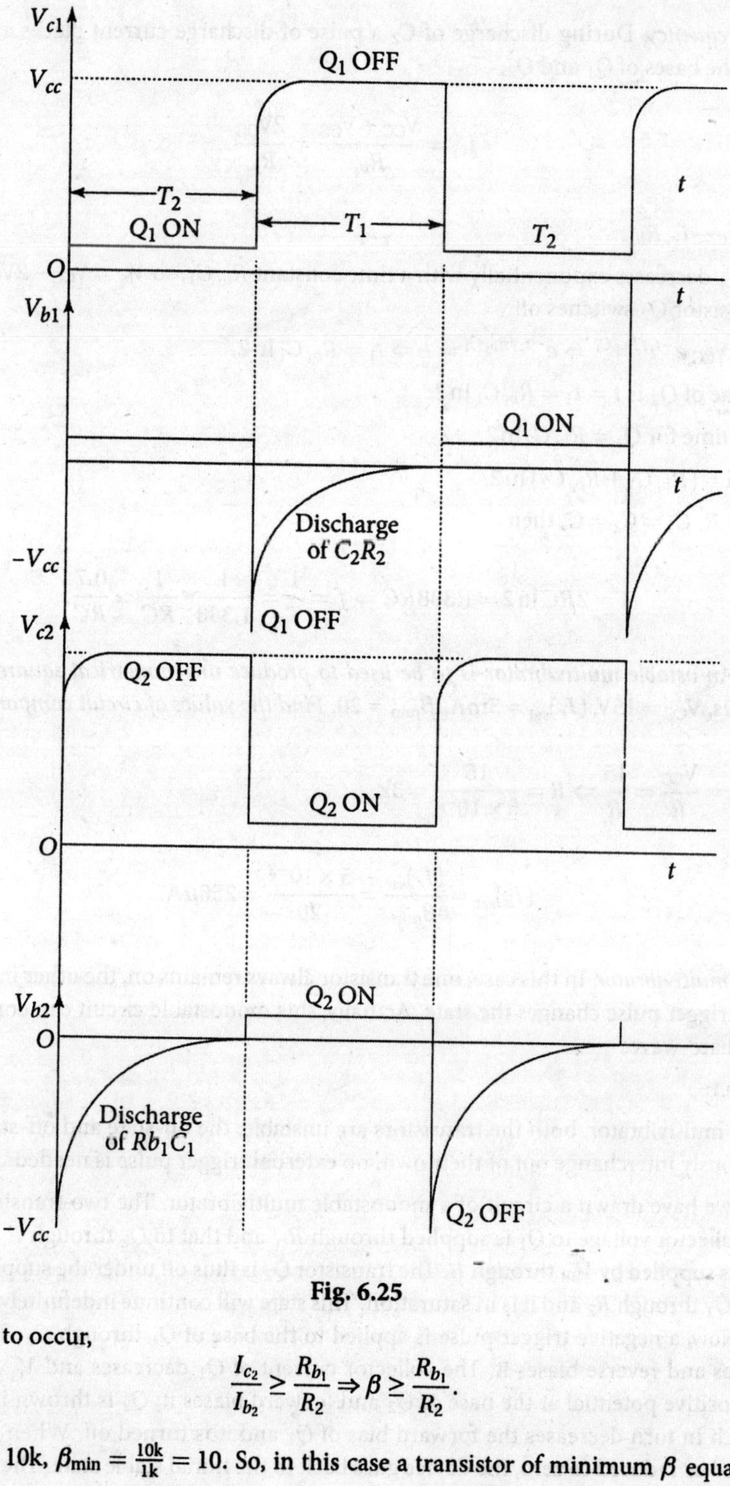

Fig. 6.25

So for saturation to occur,

$$\frac{I_{c_2}}{I_{b_2}} \geq \frac{R_{b_1}}{R_2} \Rightarrow \beta \geq \frac{R_{b_1}}{R_2}.$$

If $R_2 = 1\text{k}$, $R_{b_1} = 10\text{k}$, $\beta_{\min} = \frac{10\text{k}}{1\text{k}} = 10$. So, in this case a transistor of minimum β equal to 10 is to be used.

Calculation of frequency. During discharge of C_2 a pulse of discharge current places a negative potential equal to V_{CC} at the bases of Q_1 and Q_2,

$$i_{c_1} = \frac{V_{CC}+V_{CC}}{R_{b_1}} = \frac{2V_{CC}}{R_{b_1}}.$$

Now, $V_{b_2} = V_{CC} - i_{c_1} R_{b_1}$.

The current i_{c_1} decreases exponentially with a time constant $R_{b_1}C_1$, so $V_{b_2} = V_{CC} - 2V_{CC}\,e^{-t/R_{b_1}C_1}$. When $V_{b_2} \simeq 0$, the transistor Q_2 switches off.

$\therefore\ 0 = V_{CC} - 2V_{CC}\,e^{-t_1/R_{b_1}C_1} \Rightarrow e^{-t_1/R_{b_1}C_1} = \frac{1}{2} \Rightarrow t_1 = R_{b_1}C_1 \ln 2.$

So, the off-time of Q_2 is $t = t_1 = R_{b_1}C_1 \ln 2$.

Similarly, off-time for $Q_1 = R_{b_2}C_2 \ln 2$.

$\therefore$ total period $= (R_{b_1}C_1 + R_{b_2}C_2)\ln 2$.

If $R_{b_1} = R_{b_2} = R$, $C_1 = C_2 = C$, then

$$T = 2RC\ln 2 = 1.388RC \Rightarrow f = \frac{1}{T} = \frac{1}{1.388}\cdot\frac{1}{RC} \simeq \frac{0.7}{RC}.$$

Example 6.1. *An astable multivibrator is to be used to produce unsymmetrical square-wave of durations* $310\mu s$ *and* $250\mu s$, V_{CC} = 15V, $(I_c)_{sat}$ = 5mA, β_{min} = 20. *Find the values of circuit components.*

Solution: $(i_c)_{sat} = \dfrac{V_{CC}}{R} = \dfrac{15}{R} \Rightarrow R = \dfrac{15}{5\times 10^{-3}} = 3\text{k}.$

Now,

$$(I_b)_{sat} = \frac{(I_c)_{sat}}{\beta_{min}} = \frac{5\times 10^{-3}}{20} = 250\mu\text{A}.$$

(b) *Monostable multivibrator.* In this case, one transistor always remains on, the other in the off-state unless some external trigger pulse changes the state. Actually, this monostable circuit can convert a pulse of any shape into a square-wave.

Circuit diagram.

In an astable multivibrator, both the transistors are unstable; the on-state and off-state of the two transistors continuously interchange out of their own, no external trigger pulse is needed.

In Fig. 6.26, we have drawn a circuit of a monostable multivibrator. The two transistors Q_1 and Q_2 are identical; the collector voltage to Q_1 is supplied through R_{c_1} and that to Q_2 through R_{c_2}. The transistor Q_2 is in reverse bias supplied by V_{BB} through R. The transistor Q_2 is thus off under the supply source V_{CC} which forward biases Q_1 through R_2 and it is in saturation. This state will continue indefinitely, hence it represents a stable state. Now, a negative trigger pulse is applied to the base of Q_1 through C_2. This pulse decreases the forward bias and reverse biases it. The collector current of Q_1 decreases and V_{c_1} increases, which in turn places a positive potential at the base of Q_2 and forward biases it; Q_2 is thrown into conduction. V_{c_2} decreases, which in turn decreases the forward bias of Q_1 and it is turned off. When the external trigger pulse is removed or reduced to zero, the device goes back to the initial stable state. The waveform available at the output [Fig. 6.27].

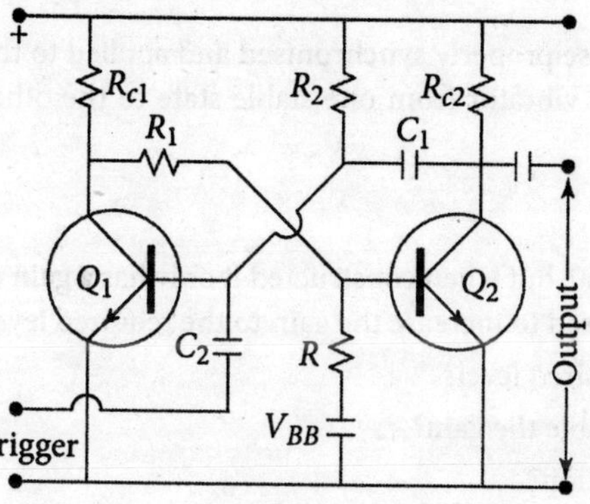

Fig. 6.26

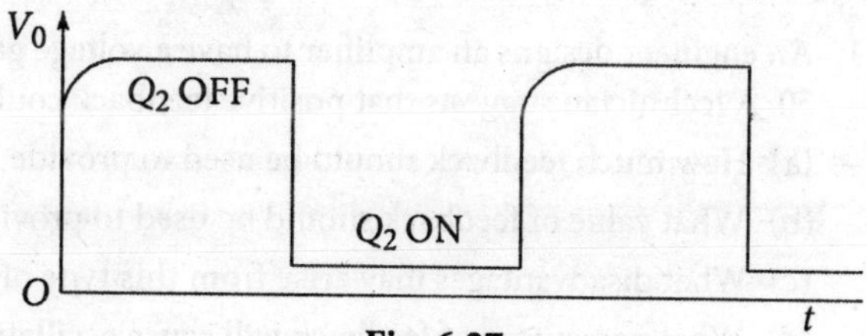

Fig. 6.27

(c) ***Bistable multivibrator.*** **In this design both the transistors have stable state, they can remain either in off-state or on-state if no external trigger is applied.**

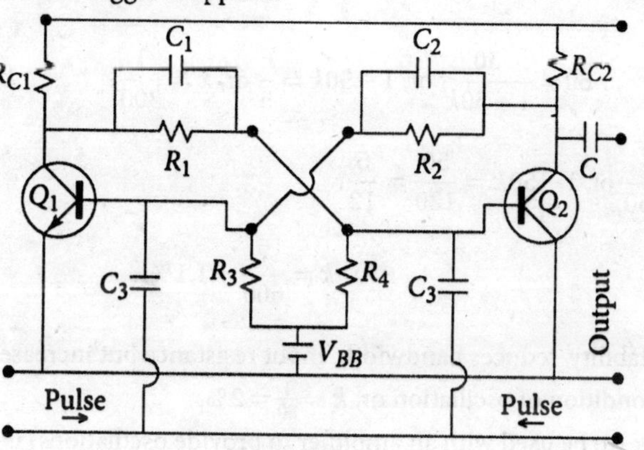

Fig. 6.28

The circuit [Fig. 6.28] shows a bistable multivibrator consisting of two identical sections, one containing transistor Q_1, load resistor R_{c_1} biased by voltage source V_{BB}, and the other section contains an identical transistor Q_2, load resistor $R_{c_2} = R_{c_1}$ biased by the same voltage source V_{BB} through R_4. Resistors R_1 and R_2 are shunted by capacitors C_1 and C_2 which control the rise and fall times of the circuit and hence, they improve switching characteristics of the circuit and help to remove distortion in the square-wave output.

Operation. As the dc supply is switched on, both of Q_1 and Q_2 will start conducting; though the transistors are alike, still there will be some differences in their characteristics, because of which there will be difference in their behaviour. Suppose, Q_1 conducts more than Q_2, then V_{c_1} decreases and this decrease is coupled to the base of Q_2 thus decreasing the base-bias, hence its collector current decreases. V_{c_2} thus increases, which in turn increases the base-bias of Q_1. Thus, in a cumulative manner the two transistors Q_1 and Q_2 will soon pass into full conduction and complete closure state respectively. The circuit will remain in this condition unless compelled by some trigger pulse to change the state.

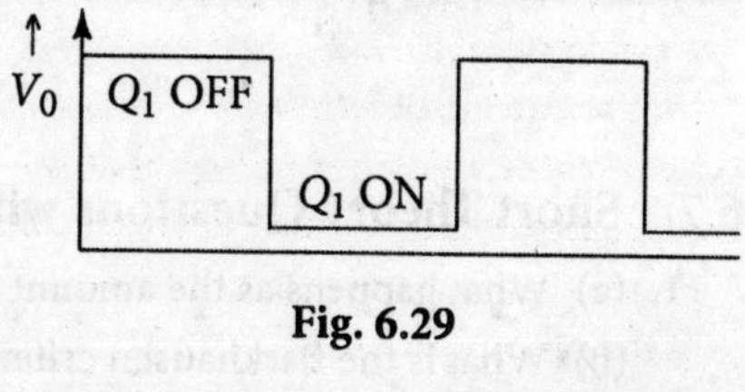

Fig. 6.29

If in the stable state of affair to the base of Q_1 which has passed into full conduction, a negative sawtooth pulse is injected and a positive sawtooth pulse injected into the base of Q_2 in cut-off, the transistor Q_1 will soon pass into non-conducting state and Q_2 will be pushed to closure in a cumulative manner. This will be another stable state of the vibrator. Thus, under trigger pulse properly synchronised and applied to the bases of Q_1 and Q_2, there will be continual transition of the vibrator from one stable state to the other producing square-wave pulses at the output [Fig. 6.29].

6.6 Numerical Problems

1. An engineer designs an amplifier to have a voltage gain of 60, but when constructed it only has a gain of 50. A technician suggests that positive feedback could be used to increase the gain to the required level.
   (a) How much feedback should be used to provide the desired level?
   (b) What value of feedback should be used to provide double the gain?
   (c) What disadvantages may arise from this type of operation?
   (d) What percentage of feedback will cause oscillation?

Solution:

(a) $A_{vf} = \frac{A}{1-kA}$.

$$\therefore 60 = \frac{50}{1-50k} \text{ or, } 1-50k = \frac{6}{5} \text{ or, } k = \frac{1}{300}. \qquad \therefore k = 0.33\,\%.$$

(b) $120 = \frac{50}{1-k\times 50}$ or, $1-50k = \frac{50}{120} = \frac{5}{12}$.

$$\therefore k = \frac{7}{600} = 1.1\,\%.$$

(c) It brings in instability, reduces bandwidth, input resistance but increases distortion and noise.

(d) $kA = 1$ is the condition of oscillation or, $k = \frac{1}{50} = 2\%$.

2. A feedback network (to be used with an amplifier to provide oscillations) is tested and found to give an output of 0.0125V with a 0.5V input.
   (a) What must be the minimum gain of the amplifier to provide oscillations?
   (b) What will be the effect on output if the amplifier's gain is twice that required by the Barkhausen criterion?
   (c) What percentage of negative feedback should be introduced to the amplifier so that linear oscillations result?

Solution:

(a) $k = \frac{0.0125}{0.5} = \frac{1}{40}$. The condition for oscillation is $kA = 1$. $\therefore A = 40$.

(b) Non-linear oscillations will occur.

(c) $A_{vf} = \frac{A}{1+kA}$.

$$\therefore k = -\frac{1}{A} = -\frac{1}{80} \text{ or, } k = -1.25\,\%.$$

6.7 Short Theory Questions with Answers

1. (a) What happens as the amount of positive feedback introduced to an amplifier is increased?
   (b) What is the Barkhausen criterion for oscillation?

(c) When are linear oscillations obtained?

Ans. (a) Gain will increase until a stage reaches when the amplifier stops amplifying and circuit starts oscillating.

(b) The criterion is $kA = 1$, where k is the feedback factor, A is the voltage gain without feedback.

(c) For linear oscillations in which the output is relatively free from distortion, kA should be slightly greater than unity.

2. How do resonant circuits of RF oscillators differ from most audio-oscillators?

Ans. In resonant circuit of RF oscillator, the oscillation frequency determined by the elements of the tank is a fixed frequency, whereas an audio-oscillator is a variable frequency oscillator.

3. What factors determine the maximum frequency of oscillation of a transistor?

Ans. As the frequency increases it is not the external circuit that determines the frequency of the oscillator but here, we have to consider the interior of the transistor so far as its capacitative effects are concerned. Using the high frequency model (see Chapter on *Amplifier*), the maximum frequency of oscillation for a given transistor, beyond which Barkhausen criterion does not hold is given by $f_{max} \simeq \sqrt{f_T/8\pi r_{bb'} C_{b'c}}$, where $r_{bb'}$ = base spreading resistance, $C_{b'c}$ = collector-base junction capacitance, f_T = current gain-bandwidth product.

So it is the (a) junction capacitance (b) junction resistance (c) bandwidth product that determine the maximum frequency of oscillation of a transistor.

4. What are the distinguishing features of (a) the Colpitts (b) the Hartley oscillators.

Ans. Both are RF oscillator using tank circuits. The Colpitts oscillator consists of two condensers in series while Hartley oscillator consists of a tapped inductance and a capacitor across it. The frequency of the Colpitts oscillator is given by

$$f_{Col} = \frac{1}{2\pi}\sqrt{\frac{1}{LC}}, C = C_1C_2/(C_1 + C_2),$$

whereas the frequency of the Hartley oscillator is given by $f_{Hart} = 1/2\pi\sqrt{1/LC}$, where L is the total inductance of the tapped coil.

6.8 Exercises

1. For the Hartley Oscillator derive an expression for oscillation frequency when M is very small.
2. For the collector tuned oscillator derive the condition for resonant frequency.
3. Distinguish between a Colpitts Oscillator and a Hartley Oscillator. What type of oscillator is the local oscillator of a radio receiver? Describe the operation of any one of the two and derive an expression for the resonant frequency.
4. A Colpitt oscillator is used as the local oscillator in an AM radio receiver to produce frequencies from 1 to 2 MHz.

   (a) What must be the inductance of the coil if the minimum capacitance obtainable is 43pF (C_1 and C_2)?

   (b) What must be the maximum value of C to produce the necessary frequencies?

   (c) What frequency would be produced if this coil were used in a Hartley oscillator circuit with C = 100pF?

Chapter 7

Field Effect Transistors

FET and MOSFET

7.1 Introduction

We will now consider another type of transistor distinctly different from the transistor that we so long discussed, a type of transistor that finds many applications as in noiseless amplifier in FM tuner, in both digital and analog circuits as switch. This new type of transistor is of two varieties, one called JFET (*Junction Field Effect Transistor*), the other called *metal oxide semiconductor* FET or more commonly called MOS-FET.

Difference between a FET *and bipolar transistor* (FET vs. BJT). The FET transistor has a few special features that distinguish between this type of transistor and the BJT transistor.

1. In an FET, current is due to only one type of carriers, the majority carriers, either electrons on holes unlike in a BJT, where both electrons and holes conduction constitutes the current. Hence, an FET is a unipolar device, whereas a BJT is a bipolar device.
2. The FET has a very high input resistance of the order of 100 MΩ for JFET and 10^{10} to $10^{15}\Omega$ for MOS-FET. So an FET is a voltage-controlled device unlike a BJT which is a current-controlled device.
3. FET is less noisy than the usual BJT.
4. Because of high input resistance, the time constant of the input circuit is large enough to enable the charge stored on the small input capacitance to remain sufficiently long for the device to be used as a storage element in digital circuits.
5. FET can dissipate high power and switch large current in several nano-seconds, hence it can be used as a high frequency, high power switch.

7.2 Junction Field Effect Transistor (JFET)

Junction Field Effect Transistor (JFET) *current direction if the gate was forward-biased.* The majority carriers, electrons in *n*-type FET, flow through the *n*-channel and out through the drain.

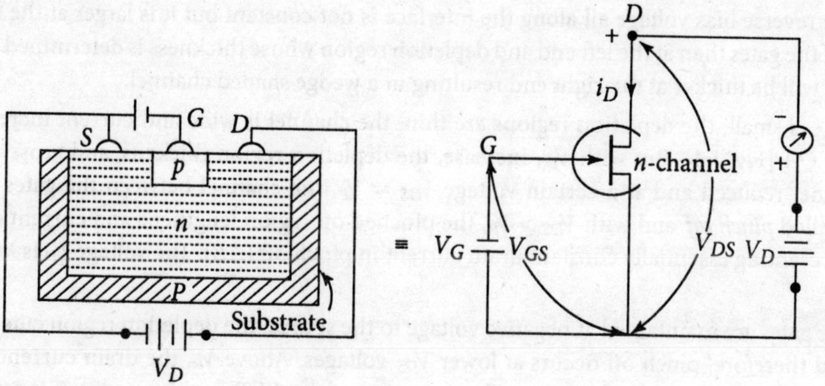

Fig. 7.1

Principle of operation of a **FET. The drain current in an FET is controlled by the narrowing of the channel caused by the gate.**

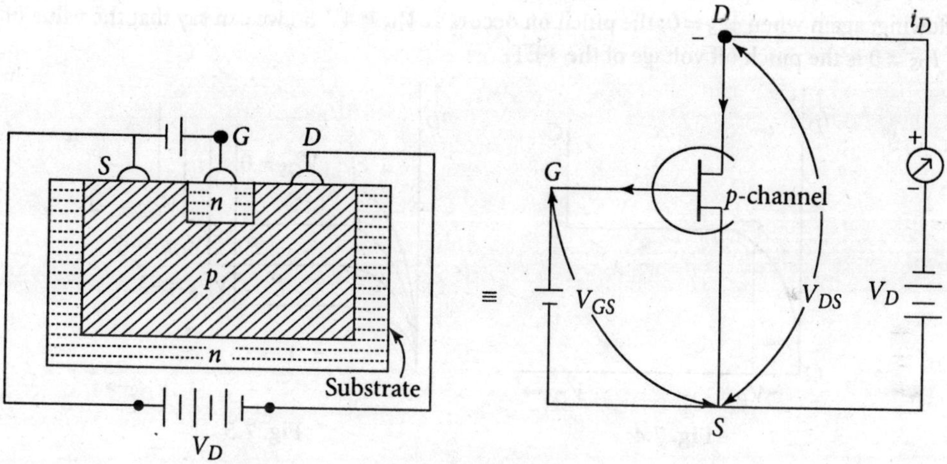

Fig. 7.2

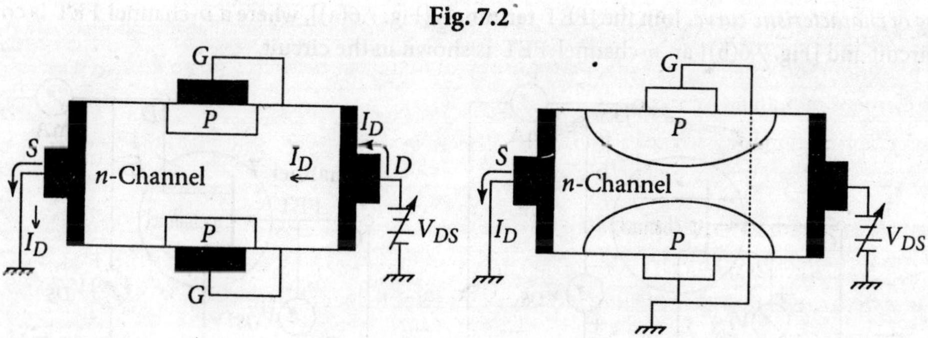

Fig. 7.3

When a voltage V_{DS} is applied between source and drain (S and D), electrons in the channel of n-type FET move from S to D. There is a voltage gradient along the length of the bar thus the gates connected to the source, the gate channel interface junctions are reverse-biased, the channel will be positive to the *p*-gates. The reverse bias voltage all along the interface is not constant but it is larger at the right end, i.e., drain end of the gates than at the left end and depletion region whose thickness is determined by the reverse bias voltage will be thicker at the right end resulting in a wedge shaped channel.

When V_{DS} is small, the depletion regions are thin, the channel is wide and current increases with V_{DS} (cf. portion *OA* [Fig. 7.4]. But with V_{DS} increase, the depletion region thickens, and cross-sectional area of the channel reduced and at a certain voltage $V_{DS} = V_P$, the channel between the gates nears closure (point *A*) called *pinch off* and with $V_{DS} > V_P$, the pinched off region lengthens and current curve shows a plateau representing essentially constant drain current in pinch off (*AB*). The voltage V_P is known as *pinch off voltage*.

When the gates are maintained at negative voltage to the source, the depletion region cuts more into the channel and therefore, pinch off occurs at lower V_{DS} voltages. Above V_P, the drain current is very much influenced by the gate source voltage V_{GS}. The operation of the FET in this condition is said to be *depletion mode* operation. Such FETs are called DEFET and here gate channel is reverse-biased. Sudden steep increase in the characteristic curve (portion *BC*) at a higher value of V_{DS} voltage is due to avalanche voltage breakdown in the depletion region between the gate and drain. This voltage is known as *breakdown voltage denoted* as BV_{DSO}. An important parameter in majority FETs is that when $V_{GS} = -4V$ the drain current stops flowing; again when $V_{GS} = 0$, the pinch off occurs at $V_{DS} = 4V$. So we can say that the value of V_{GS} for which $I_{DS} = 0$ is the pinch off voltage of the FET.

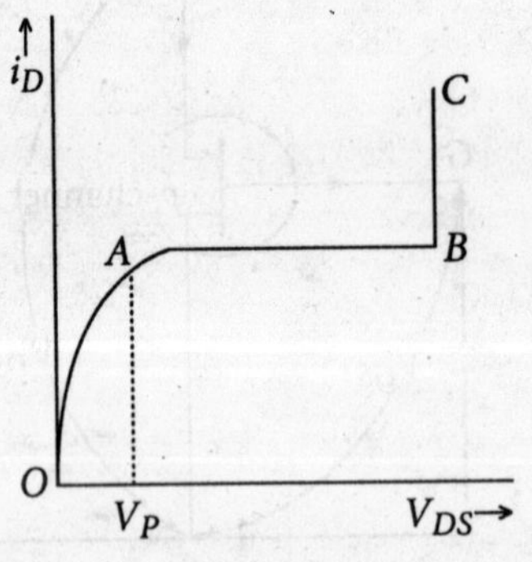

Fig. 7.4

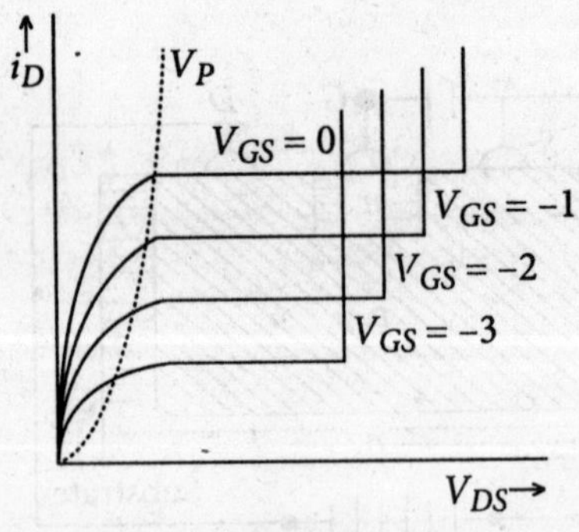

Fig. 7.5

Drawing of characteristic curve. Join the JFET terminals [Fig. 7.6(a)], where a *p*-channel FET is connected in the circuit and [Fig. 7.6(b)] an *n*-channel FET is shown in the circuit.

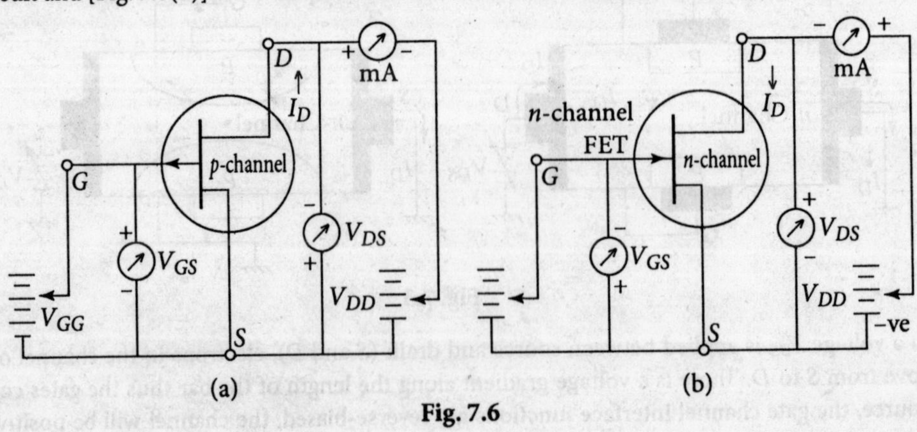

Fig. 7.6

Keep $V_{GS} = 0$ and vary V_{DS} and record the corresponding i_D. This is repeated for V_{GS} = 1V, 2V, 3V, etc. for p-FET and −1, −2, −3 volts, etc. for n-channel FET. We get a series of characteristic curves shown [Fig. 7.7(a), (b)].

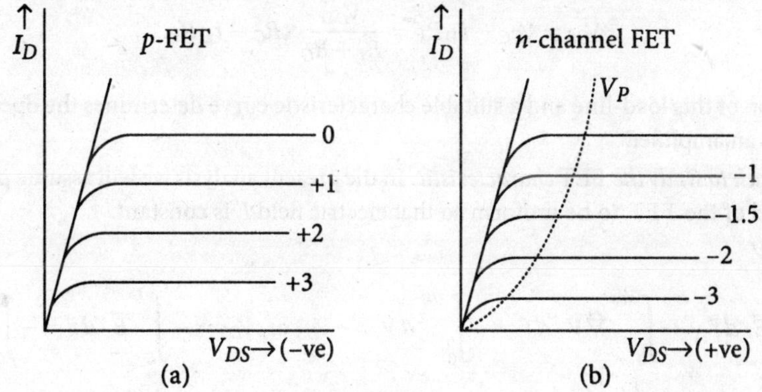

Fig. 7.7

If we consider any characteristic curve we see that it consists of two distinct portions, one portion linearly increasing, which shows that i_D is almost proportional to the voltage V_{DS} which is termed as the ohmic region and another portion, almost straight for higher V_{DS}, where i_D is practically independent of V_{DS}. There is a value of V_{DS} at which i_D levels off to a constant value and this voltage V_P is called the *pinch off voltage*. If the reverse bias of the gate increases, the pinch off occurs at a lower value of V_{DS} (cf. V_P line), the pinch off points corresponding to different V_{GS} values lie on a parabolic trace, also the maximum drain current decreases.

Drawing of load-line. Fig. 7.8 shows an actual FET circuit in common source configuration (CS) with biasing arrangement. The output circuit equation is given by

$$V_{DS} = V_{DD} - i_D(R_l + R_k). \tag{7.1}$$

When $i_D = 0$, $V_{DS} = V_{DD}$.

When $V_{DS} = 0$,

$$i_D = \frac{V_{DD}}{(R_l + R_k)}.$$

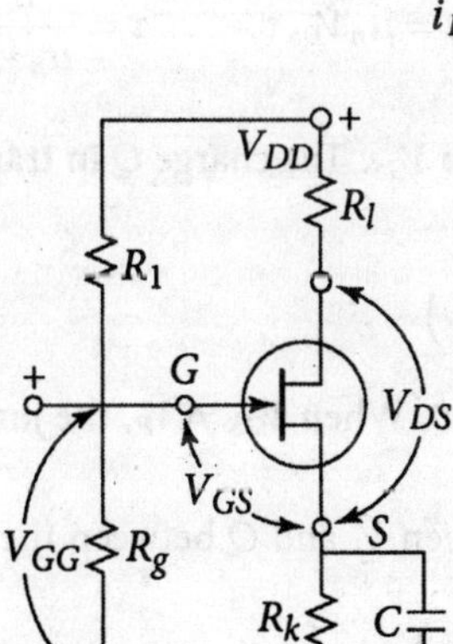

Fig. 7.8

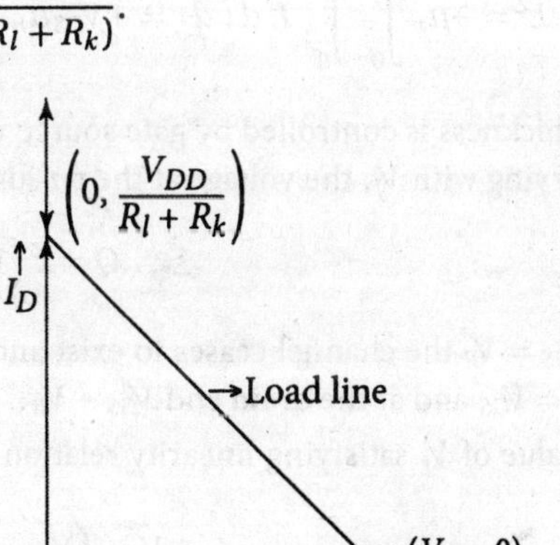

Fig. 7.9

So, the two points of the load-line are $(V_{DD},0)$ and $\left(0, \frac{V_{DD}}{R_l+R_k}\right)$. Joining these two points, we get the load-line as shown in Fig. 7.9. The bias circuit equation is given by

$$V_{GS} = V_{GC} - i_D R_k = \frac{V_{DD}}{R_l + R_G} \cdot R_G - i_D R_k. \tag{7.2}$$

The intersection of this load-line and a suitable characteristic curve determines the operating point Q of the FET used as an amplifier.

Analytical approach to study the FET *characteristic.* In the present analysis we will assume potential gradient along the channel of the FET to be uniform so that electric field E is constant.

Now, $\vec{E} = -\vec{\nabla} V$

$$\therefore \int_0^L \vec{E} \cdot d\vec{r} = -\int_0^{V_{DS}} \vec{\nabla} V \cdot d\vec{r} = -\int_0^{V_{DS}} dV = -V_{DS} \text{ or, } V_{DS} = -\int_0^L \vec{E} \cdot d\vec{r} = -\int_0^L E dr. \tag{7.3}$$

When the drain of an n-channel FET is positively biased, electrons flow from source to the drain through the channel with a drift velocity $\vec{v}_d$ given by

$$\vec{v}_d = -\mu_n \vec{E}, \tag{7.4}$$

where μ_n is the electron mobility.

Now

$$\vec{v}_d = \frac{d\vec{r}}{dt} \text{ or, } d\vec{r} = \vec{v}_d \, dt.$$

Integrating over the time interval τ, called the *average transit time* of the electrons to traverse the length L, we get

$$L = -\mu_n \int_0^\tau E dt. \tag{7.5}$$

Integrating (7.5) w.r.t. dr over the entire length, we get

$$L^2 = -\mu_n \int_0^L \int_0^\tau E dt\, dr = +V_{DS}\mu_n \int_0^\tau dt = \mu_n V_{DS} \tau. \quad \therefore \tau = \frac{L^2}{\mu_n V_{DS}}. \tag{7.6}$$

The channel thickness is controlled by gate source voltage V_{GS}. The charge Q in transit will be assumed to be linearly varying with V_j, the voltage at the pn-junction.

$$\therefore Q = k\left(V_j - V_P\right). \tag{7.7}$$

When $V_j \simeq V_{GS} = V_P$ the channel ceases to exist and $Q = 0$. When $V_{GS} \neq V_P$, the junction voltage at the source end is $V_j = V_{GS}$ and at the drain end $V_{GS} - V_{BS}$.

The average value of V_j satisfying linearity relation between V_j and Q between these extreme values is given by

$$V_j = \left(V_{GS} - \tfrac{1}{2} \cdot V_{DS}\right). \tag{7.8}$$

Putting this value,

$$Q = k\left(V_{GS} - \frac{1}{2} V_{DS} - V_P\right).$$

$$\therefore\ i_D = \frac{Q}{\tau} = \frac{k\mu_n V_{DS}}{L^2}\left(V_{GS} - \frac{1}{2}V_{DS} - V_P\right) = \frac{\mu_n k}{2L^2}\left[2(V_{GS} - V_P)V_{DS} - V_{DS}^2\right]. \tag{7.9}$$

At pinch off, $V_{DS} = (V_{GS} - V_P)$, so we get

$$i_D = \frac{\mu_n k}{2L^2}\left[(V_{GS} - V_P)^2\right] = \frac{\mu_n k V_P}{2L^2}\left(1 - \frac{V_{GS}}{V_P}\right)^2 = I_D\left(1 - \frac{V_{GS}}{V_P}\right)^2. \tag{7.10}$$

The boundary between the pinch off region and the ohmic region of the drain characteristic corresponds to the condition $V_{DS} = V_{GS} - V_P > 0$.

$\therefore$ so joining the points of such transition, we get a curve $i_D = \text{const} \times V_{DS}^2$ giving the parabolic trace for V_P.

Parameters of a FET. There are three parameters of an FET and they are

(a) *Drain resistance* (r_d): This is similar to r_P of a triode; it is defined by

$$r_d = \left(\frac{\partial V_{DS}}{\partial i_D}\right)_{V_{GS}}.$$

The low slope of the $i_D \sim V_{DS}$ characteristic curve in the pinch off region shows that r_d (which is reciprocal of slope) is large and independent of V_{DS}.

From (7.9) when $V_{DS} \ll V_{GS}$, we get

$$i_D = \frac{\mu_n k}{2L^2} \times 2(V_{GS} - V_P)V_{DS} = \left[\frac{\mu_n k}{L^2}(V_{GS} - V_P)\right]V_{DS}.$$

$$\therefore\ r_D = \frac{L^2}{\mu_n k}\cdot\frac{1}{(V_{GS} - V_P)} = \frac{L^2}{\mu_n k}\cdot V_{GS}^{-1}\frac{1}{\left(1 - \frac{V_P}{V_{GS}}\right)} \simeq \frac{L^2}{\mu_n k}V_{GS}^{-1} = \text{cons} \times V_{GS}^{-1}. \tag{7.11}$$

So r_d is independent of V_{DS}.

The input resistance is that of the gate electrode and in most of the circuit this resistance is negligible in effect.

Transconductance g_m (or g_{fs}). This parameter of an FET measures the control of i_D by the gate voltage and is defined as

$$g_m = \left(\frac{\partial i_D}{\partial V_{GS}}\right)_{V_{DS}}.$$

(7.10) predicts the transfer curve, which represents variation of i_D with V_{GS}.

$i_D = I_D\left(1 - \frac{V_{GS}}{V_P}\right)^2$, here I_D is the drain current when the gate is shorted to the source.

A typical transfer curve is shown in Fig. 7.10. From the theoretical transfer curve equation, we get differentiating i_D w.r.t. V_{GS}

$$\frac{di_D}{dV_{GS}} = -2\frac{i_D^{1/2}I_D^{1/2}}{V_P} \text{ or, } \frac{di_D}{d(-V_{GS})} = \frac{i_D^{1/2}I_D^{1/2}}{V_P/2}.$$

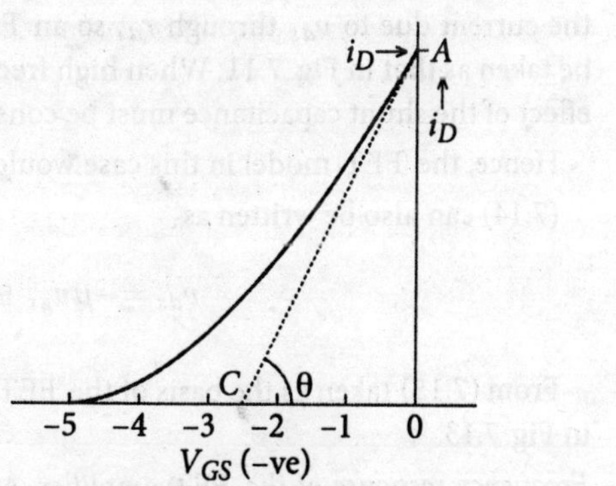

Fig. 7.10

$$\therefore \left.\frac{d i_D}{d(-V_{GS})}\right|_{i_D=I_D} = \frac{I_D}{V_P/2}. \tag{7.12}$$

$$\therefore \tan\theta = \left.\frac{d i_D}{d(-V_{GS})}\right|_{i_D=I_D} = \frac{AO}{OC} = \frac{I_D}{OC}.$$

$\therefore OC = \frac{V_P}{2}$, hence V_P, the pinch off voltage can be found out.

Amplification factor (μ). The amplification factor μ is defined as

$$\mu = -\left(\frac{\partial V_{DS}}{\partial V_{GS}}\right)_{i_D}.$$

Relationship between r_d, g_m and μ. $i_D = f(V_{GS}, V_{DS})$.

$$\therefore d i_D = \left(\frac{\partial i_D}{\partial V_{GS}}\right)_{V_{DS}} d V_{GS} + \left(\frac{\partial i_D}{\partial V_{DS}}\right)_{V_{GS}} d V_{DS}.$$

If i_D is constant, then

$$\left(\frac{\partial i_D}{\partial V_{GS}}\right)_{V_{DS}} = -\left(\frac{\partial i_D}{\partial V_{DS}}\right)_{V_{GS}} \left(\frac{\partial V_{DS}}{\partial V_{GS}}\right)_{i_D} \quad \text{or, } g_m = \frac{\mu}{r_d} \text{ or, } \mu = g_m r_d. \tag{7.13}$$

This relationship between the parameters is similar to that in case of a valve.

Small signal model. $i_D = f(V_{GS} \cdot V_{DS})$.

$$\therefore d i_D = \left(\frac{\partial i_D}{\partial V_{GS}}\right)_{V_{DS}} d V_{GS} + \left(\frac{\partial i_D}{\partial V_{DS}}\right)_{V_{GS}} d V_{DS} = g_m d V_{GS} + \frac{1}{r_d} d V_{DS}.$$

If the incremental changes $d i_D$, $d V_{GS}$, $d V_{DS}$ are represented by i_d, v_{gs}, v_{ds}, then from the above equation, we get

$$i_d = g_m v_{gs} + \frac{v_{ds}}{r_d}. \tag{7.14}$$

(7.14) forms the basis of the FET model. The drain current consists of two terms, the first term represents a current generator g_m, v_{gs}, the second term is the current due to v_{ds} through r_d, so an FET model (at low frequency) may be taken as that in Fig. 7.11. When high frequency operation is considered, the effect of the shunt capacitance must be considered.

Hence, the FET model in this case would be as shown in Fig. 7.12.

(7.14) can also be written as

$$v_{ds} = -\mu v_{gs} + i_d r_d. \tag{7.15}$$

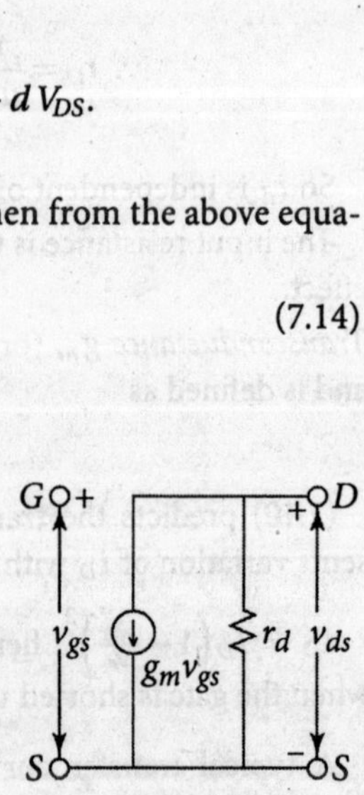

Fig. 7.11

From (7.15) taken as the basis of the FET model, voltage equivalent Thévenin's representation is shown in Fig. 7.13.

Frequency response of the FET *amplifier.* At low frequencies an FET amplifier is preferable to a bipolar transistor amplifier, because an FET has almost infinite input resistance and no feedback as in bipolar transistor through the hybrid parameter h_{re}. But at high frequencies effect of capacitances C_{gs}, C_{pd}, G_{ds} become pronounced and feedback occurs between input and output circuits. So an FET is very useful as an amplifying device in the AF range.

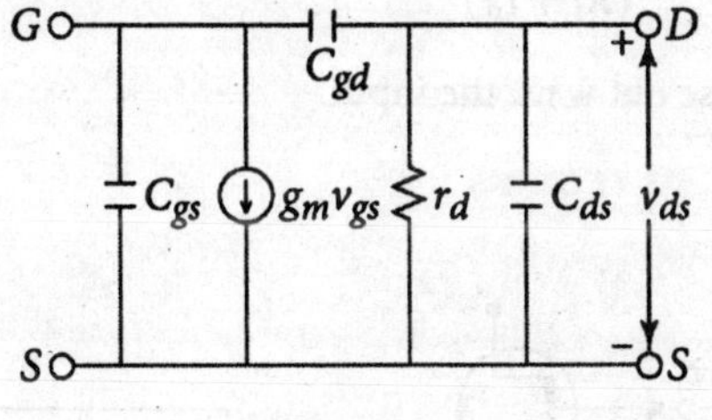

Fig. 7.12

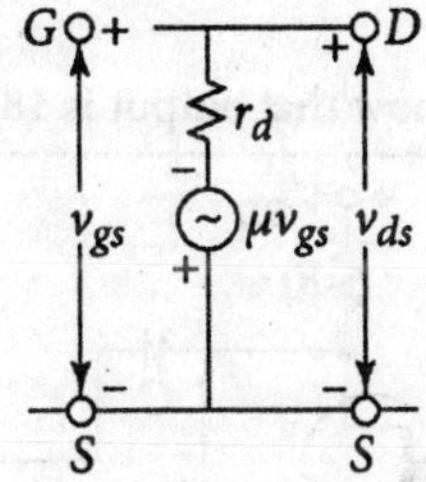

Fig. 7.13

The common source amplifier (*CS configuration*). An FET amplifier in the common source (CS) configuration is shown [Fig. 7.14(a)]. This is equivalent to common emitter (CE) configuration of a bipolar transistor. The ac equivalent of the circuit [Fig. 7.14(a)] is shown in Fig. 7.14(b). Applying Kirchhoff's law, we get

$$i_d R_l + \left(i_d - g_m v_{gs}\right) r_d + i_d R_s = 0.$$

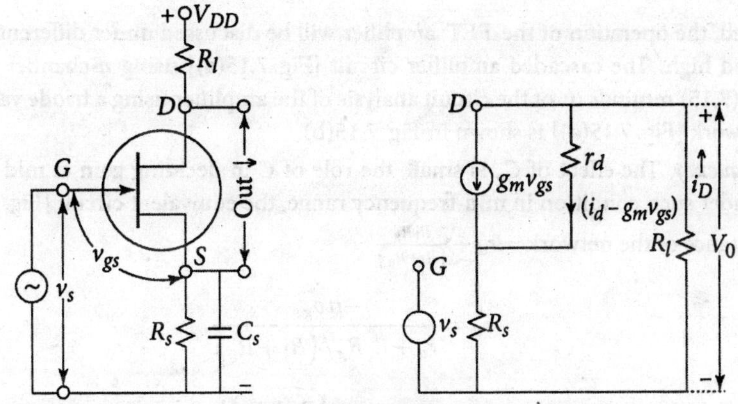

Fig. 7.14

Also, $v_{gs} = v_s - i_d R_s$. From the first equation, we get $i_d (R_l + r_d + R_s) - g_m r_d v_{gs} = 0$.

$$i_d R_s + v_{ge} = v_s.$$

$$\therefore\ i_d = \frac{\begin{vmatrix} 0 & -g_m r_d \\ v_s & 1 \end{vmatrix}}{\begin{vmatrix} R_l + r_d + R_s & -g_m r_d \\ R_s & 1 \end{vmatrix}} = \frac{r_s g_m r_d}{(R_l + r_d + R_s) + g_m r_d R_s} = \frac{r_s g_m r_d}{(r_d + R_l) + (\mu + 1) R_s}.$$

$$v_0 = -i_d R_l = -\frac{\mu r_s R_l}{(r_d + R_l) + (\mu + 1) R_s}.$$

$$\therefore\ A_v = \frac{v_0}{v_s} = \frac{-\mu R_l}{(R_l + r_d) + (\mu + 1) R_s}. \tag{7.16}$$

If the source is grounded, $R_s = 0$, in this case the voltage gain is

$$A_v = \frac{-\mu R_l}{(R_l + r_d)}. \tag{7.17}$$

(7.16) and (7.17) show that output is 180° phase out w.r.t. the input.

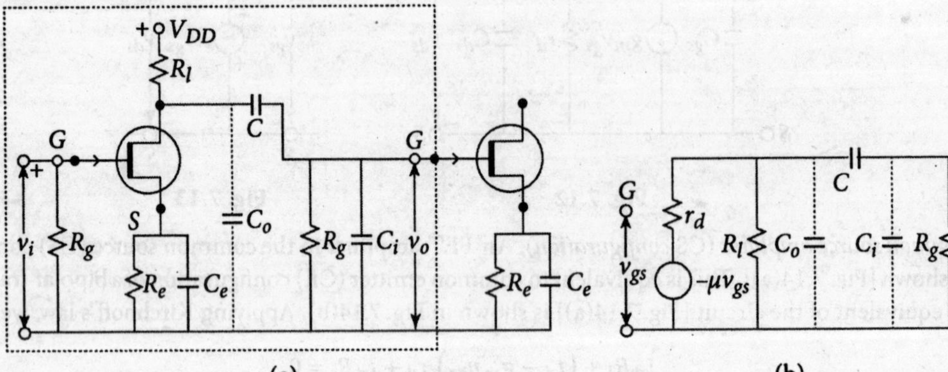

Fig. 7.15

When cascaded, the operation of the FET amplifier will be discussed under different frequency range, low, medium and high. The cascaded amplifier circuit [Fig. 7.15(a)] using n-channel FET in the RC-coupling mode. (7.15) reminds us of the circuit analysis of the amplifier using a triode valve. So, the equivalent of RC network [Fig. 7.15(a)] is shown in Fig. 7.15(b).

Gain at mid-frequency. The effect of C_o is small, the role of C in deciding gain at mid-frequency is also negligible. So under such condition in mid-frequency range, the equivalent circuit [Fig. 7.16(a)].

The total resistance of the network $= r_d + \frac{R_l R_g}{(R_l + R_g)}$.

$$\therefore\ i_d = \frac{-\mu v_{gs}}{r_d + R_l R_g / \left(R_l + R_g\right)}.$$

$$\therefore\ v_o = \frac{-\mu v_{gs} \times \left(R_g \parallel R_l\right)}{r_d + R_l R_g / \left(R_l + R_g\right)}.$$

$$\therefore\ A_{mf} = \frac{v_o}{v_{gs}} = \frac{-\mu\left(R_l + R_g\right)}{r_d R_l + r_d R_g + R_l R_g} \times \frac{R_l R_g}{\left(R_g + R_l\right)} = -\frac{\mu R_l R_g}{r_d\left(R_l + R_g\right) + R_l R_g} = -g_m R \tag{7.18}$$

where $R = \left(r_d \parallel R_l \parallel R_g\right)$.

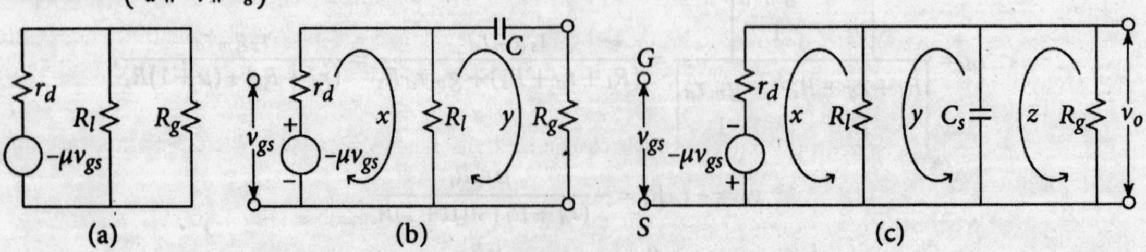

Fig. 7.16

Low-frequency response. In this range of frequency the shunting stray capacitance C_o is of no serious consequence but the effect of C needs be considered. The equivalent circuit in this case is shown [Fig. 7.16(b)].

Applying Kirchhoff's law to the current-loop carrying current x and y as shown, we get

$$-(x-y)R_l - xr_d = \mu v_{gs} \text{ or, } yR_l - x(R_l + r_d) = \mu v_{gs} \tag{7.19}$$

$$\text{and } -y\left(R_g + \frac{1}{j\omega C} + R_l\right) + xR_l = 0 \text{ or, } y\left(R_l + R_g - j/\omega C\right) - xR_l = 0. \tag{7.20}$$

Let

$$\Delta = \begin{vmatrix} R_l & -(R_l + r_d) \\ (R_l + R_g - j/\omega C) & -R_l \end{vmatrix} \text{ and } \Delta_1 = \begin{vmatrix} \mu v_{gs} & -(R_l + R_d) \\ 0 & -R_l \end{vmatrix}.$$

$$\therefore\ y = \frac{\Delta_1}{\Delta} = \frac{-\mu v_{gs} R_l}{\Delta},$$

where $\Delta = R_l R_g + r_d\left(R_g + R_l\right) - \frac{j}{\omega C}(R_l + r_d)$.

$$\therefore\ v_o = yR_g = \frac{-\mu v_{gs} R_l R_g}{R_l R_g + r_d\left(R_l + R_g\right) - \frac{j}{\omega C}(R_l + r_d)}.$$

$$\therefore\ A_{lf} = \frac{-\mu R_l R_g}{R_l R_g + r_d\left(R_l + R_g\right) - \frac{j}{\omega C}(R_l + r_d)} = \frac{-\mu R_l R_g}{R_l R_g + r_d\left(R_l + R_g\right)} \cdot \frac{1}{1 - \frac{j/\omega C(R_l + r_d)}{r_d\left(R_l + R_g\right) + R_l R_g}}$$

$$= \frac{A_{mf}}{1 - \frac{j}{\omega C} \cdot \frac{R_l + r_d}{r_d\left(R_l + R_g\right) + R_l R_g}} = \frac{A_{mf}}{1 - j\omega_1/\omega},$$

where

$$\omega_1 = \frac{R_l + r_d}{C\left\{R_l R_g + r_d\left(R_l + R_g\right)\right\}} \text{ or, } f_1 = \frac{1}{2\pi C} \cdot \frac{1}{\left[\frac{R_l r_d}{R_l + r_d} + R_g\right]} = \frac{1}{2\pi\ C\left[(R_l \parallel r_d) + R_g\right]}.$$

$$\therefore\ A_{lf} = \frac{A_{mf}}{1 - jf_1/f}.$$

$$\therefore\ \left|A_{lf}\right| = A_{mf} \cdot \frac{1}{\sqrt{1 + (f_1/f)^2}},$$

when $f = f_1$, then $|A_{lf}| = \frac{A_{mf}}{\sqrt{2}}$, this frequency is known as *lower half-power frequency.*

High-frequency range. In this frequency domain, the stray capacitances are of considerable influence while the coupling capacitor is of no concern, because it is connected in series. So, the equivalent circuit in this case is shown in Fig. 7.16(c).

Here c_s is the effective shunt capacitance, which can be written as $c_s = c_o + c_i$, where c_i = input capacitance of the next stage = $c_{gs} + \left(1 - A_{mf}\right) c_{gs}$.

Applying Kirchhoff's second law to the cyclic currents in the closed meshes [Fig. 7.16(c)], we get

$$0 + x(r_d + R_l) - yR_l = -\mu v_{gs}. \tag{7.21}$$

$$-\frac{1}{j\omega c_s}z - xR_l + y\left(R_l + \frac{1}{j\omega c_s}\right) = 0. \tag{7.22}$$

$$z\left(R_g + \frac{1}{j\omega c_s}\right) + 0 - \frac{1}{j\omega c_s} = 0. \tag{7.23}$$

Solving for z, we get $z = \frac{\Delta_1}{\Delta}$, where

$$\Delta_1 = \begin{vmatrix} -\mu v_{gs} & (r_d + R_l) & -R_l \\ 0 & -R_l & R_l + \left(\frac{1}{j\omega c_s}\right) \\ 0 & 0 & -\frac{1}{j\omega c_s} \end{vmatrix} = -\mu v_{gs}\frac{R_l}{j\omega c_s}.$$

$$\Delta = \begin{vmatrix} 0 & (r_d + R_l) & -R_l \\ -\frac{1}{j\omega c_s} & -R_l & \left(R_l + \frac{1}{j\omega c_s}\right) \\ \left(R_g + \frac{1}{j\omega c_s}\right) & 0 & -\frac{1}{j\omega c_s} \end{vmatrix} = r_d\left\{R_l R_g + \left(R_l + R_g\right)/j\omega c_s\right\} + R_l R_g/j\omega c_s.$$

$$\therefore z = -\frac{\mu v_{gs} R_l R_g}{r_d\left\{R_l R_g j\omega c_s + \left(R_l + R_g\right)\right\} + R_l R_g}.$$

$$\therefore v_o = R_g z = \frac{-\mu v_{gs} R_l R_g}{r_d\left\{R_l R_g j\omega c_s + \left(R_l + R_g\right)\right\} + R_l R_s}.$$

So, the high frequency gain is given by

$$A_{hf} = \frac{v_o}{v_{gs}} = \frac{-\mu R_l R_g}{\left(R_l + R_g\right) r_d + R_l R_g\left(1 + j\omega c_s r_d\right)} = A_{mf} \times \frac{r_d\left(R_l + R_g\right) + R_l R_g}{r_d\left(R_l + R_g\right) + R_l R_g + j\omega c_s R_l R_g r_d}$$

$$= \frac{A_{mf}}{1 + \frac{j\omega c_s R_l R_g r_d}{r_d\left(R_l + R_g\right) + R_l R_g}} = \frac{A_{mf}}{1 + j\omega c_s R}, \text{ where } R = r_d \| R_l \| R_g$$

$$= \frac{A_{mf}}{1 + j\frac{f}{f_2}}, \text{ where } f_2 = \frac{1}{2\pi R c_s}.$$

$\therefore \left|A_{hf}\right| = \dfrac{A_{mf}}{\sqrt{1 + (f/f_2)^2}}$, when $f = f_2$, $\left(A_{hf}\right) = \frac{1}{\sqrt{2}} A_{mf}$ and this frequency is known as *upper half-power frequency.*

Practical circuits with **FET** ***transistor.*** **For good recording two mikes are better than one and the supper mike mixer [Fig. 7.17] using FET transistor does this mixing after amplification.**

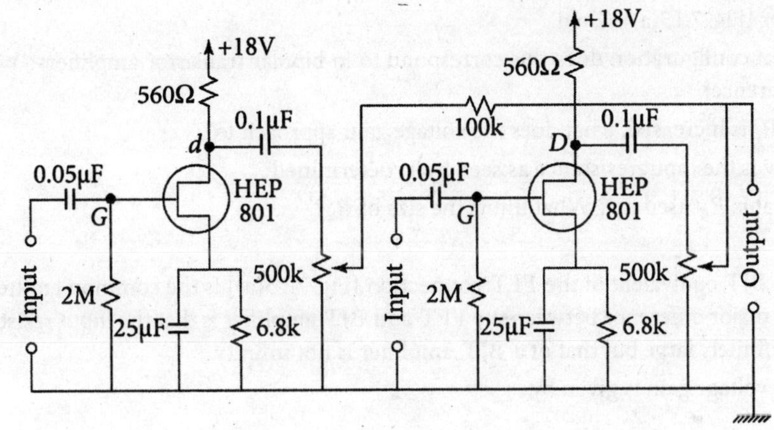

Fig. 7.17

Null detector using BFW 10.

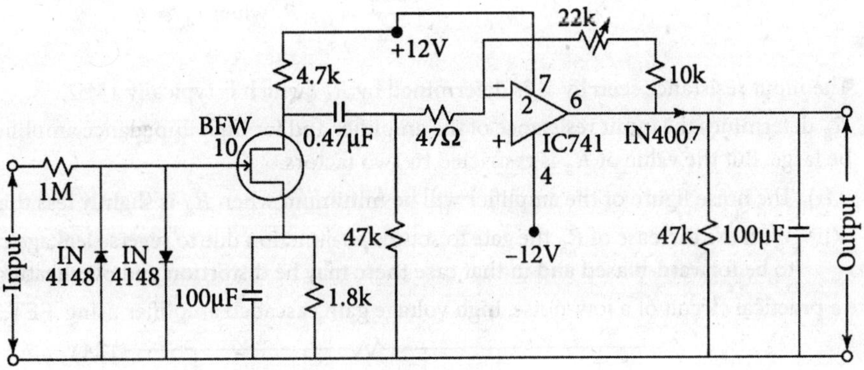

Fig. 7.18

7.3 Short Theory Questions with Answers

1. (a) What is meant by the *substrate*?
   (b) What type of material is the substrate in a *p*-channel FET?
   (c) What is the normal polarity of the gate w.r.t. the source in a *p*-channel FET?
   (d) What effect will an increased positive voltage on the gate of a *p*-channel FET have on the drain current?

Ans.

(a) ***Substrate*** is the bulk of one type of semiconductor into which is diffused another type of lightly doped semiconductor. Thus, an *n*-channel FET consists of *p*-type substrate into which a lightly doped *n*-type channel is diffused.

(b) *n*-type material forms the substrate of a *p*-channel FET.

(c) The gate electrodes are usually at positive potential w.r.t. the source in case of a *p*-channel FET.

(d) The maximum drain current decreases when the positive voltage on the gate of a *p*-channel FET is increased. The pinch off voltage also occurs at a lower value of V_{DS}.

2. Consider [Fig. 7.15(a)]. Then
   (a) What configuration does this correspond to in bipolar transistor amplifiers? What is the major difference?
   (b) As R_l is increased, what does the voltage gain approach to?
   (c) How is the input resistance as seen by V_i determined?
   (d) What is R_g used for? What limits the size of R_g?

Ans.

(a) The BJT equivalent of the FET connection [Fig. 7.15(a)] is the common emitter configuration. The major difference between the FET and BJT amplifier is that the input resistance of the FET is infinitely large but that of a BJT amplifier is not infinity.

(b) The voltage gain is given by

$$|A_v| = \frac{\mu R_l}{(R_l + r_d)} = \frac{\mu}{1+\frac{r_d}{R_l}} \to \mu \text{ when } R_l \gg r_d$$

$$\to \frac{\mu R_l}{r_d} = g_m R_l \text{ when } r_d \gg R_l.$$

(c) The input resistance seen by V_i is determined by R_g (which is typically 1MΩ).

(d) R_g determines the input resistance of the amplifier and for high impedance amplifier R_g should be large. But the value of R_g is restricted by two factors.
   (i) The noise figure of the amplifier will be minimum when R_g is slightly less than 1MΩ.
   (ii) With the increase of R_g the gate to source *pn*-junction due to reverse leakage current tends to be forward-biased and in that case there may be distortion and overheating of the gate.

3. Give a practical circuit of a low-noise, high voltage gain cascaded amplifier using FET.

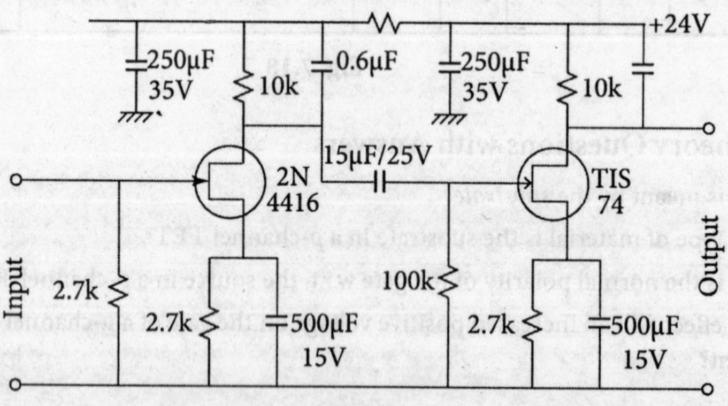

Fig. 7.19

Such a circuit can be used for a low-noise, high voltage gain ≃ 400 amplifier using FET transistors.

4. Define: (a) Channel ohmic region (b) Drain resistance and (c) Pinch-off voltage.
   Ans. See the text, cf. *Principle of operation of a* FET.

5. (a) Which capacitance is the most troublesome in an FET amplifier?
   (b) How does the total input capacitance vary with R_l?
   Ans. The input capacitance is very troublesome in an FET amplifier. Because of Miller effect, the

input capacitance increases and it is given as $c_i = c_{gs} + \left(1 - A_{mf}\right) c_{gd}$. It is the large value of c_{gd} which very much affects the high frequency operation.

Now, $|A_{mf}| = g_m R$, where $R = r_d \parallel R_l \parallel R_g$.

So c_i is affected through the factor A_{mf}.

6. Give a practical circuit of a high voltage regulator using FET.

The circuit shown in Fig. 7.20 is a practical design for a high voltage regulator producing output varying from 50V to 200V. It is a simple high voltage regulator circuit using BFW 10 as the FET element acting as a source follower and it serves as an impedance transformer for the error amplifier transistor 2N3439.

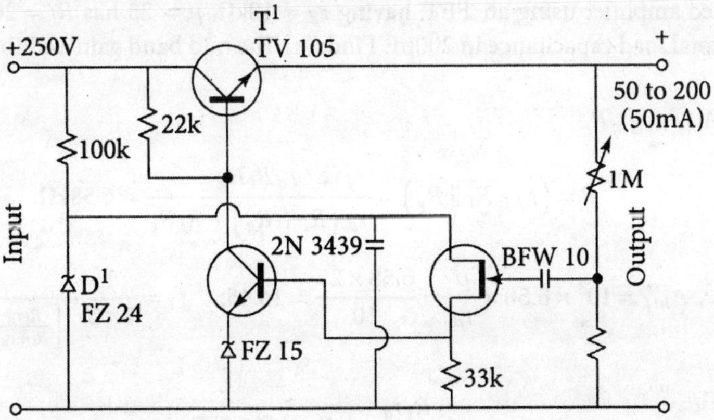

Fig. 7.20

7.4 Numerical Problems

1. An RC-coupled FET amplifier has the following parameter values: $R_l = 10\text{k}$, $R_g = 500\text{k}$, $C = 0.01\mu\text{F}$, $C_s = 150\text{pF}$, $\mu = 25$, $r_d = 10\text{k}$, $g_m = 2.5\text{mA/V}$. Find A_{mf}, f_1 and f_2.

Solution: $\left|A_{mf}\right| = g_m R$.

$$R = \left(r_d \parallel R_l \parallel R_g\right) = \frac{r_d R_l R_g}{r_d\left(R_l + R_g\right) + R_l R_g} = \frac{10 \times 10 \times 500}{10 \times 510 + 10 \times 500} = 4.95\text{k}\Omega.$$

$$\left|A_{mf}\right| = 2.5 \times 10^{-8} \times 4.95 \times 10^3 = 12.38.$$

$$f_1 = \frac{1}{2\pi C} \cdot \frac{1}{\left[(R_l \parallel r_d) + R_g\right]}. \qquad R_l \parallel r_d = \frac{10 \times 10}{20} = 5\text{k}\Omega.$$

$$f_1 = \frac{1}{2 \times 3.14 \times 0.01 \times 10^{-6}\left[5 \times 10^8 + 500 \times 10^8\right]} = 31.73\text{Hz}.$$

$$f_2 = \frac{1}{2\pi R C_s} = \frac{1}{2 \times 3.14 \times 4.95 \times 10^3 \times 150 \times 10^{-12}} = 214.5\text{Hz}.$$

2. The drain resistance and transconductance of an FET are to be determined experimentally. With $V_{GS} = 0\text{V}$ and $V_{DS} = 7\text{V}$, the drain current is 10mA. When the drain voltage is increased to 15V with V_{GS} held constant at 0V, the drain current is increased to 10.25mA.

When the gate to source voltage is now increased to -0.2V, while V_{DS} is held constant at 15V, the drain current decreases to 9.65mA. Determine: (a) Drain resistance (b) The transconductance (c) The amplification factor and (d) The type of transistor.

Solution:

(a) $r_d = \left(\dfrac{\partial V_{DS}}{\partial i_D}\right)_{V_{GS}} = \dfrac{15-7}{(10.25-10.00)\times 10^{-3}} = \dfrac{8}{0.25}\times 10^3 = 32\text{k}\Omega.$

(b) $g_m = \left(\dfrac{\partial i_D}{\partial V_{GS}}\right)_{V_{DS}} = \dfrac{0.60\times 10^{-3}}{0.2} = 3\times 10^{-3}\text{mho} = 3000\mu\text{mho}.$

(c) $\mu = g_m r_d = 96.$ (d) n-channel FET.

3. An RC-coupled amplifier using an FET having $r_d = 10\text{k}\Omega$, $\mu = 25$ has $R_l = 20\text{k}$, $C = 0.01\mu\text{F}$, $R_g = 500\text{k}$ and the total load capacitance in 200pF. Find: (a) The mid band gain and (b) The *hf* and *lf* cut-off frequencies.

Solution: (a) $A_{mf} = g_m R.$

$$R = \left(r_d \parallel R_l \parallel R_g\right) = \frac{r_d R_l R_g}{r_d\left(R_l + R_g\right) + R_l R_g} = 6.58\text{k}\Omega.$$

$$\therefore A_{mf} = 10^3 \times 6.58 \times \frac{\mu}{r_d} = \frac{6.58\times 25}{10} = 16.45; \quad f_1 = \frac{1}{2\pi C}\cdot\frac{1}{\left[\frac{R_l r_d}{R_l + r_d} + R_g\right]}.$$

Now,

$$\frac{R_l r_d}{R_l + r_d} + R_g = 506.7\text{k}\Omega.$$

$$\therefore f_1 = \frac{1}{2\pi C}\cdot\frac{1}{506.7\times 10^3} = \frac{1}{2\pi\times 0.01\times 10^{-6}\times(506.7\times 10^3)} = 31.4\text{Hz}.$$

$$f_2 = \frac{1}{2\pi C_s} = \frac{1}{2\pi\times 6.58\times 10^3\times 200\times 10^{-12}} = 121\text{kHz}.$$

Chapter 8

Overview of Field Effect Transistors (MOSFETs) and Prelude to IC Technology

8.1 JFET and MOSFET Compared

Though we are very much familiar with BJT, the field effect transistors are of more practical utility than the popular BJT.

Field effect transistors are of two varieties: (a) Junction field effect transistor — JFET (b) Metal oxide semiconductor FET — MOSFET, also called IGFET (Insulated Gate FET).

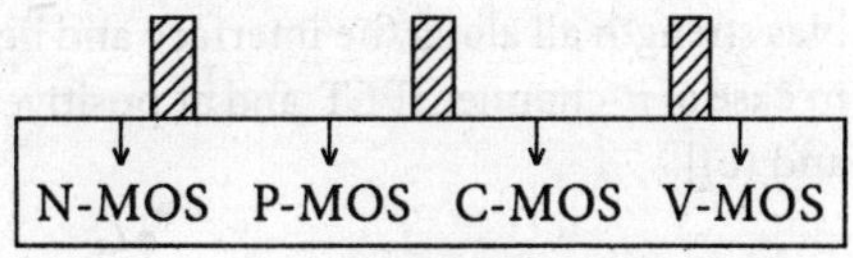

Fig. 8.1

The basic differences between JFET and MOSFET are:

1. The input impedance of a MOSFET is very high $\simeq 10^{10}$-$10^{15}\Omega$ while that of a JFET is $\simeq 100\text{M}\Omega$.
2. Both of JFET and MOSFET are used as electronic switch but MOSFETs form more perfect switch than JFETs.
3. For the same area, the MOSFETs can be more densely packed than JFETs.

For this reason in IC fabrication, MOSFETs are widely used, so MOSFETs form the core of microelectronics, which is the heart of digital electronic devices.

After its first introduction in 1958 by Jack Kilby of Texas Instruments, USA, the IC technology saw rapid revolution such that today the commercial market is flooded with various IC products in which logic circuits are fabricated on a small chip mainly of silicon, more recently of gallium arsenide (Ga-As).

An integrated circuit consists of a number of transistors, diodes, resistors, capacitors and other passive components. These components are interconnected so as to build some logic circuits. There are many types of integrated logic circuits composing the logic families. The fabrication of integration is done by advanced photographic technique, called the photolithography; finished network is so fine and small that one needs a microscope to see the connection in the circuit.

The two basic technologies for manufacturing ICs are bipolar and MOS; bipolar transistors are fabricated in the first and MOSFETs are used in the second.

Depending on the number of components or logical units fabricated on a chip, ICs are classified to belong to different levels of integration:

1. Small scale integration (SSI) for less than 10–12 logical units per chip, bipolar technology is suitable (e.g., logic gates, flip-flops).

2. Medium scale integration (MSI): For logical units numbering between 12 to 1000 per chip. Bipolar technology is suitable here also (e.g., counters, multiplexers, adders).
3. Large scale integration (LSI): For logical units numbering greater than 1000, but less than 20,000 per chip (e.g., single chip telephone system; 8-bit μ-processors).
4. Very large scale integration (VLSI): For logical units greater than 10,000, but less than 10^6 per chip (e.g., 16–32 bit μ-processor).
5. Ultra large scale integration (ULSI): For logical units ranging from 10^6 to 10×10^6 per chip (e.g., special processors).
6. Giant scale integration (GSI): For logical units $> 10 \times 10^6$ per chip.

8.2 VLSI Design about the Transistors and Wires

MOS *technology in* VLSI *family.* In VLSI family MOSFETs are commonly employed. The circuit realisation may be based on *p*-MOS, *n*-MOS, a combination of *p*- and *n*-MOS, called C-MOS and also Bi-C-MOS. So, before VLSI design consideration, we must be familiar with the construction, characteristics and working of MOSFETs.

MOSFET *transistors — n*-MOS, *p*-MOS and C-MOS: JFETs differ from MOSFETs both as regards their construction as well as their operation. In a JFET, the channel width modulation is due to the reverse bias strength all along the interface and hence it is used only with gate at negative potential w.r.t. the source in case of *n*-channel JFET and at positive potential w.r.t. the source in case of *p*-channel JFET [Fig. 8.2(a) and (b)].

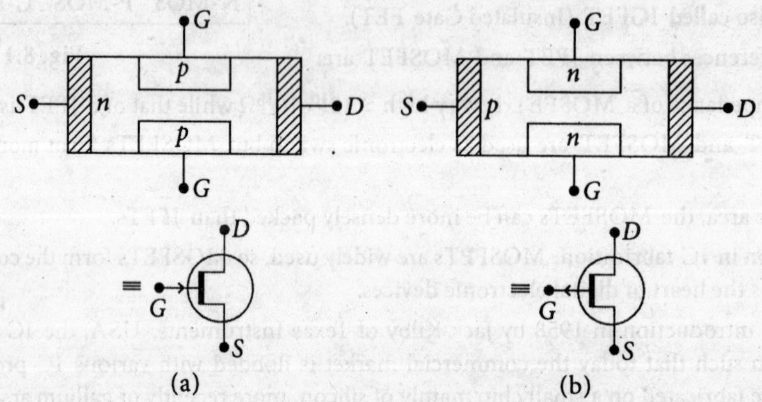

Fig. 8.2

From the construction viewpoint, an *n*-JFET or a *p*-JFET can operate only in depletion mode. But MOSFETs may be operated in depletion mode as well as in enhancement mode; of course, by special construction we manufacture exclusive D-MOS and exclusive E-MOS (E-DMOS, E-EMOS).

8.3 General MOS Structure

In order to manufacture *n*-channel MOS, the substrate is a lightly doped *p*-type material whereas for *p*-channel MOS, the substrate is a lightly doped *n*-type material [Fig. 8.3(a) and (b)].

The top of the substrate is overlaid with a layer of SiO_2, called field oxide layer, typically $0.1\mu m$ thick. Two highly doped *n*-regions (n^+) are diffused into the substrate (in *n*-channel MOS) or two highly doped *p*-regions are diffused into the substrate (in *p*-channel MOS), one forming the source, the other forming the drains. Holes are drilled through the SiO_2-layer, called the cuts for metallic contacts with S and D n^+ or p^+ regions. A lightly doped n-region (n-MOS) or p-region (p-MOS), forming the physical existence of a channel connects the source and the drain. A metal gate G lies on the insulating SiO_2 layer overlying the channel path between S and D. There is no pn-junction interface here as in JFET, rather there is a capacitor formed between the metal gate as one plate and the semiconducting channel as the other plate with SiO_2 as dielectric in between. Because of this capacitor formation, the input impedance of a MOS is tremendously large, as high as $10^{10}–10^{16}\Omega$.

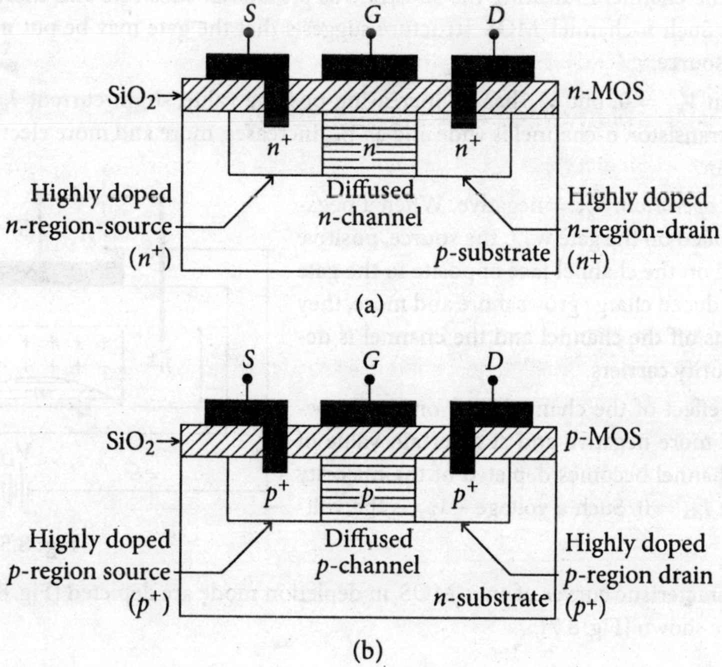

Fig. 8.3

The MOSFET transistor so formed can act in two modes: (a) Depletion mode (D-mode) (b) Enhancement mode (E-mode).

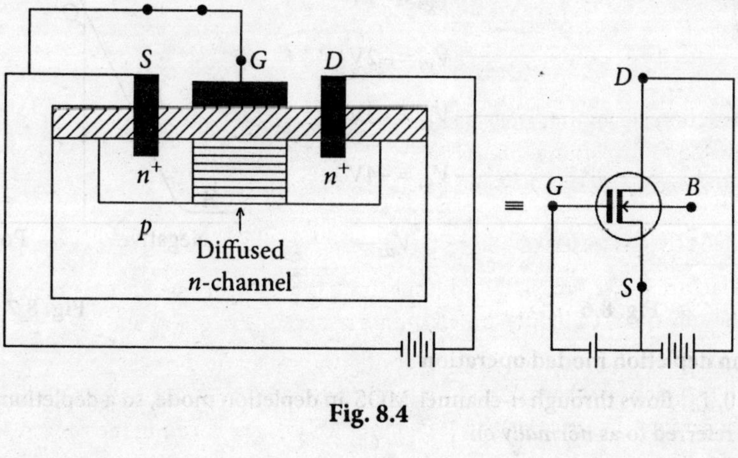

Fig. 8.4

8.4 Operation of a D-Mode MOSFET (DM MOS)

The operation of an n-MOS or a p-MOS in D-mode will be similar, the difference being in biasing arrangement and voltage on the gate (the arrangements will be opposite in nature).

Operation of n-MOS in D-mode: The construction of Fig. 8.3(a) is redrawn as Fig. 8.4 with proper biasing and the circuit symbol.

Mechanism of operation. The point B is the substrate (see the circuit symbol), the arrow is directed from the substrate to the channel indicating the substrate as p-material substrate and therefore, it represents n-channel MOS. Such n-channel MOS structure suggests that the gate may be put negative or positive voltage w.r.t. the source.

$V_{gs} = 0$: When $V_{gs} = 0$, but V_{ds} has an increasing positive value, drain current I_d increases. In this condition of the transistor, n-channel is wide and as V_{ds} increases, more and more electrons flow from the source to the drain.

Depletion moded operation: V_{gs} = negative. When a negative voltage is applied on the gate w.r.t. the source, positive charge is induced on the channel face opposite to the gate and as positive induced charge grows more and more, they push the electrons off the channel and the channel is depleted of the majority carriers.

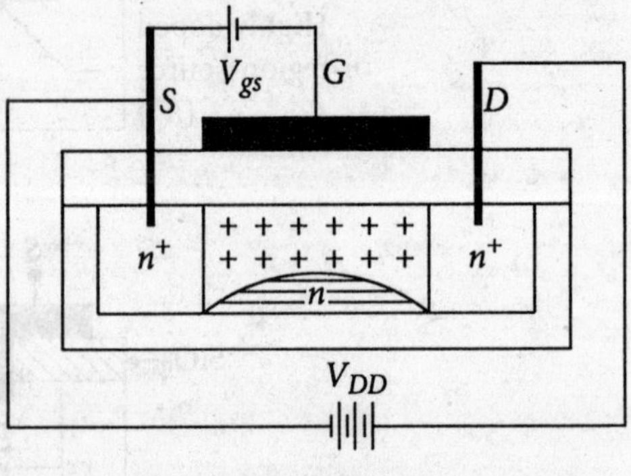

Fig. 8.5

The depletion effect of the channel goes on as V_{gs} becomes more and more negative and at a certain value of v_{gs}, the whole channel becomes depleted of the majority carriers and now $I_{ds} = 0$. Such a voltage $-V_{gs}$ is the voltage pinch-off.

The output characteristic curves of an n-MOS in depletion mode are depicted [Fig. 8.6] and its transfer characteristics are shown [Fig. 8.7].

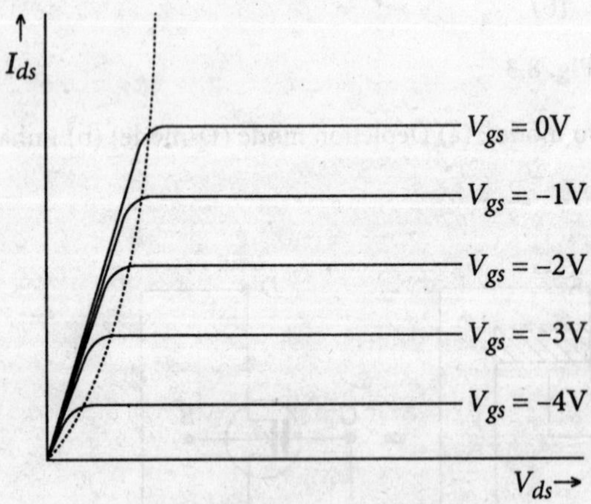

Fig. 8.6

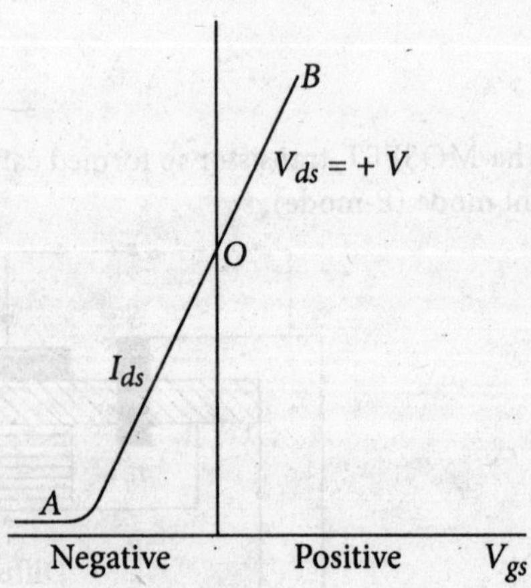

Fig. 8.7

A few remarks on depletion moded operation

1. Even if $V_{gs} = 0$, I_{ds} flows through n-channel MOS in depletion mode, so a depletion moded transistor is sometimes referred to as *normally on*.

2. The transfer characteristic shows that if V_{gs} is beyond a certain negative voltage, no I_{ds} flows; this negative voltage on the gate w.r.t. source is called the threshold voltage denoted by V_{td}.

8.5 Enhancement Moded Operation

The *n*-MOS structure as depicted [Fig. 8.4] may be operated with positive voltage on the gate w.r.t. the source.

If the gate is connected to positive voltage w.r.t. source, negative charge will grow on the opposite side and this increasing negative charge will increase the channel size, pushing away the *p*-type carriers, allowing more charge carriers and, therefore, greater channel current. The characteristic curve under this condition is shown in transfer characteristic by OB portion of AOB curve.

Though the MOSFET with the above construction may be operated in depletion and enhancement mode, the above structure is used with negative gate voltage (V_{gs} negative), hence in D-MOS, there is a physical channel connecting the source and drain.

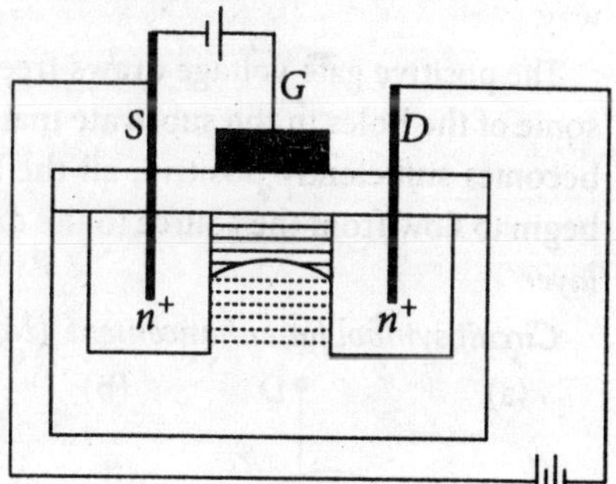

Fig. 8.8

8.6 Exclusive Enhancement MOS: (EEMOS)

In a *p*-substrate, two heavily doped *n*-regions (n^+) are diffused, one acting as source, the other as drain. On the top there is SiO_2-layer forming the insulating cover. Therefore, two cuts are there in the SiO_2-layer through which metal is allowed to flow to make contact with the two n^+ regions forming source and the drain.

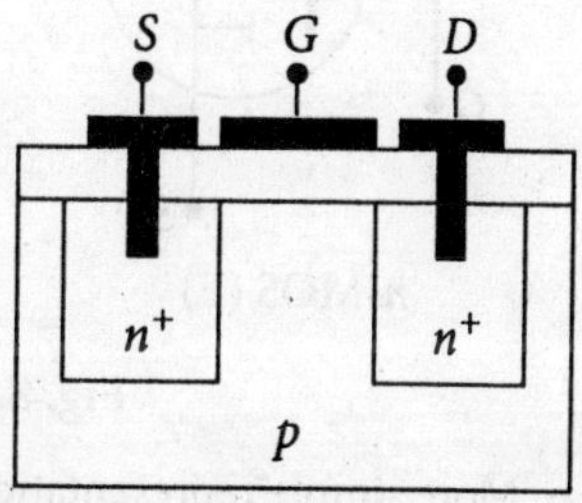

Fig. 8.9

Mechanism of operation

1. When $V_{gs} = 0$, there is no drain current ($I_{ds} = 0$), because of the formation of two diodes between the substrate (*p*-type) and the n^+ regions forming the source and the drain. So, the enhancement MOS (E-MOS) is referred to as *normally off.* So it acts as a perfect switch. This type of MOSFET is, therefore, widely used in transistor technology of VLSI design.
2. *When V_{gs} is positive*: To get drain current a positive voltage is applied on the gate w.r.t. the source.

In this construction the substrate extends all the way to the SiO_2-layer. When V_{gs} is positive the positive gate attracts free electrons from the source into the left corner of the *p*-region and with positive gate voltage of sufficient strength, there is creation of a thin layer of electrons extending from source to drain. This layer of electrons is called the *inversion layer.* The conductivity of the inversion layer increases with the increase of positive gate voltage and the drain current increases. The above facts have been elaborated with the help of diagrams [Figs. 8.8, 8.9 and 8.10].

The enhancement MOS operation can be interpreted in another way.

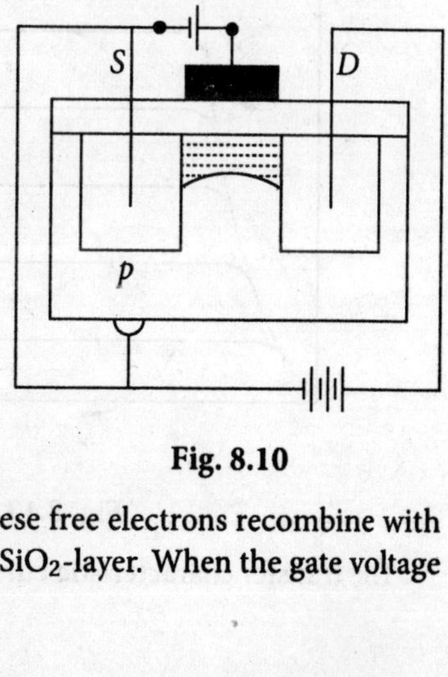

Fig. 8.10

The positive gate voltage draws free electrons into the *p*-substrate. These free electrons recombine with some of the holes in the substrate material near the lower surface of the SiO_2-layer. When the gate voltage becomes sufficiently positive, all the holes touching the SiO_2-layer are filled and then the free electrons begin to flow from the source to the drain thus with the creation of a thin layer of electrons, called the *inversion layer.*

The conductivity of the inversion layer increases with the increase of positive gate voltage and the drain current increases. The above facts have been elaborated. MOS operations can be interpreted in another way.

The positive gate voltage draws free electrons into the *p*-substrate. These free electrons recombine with some of the holes in the substrate material near the lower surface of the SiO_2-layer. When the gate voltage becomes sufficiently positive, all the holes touching the SiO_2-layer are filled and then the free electrons begin to flow from the source to the drain thus with creation of a thin layer of electron, called the *inversion layer.*

Circuit symbol for enhancement (MOS). The circuit symbol for an E-MOS is shown in Fig. 8.11.

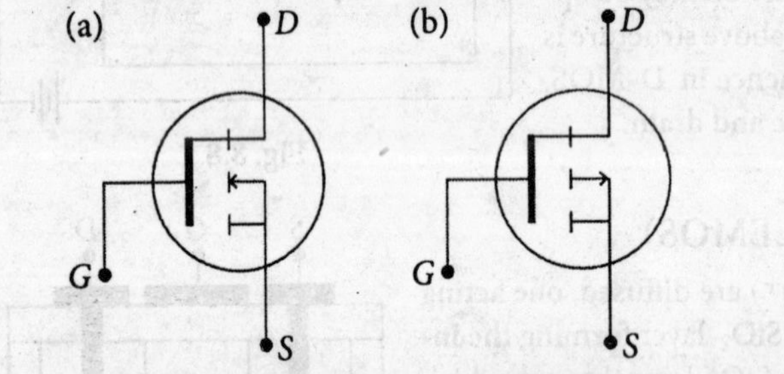

n-MOS (E)   *p*-MOS (E)

Fig. 8.11

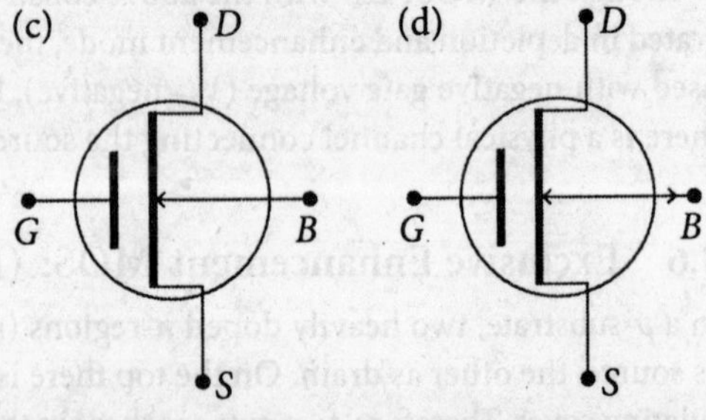

D-MOS (*n*-type)   E-MOS (*n*-type)

Fig. 8.12

More simple representation for D-MOS and E-MOS is shown [Fig. 8.12].

8.7 Characteristic Curves: Output Characteristics and Transfer Characteristics

The curves showing the relationship between I_{ds} and V_{ds} are known as *output characteristics* of the transistor [Fig. 8.12].

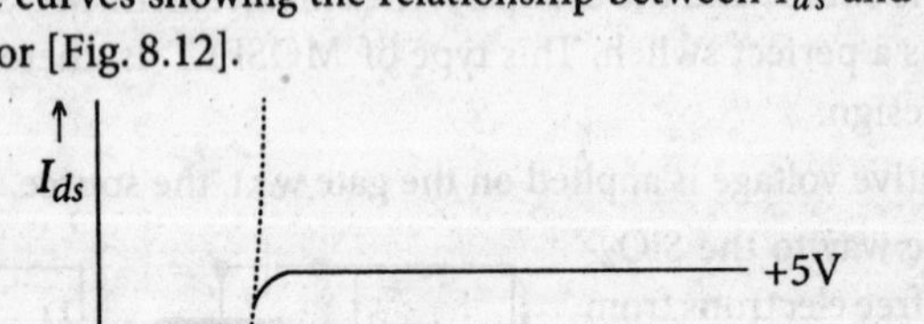

Fig. 8.13

Fig. 8.14

The transfer characteristic curve shows the relationship between I_{ds} and V_{gs} at constant V_{ds} [Fig. 8.14].

The transfer characteristic shows that the current flows through the transistor for a positive gate voltage greater than a voltage V_t, called the *threshold voltage.*

C-MOS *Construction*: (INVERTER). A popular connection in digital circuits is to connect enhancement *p*-MOS and *n*-MOS transistors to form what is known as complementary or C-MOS device [Fig. 8.15].

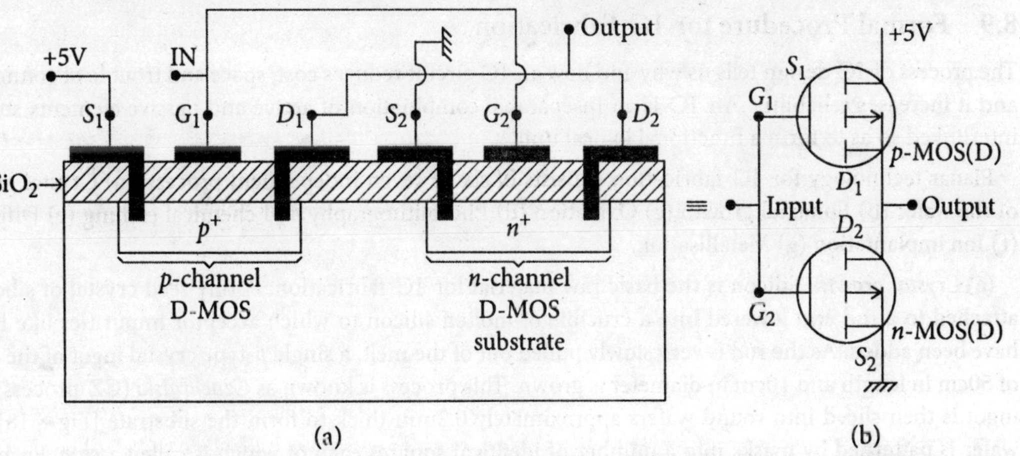

Fig. 8.15 Basic C-MOS connection.

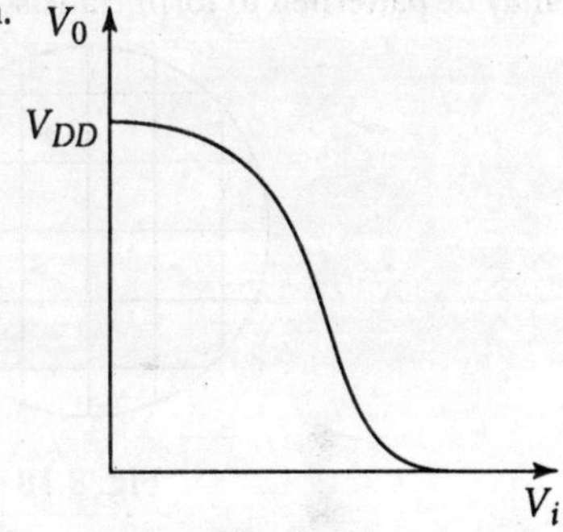

Fig. 8.16

Operation. The gates G_1 and G_2 of *p*-MOS are connected together and form the input terminal of the C-MOS device. The drains of *p*-MOS and *n*-MOS are connected together and form the output terminal of the device. The source S_1 of the *p*-MOS is connected to positive supply rail and the source S_2 is grounded. Fig. 8.15(a) shows the basic construction and Fig. 8.15(b) shows circuit symbol representation.

A positive input voltage drives the *p*-MOS OFF and *n*-MOS ON. The output will be 0 (zero). A low input voltage drives *p*-MOS ON, *n*-MOS OFF; the output rising to $+V_{DD}$. The behaviour of the device is shown by graphical plot as shown in Fig. 8.16.

8.8 The *n*-MOS Inverter

A complete range of logic circuits is based on the construction of an inverter.

A basic inverter circuit, consisting of a transistor, whose source is grounded and the drain through a resistance of some sort is connected to the positive supply rail V_{DD}. In an IC fabrication, resistors are not produced on the silicon substrate, because even a modest value of the resistance requires a large area so that some other form of load resistance is required. A depletion model MOS solves the problem of load resistance. So an inverter circuit is made as shown in Fig. 8.17.

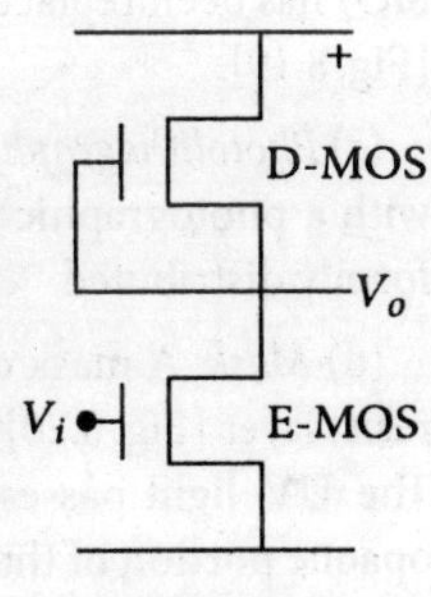

Fig. 8.17

Explanation

1. If no current is drawn from the output, I_{ds} for both D-MOS and E-MOS transistors are same.
2. For D-MOS, the gate is connected to the source and it is always ON.
3. In this case, the D-MOS device is called the pull-up (pu) transistor and E-MOS device is called the pull-down (pd) transistor.

4. When v_i is low (positive), the E-MOS is OFF, the output is high (State 1).
5. When V_i is high, the E-MOS is ON and the output is low (0-state).

Thus, the circuit behaves as an inverter and the D-MOS transistor offers the necessary resistance.

8.9 Formal Procedure for IC Fabrication

The process of IC design tells us why and how an IC circuit reduces cost, space and trouble of connection and it increases reliability. An IC is an inseparable combination of active and passive elements suitably interlinked so as to form a functional logical unit.

Planar technology for IC fabrication consists of six or seven independent process: (a) Crystal growth of the wafer (b) Epitaxial growth (c) Oxidation (d) Photolithography and chemical etching (e) Diffusion (f) Ion implantation (g) Metallisation.

(a) *Crystal growth.* Silicon is the basic raw material for IC fabrication. A tiny seed crystal of silicon is attached to a rod and lowered into a crucible of molten silicon to which acceptor impurities like Boron have been added. As the rod is very slowly pulled out of the melt, a single p-type crystal ingot of the order of 50cm in length and 10cm in diameter is grown. This process is known as *Czochralski* (CZ process). The ingot is then sliced into round wafers approximately 0.2mm thick to form the substrate [Fig. 8.18]. The wafer is patterned by masks into a number of identical squares each of which is called a *chip.* Each chip may be patterned to form transistors and wires.

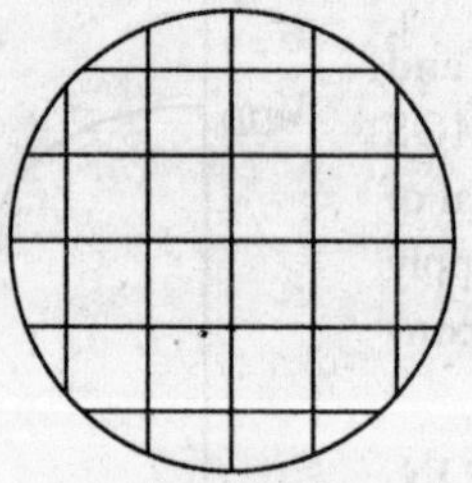

Fig. 8.18

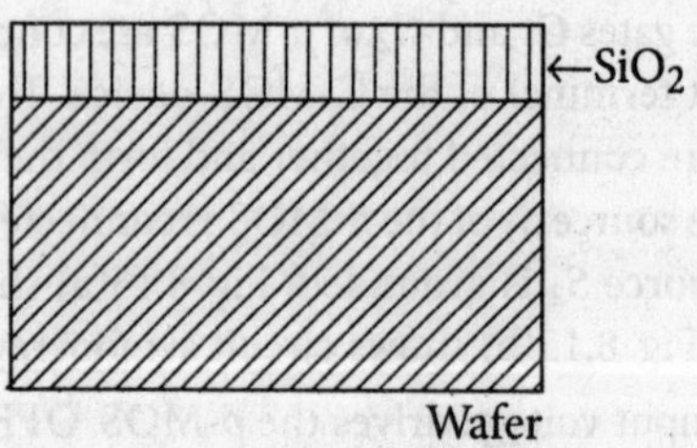

Fig. 8.19

(b) *Oxidation.* SiO_2 layer is grown over the wafer surface having thickness $\simeq 1\mu m$ (10^{-3}mm). Nowadays, SiO_2 has been replaced by silicon nitride (Si_3N_4), because Si_3N_4 is more impervious to dopants than SiO_2 [Fig. 8.19].

(c) *Photolithography.* SiO_2-layer being deposited, the wafer surface is sprayed with a photographic emulsion and it is spun such that the photo-resist is uniformly distributed.

(d) *Mask.* A mask containing the desired pattern is then placed on the photo-resist layer [Fig. 8.20] and ultraviolet light is allowed to be incident on the mask. The UV light passes through the transparent portion but it is blocked by the opaque portion of the mask.

The portion of the photo-resist layer exposed to UV light becomes polymerised (i.e., hardened), the opaque portion of the mask does not transmit the radiation and hence the photo-resist corresponding to this opaque portion remains unaffected.

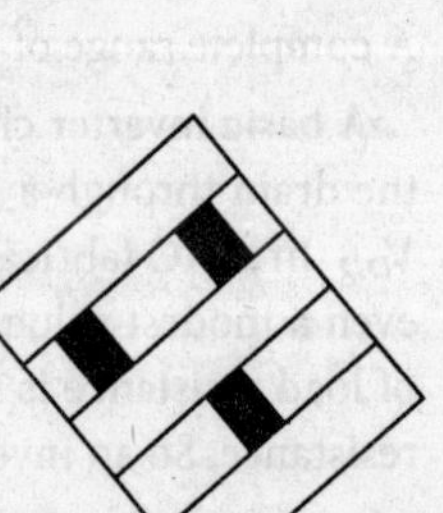

Fig. 8.20

The mask is removed and the wafer is developed by using a chemical (trichloroethylene), which dissolves the unexposed (un-polymerised) portion of the photo-resist film and leaves the surface pattern [Fig. 8.22].

1. The field oxide layer in between the photo-resist is removed by using hydrofluoric acid such that here the wafer surface is exposed.
2. The remaining photo-resist is removed and a thin layer (0.1μm) of SiO_2 is grown over the entire surface. This thin-oxide layer is also known as thinox layer.

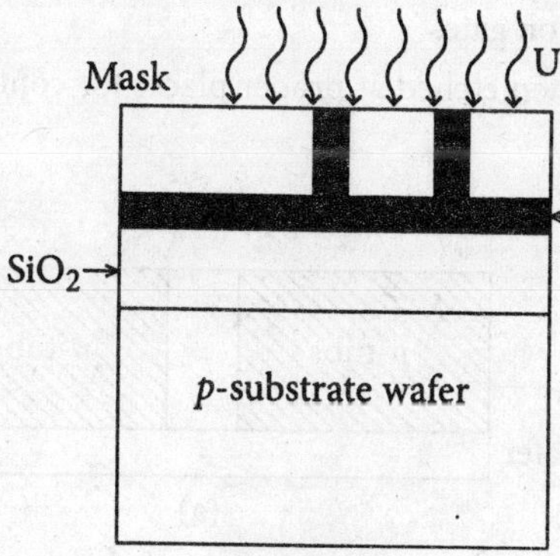

Fig. 8.21

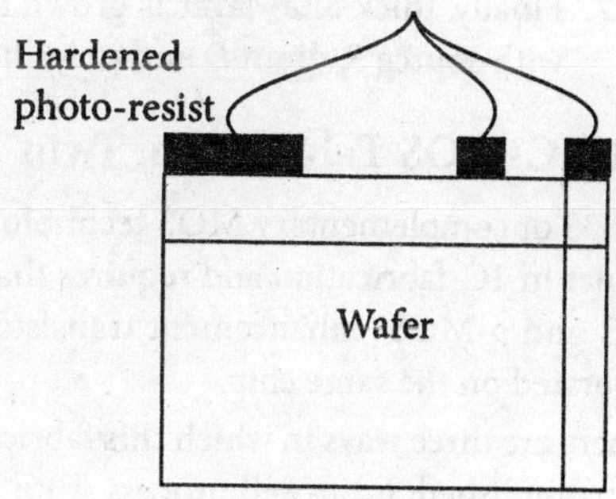

Fig. 8.22

3. Now, polysilicon is deposited on the top of the thinox layer at proper place to form the gate structure. The polysilicon layer is a heavily doped polysilicon deposited by chemical vapour deposition technique (CVD).
4. Further, photo-resist coating and masking allows the polysilicon to be patterned and then the thin-oxide is removed to expose. The area in which heavily doped n^+ region is created by ion implantation technique to form the source and the drain. The resulting structure is as shown in Fig. 8.23.

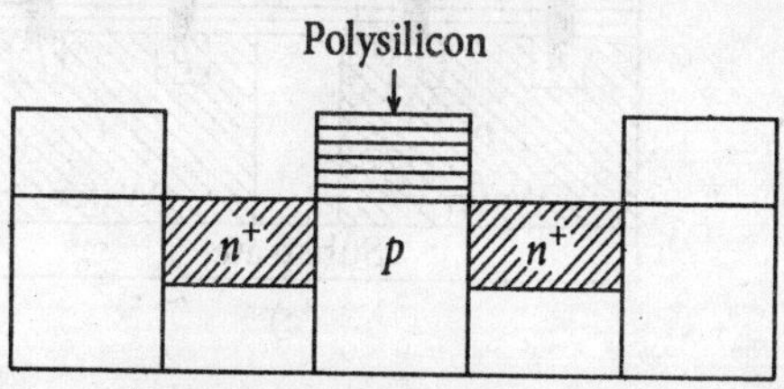

Fig. 8.23

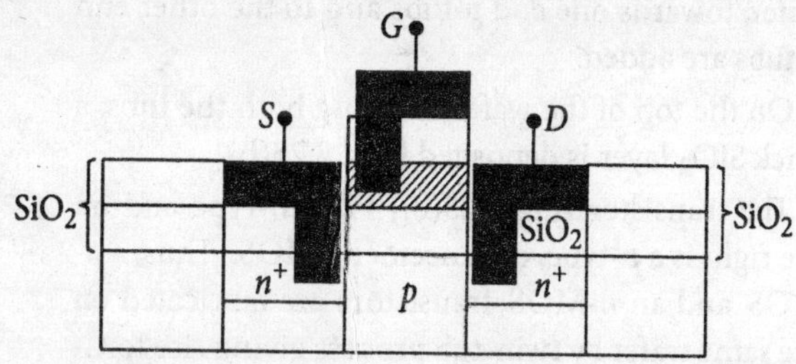

Fig. 8.24

5. Finally, thick layer of SiO_2 is grown all over again and holes are drilled through the thick oxide layer for metal contact with source, drain and gate [Fig. 8.24].

8.10 Summary of the *n*-MOS Depletion Transistor Fabrication Process

The different steps followed for the fabrication of D-MOS (*n*-type) transistor are summarised below:

Step 1. *p*-type wafer is prepared from a silicon crystal.

Step 2. A thick layer of SiO_2 (or Si_3N_4). Using photo-resist and mask 1, the thick oxide layer is patterned. The photo-resist under the transparent portion of the mask becomes hardened while that under the opaque portion remains unaffected, when UV light is incident on the mask.

Step 3. The photo-resist under transparent portion of the mask is removed by a suitable chemical, and SiO_2 beneath it is etched out to get the wafer surface exposed as window. The remaining photo-resist is then removed.

Step 4. The exposed areas are covered with a heavily doped n^+ region by ion implantation technique. One of these n^+ regions acts as the source, the other as the drain.

Step 5. A thin oxide layer is spread on the top, and polysilicon is deposited on the thinox layer to form the gate structure.

Step 6. The thinox layer is removed except under the polysilicon gate.

Step 7. Finally, thick SiO_2 layer is grown all over again and then etched at proper places for contact cuts with source S, drain D and polysilicon gate G.

8.11 C-MOS Fabrication: Twin Tub Process

C-MOS or complementary MOS technology is very popular in IC fabrication and requires that both n-MOS and p-MOS enhancement transistors be incorporated on the same chip.

There are three ways in which this fabrication can be accomplished: (a) p-well process (b) n-well process (c) Twin-Tub process.

In a p-well process, n-tubs are added to a p-wafer, whereas in n-well process, p-tubs are added to an n-wafer. But, however, the twin-tub processes use an undoped wafer, where p-tubs and n-tubs are added [Fig. 8.25].

Twin-tub process. As in Fig. 8.25(a) in a silicon wafer, towards one end p-tubs and to the other end n-tubs are added.

On the top of the wafer covering both the tubs, a thick SiO_2 layer is deposited [Fig. 8.25(b)].

The transistor [Fig. 8.25(c)] is an n-type and on the right is a p-type, enhancement MOS. Thus, a p-MOS and an n-MOS transistors are fabricated on the same wafer by twin-tub process giving rise to C-MOS technology.

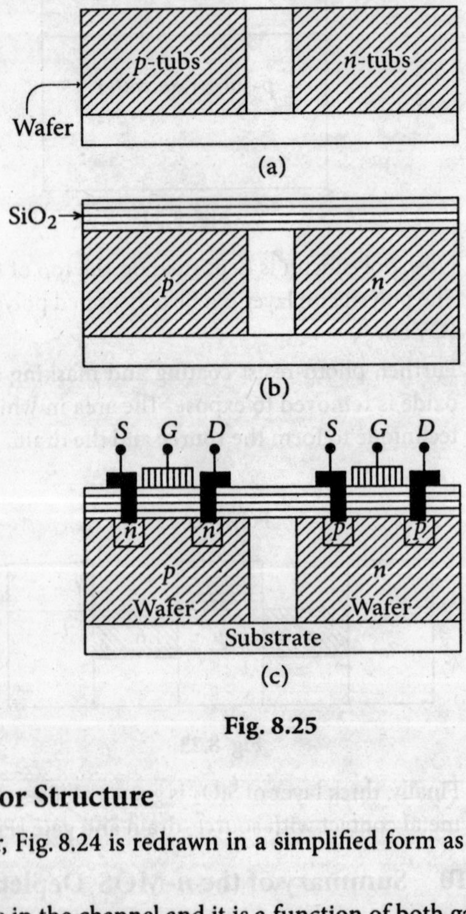

Fig. 8.25

8.12 Electrical Behaviour of n-MOS Transistor Structure

To study the electrical behaviour of an n-MOS transistor, Fig. 8.24 is redrawn in a simplified form as of Fig. 8.26.

I_{ds} is the source to drain current due to induced charge in the channel and it is a function of both gate to source voltage V_{gs} as well as source to drain voltage V_{ds}.

So,

$$I_{ds} = -I_{sd} = \frac{\text{Charge induced in the channel of length } L}{\text{Electron transit time } \tau}.$$

Again, τ is given by

$$\tau = \frac{\text{Length } L}{\text{Velocity of the electron } (v)}.$$

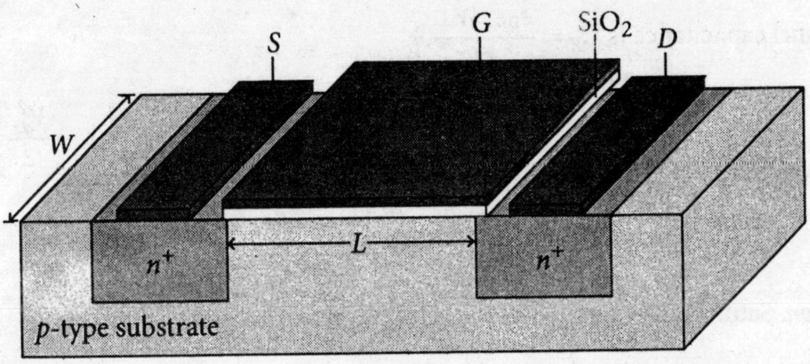

Fig. 8.26

The velocity $v=-\mu_n E$, μ_n is the electron mobility, where E is the electric field at a certain distance from the source end.

$$E=-\frac{dV}{dx}\Rightarrow -dV=E\,dx\Rightarrow \int_{V_S=0}^{V_D} -dV=\int_0^L E\,dx\Rightarrow V_D-V_s=V_{ds}=-\int_0^L E\,dx. \tag{8.1}$$

$$v=-\mu_n E\Rightarrow \frac{dx}{dt}=-\mu_n E\Rightarrow dx=-\mu_n E\,dt\Rightarrow \int_0^L dx=-\mu_n\int_0^\tau E\,dt\Rightarrow L=-\mu_n\int_0^\tau E\,dt.$$

Multiplying throughout by dx and integrating w.r.t. x, we get

$$\int_0^L L\,dx=\int_0^L\left(-\mu_n\int_0^\tau E\,dt\right)dx=-\mu_n\int_0^L\int_0^\tau E\,dx\,dt=-\mu_n\left[\int_0^L E\,dx\int_0^\tau dt\right]=\mu_n V_{ds}\times\tau$$

$$\Rightarrow \tau=\frac{L^2}{\mu_n V_{Ds}}. \tag{8.2}$$

Typical values of μ at room temperature $\mu_n=650\text{cm}^2/\text{V-sec}$
$\mu_p=240\text{cm}^2/\text{V-sec}$.

Behaviour in the resistance or non-saturated region

The charge induced in the channel is due to difference in voltage V_{gs} between gate and the channel. It is to remember that voltage along the channel varies linearly with the distance x from the source, because of IR drop in the channel, R is the channel resistance. So, the value of the voltage at any point in the channel may be taken as $(V_{ds}/2)$.

So, the charge induced in the channel is given by $Q_c=\sigma A=\epsilon_0\epsilon_r E_g\times A$.

Now, E_g = average electric field between gate and channel $=\left[\left(V_{gs}-V_t\right)-V_{ds}/2\right]/D$,
where V_t = threshold voltage, D = thickness of SiO_2 layer, $A=W\times L$, ϵ_r = relative permittivity of the SiO_2 layer, $Q_c=\frac{\epsilon_0\epsilon_r W\times L\times[(V_{gs}-V_t)-V_{ds}/2]}{D}$, $\epsilon_0=8.85\times10^{-12}\text{F/m}$, $\epsilon_r=4.0$.

$$\therefore I_{ds}=\frac{\epsilon_0\epsilon_r WL}{\tau D}\left[\left(V_{gs}-V_t\right)-\frac{V_{ds}}{2}\right]=\frac{\epsilon_0\epsilon_r\mu_n V_{ds}WL}{DL^2}\left[\left(V_{gs}-V_t\right)-\frac{V_{ds}}{2}\right]=k\frac{W}{L}\left[\left(V_{gs}-V_t\right)V_{ds}-\frac{V_{ds}^2}{2}\right],$$

where $k=\frac{\epsilon_0\epsilon_r\mu_n}{D}\Rightarrow I_{ds}=\beta\left[\left(V_{gs}-V_t\right)V_{ds}-\frac{V_{ds}^2}{2}\right]$, where $\beta=\frac{\epsilon_0\epsilon_r\mu_n}{D}\cdot\frac{W}{L}=\frac{kW}{L}$.

The gate-channel capacitance is $C_g = \dfrac{\epsilon_0 \epsilon_r WL}{D}$.

$$\therefore I_{ds} = \frac{\mu_n C_g}{L^2}\left[\left(V_{gs}-V_t\right)V_{ds}-\frac{V_{ds}^2}{2}\right] = \frac{\mu_n}{L^2}C_g\left[\left(V_{gs}-V_t\right)V_{ds}-\frac{V_{ds}^2}{2}\right]$$

$$= \frac{\mu_n C_g}{L^2}\left[\left(V_{gs}-V_t\right)V_{ds}-\frac{V_{Ds}^2}{2}\right].$$

Saturated region. Saturation begins when $V_{ds} \geq \left(V_{gs}-V_t\right)$.

$$\therefore I_{ds} = k\frac{W}{2L}\left(V_{gs}-V_t\right)^2 = \frac{\beta}{2}\left(V_{gs}-V_t\right)^2.$$

$$\beta = \frac{kW}{L} = \frac{\epsilon_0 \epsilon_r \mu_n}{D}\cdot\frac{W}{L} = \frac{\epsilon_0 \epsilon_r WL}{D}\cdot\frac{\mu_n}{L^2} = \mu_n C_g / L^2.$$

$$\therefore I_{ds} = \frac{\mu_n C_g}{2L^2}\left(V_{gs}-V_t\right)^2.$$

MOS transistor transconductance g_m and output conductance g_{ds}.

The relationship between input voltage V_{gs} and the output current I_{ds} is expressed through a parameter called *transistor transconductance.*

g_m is defined by

$$g_m = \frac{\partial I_{ds}}{\partial\left(+V_{gs}\right)} = \beta\left(V_{gs}-V_t\right) = \frac{\mu_n C_g}{L^2}\left(V_{gs}-V_t\right) = \frac{\mu_n \epsilon_0 \epsilon_r W}{DL}\left(V_{gs}-V_t\right).$$

So, g_m increases when the width W increases, but it increases the input capacitance and area occupied.

Chapter 9

Integrated Circuits IC

9.1 What is an IC?

An electronic device that can meet the need of small size, light weight, that can run on low power consumption and at the same time high degree of reliability — is an IC device.

IC circuits originated from the miniaturisation programme. The use of ICs in electronic circuits is increasing at a very fast rate and has given birth to the modern computer. It is also widely used in entertainment, industrial and medical equipments.

An integrated circuit consists of one single chip of silicon, typically 1mm × 1mm in area, containing both *active elements* and *passive elements* and their interconnections. Active elements refer to *transistors,* etc. while passive elements refer to resistances, capacitors, etc.

9.2 Constructional Features of an IC

In an IC various components of the circuit like resistance, capacitance, etc. are produced on a small common base, called a *wafer* or a chip. It is very interesting to know how resistors, capacitors, diodes, transistors are incorporated in a single wafer.

There are two types of IC (1) Linear monolithic IC (2) Non-linear IC (a) Combinational digital (b) Sequential digital circuits.

Linear IC circuit is that where the output is proportional to the input. It is called *monolithic,* because it has a uniform base, known as a *substrate.* The substrate is a wafer or a chip of *p*-type silicon. By the diffusion process appropriate substances are added to the substrate to produce resistors, capacitors diodes, transistors, etc.

In non-linear IC combinational digital circuits, the output at some instant depends in some manner (not linearly) on the input at the same instant, independent of the previous history of the input. In non-linear sequential circuits, the output depends upon the order or sequence in which the inputs are applied, so such a circuit does not forget the history and is said to possess memory.

Diffusion is the technique of introducing minute amount of impurities on the substrate by thermal process.

A slice of silicon is heated in a furnace to a temperature of about (1000–1200 °C) and vapour containing the impurities is then blown through the furnace over the slice. Some of the impurity atoms settle down, come in contact with the surface of the slice, then very slowly diffuse into the silicon.

9.3 Fabrication of an IC

The process of fabrication of an IC will tell us why and how an IC circuit reduces cost, space, trouble of interconnection and hence increase in reliability of operation. An IC is an inseparable combination of active and passive elements suitably interconnected so as to form a functional circuit. The fabrication of an IC consists of a number of steps shown in Fig. 9.1(a) through (e) and the sectional view of Fig. 9.1(d) is shown in Fig. 9.1(e).

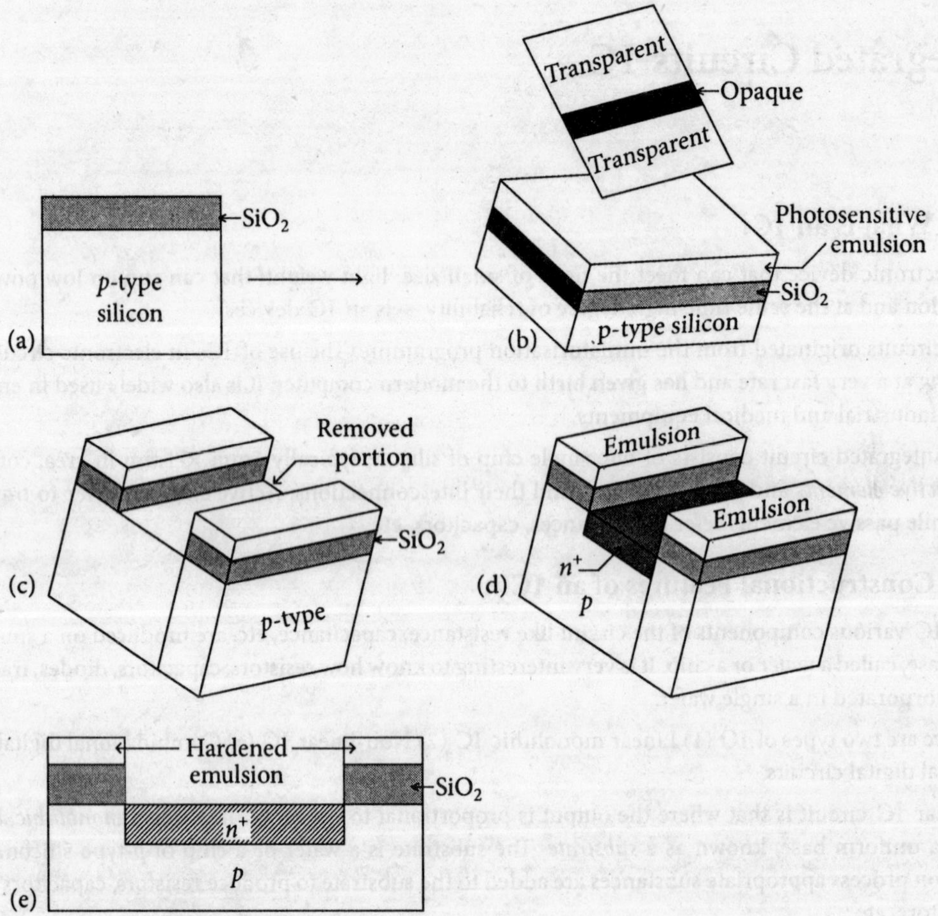

Fig. 9.1

Fabrication of an IC begins with (a) where a *p*-type silicon wafer is taken, also called substrate, it is then polished and placed in an oven having an oxygen atmosphere and heated to a temperature of 1200 °C. As a result, a layer of SiO_2 is formed on the top of the wafer, [Fig. 9.1(a)]. It is then coated with a photosensitive emulsion on which is placed a mask containing transparent and opaque portions at required points [Fig. 9.1(b)].

The masked wafer is exposed to light and the portion of the emulsion not exposed to light is removed using a solvent, also the corresponding portions of the SiO_2 coating are removed, the result is shown in Fig. 9.1(c). The wafer is then placed in a high temperature oven and exposed to an *n*-type dopant (like

As), which diffuses into the substrate. The heavy concentration of the doping material results in a highly doped *n*-region, denoted by n^+ region, [Fig. 9.1(d)], the sectional view is shown in Fig. 9.1(e). Finally, an *n*-type layer is formed above the n^+ layer bypassing a gas which contains an *n*-type impurity over the n^+ region. This *n*-type layer on the top of n^+ region is called the *epitaxial layer* and it will eventually become the collector.

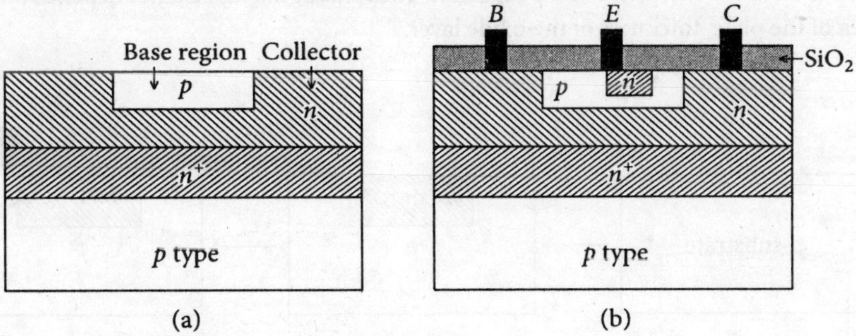

Fig. 9.2

To form the base of the transistor, a *p*-type diffusion is applied [Fig. 9.2(a)] and finally an *n*-type diffusion completes the diffusing process and this forms the emitter [Fig. 9.2(b)]. A SiO_2 protective covering is applied on the top except at the sites for base, emitter and collector, where metallic terminals are put. The interconnection of the emitter, base and collector terminals determine whether element becomes a transistor, diode, resistor or capacitor.

Each wafer or substrate is a disk of diameter about 1 inch. This is subdivided into squares or rectangles, called *dice* or *chip* of the order of 0.050 to 0.100 inch. Thus, each wafer may contain as many as 300 to 400 complete circuits. The small scale integration (SSI) and medium scale integration (MSI) involve 10 to 100 integral components per chip respectively. But, if over 1000 components are involved per chip, the integration is termed as large scale integration (LSI).

9.4 Formation of active and passive elements in IC: How a resistance is formed in an IC?

Resistor or capacitor can be constructed either from the basic transistor element or by thin film technique. We consider here the thin film technique.

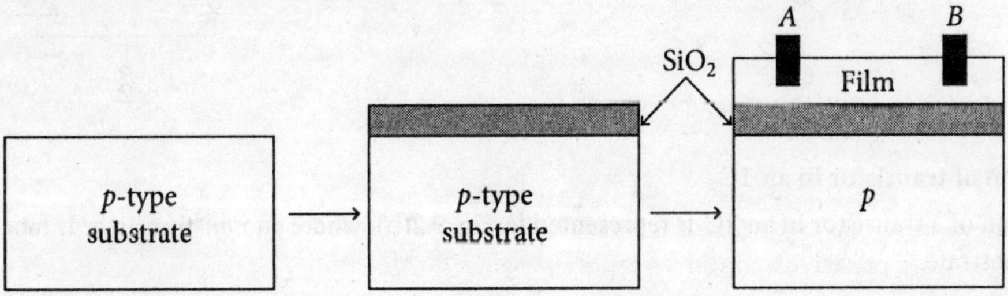

Fig. 9.3

In this process a thin film of resistive material such as nitrided tantalum, nichrome or tin oxide is deposited on the SiO_2 covering over the *p*-type substrate. Finally, an oxide layer at the top of the film serves to put terminals. The resistance so constructed depends on the material of the film, its length and cross section.

How capacitor is formed in an IC?

The formation of a capacitor in an IC is represented in Fig. 9.4. An *n*-type area is diffused into *p*-type substrate and it forms one plate of the capacitor. Then an oxide layer is applied and it becomes the dielectric. Terminal 1 is etched on it making contact with *n*-type area. Finally, a large metallic electrode is laid on the oxide layer top to form the terminal 2 of the capacitor. The value of the capacitance depends on the dielectric constant, area of the plate, thickness of the oxide layer.

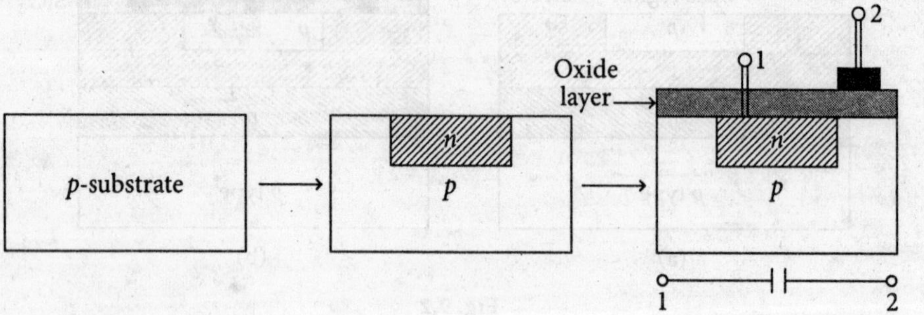

Fig. 9.4

How diodes are formed in an IC?

Fabrication of diodes in an IC is shown Fig. 9.5(a), (b) and (c)]. *n*-type region is diffused at suitable locations [Fig. 9.5(b)]. An oxide layer is then spread over the entire surface and terminals 1, 2, 3 are etched on it. The figure shows two diodes with a common cathode (which is the *p*-type material).

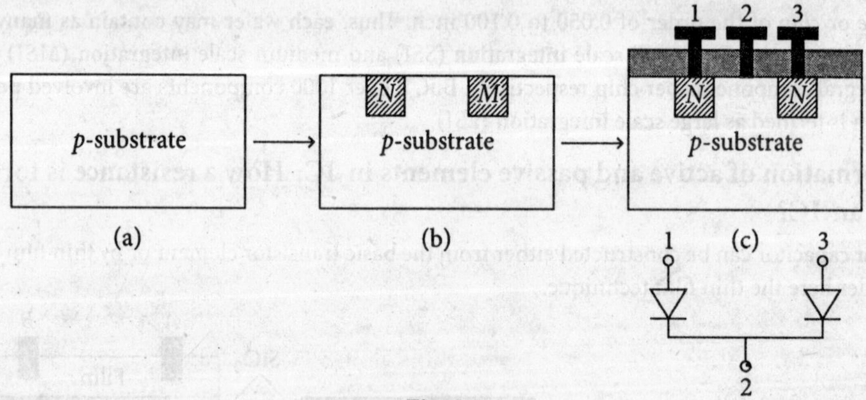

Fig. 9.5

Formation of transistor in an IC

Fabrication of a transistor in an IC is represented in Fig. 9.2(b), where an *npn*-transistor is fabricated on a *p*-type substrate.

Formation of two transistors on a single wafer

Fig. 9.6 shows two transistors formed on the same wafer, a *p*-type substrate, the discrete equivalent being shown in Fig. 9.6(b). r_{SC} represents the series collector resistance, which in presence of a highly doped n^+ region, shown in Fig. 9.2(b), takes up typical value less than 1Ω, whereas in absence of n^+ region as shown in Fig. 9.6(a) attains a value between 6 to 15Ω. This increase in r_{SC} results in the increase of the saturation voltage of the transistor.

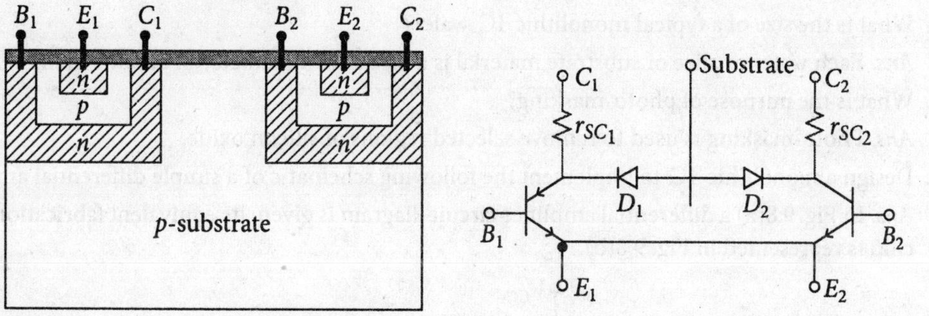

(a) (b)

Fig. 9.6

In Fig. 9.6 two transistors (C_1, B_1, E_1), (C_2, B_2, E_2) are separated by a p-type substrate. The collectors C_1, C_2 and the substrate form pn-junction diodes. The substrate when connected to the most negative potential, the diodes act as if in reverse bias and offer a high impedance thus isolation between the two transistors is effected.

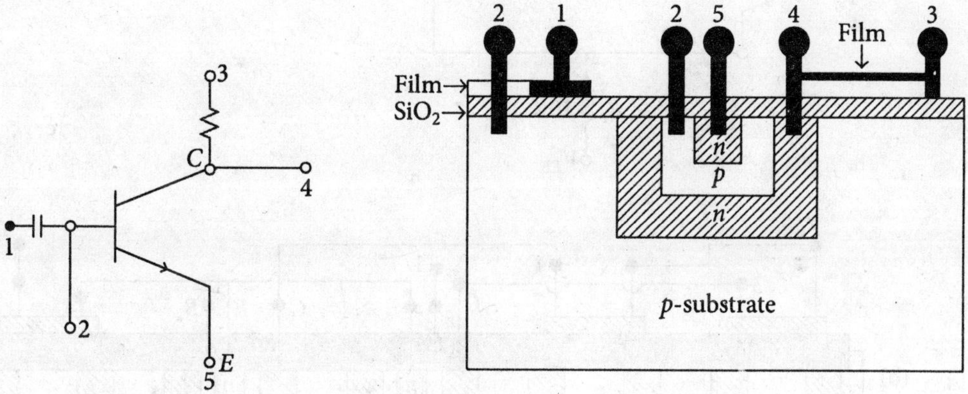

(a) (b)

Fig. 9.7

Suppose, a circuit shown in Fig. 9.7(a) is to be incorporated in an IC chip. How to fabricate?

The IC fabrication is shown in Fig. 9.7 with the terminals. It is a monolithic IC consisting of one transistor, one condenser and one resistance. Thus, on a single wafer large number of discrete components can be fabricated so as to serve as different functional units. One functional unit may be an audio-amplifier, another functional unit may be an oscillator, etc. Thus, IC BEL 700 widely used nowadays in 2 band or 3 band solid state radio receiver is a monolithic linear IC consisting of two sections, one high frequency section, the other audio section. This IC is a fabrication of 25 transistors, 11 resistors and 18 diodes.

9.5 Short Theory Questions and Answers

1. What do you mean by the term microelectronics?

   Ans. The term microelectronics refers to the integrated circuits, which brought reduction in size of electronic equipment.

2. What is the size of a typical monolithic IC wafer?

   Ans. Each wafer or slice of substrate material is a disc with a diameter of about 1 inch.

3. What is the purpose of photo-masking?

   Ans. Photo-masking is used to remove selected regions of silicon oxide.

4. Design a monolithic IC to implement the following schematic of a simple differential amplifier.

   Ans. In Fig. 9.8(a) a differential amplifier circuit diagram is given, its equivalent fabrication on an IC chip is represented in Fig. 9.8(b).

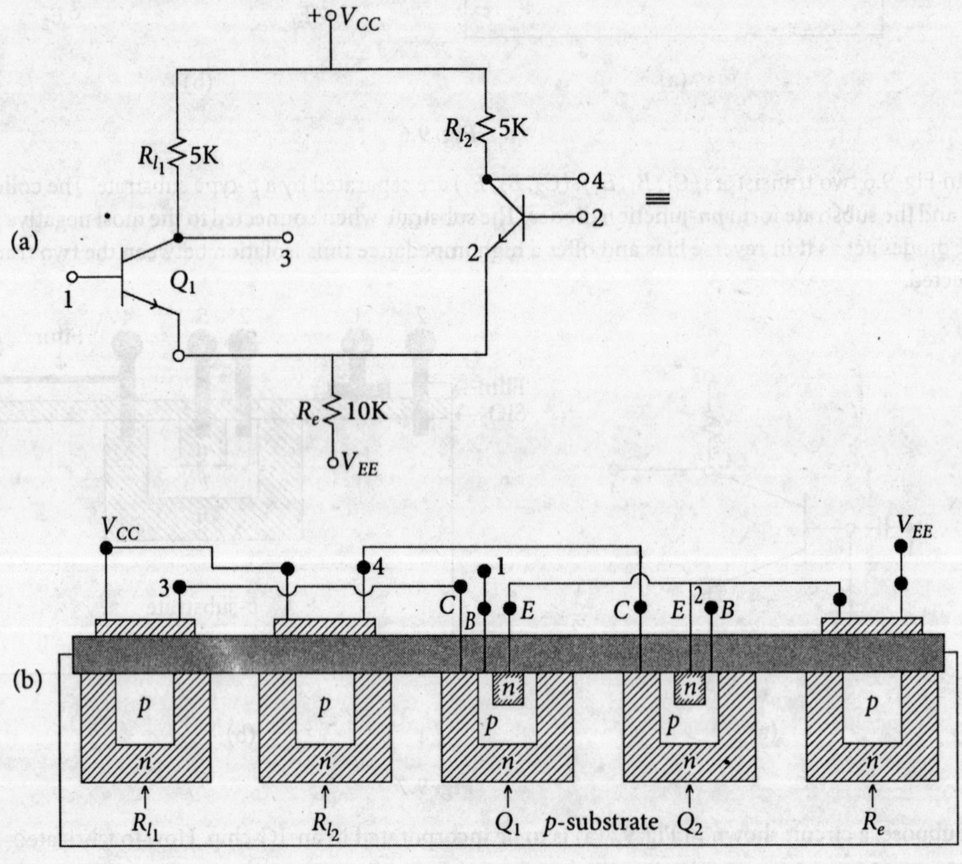

Fig. 9.8

5. Design a monolithic IC to implement the schematic shown in Fig. 9.9(a).

   Ans. In Fig. 9.9(b), the IC fabrication of the circuit shown in Fig. 9.9(a) is schematically represented.

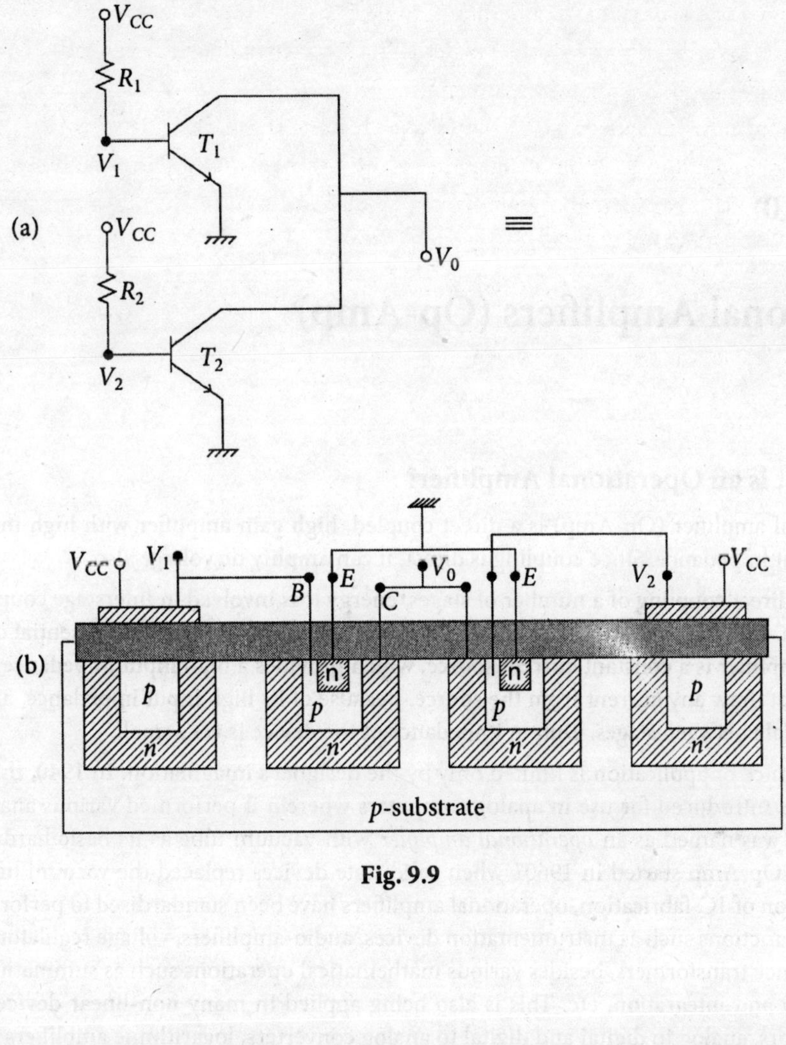
V_CC
R_1
T_1
V_1
(a)
V_CC
V_0
≡
R_2
T_2
V_2
V_CC
V_1
B
E
C
V_0
E
V_2
V_CC
(b)
n
n
p
p
p
p
n
n
n
n
p-substrate

Fig. 9.9

Chapter 10

Operational Amplifiers (Op-Amp)

10.1 What is an Operational Amplifier?

An operational amplifier (Op-Amp) is a direct coupled, high gain amplifier with high input impedance and low output impedance. Since coupling is direct, it can amplify dc voltage also.

Because of direct coupling of a number of stages, energy loss involved in interstage coupling is reduced to a minimum. Basically it is a differential amplifier forming its heart. The most essential component of a differential amplifier is a constant current source, which provides a high input impedance and hence this device does not draw any current from the source. Because of its high input impedance, an Op-Amp can be used to isolate between stages. Output impedance of the device is very small.

Its performance or application is limited only by the designer's imagination. In 1940, this highly versatile device was introduced for use in analog computers wherein it performed various analog operations. This is why it was named as an *operational amplifier* with vacuum tube as its basic hardware. The wide acceptance of Op-Amp started in 1960's when solid state devices replaced the vacuum tubes. Later with the introduction of IC fabrication, operational amplifiers have been standardised to perform a wide range of electronic functions such as instrumentation devices, audio-amplifiers, voltage regulators, signal generators, impedance transformers, besides various mathematical operations such as summation, subtraction, differentiation and integration, etc. This is also being applied in many non-linear devices such as voltage comparators, analog to digital and digital to analog converters, logarithmic amplifiers and non-linear function generators.

In addition to its being a high gain amplifier, an operational amplifier has two floating input terminals and an ideal Op-Amp has the following properties.

1. Input resistance $R_i = \infty$.
2. Output resistance $R_0 = 0$.
3. Voltage gain $A_v = \infty$.
4. Bandwidth = 0 to ∞.
5. Perfect balance, i.e., $V_0 = 0$, when $V_1 = V_2$, where V_1 and V_2 are input voltages to the two-input terminals and V_0 is the output voltage.
6. No temperature dependence.

However, practical Op-Amps have properties that deviate from the ideal properties and the deviation sets a criterion for determining the quality of the Op-Amps.

10.2 Symbolic representation of an Op-Amp

The symbol for an Op-Amp (IC741) [Fig. 10.1], where I denotes the inverting input terminal (Pin 2) meaning that any input applied at this terminal would be inverted at the output (Pin 6), while non-inverting input terminal NI (Pin 3) will give rise to output in phase with the input. These facts are shown schematically in Fig. 10.2(a) and (b).

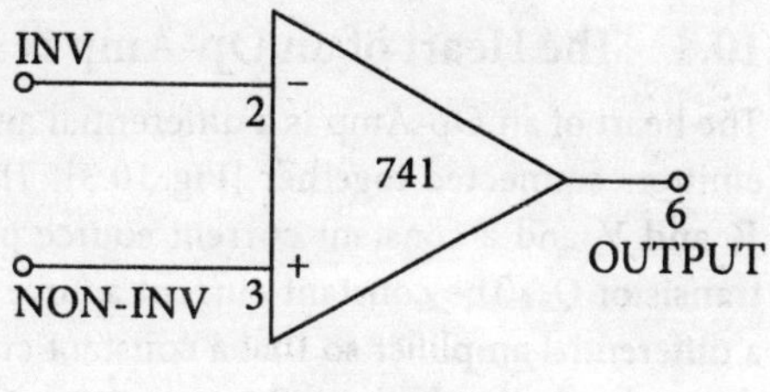

Fig. 10.1

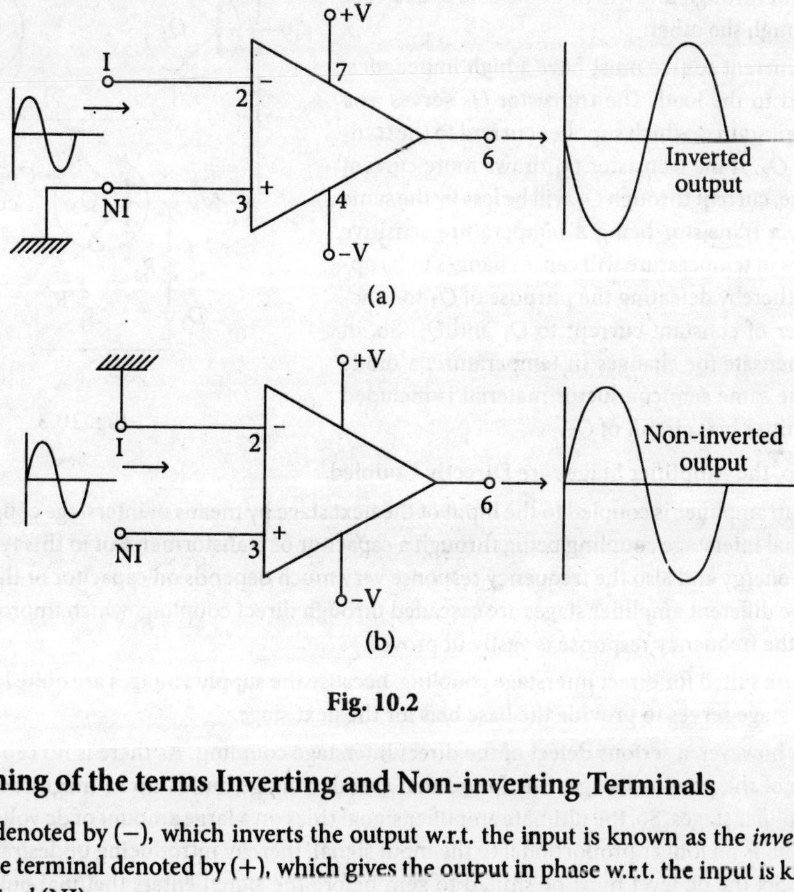

Fig. 10.2

10.3 Meaning of the terms Inverting and Non-inverting Terminals

The terminal denoted by (–), which inverts the output w.r.t. the input is known as the *inverting terminal* (I), whereas the terminal denoted by (+), which gives the output in phase w.r.t. the input is known as *Non-inverting terminal* (NI).

A single stage common emitter configuration we have seen to produce a phase shift of 180°, but again if another similar stage is added, the output is in phase w.r.t. the input. If a third amplifier stage is used, the ultimate output again becomes inverted w.r.t. the input.

Thus, we see that a system of even numbers (2, 4, 6, etc.) of common emitter configuration produces non-inverted output, whereas a system of odd number (1, 3, 5, etc.) of common emitter stages produces an inverted (180° out of phase) output w.r.t. the input. Thus, the inverted terminal (I) includes an odd number of amplifying stages and the non-inverting terminal (NI) includes an even number of amplifying stages. There is also possibility of having non-inversion terminal only if common collector (emitter follower) amplifier configuration is used irrespective of the number of stages involved.

10.4 The Heart of an Op-Amp is a Differential Amplifier

The heart of an Op-Amp is a differential amplifier, which consists of two transistors Q_1 and Q_2 with their emitters connected together [Fig. 10.3]. There are two-input terminals e_1 and e_2 and output terminals V_1 and V_2 and a constant current source provided by the transistor Q_3. The constant current source is essential for a differential amplifier so that a constant current is always drawn by the loads (here the transistors Q_1 and Q_2), increase in current through one will be at the cost of decrease of current through the other.

A constant current source must have a high impedance, high compared to the load. The transistor Q_3 serves as a constant current source, which supplies current to the transistors Q_1 and Q_2. If the transistor Q_1 draws more current from the source, current through Q_2 will be less by the same amount. Now, a transistor being a temperature sensitive device, changes in temperature will cause changes in its operating point, thereby defeating the purpose of Q_3 to function as supplier of constant current to Q_1 and Q_2. So, in order to compensate for changes in temperature, a diode D_1 made of the same semiconductor material is included in the base-emitter bias circuit of Q_3,

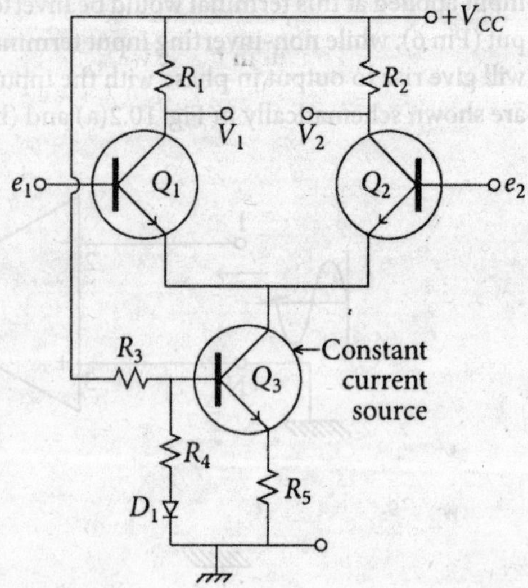

Fig. 10.3

In an Op-Amp, the Amplifier Stages are Directly Coupled

The output of an amplifier is coupled to the input of the next stage by means of interstage coupling network, the conventional interstage coupling being through a capacitor or transformer. But in this type of coupling there is loss of energy and also the frequency response very much depends on capacitor or the inductor. So, in Op-Amp the different amplifier stages are cascaded through direct coupling, which improves the overall gain and also the frequency response is vastly improved.

Transistors are suited for direct interstage coupling, because the supply voltages are quite low. The output voltage of one stage serves to provide the base bias for the next stage.

But there is, however, a serious defect of the direct interstage coupling. As there is no capacitor to block the dc voltage of the previous stage, it is found that this dc voltage builds up in magnitude through the successive amplifier stages. So, the ultimate amplifier signal rides on a large amount of dc voltage, as a result, the signal output is no longer proportional to the input signal, thereby introducing undesirable distortions in the output. So, the dc level must be shifted to zero before the signal enters the final output stage. The process by which the dc voltage is eliminated from the ac signal is called *dc level shifting*. This shifting is effected through either of the following: (i) Emitter follower stages (ii) *pnp* transistor if the prior amplifier stages employ *npn* transistors (iii) Zener diodes (iv) Forward biased diodes.

10.5 Basic Requirement of a Constant Current Source

A source having high internal impedance compared to the load impedance is capable of supplying constant current to the load. A constant current source I_s of internal impedance Z_i connected in parallel with the source [Fig. 10.4], Z_l being the load. The equivalent impedance of the network is given by

$$Z_e = \frac{Z_i Z_l}{Z_i + Z_l}.$$

$$\therefore\ V_{AB} = I_s \times \frac{Z_i Z_l}{Z_i + Z_l}. \qquad (10.1)$$

$$\therefore\ I_l = \frac{V_{AB}}{Z_l} = I_s \frac{Z_i}{Z_i + Z_l}.$$

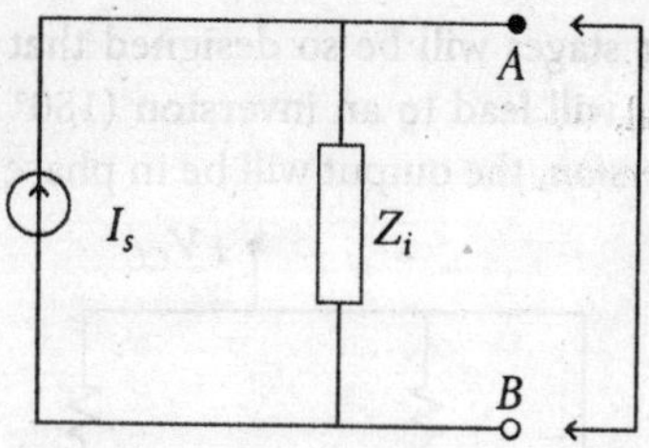

Fig. 10.4

So, if $Z_i \gg Z_l$, $I_l = I_s$, hence it functions as a constant current source. So, the collector impedance of the transistor Q_3 is designed to be much higher value than the impedance of the differential amplifier comprising Q_1 and Q_2.

10.6 Modes of Input Supply: Inverting and Non-inverting Input Terminals

Refer to Fig. 10.3, we will discuss here the various ways in which the input signals may be supplied to the differential amplifier and the nature of the outputs obtained.

The input signals can be applied in four possible ways to the differential amplifier.

1. Between the terminal e_1 and the earth.
2. Between the terminal e_2 and the earth.
3. Between the input terminals e_1 and e_2.
4. Between e_1 and e_2 shorted and the earth.

Nature of the output when signal is applied between e_2 and the earth. When the signal is applied between e_1 and the ground, as the input goes positive, the emitter-collector current of Q_1 increases, causing a greater potential drop across R_1, and so the available voltage at the output terminal V_1 goes less positive, i.e., negative going voltage. Thus, Q_1 produces a negative going voltage output when its input signal goes positive. Similarly, when the input signal is negative going, the output at V_1 is positive going. So it inverts the signal. Since current is drawn from the constant current source, increase of current through Q_1 results in decrease in current through Q_2 and hence the output at V_2 is in phase with the input. In other words, Q_1 inverts the input whereas Q_2 produces an output in phase with the input at e_1.

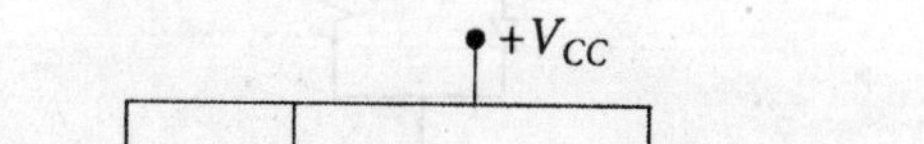

Fig. 10.5

Nature of the output when signal is applied between e_2 and the earth. When the signal is applied between e_2 and ground, as the signal is positive going, the voltage at the terminal V_2 is negative going but Q_2 being constant current source, the current drawn by Q_1 decreases and hence, the voltage at V_1 is also positive going. Thus, the output at terminal V_2 is inverted while the output at V_1 is non-inverted. So, we find from Figs. 10.5 and 10.6 that either of the input terminals e_1 and e_2 of the differential amplifier is capable of producing inverting or non-inverting outputs. But in an Op-Amp two terminals are permanently labelled *inverting* denoted by the symbol (−) and *non-inverting* denoted by the symbol (+) as shown in Fig. 10.16. A signal applied between Pin 2 and the earth with Pin 3 grounded, the output obtained is inverted, whereas for input between Pin 3 and ground produces non-inverted output. It means, therefore, that the chain of amplifier stages will be so designed that an input signal applied to one specific terminal, say, inverting terminal, will lead to an inversion (180° out of phase) while when applied to the other terminal, called *non-inversion,* the output will be in phase w.r.t. the input.

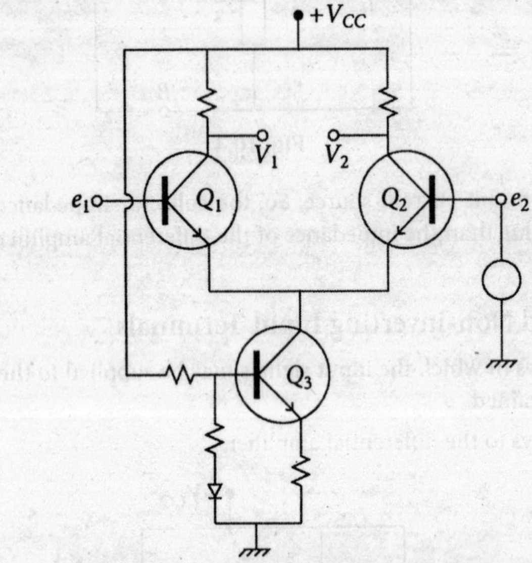

Fig. 10.6

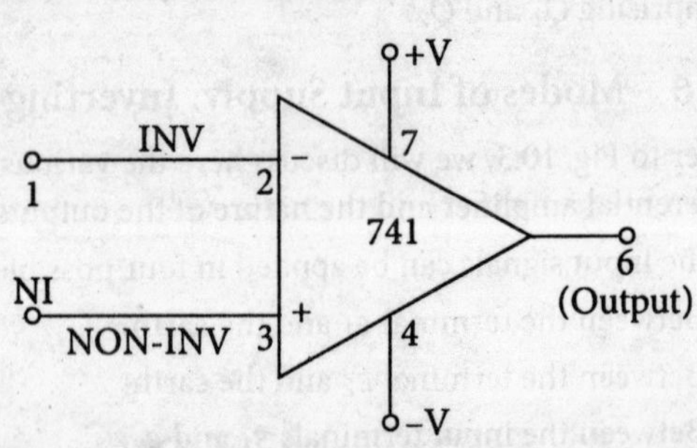

Fig. 10.7

10.7 Gain of the Op-Amp

An Op-Amp is a negative feedback amplifier so as to stabilise the gain. By gain we mean the ratio of the output voltage to the input voltage. We have introduced two types of gains, one called *open loop gain,* i.e., gain of the Op-Amp without feedback and the other, closed-loop gain, i.e., gain when feedbacking is done. Because of negative feedback though gain reduces, its value may be as high as 10^3 to 10^6.

Op-*amp with negative feedback: Formation of virtual ground.* In Fig. 10.8 we have shown an inverting configuration of an Op-Amp with negative feedback. The signal V_i is applied to the (–) inverting terminal through R_1. Then according to the definition of gain, we can write,

$$A = -\frac{V_0}{V_1}. \qquad (10.2)$$

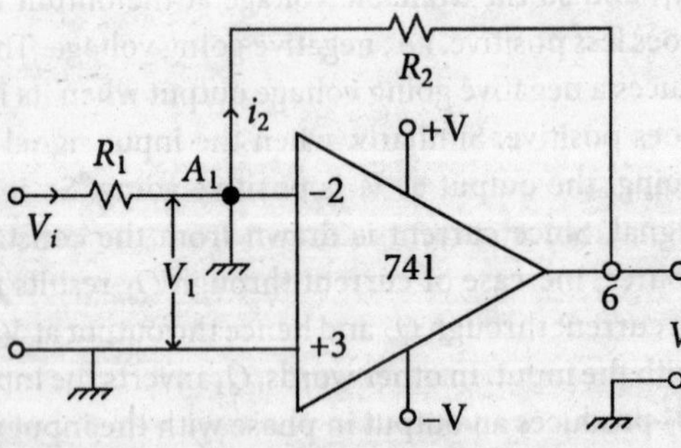

Fig. 10.8

The negative sign indicates inversion, i.e., 180° phase shift. We add a voltage feedback resistance R_2 to lower the output and stabilise the gain. Since the gain A is very large, the input V_1 is so small that the terminal A_1 is virtually at ground potential shown by line at A_1 to ground. If $V_0 = 10\text{V}$, $A = 10^4$, then the voltage at A_1 required is 1mV. With high input impedance and negligible applied voltage, essentially zero current enters the gain element at A_1. So we can write,

$$i_1 = i_2 \text{ or, } \frac{V_i - V_1}{R_1} = \frac{V_1 - V_0}{R_2} \text{ or, } \frac{V_i - V_1}{V_1 - V_0} = \frac{R_1}{R_2}. \qquad (10.3)$$

Dividing by V_0 and introducing the definitions of open loop gain (A) and closed-loop gain $\left(A_f\right)$, we get

$$\frac{\frac{1}{A_f}+\frac{1}{A}}{-\frac{1}{A}-1}=\frac{R_1}{R_2}. \qquad \therefore A_f=-\frac{R_2}{R_1}\times\frac{1}{1+\frac{1}{A}\left[1+\frac{R_2}{R_1}\right]}. \tag{10.4}$$

If we put $\beta=\dfrac{R_1}{R_1+R_2}$, called feedback factor, we get

$$A_f=-\frac{R_2}{R_1}\left(1-\frac{1}{1+A\beta}\right).$$

When A is very large, (10.4) gives

$$A_f=-\frac{R_2}{R_1}. \tag{10.5}$$

Formation of virtual ground at A_1 leads to the following results:

1. $V_i-i_1R_1=0$, or, $V_i=i_1R_1$.

   It means, therefore, the p.d. across R_1 is equal to the input V_i.
2. $V_0-R_2i_2=V_1=0$ or, $V_0=R_2i_2$.

   It means that the voltage drop across the resistor R_2 is equal to the output voltage V_0.
3. The input current and feedback current are almost of the same magnitude and because of negative feedback, the feedback current and input current meet at A_1 in anti-phase and tend to cancel each other thus forming a state of almost zero potential at A_1. This is what is known as *virtual grounding.*
4. The difference in potential between (−) terminal and (+) terminal is also equal to zero, because (−) terminal is virtually at ground potential and (+) terminal is de facto grounded.

From (10.5), it is evident that with the application of negative feedback, the gain of the inverting Op-Amp is determined entirely by the ratio of the feedback resistor R_2 to the input resistor R_1.

The non-inverting Op-*amp.* In Fig. 10.9 the input V_i is applied between the (+) terminal and ground. As in the case of inverting amplifier, the input to the amplifier here also must be limited to a small value, since the gain is very large. When the input is positive going, a large positive going output is obtained (non-inverting). This output forces a current through R_2 and R_1 to earth. The potential of the junction of R_2 and R_1 is applied to the (−) terminal. Since this potential is positive going, its inversion by the amplifier in the output will reduce the output V_0 and stabilise the amplifier operation. Since the feedback potential at (−) terminal is positive going when the input in the non-inverting terminal is also positive going, almost zero p.d. exists between the (+) terminal and (−) terminal. It, in other words, means that the feedback tends to take the (−) terminal to the same potential as that which exists in the (+) terminal.

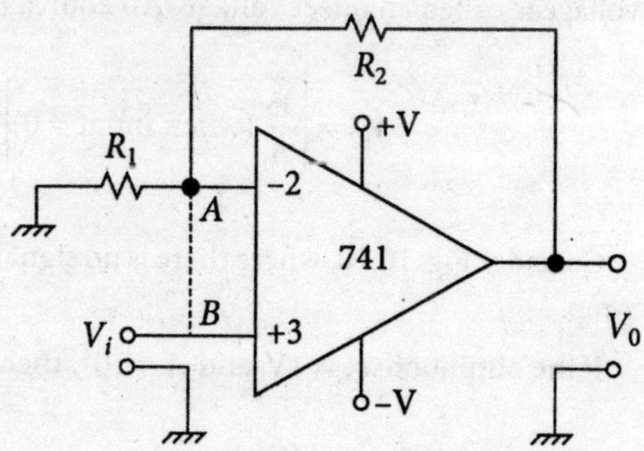

Fig. 10.9

Here,

$$V_f=\frac{V_0R_1}{(R_1+R_2)} \Rightarrow \beta=\frac{V_f}{V_0}=\frac{R_1}{R_1+R_2}.$$

∴ differential input to the amplifier is given by $\left(V_i - V_f\right)$.

$$\therefore A = \frac{V_0}{\left(V_i - V_f\right)} \Rightarrow V_0 = A\left(V_i - V_f\right).$$

Now, $V_0/V_i = A_f$, so dividing by V_0, we get

$$1 = \left(\frac{V_i}{V_0} - \frac{R_1}{R_1+R_2}\right) = A\left(\frac{1}{A_f} - \beta\right)$$

$$\Rightarrow \frac{1}{A_f} - \beta = \frac{1}{A}, \text{ since } A \to \text{large, } A_f = \frac{1}{\beta} = \frac{R_1+R_2}{R_1} = 1 + \frac{R_2}{R_1}.$$

Then we get

$$A_f = \frac{R_1+R_2}{R_1} = 1 + \frac{R_2}{R_1}. \tag{10.6}$$

10.8 What is the cause of the high gain of the Op-Amp?

The use of at least two stages of differential amplifier, direct interstage coupling and Darlington-pair amplifiers are some of the factors contributing to the high gain of the Op-Amp. In the present day IC circuitry, the extremely high gain of the Op-Amp is largely due to the development of super-β transistor, bipolar transistor with current gain approaching 10^4. More recently, a new type of high performance Op-Amp has come out based on the bipolar and FET technologies. Such Op-Amp, called Bi-FET Op-Amp, uses FET input transistors and bipolar transistors on a common monolithic chip.

The latest technology giving rise to MOSFET has been used as the input element of the latest Op-Amps, which are characterised by (i) very high gain (ii) very high impedance (iii) wide frequency response and lastly (iv) high *slew rate,* which means ability to swing output voltage very rapidly.

10.9 Errors (Offset Voltage or Current) generated by the Op-Amp

In an ideal Op-Amp when the input voltages are zero, the output voltage should be zero. But because of the inherent imbalance of the circuit, an output voltage is obtained even in absence of any input voltage. Such voltage is called an offset voltage. An equivalent input offset voltage is defined as

$$V_{io} = \left|\frac{V_0}{A} \text{ (when input = 0)}\right|.$$

Consider Fig. 10.10, where there is no signal applied to the Op-Amp.

If the output offset is 1V, and $A = 10^4$, then

$$V_{io} = \left(\frac{V_0}{A}\right) = \frac{1}{10^4} = 0.1\text{mV}.$$

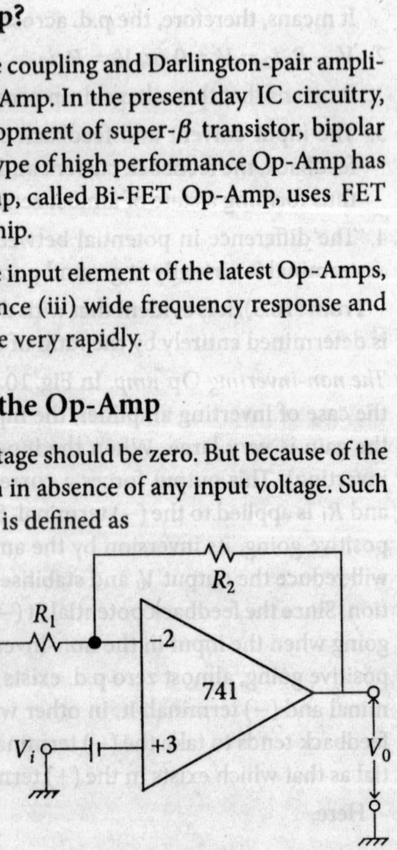

Fig. 10.10

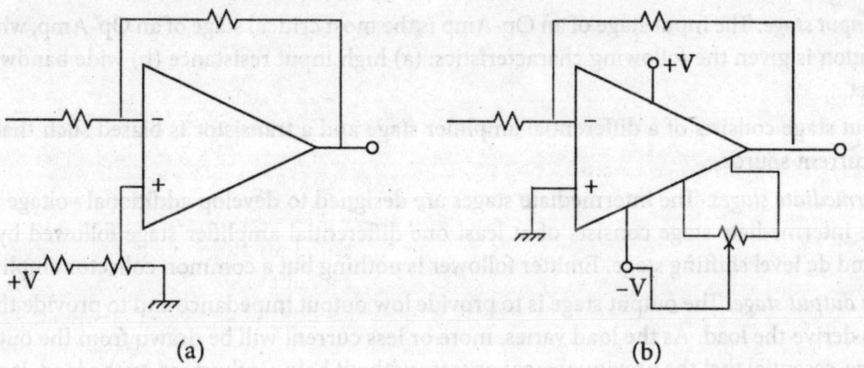

Fig. 10.11

So applying a correcting voltage V_{io} with polarity shown in the (+) terminal, offset correction is effected. The input offset can be calculated w.r.t. Fig. 10.10 as

$$V_{io} = \frac{R_1}{R_1 + R_2} V_0 \text{ (when inputs = 0).}$$

This nulling voltage may be added externally or internally as shown in Fig. 10.11(a) and (b) respectively.

Current offset. Even in absence of any voltage at the input ports of the Op-Amp, currents I_1 and I_2 [Fig. 10.12] enter the input terminals and they are not equal but they are small. If the output impedance is small, the output terminal of R_2 is at ground for dc effectively. Theoretically, it appears that due to the symmetry of the differential amplifier stage, these two bias currents should be same. But in practice, however, they are never exactly equal. This implies that a resulting *offset current* $I_{io} = I_1 \sim I_2$ would always flow causing a non-zero output voltage from the amplifier without any input. This can be, of course, balanced or nullified by applying a small potential, called the input *offset* voltage V_{io}, at any one of the internal offset balance terminals of the Op-Amp.

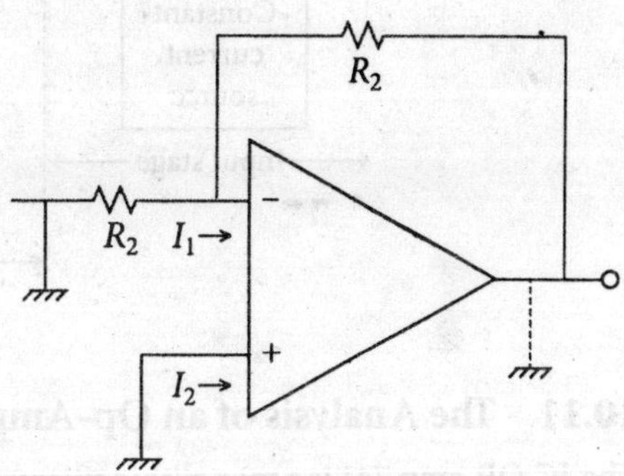

Fig. 10.12

Definitions

Input offset voltage. This is defined as the dc input voltage V_{io} that must be provided between the two-input terminals of the Op-Amp to get zero output voltage.

Input offset current. This is the difference between the separate bias currents entering the terminals of a balanced amplifier.

Input bias current. The input bias current is one-half of the sum of the separate bias currents entering the two-input terminals of a balanced amplifier, i.e.,

$$I_B = \frac{I_1 + I_2}{2}, \text{ when } V_0 = 0.$$

10.10 Block Diagram of an Op-Amp

The block diagram of a typical Op-Amp is shown in Fig. 10.13. Broadly speaking an Op-Amp consists of three parts: (i) the input stage (ii) a number of intermediate stages (iii) the output stage.

(i) *The input stage.* The input stage of an Op-Amp is the most critical stage of an Op-Amp, which by careful fabrication is given the following characteristics: (a) high input resistance (b) wide bandwidth (c) low input offset.

The input stage consists of a differential amplifier stage and a transistor is biased such that it acts as a constant current source.

(ii) *Intermediate stages.* The intermediate stages are designed to develop additional voltage and current gains. The intermediate stage consists of at least one differential amplifier stage followed by an emitter follower and dc level shifting stage. Emitter follower is nothing but a common collector amplifier.

(iii) *The output stage.* The output stage is to provide low output impedance and to provide the necessary current to derive the load. As the load varies, more or less current will be drawn from the output stage. It is, therefore, essential that the previous stages operate without being influenced by the load. It requires that the output stage must isolate the previous stages from load variations. So, the input resistance of this stage must be high, current gain high but low output resistance.

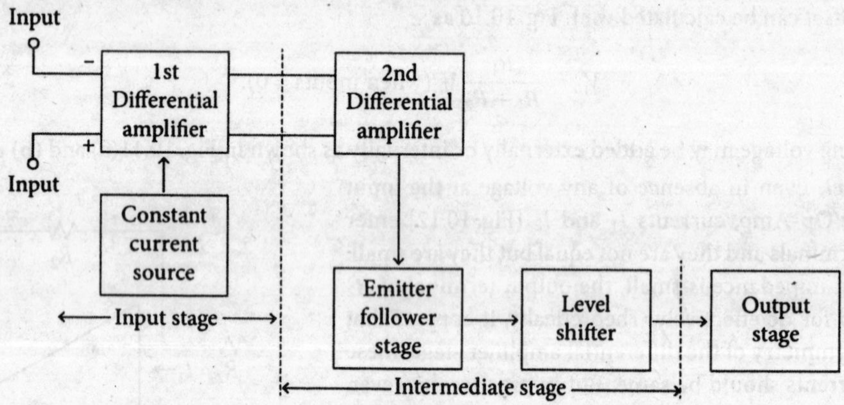

Fig. 10.13

10.11 The Analysis of an Op-Amp IC 741

The IC OP-amp 741 is a monolithic silicon integrated amplifier consisting of 20 transistors connected so as to form a differential amplifier, level shifting stage; it also provides output short-circuit protection, internal phase compensation. The schematic of IC 741 is shown in Fig. 10.14.

IC 741 is fully protected against damage from shorting occurring at either the input or output. Q_1, Q_2 are transistors in common collector mode while Q_3, Q_4 are common-base-mode forming a CC–CB cascade differential pre-amplifier. The output of the difference amplifier is fed to a Darlington connected CE intermediate amplifier formed by transistors Q_{18} and Q_{17}. The Darlington amplifier has a voltage gain of about 1000 and drives the class — AB complementary emitter-follower output amplifiers formed by transistors Q_{14}, Q_{16}. The transistors Q_{15} and Q_{19} provide the overload protection.

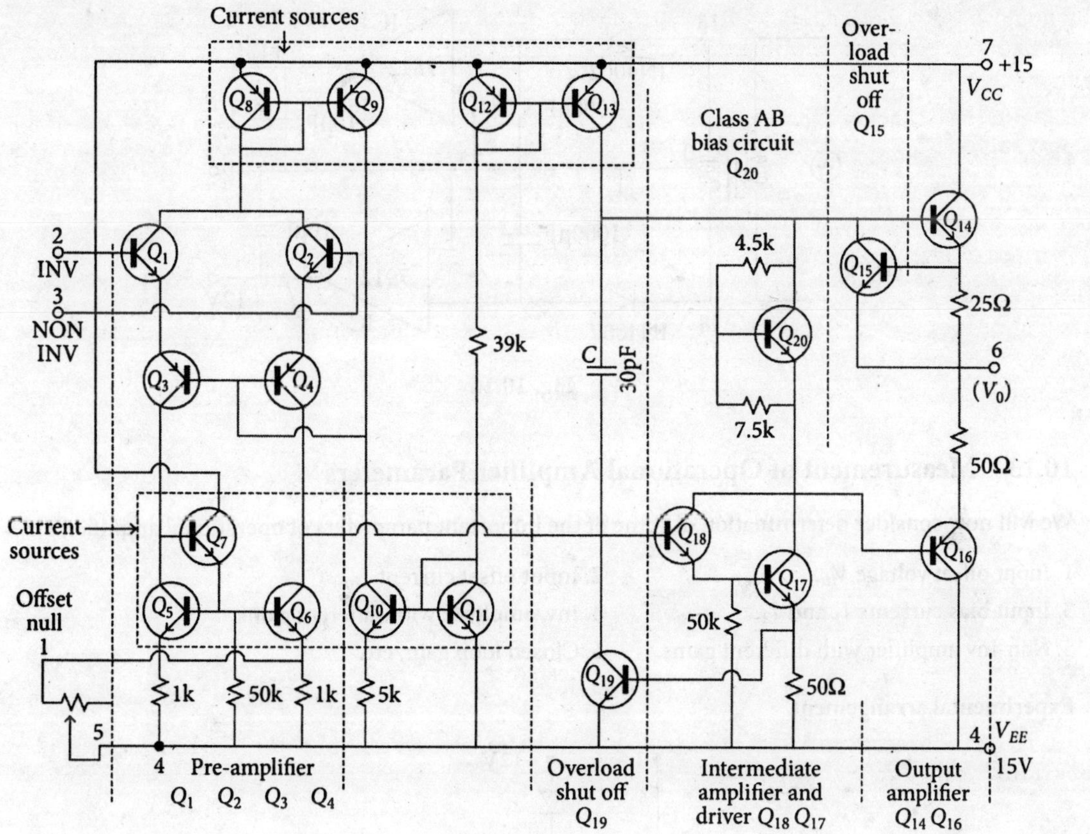

Fig. 10.14

The input offset of the amplifier is nullified by connecting a 10k potentiometer to the offset null terminals. This allows the collector currents of transistors Q_3 and Q_4 to have equal values. The circuit symbol is shown in Fig. 10.15.

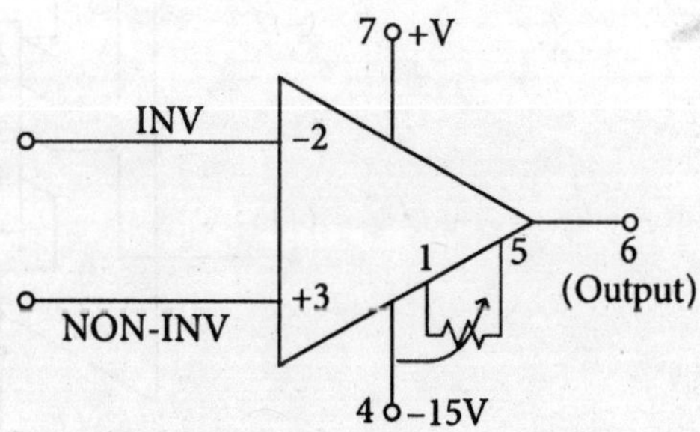

Fig. 10.15

The amplifier can be used in a wide variety of applications. Of the wide range of applications we will be considering only a few of them. The built-in compensation and built-in protection are some of the assets of Op-Amp 741, a general purpose IC amplifier.

10.12 Power Supply for an Op-Amp

The Op-Amp requires power supplies, one positive, the other negative. So a dual polarity power supply facilitates application of bias to the circuits. In Fig. 10.16 we have shown a suitable dual power supply using IC 7812 and IC 7912.

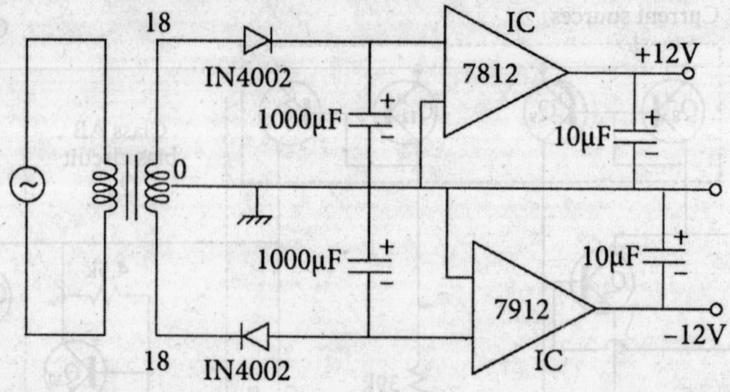

Fig. 10.16

10.13 Measurement of Operational Amplifier Parameters

We will now consider determination of some of the important parameters of operational amplifier, viz.,

1. Input offset voltage V_{io}.
2. Input offset current I_{io}.
3. Input bias currents I_1 and I_2.
4. Inv. amplifier with different gains.
5. Non-inv amplifier with different gains.
6. Closed loop gain, etc.

Experimental arrangement

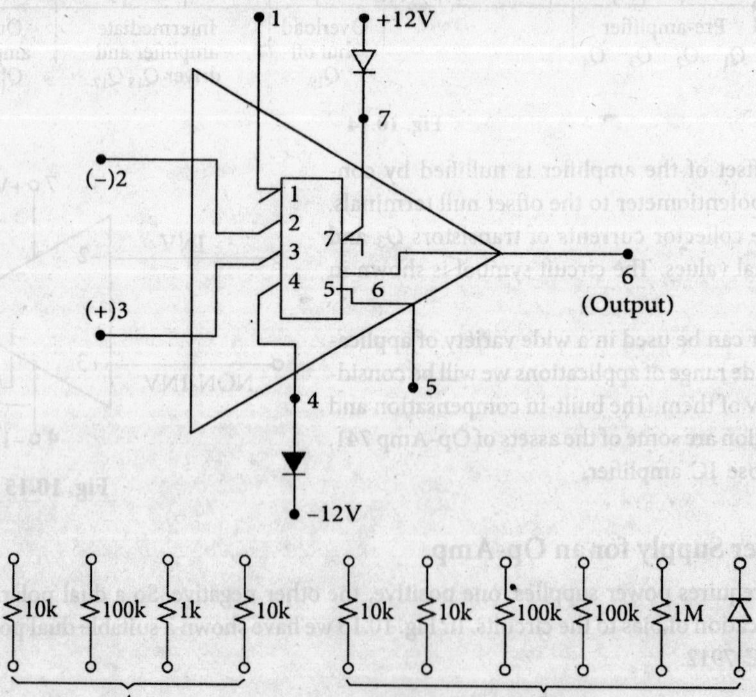

Fig. 10.17

Measurement of input offset voltage V_{io} : The circuit is shown in Fig. 10.18.

$$V_0 = -\left(\frac{R_1 + R_2}{R_1}\right) V_{io}. \tag{10.7}$$

The negative sign indicates inverting amplifier mode. Take R_2 values from B arrangement and R_1 from A arrangement.

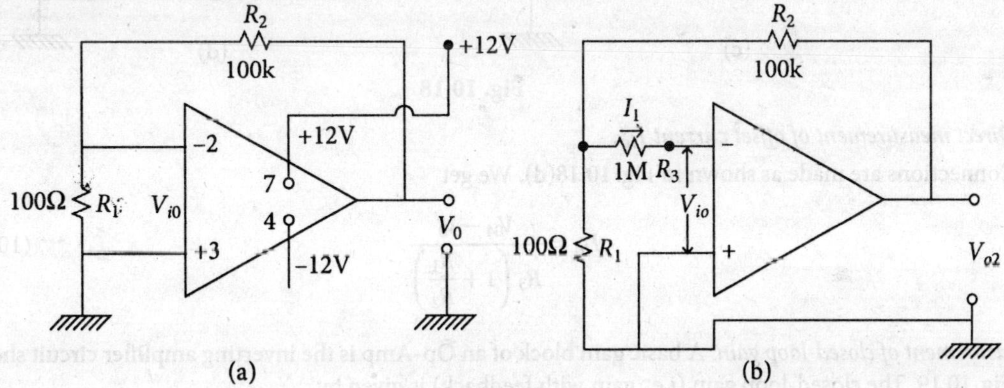

Fig. 10.18

Measurement of input bias current I_1 *at inverting input.* Connect as shown in Fig. 10.18(b) taking resistances R_1, R_2 and R_3 from the arrangements A and B.

Now,

$$V_{io} = -\frac{V_{a_2}}{R_1 + R_2} \times R_1 + I_1 R_3.$$

Again,

$$V_{io} = -\frac{R_1}{R_1 + R_2} V_0.$$

$$\therefore\ I_1 R_3 - \frac{V_{02} R_1}{R_1 + R_2} = -\frac{R_1}{R_1 + R_2} V_0 \text{ or, } I_1 = \frac{V_{02} - V_0}{R_3(1 + R_2/R_1)}. \tag{10.8}$$

Measurement of input bias current I_2 *at the non-inverting terminal.* Connection is made as shown in Fig. 10.18(c) and (d) introducing R_1, R_2 and R_3 from the arrangements A and B.

$$\frac{-V_{03} R_1}{(R_1 + R_2)} + I_2 R_2 = V_{io}.$$

Now,

$$V_{io} = \frac{R_1}{R_1 + R_2} V_0.$$

$$\therefore\ I_2 R_3 = \frac{R_1}{R_1 + R_2} V_0 \text{ or, } I_2 = \frac{V_{03} - V_0}{R_3(1 + R_2/R_3)}. \tag{10.9}$$

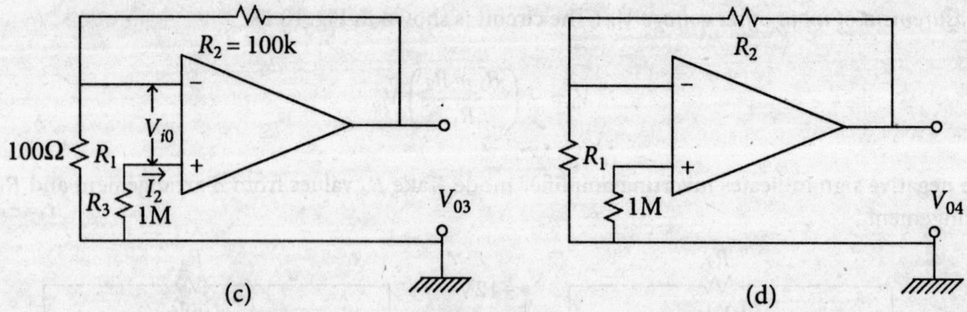

Fig. 10.18

Direct measurement of offset current I_{io}.

Connections are made as shown in Fig. 10.18(d). We get

$$I_{io} = \frac{V_{04} - V_0}{R_3\left(1 + \frac{R_1}{R_2}\right)}. \tag{10.10}$$

Measurement of closed-loop gain. A basic gain block of an Op-Amp is the inverting amplifier circuit shown in Fig. 10.19. The closed-loop gain (i.e., gain with feedback) is given by

$$A_f = -\frac{R_2}{R_1}\left(1 - \frac{1}{A\beta}\right),$$

where $\beta = \dfrac{R_1}{R_1 + R_2}$, called *feedback factor.*

If $AB \gg 1$, we have, $A_f = -R_2/R_1$. In a practical amplifier, the offset error minimisation is done through a resistor R_3 connected between NI terminal and ground. R_3 should be chosen as

$$R_3 = \frac{R_1 R_2}{R_1 + R_2}.$$

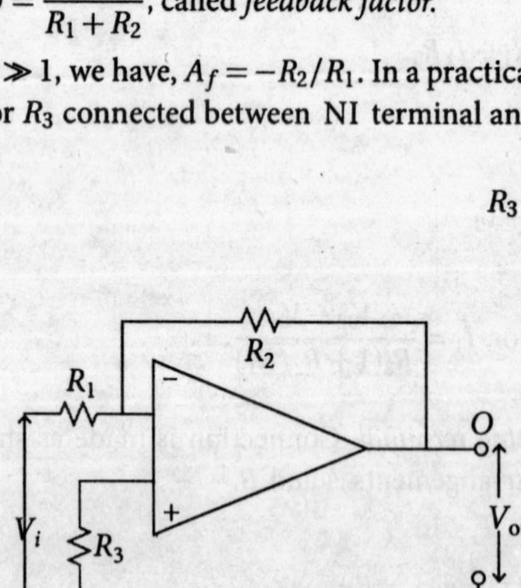

Fig. 10.19

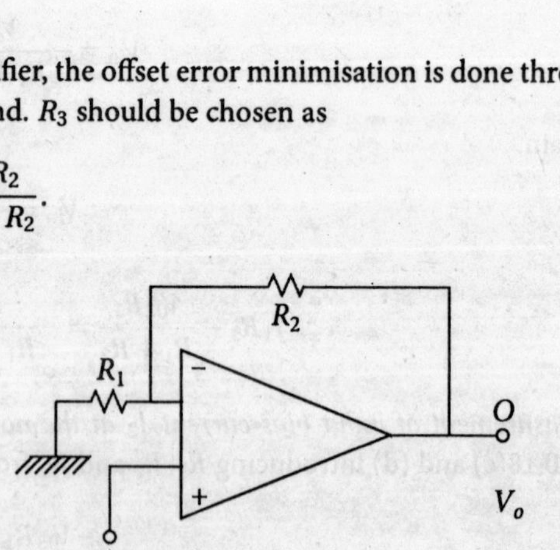

Fig. 10.20

Non-inverting amplifier. In case of non-inverting mode, the closed-loop gain is given by

$$A'_f = \frac{V_0}{V_i} = \frac{A}{1 + A\beta}.$$

A is the gain, called *open loop gain,* and β is the feedback factor.

When A is very large, $A'_f = \frac{1}{\beta}$.

In addition, if $R_1 = \infty$ or $R_2 = 0$, $A'_f = 1$ and the amplifier acts as a *voltage follower.*

For supplying the input voltages both positive and negative, to the inverting or non-inverting amplifier, we use the resistive network as shown in Fig. 10.21.

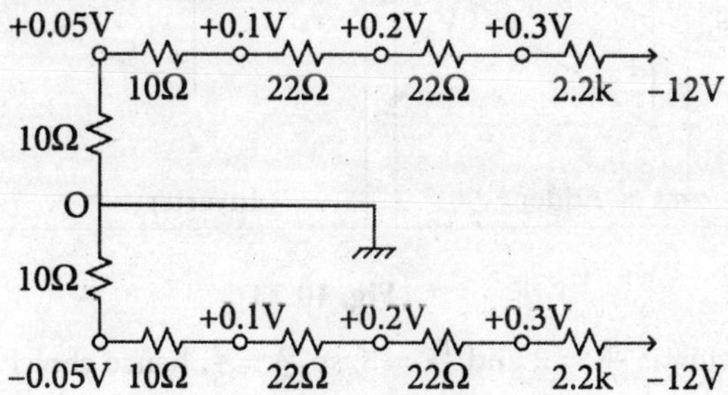

Fig. 10.21

10.14 Applications of Op-Amp

1. Unit gain buffer or voltage follower.
2. Adder or summing amplifier and difference amplifier.
3. Log amplifier.
4. Integrator.
5. Differentiator.
6. Null detector.
7. Multivibrator.
8. Wien's bridge oscillator.
9. Solutions of linear equations.
10. Simulation of inductance.
11. Voltage comparator.
12. Schmitt trigger.
13. Current to voltage amplifier, etc.

1. *Unit gain buffer.* This is a non-inverting mode Op-Amp already discussed in previous section [Fig. 10.20]. Whatever, may be the input $V_i(\pm 0.3, \pm 0.1\text{V}$, etc.).

2. *Adder.* As shown in Fig. 10.22, voltages V_1, V_2, V_3 are applied at the inverting terminal and the terminal is virtually at ground potential, so the currents i_1, i_2 and i_3 flow through the respective input resistors uninfluenced by each other. Here,

$$i = i_1 + i_2 + i_3 = \frac{V_1}{r_1} + \frac{V_2}{r_2} + \frac{V_3}{r_3} \text{ and } V_0 = -iR_2.$$

$$\therefore\ V_0 = -R_0\left(\frac{V_1}{r_1} + \frac{V_2}{r_2} + \frac{V_3}{r_3}\right). \qquad (10.11)$$

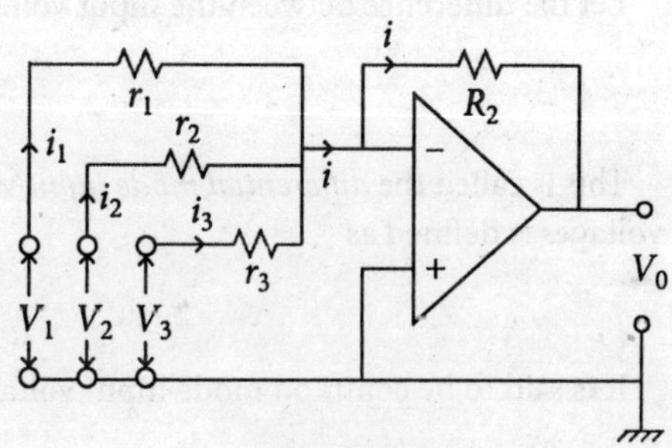

Fig. 10.22

By making $r_1 = r_2 = r_3 = R_3$, we get $V_0 = -(V_1 + V_2 + V_3)$ thus it acts as an adder. If $r_1 = r_2 = r_3 = R$ and $R_2/R = \frac{1}{3}$, we get

$$V_0 = -\frac{V_1 + V_2 + V_3}{3},$$

i.e., average of the inputs. If the minimum input voltage is comparable to the input offset, offset null-adjustment is first to be made. If $r_1 \neq r_2 \neq r_3$, then (10.11) gives us the output as sum of $\frac{R_2}{r_1}V_1$, $\frac{R_2}{r_2}V_2$ and $\frac{R_2}{r_3}V_3$, where $\frac{R_2}{r_1}$, etc. will be called a *scale factor.*

When we want an output to be given by $V_0 = 2V_1 + 5V_2$, the Op-Amp design will be as shown in Fig. 10.23.

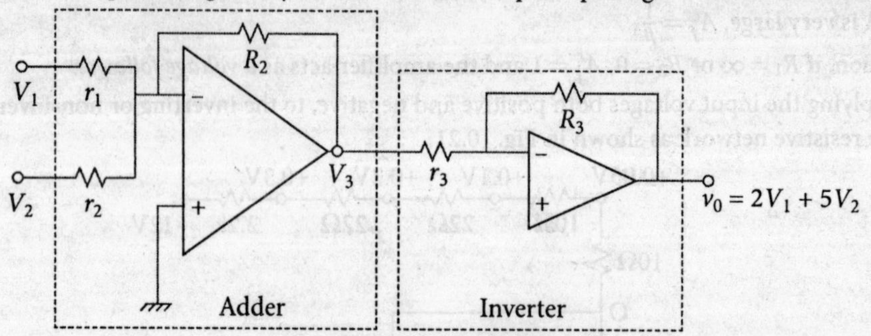

Fig. 10.23

We choose the scale factor as $\frac{R_2}{r_1} = 2$ and $\frac{R_2}{r_2} = 5$, so $\frac{r_2}{r_1} = \frac{2}{5}$, hence choice may be made like $r_1 = 5\text{k}\Omega$, $r_2 = 2\text{k}\Omega$ and $R_2 = 10\text{k}\Omega$.

$$\therefore\ V_s = -10\left(\frac{V_1}{5} + \frac{V_2}{2}\right) = -(2V_1 + 5V_2).$$

This V_s is then applied to the inverting terminal via r_3 and making $R_3 = r_3$, we get unit gain buffer inverter producing ultimate output $V_0 = 2V_1 + 5V_2$.

Difference amplifier. Just as sum of two or more input voltages is obtained through a summing amplifier, so also sometimes it is necessary to amplify the difference of two-input voltages. This is achieved by the difference amplifier shown in Fig. 10.24.

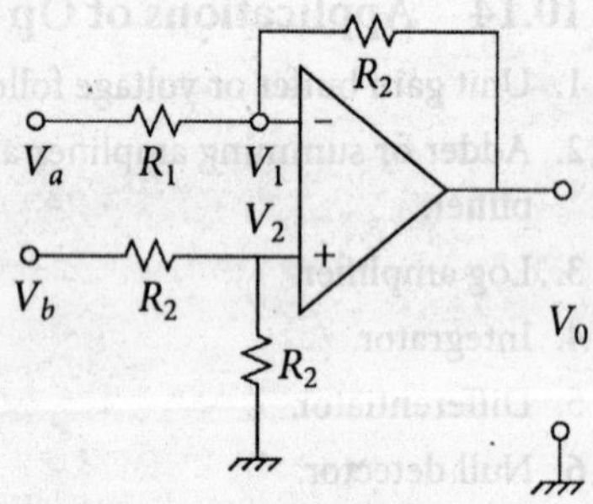

Fig. 10.24

In addition to the output voltage proportional to the difference of the inputs, a practical Op-Amp also produces in the output a small term proportional to the common mode voltage $(V_a + V_b)/2$.

Let the difference between the input voltages be represented by

$$\Delta V_d = V_2 - V_1. \tag{10.12}$$

This is called the *differential mode input voltage.* Another input voltage, which is the average of the input voltages is defined as

$$\Delta V_a = \frac{V_1 + V_2}{2}. \tag{10.13}$$

It is said to be common mode input voltage. Solving (10.12) and (10.13), we can write,

$$V_1 = \Delta V_d - \frac{\Delta V_a}{2} \text{ and } V_2 = \Delta V_a - \frac{\Delta V_d}{2}. \tag{10.14}$$

From the above set it is evident that each of the input voltages can be represented in terms of common mode input voltage as well as differential mode input voltage. So from the circuit shown in Fig. 10.24 we can write,

$$V_0 = -A_d(V_2 - V_1) - A_{CM}\frac{V_1 + V_2}{2}, \tag{10.15}$$

where A_a = gain of the differential mode and A_{CM} = gain of the common mode.

Now,

$$\left.\begin{aligned} V_2 &= \frac{R_2}{R_1+R_2} V_b \\ \text{and } V_1 &= \frac{R_2}{R_1+R_2} V_a + \frac{R_1}{R_1+R_2} V_0 \end{aligned}\right\}. \tag{10.16}$$

Putting these values of V_1 and V_2 in (10.15), we get

$$V_0 = \frac{-A_d \dfrac{R_1}{R_1+R_2}(V_b - V_a) - \dfrac{A_{CM} R_1}{R_1+R_2}\left(\dfrac{V_b+V_a}{2}\right)}{1 - \dfrac{A_d R_1}{R_1+R_2} + \dfrac{A_{CM} R_1}{2(R_1+R_2)}}.$$

Since $A_a \gg A_{CM}$ and $A_a \gg 1 + \frac{R_2}{R_1}$, the above equation reduces to

$$V_0 = \frac{R_2}{R_1}(V_d - V_a) + \frac{A_{CM}}{A_d}\left(\frac{V_a+V_b}{2}\right). \tag{10.17}$$

The first term in (10.17) represents the part of output proportional to the difference of inputs $(V_b - V_a)$ and the second term represents the part proportional to the common mode voltage input. This term is a small quantity, since $A_a \simeq 10$, $A_{CM} \simeq 1$.

$$\therefore\ V_0 = \frac{R_2}{R_1}(V_d - V_a). \tag{10.18}$$

The scale factor is R_2/R_1 and depending on the value of the scale factor the output will be magnified version of the difference of the input voltages.

3. *Log amplifier.* Logarithmic amplifier is one whose output is proportional to the logarithm of the input voltage. The log function generator is a transistor Q shown in the circuit design [Fig. 10.25].

Here

$$i = \frac{V_i}{R_1} = i_c.$$

Now, the collector current is given by $i_c = I_o\, e^{V_{BE}/V_T}$, where $V_T = kT/q_e$, called *volt-equivalent* of temperature and q_e is the electronic charge.

$$\therefore\ \frac{V_1}{R_1} = I_0\, e^{V_{BE}/V_T} \text{ or, } \frac{V_i}{R_1 I_0} = e^{V_{BE}/V_T}.$$

$$\therefore\ \frac{V_{BE}}{V_T} = \ln\frac{V_i}{I_0 R_1} \text{ or, } V_{BE} = V_T \ln\frac{V_i}{I_0 R_1}.$$

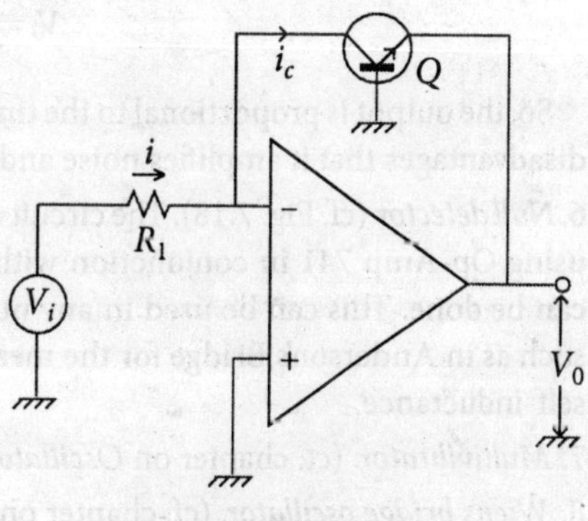

Fig. 10.25

Now $V_{BE} = -V_E = -V_e$, since the base is at zero potential.

$$\therefore\ V_0 = -V_T \ln\frac{V_i}{I_0 R_1}. \tag{10.19}$$

(10.19) shows that V_0 is proportional to the logarithm of V_i. The scale factor is $I_0 R_1 = a$ (say), then we get a function generated at the output $\ln a V_i$. However, it is to be noted that the output function is temperature dependent, so temperature compensated logarithmic amplifier is in use.

4. *Integrator* Op-Amp. Here the feedback element is a condenser C as shown in Fig. 10.26. This circuit provides output proportional to the integral of the input function.

Here,

$$I_R = \frac{V_i}{R} = I_0 = -C\frac{dV_0}{dt}.$$

Hence,

$$V_0 = -\frac{1}{CR}\int_0^t V_i(t)\,dt. \qquad (10.20)$$

Thus, the output is proportional to the integral of the input, the scale factor being $1/CR$.

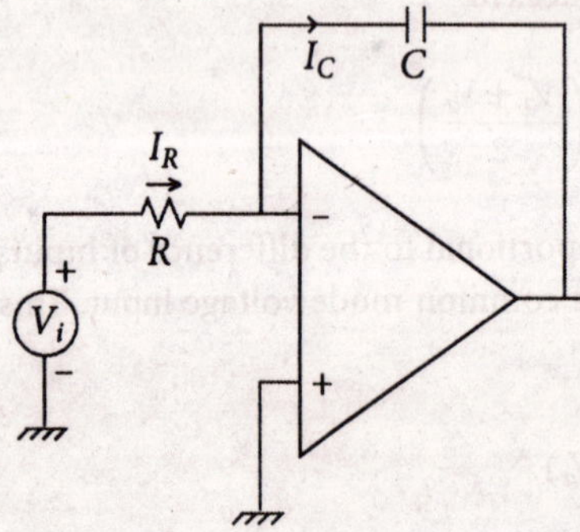

Fig. 10.26

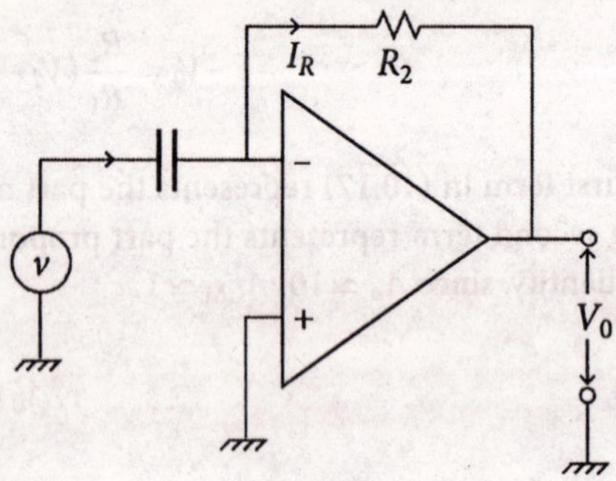

Fig. 10.27

5. *Differentiating* Op-Amp. A differentiator circuit is shown in Fig. 10.27, where C is in the inverting input circuit.

Here,

$$V_0 = -R_2 I_R = -R_2 I_c = -R_2 C\frac{dv}{dt}. \qquad (10.21)$$

So, the output is proportional to the time derivative of the input. The Op-Amp as the differentiator has the disadvantages that it amplifies noise and ripple present at the input end and that it gives rise to instability.

6. *Null detector* (cf. Fig. 7.18). The circuit of the null-detector using Op-Amp 741 in conjunction with FET BFW 10 can be done. This can be used in any null measurement such as in Anderson's Bridge for the measurement of the self-inductance.

7. *Multivibrator.* (cf. chapter on *Oscillator*).

8. *Wien's bridge oscillator.* (cf. chapter on *Oscillator*).

9. *Solutions of linear equations.* (a) Simultaneous algebraic equation

$$2x + 5y = 10$$
$$2x - y = 8.$$

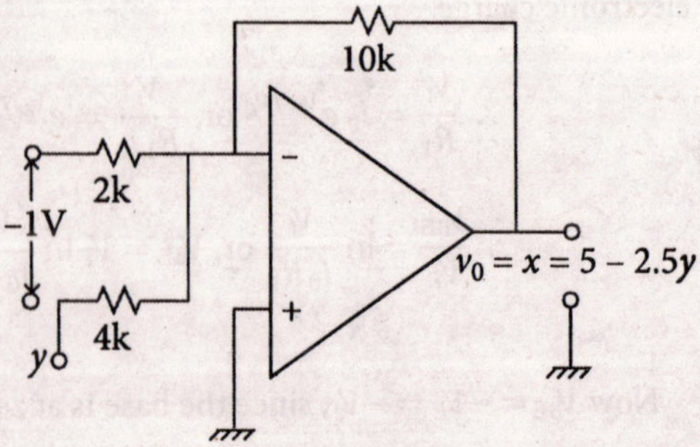

Fig. 10.28

From the first equation we get

$$x = \frac{10}{2} - \frac{5}{2}y = -\left(-\frac{10}{2} + \frac{5}{2}y\right).$$

$$v_0 = -R_2\left(\frac{V_1}{r_1} + \frac{V_2}{r_2}\right) = -\left(\frac{-1}{2k} \times 10k + \frac{y}{4k} \times 10k\right) = 5 - 2.5y.$$

From the second equation, we get

$$y = 2x - 8 = -(8 - 2x).$$

$$\therefore\ v_0 = -R_2\left(\frac{V_1}{r_1} + \frac{V_2}{r_2}\right) = -\left(\frac{2k}{1k} \times 4 + -x \times \frac{2k}{1k}\right)$$

$$= -(8 - 2x) = y \text{ (see circuit Fig. 10.29).}$$

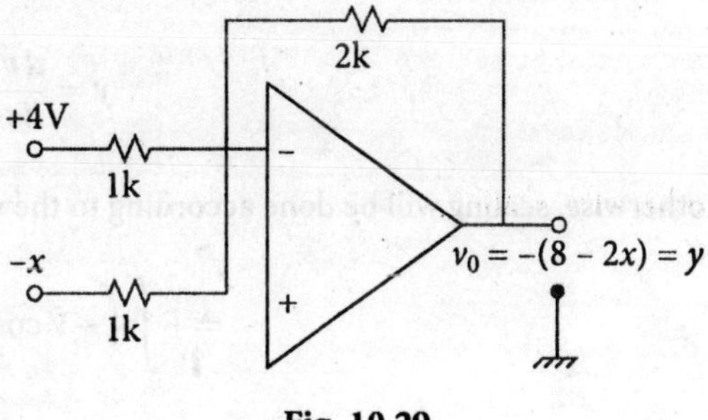

Fig. 10.29

Voltage representing $-x$ may be obtained by inverting x generated w.r.t. circuit (10.28). The complete analog computer circuit is shown in Fig. 10.30.

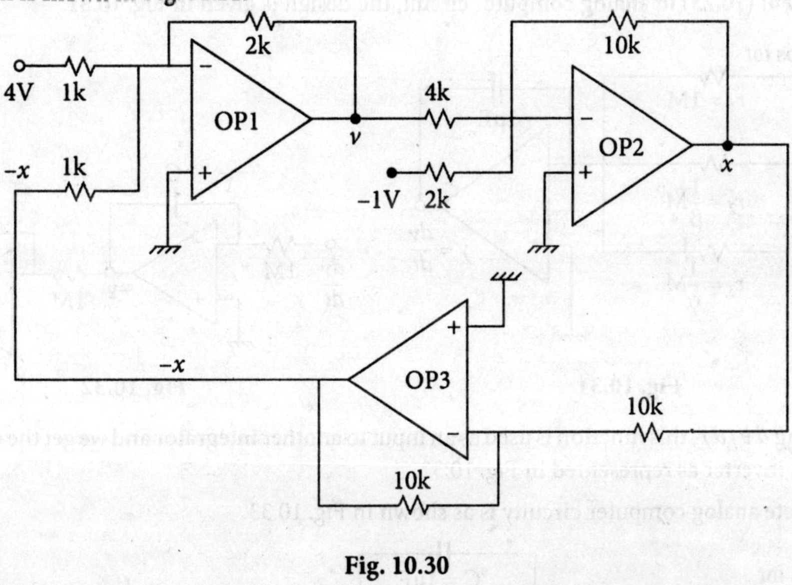

Fig. 10.30

The composite Op-Amp circuit giving solution of the simultaneous equation is represented in Fig. 10.30 and a dc voltmeter at y and x will read out the solution.

(b) *Solution of a differential equation.* For solution of a second order differential equation we would require adder integrator and proper scaling, all these processes can be accomplished by using Op-Amps. Let us take a differential equation as

$$\frac{d^2v}{dt^2} + \alpha\frac{dv}{dt} + \beta v = \hat{V}\cos\omega t.$$

We can write,

$$\frac{d^2v}{dt^2} = \left(\hat{V}\cos\omega t - \alpha\frac{dv}{dt} - \beta v\right) = -\left(-\hat{V}\cos\omega t + \alpha\frac{dv}{dt} + \beta v\right). \tag{10.22}$$

An integrator would give us an output function $y = dv/dt$.

$$\therefore\ y = \frac{dv}{dt} = \frac{1}{CR}\int -\frac{d^2v}{dt^2}\,dt,$$

remembering that an Op-Amp produces an inversion. If we take $CR = 1$, the scaling factor being unity, we get the output function as an integral of the input, i.e.,

$$y = \frac{dv}{dt} = \int -\frac{d^2v}{dt^2}\,dt,$$

otherwise, scaling will be done according to the weight factor of the inputs

$$= +\int\left(-\hat{V}\cos\omega t + \alpha\frac{dv}{dt} + \beta v\right)dt. \tag{10.23}$$

To implement (10.23) in analog computer circuit, the design is given in Fig. 10.31.

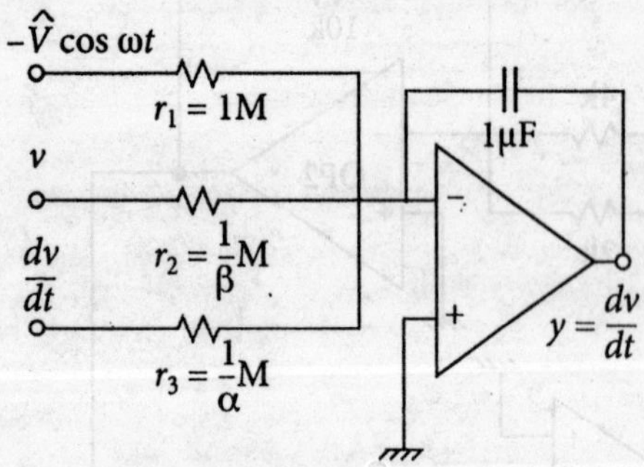

Fig. 10.31

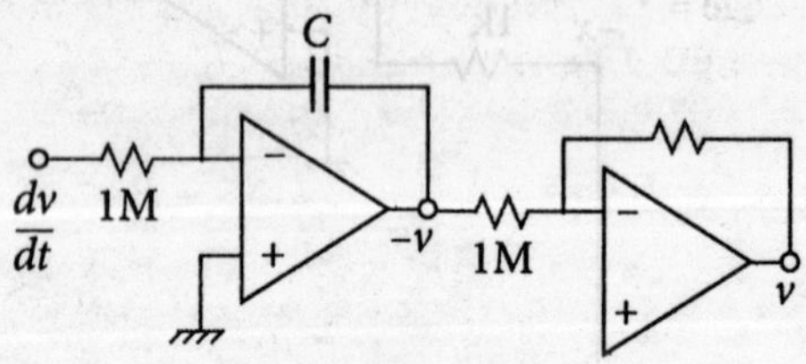

Fig. 10.32

After getting dv/dt, this function is used as an input to another integrator and we get the output function v through an inverter as represented in Fig. 10.32.

The complete analog computer circuity is as shown in Fig. 10.33.

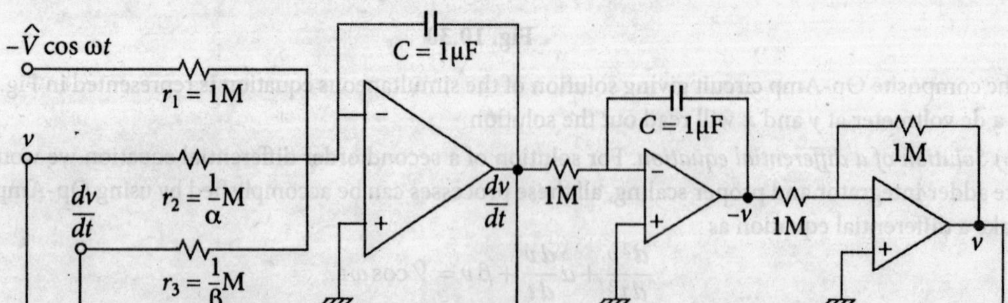

Fig. 10.33

10.15 Some Practical Project Circuits using Op-Amp 741

Of numerous practical projects using IC 741, we are giving below a few of them.

(1) *Voltage regulator.*

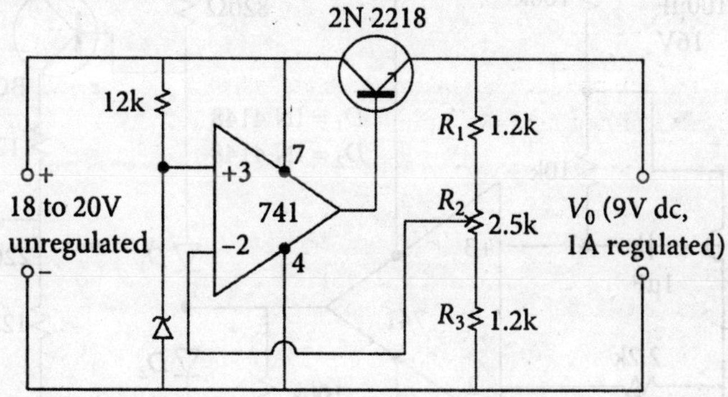

Fig. 10.34

(2) *Battery voltage state indicator.* The circuit may be used to indicate the state of the 12V car battery by the visible display of the LEDs D_1, D_2, D_3. Suppose, we want to know the condition of our car battery. R_1 is adjusted so that the potential between inverting terminal (2) of the first IC and ground is +13V. Adjust R_2 so that the potential of the non-inverting terminal 3 is +11V. The supply voltage must be at least −15V. The battery to be tested is connected with its positive polarity to the point E and negative to ground. If the test voltage = 12V, only D_3 will glow and not the LED D_1 and D_2. But when the test voltage is less than 11V, the LED D_2 will glow and not D_1 and D_3.

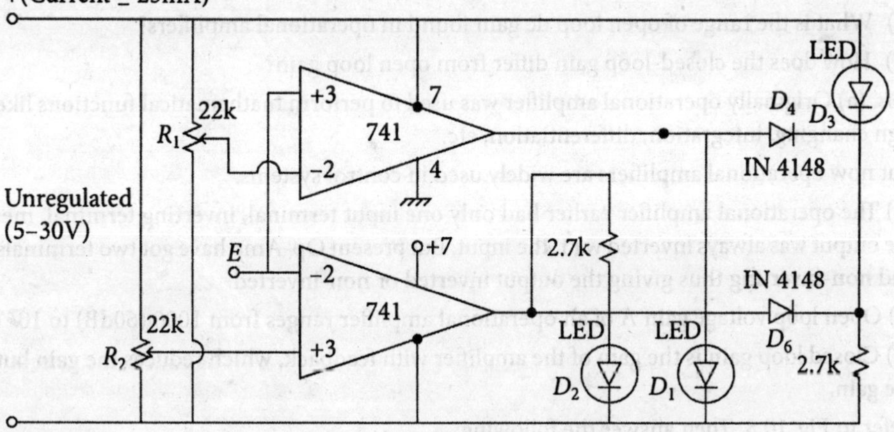

Fig. 10.35

If the test voltage > 13V, D_1 will glow and not D_2 and D_3.

(3) *Audio-power amplifier.*

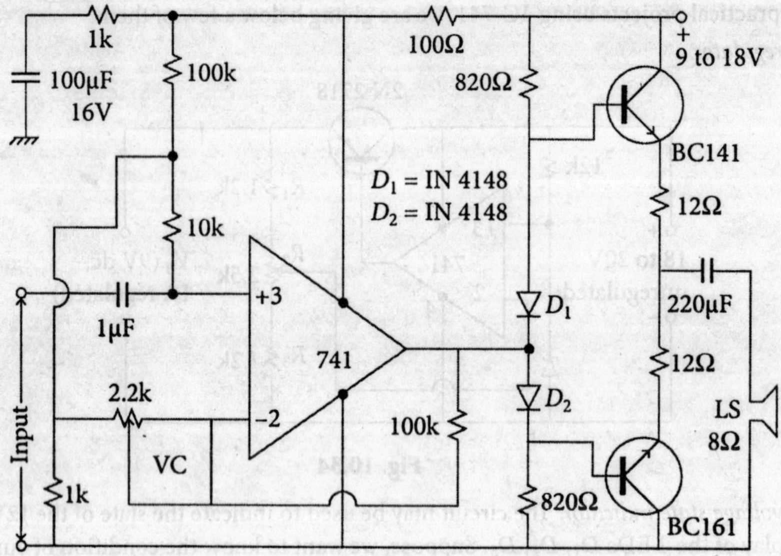

Fig. 10.36

10.16 Short Theory Questions with Answers

1. (a) What was the operational amplifier originally used for?
   (b) How does it differ from the present differential operational amplifiers?
   (c) What is the range of open loop dc gain found in operational amplifiers?
   (d) How does the closed-loop gain differ from open loop gain?

   Ans. (a) Originally operational amplifier was used to perform mathematical functions like addition, sign changing, integration, differentiation, etc.

   But now operational amplifiers are widely used in control systems.

   (b) The operational amplifier earlier had only one input terminal, inverting terminal, meaning that the output was always inverted w.r.t. the input. But present Op-Amp have got two terminals inverting and non-inverting thus giving the output inverted or non-inverted.

   (c) Open loop voltage gain A of an operational amplifier ranges from 1000 (60dB) to 10^7 (140dB).

   (d) Closed loop gain is the gain of the amplifier with feedback, which reduces the gain but stabilises the gain.

2. Refer to Fig. 10.8. Then answer the following:
   (a) Which point may be thought of as virtual ground?
   (b) What does this imply about the input resistance as seen by v_i?
   (c) What is meant by the loop gain of the circuit?
   (d) How does this affect the closed-loop output impedance?

   Ans. (a) Fig. 10.8 is an inverting operational amplifier. According to the definition of gain, $A = -V_0/V_i$.

   Since A is very large, it implies $V_i \simeq 0$, hence the point A_1 is virtually at ground potential.

(b) The input impedance r_i is given by $r_i = v_i/i_1$.
Now $v_i = R_1 i_1 - V_1$, where V_1 is the difference of potential between virtual ground and actual ground and hence $V_1 \simeq 0$.
$\therefore\ v_i = R_1 i_1. \therefore\ R_i = R_1$.
(c) The loop gain is given by (closed-loop gain) $A_f = -V_0/V_i$ and by calculation we have shown that

$$A_f = -\frac{R_2}{R_1}\left(1 - \frac{1}{1+A\beta}\right),$$

where $A = -V_0/V_1$, called open loop gain.
(d) The output impedance of the Op-Amp is found by setting $v_i = 0$ and inserting a test voltage source at the output terminal and measuring the current drawn from the test source. Then the output impedance (resistance) is given by

$$R_{\text{out}} = \frac{R_0}{1 + \dfrac{AR_1}{R_1 + R_2}} = \frac{R_0}{1+A\beta},$$

where R_0 is the open loop output resistance.
Since $A \to$ large value, $R_{\text{out}} \to$ small value.

3. How does the closed-loop voltage gain in a non-inverting amplifier compare with that in an inverting amplifier for the same values of resistors?
Ans. The voltage output of a non-inverting amplifier is in phase with the input but in case of inverting amplifier it is 180° out of phase. The closed-loop voltage gains in INV and NON-INV configurations are respectively given by

$$A_f = -\frac{R_2}{R_1}\left(1 - \frac{1}{1+A\beta}\right) \simeq -\frac{R_2}{R_1}; \quad A'_f = \frac{A}{1+A\beta} \simeq \frac{R_1 + R_2}{R_1}.$$

4. (a) What is the purpose of a differential amplifier?
(b) What is the meaning of common mode input voltage?
(c) What is the difference between the common mode gain and the differential gain?
Ans. (a) The purpose of a differential amplifier is to amplify the difference between two-input voltage levels and reject the voltage common to both.
(b) Common mode input voltage refers to the average of the input voltages V_1 and V_2, i.e.,

$$\Delta V_a = \frac{V_1 + V_2}{2}.$$

The differential mode input voltage is given by $\Delta v_d = (V_2 - V_1)$.
Solving,

$$V_1 = \Delta V_a - \frac{\Delta v_d}{2}; \quad V_2 = \Delta V_a + \frac{\Delta v_d}{2}.$$

So, the input voltages can be represented in terms of common mode input voltage as well as differential mode input voltage. If A_d represents the differential mode gain and A_{CM}, the common mode gain, then

$$V_0 = \frac{R_2}{R_1}(V_b - V_a) + \frac{A_{\text{CM}}}{A_d}\left(\frac{V_a + V_b}{2}\right).$$

Now, $A_d \simeq 10^5$, $A_{CM} \simeq 1$.

5. **Draw an analog computer circuit to solve the following equations:**

$$X - 4Y = 6$$
$$3X + 2Y = 5.$$

Solution: $X = 6 + 4Y = -(-6 - 4Y)$.
$Y = \frac{5}{2} - \frac{3}{2}X = -\left(-\frac{5}{2} + \frac{3}{2}X\right)$.

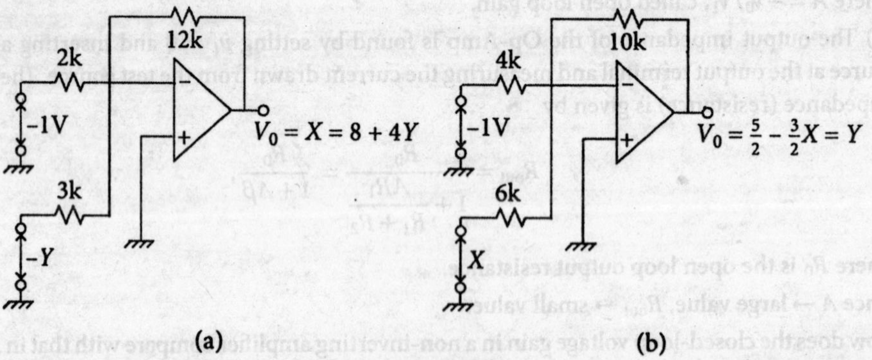

Fig. 10.37

So, the complete analog computer circuit is shown in Fig. 10.38.

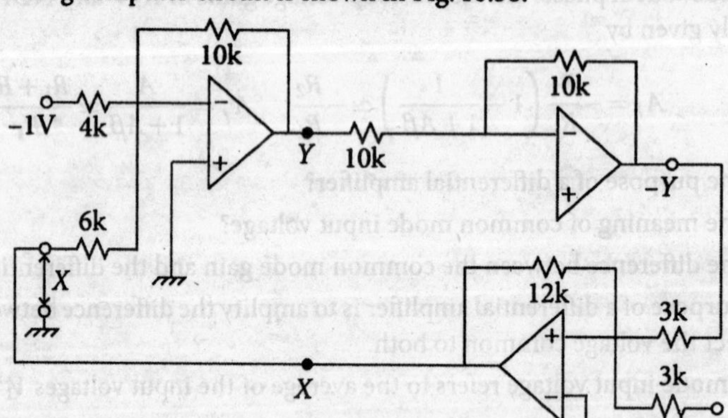

Fig. 10.38

10.17 Numerical Problems

1. The summing amplifier has $R_2 = 10k$ and $r_1 = 10k, r_2 = 2.2k$ and $r_3 = 3.3k$. If $V_1 = 6V, V_2 = -3V$ and $V_3 = -0.75V$, what is V_0?

Solution: $V_0 = -R_2\left(\frac{V_1}{r_1} + \frac{V_2}{r_2} + \frac{V_3}{r_3}\right) = -10\left(\frac{6}{10} - \frac{3}{2.2} - \frac{0.75}{3.3}\right) = -6 + \frac{30}{2.2} + \frac{7.5}{3.3} = 9.91V.$

2. An Op-Amp has $R_i = 100k, A = 100{,}000, R_0 = 100\Omega$. If $R_1 = 1k\Omega, R_2 = 50k\Omega$. Calculate (a) A_f (b) r_i (c) r_0.

Solution: (a) $A_f = -\frac{R_2}{R_1} = -50.$

(b) $r_i = R_1 + \frac{R_2 + R_0}{1+A} = 1000 + \frac{50{,}000 + 100}{1 + 100{,}000} = 1\text{k}.$

(c) $r_0 =$ output impedance $= \frac{R_0}{1 + \frac{R_1 A}{R_1 + R_2}} = 0.05\Omega.$

3. The common mode rejection ratio is defined as

$$\text{CMRR} = \frac{A_d}{A_{CM}} = \frac{\text{Differential gain}}{\text{Common mode gain}}.$$

In an Op-Amp the CMR is 55dB and $A_d = 1200$. What are A_{CM} and CMRR?

Solution: CMR = 20 log CMRR = $20\log A_d - 20\log A_{CM} = 20\log 1200 - 20\log A_{CM}$.

$\therefore$ $55 = 61.6 - 20\log A_{CM}$ or, $\log A_{CM} = 0.329$. $\therefore$ $A_{CM} = 2.14$.

$\therefore$ $\text{CMRR} = \frac{A_d}{A_{CM}} = \frac{1200}{2.14} = 561.$

4. In the differential amplifier of Fig. 10.24, let V_1 = 1.3mV and V_2 = 1.5mV, find the total output. $A_d = 1200$, $A_{CM} = 2.14$.

Solution: The output is given by

$$V_0 = -A_d(V_2 - V_1) - A_{CM}\frac{V_1 + V_2}{2} = -1200(1.5 - 1.3) \times 10^{-3} - 2.14 \times \frac{(1.5 + 1.3)}{2} \times 10^{-8} = -0.243\text{V}.$$

5. An IC OP-amp with a differential open loop gain of 20,000 has a common mode rejection of 80dB. Then

(a) What is its CMRR?

(b) What is its common mode gain?

(c) What is the output from the amplifier if $V_1 = 0.16$mV and $V_2 = 0.18$mV?

Solution: $\text{CMRR} = \frac{A_d}{A_{CM}}.$

CMR = 20 log CMRR = $20\log A_d - 20\log A_{CM}$ or $80 = 20\log 20{,}000 - 20\log A_{CM} = 86.02 - 20\log A_{CM}$.

$\therefore$ $\log A_{CM} = 2.$

(a) $\text{CMRR} = \frac{20{,}000}{2} = 10{,}000.$

(b) $A_{CM} = 2.$

(c) $V_0 = -A_d(V_2 - V_1) - A_{CM}\left(\frac{V_1 + V_2}{2}\right) = -20{,}000(0.18 - 0.16) - 2 \times \frac{0.18 + 0.16}{2} = -0.40034\text{V}.$

10.18 What is a Logic Gate and Logic Function?

The basic building blocks of digital instruments and computers consist of logic switching circuits. A switch has only two distinct states, open or close, off or on. A number of switches can be used to perform a variety of circuit actions. Such a combination of switches, which passes a signal to the output terminal for a predetermined combination of inputs is called a *gate* or logic circuit. The interconnections of the switches are designed to implement the laws of *logic* and are manipulated so as to perform desired calculations. The mathematical technique by which logical analysis of systems containing a large number of switches is accomplished is known as *Boolean Algebra,* characterised by a set of rules similar to those of ordinary algebra but it is based on binary system, i.e., on two possible states (ON or OFF) and not on 10 states possible in decimal system.

There are three basic logic functions performed by the components of a digital computer — they are the NOT, AND and OR functions. The devices by which these functions are performed are respectively termed as NOT-gate, AND-gate and OR-gate. Before we take up the discussion about these gates, let us have a discourse on Binary system, decimal system and their interconversion and also let us have a bird's eye view of the Boolean algebra for better understanding of the contents of this section.

10.19 Binary Number System and Decimal Number System

In binary system two possible states are there, 0 and 1, for example when a switch is closed, the state is referred to as state 1, when the switch is open, the state is referred to as 0. But in decimal number system, 10 numerals are required 0, 1, 2, 3, ..., 9. So, if a digital instrument were to count in decimal numbers with transistors, they would have to recognise 10 states, whereas in binary system, using conduction and cut-off as the operating states, only two states are to be recognised. So binary system is used in all digital instruments.

In decimal number system, the number 1985 is represented by $1\times10^3+9\times10^2+8\times10^1+5\times10^0$.

Here 10 is the arithmetic base or radix and each position to the left or right of the decimal point represents an increased or decreased weight as a power of 10. Thus, the individual digits in the number are the coefficients of the different powers of 10 (ten). But in binary system the base or radix is 2 and the position to left or right of the binary point represents an increased or decreased weight as a power of 2. In binary system the representation of 1985 is carried out as follows.

$\frac{1985}{2}=992$ Remainder 1 Least significant digit

$\frac{992}{2}=496$ " 0

$\frac{496}{2}=248$ " 0

$\frac{248}{2}=124$ " 0

$\frac{124}{2}=62$ " 0

$\frac{62}{2}=31$ " 0

$\frac{31}{2}=15$ " 1

$\frac{15}{2}=7$ " 1

$\frac{7}{2}=3$ " 1

$\frac{3}{2}=1$ " 1

$\frac{1}{2}=0$ " 1 Most significant digit.

So, the binary equivalent of 1985 is 11111000001. The digit in binary is termed as *bit*, so the most significant bit (MSB) is on the left and the least significant bit (LSB) is to the right. The decimal and binary representation of number 1985 is shown.

$$1024+512+256+128+64+0+0+0+0+0+1=1985$$

Binary: $1\times2^{10}+1\times2^9+1\times2^8+1\times2^7+1\times2^6+0\times2^5+0\times2^4+0\times2^3+0\times2^2+0\times2^1+1\times2^0=11111000001$.

Thus, when we write a number in binary system with base 2, the coefficient of the different powers of 2 give the binary representation of the number and the different powers of terms added together give the number in decimal system. The following table gives the binary representation of the different decimal numbers.

Decimal number	Binary equivalent	Binary representation	Remark
0	0	0	The table shows that when any number is expressed in terms of different powers of 2, the coefficients of the different powers would give the binary representation and the terms of different powers of 2 added together give the number in decimal system.
1	1×2^0	1	
2	$1\times2+0\times2^0$	10	
3	$1\times2^1+1\times2^0$	11	
4	$1\times2^2+0\times2^1+0\times2^0$	100	
8	$1\times2^3+0\times3^2+0\times2^1+0\times2^0$	1000	
9	$1\times2^3+0\times2^2+0\times2^1+1\times2^0$	1001	

In case, the decimal fraction written as 0.812 when converted to binary fraction will read as:

	Binary	
$0.812\times2=1.624$	1	Most significant bit (MSB)
$0.624\times2=1.248$	1	
$0.248\times2=0.496$	0	
$0.496\times2=0.992$	1	Least significant bit (LSB).

Therefore, binary equivalent of 0.812 is 0.1101. Conversely, if a binary fraction 0.1101 is given, to get the decimal we proceed as under:

$0.1101=1\times2^{-1}+1\times2^{-2}+0\times2^{-3}+1\times2^{-4}=0.500+0.250+0+0.0625=0.812$.

Thus, let us frame the rule of conversion.

Decimal to Binary

Rule 1. A number given in decimal system is to be expressed as the sum of ascending powers of 2 and then collect the coefficients of the different powers. Thus, 11 (eleven) $=1\times2^3+0\times2^2+1\times2^1+1\times2^0$ thus the binary representation of the number eleven is 1011 (read as *one zero one one*).

If the number be a fraction, express it in descending powers of 2, then collect the coefficients for binary representation.

Thus, $0.812=1\times2^{-1}+0\times2^{-2}+0\times2^{-3}+0\times2^{-4}$ thus binary equivalent is 0.1101.

Rule 2. This rule is of more operational importance than the first rule. The given number is divided progressively by 2 and the first remainder 0 or 1 is the least significant bit (LSB) and the last remainder when the quotient is zero is most significant bit (MSB). See the illustration as shown earlier.

In case of decimal fraction, multiply the number successively by 2, recording each time a *carry* in the integer position. The carriers in the forward order give the binary representation. Thus, take decimal fraction 0.65.

$0.65 \times 2 = 1.30 = 0.30$ with a carry of 1

$0.30 \times 2 = 0.60$ „ „ „ „ 0

$0.60 \times 2 = 1.20 = 0.20$ „ „ „ „ 1

$0.20 \times 2 = 0.40 = 0.40$ „ „ „ „ 0

$0.40 \times 2 = 0.80 = 0.80$ „ „ „ „ 0

$0.80 \times 2 = 1.60 = 0.60$ „ „ „ „ 1

Thus, the binary equivalent is 0.101001.

Binary System of Operations

Binary Addition

The binary addition is accomplished through the following four rules.

1. $0+0=0$, nothing added to nothing gives nothing.
2. $1+0=1$, one added to nothing gives one.
3. $0+1=1$, same meaning.
4. $1+1=10$, one added to one gives 2 (decimal equivalent of 10).

Thus, consider the decimal addition and corresponding binary equivalent operation.

Decimal addition	Binary addition
5	101
6	110
11 (eleven)	$1011 \equiv 1 \times 2^8 + 0 \times 2^2 + 1 \times 2^1 + 1 \times 2^0$ = 11 (eleven)

In finding the addition we have used the above four rules thus in the first column 1 + 0 = 1, the rule (b) is applied, the second column 0 + 1 = 0, the rule (c) is applied and in the last column 1 + 1 = 10, the rule (d) is applied.

Binary subtraction

The subtraction in binary system is also subject to four rules

1. $0-0=0$, nothing subtracted from nothing yields noting.
2. $1-0=1$, nothing subtracted from one yields one.
3. $1-1=0$, one subtracted from one yields nothing.
4. $10-1=1$, one subtracted from one zero (i.e., two) yields one.

Thus 10011 (= 19)
 10001 (= 17)
 00010 (= 2)

Column-wise subtraction in the above example is according to the rules of subtraction.

Remember if a larger binary number is subtracted from a smaller one, we are to subtract the smaller magnitude from the larger one and the sign of the larger magnitude is prefixed before the result.

Thus, 10001 (17)
 10011 (19)
 −00010 (−2)

Binary multiplication

Multiplication in binary system also obeys four rules.

1. $0 \times 0 = 0$, nothing multiplied by nothing yields nothing.
2. $0 \times 1 = 0$, nothing multiplied by one yields nothing.
3. $1 \times 0 = 0$, one multiplied by nothing yields nothing.
4. $1 \times 1 = 1$, one multiplied by one yields one.

Thus, 10011 (19)

$$\begin{array}{r} 10011 \quad (19) \\ \times 101 \quad (5) \\ \hline 10011 \\ 00000 \\ 10011 \\ \hline 1011111 \end{array}$$

$(= 1 \times 2^6 + 0 \times 2^5 + 1 \times 2^4 + 1 \times 2^3 + 1 \times 2^2 + 1 \times 2^1 + 1 \times 2^0 = 95)$

Binary division

Binary division can be accomplished in the same way as the decimal division.

Thus, 101) 101111(10011

$$\begin{array}{l} 101 \\ \hline 111 \\ 101 \\ \hline 0101 \\ 0101 \\ \hline 0000 \end{array}$$

Thus, quotient is $10011 = 1 \times 2^4 + 0 \times 2^3 + 0 \times 2^2 + 1 \times 2^1 + 1 \times 2^0 = 19$.

The divisor is 5, the dividend is 95, so the result of division in decimal system is 19, which is found also in binary system.

In a digital system, voltages representing analog quantities such as speed, temperature, pressure, etc. must be converted to binary. Such a process is called analog to digital conversion. Finally, when the signal has been processed, digital to analog conversion is needed to display the information in readable form.

Boolean algebra. This is the language in which the digital instruments and computer system speak. This algebra allows only two states for its variables — true or false or 1 and 0. Thus, this algebra is one of words of ideas, not of numbers.

Only three basic operations are permitted on the variables, they are:

1. The AND operation, denoted by a dot or cross thus $A \cdot B$ reads A AND B.
2. The OR operation denoted by the symbol + thus $A + B$ reads A OR B.
3. The NOT operation as complementation or negation of the variables, denoted by a bar or prime on the variable thus $\bar{A}$ or A' as complement of A.

Project 1: Medium wave broadcasting circuit

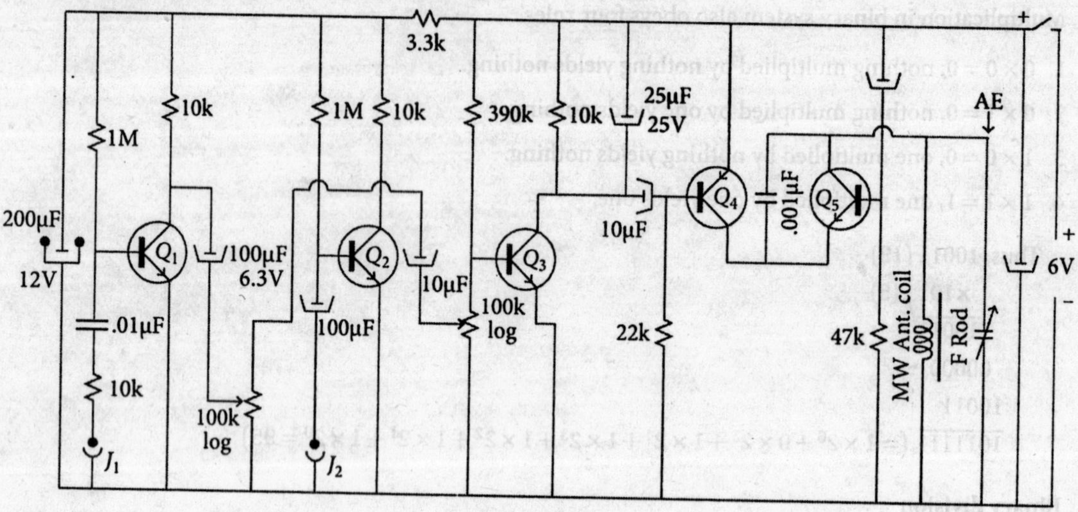

Fig. 10.39

$Q_1, Q_2, Q_3\, BC148B : Q_4$, is ac125 and Q_5 is 2N 483.

Project 2: Touch switch

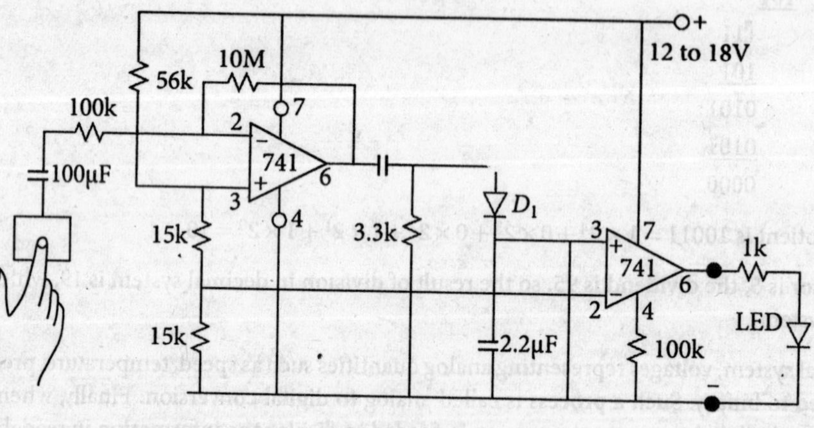

Fig. 10.40

2 IC 741, D_1 = 1N 4148, LED = TIL 209.

Project 3: A sawtooth generator

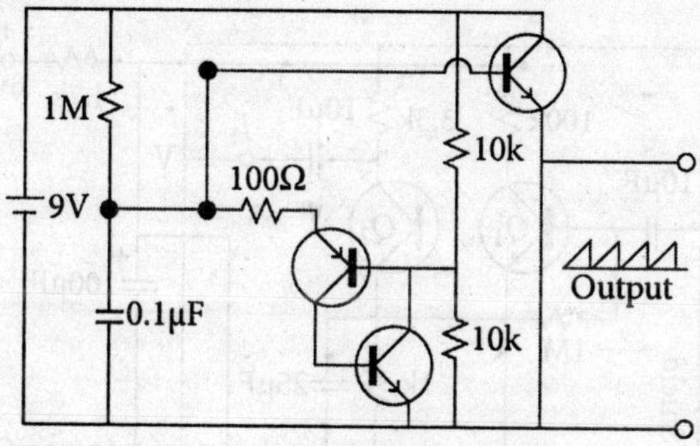

Fig. 10.41

Project 4: Transformer less 1.5 watt amplifier

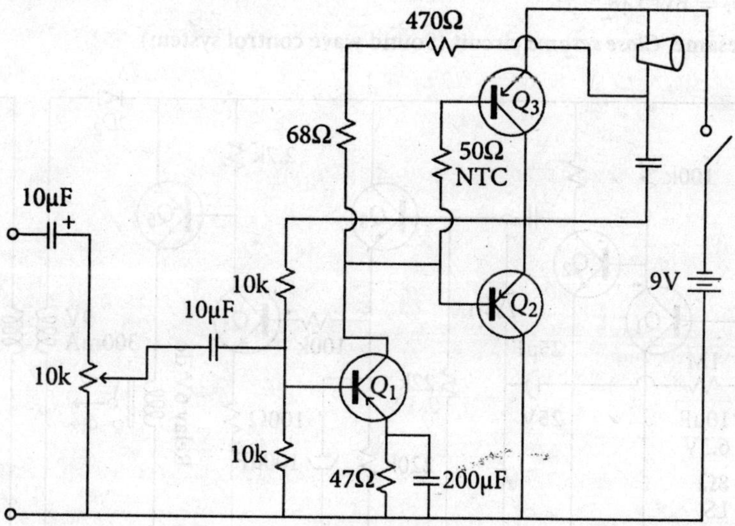

Fig. 10.42

Project 5: A high-gain pre-amplifier

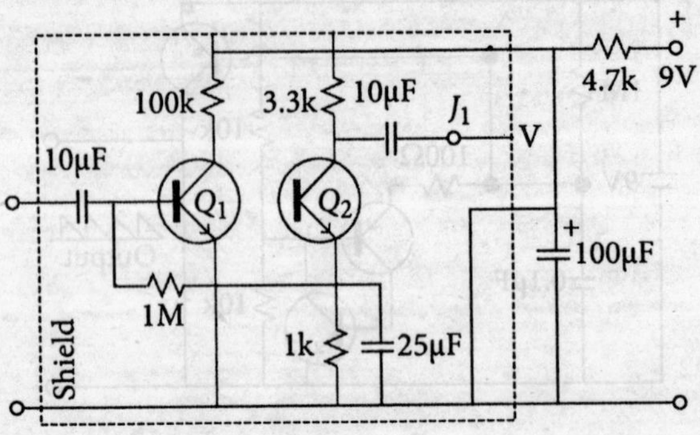

Fig. 10.43

Q_1 = BC 148, Q_2 = BC 148.

Project 6: *Open* sesame, *Close sesame* circuit (Sound wave control system)

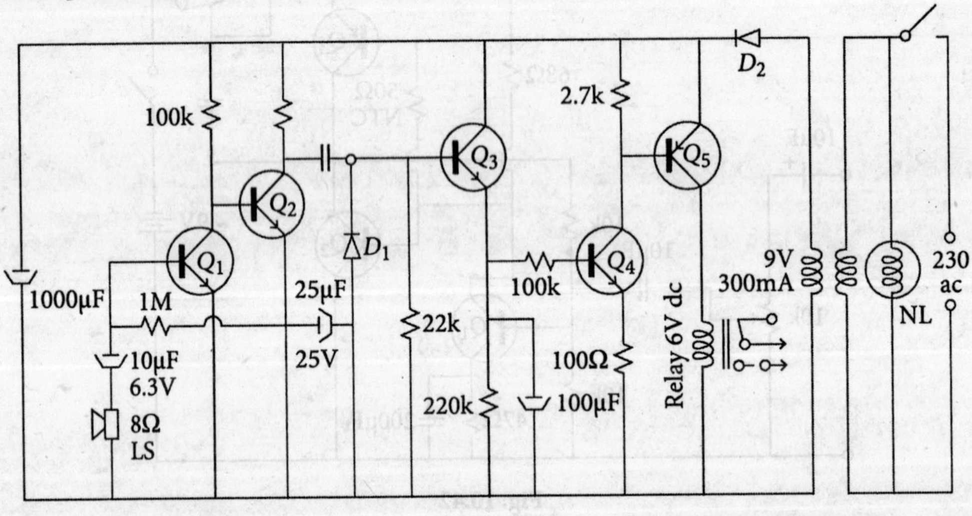

Fig. 10.44

Q_1 = BC 148B, Q_2 = BC 184, Q_3 = BC 149B, Q_4 = BC 148B, Q_5 = ac 126, D_1 = 1N 4001.

Project 7: Electronic puzzle

Here is a project, which will give a lot of entertainment as well as an aid to understand the logic circuits. A person sets the switches from S_1 to S_{14} and the second person is asked to switch on from S_{15} to S_{20} only four times to set the LED glow. There are millions of possible combinations to set the LED glow.

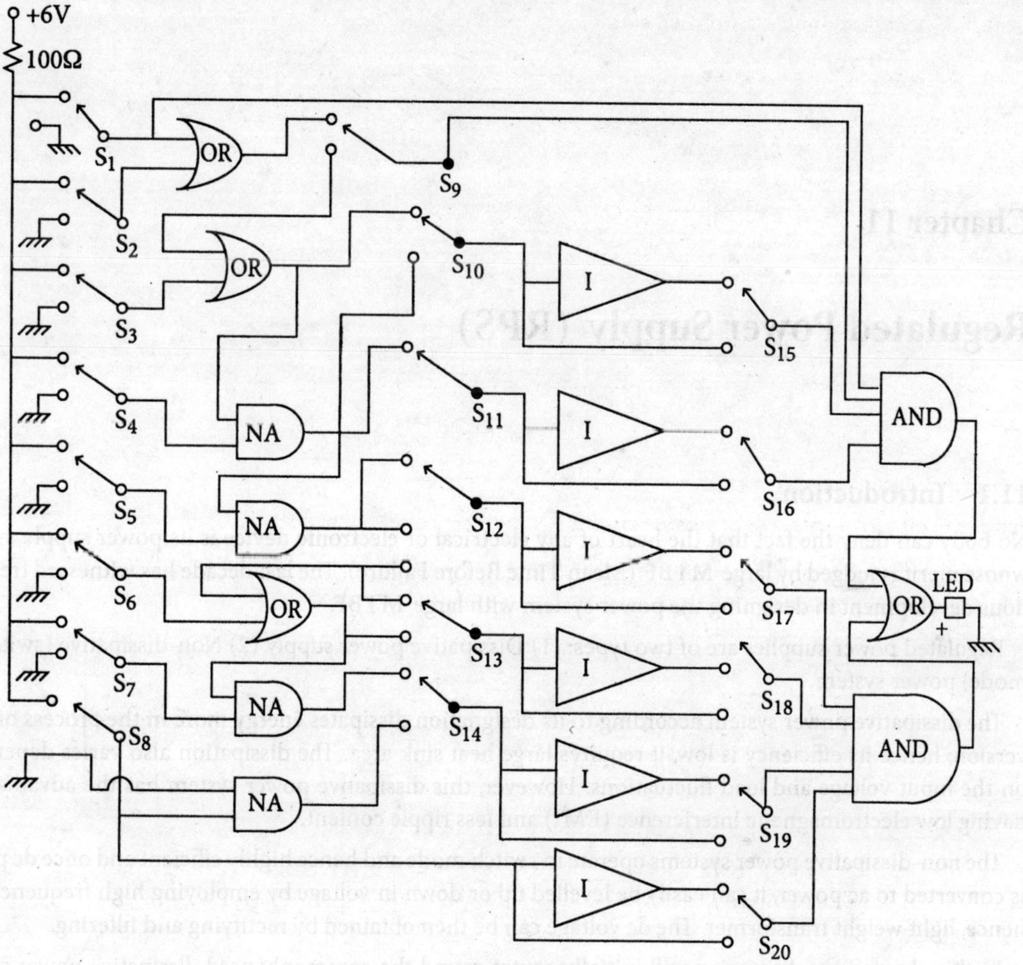

Fig. 10.45

For the six inverters use IC 7404.

For the AND use IC 7408 (quad 2-input AND).

For the NAND use IC 7400 (quad 2-input NAND).

Also IC 7421 (Dual 4 input AND-gates) is used.

Project 8: Null-detector

This is an electronic system used for detecting the condition of balance thus doing away with the headphone. This can be used in the Wheatstone's bridge experiment or in Anderson's bridge for the determination of self-inductance of a coil. The circuit is given earlier in the chapter on FET and MOSFET.

Chapter 11

Regulated Power Supply (RPS)

11.1 Introduction

No body can deny the fact that the heart of any electrical or electronic device is its power supply system whose merit is judged by large MTBF (Mean Time Before Failure). The last decade has witnessed tremendous development in designing the power system with large MTBF.

Regulated power supplies are of two types: (1) Dissipative power supply (2) Non-dissipative (switching mode) power system.

The dissipative power system according to its designation dissipates energy more in the process of conversion, hence its efficiency is low, it requires large heat sink area. The dissipation also varies depending on the input voltage and load fluctuations. However, this dissipative power system has the advantage in having low electromagnetic interference (EMI) and less ripple content.

The non-dissipative power systems operate in switch mode and hence highly efficient and once dc power is converted to ac power, it can easily be levelled up or down in voltage by employing high frequency and hence, light weight transformer. The dc voltage can be then obtained by rectifying and filtering.

In this chapter our discussion will initially centre round the commonly used dissipative power supply systems and finally wind up with switching mode power supply system (SMPS).

1. *Dissipative power supply system: Voltage regulator.* All electronic devices whether analog or digital require dc operating voltages and currents. Electronic devices using BJTs and FETs require ±18V dc and the TTL-gates in digital circuits run on ±5V. Such voltages may be available from battery but electronic devices run from battery will be prohibitively costly. So we have to supply electrical energy from the ac mains after suitable conversion into dc.

Basically, any power supply converting ac voltage into dc voltage consists of three essential components.

- *A transformer.* The primary of a transformer includes the ac source and depending on the number of secondary turns, ac voltage greater than or lesser than the primary input voltage is developed in the secondary. Thus, the transformer transforms electrical energy from the primary to the secondary at the rated voltage and current.

- *A rectifier.* The ac developed in the secondary due to ac input in the primary according to our specific requirement is rectified by a suitable solid state diode. Thus, a dc voltage is available.

- *A filter circuit is added.* The output of the rectifier is not steady but unidirectional. It contains enough of fluctuation. To suppress the fluctuation, a filter circuit is added. There is a great variety of filter circuits,

CR filter, LC filter or a π-section filter, but whatever may be the type, a filter tends to smoothen out the rectified output.

This trio-combination of transformer-rectifier-filter forms what is known as *ordinary power supply.* But the output of the ordinary power supply may vary due to variation of ac mains, called line variation or may be due to variation of load current, called load variation. In order to save the output of the power supply against either line variation or load variation, some special circuitry is to be added to the ordinary supply system. Such a circuit maintaining constancy of the voltage output against line or load variation is called a *regulatory circuits* and the supply with modification is termed as *regulated power supply.* The whole sequence of a regulated power supply is shown in Fig. 11.1 as a schematic diagram.

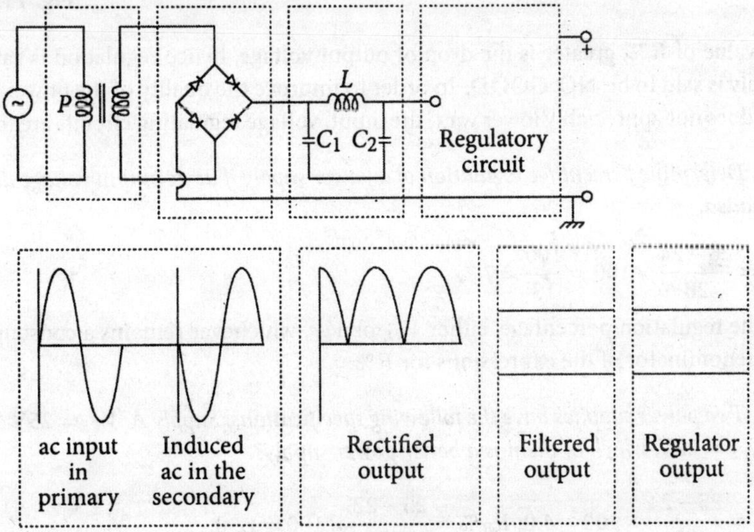

Fig. 11.1

2. *Why to consider a regulated power supply*? A regulated power supply is one, which keeps the output voltage constant against the line variation or load variation.

Variation of line voltage. ac line voltage is permitted to vary by ±6%, which means the mains voltage may vary between 160V to 240V.

Due to this variation, the dc output of the ordinary supply changes in consequence of which the electronic device does not work satisfactorily. So, a voltage regulator circuit is needed and that circuit will keep the output constant against line variation.

Fluctuation due to load variation. Ordinary power supply has internal resistance of value greater than 30Ω or so. So, there is drop of voltage output when a load is connected to it. If the load draws a current of 100mA, the output of the supply drops out by $30 \times 100 \times 10^{-3} = 3$V as internal drop. So, a regulated power supply is to be added next to the ordinary power supply so that in spite of load variation, the output of the supply remains sensibly constant.

3. *Voltage regulation per cent:* (*R* %). The quality of a voltage regulator is expressed through what we call voltage regulation per cent.

Let V_{NL} = voltage output under no-load, V_{FL} = voltage output under full-load.

$$\text{Then, } R\% = \frac{V_{NL} - V_{FL}}{V_{FL}(\text{or, } V_{NL})} \times 100.$$

If for a power supply, $V_{NL} = V_{FL}$, $R\% = 0\%$, it means that there is no change in the output voltage when loaded and when unloaded. If V_{NL} is appreciably different from V_{FL}, the power supply departs from ideal behaviour. In Fig. 11.2 we have shown the graphical plot of the behaviour of an ideal power supply and practical power supply.

In the adjoining figure, $V_{NL} = 9V$, $V_{FL} = 8V$.

$$R\% = \frac{9-8}{9} \times 100 = 11\%.$$

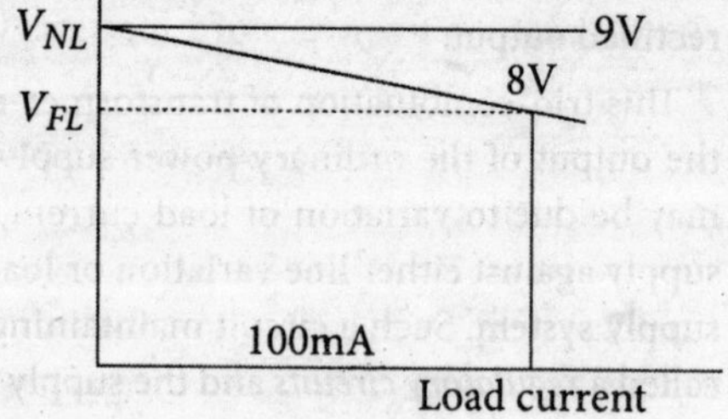

Fig. 11.2

Greater the value of $R\%$, greater is the drop of output voltage, hence regulation is said to be poor and the power supply is said to be NO GOOD. In order to improve the quality of the power supply so that the output voltage does not appreciably lower w.r.t. the input, voltage regulating circuits are to be incorporated.

Example 11.1. *Determine percentage regulation of a power supply if its dc output voltage decreases from 26V to 24V when loaded.*

Solution: $R\% = \frac{26-24}{26} \times 100 = \frac{100}{13} = 7.7.$

In defining the regulation percentage, either V_{NL} or V_{FL}, whichever remains a constant, should be considered in the denominator of the expressions for $R\%$.

Example 11.2. *Two power supplies have the following specifications: Supply A: $V_{NL} = 25V$ $V_{FL} = 24V$, Supply B: $V_{NL} = 25V$ $V_{FL} = 22V$, which of them is a better power supply?*

Solution: $R_A\% = \frac{25-24}{25} \times 100 = 4.0$; $R_B\% = \frac{25-22}{25} \times 100 = 12.0.$

Supply *A* is a better power supply.

4. *Classes of regulated supply: Types of voltage regulators.* These are two classes of regulated power supply.

- *Low voltage regulator.* Voltage regulator for regulating voltage less than 50V is called low voltage regulator.

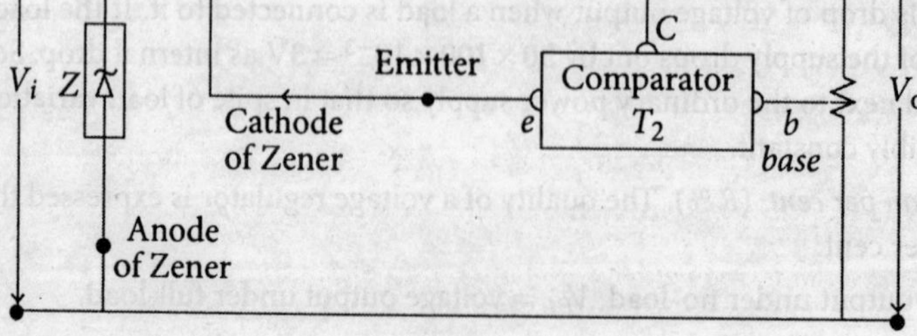

Fig.11.3

- *High voltage regulators* are meant for regulating voltages > 50V.

There are again two types of voltage regulators: (a) series voltage regulator (b) parallel or shunt voltage regulator.

(a) *Series voltage regulator.* In this type of voltage regulator, the regulating element, a simple resistor or a transistor, is in series with the load. The main drawback with this type is its large amount of dissipation in the series transistor (also called *pass transistor*). Such type is represented in block diagram as in Fig. 11.3.

(b) *Shunt regulator.* In this type the regulating element, say a transistor, is an in parallel with the load. If the regulating element be a Zener diode, it is to be connected in parallel with the load.

5. *Regulated power supplies frequently used.*

(a) *Zener controlled regulator*[1]*: Shunt type.* The simplest regulating element is a Zener diode operated at its breakdown region and the Zener is connected in parallel to the load, which means, it is shunt regulator. The $v-i$ characteristics of the Zener diode show that when it is operated in the breakdown region, the voltage across it remains constant. The regulator circuit is shown in Fig. 11.4. The Zener is so chosen that its V_z value is less than V_i. The input current I through the ballast resistor R_s is given by

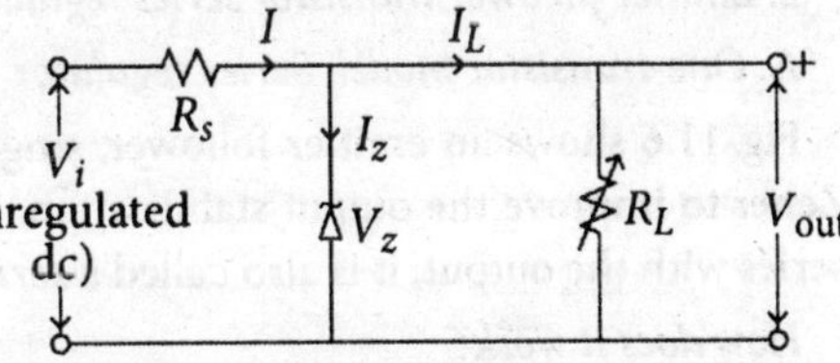

$$I = \frac{V_i - V_{out}}{R_s} \Rightarrow \boxed{V_{out} = V_i - IR_s}. \qquad (11.1)$$

Fig. 11.4

Now, $I = I_L + I_z$, it is show that for constancy of I, load current increases at the cost of decrease in Zener current and vice versa.

Any change in V_i is compensated by opposite change in voltage across R_s to keep the output voltage V_{out} constant. If the load current I_L is going to change, it is compensated by opposite change in the Zener current I_z so as to keep the output voltage constant.

Calculation of the ballast resistor R_s. To find the Thévenin's equivalent of the circuit in Fig. 11.4, disconnect the Zener from the circuit and find the open-circuited voltage $V_{oc} = V_{Th}$, and also find R_{Th},

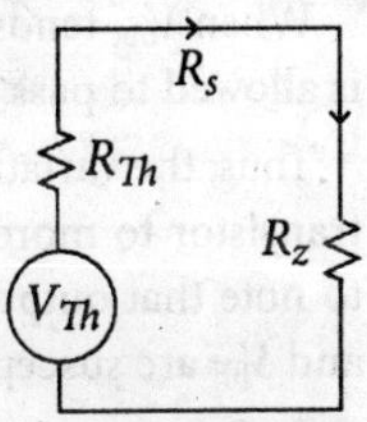

$$V_{Th} = \frac{V_i}{R_s + R_L} \times R_L; \; R_{Th} = R_s \| R_L = \frac{R_s R_L}{(R_s + R_L)}; \; I_z = \frac{V_{Th}}{(R_{Th} + R_z)},$$

where $R_z = Z_z =$ Zener impedance.

$$\Rightarrow I_z = \frac{V_i R_L}{(R_s + R_L)} / (R_{Th} + R_z) = \frac{V_i R_L}{R_s R_L + R_z (R_s + R_L)}.$$

Fig. 11.5

Now, applying KVL, we can write:

$$-IR_s - V_0 + V_i = 0 \Rightarrow I = \frac{V_i - V_0}{R_s} \Rightarrow R_s = \frac{V_i - V_0}{I} = \frac{V_i - V_0}{(I_z)_{min} + (I_L)_{max}}$$

$(I_z)_{min}$ is empirically taken as 10 % of maximum load current in case of shunt regulator.

Example 11.3. *A 7.2V Zener diode is used in the regulator circuit of* Fig. 11.4 *and the load current is to vary from* 12mA *to* 100mA. *Find the value of R_s required to maintain this load current if the supply voltage* $V_i = 12$V.

---

[1] For details see *Theory of Electricity and Magnetism* by the author; Scitech Publication, Chennai.

Solution: The value of R_s determines the extent of load current variation to be accommodated by the device. According to the above formulation

$$R_s = \frac{(V_i - V_0)}{(I_z)_{min} + (I_L)_{max}} = \frac{12 - 7.2}{0.1 \times (I_L)_{max} + (I_z)_{max}} = \frac{4.8}{(1+0.01)\times(I_L)_{max}} = \frac{4.8}{1.1 \times 100 \times 10^{-3}} = 43.5\Omega.$$

With this value of R_s thus once set, as the load current decreases, the Zener current increases but their sum remaining constant at 110mA.

If the load accordingly becomes open, then $I_L = 0, I_z$ becomes equal to 110mA. So, the wattage of the diode requires to be voltage $= (110 \times 10^{-3}) \times 7.2 = 0.8$W.

2. *Emitter follower transistor series regulator.*

A. *One-transistor model: Series regulator*

Fig. 11.6 shows an emitter follower, single transistor model with Zener to improve the output stabilisation. Since the transistor is in series with the output, it is also called a *series regulator.*

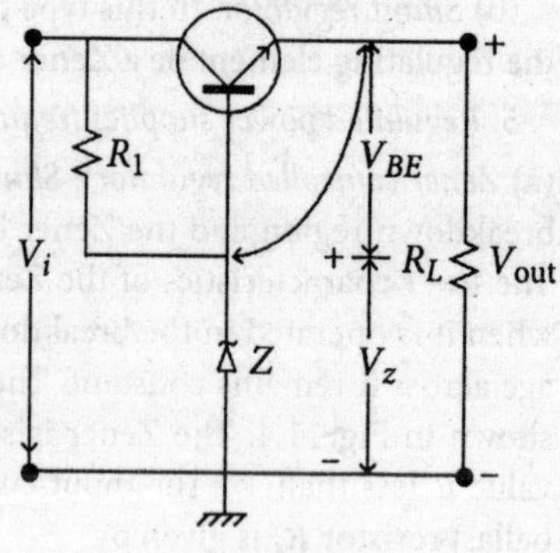

Fig. 11.6

How does it work?

Applying KVL to the output circuit we can write,

$$-V_{out} + V_z - V_{BE} = 0 \Rightarrow V_{out} = V_z - V_{BE}. \qquad (11.2)$$

As V_{out} tends to increase due to load current variation, V_{BE} decreases, because V_z is remaining constant as long as the Zener is restricted to operate in the breakdown region. When V_{BE} decreases, the transistor is thrown to less conducting state passing lesser amount of current so as to reduce the collector-emitter voltage (output voltage).

When V_{out} tends to fall, V_{BE} increases and the transistor is pushed to more conducting state, more current is allowed to pass through so as to increase the collector-emitter voltage (output voltage).

Thus, the variation of the output voltage is compensated by varying the emitter-base voltage pushing the transistor to more or less conducting state. It is, however, to note that output compensated is not 100 %, because V_z and V_{BE} are susceptible to temperature.

B. *One-transistor model: Shunt regulator.* The circuit in Fig. 11.7 is the shunt version of single transistor regulator model, which differs from the previous circuit in the disposition of the resistor R_s and the Zener.

Applying KVL, we get in Fig. 11.7

$$-V_0 + V_{BE} + V_z = 0 \Rightarrow V_0 = V_z + V_{BE}. \qquad (11.3)$$

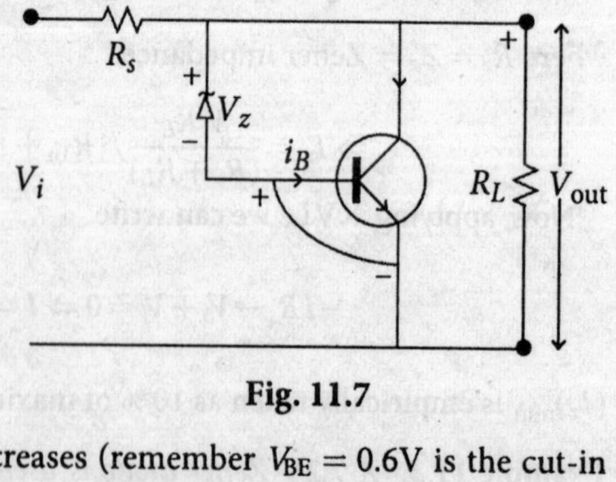

Fig. 11.7

It is evident from (11.3) that V_0 tends to increase if V_{BE} increases (remember $V_{BE} = 0.6$V is the cut-in voltage). Increase in V_{BE} causes more current to be drawn from the supply, more voltage drop across R_s so as to compensate the attempted increase in the output voltage. Similarly, when the voltage V_0 tends to decrease, V_{BE} decreases drawing lesser amount of current from the supply thus dropping lesser voltage across R_s so as to compensate for the decrease in the output.

C. *Feedback regulator: Two-transistor model.* The circuit shown in Fig. 11.8 is a two-transistor model feedback regulator.

The transistor Q_1 is the usual pass transistor in the emitter follower mode. The resistor R_3 connecting Q_1-base to the input dc voltage serves to supply the base bias to Q_1; the resistor R_4 is so chosen that the Zener is in the breakdown region so that the voltage V_z is remaining constant against source voltage or load current variation. The potential divider through R_1 and R_2 samples out the output voltage and feeds back a voltage V_F to the base of Q_2 and it controls the collector current of Q_2. The larger the value of V_F, the greater is the collector current of Q_2.

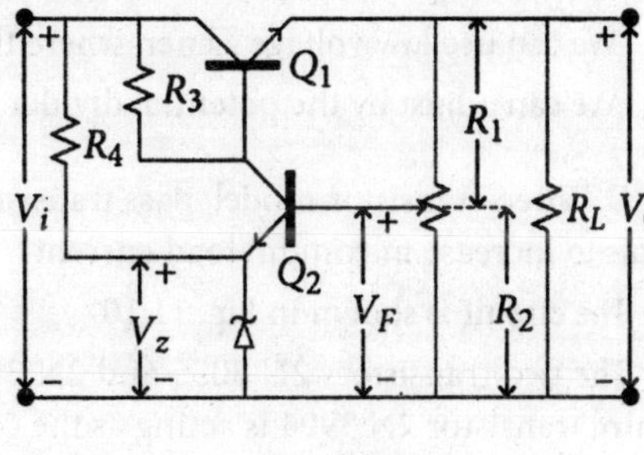

Fig. 11.8

Input voltage V_i may be taken to be the output of an ordinary power supply having ripple content of 10 % of the unregulated dc voltage (peak-to-peak). This regulator output V_0 is almost perfectly constant against the variation of both the input voltage as well as load current. If the output voltage tends to change, the feedbacking to the base of the control transistor Q_2 produces an error voltage that compensates for the attempted change. Why and how?

Explanation. Suppose, V_0 tends to rise, either due to rise in V_i or in the load current, the feedback voltage V_F increases drawing more collector current for the control transistor Q_2 through R_3 with consequent decrease in base voltage of the pass transistor Q_1. The resulting lesser base voltage of Q_1 lowers the conduction through it and hence, the output V_0 is reduced.

Similarly, if the output tends to drop, the base bias of Q_2 lowers, its collector current through R_3 reduces, which in turn produces greater base bias to the pass transistor Q_1. The voltage drop in the output is thus compensated.

Mathematical viewpoint. For the sake of analysis, the circuit in Fig. 11.8 is equivalently redrawn in Fig. 11.9.

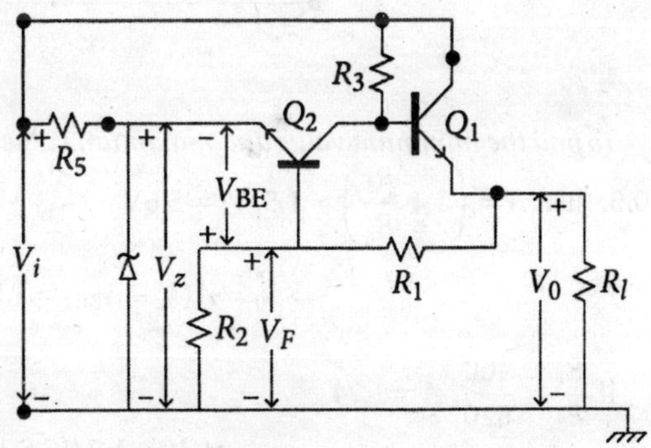

Fig. 11.9

The output voltage V_0 samples out a voltage V_F, which is fed back to the emitter of Q_2. We can write,

$$V_F = \frac{V_0}{R_1 + R_2} \times R_2 \Rightarrow \beta = \text{feedback factor} = \frac{V_F}{V_0} = \frac{R_2}{R_1 + R_2}.$$

The closed-loop voltage gain

$$A = \frac{1}{\beta} = 1 + \frac{R_1}{R_2}.$$

Applying KVL to the input side of the control transistor Q_2, we easily get

$$V_F - V_{BE} - V_z = 0 \Rightarrow V_F = V_{BE} + V_z \Rightarrow \beta V_0 = V_{BE} + V_z \Rightarrow V_0 = \frac{V_{BE} + V_z}{\beta}. \quad (11.4)$$

$$\therefore A = \frac{1}{\beta} = \frac{V_0}{(V_{BE} + V_z)}.$$

$$\therefore V_0 = A(V_{BE} + V_z),$$

where V_0 = regulated output, A = closed-loop gain = $1 + R_1/R_2$, V_z = Zener breakdown voltage, V_{BE} = base-emitter voltage of Q_2 (control transistor).

Remarks.

1. V_0 remains practically constant against line variation and load variation.
2. We can use low voltage Zener, where the temperature coefficient approaches zero.
3. We can adjust by the potential divider rheostat to the exact value required in a particular purpose.

D. Three-transistor model, pass transistor is of high current bearing capacity and forming a Darlington pair to increase maximum load current.

The circuit is shown in Fig. 11.10.

The two transistors 2N 3055 and 2N 3904 form a Darlington pair for increasing the load current. The third transistor 2N 3904 is acting as the control transistor in the series regulator.

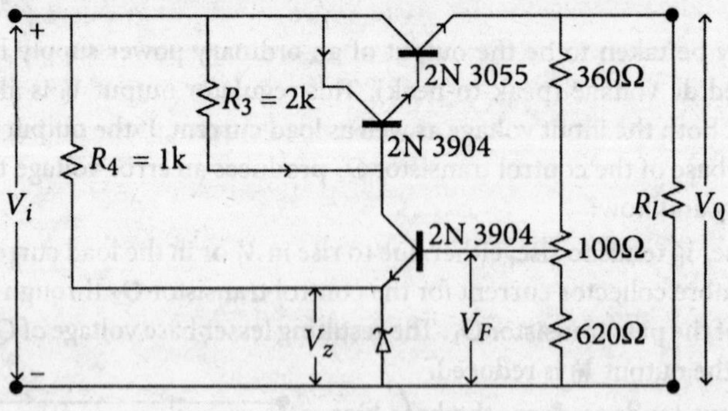

Fig. 11.10

To find the minimum value and maximum value of V_0. Depending on the position of the wiper, let $R_1/R_2 = 0.5$, then $A = \left(1 + \dfrac{R_1}{R_2}\right) = 1.5$, $V_z = 5.6\text{V}$

$$\Rightarrow V_0 = A(V_z - V_{BE}) = 1.5(5.6 + 0.7) = 1.5 \times 6.3 = 9.45\text{V}.$$

If $\dfrac{R_1}{R_2} = \dfrac{460}{620}$, $A = 1.74$

$$\Rightarrow V_0 = 1.74(5.6 + 0.7) = 1.74 \times 6.3 = 11\text{V}.$$

6. *Line regulation and load regulation revisited.*

The present discussion is held w.r.t. the Zener regulating circuit [Fig. 11.4].

Load regulation. The regulating characteristic of a Zener is expressed through two quantities (a) load regulation (b) line regulation.

If the load resistance varies, load current also varies at a constant supply voltage resulting in variation of output voltage V_0. So we have introduced a term, called *load regulation*, as

$$\text{Load regulation} = \lim \left(\frac{\delta V_0}{\delta I_0}\right)_{V_i} = \left(\frac{\partial V_0}{\partial I_0}\right)_{V_i}.$$

Such a variation is shown in Fig. 11.11.

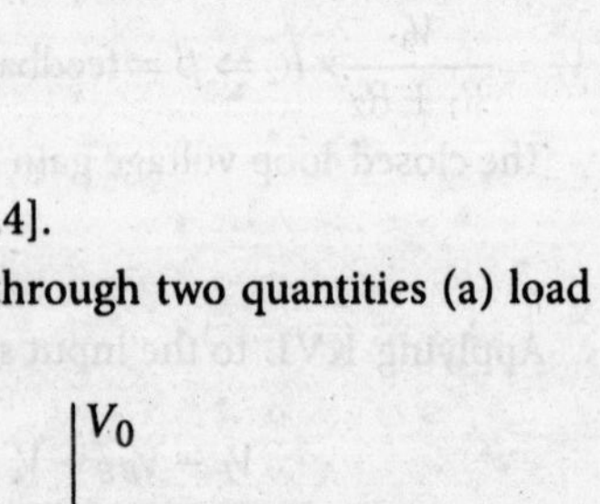

Fig. 11.11

(b) *Line regulation.* The rate of variation of output voltage V_0 with input voltage V_i is termed as line regulation.

So, line regulation = $\left(\frac{\partial V_0}{\partial V_i}\right)_{I_L}$.

11.2 Short Theory Questions with Answers

1. What do you mean by a series regulator and a shunt regulator?

   Ans. If the regulating element is connected in series with the load, it is called *series regulator.* If it is in parallel to the load, it is called a *shunt regulator.*

2. With respect to circuit in Fig. 11.10, what would occur to the output voltage if (a) 2N 3055 is shorted (b) 2N 3055 is open?

   Ans. (a) Current through the load increases increasing the output voltage.

   (b) Load current decreases and as a result, the output voltage decreases.

3. What is the purpose served by the Zener?

   Ans. The Zener serves supply, a reference voltage.

4. What do you mean by load regulation and load regulation percentage?

   Ans. Load regulation, also called *load effect,* is defined as the change in the regulated output voltage when the load current changes from minimum to maximum. $R = V_{NL} - V_{FL}$.

   Regulation $R\%$ is defined as the ratio of regulation to the no-load voltage output.

$$R\% = \frac{V_{NL} - V_{FL}}{V_{NL}} \times 100.$$

5. If no-load voltage is 10V and the full voltage is 9.9V, then what is the per cent load regulation?

   Ans. $R\% = \frac{10 - 9.9}{10} \times 100 = 1\,\%$.

6. What is source regulation? What is source regulation per cent?

   Ans. Source regulation or source effect or also called line regulation is defined as the change in the regulated output voltage for a specified range of line voltage.

   Thus, if the line voltage varies by 10 %, then for a line voltage 100V ±10 %, it will be 90V to 110V. For this specified range of line voltage variation, if the load voltage changes from 10V to 9.8V, then the source regulation (SR) = 10 − 9.8 = 0.2V.

$$\text{SR}\,\% = \frac{0.2}{10} = 0.02.$$

7. In Fig. 11.8, what would happen if the Zener is shorted?

   Ans. More current drains through the resistor R_4 to ground through the shorted Zener, so output voltage decreases.

11.3 Numerical Problems

1. Find the dissipation in the pass transistor in regulator circuit of Fig. 11.8, where $V_0 = 10$V, $V_i = 20$V, load current $I_l = 2$A.

   Solution: The power dissipation in the pass transistor 2N 3055 is $P_{dc} = (20 - 10) \times 2 = 20$W.

2. With respect to Fig. 11.10, find the following quantities: (a) current through 2kΩ resistor. (b) current through 1kΩ resistor. (c) Current through the Zener. Output voltage 10V, load resistance = 5Ω, β_{dc} (2N 3055) = 50, β_{dc} (2N 3904) = 100, load resistance = 1080Ω, $V_z = 5.6V$, $V_i = 20V$.

Solution: $I_L = \frac{10}{5} = 2A.$

Current through the potential divider $= \frac{10}{1080} = 9.26mA.$

So, the total current through the pass transistor = 2A (practically).

(a) The potential at the base of 2N 3904 is $(2 \times 0.7) = 1.4V$ higher than the output. So, it is $(10 + 1.4) = 11.4V$.

$\therefore$ the current through 2kΩ resistor $= \frac{20 - 11.4}{2 \times 10^3} = 4.3mA.$

(b) The current through 1kΩ resistor is $= \frac{20 - 5.6}{2 \times 10^6} = 14.4mA.$

(c) Collector current for 2N 3055 = 2A ⇒ its base current $= \frac{2}{50} = 0.04A.$

The current through 2kΩ resistor is divided into two parts: (i) one goes to supply the base current for 2N 3904 of the Darlington pair, the other (ii) for the collector current of the control transistor 2N 3904.

$\therefore$ the base current of the lower 2N 3904 is $= \frac{40mA}{100} = 0.4mA.$

Collector current for the lower 2N 3904 = 4.3mA – 0.4mA = 3.9mA.

So, the current through the Zener = 14.4mA + 3.9mA = 18.3mA.

3. In the above example find out the power dissipation is 2N 3055.

Solution: The total current passing through 2N 3055 is 2A (practically), so the power dissipation is 2N 3055 is $(20 - 10) \times 2 = 20W$.

4. In the above example find the required wattage of the Zener. $V_z = 5.6V$.

Solution: $I_z = 18.3 \times 10^{-3}A$. $\therefore$ $P_z = 18.3 \times 10^{-3} \times 5.6 = 0.1W$.

7. *Series regulator with short-circuit protection.*

If accidentally, the load terminals get shorted, tremendously large amount of current passes through the pass transistor and it is likely to be burnt. It is because in this case, practically whole of the input voltage is impressed upon the pass transistor. Shorting of the load has also effect on the rectifier diodes over and above damaging the pass transistor. So every regulating device has short-circuit protection artifice incorporated.

Short-circuit protection technique may be of two types: (a) simple limiting (b) Fold back limiting.

(i) *Simple current limiter.* This current limiting is simple in design but it has the disadvantage of having large amount of dissipation (110W to 120W). This problem of dissipation is reduced to a large extent by fold back limiting.

(b) *Fold back limiting device.* If a constant current passes through the load, then when the load is short-circuited, whole of the input voltage is impressed on the pass transistor, hence the pass transistor is to dissipate large amount of power. Suppose, in a series regulator with simple limiting device, the input $V_i =$ 20V, a constant current of 6A flows through the load with nominal voltage of 10V. Power dissipation of the pass transistor under normal condition is $P = (20 - 10) \times 6 = 60W$. This dissipation shoots up to $P_{sh} = 20 \times 6 = 120W$. So, simple current limiter is not a suitable solution.

Refer to Fig. 11.12, where the essential parts of the short-circuit protection is shown with fold back limiting device.

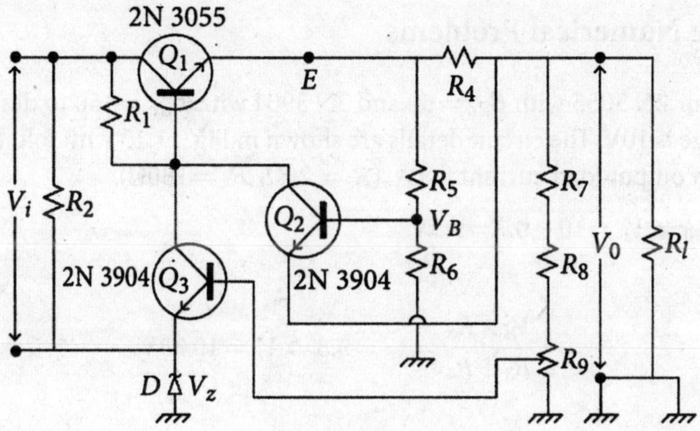

Fig. 11.12

R_4 is called current sensing resistance and the load current I flows through R_4 producing a voltage drop of IR_4. This means a voltage of $(IR_4 + V_0)$ is fed to the voltage divider consisting of R_5 and R_6. The control voltage equal to the drop over R_6 is fed to the base of Q_2. This voltage is given by

$$V_B = \frac{(IR_4 + V_0)R_6}{(R_5 + R_6)} = K(IR_4 + V_0) = V_0 + V_{BE},$$

where K = feedback fraction of the voltage divider.

$$\Rightarrow V_{BE} = K(IR_4 + V_0) - V_0 = K(IR_4) + (K-1)V_0$$

$$\Rightarrow \quad I = \frac{V_{BE} - (K-1)V_0}{KR_4} = \frac{V_{BE} + (1-K)V_0}{KR_4}. \tag{11.5}$$

In case of accidental shorting of the output terminals, $V_0 = 0$ and $I = I_{SL} \Rightarrow I_{SL} = \dfrac{V_{BE}}{KR_4}$.

So, when the terminals are not shorted, the load current is

$$I_L = \frac{V_{BE}}{KR_4} + \frac{(1-K)V_0}{KR_4} = I_{SL} + \frac{(1-K)V_0}{KR_4}. \tag{11.6}$$

Illustration. If for a series regulator $V_0 = 10\text{V}$, $R_4 = 1\Omega$, $R_5 = 20\Omega$, $R_6 = 180\Omega$, find the current passing through the load and also the current when load terminals are shorted.

Here, $K = \dfrac{R_6}{R_5 + R_6} = \dfrac{180}{180+20} = 0.9.$

$$\therefore\ I_L = \frac{V_{BE}}{KR_4} + \frac{(1-K)}{KR_4}V_0 = \frac{0.7}{0.9\times 1} + \frac{(1-0.9)\times 10}{0.9\times 1} = 1.89\text{A}.$$

$$I_{SL} = \frac{0.7}{0.9\times 1} = 0.778\text{A}.$$

Under normal working condition, the load current is 1.89A, which drops to 0.778A when the load terminals are shorted. This explains the functioning of the circuit as protection against short-circuiting the load terminals.

11.4 Challenging Numerical Problems

1. Use a Darlington pair 2N 3055 with $\beta_{dc} = 50$ and 2N 3904 with $\beta_{dc} = 150$ to design a voltage regulator whose output voltage is 10V. The circuit details are shown in Fig. 11.13 with fold back limiting. Find the Zener current when output load current is 1A. ($R_3 = 20\Omega$, $R_4 = 180\Omega$).

   [*Hints.* $V_B + V_{BE} = V_0 \Rightarrow V_B = 10 - 0.7 = 9.3\text{V}$

$$\frac{V_E \times R_4}{R_3 + R_4} = V_B = 9.3 \Rightarrow V_E = 10.33\text{V}.$$

   $V_G = V_E + 2V_{BE} = 11.73\text{V}$]

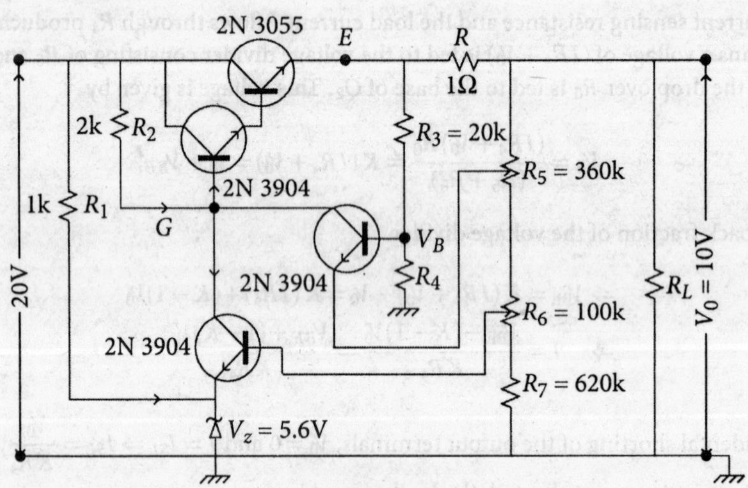

Fig. 11.13

2. Find the power dissipation in the pass transistor when the load terminals are shorted.

   [*Hints.* $I_{SL} = \dfrac{V_{BE}}{KR} = \dfrac{0.7}{0.9 \times 1} = 0.778\text{A} \Rightarrow V_E = 0.778 \times 1 = 0.778\text{V}.$

   $\therefore V_{CE} = 20 - 0.778 = 19.2\text{V} \Rightarrow P_D = 19.2 \times 0.778 = 14.9\text{W}$]

3. Design a series regulated power supply to provide a nominal output of 15V at a maximum load current of 150mA. The input voltage is 35V with internal resistance of 10Ω. Calculate (a) closed-loop voltage gain (b) power dissipation, given for pass transistor: $h_{fe} = 200$, $h_{ie} = 1.2\text{k}$, $V_{BE} = 0.7\text{V}$, $i_c = 10\text{mA}$.

   Feedback transistor: $h_{fe} = 120$, $h_{ie} = 220\Omega$.

   Zener: 120mW at 5.6V.

4. In Fig. 11.14 we have shown a series regulator circuit with the circuit details. The transistor Q_1 is a pass transistor 2N 3055; Q_2 is 2N 3904, which is serving as the difference amplifiers; the Zener diode is having $V_z = 5.6\text{V}$. Show that $V_0 = V_R + V_{BE2} + \dfrac{R_1}{R_1 + R_2} V_0$.

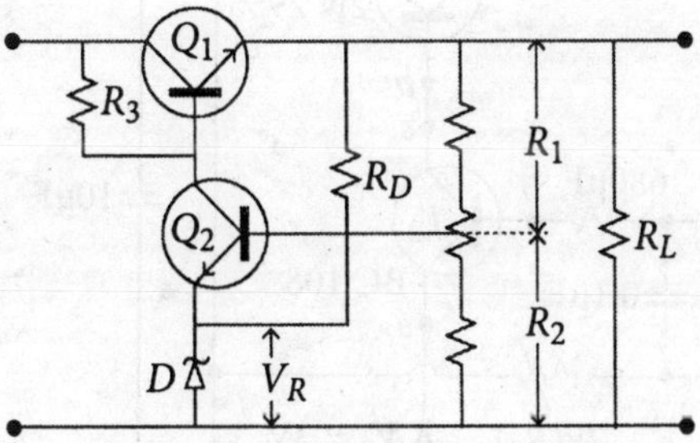

Fig. 11.14

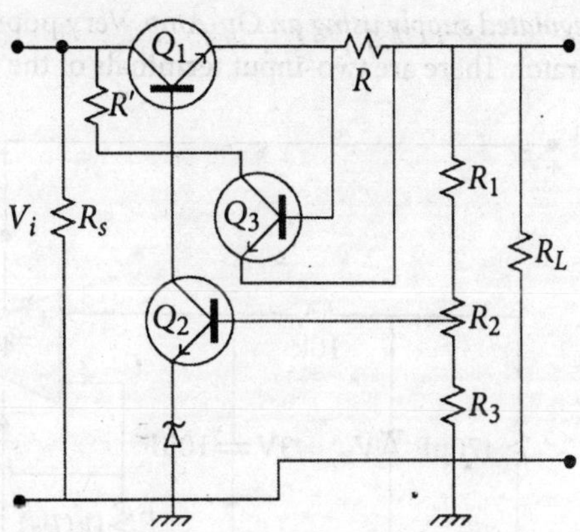

Fig. 11.15

5. The circuit shown in Fig. 11.15 is a series regulator with short-circuit protection. Q_1 is a pass transistor while Q_2 is the control transistor serving as a difference amplifier.
   Find the values of R', R_s, R_2, R_3 under given specifications as below.

   Nominal output voltage 15V at a maximum load current of 150mA. The input voltage $V_i = 25$V, with its internal resistance of 10Ω. Also, calculate (a) closed-loop gain (b) power dissipation in the pass transistor.

   Pass transistor: $h_{fe} = 200$, $h_{ie} = 1.2$k, $V_{BE} = 0.7$V, $I_e = 10$mA.

   Feedback transistor: $h_{fe} = 120$, $h_{ie} = 220Ω$.

   Zener: Wattage: 120mW at 5.6V.

11.5 Theory Questions Exercises

1. Give the block diagram of a regulated power supply. What is the need for a regulated power supply?
2. Explain with a neat diagram the working principle of transistor series emitter follower regulator.
3. Use a Zener diode as a regulating element. Draw a neat diagram of a Zener regulator and explain the functioning of the device.
4. Draw a neat diagram of a one transistor model series regulator. Explain how it works.
5. What is the meaning of voltage regulation? How do you define regulation percentage?
6. What is the meaning of an ordinary power supply? In case of an ordinary power supply with a nominal output of 6V, it is found that when a load is connected to it, the output voltage drops to 4V, why?
7. Describe a voltage regulator, which survives even when the output terminals are accidentally shorted.
8. What is fold back limiting? Draw a neat diagram of a series regulator with fold back limiting.
9. Design a simple emitter follower regulator to give an output voltage of 20V and a maximum load current of 1A using a silicon transistor with $h_{FE} = 50$, $V_{BE} = 0.6$V at $I_E = 1$A along with a suitable reference diode if $R_s = 2.4Ω$.
10. Use a Darlington pair as the pass element and with a fold back limiting, draw a neat diagram of series emitter follower voltage regulator explaining the functioning of the different parts.
11. Develop the mathematical formulation of the voltage with short-circuit protection as shown in Fig. 11.13 in presence of R and in absence of R.

8. *Regulated supply using an Op-Amp.* Very popular Op-Amp like IC 741 may be conveniently used as a comparator. There are two-input terminals of the Op-Amp..

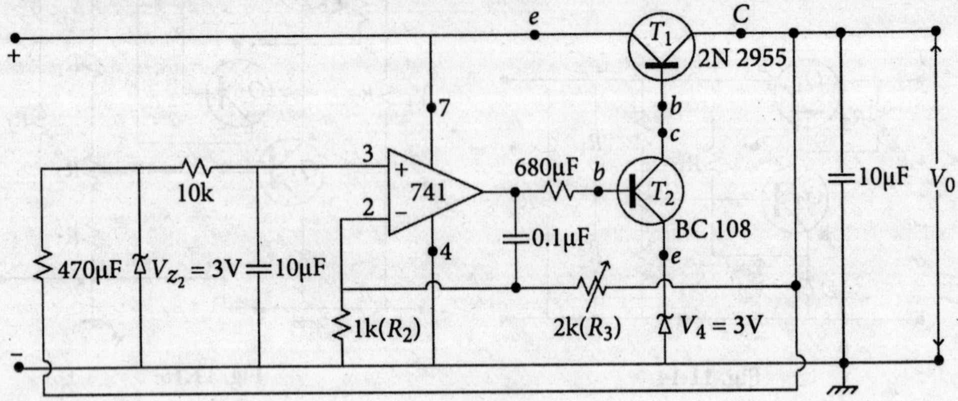

Fig. 11.16

1. One is the non-inverting terminals 3 (indicated by positive sign), the whole of Zener z_2 voltage is applied on to it.
2. The second is the inverting terminal 2 (indicated by negative sign).

The whole of Zener voltage V_{z_2} is applied to the non-inverting terminal and the potential divider consisting of R_3 and R_2 samples out from the output voltage to the inverting terminal. Input positive voltage to the non-inverting terminal is usually greater than the input voltage (negative) to the inverting terminal. The difference between them, called *error voltage* is positive and that is amplified by the Op-Amp so as to push the transistor Q_2 into conduction. The error voltage amplified supplies the necessary base bias and as it is becoming more and more, the transistor Q_2 is pushed to more and more conducting state and the output becomes also more and more.

Working of the regulator explained. As the output voltage tends to increase, the negative terminal feedback voltage increases, the error voltage decreases and Q_2 conducts less, the increase in the output is compensated. On the other hand, if the output voltage decreases, the feedback to the negative terminal decreases, the error voltage increases, the transistor Q_2 is now pushed to more conducting state. The output increases so as to compensate the drop in the output.

Another simpler type of regulator using an Op-Amp

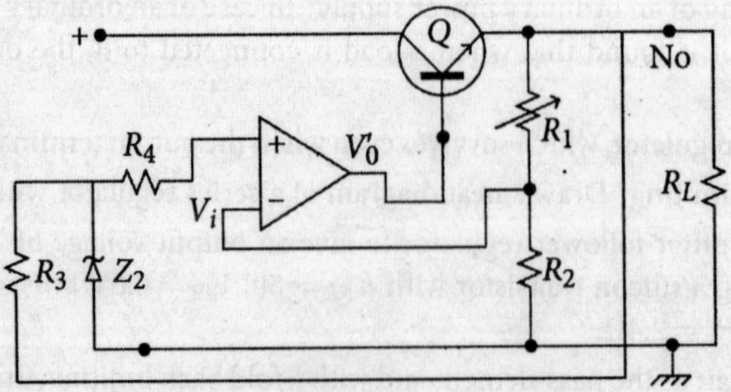

Fig. 11.17

Fig. 11.17 shows an almost similar circuit, similar to circuit shown in Fig. 11.16 using an Op-Amp.

9. IC *Regulators.*

We have discussed in the last few sections how we can make regulated power supplies using discrete components including Op-Amp also. To make the job of power supply design more easy, a number of readymade ICs are available in the market. Some of them are fixed voltage regulator ICs having only three terminals — input, ground and output, while others are also there having multiple terminals with wider involvements.

Three terminal voltage regulator ICs: Exploded view of the internal to such an IC is shown schematically in Fig. 11.18.

Inside the clip there is a pass transistor capable of handling load current greater than 1.5A (with proper heat sink), provision for thermal shutdown and limiting. Thermal shutdown refers to the mechanism, which shuts down the device when the internal temperature exceeds the danger level of 175°C.

We will consider here only the common type of fixed or variable voltage IC regulators (a) LM 78 or 79 series (b) CA series.

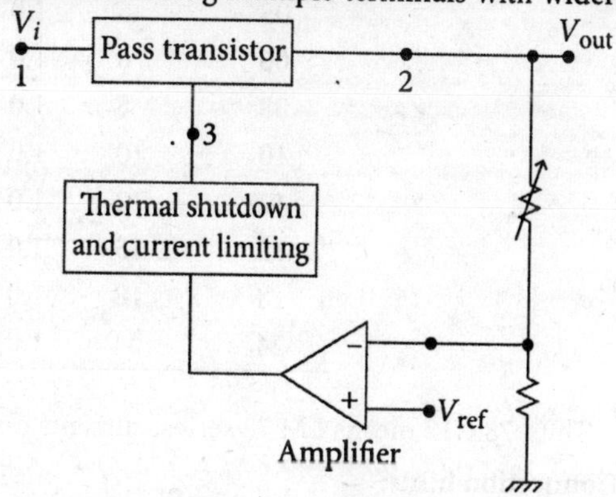

Fig. 11.18

(a) LM 78 series is used for positive voltage supply and LM 79 series for negative voltage supply.

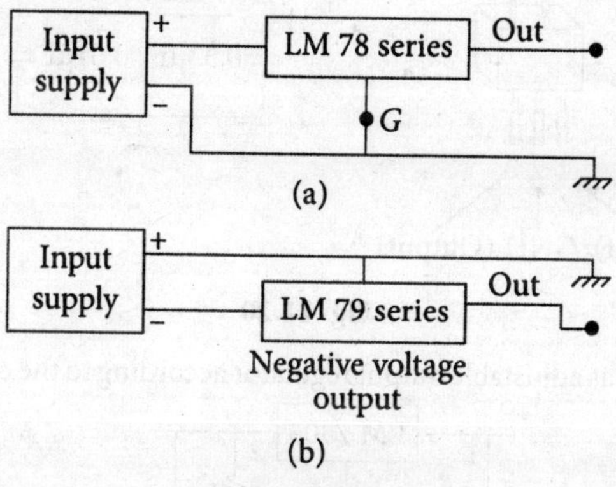

Fig. 11.19

Fig. 11.19 (a) shows positive output and (b) shows negative output supply.

LM 78 series (for positive voltage)

XX (Voltage)	V(out) (V)	Current (I_0) C	L	M	Type	Input (v) Min	Max
05	5	1.0	0.1	0.5	7805	7.5	20
06	6	1.0	0.1	0.5	7806	8.6	21
08	8	1.0	0.1	0.5	7808	10.6	23
10	10	1.0	0.1	0.5	7810	12.7	25
12	12	1.0	0.1	0.5	7812	14.8	27
15	15	1.0	0.1	0.5	7815	18	30
18	18	1.0	0.1	0.5	7818	21	33
24	24	1.0	0.1	0.5	7824	27.3	38

Thus, 78 C12 means LM 78 series, currents output 1A, volt output 12, regulator three-terminal IC.

Connection hints:

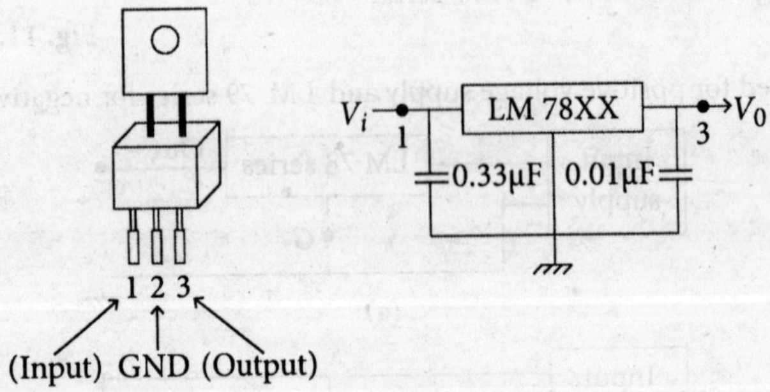

Fig. 11.20

LM 78 XX can also be used as adjustable output regulator according to the connection shown in Fig. 11.21.

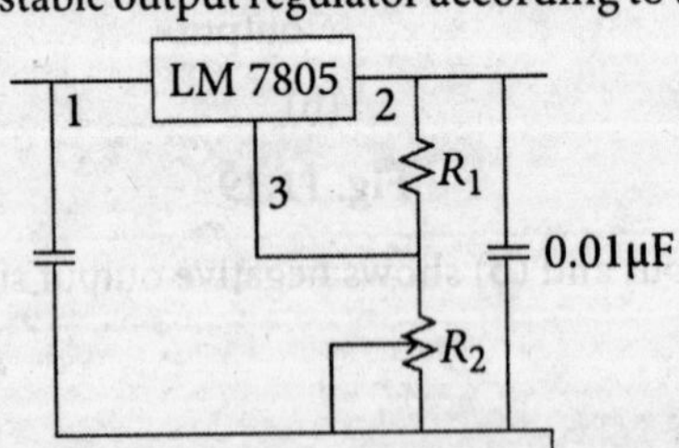

Fig. 11.21

79 XX voltage regulator IC series is meant for negative voltage output, otherwise, similar to positive series. Thus,

LM 79 series (for negative voltage)

XX (indicator of voltage)	V_0	Current I_0(A) *C*	*L*	*M*	Type	V_i(V) Min	Max
05	−5.0	1.5	0.1	0.5	7905	−20	−7.5
05.2	−5.2	1.5	—	—	7905.2	−20.5	−7.5
06	−6.0	1.5	—	0.5	7906	−21.0	−8
08	−8.0	1.5	—	0.5	7908	−23.0	−10.5
09	−9.0	1.5	—	—	7909	−24.0	−11.5
12	−12.0	1.5	0.1	0.5	7912	−27.0	−14.5
15	−15.0	1.5	0.1	0.5	7915	−30.0	−18.0
18	−18.0	1.5	0.1	—	7918	−33.0	−21.5
24	−24.0	1.5	0.1	0.5	7924	−38	−24

Connection hints

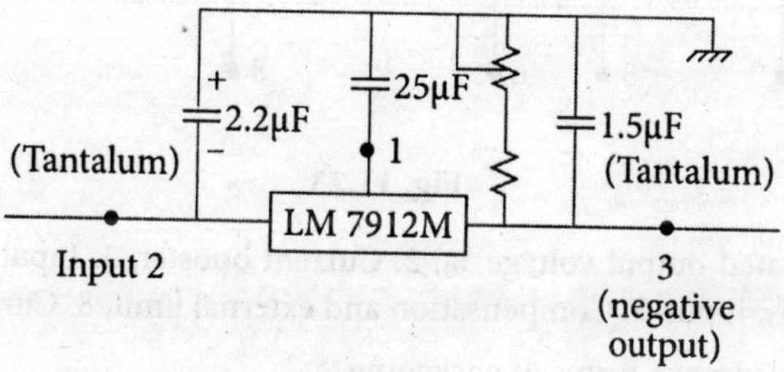

Fig. 11.22

Multiple terminal IC regulator CA series

CA 3085 series regulators: Regulated voltage from 1.8V to 46V with current rating up to 100mA.

CA 3085 and CA 3085A are silicon monolithic integrated circuits designed for voltage regulators at output voltages ranging from 1.7V to 46V with current rating at 100mA.

Interior of the device is shown is block diagram [Fig. 11.23].

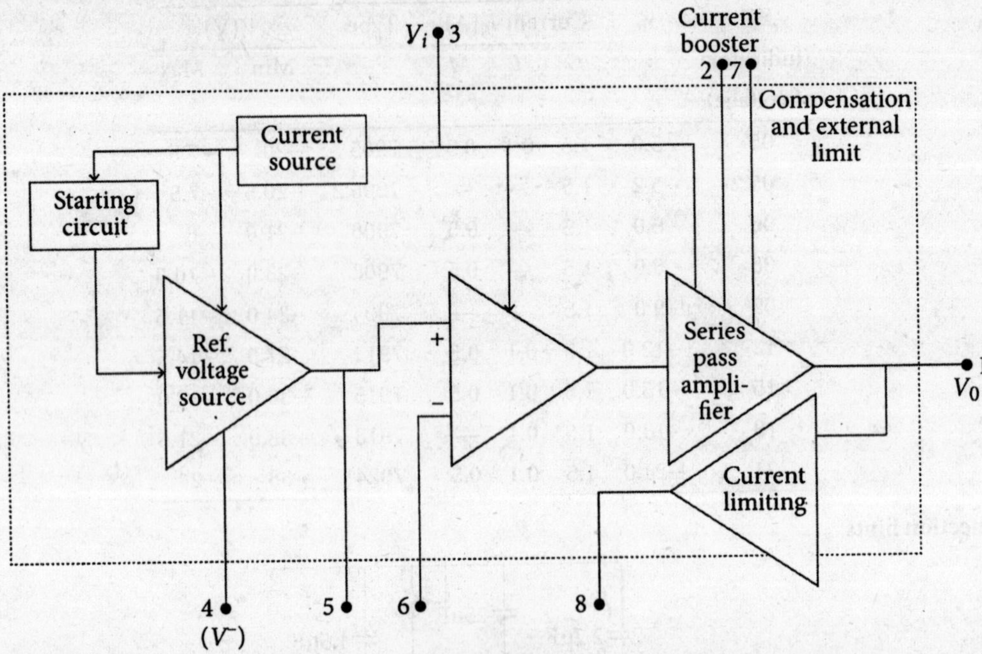

Fig. 11.23

Pin description: 1. Regulated output voltage V_0, 2. Current booster, 3. Input unregulated voltage V_i, 4. Substrate (V^-), 5. V_{ref}, 6. Inv. input, 7. Compensation and external limit, 8. Current limiting (Fold back).

CA 3085 series is available in two forms of packaging.

One packaging is in 8-lead TO 5 packaging shown in diagram (11.24).

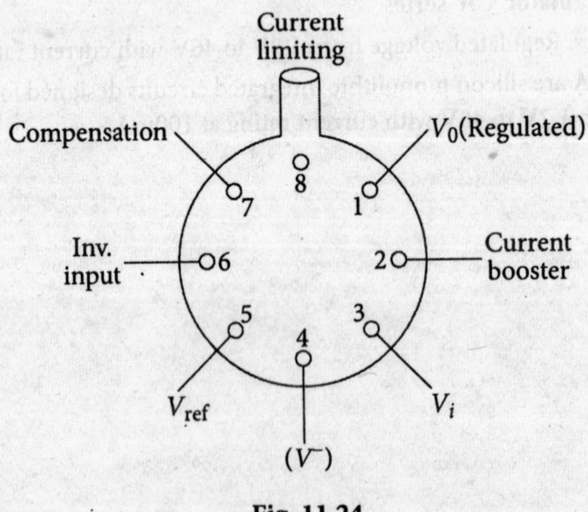

Fig. 11.24

TO 5 dual-in-line (DIP). It is an eight pin IC; the pins are arranged in lines, 4 on the left flange, 4 on the right [Fig. 13.25].

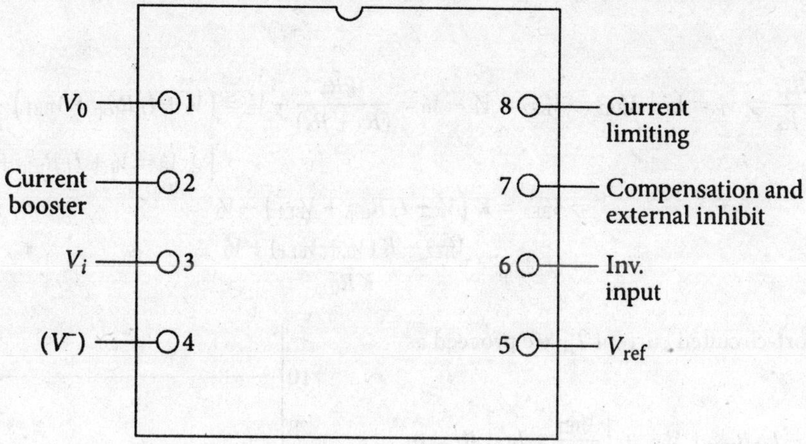

Fig. 11.25

Application hints for CA 3085

Refer to Fig. 11.23. The circuit fundamentally consists of a frequency compensated error amplifier, which compares the sampled output voltage with an internally generated reference voltage V_{ref}.

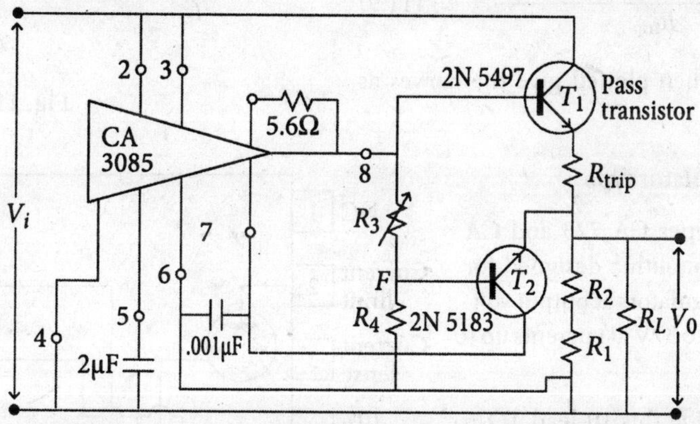

Fig. 11.26

The circuit [Fig. 11.26] is designed for high-output-current voltage regulator using auxiliary transistor for fold-back current limiting using IC CA 3085.

At low load current, T_2 is effectively in reverse bias by a small amount controlled by the potential divider resistors R_3 and R_4. When the load current, the drop across R_{trip} increases and the base bias is raised to T_1 and T_2 starts conducting. With T_2 becoming more and more conductive, it diverts base current from transistors inside CA 3085 and thus reduces the base drive to the transistor T_1. The net effect is to reduce the output voltage.

To find the current limiting point I_{trip}, we calculate as described below:

At low load current (voltage across R_{trip} small): We can write w.r.t. Fig. 11.26.

Voltage at the terminal δ, is given by $V_8 - V_{BE1} = V_0 \Rightarrow V_{BE1} = V_8 - V_0$.

Now

$$V_F = \frac{V_8 \cdot R_4}{R_3 + R_4} \Rightarrow V_F = V_0 + V_{BE2} \Rightarrow V_{BE2} = V_F - V_0 = \frac{V_8 R_4}{(R_3 + R_4)} - V_0 = \left(V_0 + I_L R_{trip} + V_{BE1}\right)\frac{R_4}{R_3 + R_4} - V_0$$

$$\left[\because\ V_8 = V_0 + I_L R_{trip} + V_{BE1}\right]$$

$$\Rightarrow V_{BE2} = K\left(V_0 + I_L R_{trip} + V_{BE1}\right) - V_0$$

$$\Rightarrow (I_L)_{trip} = \frac{V_{BE2} - K(V_0 + V_{BE1}) + V_0}{K R_{trip}}. \qquad (11.7)$$

To find short-circuited current I_{sc} we proceed as below:

$$-V_{BE1} - I_{sc} R_{trip} + V_{BE2} + \left[\frac{V_{BE2}}{R_2} + I_{B2}\right] R_1 = 0 \qquad (11.8)$$

$$\Rightarrow V_{BE2} + \left[\frac{V_{BE2}}{R_2} + I_{B2}\right] R_1 = V_{BE1} + I_{sc} R_{trip}$$

$$\Rightarrow I_{sc} = \frac{V_{BE2} + \left[\frac{V_{BE2}}{R_2} + I_{B2}\right] R_1 - V_{BE1}}{R_{trip}}. \qquad (11.9)$$

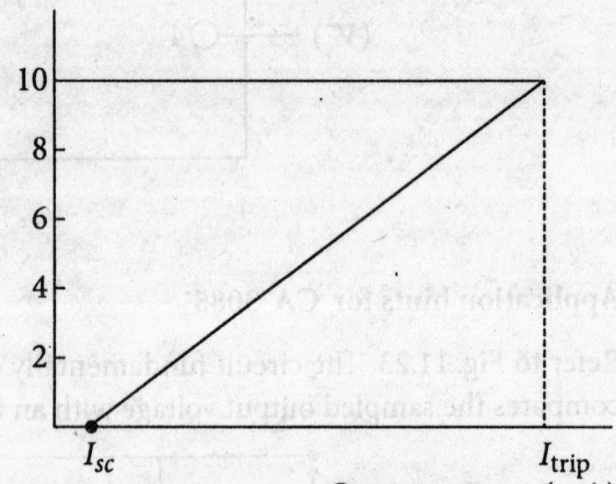

Fig. 11.27

(11.7) and (11.8) when plotted give the curves as shown in Fig. 11.27.

CA 723 voltage regulator type

Integrated circuit types CA 723 and CA 723C are silicon monolithic designed for serving as voltage regulators at output voltages ranging from 2 to 37V at currents up to 150mA.

What is there inside this 10-lead TO-5 ceramic package (T suffix) and 14-lead dual in line plastic package (E suffix) IC?

Each type includes temperature compensated reference amplifier, an error amplifier, a power series-pass transistor and a current limiting circuit. They also provide independently accessible inputs for adjustable current limiting and also remote shutdown [Fig. 11.28].

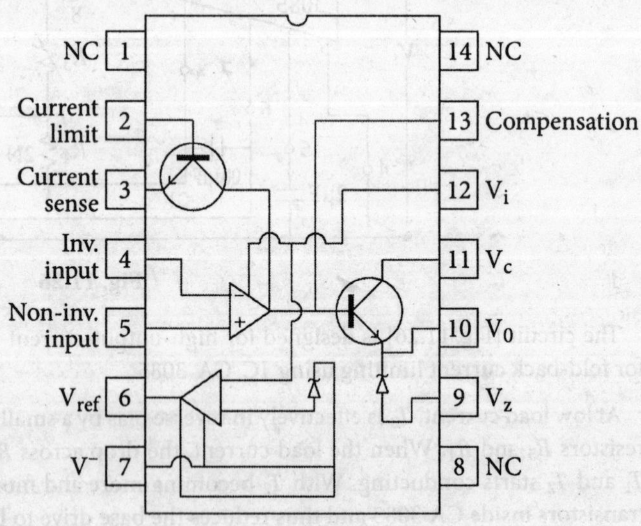

Fig. 11.28

Application hints for CA723: Fig. 11.29 shows a voltage regulator using CA723 to provide voltage output ranging from 7 to 37V. The basic circuit on which this IC works is shown in Fig. 11.29.

CL = current limiting terminal, CS = current sense terminal.

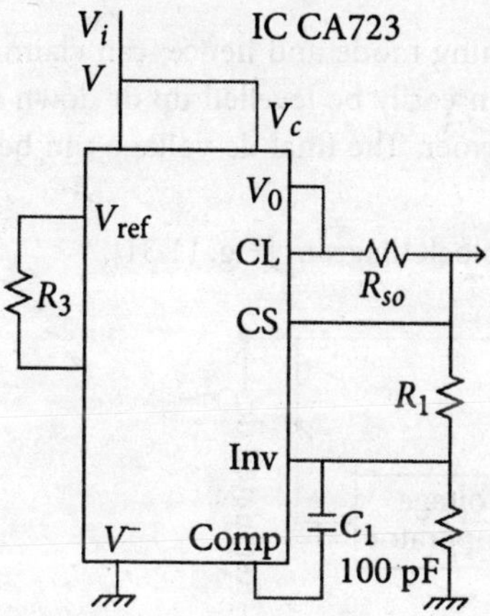

Fig. 11.29

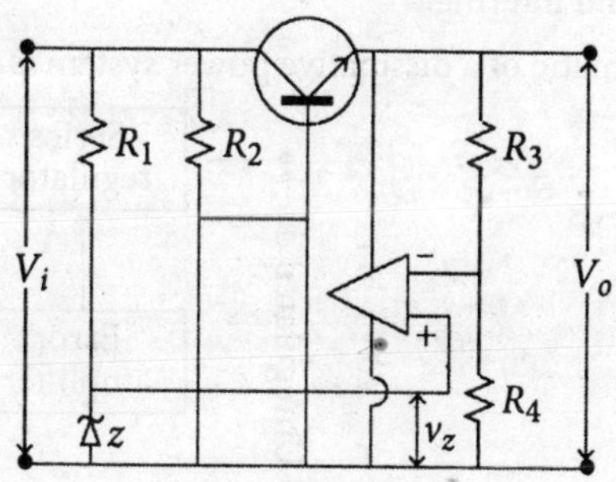

Fig. 11.30

The potential divider arrangement samples out a certain voltage from the output and is compared by the comparator. Op-Amp with a reference voltage V_z provided by the Zener is applied to the non-inverting terminal, while the sampled voltage is applied to the inverting terminal. Usually, $v_z >$ sample voltage, their difference, called *error voltage*, is positive. This error signal is amplified by the error amplifier so as to control the conductivity of the pass transistor to reduce the error voltage at the input of the comparator. The system by adjusting the conductivity of the pass transistor due to increase or decrease of the output of the error amplifier thus changing the base drive of the pass transistor, maintains a steady level of the output voltage.

$$\frac{V_{\text{out}} \times R_4}{(R_3 + R_4)} = V_F = V_z, \text{ error voltage is small}$$

$$\Rightarrow V_{\text{out}} = \left(\frac{R_3 + R_4}{R_4}\right) V_z = \left(1 + \frac{R_3}{R_4}\right) V_z.$$

11.6 SMPS — revolution in Power Electronics

Power electronics is entirely devoted to switching mode power (SMP) conversion for efficiency, control and regulation of electrical energy. This type of power supply is in increasing demand particularly in computer system requiring low voltage, high current source with low output ripple and fast transient response. Now-a-days SMPS are recommended to be used in CTV, dc-to-dc single and multiple output power supplies; bidirectional power supplies; dc-to-dc inverters and also in robotics.

SMPS is the fruit of labour for long years of active search for a suitable power supply meeting the requirements for highly reliable, efficient and light-weight electrical power system to be installed in spacecraft. Nobody can deny the fact that the heart of any electrical or electronic device is its power supply system whose merit is judged by large MTBF (Mean Time Before Failure). The last decade has witnessed tremendous development in designing the power system with large MTBF. Such a design is the design of SMPS with different forms and features.

Regulated power supplies are of two types: (i) Dissipative (linear) power supply (ii) Non-dissipative (switching mode) power system. The dissipative power system, as the name itself suggest, dissipates energy more in the process of conversion, hence its efficiency is low, it requires large heat sink area. The dissipation also varies depending on the input voltage and load fluctuations. However, this dissipative power systems has advantage in having low electromagnetic inference (EMI) and less ripple content.

The non-dissipative power system operates in the switching mode and hence, can claim to be highly efficient and once dc power is converted to ac power, it can easily be levelled up or down in voltage by employing high frequency and hence, light weight transformer. The final dc voltage can be obtained by rectifying and filtering.

The schematic of a dissipative power system is shown in block diagram [Fig. 11.31].

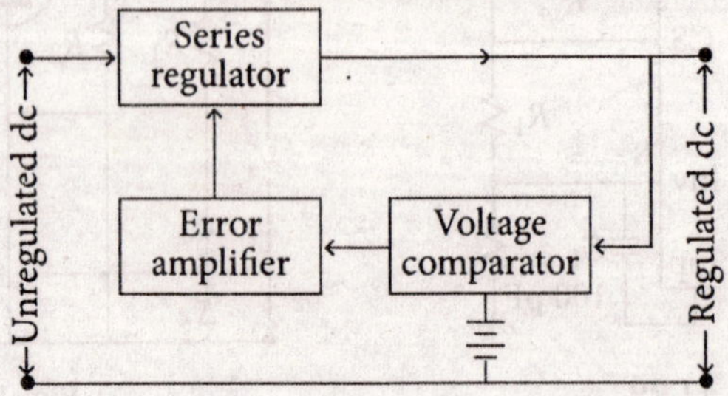

Fig. 11.31

The conventional practical circuit using the above scheme is shown in the diagram of Fig. 11.32 with explanation given below.

The pass transistor T_1 is the series regulator and the transistor T_2 serves as the voltage comparator and error amplifier. The reference voltage V_{ref} is provided by the Zener diode ZDI, the cathode receiving the voltage from the output through the dropping resistor R_2 thus the emitter of T_2 is maintained at a fixed value which is the Zener Break-down voltage.

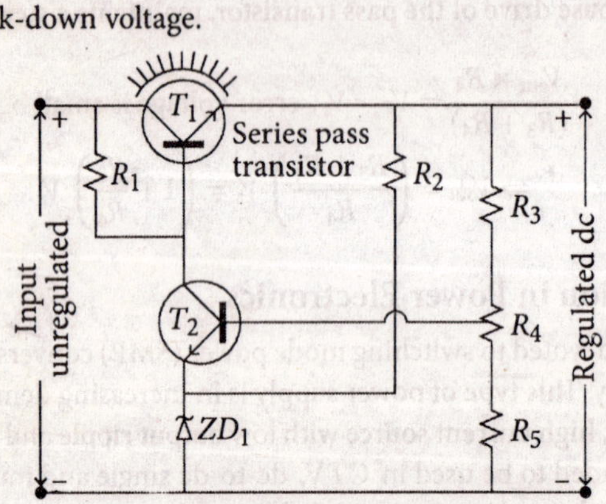

Fig. 11.32

A part of the output voltage is fed to the base of T_2 by the potential divider arrangement R_3, R_5 and the pre-set R_4. The transistor T_2 compares the output voltage with the reference voltage and differences amplified and fed to the base of the pass transistor T_1 for controlling its resistance.

Suppose, the circuit at some instant has attained a steady state giving steady output, the value of which can be set as desired by the preset R_4; if now there is variation of output voltage due to some reason or other, the regulating transistor varying its resistance components for any change. If there is sudden increase in the output (may be due to increase in the input or decrease in the load current), the feedback voltage to the base of T_2 increases, hence the base bias increases which in turn increases the collector current of this transistor. Since this collector current is drawn through the series transistor T_1, the base current of T_1 decreases in the output. Thus, the pass transistor T_1 by controlling the resistance offered by it, levels off any variation in the output. In this power system, therefore, the pass transistor T_1 has to withstand a large or small current passing through it depending on the load-current demand. So there is energy dissipation in this transistor, hence the pass transistor ought to be a power transistor fitted with proper heat sink.

SMPS, on the other hand, is less dissipative, highly efficient, smaller in size, lighter in weight. So many good qualities are inherent in the design of SMPS.

In SMPS transistors are used as rapidly opening and closing switches to regulate the output voltage. This switching action can be performed with sufficient efficiency by a transistor. So an SMPS ought to be highly efficient. Since in this case the pass transistor operates in ON-state or OFF-state, the dissipation here is minimum. The regulation is achieved by controlling the duty ratio of the pass transistor.

There are three types of such regulators (i) buck-type (ii) boost-type (iii) buck-boost-type of regulators.

The basics of the buck-type and boost-type switching converts are explained with reference to Fig. 11.33 and Fig. 11.34 respectively. Fig. 11.33 is the step-up switching converter, while Fig. 11.34 is the step-down switching converter.

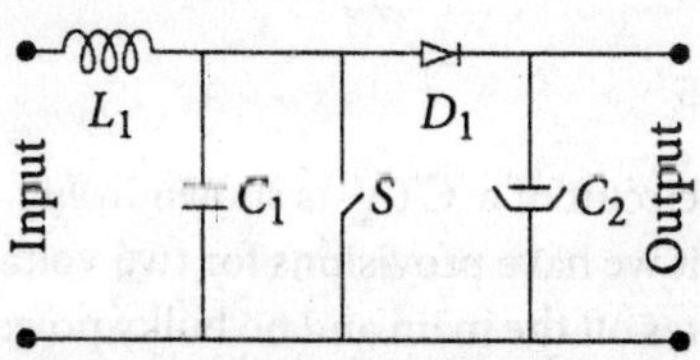

Fig. 11.33

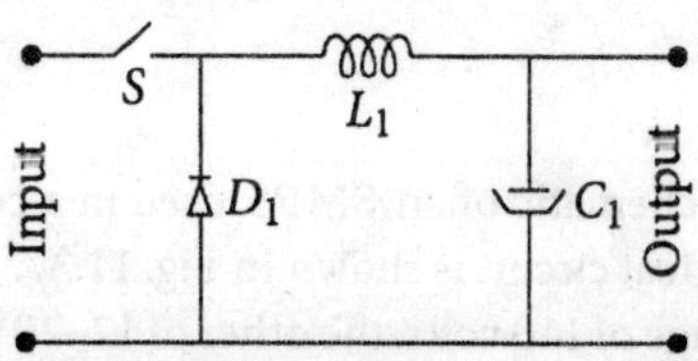

Fig. 11.34

Fig. 11.33 is a step-up switching converter where the output voltage is greater than the input voltage. This is also called a buck-type regulator. The switch S shown here which opens or closes rapidly is done by transistor switching. When S is closed, the voltage across C_1 is zero, the diode is non-conducting. The current flows through the inductance and energy is stored in the coil. As the switch opens, the magnetic field of the coil collapses and the energy stored is released charging the condenser C_1, the voltage across it rising rapidly and when it exceeds the output, the diode D_1 fires and the condenser C_2 gets charged. Due to rapid switching action performed through a transistor, the output voltage varies only slightly. Fig. 11.33 shows the basics of a step-up converter.

The circuit in Fig. 11.34 is a slight variation of the previous one. Here when the switch closes, the current passes through the inductor L_1 and energy is supplied to the load, coil L_1 and the condenser C_1. When the switch opens, the output current is supplied by the stored energy in the coil and the condenser. The output voltage in this case will be less than the input. It is step-down converter. The output voltage in both types of converters can be adjusted by varying the ON-period of the switch (longer → higher, shorter → lower).

Based on the above model, the schematic of a switching regulator design will be as shown in Fig. 11.35 in block representation.

Compare the circuit drawn in Fig. 11.35 with the basic circuit in Fig. 11.34; we have shown here a step-down switching converter. The transistor T_1 acts as a switch which opens or closes rapidly. The output depends on the duty cycle (ON-OFF period) of the switching transistor. The circuit consists of a transistor T_1, free wheeling diode D_1, coil L_1, condenser C_1, voltage comparator and the ramp generator.

The ramp generator is a generator of square-wave pulses of high frequency in the range of 15 of 20 kHz. These high frequency pulses are fed to the base of the transistor T_1 closing or opening the circuit at a high frequency. The output voltage value may be adjusted by properly adjusting the duty cycle of the switching transistor. This output voltage is compared to a reference voltage (V_{ref}) provided by the Zener diode and if the output suddenly changes an error voltage is produced. This error voltage is given to the ramp generator and the duty cycle of the transistor is adjusted (on increasing or decreasing the ON-time) so that the change is nullified.

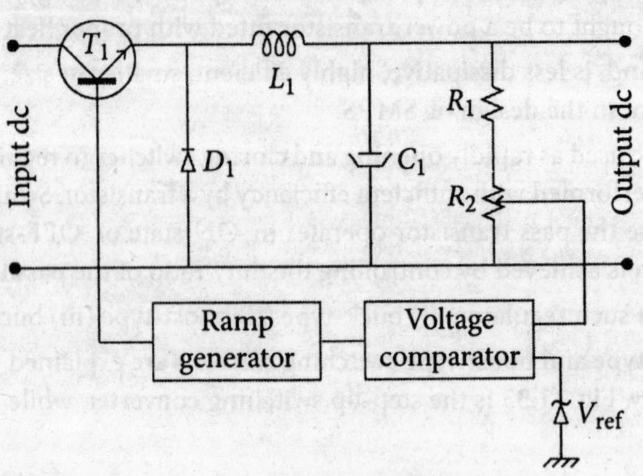

Fig. 11.35

The full schematic of an SMPS used in a conventional circuit of a CTV is shown in Fig. 11.36 and the actual practical circuit is shown in Fig. 11.37. In this circuit we have provisions for two voltages, one main supply voltage of 110 volts, the other of 12–20V. This operates off the main and no bulky power transformer is needed.

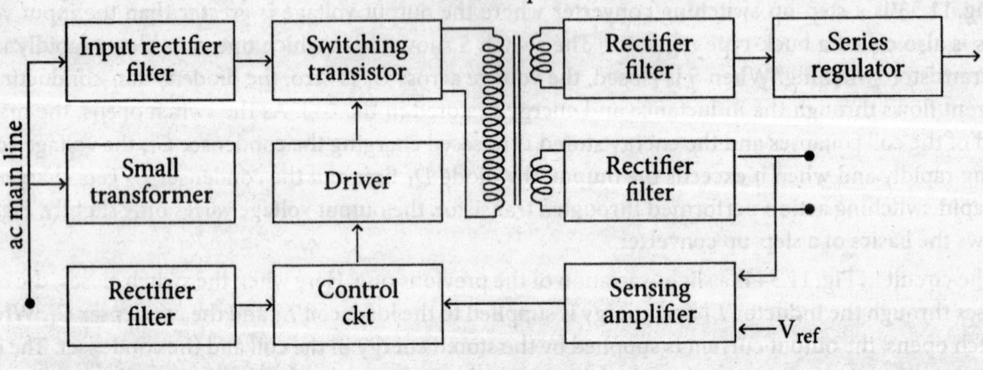

Fig. 11.36

The main ac is rectified and filtered. The unregulated dc voltage from the rectifier filter section is interrupted by making rapid make and break by the switching transistor at a high frequency. The interrupted dc is applied to the primary of the SMPS transformer T_r and the voltage in the secondary is rectified and filtered. A part of the voltage output is fedback to the sensing amplifier which compares it with a reference voltage (V_{ref}) provided by the Zener diode of proper Zener voltage. If there is any difference, an error voltage is produced and it is given to the control circuit which by adjusting the duty cycle of the switching transistor (if the output increases, the control circuit reduces the ON-time and vice versa) via the driver stage stabilises the output.

The Fig. 11.37 shows the essence of practical SMPS which is more or less universally employed in all CTVs with slight changes here and there.

The circuit shown provides two outputs of voltages + 110V and the other of 20V dc.

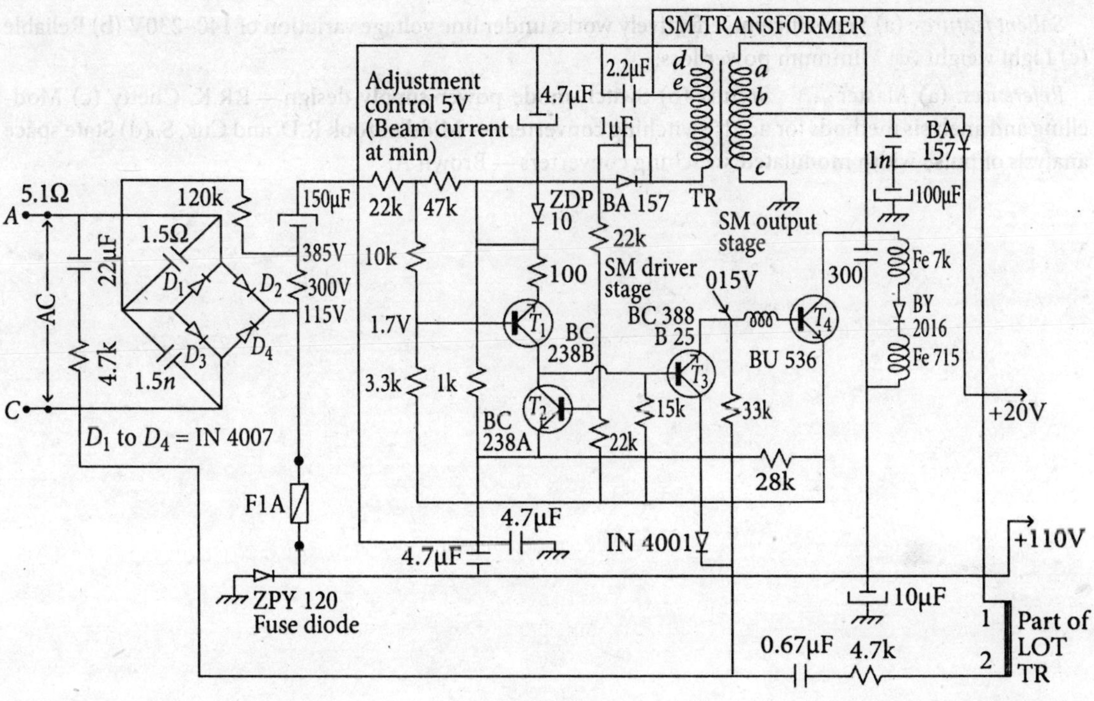

Fig. 11.37

Four transistors T_1, T_2, T_3 and T_4 are used here; the first transistor T_1 (BC 238B) is a regulator transistor; the second T_2 is BC 238A, it acts as an electronic fuse for protection of the circuit. The third transistor T_3 (BC 238 25) acts as the driver feeding the control voltage to the switching transistor T_3 (BC 238 25) acts as the driver feeding the control voltage to the switching transistor T_4 (BU 536).

As the ac main is turned on, the output transistor gets the base voltage through resistor 5.1Ω, capacitor 22μF, resistor 4.7k, 33Ω and then finally through coil L_1.

The transistor T_4 is immediately switched on. As it is on, the SMPS transformer starts functioning. The SMPS transformer primary has the tappings a, b, c while the tappings d, e, f belong to the secondary. The voltage induced between the tappings d and e is fed to the base of T_4 via the LOT (a portion of the EHT of the CTV), resistor 4.7Ω, capacitor 0.47μF, resistor 33Ω and coil L_1. Thus, oscillation starts. In this circuit the oscillation frequency is 15,625 cycles/s. Between d and f of the SMPS transformer a diode BA157 is connected producing a negative voltage of 11.9V, filtered by the capacitor 4.7μF placed across it. This negative voltage is applied to the base of T_1 (BC 238B) through 4.7k potentiometer and resistor 10k. A Zener ZDP10 of 10V breakdown voltage is connected in the emitter of this transistor, the anode of the diode is connected to the −11.9V supply line. As soon as the transistor T_4 (BU536) is thrown into action a voltage is developed across the 22Ω resistor connected in its emitter. This voltage is feed to the base of T_1 (BC 238D) through 3.3k. A negative going sawtooth pulse is available at the collector of T_1 which is connected directly to the base of T_3 (BC 238B 25). T_3 being a *pnp* transistor it is switched on, shorting thereby the base and emitter of T_4 to which it is connected. Thus, the transistor T_4 is switched off.

Since the switching is done by transistor, no big power transformer is needed. This circuit can work at very low voltage, low power loss, no big filter capacitor is needed in the SMPS circuit.

Salient features: (a) Very efficient, effectively works under line voltage variation of 140–230V (b) Reliable (c) Light weight (d) Minimum power loss.

References: (a) Master TV circuits (b) Switch-mode power supply design — P.R.K. Chetty (c) Modelling and analysis methods for ac-dc switching converters — Middlebrook R.D. and Cuk, S. (d) State space analysis of pulse width modulated switching converters — Brown A.

Chapter 12

IC 555 (Bipolar Version) and IC 7555 (C-MOS Version) Timer: IC 556 — the Dual Version of IC 555

12.1 Introduction

IC 555 or IC 7555 is widely used as a timer IC serving various purposes so much so that it has become an industry standard.

Basically it is a combination of a relaxation oscillator, two comparators, an RS flip-flop and a discharge transistor. In order to understand the working of the device, let us first see what is a flip-flop and how it works.

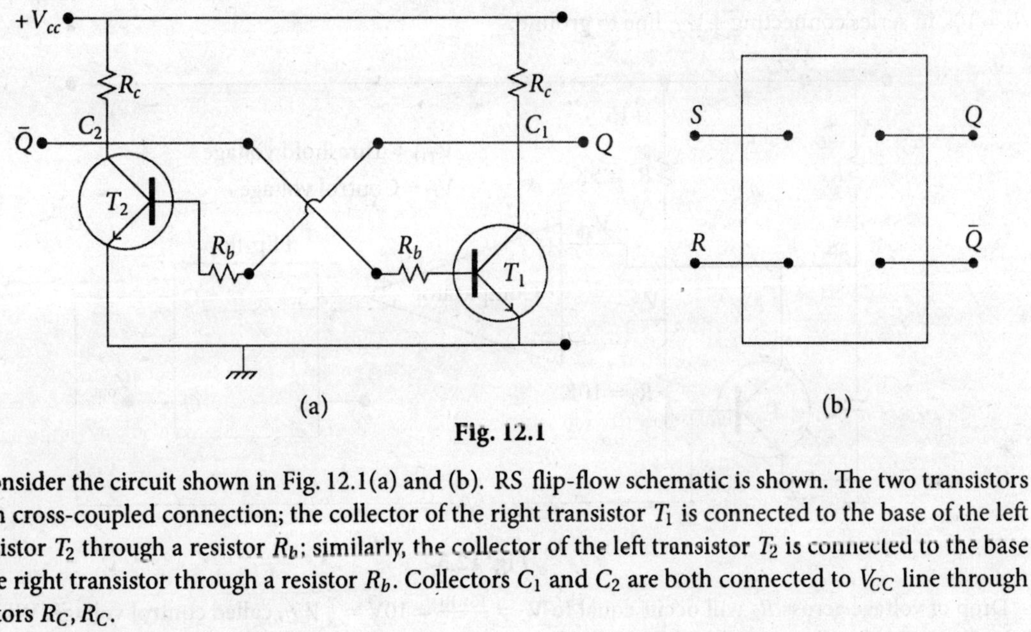

Fig. 12.1

Consider the circuit shown in Fig. 12.1(a) and (b). RS flip-flow schematic is shown. The two transistors are in cross-coupled connection; the collector of the right transistor T_1 is connected to the base of the left transistor T_2 through a resistor R_b; similarly, the collector of the left transistor T_2 is connected to the base of the right transistor through a resistor R_b. Collectors C_1 and C_2 are both connected to V_{CC} line through resistors R_C, R_C.

12.2 How does the flip-flop work?

Even assuming the two transistors to be of same make, the two cannot be exactly identical in behaviour, one of them conducting relatively more than the other. Let the right transistor conduct more at the moment the power is on. It will reach faster the full conduction state than the other. So the right transistor collector will soon attain ground potential and base bias to the other transistor will be nearly zero. Thus, collector voltage of the right transistor will be zero and the collector voltage of the other transistor will be high. This is stated as $Q = 0, \bar{Q} = 1$. After this situation, the table turns. T_2 now passes into full conduction state rendering its collector voltage zero ($\bar{Q} = 0$), so no base bias is supplied to the transistor T_1, which is held off, so the potential at the collector of T_1 is high ($Q = 1$). It is evident from this discussion that the outputs Q and $\bar{Q}$ are complementary.

Truth table

Q	$\bar{Q}$
0	1
1	0

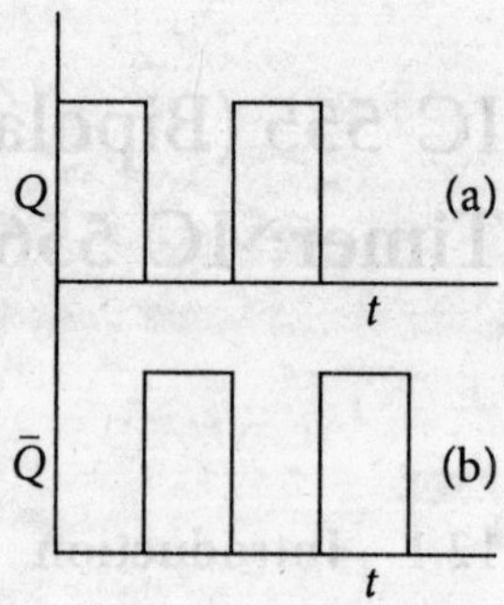

Fig. 12.2

This complementary behaviour is exhibited in the table above.

Graphical representation of the behaviour of the two transistors T_1 and T_2 is shown in Fig. 12.2(a) and (b) respectively. The diagram shows that Q and $\bar{Q}$ are complementary.

12.3 How does the concept of time come in? At regular intervals of time, the flip-flop sets and resets?

Consider Fig. 12.3, where R_C and C are the timing elements; there is a resistive network containing $R_1 = 5\text{k}$, $R_2 = 10\text{k}$ in series connecting $+V_{CC}$ line to ground.

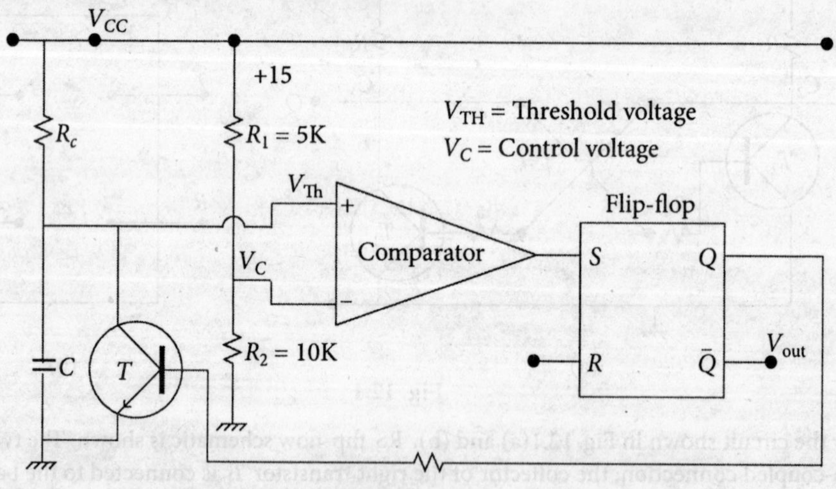

Fig. 12.3

Drop of voltage across R_2 will occur equal to $V_C = \frac{15\times10}{15} = 10\text{V} = \frac{2}{3}V_{CC}$, called control voltage. When Q is high, the RS flip-flop is said to set. This high output Q saturates the transistor T, and the capacitor C is shorted and the process of charging stops. If the capacitor is initially charged, it discharges through T in saturation.

Thus, the voltage to the non-inverting terminal of the comparator Op-Amp is now zero. The voltage to the non-inverting terminal is called threshold voltage (V_{Th}). Voltage to the inverting terminal is always equal to the voltage across R_2 and this voltage is the control voltage (V_C), which is in the present case equal to 10V $\left(=\frac{2}{3}V_{CC}\right)$.

In the circuit shown in Fig. 12.3, we have denoted V_{Th} = threshold voltage, V_C = control voltage.

R_C and C are the timing elements; R_1 and R_2 are in this case taken in the ratio 1 : 2 so that the potential divider consisting of R_1 and R_2 supplies a constant voltage $V_C = \frac{2}{3}V_{CC}$ to the inverting terminal of the comparator Op-Amp.

T = discharge transistor

The Q output is either low or high depending on which transistor is saturated. When Q is high, the flip-flop is said to set and it is reset when it is low. The two transistors T_1 and T_2 [Fig. 12.1(a)] latch such as to give Q output low and when low, the discharge transistor T is off and charging of the capacitor C goes on until the voltage across the capacitor denoted by V_{Th}, called the threshold voltage, is attained. When this V_{Th} slightly exceeds the control voltage V_c provided by the potential divider resistive network and applied to the inverting terminal, the circuit latches to give Q output high.

When it is high, the discharges transistor T is in saturation and the capacitor discharge to reset the device. Thus, we get setting and resetting of the flip-flop device.

The output $\bar{Q}$ is complementary to the output Q and similar arguments can be offered for setting and resetting of the device. So, if the output $\bar{Q}$ is externally available and Q is inside the chip, $\bar{Q}$ high means Q low, the charging is ON; when $\bar{Q}$ is low, Q is high and the capacitor discharges to reset the device.

What does mathematics say?

Starting from uncharged condition of the capacitor, time to reach $\frac{2}{3}V_{CC}$ value is given by t_1, called 'ON time' of the device and time to discharge the capacitor down to $\frac{1}{3}V_{CC}$ is termed as the 'OFF time' of the device. So the period of the device $T = t_1 + t_2$.

So the problem with us is a physical problem of charging and discharging of a capacitor through a resistor between $\frac{1}{3}CV_{CC}$ and $\frac{2}{3}CV_{CC}$ amount of charge.

To find the charging time t_1 (In case of Monostable mode). [1]

According to theory, $q = q_0\left(1 - e^{-t/CR}\right)$, where q is the charge on the capacitor at time instant t, q_0 is the charge that can be stored in the capacitor.

$$q_1 = \tfrac{1}{3}CV_{CC},\ q_2 = \tfrac{2}{3}CV_{CC}.$$

So,

$$\tfrac{2}{3}CV_{CC} = CV_{CC}\left(1 - e^{-t/CR}\right).$$

Now, $t = t_1$ when $V = \frac{2}{3}V_{CC}$, hence,

$$\tfrac{1}{3}V_{CC} = \tfrac{2}{3}V_{CC}\left(1 - e^{-t_1/CR_C}\right) \Rightarrow e^{-t_1/CR_C} = \tfrac{1}{2}$$

$$\Rightarrow t_1 = CR_C \ln 2 = 0.693CR_C$$

$$\tfrac{1}{3}V_{CC} = V_{CC}\left(1 - e^{-t_1'/CR_C}\right) \Rightarrow e^{-t_1''/CR_C} = \tfrac{2}{3}$$

$$\Rightarrow t_1' = CR_C \ln \tfrac{3}{2} = CR_C \times 0.405.$$

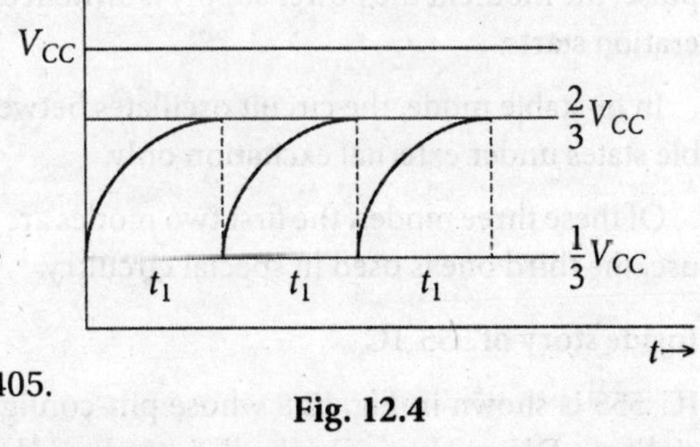

Fig. 12.4

So the time required for the capacitor to rise to $\frac{2}{3}V_{CC}$ voltage from $\frac{1}{3}V_{CC}$ is $\left(t_1 + t_1'\right) = 1.1CR_C$ = periodic time. For alternative derivation, see the solution in pp. 334 given as Mathematical analysis.

[1]For further details, refer to *Theory of Electricity and Magnetism* by the author, Scitech Publication, Chennai.

To find the decay time t_2

The time taken by the capacitor to drop to $\frac{1}{3}V_{CC}$ voltage from $\frac{2}{3}V_{CC}$ is the OFF time t_2 of the device.

The discharging of the capacitor is given by $q = q_0\, e^{-t/CR}$, here $t = t_2$.

When

$$V = \tfrac{1}{3}V_{CC} \Rightarrow C \times \tfrac{1}{3}V_{CC} = C \times \tfrac{2}{3}V_{CC}\, e^{-t_2/CR_C} \Rightarrow t_2 = CR_C \ln 2 = 0.693CR_C. \qquad (12.1)$$

The discharge of the capacitor is graphically represented by in Fig. 12.5.

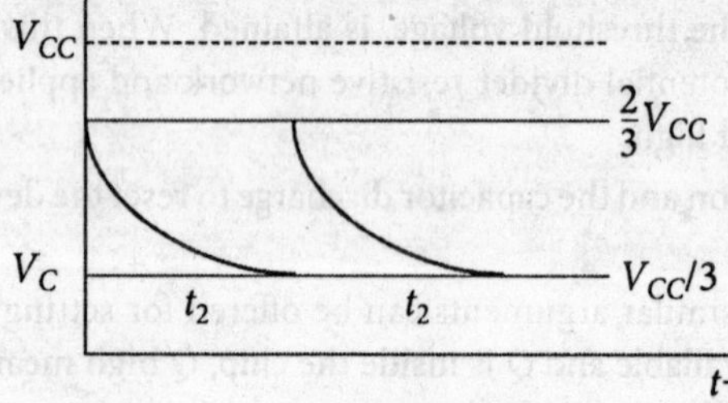

Fig. 12.5

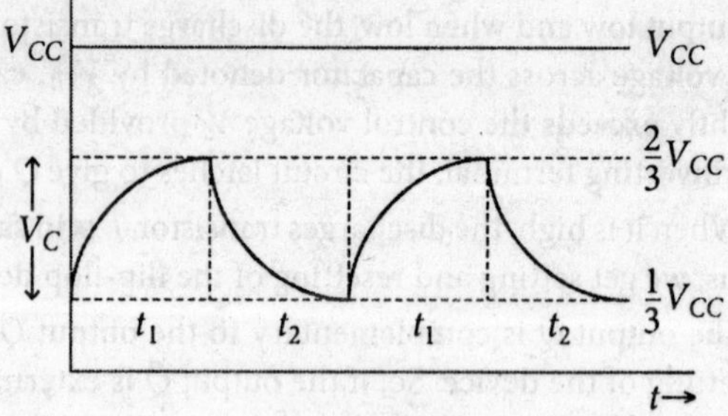

Fig. 12.6

The combined effect is graphically represented in Fig. 12.6. So the device oscillates between two voltages $\frac{1}{3}V_{CC}$ and $\frac{2}{3}V_{CC}$.

Know your timer IC

IC 555 is shown in Fig. 12.7 and IC 7555 which is the C-MOS version of 555 has identical pin-configuration.

There are three modes of connection of IC 555 in general: (a) monostable mode (b) astable mode (free running) (c) bi-stable mode.

In monostable (one shot) mode, external trigger-pulse is needed to initiate the operation.

In astable mode there is no need of any external trigger pulse; the moment the power supply is switched on, the operation starts.

In bi-stable mode, the circuit oscillates between two stable states under external excitation only.

Of these three modes, the first two modes are of frequent use, the third one is used in special circuitry.

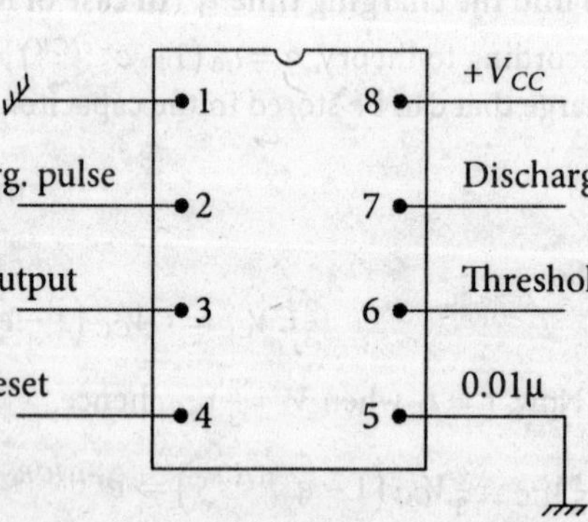

Fig.12.7

Inside story of 555 IC

IC 555 is shown in Fig. 12.8 whose pin-configuration is exhibited in Fig. 12.7. It is usually housed in an eight pin DIP package. Originally introduced by Signatics using bipolar technology, it is now available from several manufacturers in both bipolar and C-MOS technology. A series connection of three resistors, 5kΩ each, set the reference level inputs to the two comparators C_1 and C_2 at $\frac{2}{3}V_{CC}$ and $\frac{1}{3}V_{CC}$ respectively, the outputs of these comparators setting or resetting the flip-flop unit.

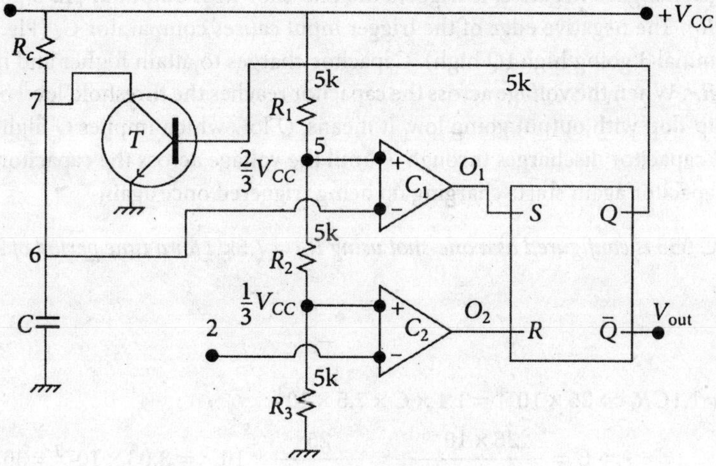

Fig. 12.8

Know the important terminals as numbered below:

Terminals	Identification
1	Substrate (grounded)
2	Trigger to be applied
3	V_{out}($\bar{Q}$ of Fig. 12.1)
4	Reset, to avoid false resetting pin 4 is usually connected to V_{CC} line
5	Control voltage
6	Threshold voltage input
7	Collector of discharge transistor connected
8	Connected to voltage line.

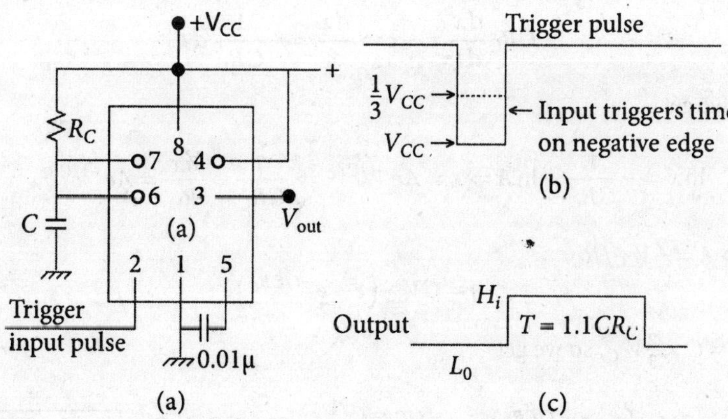

Fig. 12.9

Monostable mode. This is also called *one shot* or monostable mode where an external trigger pulse initiates the operation of the timer. Fig. 12.9 shows the connections of the pins in this mode of operation. When the trigger input signal goes negative, it triggers the one-shot with output at pin 3 going high for a time period $T = 1.1CR_C$. The negative edge of the trigger input causes comparator C_2 [Fig. 12.8] to trigger the flip-flop with terminal 3 going high ($\bar{Q}$ high). Capacitor charges to attain higher and higher voltage value through resistor R_C. When the voltage across the capacitor reaches the threshold level of $\frac{2}{3}V_{CC}$, comparator C_1 triggers the flip-flop with output going low. It means, $\bar{Q}$ low which implies Q high and transistor T is in saturation, the capacitor discharges through T until the voltage across the capacitor falls slightly below $\frac{1}{3}V_{CC}$ when the capacitor again starts charging up being triggered once again.

Example 12.1. IC 555 *is configured as a one-shot using* $R_C = 7.5\text{k}\Omega$ *for a time period of* $25\mu s$. *Find the value of C needed.*

Solution:

$$T = 1.1CR_c \Rightarrow 25\times10^{-6} = 1.1\times C\times 7.5\times10^3$$

$$\Rightarrow C = \frac{25\times10^{-6}}{1.1\times7.5\times10^3} = \frac{25}{1.1\times7.5}\times10^{-9} = 3.03\times10^{-9} = 3030\text{pF}.$$

Mathematical analysis. In this mode of operation when the trigger pulse going negative has a value less than $\frac{1}{3}V_{CC}$, the lower comparator C_2 has a high output and resets the flip-flop. The output Q becomes high ($\bar{Q}$ low) and cuts off the transistor to allow the capacitor to charge. As soon as the capacitor voltage exceeds $\frac{2}{3}V_{CC}$, the transistor turns on and discharges the capacitor very quickly. In Fig. 12.9, we have shown the trigger pulse (b) having a narrow width and depth of quiescent value V_{CC}.

The differential equation for the charge growth is written as

$$R_C i + \frac{q}{C} = V_{CC} \Rightarrow R_C\frac{dq}{dt} + \frac{q}{C} = V_{CC} \Rightarrow \frac{dq}{dt} + \frac{q}{CR_C} = \frac{V_{CC}}{R_C} \Rightarrow \frac{dq}{dt} = \frac{V_{CC}}{R_C} - \frac{q}{CR_c} = -\left(\frac{q}{CR_c} - \frac{V_{CC}}{R_C}\right).$$

Set,

$$\frac{q}{CR_C} - \frac{V_{CC}}{R_C} = x \Rightarrow \frac{dq}{dt} = CR_C\frac{dx}{dt}.$$

$$\therefore\ CR_C\frac{dx}{dt} = -x \Rightarrow \frac{dx}{x} = -\frac{1}{CR_C}dt.$$

On integration, we get

$$\ln x = -\frac{t}{CR_C} + \ln A \Rightarrow x = Ae^{-t/CR_C} \Rightarrow \frac{q}{CR_C} - \frac{V_{CC}}{R_C} = Ae^{-t/CR_C}.$$

At $t = 0, q = 0 \Rightarrow A = -V_{CC}/R_C$.

$$\therefore\ q = CV_{CC}\left(1 - e^{-t/CR_C}\right). \tag{12.2}$$

Let $t = T$ and $q = C\times\frac{2}{3}V_{CC}$, so we get

$$C\times\frac{2}{3}V_{CC} = CV_{CC}\left(1 - e^{-T/CR_C}\right) \Rightarrow e^{-T/CR_C} = \frac{1}{3} \Rightarrow T = CR_C\ln 3 \Rightarrow \boxed{T = 1.1CR_C}. \tag{12.3}$$

Equation (12.3) determines the output pulse width.

The different voltage levels and waveform in case of monostable timer are shown in Fig. 12.10.

	UC	LC
Control voltage	$\frac{2}{3}V_{CC}$	$\frac{1}{3}V_{CC}$

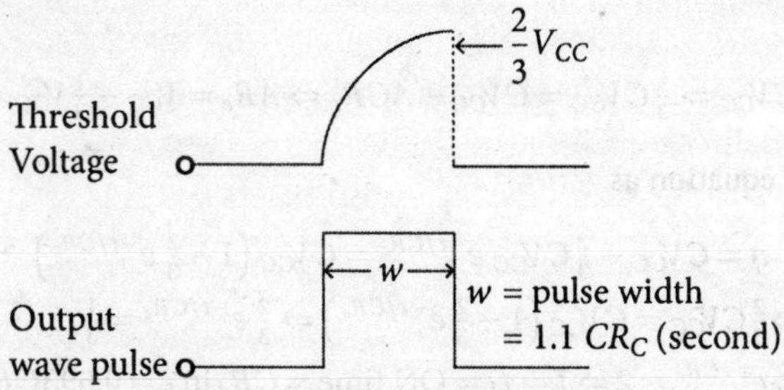

Fig. 12.10

Astable operational mode

For most practical purposes, a 555 IC is configured in astable mode. The pin connection in this mode is shown in Fig. 12.11.

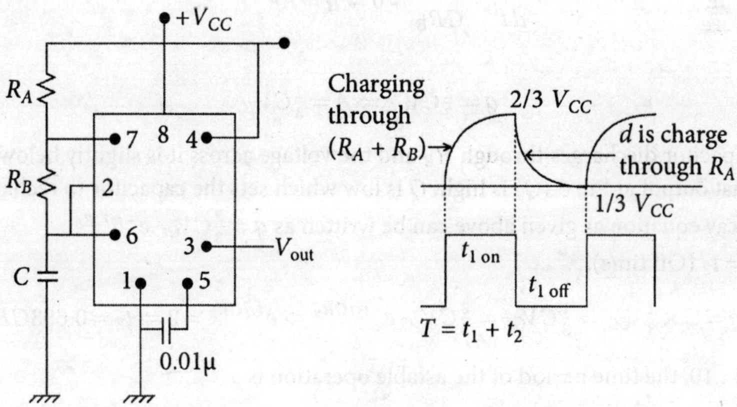

Fig. 12.11

Capacitor C charges through $(R_A + R_B)$ until the voltage across the capacitor slightly exceeds $\frac{2}{3}V_{CC}$ which is the threshold voltage at Pin 6. Thus, attainment of this threshold voltage drives the comparator C1 [Fig. 12.8] to trigger the flip-flop so that the output at Pin 3 ($\bar{Q}$) goes low; Q is high, the transistor T sets to discharge the capacitor through R_B. The discharge will go on until the voltage falls below $\frac{1}{3}V_{CC}$ when the output at Pin 3 ($\bar{Q}$) goes high; Q is low and the transistor resets to allow the capacitor to charge.

Thus, charging of the capacitor takes place through $(R_A + R_B)$ but discharge takes place through the resistor R_B only.

How would you look into the problem mathematically?

Charge growth on the capacitor plates. According to the theory of growth of charge on the capacitor plates[2],

$$Ri + \frac{q}{C} = V_{CC} \Rightarrow (R_A + R_B)\,i + \frac{q}{C} = V_{CC} \Rightarrow R_e\frac{dq}{dt} + \frac{q}{C} = V_{CC} \Rightarrow \frac{dq}{dt} = \frac{V_{CC}}{Re} - \frac{q}{CR_e}.$$

[2]Theory of *Electricity and Magnetism* by the author, Scitech Publication, Chennai.

In this case, the charge oscillates between $C \times \frac{1}{3} V_{CC}$ and $C \times \frac{2}{3} V_{CC}$. On solution, we get $q = CV_{CC} - ACR_e\, e^{-t/CR_e}$, where $R_e = R_A + R_B$.

When $t = 0$,

$$q = \tfrac{1}{3} CV_{CC} \Rightarrow \tfrac{1}{3} CV_{CC} = CV_{CC} - ACR_e \Rightarrow AR_e = V_{CC} - \tfrac{1}{3} V_{CC} = \tfrac{2}{3} V_{CC}.$$

So, we get the charging equation as

$$q = CV_{CC} - \tfrac{2}{3} CV_{CC}\, e^{-t/CR_e} = CV_{CC}\left(1 - \tfrac{2}{3} e^{-t/CR_e}\right) \qquad (12.4)$$

$$\Rightarrow \tfrac{2}{3} CV_{CC} = CV_{CC}\left(1 - \tfrac{2}{3} e^{-t/CR_e}\right) \Rightarrow \tfrac{2}{3} e^{-t/CR_e} = \tfrac{1}{3}$$

$$\Rightarrow e^{t-CR_e} = 2 \Rightarrow t = t_1 = \text{ON time} = CR_e \ln 2 = 0.693CR_e, \qquad (12.5)$$

where $R_e = R_A + R_B$.

To find t_2 (OFF-time)

The differential equation representing the discharge of the capacitor through resistor R_B is given as

$$\frac{dq}{dt} + \frac{q}{CR_B} = 0 \Rightarrow q = Ae^{-t/CR_B}.$$

At $t = 0$,

$$q = \tfrac{2}{3} CV_{CC} \Rightarrow A = \tfrac{2}{3} CV_{CC}.$$

Now, as the capacitor discharges through R_B and the voltage across it is slightly below $\frac{1}{3} V_{CC}$, the circuit rejuvenates so that output at Pin 3 ($\bar{Q}$) is high; Q is low which sets the capacitor to recharge.

The charge decay equation as given above can be written as $q = \frac{2}{3} CV_{CC}\, e^{-t/CR_B}$.

Now, when $t = t_2$ (Off time),

$$q = C \times \tfrac{1}{3} V_{CC} \Rightarrow \tfrac{1}{3} CV_{CC} = \tfrac{2}{3} CV_{CC}\, e^{-t_2/CR_B} \Rightarrow e^{t_2/CR_B} = 2 \Rightarrow t_2 = 0.693CR_B$$

So, from Fig. 12.10, the time period of the astable operation is

$$T = t_1 + t_2 = 0.693(R_A + R_B)C + 0.693R_BC$$

$$\Rightarrow \boxed{T = 0.693(R_1 + 2R_B)\,C} \qquad (12.6)$$

$$\Rightarrow \boxed{f = \frac{1}{T} = \frac{1}{0.693(R_A + 2R_B)C} = \frac{1.44}{(R_A + 2R_B)C}} \qquad (12.7)$$

12.4 Solved Theory and Numerical Problems

1. What do you mean by duty cycle? What is its implication?

Ans. Duty cycle D is defined as ratio of the t_{ON} to the sum of $(t_{ON} + t_{OFF})$.

$$\therefore D = \frac{t_1}{t_1 + t_2} \times 100\%.$$

If D is greater, it implies greater asymmetry in the output waveform. Take $t_1 = w =$ pulse width $= 2 \times 10^{-3}$ s, $T = 2.5$ m/s, then

$$D = \frac{w}{T} \times 100 = \frac{2 \times 10^{-3}}{2.5 \times 10^{-3}} \times 100 = 80\%.$$

In terms of R_A and R_B,

$$D=\frac{R_A+R_B}{R_A+2R_B}.$$

If $R_B \gg R_A, D=\frac{1}{2}\times 100=50\%$.

2. A 555-timer is connected for monostable operation. If $R_C=10\,k\Omega, C=0.022\mu F$, find the width of the output pulse.

   Solution: $w=1.1R_C C=1.1\times 10\times 10^3\times 0.022\times 10^{-6}=11\times 0.022\times 10^{-3}=0.24\,ms$.

3. Explain physically the working of a 555-timer.

   Ans. Refer to Fig. 12.8. There are two comparators C_1 and C_2 and the fixed inputs to the non-inverting terminals are $\frac{2}{3}V_{CC}$ and $\frac{1}{3}V_{CC}$ respectively, where V_{CC} is the supply voltage (5V–15V). The inputs to the inverting terminals of the two comparators are the voltages across the capacitor C and the trigger voltage at Pin 2 respectively. The fixed voltages to the non-inverting terminals are known as control voltage. When the threshold is slightly greater than the control voltage $\frac{2}{3}V_{CC}$, Q is high ($\bar{Q}$ low), the flip-flop is said to set when the capacitor discharges through the transistor T. The capacitor clamps to ground potential and now the output of the flip-flop Q is low ($\bar{Q}$ high); the transistor T opens and the flip-flop resets so that the capacitor once again charges to $\frac{2}{3}V_{CC}$.

4. Bring out the differences between monostable operation and astable operation.

   Ans. In monostable operation, trigger pulse is needed to initiate operation but in astable operation there is no triggering for initiating the operation of the timer.

   In monostable timer the discharge of the capacitor takes place almost through zero resistance which means, the time constant of the discharge circuit is zero. So, as Q output of the flip-flop becomes high, the discharge transistor very quickly passes into saturation, so that the collector of the transistor is grounded. This is the setting of the flip-flop. So, in monostable timer, threshold voltage swings between zero and $\frac{2}{3}V_{CC}$.

   In astable timer, charging of the capacitor is through the resistance sum of the two resistors R_A and R_B; discharge of the capacitor takes place through a resistor R_B. So charging and discharging of the capacitor occur between $\frac{1}{3}V_{CC}$ and $\frac{2}{3}V_{CC}$.

   Pulse width in case of monostable timer is $T=1.1CR_C$.

   Pulse width in astable timer is $T=0.693(R_A+2R_B)$.

5. Determine the frequency and output waveform of an astable timer, where $R_A=7.5\,k\Omega$, $R_B=7.5\,k\Omega$, $C=0.1\mu F$.

   Solution: t_1 (ON time) $=0.693(R_A+R_B)C=0.693(7.5+7.5)\times 10^3\times 0.1\times 10^{-6}=0.693\times 2\times 7.5\times 10^{-3}\times 0.1=1.04\,ms$.

   t_2 (OFF time) $=0.693R_B C=0.693\times 7.5\times 0.1\times 10^{-3}=0.52\,ms$.

$$f=\frac{1.44}{(R_A+2R_B)C}=\frac{1.44}{(3\times 7.5)\times 0.1}\times 10^3=640\,Hz.$$

6. Fig. 12.12 shows a practical circuit used in making an IR-transmitter. The timer used is NE 555. Find the transmitted frequency.

   Solution:

$$f=\frac{1.44}{(R_A+2R_B)C}=\frac{1.44}{(0.22+3.6)\times 10^3\times 0.01\times 10^{-6}}=36.69\times 10^3=37\,kHz.$$

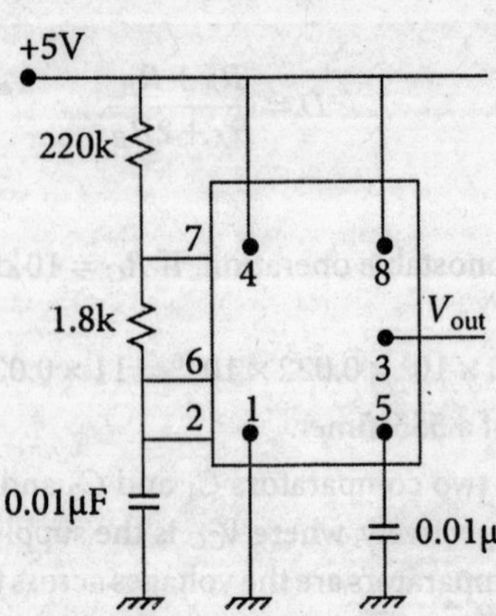

Fig. 12.12

12.5 Theory Questions Exercises

1. What is a 555-timer? Explain the principle of RS flip-flop.
2. How does the concept of time come in the function of IC 555?
3. Explain neatly how a 555-timer works. Explain the terms: Threshold Voltage, Control Voltage, Comparators.
4. What are the differences, from operational point of view, between a monostable timer and an astable timer?
5. In an astable timer, the threshold voltage is considered from $\frac{1}{3}V_{CC}$ to $\frac{2}{3}V_{CC}$ while in case of a monostable timer it is taken from zero to $\frac{2}{3}V_{CC}$. Why?
6. What is the meaning of duty cycle D of an astable timer? Under what condition it is 50 %.
7. Work out the mathematical theory of a monostable timer. Why is it called *one shot* timer?
8. Establish mathematically that time period of an astable timer is given by $T = 0.693(R_A + 2R_BC)$. Discuss duty cycle of the timer in this connection.

Chapter 13

Power Control Devices and Circuits

13.1 Silicon Controlled Rectifier

Power control circuit is defined as a circuit that can control distribution or levels of ac or dc power sources. For controlling automatically or otherwise the brilliance of a lamp, the speed of a motor, temperature of a heating device or loudness of audio-signals, a power control circuit is used.

There is a large number of elements which are used in power control applications starting from simple mechanical switches or electromechanical switches like relays and solenoids meant for low speed power control to solid state elements like transistors, FETs; C-MOS multiplexers; IGBT technology; SCR (Thyristors), TRIACs, DIACs or Power ICs used as high speed power controllers or magnitude controllers. Power controllers devices can be made to control power ranging from a fraction of a milliwatt to several kilowatts.

The silicon controlled rectifier (SCR) is the solid state equivalent of a thyratron and is very important in the power control device family as a circuit element. It does a great variety of functions such as rectification, inversion and regulation of power. The power control family consists of members, other than SCR, like Triacs, Diacs, SCS, etc. Gone are the days when only a variable resistor was available for control of power with tremendous amount of power loss, thanks to General Electric Company for their pioneering work on the development of SCR.

The first SCR came in 1957 as a small 35A, 200V device. Could any one think at that time that this small device will bring in a revolution in power electronics? Since then this small device has proliferated in so many different types, sizes, ratings and applications that now nobody can dare think of power control without an SCR. The SCRs are used to control power from a low watt to a very high value of power of the order of a few mega watts. They have gone out into space and deep into earth as an important component in an electronic device.

The list of applications of an SCR goes on increasing, but some of them include control of induction heating equipment at high frequency of a few kHz, ac and dc motor speed control, lighting and temperature control, welding, electromechanical refining, laser power supplies, traction and pulse modulation, etc.

The SCR is the best known member in the thyristor family. Some other members belonging to it are SCS (silicon controlled switch), LASCR (light activated silicon controlled rectifier), CSCR (complimentary SCR); GTO (gate turn off switch); unijunction transistor and the programmable unijunction transistor (PUT).

13.2 Learning about an SCR

Before applying it to do our desired task, it requires to know the architecture of the device.

An SCR is a three *pn*-junction semiconductor device consisting of the array *pn-pn* made of silicon wafer or substrate. The schematic of the construction with its circuit symbol is shown in Fig. 13.1(a).

In one type of construction, a region of *n*-type is diffused into a silicon *pnp* pellet as shown in Fig. 13.1(b).

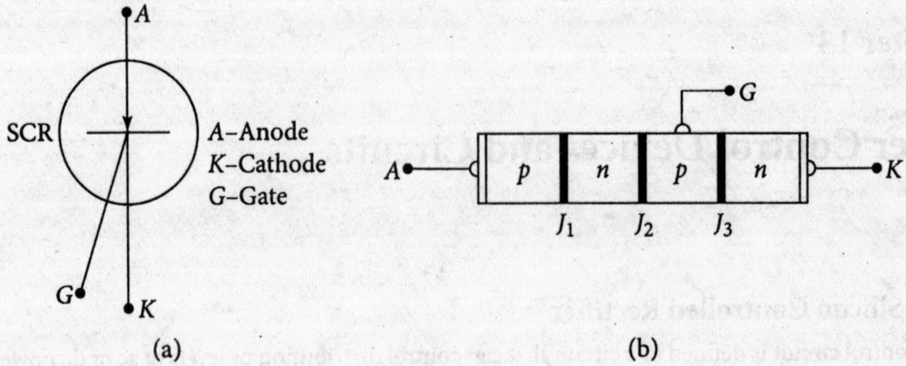

Fig. 13.1

The terminal in contact with *p* is the anode *A*, the contact terminal to the *n*-block is the cathode *K*; the gate terminal *G* is welded into the second *p*-block or the *p*-block where *n*-diffusion has been made. Silicon is used here as the substrate, because it has a small value of leakage current, so when it is off, the SCR acts as a perfect switch.

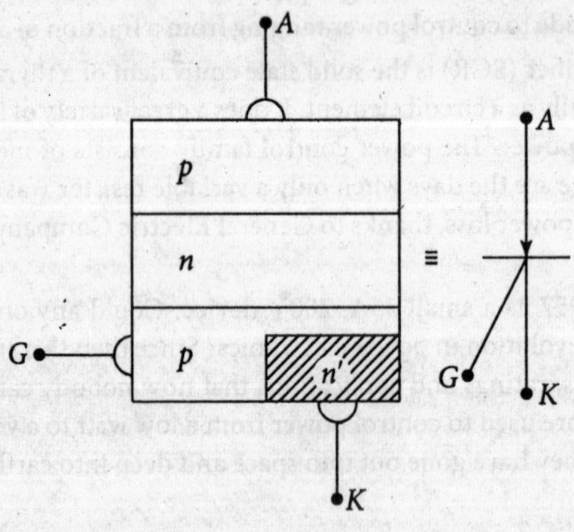

Fig. 13.2

13.3 Operation of an SCR

1. *Anode is kept at positive voltage w.r.t. the cathode but no gate voltage.* The anode is at positive potential w.r.t. the cathode and the gate voltage is zero. The junctions J_1 and J_3 are forward-biased while J_2 is in reverse bias. So no current flows through the SCR and the SCR is said to be in *off-state*. But if positive voltage on the anode is increased, there is a certain voltage V_{FBO} between the anode and the cathode when the junction J_2 breaks down and the SCR is thrown into conduction. This voltage between the positive anode and cathode for which the SCR switches from OFF-state to ON-state even in absence of any gate voltage is known as forward breakover voltage (V_{FBO}).

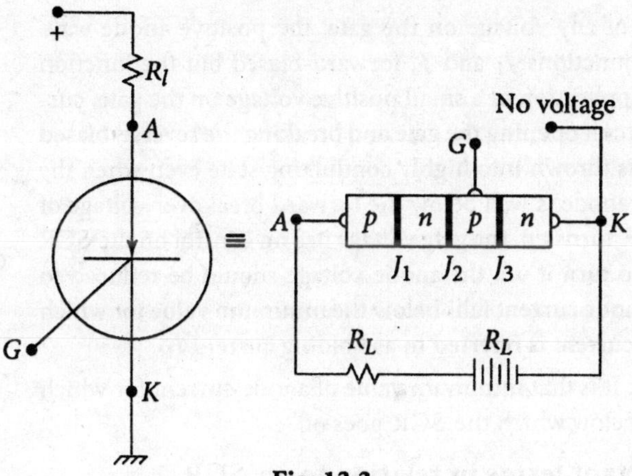

Fig. 13.3

So Anode	Cathode	Gate	State
$+V$	(−) (or zero)	0	OFF
$-V_{FBO}$	(−)	0	ON

Forward breakover voltage V_{FBO} is that positive voltage on the anode for which the SCR switches on even in absence of any gate voltage.

2. *When the anode is negative w.r.t. the cathode, gate at zero voltage.* In this case, the junctions J_1 and J_3 are in reverse bias while J_2 is in forward bias. The SCR does not conduct, it is OFF. If now the negative voltage on the anode is increased, at a certain value of the negative voltage, the junctions J_1 and J_3 are forced to break down and the SCR is thrown into high conduction state. This negative value of the anode for which the reverse-biased junctions J_1 and J_3 break is referred to as reverse breakdown voltage V_{RBV} of the SCR.

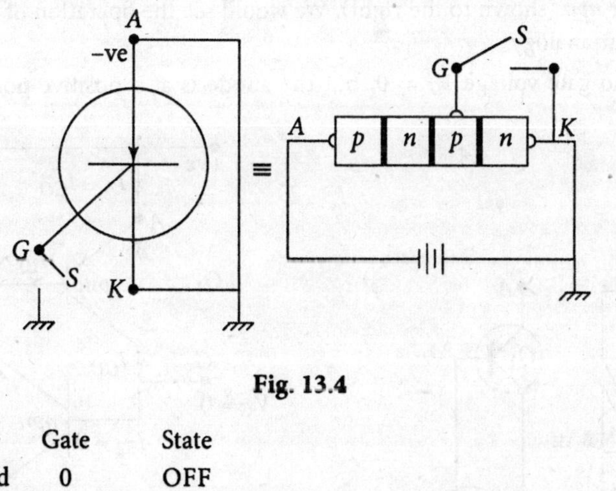

Fig. 13.4

So Anode	Cathode	Gate	State
$-V$	Grounded	0	OFF
$-V_{RBV}$	Grounded	0	Heavily ON

What is the role of the gate voltage V_G?

The voltage on the gate has far reaching consequences on the functioning of the SCR. Start from a small positive voltage on the gate [Fig. 13.5].

Prior to application of any voltage on the gate, the positive anode w.r.t. the cathode made the junctions J_1 and J_3 forward-biased but the junction J_2 reverse-biased. On application of a small positive voltage on the gate, current flows in the gate circuit opening the gate and breaking the reverse-biased junction J_2. The SCR is thrown into highly conducting state even when the positive voltage on the anode is well below the forward breakover voltage of the SCR. Once the SCR turns on, the gate voltage has no control on the SCR functioning. In order to turn it off, the anode voltage should be reduced to zero value so that the anode current falls below the minimum value for which conduction is on. This current is referred to as holding current I_H.

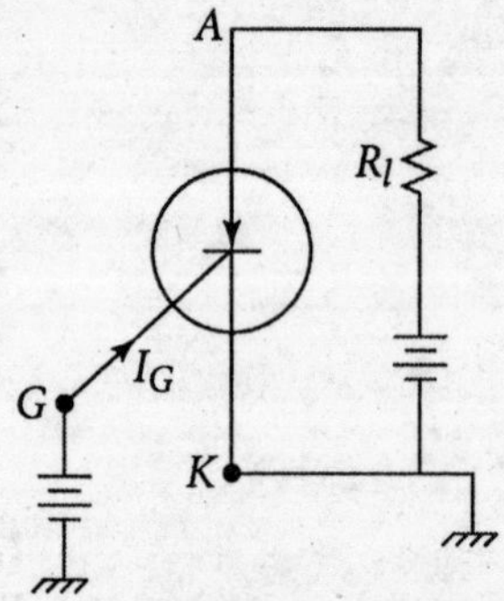

Fig. 13.5

Holding current I_H : It is that minimum value of anode current for which the SCR turns on but below which the SCR goes off.

13.4 Some important terms in relation to an SCR

1. *Break over voltage* (V_{FBO}). In absence of any gate current ($V_G = 0$), the voltage on the anode positive w.r.t. cathode required for an SCR to turn-on is called the *breakover voltage* (V_{FBO}).

2. *Reverse breakdown voltage* (V_{RBV}). When the anode is at a negative voltage w.r.t. the cathode, with gate voltage $V_G = 0$, the reverse voltage at which the SCR breaks and turns on, is referred to as reverse breakdown voltage (V_{RBV}).

3. *Holding current* I_H. This refers to the anode current at which the SCR will turn-on and below which it will not sustain conduction or the SCR turns off.

13.5 Different facts about an SCR explained

An SCR which is a combination of four semiconductor blocks *pn-pn* may be looked upon as consisting of two systems of three blocks [Fig. 13.6]. So an SCR may be taken as a combination of two transistors, one *pnp* and the other *npn* (shown to the right). We would see the operation of an SCR in terms of two transistor-combination-analogy.

Suppose, there is no gate voltage $V_G = 0$, but the anode is at a positive potential w.r.t. the cathode [Fig. 13.7].

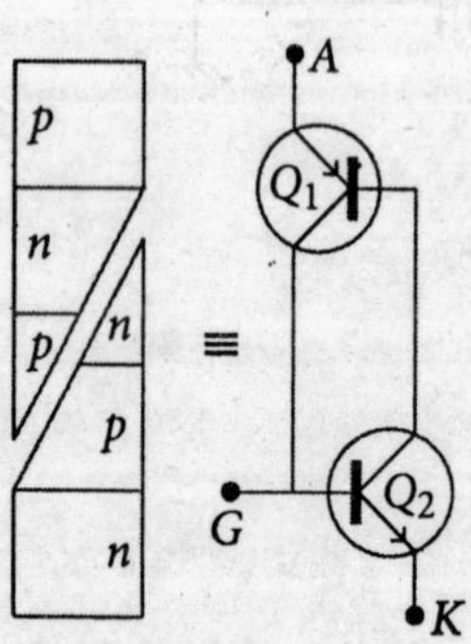

Fig. 13.6

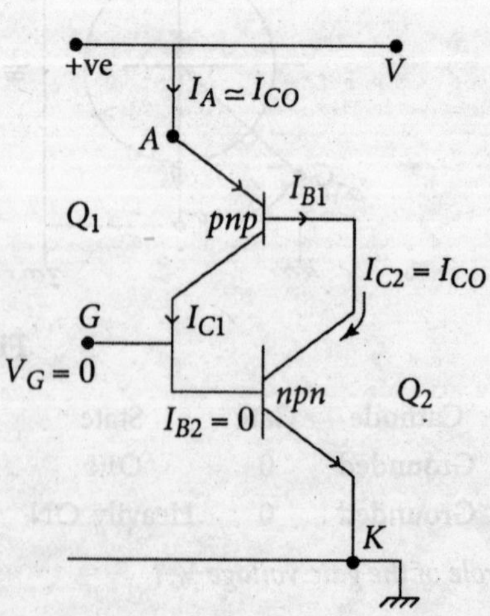

Fig. 13.7

Since $V_G = 0$, the gate terminal may be supposed to be grounded, hence the *pnp* transistor Q_1 is ON, but the *npn* transistor Q_2 is in cut-off position, because $I_{B2} = 0$. So $I_{B1} = I_{C2} = I_{CO}$ will be representing the anode current of the SCR. This current is of the order of nano ampere (10^{-9} A) which is much lower than the holding current, hence the SCR is in OFF-state. So, if the voltage V is below the forward breakover voltage (V_{BFO}) of the SCR, it remains OFF offering high impedance between the anode and the cathode. This situation is shown in Fig. 13.8 equivalent to latch open. But if the positive applied voltage $V > V_{FBO}$, the forward breakover voltage, an anode current $I_A > I_H$ is forced past through the device and the SCR is ON (latch closes).

V_G is kept slightly positive [Fig. 13.9]. Even when the applied voltage V is much below V_{FBO}, a small positive voltage to the gate makes the SCR fire [Fig. 13.9].

Fig. 13.8

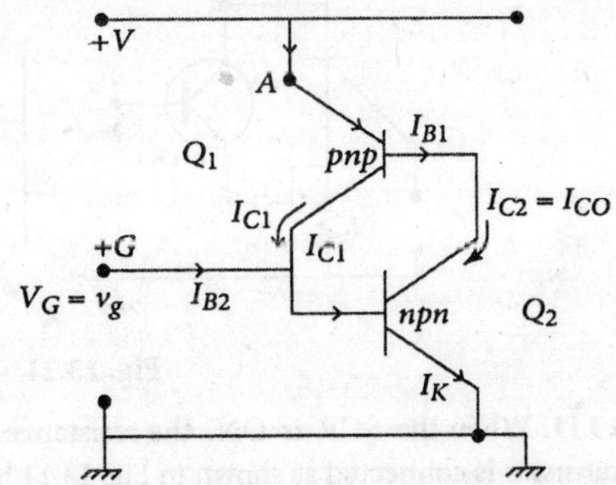

Fig. 13.9

If $V_G = v_g$ is sufficient to turn the transistor Q_2 ON, its collector current rises, which implies the base current of Q_1 to increase and it also turns on. As Q_1 turns on, its collector current I_{C1} increases which in turn increases the base current of Q_2, thus increasing its collector current further. So, once the action sets in, both of Q_1 and Q_2 are in a cumulative manner pushed to heavy conducting state in a very short time, called turn-on time which typically may vary from 0.1 to 1μs. However, high power SCRs may have turn-on time ranging from 10 to 25μs. As both of Q_1 and Q_2 are in deep conduction state, the anode to cathode resistance (V/I_A) becomes small and the SCR is said to be fully ON, the latch is said to be closed [Fig. 13.10].

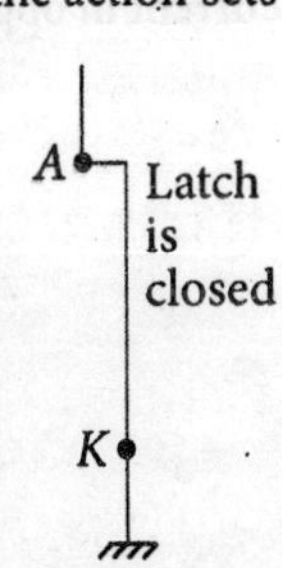

Fig. 13.10

The gate voltage initiates the process of conduction but once the anode current floods through the SCR forcing the device to be ON, the anode current does no longer obey the dictates of the gate voltage; so even reducing the gate voltage to zero value we cannot make the SCR off. In the ON-state when Q_1 and Q_2 are in full conduction, the anode current is fully determined by the circuit resistance.

13.6 How and when the SCR can be made to stop firing?

To initiate the process of firing of an SCR gate voltage is very much important but once the SCR is ON, the gate voltage loses all its control over the operation; even when it is reduced to zero value, the firing of the SCR does not stop. Then how to stop SCR from firing?

In some cases, an SCR can be turned off by making gate voltage negative. But this is not the general mechanism to make an SCR inoperative from turn-on condition.

There are in general two methods for stopping an SCR after it is triggered.

1. *Cutting the voltage source connection: Anode current interruption method.* As we know that when the SCR is ON, the anode current is determined by external circuit resistance. So, if the circuit resistance is so adjusted or the voltage source is momentarily cut-off, the anode current decreases and when it falls below the threshold value, the holding current I_H, the SCR is cut off. This method is known as anode current interruption method.

2. *Forced commutation technique.* In this method a current is sent through the SCR in opposite direction to the anode current I_A so that the current through the SCR falls below the holding current and the SCR stops functioning.

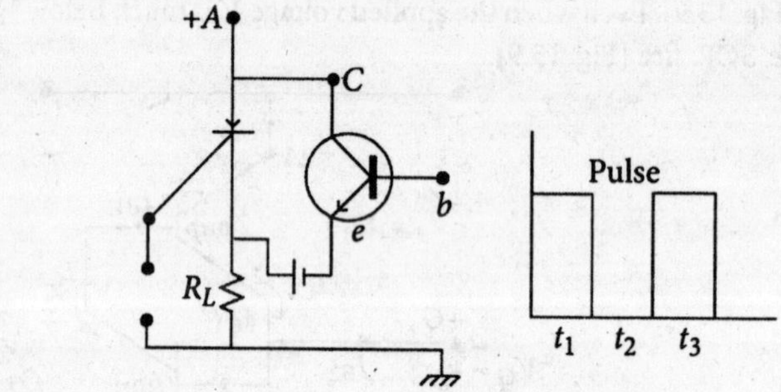

Fig. 13.11

Refer to Fig. 13.11. When the SCR is ON, the resistance between the anode and the cathode is very small. An *npn* transistor is connected as shown in Fig. 13.11 having a dc voltage source in the emitter.

By means of a pulse generator, the voltage pulse shown to the right top of Fig. 13.11 is generated and it is impressed on the base of the transistor. When the voltage pulse is of zero value, the base bias is absent and transistor is cut-off. The SCR maintains its ON-state. When the pulse voltage appears at the base, the transistor is triggered into conduction and the full voltage of the dc source appears across the SCR sending current in opposition to the anode current I_A thus the SCR is put off.

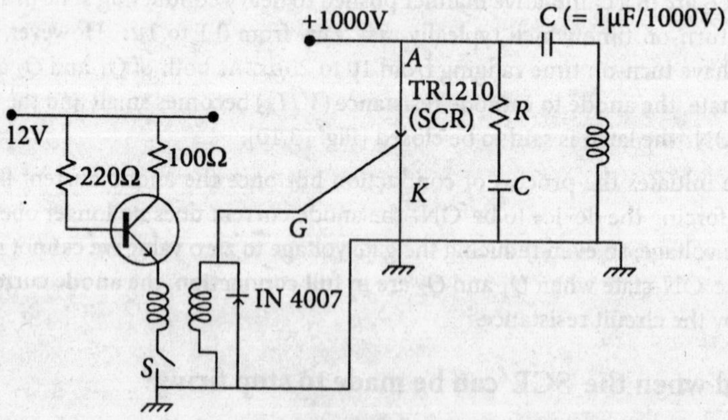

Fig. 13.12

3. *A practical circuit for commutation of the* SCR. The circuit is meant for charging and discharging of a condenser. The condenser is charged to 1000V by a dc voltage source through the inductance *L*. The subsidiary circuit consists of an *npn* transistor powered by a separate dc voltage source (12V) and the secondary of a small step-down transformer is connected in the emitter part in conjunction with a make and break switch S. During break of the circuit, a certain voltage pulse appears at the gate *G* of the SCR and

it fires at once. The condenser C discharges through the inductor. After each discharge, the anode voltage becomes zero or even negative, because of the inductance L and thus the SCR is put off. So the charging and discharging action of the condenser goes on due to conduction (ON) or no conduction (OFF)-state of the SCR.

13.7 SCR Characteristics

The essential features of an SCR characteristics are shown in Fig. 13.13. By characteristics we mean relationship between anode current and anode voltage taking gate current as a parameter ($I_g = 0, I_{g1}, I_{g2}$, etc.)

The characteristics V vs. I_A are drawn at $I_g = 0, I_g = I_{g1}, I_g = I_{g2}$. The characteristics curves show a peculiar feature of possessing a flat portion like AB (for $I_g = 0$). Current corresponding to this flat portion for any value of I_g is the threshold current, called the holding current I_H, for that particular value of I_g. It is also evident from Fig. 13.13 that the threshold current for $I_g = 0$ is maximum, the flat portion decreases with increase in the value of I_g and culminates to a point at a certain value of I_g when the SCR switches on to heavy conduction state almost immediately. Another fact becomes evident from the plot that when the breakover voltage is lesser, greater the value of I_g.

1. *A few important aspects of the characteristic curve at* $I_g = 0$. Consider the characteristic curves shown in Fig. 13.13 and in particular the characteristic curve at $I_g = 0$. In this case, the junction J_2 is in reverse bias [Fig. 13.3], so when there is no gate current, reverse saturation current I_{CO} flows for small positive voltage between anode and the cathode. So the SCR remains off. But when $V > (V_{FBO})_0$ which corresponds to the point A, the junction J_2 breaks and current equal to the holding current flews, the SCR is ON and passes soon into heavy conduction. This $(V_{FBO})_0$ where the holding current (maximum possible current) flows for $I_g = 0$ is the breakover voltage at zero value of gate current.

When $I_g \neq 0$. As some positive pulse is applied to the gate, some gate current flows, the junction J_2 is broken and current equal to the holding current is attained for lesser value of the applied voltage. The SCR is said to be ON and in a regeneration process soon passes into highly conducting state.

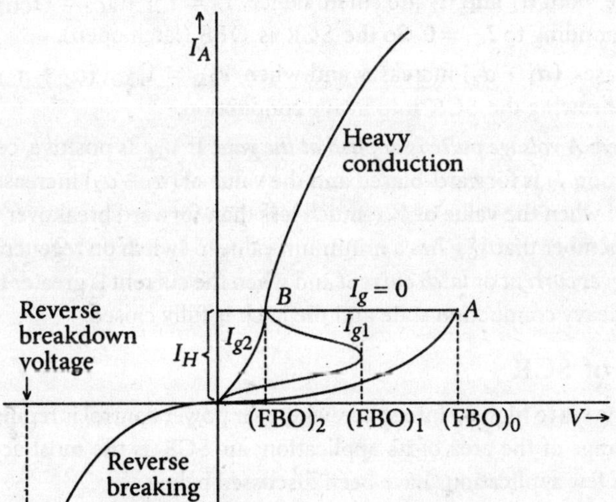

Fig. 13.13

13.8 Mathematical Viewpoint

The facts about an SCR as stated above can be mathematically established by suitable formulation.

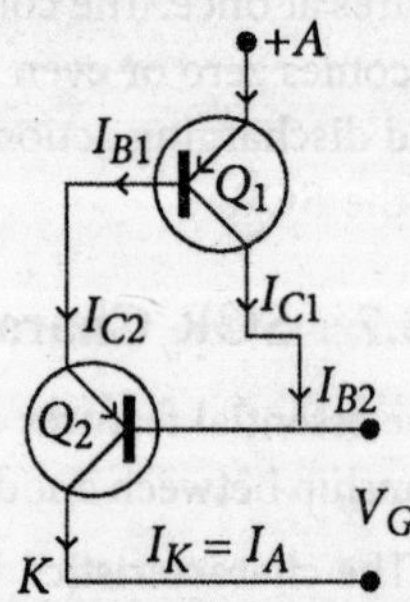

Fig. 13.14

Assuming the ON-state of the SCR, we get $I_A = I_{C1} + I_{C2}$.

If α_1 and α_2 be current gains of the transistors Q_1 and Q_2, we can write:

$$(1-\alpha_1)I_{C1} = \alpha_1 I_{B1} + (I_{CO})_1$$
$$(1-\alpha_2)I_{C2} = \alpha_2 I_{B2} + (I_{CO})_2,$$

where $(I_{CO})_1$ and $(I_{CO})_2$ are the leakage currents of Q_1 and Q_2 respectively. From Fig. 13.15, $I_{C1} = I_{B2}, I_{C2} = I_{B1}$.

Using these equations, we at once get

$$(1-\alpha_1)I_{C_1} = \alpha_1 I_{c2} + (I_{CO})_1$$
$$(1-\alpha_2)I_{C_2} = \alpha_2 I_{c1} + (I_{CO})_2$$
$$\Rightarrow (1-\alpha_1)I_{C1} + (1-\alpha_2)I_{C2} = \alpha_1 I_{C2} + \alpha_2 I_{C1} + (I_{CO})_1 + (I_{CO})_2$$
$$\Rightarrow (1-\alpha_1-\alpha_2)I_{C_1} + (1-\alpha_1-\alpha_2)I_{C_2} = (I_{CO})_1 + (I_{CO})_2$$
$$\Rightarrow (1-\alpha_1-\alpha_2)(I_{C1}+I_{C2}) = (I_{CO})_1 + (I_{CO})_2$$
$$\Rightarrow I_A = \frac{(I_{CO})_1 + (I_{CO})_2}{1-\alpha_1-\alpha_2} = \frac{(I_{CO})_1 + (I_{CO})_2}{(1-\overline{\alpha_1+\alpha_2})}. \qquad (13.1)$$

Cases of interest

Gate is open: $V_{GK} = 0$: The gate may now be supposed to be connected to ground. The *pnp* transistor Q_1 conducts but not heavily, because $I_{B1} \simeq 0 \simeq I_{C2}, I_{C1}$ is also feeble, so $I_A = I_{C1} + I_{C2}$ = sum of the leakage currents. In such a case, both α_1 and α_2 are small, hence, $I_A = I_{C1} + I_{C2} = (I_{CO})_1 + (I_{CO})_2$, less than the holding current corresponding to $I_{C1} = 0$. So the SCR is OFF (latch open).

But when V_{GK} increases, $(\alpha_1 + \alpha_2)$ increases and when $V_{BK} = V_{FBO}$, $(\alpha_1 + \alpha_2) \to 1$ and $I_A \to \infty$ and, therefore, latch closes throwing the SCR into heavy conduction.

When the gate is closed: A voltage pulse is applied at the gate. If V_{GK} is positive, certain gate current flows. If $V_{GK} > 0.7$V, the junction J_2 is forward-biased and the value of $(\alpha_1 + \alpha_2)$ increases and tends to assume a value equal to unit even when the value of V_{AK} much less than forward breakover voltage (V_{FBO}). The SCR is turned on. It is to remember that V_{GK} has a minimum value to switch on regeneration; the corresponding current is called the *trigger current* or *latch current* and when the current is greater than the holding current, the SCR is pushed to heavy conduction state and the latch is fully closed.

13.9 Application of SCR

From deep down into the sea to high up into space, wherever power control is required, an SCR is inevitably used. In the wide coverage of the area of its application, an SCR is the most acceptable member of the thyristor family. Only a few applications have been discussed below.

1. *Static switch.* An SCR can serve as an efficient electronic switch having noiseless operation and at the same time high speed operation of the order of 10^9 ON-OFF per sec.

Any SCR has only two states ON and OFF and nothing in between. When the gate is open, the SCR is in OFF-state, the switch is open. When the latching current or triggering current is supplied to the device through the gate, the SCR switches ON and it remains in the ON-state even when the gate is now opened. To switch the SCR off, the anode current is to be reduced below the holding current.

Thus, the SCR can act as a static switch opening or closing a circuit like a mechanical switch or an electromagnetic switch like a relay but here, in case of an SCR, there is no moving part, hence, it acts as long lasting switch.

2. *Phase-controlled rectifier.* In case of an ac voltage, the positive half is during the time $T/2$, the negative half is also $T/2$ giving the RMS value $V_R = \hat{V}/\sqrt{2}$. This RMS value can be varied by half-wave rectification or full-wave rectification. In half-wave rectification, we get

$$I_{\text{rms}}^2 = \frac{1}{T}\int_0^T i^2\,dt = \frac{1}{T}\int_0^{T/2} \hat{I}^2 \sin^2 \omega t\,dt + \frac{1}{T}\int_{T/2}^{T} 0\,dt = \hat{I}^2/4 \Rightarrow I_{\text{rms}} = \hat{I}/2.$$

In full-wave rectification, the RMS value is given by

$$I_{\text{rms}}^2 = \frac{1}{T}\int_0^T i^2\,dt = \frac{1}{T}\int_0^T \hat{I}^2 \sin^2 \omega t\,dt = \frac{\hat{I}^2}{T}\cdot\frac{T}{2} = \frac{\hat{I}^2}{2} \Rightarrow I_{\text{rms}} = \frac{\hat{I}}{\sqrt{2}}.$$

These RMS values may further be controlled by varying the conduction angle by means of an SCR. The normal conduction angle is from 0 to 180° for half-wave rectification and from 0 to 360° for full-wave rectification.

Phase-controlled half-wave rectifier through an SCR. By means of an SCR, the rectified output may be made to be available during 0° to 180° or from θ° to 180°. During positive half-cycle when A is positive, the diode in the gate is forward-biased and the gate current flows, its value can be controlled by R_1.

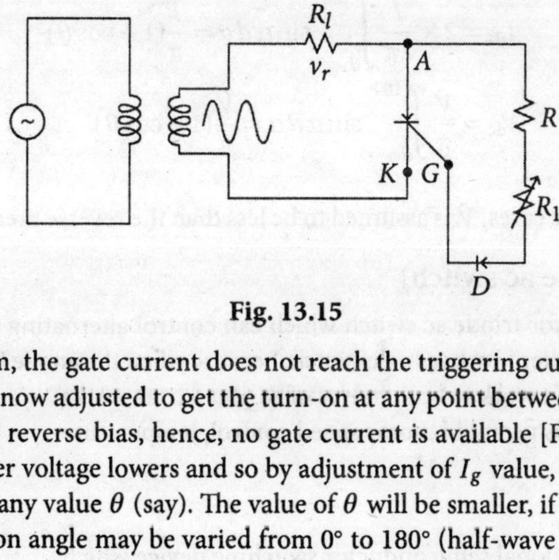

Fig. 13.15

Setting R_1 to its maximum, the gate current does not reach the triggering current value during positive half-cycle of the input. R_1 is now adjusted to get the turn-on at any point between 0° to 90° during positive half-cycle, the diode D is in reverse bias, hence, no gate current is available [Fig. 13.16]. It is to note that as I_g increases, the breakover voltage lowers and so by adjustment of I_g value, the turn-on can be made at any value θ (say). The value of θ will be smaller, if I_g is larger, so the conduction angle may be varied from 0° to 180° (half-wave rectifier) or from θ to 180°.

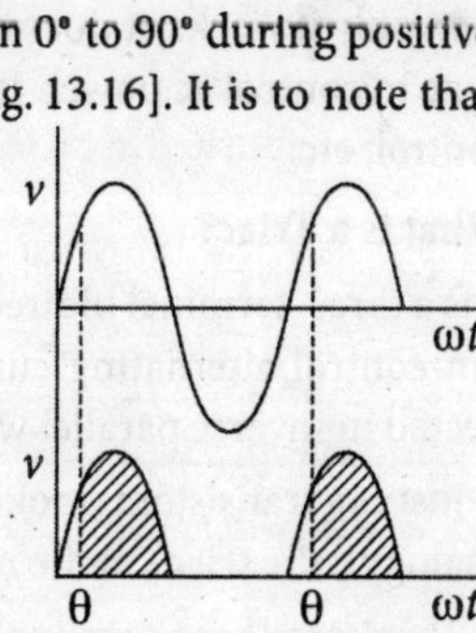

Fig. 13.16

Mathematical exposition:

$$I_{\text{dc}} = \frac{1}{2\pi}\int_0^{2\pi} i\,d\theta = \frac{\hat{I}}{2\pi}\left[\int_\theta^{180°} \sin\alpha\,d\alpha + 0\right] = \frac{\hat{I}}{2\pi}(1+\cos\theta)$$

$$V_{\text{dc}} = \frac{\hat{V}}{2\pi}(1+\cos\theta),\ \hat{V} = R_l\hat{I}.$$

Full-wave phase-controlled rectifier. For full-wave phase-controlled rectification, two SCRs are connected to the centre-tapped secondary; the cathodes of the SCRs are connected together. During positive half-cycle of the induced voltage, the firing angle of the SCR1 is controlled by R_1 between θ to 180° when the second SCR2 is off. During the other half-cycle of the induced voltage, SCR2 is fired at θ and conducts between θ and 180°. Thus, the voltage that appears across the load R_l is as shown in Fig. 13.18.

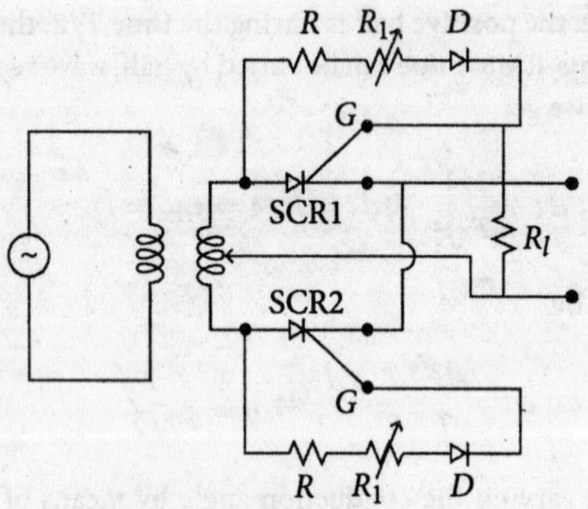

Fig. 13.17

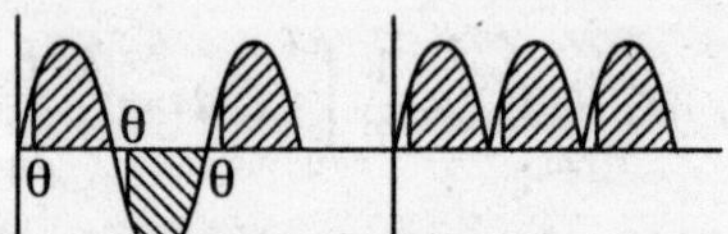

Fig. 13.18

Mathematical analysis. The first SCR1 conducts between θ to 180°, the second SCR2 between $(180+\theta)$ to 360°. The average values of current and voltage are given by

$$I_{dc} = 2\times\frac{\hat{I}}{2\pi}\int_{\theta}^{180^\circ}\sin\alpha\, d\alpha = \frac{\hat{I}}{\pi}(1+\cos\theta)$$

$$V_{dc} = \frac{\hat{V}}{\pi}\int_{\theta}^{180^\circ}\sin\alpha\, d\alpha = \frac{\hat{V}}{\pi}(1+\cos\theta).$$

Note. Remember, in all these cases, $\hat{V}$ is assumed to be less than the reverse breakdown voltage of the SCR.

13.10 TRIAC (Triode ac switch)

TRIAC is the abbreviation for triode ac switch which can control alternating current in a load. An SCR, which we have discussed earlier, works as a dc switch, because when the anode is positive w.r.t. the cathode, then only it works, it does not work in both halves of ac. But a triac works in ac to control ac power. It is useful in motor speed control, temperature control, etc.

What is a Triac?

It is a three-terminal bidirectional semiconductor switching device which can control alternating current in a load. It consists of two SCRs connected in inverse parallel with the gates connected as shown in Fig. 13.19.

Just as a transistor is looked upon as a combination of two diodes (diode-analogy), the triac can be represented by two-SCR analogy.

There are three terminals (1) MT1, junction of A_1 and K_2 (2) MT2, junction of A_2 and K_1 (3) G, the common gate, the junction of G_1, G_2.

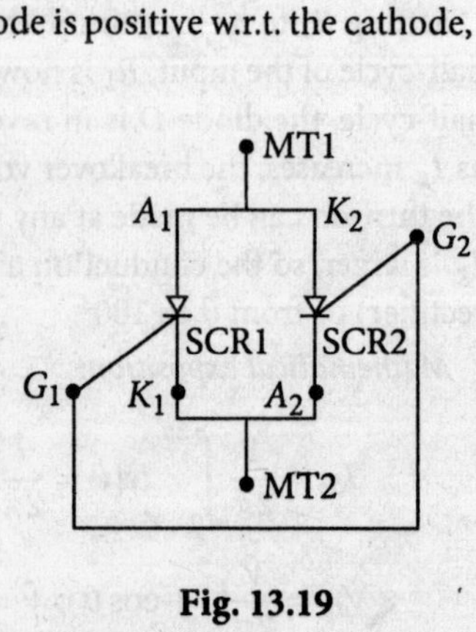

Fig. 13.19

Constructional basics:

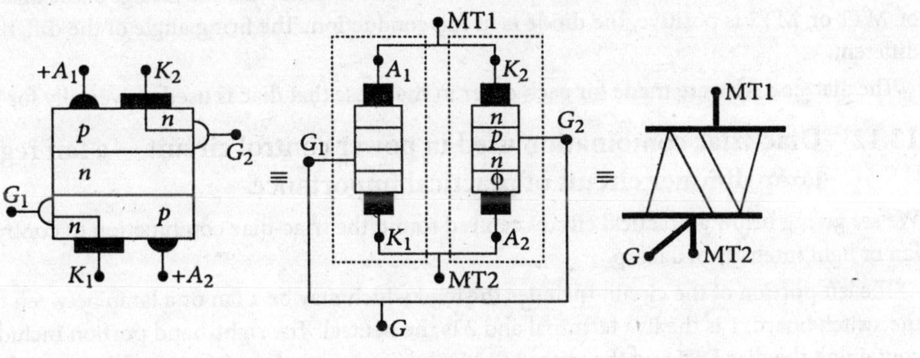

Fig. 13.20

Fig. 13.20 shows the constructional basics of the bidirectional current-control device in the name of a triac. It is in reality a parallel combination of SCRs; one SCR is formed with *pnpn* ($A_1 G_1 K_1$) and the other with *npnp* ($A_2 G_2 K_2$) [Fig. 3.20(b)]. Anode of one A_1 is connected to the cathode K_2 of the other. Similarly, the A_2 of one is connected to the cathode K_1 of the other. The two gates are connected together to form one common terminal. The circuit symbol is shown to the extreme right [Fig. 3.20(c)].

Simple triac circuit: *Lamp Dimmer*. Phase-control of power by a triac: The load may be a lamp which may be dimly lit or heavily lit by controlling the gate current by a variable resistor *R*. When the resistance is too high, the gate current may not be sufficient to turn the SCR on. But by controlling the gate current, the firing angle of the SCR may be changed. During one half-cycle of the input, one SCR is in ON-state, the other in the OFF-state. During the other half, the SCR which was off in the first half-cycle now switches on, so that lamp glows during both halves and the power available across the lamp may be varied by phase-controlling the ac input.

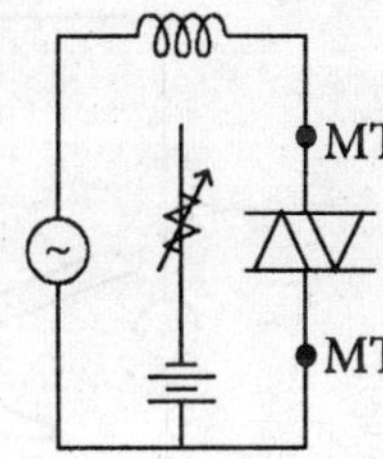

Fig. 13.21

13.11 DIAC (Diode ac switch)

It is a two-terminal bidirectional ac switching device. It is a pair of four-layer diodes (Shockley diode) connected in reverse parallel equivalent to latches which open or close simultaneously.

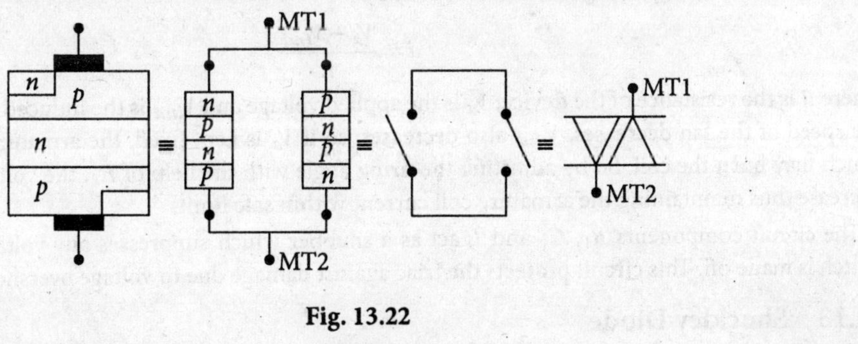

Fig. 13.22

Constructional basics: In Fig. 13.22, the constructional basics are shown having two main terminals MT1 and MT2. It may be looked upon as a triac without the gate. So here there is no triggering such that dirac will switch on when the voltage on MT1 and MT2 exceeds the breakover voltage of the diac. So, when either of MT1 or MT2 is positive, the diode is in full conduction. The firing angle of the different diacs may be different.

The diac and triac are made for each other in the sense that diac is used universally for firing a triac.

13.12 Diac-triac combination used in power control circuit — a fan regulator or a lamp dimmer circuit of practical importance

We are giving below a practical circuit centred round the triac-diac combination to control the speed of a fan or light intensity of a lamp.

The left portion of the circuit includes the load which may be a fan or a lamp between two terminals in the switch board. 1 is the live terminal and 2 is the neutral. The right-hand portion includes the regulator containing the diac D84 and the triac BT136 made for each other.

Working of the device. The diac used in the circuit is a matching pair for firing the triac BTC136.

As the switch S is closed, the capacitors C_2 and C_3 are charged during both halves of the input; the rate of charging is controlled by the variable resistor R_2. When the voltage across C_3 exceeds the breakover voltage of the diac, it is thrown into heavy conduction state. The condenser C_3 thus discharges through the gate of the triac which gets triggered and the load circuit is closed. The voltage and hence, the power delivered to the load depends on the firing angle of the triac which is again controlled by the variable resistor R_2.

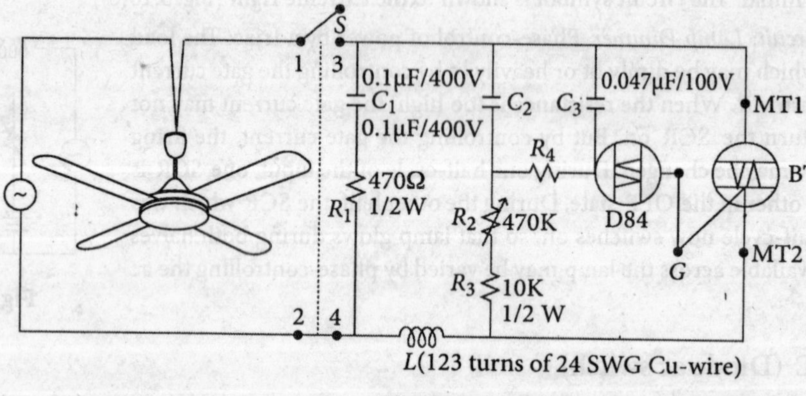

Fig. 13.23

The current flowing through the load is given by

$$i = \frac{V_a - V_{ind}}{R},$$

where R is the resistance of the device; V_a is the applied voltage and V_{ind} is the induced back emf. So, when the speed of the fan decreases, V_{ind} also decreases; so if V_a is kept fixed, the armature current increases which may burn the coil. So by adjusting the firing angle with the help of R_2, the voltage is also made to decrease thus maintaining the armature coil current within safe limit.

The circuit components R_1, C_1 and L act as a snubber which suppresses any voltage surge when the switch is made off. This circuit protects the triac against damage due to voltage overshoot.

13.13 Shockley Diode

It consists of four layers of doped semiconductors consisting of an array of *pnpn*, hence often called *pnpn* diode or Shockley diode [Fig. 13.24].

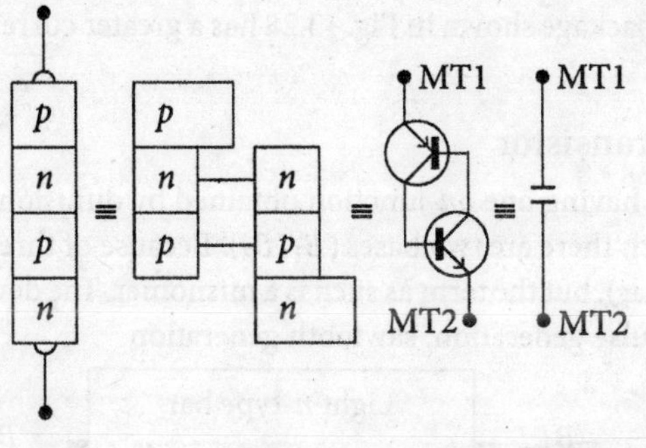

(a) (b) (c) (d)

Fig. 13.24

The basic structure is shown in Fig. 13.24 where we have shown it as a combination of four layers of semiconducting materials in the form of an array *pnpn* [Fig. 13.24(a)]; it is shown broken into two halves equivalently [Fig. 13.24(b)] and ultimately as a combination of two transistors (*pnp*) and (*npn*) in latch [Fig. 13.24(c)]. The circuit symbol is, however, shown in Fig. 13.24(d). Only two terminals are there, MT1 and MT2 and no gate terminal. So there is no gate-trigger-input to the device. The only way to set it is to apply a voltage greater than the breakover voltage of the Shockley diode. It is open when the current is less than the holding current.

Working of the diode. When the positive half-cycle of the ac input is on MT1, both the transistors are latched if the voltage exceeds the breakover voltage; but it is open if current flowing through it is less than the holding current. The device remains off in the negative half-cycle, because the peak ac voltage is less than the reverse breakdown voltage. So during positive half-cycle of the ac, the latch is closed throwing the Shockley diode into deep conduction but during the negative half-cycle the latch is open.

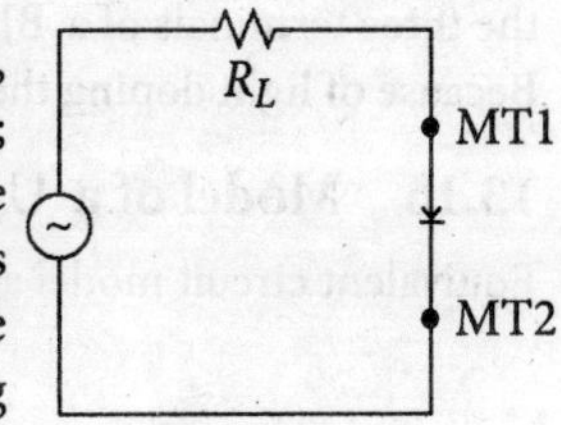

Fig. 13.25

13.14 Different SCR Packages

There are at least three distinct package for an SCR.

1. *Discrete plastic package.* In Fig. 13.26 a plastic packaged SCR is shown. Such plastic packaged SCR has a normal rating as 25A, 1000V.

2. *Stud base package.* This type of packaging is shown in Fig. 13.27 suitable for SCR having rating as 5–150A and voltage up to 3000V.

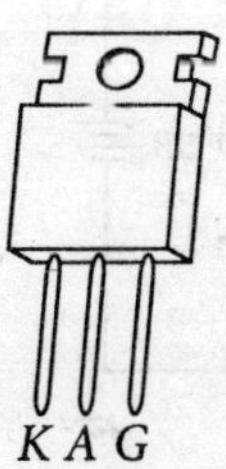

Fig. 13.26

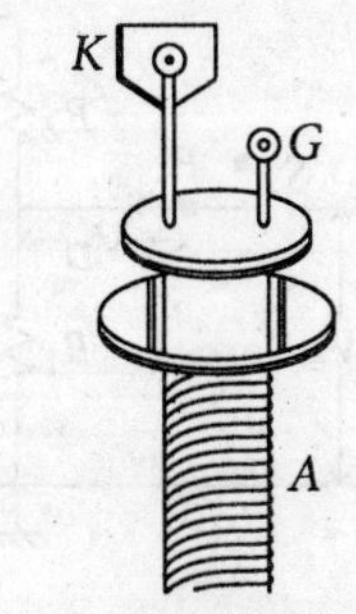

Fig. 13.27

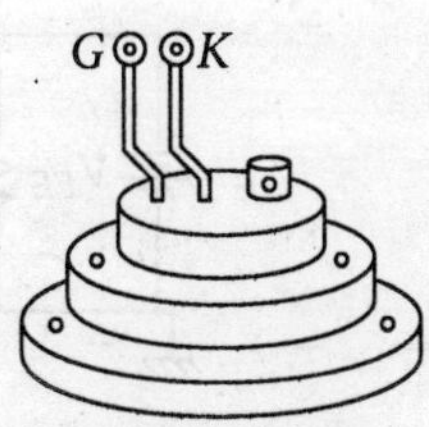

Fig. 13.28

3. *Flat base package.* This package shown in Fig. 13.28 has a greater current handling capability 10–400A, voltage 3000V.

13.15 Uni-junction transistor

It is a three-terminal device having one *pn*-junction obtained by diffusion of one *p*-type region into an *n*-type bar, at two ends of which there are two bases (B_1, B_2). Because of three terminals and external look, it resembles a transistor (bipolar), but the term as such is a misnomer. The device is very suitable for switching or triggering an SCR, for pulse generation, sawtooth generation.

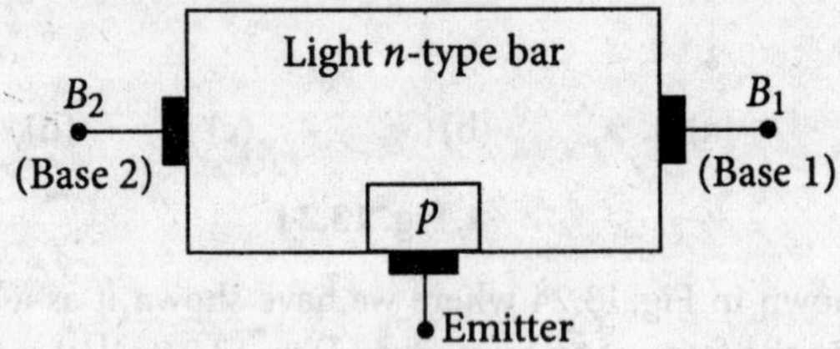

Fig. 13.29

Constructional basics. A lightly doped *n*-bar has at its two ends two terminals B_1 and B_2, called the base terminals [Fig. 13.29] and a heavily doped *p*-regions diffused into the bar almost at the middle forming what is known as the emitter (E). The two base terminals B_1, B_2 and the third terminal emitter are equivalent to the three terminals of a BJT, but since there is only one *pn*-junction, it is not a transistor device in reality. Because of light doping there is a large resistance between the two bases, ranging from 5 to 10kΩ.

13.16 Model of a UJT

Equivalent circuit model and its circuit symbol are shown in Fig. 13.30(a) and (b) respectively.

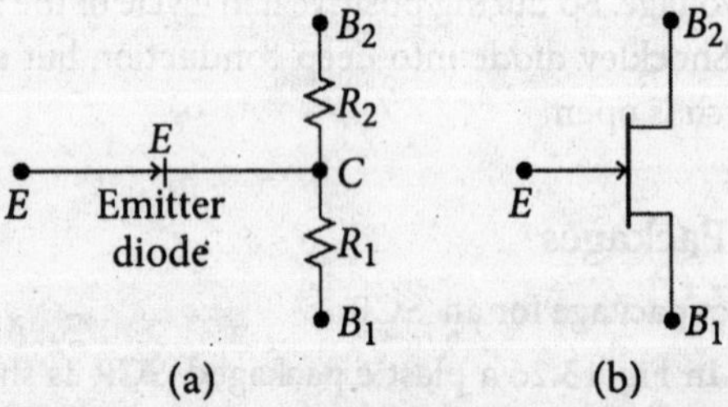

Fig. 13.30

Usually, B_1 is grounded and positive voltage is applied at the base B_2. The situation is shown in Fig. 13.31.

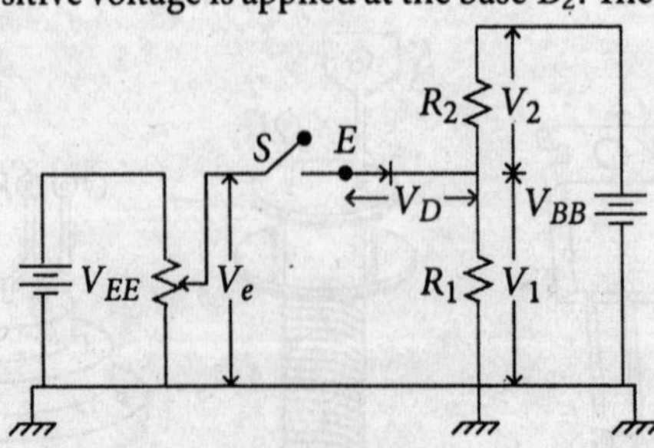

Fig. 13.31

The constructional features of a UJT speak out a few important terms w.r.t. it.

The emitter-diode cathode end C divides the resistance between the two bases as R_1 and R_2 where the voltage drop across R_1 is V_1 and across R_2 is V_2.

$$R_{BB} = R_1 + R_2.$$

When a voltage V_{BB} is applied between the two bases, then

$$V_1 = \frac{V_{BB}}{R_1 + R_2} \cdot R_1 = \eta V_{BB},$$

where

$$\eta = \frac{R_1}{R_1 + R_2} = \frac{V_1}{V_{BB}},$$

called the *Stand-off ratio,* is a constant provided V_1 remains constant. It is a constant when the switch S is open, but when it is closed and the device works as a UJT, it is not a constant. η ranges between 0.5 to 0.8 depending on the value of the emitter current.

Now, the switch S is closed so that a positive voltage V_e is applied at the anode end of the emitter diode.

$$V_e = V_1 + V_D = \eta V_{BB} + 0.7.$$

The voltage source V_{BB} places a voltage gradient along the bar which maintains the diode D in reverse bias as long as $V_e < V_1$. As V_e is increased and becomes greater then V_1, the diode is thrown into forward bias.

When in forward bias and the bias voltage exceeds V_1 by V_D, p-type emitter injects holes into the n-bar. These holes are repelled by the positive base B_2 and attracted by the base B_1 which is maintained at zero potential. Thus, the concentration of holes goes on increasing as the forward bias voltage V_e is progressively increased. This increasing hole concentration decreases the resistance of the base region between the emitter and base B_1. Emitter current I_E thus increases.

13.17 Characteristic Curve of a UJT

The relationship between V_e and I_E at a constant V_{BB} is called the emitter characteristic of a UJT. Such a characteristic curve is shown in Fig. 13.32 as ABCDEF.

As in case of BJT, here also the curve shows three distinct regions.

Cut off region. The region in shade shown in Fig. 13.32, where the emitter diode voltage V_e satisfies

$$\begin{aligned} V_e &< (V_1 + V_D) \\ &< (\eta V_{BB} + V_D) \end{aligned}$$

is called the cut-off region. $V_1 = \eta V_{BB}$ and V_D the sum of these two is called the threshold voltage such that for $V_e < (\eta V_B + V_D)$, the UJT is in cut-off when only a small reverse alteration current flow through the device. The point B of the curve, where the curve maximises, is known as the *peak point* and its corresponding voltage is the peak voltage V_p.

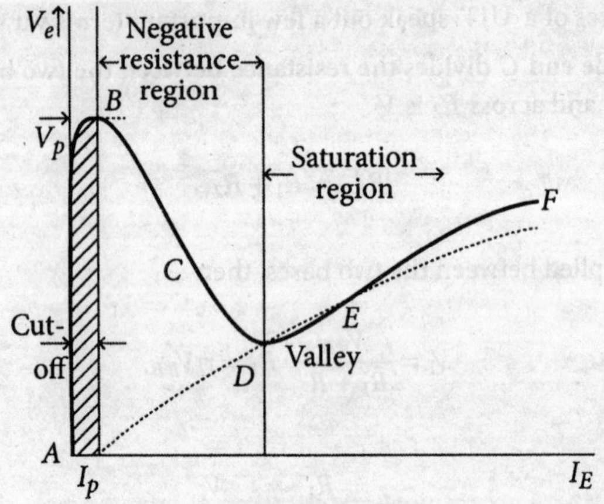

Fig. 13.32

Example 13.1. *A UJT has* $\eta = 0.63$. *What is ideal turn-on voltage? If* $V_D = 0.7$V, *what is the actual turn-on voltage V?* [Fig. 13.33].

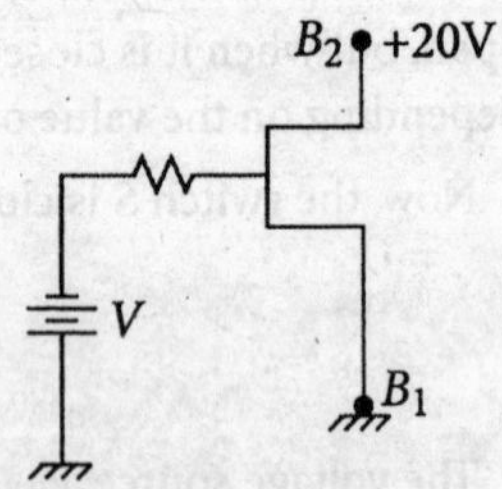

Fig. 13.33

Solution: The ideal turn-on value of V is $V = V_1 = \eta V_{BB} = 0.63 \times 20 = 12.6$V.

Actual turn-on voltage is $V = V_1 = \eta V_{BB} + 0.7 = (12.6 + 0.7) = 13.3$V.

Negative resistance region: The existence of negative resistance region in UJT characteristic reminds us of a similar situation as in tetrode valve.

As we pass across the cut-off region through the point B of the characteristic curve, from the condition $V_e < (\eta V_{BB} + V_D)$ to $V_e > (\eta V_{BB} + V_D)$, the emitter diode becomes more and more forward-biased, the unijunction transistor passes into deep conduction state, called negative resistance region. This region signifies a sudden drop in the emitter voltage (see the portion BC of the characteristic curve). As the forward bias increases, more and more holes are injected into the base region between the emitter and base B_1, causing sharp drop in resistance until the *valley point C* is reached where the curve has its lowest point. Beyond this point the curve shows ascent again, the concentration of the holes attains almost a steady value. It is the saturation region of the UJT.

Saturation region. It is the region of the curve beyond the valley point C, where the emitter voltage remains practically constant with increasing current. Here the conducting path in the emitter-base B_1 section is saturated with the carriers.

13.18 Applications of UJT

UJT *triggered* SCR *firing.* The UJT, because of its special property that when in conduction it passes deep into it, it is conveniently utilised to trigger an SCR. The circuit is shown in Fig. 13.34.

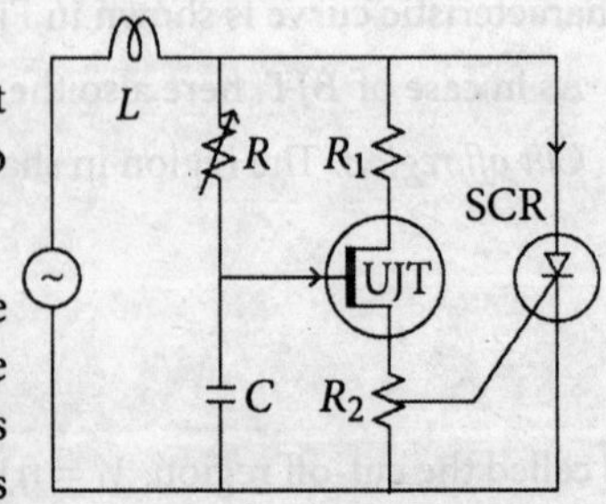

Fig. 13.34

Make a circuit as shown with a lamp L as the load. During the positive half-cycle of the ac input, the capacitor charges up through the variable resistance R. By adjustment of R, the voltage across the capacitor becomes slightly greater than V_p of the UJT when the unijunction transistor passes into deep conduction causing a drop across R_2 which is sufficient to trigger the SCR. The lamp glows to brightness. During the negative half-cycle, the capacitor discharges through the emitter diode of the UJT to reset the device. The frequency of charge and discharge of the capacitor is large enough so that the lamp appears to be lit continuously. So the voltage output of the UJT which is also the triggering pulse for the SCR firing is a sawtooth type of voltage whose frequency is calculated as below.

Calculation of the frequency of the output voltage of the UJT

Let $\langle V\rangle$ be the average value of the ac input. Then $V_c = \langle V\rangle(1-e^{-t/CR})$ is the voltage across the capacitor at any instant. The capacitor discharges when $V_C = V_P = \eta\langle V\rangle$.

$$\therefore \eta\langle V\rangle = \langle V\rangle(1-e^{-t/CR}) \Rightarrow \eta = 1-e^{-t/CR} \Rightarrow e^{-t/CR} = 1-\eta = t = CR\ln\left(\frac{1}{1-\eta}\right).$$

So, the frequency of the sawtooth pulse is

$$\boxed{f=\frac{1}{t}=\frac{1}{CR\ln\left(\frac{1}{1-\eta}\right)}}.$$

13.19 Short Theory Questions with Answers

1. What do you mean by power electronics?

   Ans. The branch of electronics that deals with power generation and control is referred to as power electronics.

2. An SCR is made of silicon and not of Germanium. Why?

   Ans. The leakage current is smaller for silicon than for Germanium so that an SCR made of silicon acts as a more perfect switch.

3. An SCR is not a bidirectional device. Why?

   Ans. An SCR works when the anode is positive w.r.t. the cathode and not when the anode is negative. So it is a bidirectional device.

4. What are the available packages for an SCR?

   Ans. There are three common packages, although there are in all five types of packages. They are: (a) Discrete plastic package (b) Stud base package (c) Flat base package.

5. Draw an automobile ignition system using a UJT triggered SCR device.

   Ans. The ignition circuit is shown in Fig. 13.35, where a UJT triggers an SCR.

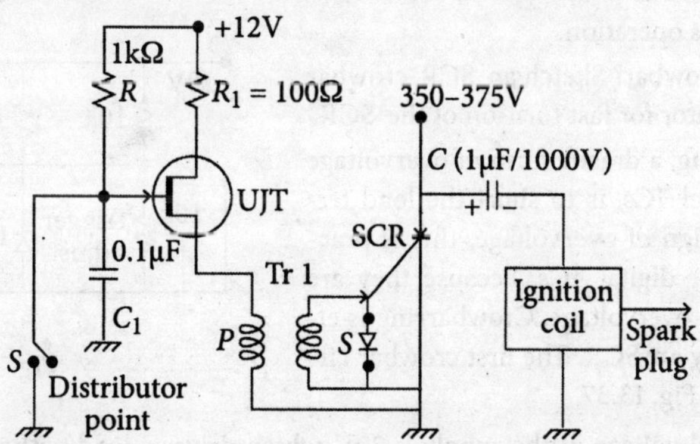

Fig. 13.35

The capacitor C_1 is charged as the distributor point opens and is discharged when it closes. When the voltage across the capacitor rises to a value slightly greater than V_p of the UJT, the UJT passes into heavy conduction state and a current pulse passes through the primary of the transformer Tr 1. Trigger voltage appears across the secondary and the SCR at once fires. The capacitor C was previously charged to a high voltage dc (350–375V) via the primary of the ignition coil. The moment the SCR fires, the output capacitor discharges through the ignition coil and dumps all its energy through the coil and a high voltage pulse drives one of the spark plugs (usually four) as determined by the rotor cam in the distributor.

6. What is a triac? Cite one application of this device.

Ans. A triac is a bidirectional device. It consists of two SCRs connected in inverse parallel with the two gates forming one common terminal. It can control alternating current in a load.

A triac can be used in a light dimmer circuit or a fan speed regulator circuit.

7. Sketch the two-transistor equivalent circuit of an SCR and explain briefly its operation.

Ans. An SCR may be looked upon as a combination of two transistors, one *pnp* and the other *npn*, because basically it consists of an array of four semiconductor blocks *pnpn* [Fig. 3.36(a), (b), (c) and (d)].

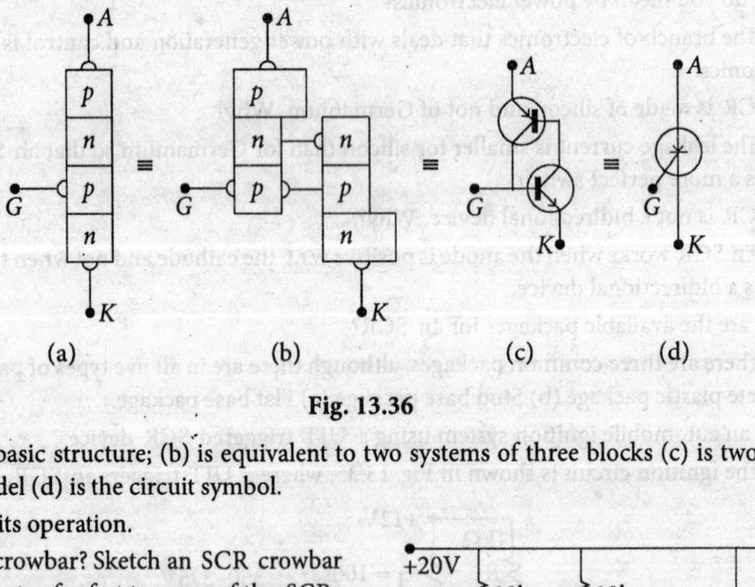

Fig. 13.36

(a) shows the basic structure; (b) is equivalent to two systems of three blocks (c) is two-transistor equivalent model (d) is the circuit symbol.

See the text of its operation.

8. What is SCR crowbar? Sketch an SCR crowbar using a comparator for fast turn-on of the SCR.

Ans. Crowbarring, a drastic form of overvoltage protection digital ICs, is to short the load terminals at first sign of overvoltage, this is practically useful for digital ICs, because they are very sensitive to overvoltage. Crowbarring is effectively done by an SCR. The first crowbar circuit is shown in Fig. 13.37.

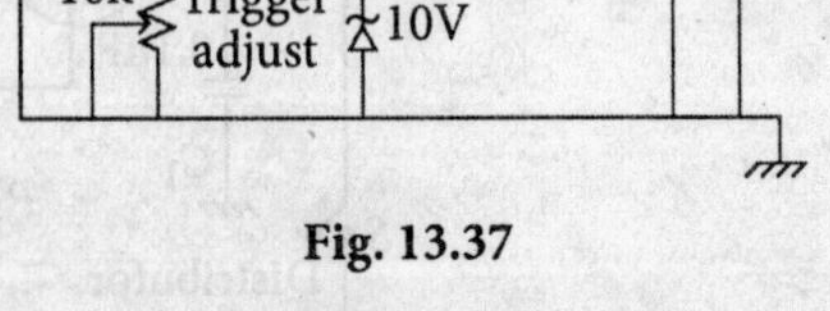

Fig. 13.37

Suppose, normal voltage of the supply is 20V, which drives a load without any danger. The Zener diode $V_z = 10V$ supplies an inverting input equal to 10V and by the trigger adjust resistor, a voltage slightly less than 10V is applied to the non-inverting terminal. Thus, the error voltage is negative and the output of the comparator Op-Amp is negative, SCR opens and has no effect on the load. But as the supply voltage tends to rise, the voltage input to the non-inverting terminal is greater than 10V thus the error voltage becomes positive and so also the output of the comparator. The SCR with positive gate fires shorting the supply thus saving the life of expensive digital ICs from overvoltage by crowbarring.

9. Is an SCR voltage-controlled or a current-controlled device?

   Ans. An SCR is a current-controlled device, because it is set to operate only when the gate current attains a value equal to holding current.

10. What is a diac?

    Ans. A diac is a pair of four-layer diodes in parallel as shown in Fig. 13.22, bidirectional in operation, hence can control alternating current.

11. Explain the term *intrinsic stand-off ratio* w.r.t. a UJT.

    Ans. The resistance offered between two base regions of the UJT, one between B_1 and the cathode of the emitter diode and the other between B_2 and the cathode, are R_1 and R_2 respectively. When the UJT is not conducting, $R_{BB} = R_1 + R_2$.

    But when a supply voltage V is applied between the two base terminals B_1 and B_2, we get the voltage distribution as

$$V_1 = \frac{VR_1}{R_1 + R_2} = \frac{R_1}{R_{BB}}V; \; V_2 = \frac{R_2}{R_{BB}}V.$$

Thus, we can write, $V_1 = \frac{R_1}{R_{BB}}V = \eta V$, this factor which varies between 0.5 to 0.8 is called the intrinsic stand-off ratio. It is nothing more than potential divider factor. Suppose, $V = 20\text{V}$ and $\eta = 0.65$, then $V_1 = 0.65 \times 20 = 13\text{V}$.

If V_1, the stand-off voltage, is less than 13V, the emitter diode is reverse-biased for all emitter voltages $V_e < 13\text{V}$.

13.20 Solved Numerical Problems

1. The η value of the UJT shown in Fig. 13.38 is 0.85. What is the ideal emitter current?

Solution: Stand-off voltage $= 0.85 \times 10 = 8.5\text{V}$.

The input voltage 20V is enough to overcome the stand-off through 400Ω. So emitter current

$$I_E = \frac{20}{400} = 50\text{mA}.$$

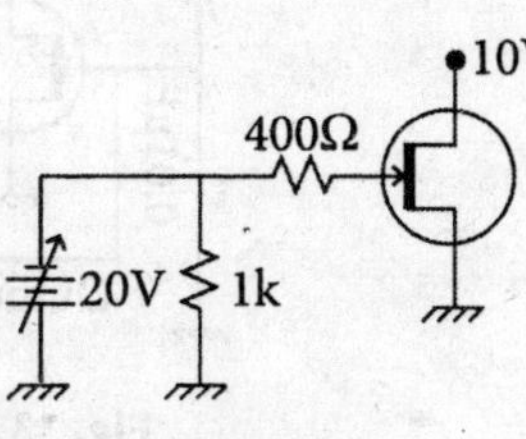

Fig. 13.38

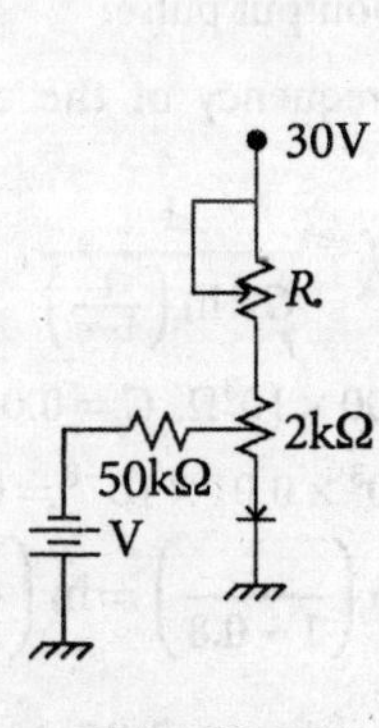

Fig. 13.39

2. The SCR shown in Fig. 13.39 has a trigger current of 0.1mA. If we allow 0.8V for the gate voltage, what is the value of V that turns on the SCR?

Solution: $0.1\times10^{-3}=\dfrac{V-0.8}{50\times10^{3}}\Rightarrow V-0.8=0.1\times50=5\Rightarrow N=5.8\text{V}.$

3. What is the maximum current through the 50Ω resistor [Fig. 13.40] after the diode latches? Allow 1V across the latched diode.

Solution: $I_{\max}=\dfrac{V_e-1}{50}=\dfrac{9-1}{50}=0.16\text{A}.$

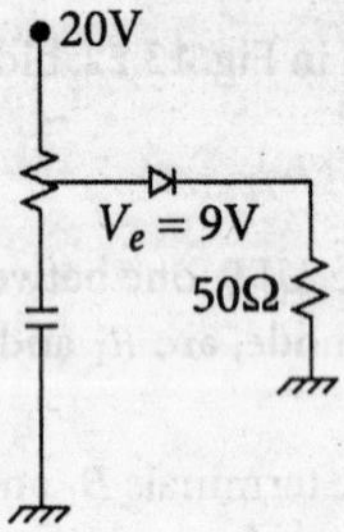

Fig. 13.40

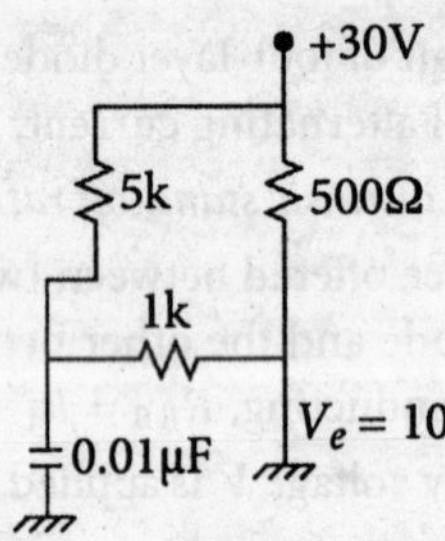

Fig. 13.41

4. The Fig. 13.41 shows an SCR circuit triggered by a Shockley diode having breakover voltage of 10V. The SCR has a trigger current of 0.1mA and trigger voltage 0.8V. If the four-layer diode has a forward drop of approximately 0.7V, what is the current through the diode just after it breaks over? What is the current through 500Ω after the SCR turns on?

Solution: $I_E=\dfrac{V_e-0.7}{1\times10^{3}}\Rightarrow I_E=\dfrac{10-0.7}{1\times10^{-3}}=\dfrac{9.3}{10^{3}}=9.3\text{mA}.$

When the SCR turns on, 30V is shorted through 50Ω to the ground, so the current through 50Ω is given by

$$i=\frac{30}{500}=0.60\text{A}=60\text{mA}.$$

5. The circuit shown in Fig. 13.42 is used to generate sawtooth pulse. $\eta=0.8$ for the UJT. Calculate the frequency of the output pulse.

Solution: The frequency of the output pulse is given by

$$f=\frac{1}{CR\ln\left(\frac{1}{1-\eta}\right)}.$$

$R=500\times10^{3}\Omega,\ C=0.01\mu\text{F}.$

$CR=50\times10^{3}\times0.01\times10^{-6}=0.5\times10^{-3}\text{s}.$

$\ln\dfrac{1}{(1-\eta)}=\ln\left(\dfrac{1}{1-0.8}\right)=\ln\left(\dfrac{1}{0.2}\right)=1.60.$

$\therefore\ f=\dfrac{1}{0.5\times10^{-3}\times1.6}=1.25\times10^{3}=1.25\text{kHz}.$

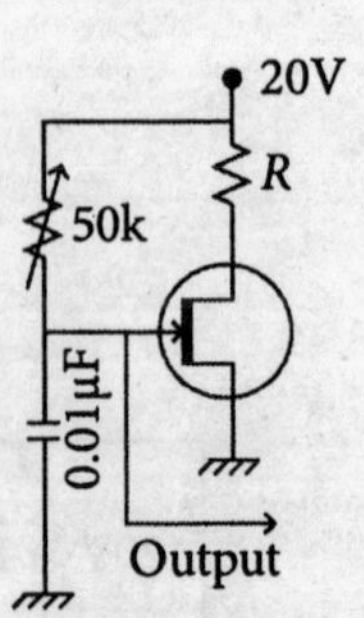

Fig. 13.42

6. The data sheet of an SCR gives the following record: *Breakover voltage* $V_{BFO} = 500\text{V}$, $I_{tgc} = 10\text{mA}$ = *triggering current*, I_H = *holding current* = 12mA. What inference you can draw from these data?

Ans. Breakover voltage is the positive voltage on the anode of the SCR which will set the SCR conduct heavily even in absence of any gate voltage. If the voltage on the anode is less than 500V, the SCR will not conduct, it is open.

Trigger current $I_{trg} = 10\text{mA}$. When the anode voltage is less than 500V, the gate must be supplied with 10mA of current just to trigger the SCR into actions. If the gate current is less than 10mA, the SCR will not fire at all for any anode voltage below 500V.

Holding current $I_H = 12\text{mA}$. Once the SCR is switched on by supplying gate with current called the triggering current of 10mA, it will not stop even if this current is removed. In order to stop the SCR, the anode forward voltage should be reduced such that $I_A = I_H$ falls below 12mA.

7. An SCR half-wave rectifier is adjusted for gate current of 2mA. The SCR conducts at 125V with a gate current of 2mA. If the input voltage is given as $v = 250\sin\alpha$, find (a) the firing angle (b) conduction angle (c) average value of voltage (d) average value of current for a load resistance $R_L = 50\Omega$.

Solution: (a) $125 = 250\sin\alpha \Rightarrow \alpha = \text{firing angle} = \sin^{-1}\left(\frac{125}{250}\right) = 30°$.

(b) Conduction angle $\theta = 180 - 30° = 150°$.

(c) $v_{de} = \frac{\hat{V}}{2\pi}(1 - \cos\alpha) = \frac{250}{2\pi}(1 + \cos\alpha) = 74.3°$.

(d) $I_{dc} = \frac{V_{dc}}{R_L} = \frac{74.3}{50} = 1.49\text{A}$.

8. The UJT shown in Fig. 13.43 has $\eta = 0.75$. Ideally at what value of V, the UJT turns on? If we allow 0.7V across the emitter diode, at what value of V, it turns on?

Solution: The turn-on voltage (ideally) is $V_1 = \eta V_{BB} = 0.75 \times 20 = 15.0\text{V}$.
Taking 0.7V as the voltage drops across the diode, the turn-on voltage is $(15 + 0.7) = 15.07\text{V}$.

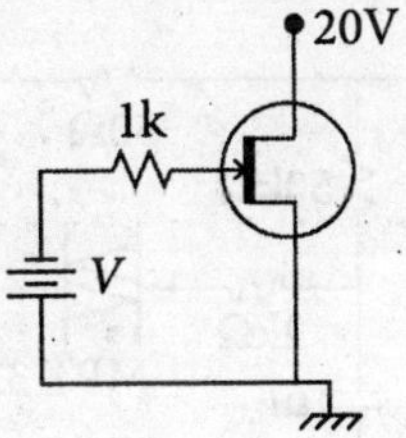

Fig. 13.43

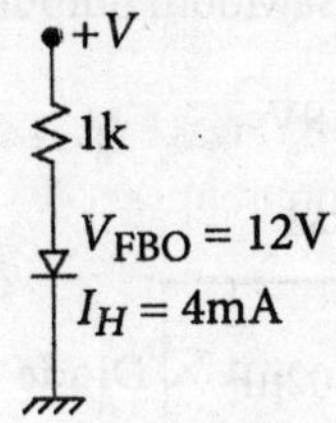

Fig. 13.44

9. What is the value of V for the Shockley diode to be in conduction if (a) diode drop is neglected? (b) diode of 0.7V is considered?

Solution: (a) Ideally, $\frac{V - 12}{1 \times 10^3} = 4 \times 10^{-3} \Rightarrow V = 16\text{V}$.

(b) Taking 0.7V as the drop across the diode, the value of $V = (16 + 0.7) = 16.7\text{V}$.

13.21 Numerical Problems Exercises

1. Find the output pulse frequency of the relaxation oscillator using the UJT in the circuit shown in Fig. 13.45.
2. Calculate the pulse repetition frequency of the relaxation oscillator if $R = 20\text{k}$, $C = 0.1\mu\text{F}$, $\eta = 0.7$.
3. A UJT has $\eta = 0.65$, inter base resistance $R_{BB} = 8\text{k}\Omega$. Find the values of R_1 and R_2.

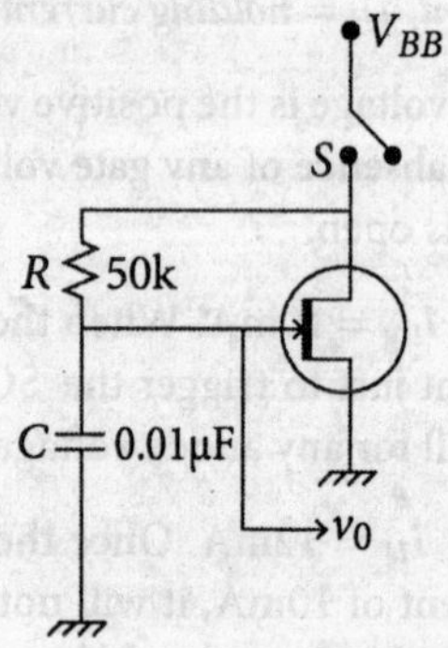

Fig. 13.45

4. We have shown in Fig. 13.46 a diac-triggered triac circuit using MPT 32 (diac). MPT 32 has a breakover when the capacitor voltage reaches 32V. It takes exactly one time constant for the capacitor to reach this voltage. How long after the switch is closed will the triac turn-on? What is the ideal value of the gate current when the diac breaks over? What is the load current after the triac has closed?

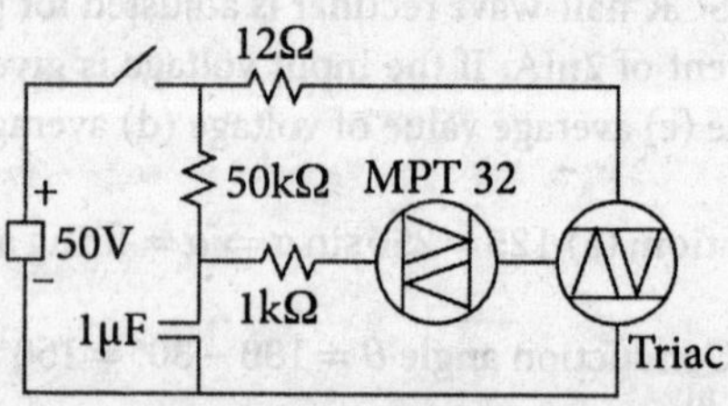

Fig. 13.46

5. The valley current of a UJT in circuit Fig. 13.43 is 2mA. If the UJT is latched, we have to reduce V to get low current drop out. Allowing 0.7V across the emitter diode, what is the value of V that just opens the UJT?
6. The valley current of a 2N 4871 is 7mA and the emitter voltage is 1V at this point. At what value of emitter supply voltage does the UJT open?
7. When the supply voltage is 19V [Fig. 13.47] it takes the capacitor exactly one time constant to charge to 12V, the breakover voltage of the diode. If we neglect the voltage across the diode conducting, what is the frequency of the sawtooth output?

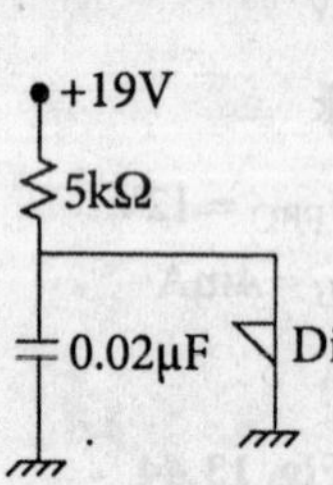

Fig. 13.47

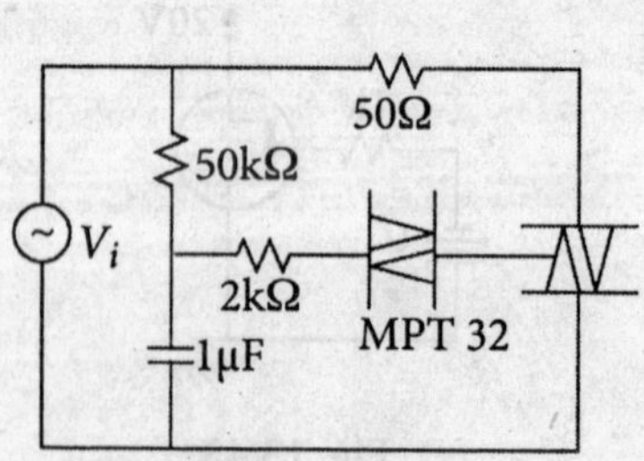

Fig. 13.48

8. The frequency of the input square-wave voltage V_i has a peak-to-peak value equal to 50V. It takes exactly one time constant for the capacitor to reach the breakover voltage of the diac. If the MPT 32 breaks over at 30V, what is the ideal value of the gate current at the instant the diac breaks over? Find the ideal load current.

9. If the UJT of Fig. 13.49 has $\eta = 0.75$, what are the maximum and minimum frequencies of the output pulse?

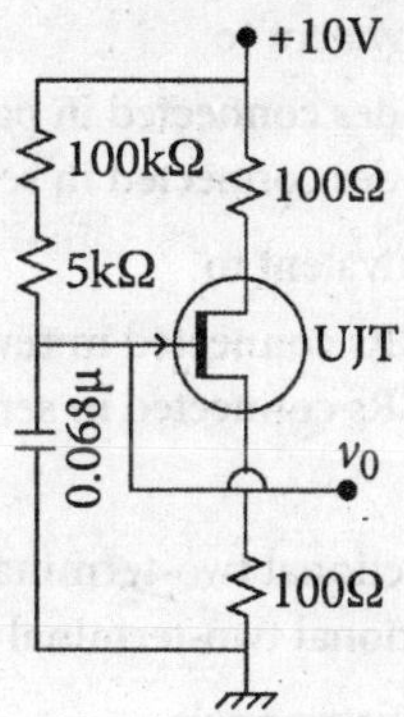

Fig. 13.49

13.22 Theory Questions

1. What are the uses of a UJT? (Relaxation oscillator, overvoltage detector, for triggering as SCR, capacity discharge automobile ignition).
2. Explain the construction and operation of an SCR. Draw a neat diagram of the I-V characteristics of an SCR.
3. What are the similarities and differences between an SCR and a triac? Draw the I-V characteristics of a triac.
4. Sketch the two-transistor equivalent circuit of an SCR and explain its operation based on this two-transistor model.
5. Is UJT really a transistor? If not, what are the differences between a UJT and an ordinary BJT?
6. Is UJT a unipolar device? Make a critical survey. Is it a current-controlled or voltage-controlled device?
7. An SCR acts as a perfect switch. Explain.
8. How can you use a UJT to generate a sawtooth wave. Estimate the frequency of the wave pulse.

13.23 MCQ Type

1. An SCR is not made of Germanium, because
   (a) it has poor conducting properties
   (b) it has larger leakage current
   (c) it is not easily available.

2. An SCR is
   (a) current-triggered
   (b) voltage-triggered
   (c) power-triggered.

3. The maximum anode current, gate being open, at which an SCR is turned off from ON condition is
   (a) cut-off current
   (b) switch-off current
   (c) holding current.

4. A UJT contains
   (a) four *pn*-junctions
   (b) three *pn*-junctions
   (c) two *pn*-junctions.

5. A diac is equivalent to
   (a) two diodes connected in parallel
   (b) two diodes connected in series
   (c) two four-layer diodes connected in reverse parallel.
6. A triac is equivalent to
   (a) two SCRs connected in reverse parallel
   (b) two SCRs connected in series
   (c) two diodes connected in reverse parallel.
7. A diac is a
   (a) unidirectional two-terminal device
   (b) bidirectional two-terminal device
   (c) unidirectional three-terminal device.
8. In a diac the firing angle
   (a) is fixed
   (b) can be changed by changing the supply voltage
   (c) can be changed by changing the gate current.

Part III

Digital Electronics

Chapter 1

Analog circuits and Digital circuits

Analog implies continuity, digital signifies discreteness. Analog circuits and systems — we have discussed in the previous few chapters such that the output of them is continuous over a certain range, maybe of any value but it exists for all instants of time. The solid state circuit elements like diodes, transistors FETs and MOSFETs are made to operate in such a way in analog circuit that a continuous output is available. It means, the elements operate in the active region of their characteristic curves. Electronic circuits designed for amplifying or processing such signals are called analog circuits and the systems as analog systems. On the other hand, digital circuits are designed to operate in such as way as to produce binary output, high or low. The discrete circuit elements here are made to operate as switches so that they are biased to operate between saturation region and cut-off region of their characteristic curves. A common example of a digital system is a calculator where the input electrical signals have only two discrete values or levels, high and low. Actual value of the signal is unimportant. Such a signal is termed as a digital signal. Digital circuits process these signals. A system developed for digital signal is known as a digital system.

A signal which is discrete w.r.t. time and also discrete in amplitude is a digital signal. The output of a digital system has only two values, either a low voltage or a high voltage; low state indicated by 0 (zero) and high state by 1 (one); if low state be taken as being represented by 0 and high state as 1, the logic is referred to as positive logic. If, on the other hand, 1 is taken as to represent low state and 0 as high state, the system logic is referred to as negative logic. It might be that in some logic system, a signal of value between 0 and 1V is recognised as binary 0 and that between 3 and 5V as binary 1.

Actually, however, digital processing was recognised as a separate field by itself towards the beginning of 1960s and since then significant developments and breakthrough in this theory and practice took place. A very important point to remember is that an analog signal exists for all instants of time requiring energy consumption as a continuous function of time but in digital processing, the signal exists only for discrete and desired instants. Hence, here the energy consumption function is also discrete time function.

Digital communication system has been introduced as replacement of Analog Communication system, because of several reasons:

1. The digital communication system is very much immune to channel noise which offsets the relative simplicity of analog communication system.
2. The digital communication system is highly reliable, has high fidelity, security and low error rate.
3. The digital communication system permits the use of computers and LSI circuits; it can accommodate various sources of information in a flexible and compatible manner.

4. The most important point in favour of the digital communication system is that the rate of data transfer volume is much higher than that in analog communication.
5. In case of digital signal, it can be stored in magnetic tape or disk without any further loss of fidelity beyond that introduced by A/D conversion process. As a consequence, digital signals become transportable and can be processed off-line in a remote laboratory.
6. It is usually very difficult to perform some mathematical operation on signals in analog form but these operations can be done very comfortably on digital signals.
7. Digital signal processing in some cases becomes cheaper than the corresponding counterpart, maybe due to the fact that the digital hardware is cheaper.

As a result of these advantages, the digital signal processing has been applied in practical systems covering a wide range of disciplines as in speech processing; transmission in telephone channel; in image processing and transmission; in seismology and geo-physical prospecting, etc.

Chapter 2

Oft-Quoted Number Systems: Data Representation and the Decimal and Binary Number Systems

2.1 Introduction

Data refers to some facts obtained from different sources — it may be organised or may be just raw. If data is refined and organised, it becomes an information meaningful to the person receiving and dealing with it. The transformation technique from data into information is referred to as data processing.

In computers we have basically two types of data to feed: (a) characters (b) numbers.

In the present chapter and that follows next, we will be dealing with a number system meaningful to and understood by the machine.

Data refers to some facts and can be organised or raw. Organised data is known as *information.* The transformation technique from data to information is known as *data processing.*

2.2 What are the different number systems and their conversions?

At least there are four different types of number systems: (a) decimal (b) binary (c) octal (d) hexadecimal.

Of these number systems, the decimal number system is very much familiar to us and it uses ten Arabic numerals symbolised by 0, 1, 2, 3, 4, 5, 6, 7, 8, 9, which we call as digits. So to understand the language of decimal system we must recognise 10 states (not a matter of joke!). But the computer or the digital instruments made of solid state or semiconductor elements cannot recognise the ten states, they can recognise only two states ON or OFF, TRUE or FALSE, YES or NO.

So the language understood by these machines is only binary and so binary number system (base 2) and the binary algebra (Boolean algebra) were formed. A part from the two number systems, we have also octal (base 8) and hexadecimal (base 16) systems for compact representation of a big number in other systems, mainly in binary representation.

The computer (or any digital instruments) made of semiconductor elements can understand only two states 1 (ON) or 0 (OFF).

2.2.1 What do you mean by radix or base of a number system?

The maximum number of distinct digits, which can be used in a number system, denotes the radix or base of the number system. Thus, in:

1. *Decimal number system,* the radix or base is 10, because there are ten distinct digits (0 to 9) used in this system.

2. *Binary number system,* the radix or base 2, because only two binary digits 0 and 1 are used. The digits are also called bits.
3. *Octal number system,* the radix or base is 8, because digits 0, 1, 2, 3, 4, 5, 6, and 7 are used.
4. *Hexadecimal number system,* the radix or base is 16, because it uses 16 symbols 0 to 9 and A, B, C, D, E and F.

Base/radix denotes the maximum number that can be used in a system.
Base of decimal number system is 10 (0–9).
Base of binary number system is 2 (0 and 1).
Base of octal number system is 8 (0–7).
Base of hexadecimal number system is 16 (0–9 and A, B, C, D, E and F).

2.2.2 What is the meaning of positional number system and non-positional number system?

In the positional number system the position of a digit indicates the value in the number. Thus, in the decimal number system, the digit on the rightmost position is called the least significant digit (LSD) and the digit on the leftmost position is called the most significant digit (MSD). So this number system is a positional number system with 10 as base.

Take an example: 195.

The last digit is 5 in the unit position.

The digit 9 is in the tenth position.

The digit 1 is in the hundredth position.

Thus, we have $195 = 1 \times 10^2 + 9 \times 10^1 + 5 \times 10^0$

(MSD) (LSD).

In non-positional number system there is no base, hence no question of LSD or MSD in this system. Thus, in roman symbol I means one, II means two, etc., so in this symbolic number system, the position of the symbol does not carry any added significance. It is non-positional number system.

Decimal number system is a positional number system where the position of a digit indicates the value in the number. Digit on the rightmost position is called the least significant digit (LSD) and the digit on the leftmost position is called the most significant digit (MSD).
Roman symbol is an example of non-positional number system.

2.3 Details of decimal number system representation

This number system is a positional number system with base 10 attributing significance to the position of the digits in the number.

For example, 8 7 6 5

5 → 5×10^0
6 → 6×10^1
7 → 7×10^2
8 → 8×10^3

Hence, $8765 = 8 \times 10^3 + 7 \times 10^2 + 6 \times 10^1 + 5 \times 10^0$
$= 8000 + 700 + 60 + 5.$

So we can see that the value of the digit in the number depends on the position of the digit.

2.4 Details of the binary number system

This is also a positional number system with radix 2. The binary digits are 1 and 0, also called *binary bits.* A binary number consists of a string of binary bits 1 and 0 in different combinations. Thus, a binary representation of a number is 1 0 1 1 0 (say).

Binary bits:	1	0	1	1	0
Positional value:	2^4	2^3	2^2	2^1	2^0
	(MSB)				(LSB)
Products	1×2^4	0×2^3	1×2^2	1×2^1	0×2^0.

So the sum of the products is 22.

Hence, $10110_2 \equiv 22_{10}$.

2.4.1 Scheme of conversion of decimal system to binary system

Suppose, a number 22 in decimal system is to be expressed in binary system. Divide the number successively by 2 and find the quotient and the remainder. Thus,

	Quotient	Remainder	Remark
22/2	11	0	LSB ↑
11/2	5	1	
5/2	2	1	
2/2	1	0	
1/2	0	1	MSB

So, $22_{10} \equiv 10110_2$.

Any binary number can be expressed in terms of different powers of 2. The coefficients of the different powers would give the binary representation and the values of different powers of 2 added together gives the equivalent decimal representation.

Look at the following table (Table 1).

Decimal number	Binary equivalent	Binary representation	Remark
0	0	0	
1	1×2^1	1	
2	$1\times2^1+0\times2^0$	10	
3	$1\times2^1+1\times2^0$	11	The statement
4	$1\times2^2+0\times2^1+0\times2^0$	100	in the above
5	$1\times2^2+0\times2^1+1\times2^0$	101	note follows
6	$1\times2^2+1\times2^1+0\times2^0$	110	
7	$1\times2^2+1\times2^1+1\times2^0$	111	
8	$1\times2^3+0\times2^2+0\times2^1+0\times2^0$	1000	
9	$1\times2^3+0\times2^2+0\times2^1+1\times2^0$	1001	

2.4.2 Decimal fraction

A decimal fraction written as 0.812 can be expressed as power series in 10 as below:

$$0.812=8\times10^{-1}+1\times10^{-2}+2\times10^{-3}.$$

So, if the number be a decimal fraction, the first digit after decimal point is MSD and the last digit is LSD.

2.4.3 Conversion of decimal fraction to binary system

Multiply the decimal fraction successively by 2, write the result, take integral part as the first carry (MSD), multiply the fraction again by 2, you will get fraction and integral part as the second carry; in this way, the last carry is the LSD. Thus, you have

Decimal fraction	Product	Fraction	Carry	
0.812 × 2	1.624	0.624	1	(MSD)
0.624 × 2	1.248	0.248	1	
0.248 × 2	0.496	0.496	0	
0.496 × 2	0.992	—	1	↓

So the binary equivalent of the decimal fraction is 0.1101. If we consider reverse conversion, we would get $0.1101_2 = 1\times2^{-1}+1\times2^{-2}+0\times2^{-3}+1\times2^{-4}$.

Adding the values of the different powers of 2, we get $2^{-1}+2^{-2}+2^{-4}=0.5+0.250+0.0625=0.8125$.

So, $0.1101_2 \equiv 0.8125_{10}$.

2.4.4 Number in decimal system having integral part as well as fractional part

Any number in decimal system containing integral part as well as fractional part may be taken as 512.812.

In decimal system expressed as power series in 10, we can write:

$$512.812 = 5\times10^2+1\times10^1+2\times10^0 \;\Big|\cdot\Big|\; 8\times10^{-1}+1\times10^{-2}+2\times10^{-3}$$
$$= 500+10+2 \;\Big|\cdot\Big|\; 0.8 \;+\; 0.01 \;+\; 0.002$$

> To convert a decimal number to binary divide the decimal number successively by 2, and place the remainder side by side. Then write the remainders starting from the down and process upward to get the equivalent binary number.

2.4.5 Conversion from decimal system to binary

For integral part: Successive division by 2: To convert the decimal number 512.812 into binary system, we first take the integral part.

	512	Remainder	Remarks
2	512	0 ↑	LSB
2	256	0	
2	128	0	
2	64	0	
2	32	0	
2	16	0	
2	8	0	
2	4	0	
2	2	0	
2	1	1	MSB

$512_{10} = 1000000000_2 = 1 \times 2^9 + 0 \times 2^8 + 0 \times 2^7 + 0 \times 2^6 + 0 \times 2^5 + 0 \times 2^4 + 0 \times 2^3 + 0 \times 2^2 + 0 \times 2^1 + 0 \times 2^0$.

For fractional part: Successive multiplication by 2.

	Product	Fraction	Carry	
0.812×2	1.624	0.624	1	MSD
0.624×2	1.248	0.248	1	
0.248×2	0.496	0.496	0	
0.496×2	0.992	—	1 ↓	

Hence, $812_{10} = 0.1101_2$ and we have $512.812_{10} = 1000000000.1101_2$.

2.5 Different transformation techniques for conversion from decimal to binary and vice versa

Rule 1. A number given in decimal system is to be expressed as the sum of ascending powers of 2 and then collect the coefficients of the different powers of 2.

For example, 11 (Eleven) $= 1 \times 2^3 + 0 \times 2^2 + 1 \times 2^1 + 1 \times 2^0$.

Collecting the coefficient of different powers of 2, we get 1011.

Thus, $11_{10} = 1011_2$.

In case of fraction, express it as ascending negative powers in 2.

Thus, $0.812_{10} = (0.1101)_2$.

Rule 2. This rule is of more operational importance than the first rule. The given number is divided progressively by 2 and the first remainder 0 or 1 is the least significant digit and the last remainder, when the quotient is zero is the most significant digit. Thus,

Divisor	Number	Remainder	
2	11		
2	5	1 ↑	LSB
2	2	1	
2	1	0	
	0	1	MSB

So, $11_{10} = 1011_2$.

In case of fraction, multiply the number successively by 2, record each time a carry in the integer position. The carry in the forward order gives the binary representation.

Thus,

	Fraction	Carry	
0.65×2	0.30	1	MSB
0.30×2	0.60	0	
0.60×2	0.20	1	
0.20×2	0.40	0	Repeating
0.40×2	0.80	0	
0.80×2	0.60	1 ↓	LSB

So, $0.65_{10} = 0.101001$.

> – Express the given decimal number in sum of ascending powers of 2 and then write the coefficient of the different powers of 2 to get the equivalent number in binary.
>
> – Divide the given decimal number successively by 2 and write the remainder side by side to get the equivalent binary number.

Rule 3. Conversion of a decimal number to binary by successive subtraction method.

Step 1. Find the nearest number to the given number in terms of non-negative integer power of 2. Express the integer power of two in binary. Subtract the nearest from the given number and find again the nearest number to this difference in terms of integer power in 2.

Step 2. Repeat until the difference reduces to zero.

Let us take an example:

Decimal fraction	Nearest no. in terms of power in 2	Binary expression
49		
−32	2^5	100000
17		
−16	2^4	10000
1	2^0	1
−1		
0		Sum = 110001

In case of a decimal fraction, the conversion to binary is also possible by successive subtraction method. the method is illustrated in the following example:

> – For conversion of decimal number neither too small nor too large, this successive subtraction method may be applied.

Decimal fraction	Nearest no. in terms of negative integer power in 2	Binary expression
0.812		
−0.500	2^{-1}	0.1
0.312		
−0.25	2^{-2}	0.01
0.062	2^{-4}	0.0001
−0.062		
0		Sum = 0.1101

2.6 Conversion methods of binary to decimal

Method 1. Expansion as a positive integer power series in 2.

Sum the value of different powers of 2 and that sum gives the decimal equivalent of the binary.

Illustration. Convert 110101 to equivalent decimal number system.

The binary number consists of 6 bits, so the highest power of 2 will be 5.

Thus, 110101 will be: 1 1 0 1 0 1

$$1\times 2^5+1\times 2^4+0\times 2^3+1\times 2^2+0\times 2^1+1\times 2^0$$

The sum of the values of the different powers of 2 is: $2^5+2^4+2^2+2^0=32+16+4+1=53$.

So $110101_2=53_{10}$.

Putting this in the box form, we can write:

1	1	0	1	0	1	
2^5	2^4	2^3	2^2	2^1	2^0	
32	16	8	4	0	1	= 53

Method 2. Double summing method: Binary 1 1 0 1 0 1

MSD

Double the MSD and add to the next digit to the right and so on		2	6	12	26	52
		+1	+0	+1	+0	+1
	2	3	6	13	26	53

In case of decimal fraction: Method 1 is called *half-summing method.*

Here the LSD is divided by 2 and added to the next digit to the left and so on until all the digits are considered.

Thus, take the binary fraction 0.1101:

Binary fraction	1	1	0	1 (LSB) ÷ 2
0.812	0.625	0.25	0.5	
	+1	+1	+0	
	1.625	1.25	0.5	

So we have decimal fraction 0.812. Hence, $0.1101_2 = 0.812_{10}$.

2.7 Short Theory Questions with Answers

1. Compare positional and non-positional number systems.

   Ans. A positional number system is one, where the position of the bit or digit determines its weight or value in terms of the power to the base of the system. Thus, decimal number system, binary number system, etc.

   A number system which has no radix or base and uses symbols is called non-positional number system. Thus, roman symbol I and stands for one, II stands for two irrespective of position.

2. What is the base or radix representation of a number?

   Ans. In every positional number system, there is a base or radix whose different powers indicate the positional values of the digits. Thus, in decimal number system, $53 = 5 \times 10^1 + 3 \times 10^0$.

3. What is data processing?

   Ans. The technique of transformation of unorganised data into organised or refined form, called *information,* is referred to as data processing.

4. How do we double a binary number?

   Ans. A binary number can be doubled by simply placing 0 (zero) to the right of the LSD of the number. Thus $101 = 1 \times 2^2 + 0 \times 2^1 + 1 \times 2^0 = 5$

   $$1010 = 1 \times 2^3 + 0 \times 2^2 + 1 \times 2^1 + 0 \times 2^0 = 10$$

   $$10100 = 1 \times 2^4 + 0 \times 2^3 + 1 \times 2^2 + 0 \times 2^1 + 0 \times 2^0 = 20.$$

5. What is a bit and a byte?

   Ans. Bit is a binary digit and may be 0 or 1 and byte is a string of 8 bits.

6. What do you mean by LSD (LSB) and MSD (MSB)?

   Ans. These two terms are related to the positional system, decimal or binary. In decimal system, the last digit to the right is the least significant digit (LSD) and the leftmost digit is the most significant digit (MSD). In binary system also, the bit to the extreme right is LSB and the bit to the extreme left is the LSB.

   Remember: (1) 8 bits make 1 byte (2) 4 bits make 1 nibble (half-byte).

7. Name the different methods of conversion of a decimal number into a binary number.

   Ans. **Method 1.** *Expansion method*: Write the number as a power series in 2, collect the coefficient of different powers.

   Method 2. *Successive division by 2*: A decimal number is to be successively divided by 2, each time writing the remainder. The first remainder is the LSB and the least remainder when the quotient is zero, is the MSB.

Method 3. *Successive subtraction method*: Nearest value to the number in terms of integer power of 2; write its binary representation. Subtract this number from the original number and then find again the nearest value in terms of integer power of 2 and write its binary representation. The process is repeated until the difference reduces to zero. Sum the binary representation.

Method 4. *Double summation method*: Double the MSD and add to the next digit to the right.

8. Show by different available methods how you can convert decimal number 53 into its binary form.

Ans. By method 1: *Expansion method*: $53 = 1\times 2^5 + 1\times 2^4 + 0\times 2^3 + 1\times 2^2 + 0\times 2^1 + 1\times 2^0$

32 16 0 4 0 1.

Hence, binary 110101.

By method 2: *Successive division by* 2:

So, $53_{10} = 110101_2$.

Divisor	Quotient	Remainder	
2	53		
2	26	1	LSB
2	13	0	
2	6	1	
2	3	0	
2	1	1	
	0	1	MSB

Method 3: Successive subtraction:

$$
\begin{array}{rll}
53 & & 100000 \\
-32 & 2^5 & \\
\hline
21 & & 10000 \\
-16 & 2^4 & \\
\hline
5 & & 100 \\
-4 & 2^2 & \\
\hline
1 & & 1 \\
-1 & 2^0 & \\
\hline
0 & \text{Sum} = & \overline{110101}
\end{array}
$$

9. Convert 53.75 into binary form by available techniques.

Ans. For the integral part any of the following methods like:

Method 1: Expansion method, Method 2: Division by 2, Method 3: Successive subtraction may be applied.

So by any method $53_{10} = 110101_2$ can be found.

For fraction: Method 1: *Expansion method*: $0.75 = 1\times 2^{-1} + 1\times 2^{-2}$

$0.5 + 0.25$

So, $0.75_{10} = 0.11_2$.

Hence, we have $53.75_{10} = 110101.11_2$.

Method 2: *Successive multiplication by* 2:

Fraction	Integral	(Carry)	
0.75 × 2	0.50	1	LSB
0.50 × 2	0.00	1	

So $0.75_{10} = 0.11_2$.

Method 3: *Successive subtraction method*:

		Binary
0.75	2^{-1}	0.1
−0.5		
0.25	2^{-2}	0.01
−0.25		
0	Sum =	$\overline{0.11}$

10. Name the different methods by which to transform a binary form into its equivalent decimal form.

Ans. For integral part you can do any one of the following:

Method 1: Expansion method: Power series in 2.

Method 2: Double summation method.

For fractional part: Method 1: Expansion method, Method 2: Half-summing method.

11. Use the different methods as in Q.10 for conversion from binary form to decimal form.

Ans. Binary form = 110101.11.

(i) Expansion method presented in the box form:

1	1	0	1	0	1	·	1	1	
2^5	2^4	2^3	2^2	2^1	2^0		2^{-1}	2^{-2}	Product
32	16	0	4	0	1		0.5	0.25	

Sum 32 + 16 + 4 + 1 + = 53 ↑ Sum = 0.75

Binary point

(ii) Double summing for integral part:

Binary	1	1	0	1	0	1	
MSB		2	6	12	26	52	
×2		3	6	13	26	53	(Sum)

So, $110101_2 = 53_{10}$.

For fractional part, half-summing method is applied: 0.1 1(LSB) ÷ 2

0.5

1.5 0.75

Thus, $0.11_2 = 0.75_{10}$

So, $110101.11_2 = 53.75_{10}$.

12. Convert 37_{10} to the binary using all possible conversion methods.

Ans. (a) By expansion method: $37 = 1\times 2^5 + 0\times 2^4 + 0\times 2^3 + 1\times 2^2 + 0\times 2^1 + 1\times 2^0$.

The coefficients of different powers of 2 give the binary representation.

Thus, $37_{10} = 100101_2$.

(b) By successive division by 2:

		Remainder	
2	37		
2	18	1	↑ LSB
2	9	0	
2	4	1	
2	2	0	
2	1	0	
	0	1	MSB

Thus, $37_{10} = 100101_2$.

(c) By successive subtraction:

		Binary
37	2^5	100000
−32		
5	2^2	100
−4		
1	2^0	1
−1		
0		
	Sum =	100101

Thus, $37_{10} = 100101_2$.

13. Convert decimal fraction 0.8125 into binary form by all possible methods.

Ans. (i) By expansion method: $0.8125 = 1 \times 2^{-1} + 1 \times 2^{-2} + 0 \times 2^{-3} + 1 \times 2^{-4}$

Product $0.5 + 0.25 + 0 + 0.0625 = 0.8125$

Thus, $0.8125_{10} = 0.1101_2$.

(ii) Successive multiplication by 2:

	Fraction	Carry	
0.8125 × 2	0.625	1	MSB
0.625 × 2	0.250	1	
0.250 × 2	0.500	0	
0.500 × 2	0	1	LSB

Thus, $0.8125_{10} = 0.1101_2$.

(iii) Successive subtraction method:

		Binary
0.8125	2^{-1}	0.1
−0.5		
0.3125	2^{-2}	0.01
−0.25		
0.0625	2^{-4}	0.0001
0.0625		
0	Sum =	0.1101

Thus, $0.8125_{10} = 0.1101_2$.

14. Convert binary number to decimal number by all possible methods.

Ans. Take the binary number as 100101.

(i) By expansion method:

$100101 = 1 \times 2^5 + 0 \times 2^4 + 0 \times 2^3 + 1 \times 2^2 + 0 \times 2^1 + 1 \times 2^0$

Product 32 0 0 4 0 1

The sum of different power terms = 32 + 4 + 1 = 37.

Thus, $100101_2 = 37_{10}$.

(ii) Double summing method:

	1	0	0	1	0	1	
	MSB	+	+	+	+	+	
	×2	2	4	8	18	36	
		2	4	9	18	37	(decimal number)

Thus, $100101_2 = 37_{10}$.

15. Convert binary fraction into decimal by all possible methods.

Ans. Let us take the binary fraction as 0.1011.

(i) By expansion method: $0.1011 = 1 \times 2^{-1} + 0 \times 2^{-2} + 1 \times 2^{-3} + 1 \times 2^{-4}$

$$= 0.5 + 0 + 0.125 + 0.0625$$

$$= 0.6875$$

Thus, $0.1011_2 = 0.6875_{10}$.

(ii) Half-summing method:

1	0	1	1	
0.6875	0.375	0.75	0.5	(LSB ÷ 2)
1.375	0.75	1.5		

Thus, $0.1011_2 = 0.6875_{10}$.

2.8 True or False Statement

1. The base of a binary number system is one. [False, correct answer 2]
2. 1KB indicates 100 bits. [True]
3. A string of 8 bits is called byte. [True]
4. MSD carries the maximum value in a number. [True]
5. When an integer is represented in binary, the LSD is always 1. [True]
6. Successive subtraction method is not good for large decimal integer or very small decimal fraction. [True]
7. Double summing method or half-summing method for conversion of a binary representation into decimal is not suitable for binary number of large bits. [False]
8. The expansion method can be better represented by the box form. [True]
9. Non-positional number system is very much in common use. [False]
10. Binary number system is only known to the computer or any digital instrument. [True]

2.9 Questions Exercises

1. Why is binary system known to the digital instrument and not the decimal system?
2. Is decimal number system a positional number system? Discuss.
3. Name a few methods for conversion from integral decimal number to binary system. Comment on the advantage of any of the conversion methods.
4. Name the different methods for conversion of decimal fraction into binary form. Apply any one method to convert 0.875 into binary equivalent.
5. Enumerate the different methods of conversion of binary representation into its decimal equivalent.
6. Explain the double summing method with reference to a specific example.

Chapter 3

Octal and Hexadecimal Number Systems — Their Conversion to Decimal and Binary Systems and vice versa

3.1 Introduction

In Chapter 2, we have discussed about decimal and binary number systems — both of which are positional number systems — the scheme of mutual conversion. The binary representation of a large number is a long chain of 1's and 0's and sometimes it becomes unwieldy long. In view of this difficulty we have introduced two more number systems, one octal; the other hexadecimal for compact representation of a large number. In the present chapter we will deal with these two number systems, formulation and their conversion into decimal and binary equivalents and vice versa.

3.2 Octal Number System

In this number system eight Arabic numerals, 0, 1, 2, ..., 7 are used. So the base or radix of this number system is $8(=2^3)$. Since the base is a power of 2, it is not at all difficult to convert a binary into its octal equivalent and vice versa.

Scheme of conversion. Since the base of the octal system is 8, which is 2 raised to the power 3, we group three bits in a binary representation starting from the least significant bit (LSB). These three-bit-groups are then replaced by their octal equivalents shown in the Table 1 for ready reference.

Table 1

Octal	Binary	Octal	Binary
0	(000)	4	(100)
1	(001)	5	(101)
2	(010)	6	(110)
3	(011)	7	(111)

> The base/radix of octal number system is 8 since eight Arabic numerals (0–7) are used, just as in decimal number system, base/radix is 10 as ten Arabic numerals (0–9) are used.

3.3 Conversion of Binary to Octal and vice versa, also its Decimal equivalent

The method of conversion consists in grouping of three terms from the least significant digit and replacement of the group by its octal equivalent.

Example 3.1. *Convert the binary number* 001100111001 *into its octal and decimal equivalents.*

$$\text{Binary:} \quad 0\,0\,1\ 1\,0\,0\ 1\,1\,1\ 0\,0\,1 \equiv (1471)_8 \equiv (825)_{10}$$

$$\equiv (001)\,(100)\,(111)\,(001)$$

LSB.

Octal: 1 4 7 1

$(1\times 8^3)+(4\times 8^2)+(7\times 8^1)+(1\times 8^0)$

Decimal: $512+256+56+1=825$.

To convert from binary to decimal, the process becomes faster if it is first converted to octal and then from octal to decimal.

Example 3.2. *Convert* 1111010_2 *into its octal equivalent.*

Binary: 1 111 010 $\equiv (172)_8 \equiv (122)_{10}$

= (001) (111) (010)

LSB.

Octal: $172=(172)_8$.

Decimal: $1\times 8^2+7\times 8^1+2\times 8^0=64+56+2=122$.

So, $(1\ 111\ 010)_2=(172)_8=(122)_{10}$.

Example 3.3. *Convert* $(6235.42)_8$ *to binary.*

Octal: 6235.42.

Express each digit of the octal representation into three-bit binary form w.r.t. binary point on both sides. Thus,

Octal: 6 2 3 5 . 4 2

(110) (010) (011) (101) (100) (010)

Binary: $(110\ 010\ 011\ 101\ .\ 100\ 010)_2$.

Decimal: $6\times 8^3+2\times 8^2+3\times 8^1+5\times 8^0+4\times 8^{-1}+2\times 8^{-2}=3072+128+24+5+(0.5+0.031)$

$=3229.531$.

The full conversion scheme is represented schematically as below: *Scheme* 1. Binary → Octal → Decimal *Scheme* 2. Binary ← Octal ← (Decimal)

Illustration:

Scheme 1: Convert $(110111)_2$ to the octal form and then to its decimal equivalent.

Binary: 1 1 0 1 1 1

≡ (110) (111)

LSB.

Octal: 6 7.

Decimal: $6\times 8^1+7\times 8^0=55$.

Thus, $(110\ 111)_2=(67)_8=(55)_{10}$.

Scheme 2: Convert $(55)_{10}$ to octal and then to binary.

Decimal: 55	Divisor	Dividend	Quotient	Remainder	
	8	55			
	8	6	6	7	(LSD)
		0	0	6	(MSD)

$(55)_{10}=(67)_8$.

Octal: 6 7.

Binary: (110) (111) = $(110\ 111)_2$.

Example 3.4. *Convert the decimal number* 119 *into octal and then to binary form.*

Decimal number: 119

Divide successively by 8.

		Quotient	Remainder	
8	119			
8	14	14	7 ↑	(LSD)
8	1	1	6	
8	0	0	1	(MSD)

1 6 7

(001) (110) (111)

∴ Octal: 167.

Binary: $(001\ 110\ 111)_2$.

Example 3.5. *Convert binary* 001110111 *into octal and then to decimal.*

Binary: 001 110 111

(LSD)

(001) (110) (111).

Octal: 1 6 7.

Decimal: $1 \times 8^2 + 6 \times 8^1 + 7 \times 8^0 = 64 + 48 + 7 = 119$.

Hexadecimal system — its conversion into other systems and vice versa

Hexadecimal number system has base or radix equal to 16(= 2^4) with 10 numerals 0–9 as in decimal number system and 6 symbols like A B C D E F to represent decimal numbers 10, 11, 12, 13, 14 and 15 respectively. So to convert a binary into hexadecimal equivalent, the binary bits are to be grouped into four starting from the LSD.

The following table shows hexadecimal and its binary representation:

Hexadecimal	Binary	Hexadecimal	Binary	Hexadecimal	Binary	Hexadecimal	Binary
0	0000	4	0100	8	1000	C (12)	1100
1	0001	5	0101	9	1001	D (13)	1101
2	0010	6	0110	A (10)	1010	E (14)	1110
3	0011	7	0111	B (11)	1011	F (15)	1111

Illustration. Convert binary 10101010110111 into hexadecimal and vice versa.

Binary: 0010 1010 1011 0111

(LSD).

Arranged as a group of 4 binary bits from LSD, we have (1010) (1010) (1011) (0111).

Hexadecimal: 2 A B 7.

Thus, we have $(10101010110111)_2 = (2AB7)_{16}$.

Decimal: $2 \times 16^3 + 10 \times 16^2 + 11 \times 16^1 + 7 \times 16^0 = 8192 + 2560 + 176 + 7 = 108935$.

Thus, $(101010110111)_2 = (2AB7)_{16} = (10935)_{10}$.

> Hexadecimal number system has base/radix equal to 16 (0–9, A, B, C, D, E, F). To convert a binary into hexadecimal equivalent, the binary bits are to be grouped into four starting from the LSD.

Example 3.6. *Follow the sequence of conversion as indicated below: Binary* 10101010110111 *→ Octal → Decimal → Hexadecimal.*

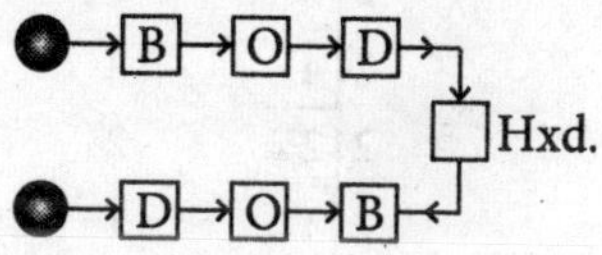

Binary: 10 101 010 110 111
(LSD)

(010) (101) (010) (110) (111) → Grouping of 3 bits.

Octal: 2 5 2 6 7.

Decimal: $2 \times 8^4 + 5 \times 8^3 + 2 \times 8^2 + 6 \times 8^1 + 7 \times 8^0 = 8192 + 2560 + 128 + 48 + 7 = 10935.$

Hexadecimal: Divide by 16.

	Quotient	Remainder		
16 | 10935				
16 | 683	683	7	(LSD)	7
16 | 42	42	11		B
16 | 2	2	10		A
	0	2	(MSD)	2

Hence, we have $(1010101011011)_2 = (25267)_8 = (10935)_{10} = (2AB7)_{16}$.

Given the equivalent forms of decimal, binary, octal and hexadecimal numbers for ready reference.

Decimal to binary:

Decimal	Binary	Decimal	Binary	Decimal	Binary
0	0	6	110	12	1100
1	1	7	111	13	1101
2	10	8	1000	14	1110
3	11	9	1001	15	1111
4	100	10	1010	16	10000
5	101	11	1011	17	10001

Octal to binary:

Octal	Binary	Octal	Binary	Octal	Binary	Octal	Binary
0	000	2	010	4	100	6	110
1	001	3	011	5	101	7	111

Hexadecimal to binary:

Octal	Binary	Octal	Binary	Octal	Binary	Octal	Binary
0	0000	4	0100	8	1000	C	1100
1	0001	5	0101	9	1001	D	1101
2	0010	6	0110	A	1010	E	1110
3	0011	7	0111	B	1011	F	1111

Example 3.7. *Convert decimal* 36.4687 *to binary and octal form.*

For integral part:

Division	Quotient	Remainder	
2 | 36			
2 | 18	18	0 ↑	(LSD)
2 | 9	9	0	
2 | 4	4	1	
2 | 2	2	0	
2 | 1	1	0	
0	0	1	(MSD)

Thus, $(36)_{10} = (100100)_2$.

For fractional part, multiply by 2:

	Fraction	Integer	
0.4687×2	0.9374	0	MSD
0.9374×2	0.8748	1	
0.8748×2	0.7496	1	
0.7496×2	0.4992	1	
0.4992×2	0	1 ↓	LSD

Hence, $(0.4687)_{10} = (0.01111)_2$.

So, $(36.4687)_{10} = (100100.01111)_2$.

Binary integral part: 100 100 → grouping of 3
(LSD)

Octal: 4 4.

Hence, $(36)_{10} = (44)_8$.

For binary fraction:

Grouping of 3 from binary point: 011 111
(011) (110)

Octal: 3 6

Thus, $(0.01111)_2 = (0.36)_8$.

So, $(36.4687)_{10} = (100100.01111)_2 = (44.36)_8$.

Example 3.8. *Complete the conversion process:* $(437)_8$ *→ Binary → Decimal.*

Octal: 4 3 7

Binary: (100) (011) (111) (Equivalent three bits in binary)

Thus, $(437)_8 = (100011111)_2$.

Binary: 1 0 0 0 1 1 1 1 1

Decimal: $1 \times 2^8 + 0 \times 2^7 + 0 \times 2^6 + 0 \times 2^5 + 0 \times 2^4 + 1 \times 2^3 + 1 \times 2^2 + 1 \times 2^1 + 1 \times 2^0$

$= 256 + 16 + 8 + 4 + 2 + 1 = 287.$

Hence, we have $(437)_8 = (100011111)_2 = (287)_{10}$.

Example 3.9. *Convert* $(A3C)_{16}$ *to decimal system.*

Hexadecimal: A 3 C

Binary: (1010) (0011) (1100)

So binary equivalent = $(101000111100)_2$.

Binary: 101000111100.

Decimal: $1\times2^{11}+0\times2^{10}+1\times2^9+0\times2^8+1\times2^7+1\times2^6+1\times2^5+1\times2^4+1\times2^3$
$+1\times2^2+1\times2^1+1\times2^0$
$=2048+512+32+16+8+4=2620.$

Alternative method of conversion: Direct conversion Hex

$\square \rightarrow \boxed{D}$

Hexadecimal: A 3 C

Decimal: $(10\times16^2)+(3\times16^1)+(12\times16^0)=2620.$

Example 3.10. *Convert* 1111110101.01011 *into hexadecimal system.*

Direct conversion: 11 1111 0101 0101 1

(Grouping the bits in groups of 4) ↓ ↓ ↓ ↓ ↓

Hence (0011) (1111) (0101) (0101) (1000)

3 F 5 5 8

$(1111110101.01011)_2=(3F5.58)_{16}$.

Alternative method:

Binary: 1111110101.01011.

Decimal: $(1\times2^9)+(1\times2^8)+(1\times2^7)+(1\times2^6)+(1\times2^5)+(1\times2^4)+(0\times2^3)+(1\times2^2)+(0\times2^1)$
$+(1\times2^0)+(0\times2^{-1})+(1\times2^{-2})+(0\times2^{-3})+(1\times2^{-4})+(1\times2^{-5})$
$=(512+256+128+64+32+16+4+1)\cdot(0.25+0.0625+0.03125)$
$=1013.34375.$

For integral part:

Divisor	Dividend	Quotient	Remainder
16	1013		
16	63	63	5 ↑
16	3	3	15
	0	0	3

$(1013)_{10}=(3F5)_{16}$

	Fraction	Integer
0.34375 × 16	0.5	5 ↓
0.5 × 16	0	8

Thus, $(0.34375)_{10}=(0.58)_{16}$.

So, we have $(1013.34375)_{10}=(3F5.58)_{16}$.

Example 3.11. *Convert* 1BF.25 *from hexadecimal to decimal.*

Hexadecimal: 1 B F . 2 5

Binary: (0001) (1011) (1111) . (0010) (0101).

Decimal: $(1\times2^8)+(1\times2^7)+(1\times2^5)+(1\times2^4)+(1\times2^3)+(1\times2^2)+(1\times2^1)$
$+(1\times2^0).(1\times2^{-3})+(1\times2^{-6})+(1\times2^{-8})$
$=(256+128+32+16+8+4+2+1).(0.125+0.0156+0.0039)=(447.1445)_{10}.$

Reverse transit:

Divisor	Dividend	Quotient	Remainder	
16	447			
16	27	27	15 ↑	LSD
16	1	1	11	
	0	0	1	MSD

$(447)_{10} = 1BF.$

	Fraction	Integer
0.1445×16	0.312	2 ↓
0.312×16	0	5

$(447.1445)_{10} \rightarrow (1BF.25)_{16}.$

Direct conversion

Hexadecimal: $1\,B\,F\,.\,2\,5 = 1\times 16^2 + 11\times 16^1 + 15\times 16^0 + 2\times 16^{-1} + 5\times 16^{-2}$

$$= 447 + (0.125 + 0.01953)$$

$$= (447.1445)_{10}.$$

Reverse conversion

Divisor	Dividend	Quotient	Remainder
16	447		
16	27	27	15 ↑
16	1	1	11
	0	0	1

$(447)_{10} = (1BF)_{16}.$

	Fraction	Integer
0.1445×16	.312	2 ↓
0.312×16	0	5

Thus, we get $(447.1445)_{10} = (1BF.25)_{16}.$

Example 3.12 (Binary Algebra). *The language that the digital instruments covering wide range from computer to camera understand is the binary language in terms of* 0 *and* 1. *The mathematics of the language of digital electronics was first developed by George Boole* (1815–18) *but his Mathematical Analysis of Logic remained practically obscure for about a century after which Claude Shannon discovered the hidden treasure of potentiality in the island of Boole's logic of thought in the form of Boolean Algebra.*

Shannon first saw one-to-one correspondence between Boolean Algebra and the switching systems and ushered in a new era in the field of electronics, called Digital electronics. In this section and that follows next, we would develop the algebra of the digital instruments and implementation of the *laws of thought* by actual electronic switches (formed by valves or transistors) forming what is known as *logic gates.*

Basic algebraic functions include addition, subtraction, multiplication and division. Of course, these four algebraic operations can be reduced to one single operation and that is addition. Thus, subtraction of two numbers x and y may be looked upon as addition of x and $-y$. Multiplication is nothing but repeated addition, and division is nothing but repeated subtraction. Thus, all the algebraic operations need not be implemented by separate hardware but only the hardware performing addition is sufficient. This reduces the size and volume of the machine as well. So all types of algebraic operations are performed by addition only.

> All the four algebraic operations, viz., addition, subtraction, multiplication and division can be reduced to one single operation and that is addition.

Example 3.13 (Binary addition). *The addition of binary number is accomplished through four rules.*

Illustration	Rule
$0+0=0$	Nothing added to nothing gives nothing.
$1+0=1$	One added to nothing gives one.
$0+1=1$	Nothing added to one gives one.
$1+1=10$	One added to one gives 2, decimally equivalent to 10.

Thus, we have

Decimal	Binary
5	101
+6	110
11 (eleven)	$1011 = 1\times2^3+0\times2^2+1\times2^1+1\times2^0$
	$= 11$ (eleven)

> Note that in the binary addition we have simply followed the four rules: $1+0=1$ (rule 2), $0+1=1$ (rule 2 or 3) and $1+1=10$ (rule 4).

Example 3.14. *Perform the following additions of binary numbers and verify by equivalent decimal addition.*

Binary	Decimal
110101	$1\times2^5+1\times2^4+0\times2^3+1\times2^2+1\times2^1+1\times2^0=53$
+100111	$1\times2^5+1\times2^4+0\times2^3+1\times2^2+1\times2^1+1\times2^0=39$
1011100	$1\times2^6+1\times2^4+1\times2^3+1\times2^2=92$

Binary	Decimal	
0101	$1\times2^2+1\times2^0$	$=5$
+0100	1×2^2	$=4$
1001		9

$=1\times2^3+1\times2^0=9.$

Binary	Decimal
110.01101	6.40625
+11.1010	3.625
1010.00001	10.03125

$=1\times2^3+1\times2^1+(1\times2^{-5})=10.03125.$

Binary	Decimal	
111.01	$1\times2^2+1\times2^1+1\times2^0+(2^{-2})$	$=7.25$
10.11	$1\times2^1+(1\times2^{-1}+1\times2^2+1\times2^{-3})$	$=2.875$
1010.001		10.125

$=1\times2^3+1\times2^1+(1\times2^{-3})=10.125.$

Binary	Decimal	
0.1011	$1\times2^{-1}+1\times2^{-3}+1\times2^{-4}$	$=0.6875$
0.1101	$1\times2^{-1}+1\times2^{-2}+1\times2^{-4}$	$=0.8125$
1.1000		1.500

$=1\times2^0+(1\times2^{-1})=1.5.$

Example 3.15 (Binary subtraction).

Illustration	Rule
$0-0=0$	Nothing subtracted from nothing yields nothing.
$1-0=1$	Nothing subtracted from one yields one.
$1-1=0$	One subtracted from one yields nothing.
$10-1=1$	One subtracted from one zero yields one.

Illustrative examples

Find the difference between 10011 and 10001.

Solution: Verification by decimal subtraction

Binary	Decimal
10011	19
10001	−17
00010	2

$1 \times 2^1 = 2$ (verified)

> ***Remember.*** **When a larger number is subtracted from a smaller number, subtract the smaller number from the larger one and the sign of the larger is to prefixed with the result.**
>
> **Thus, when 10011 is to be subtracted from 10001, proceed as follows:**
>
> | | | Decimal |
> |---|---|---|
> | 10001 (17) | 10011 | 17 |
> | 10011 (19) ≡ | 10001 | 19 |
> | | −00010 | −2 |
> | | = −2 | |

Perform the subtraction:

Binary	Decimal
11011	27
−1001	−9
10010	18

= 18 (decimal)

Perform the subtraction:

Binary		Decimal
10111.1101	→	23.8125
−10001.01	→	−17.25
00110.1001		6.5625

$= 1 \times 2^2 + 1 \times 2^1 + (1 \times 2^{-1} + 1 \times 2^{-4})$

$= 6 + (0.5 + 0.0625) = 6.5625.$

Example 3.16 (Binary multiplication). ***Though binary multiplication may be realised through repeated addition of the same number, there are, however, rules for binary multiplication exactly similar to decimal number system multiplication.***

Illustration	Rule
$0 \times 0 = 0$	Nothing multiplied by nothing yields nothing.
$0 \times 1 = 0$	Nothing multiplied by one yields nothing.
$1 \times 0 = 0$	One multiplied by nothing yields nothing.
$1 \times 1 = 1$	One multiplied by one yields one.

For example,

Binary		Decimal
11011		27
×1011		×11
11011	P_1 11011	27
11011	$+P_2$ 11011	27
00000	$S_1 = 1010001$	297
11011	P_3 00000	
100101001	$+P_4$ 11011	
	$S_2 = 110110$	

$S_1 =$ 1010001
$S_2 =$ 110110

100101001 $= 1 \times 2^8 + 1 \times 2^5 + 1 \times 2^3 + 1 \times 2^0 = 256 + 32 + 8 + 1 = 297.$

Note. Add the first two partial products $P_1 + P_2 = S_1$, then $P_3 + P_4 = S_2$, then $S_1 + S_2 =$ final result.

Example 3.17 (Binary division). *Binary division is based on the following four rules:*

Illustration	Rule
$0 \div 1 = 0$	Nothing divided by one gives nothing.
$1 \div 1 = 1$	One divided by one gives one.
$1 \div 0$	One divided by nothing is meaningless.
$0 \div 0$	Nothing divided by nothing is also meaningless.

Binary division can be done in the same way as the division in decimal system. Thus, if we want to divide 1011111 by 101, we proceed as below:

```
101) 1011111 (10011 ← Quotient
     101
     ---
       111
       101
       ---
       0101
       0101
       ----
       0000 ← Remainder
```

Decimal equivalent of the dividend: $1011111 = 1 \times 2^6 + 0 \times 2^5 + 1 \times 2^4 + 1 \times 2^3 + 1 \times 2^2 + 1 \times 2^1 + 1 \times 2^0$

$= 64 + 16 + 8 + 4 + 2 + 1 = 95.$

Divisor: $101 = 1 \times 2^2 + 0 \times 2^1 + 1 \times 2^0 = 5.$

Quotient: $10011 = 1 \times 2^4 + 0 \times 2^3 + 0 \times 2^2 + 1 \times 2^1 + 1 \times 2^0 = 19.$

So the result of decimal system division is:

```
5) 95 (19
   5
   --
   45
   45
   --
    0
```

Hence, the result is verified.

Example 3.18 (Subtraction Method in terms of Complements of Numbers). *The subtraction method in terms of complements of numbers in all positional number systems can be of two types:* (a) *Radix-minus-one complement method* (b) *Radix complement method.*

What is complemented of a number in decimal number system?

In decimal number system the radix is 10, so we often talk about 9's (10 – 1) or 10's complement of a number x. The highest number in unit digit is 9, two digits is 99; three digits 999, etc. in terms of 9. So the 9's complement of x is $(9-x)$, when x is 1 digit; $(99-x'x)$ when $x'x$ is of two digits, etc.

Thus, the 9's complement of the numbers 4,65,825.

$$\begin{array}{ccc} 9 & 99 & 999 \\ \underline{-4} & \underline{65} & \underline{825} \\ 5 & 34 & 174 \end{array}$$

So the 9's complement of the numbers 4,65 and 825 are 5,34 and 174 respectively.

Similarly, 10's complement of a number x is $(10-x)$, when x is a unidigit number, $(100-x)$ when the number is a two-digit number, etc.

Thus, the 10's complements of 6,15,120 are (10 – 6), (100 – 15) and (1000 – 120) respectively.

10's complement of a number may also be found by first finding its 9's complement, then adding 1 to the LSD. For example, 10's complemented of 4 is (10 – 4) = 6; also 9's complement of 4 is 5, then adding 1 we get 6.

Example 3.19 (Subtraction of two decimal numbers in terms of complements). *There are two complement methods to find the difference between two decimal numbers $(x-y)$. They are:* (a) *Radix-complement method* (b) *Radix minus 1 complement method.*

Radix complement method: The radix of a decimal number is ten (10), so this method is 10's complement method. This method is outline as below:

Suppose, we are to find the difference between x and y, written as $(x-y)$.

Step 1. Find the 10's complement of y.

Step 2. Adding the complement with x.

Step 3. If there is a carry in operation (2) discard it to obtain the difference $(x-y)$. If there is no carry, re-complement and then attach a negative sign to the re-complement and that will give us the difference $(x-y)$.

Example 3.20. $x = 25$, $y = 75$ (*by 10's complemented method*).

$$\begin{array}{l} \text{So } 25 \\ \underline{-75} \\ -50 \end{array}$$

10's complement of 75 = 100 – 75 = 25.

Add the complement to x, we get

$$\begin{array}{r} 25 \\ \underline{+25} \\ \text{(No carry) } 50 \end{array}$$

Since there is no carry, we complement 50 and that (100 – 50) = 50 and attach a negative sign. So the result of (25 – 75) is –50.

Schematic representation: 10's complement of 75 = 25

+x = 25

No carry: 50

Re-complement 50 = 50

Hence, result = −50.

Example 3.21. $x = 368$, $y = 279$ (*by 10's complement method*).

$x - y$: 368   10's complement of 279 = 1000 − 279 = 721

−279   Add to x + 368

89 (by conventional method)   Carry 1 ← [1] 089

Discarding carry, we have result = 89.

Example 3.22. $x = 279$, $y = 368$ (*by 10's complement method*).

Conventional method: 279

−368

−89

10's complement method: 10's complement of 368 = (1000 − 368)

= 632

Add x to the complement = 632

+279

No carry 911

Re-complement = 1000 − 911 = 89.

Attach a negative sign to re-complement.

Hence, result = −89.

Radix minus 1 *complement method:* This method of subtraction is also called 9's complement method. It is similar to 10's complement method. So to obtain $(x - y)$, we have

Step 1. Find 9's complement of y.

Step 2. Add the complement to x.

Step 3. If there is a carry 1 in operation (2), add it to the sum (y and 9's complement of y). If there is no carry re-complement the sum and then attach a negative sign to the re-complement to obtain the difference $(x - y)$.

We illustrate the second method to find the difference $(x - y)$ in the examples cited above.

Example 3.23. $x = 25$, $y = 75$.

Conventional method: 25   9's complement method: 9's complement of 75 = (99 − 75) = 24

−75   Add x to the complement + = 25

−50   (No carry) 49

Re-complement 49 = (99 − 49) = 50

Attach a negative sign to it = −50

Result = −50

Example 3.24. $x = 368, y = 279.$

Conventional method:

$$\begin{array}{r} 368 \\ -279 \\ \hline (x-y) = 89 \end{array}$$

9's complement method: 9's complement of 279 = (999 − 279) = 720

Add x to it +368

Carry 1 ← [1] 088

Add 1 to 088 = 89

Result = 89.

Example 3.25. $x = 279, y = 368.$

Conventional method:

$$\begin{array}{r} 279 \\ -368 \\ \hline (x-y) = -89 \end{array}$$

9's complement method: 9's complement of 368 = (999 − 368) = 631

Add x to the complement +279

(No carry) 910

Re-complement (999 − 910) = 89

Attach negative sign = −89

Result = −89.

Example 3.26 (Binary subtraction in terms of complement method). ***The binary number system is based on radix two (2) and so here we will have 1's compliment (radix −1) and 2's complement (radix-complement) of a binary number.***

1's complement of a binary number is obtained by subtracting each bit from 1.

2's complement of a binary number is obtained by adding 1 to the 1's complement of the given number. Thus, we get

1's complement of 1 and 0 as:

$$\begin{array}{r} 1 \\ -1 \\ \hline 0 \end{array} \qquad \begin{array}{r} 1 \\ -0 \\ \hline 1 \end{array}$$

2's complemented of the same binary bit is:

$$\begin{array}{r} 0 \\ +1 \\ \hline 1 \end{array} \qquad \begin{array}{r} 1 \\ +1 \\ \hline 10 \end{array}$$

So the rule of getting 1's complement is to replace 1 by 0 and 0 by 1 in a binary number.

Example 3.27. $x = 1100, y = 1010,$ *find* $(x - y).$

Conventional method:

$$\begin{array}{r} 1100 \\ -1010 \\ \hline 0010 \end{array}$$

1's complement method:

$y = 1010$

1's complement of $y = 0101$

Add x +1100

Carry 1 ← [1] 0001

Add to the sum +1

Result 0010

So, 1's complement method of binary subtraction method is radix minus 1 complement method.

***2's complement method of binary subtraction*: It is the radix-complement method.**

Example 3.28. $x = 10110, y = 10010,$ *find* $(x - y).$

2's complement method: 1's complement of $y = 01101$

2's complement +1

01110

Add x +10110

Carry 1 ← [1] 00100

Discard carry, hence the result = 00100.

Example 3.29. $x = 1100$, $y = 1010$, *find* $(x - y)$.

Conventional method: 1100
−1010
0010

1's complement method: $y = 1010$
1's complement of $y = 0101$
Add x + 1100
Carry 1 ← [1] 0001
Add it to the sum +1
Result = 0010

So, 1's complement method of binary subtraction method is radix minus 1 complement method.

2's complement method of binary subtraction: It is the radix-complement method.

Example 3.30. $x = 10110$, $y = 10010$, *find* $(x - y)$.

2's complemented method: 1's complement of $y = 01101$
2's complement +1
01110
Add x + 10110
Carry 1 ← [1] 00100

Discard carry, hence the result = 00100.

Example 3.31. $x = 1001$, $y = 1101$ *find* $(x - y)$.

2's complement method: 1's complement of $y = 0010$
2's complement +1
0011
Add x + 1001
(No carry) 1100

1's re-complement of the sum: 0011
+1
2's re-complement 0100

Add a negative sign and hence we get the result = −0100.

Example 3.32 (Complements in octal system). *In octal number system, the radix is eight (8), so there are two types of complements, 7's complement (radix-minus-one) and 8's complement (radix complement). So subtraction of two octal numbers may be obtained through 7's complement or 8's complement. The methods are similar to 9's complement or 10's complement method used in decimal number system.*

Example 3.33. $x = (37)_8$, $y = (16)_8$, *find* $(x - y)$.

7's complement method:

Conventional decimal equivalent:

$(37)_8 = 3 \times 8^1 + 7 \times 8^0 = (31)_{10}$

$(16)_8 = 1 \times 8^1 + 6 \times 8^0 = (14)_{10}$

Hence, difference $= (17)_{10} = (21)_8$.

7's complement method:

7's complement of $y = (77 - 16) = 61$
Added x to the complement + 37
Carry 1 ← [1] 20
Add carry to the LSD 1
Result = 21

8's complement method:

$(37)_8$
$-(16)_8$
$\overline{(21)_8}$.

8's complement of $y = $ (7's complement of y) + 1
$= (77 - 16) + 1 = 62$
Add x to 8's complement: + 37
Carry 1 ← [1] 21
Discard carry, hence result = 21.

Example 3.34. $x = (45)_8$ *and* $y = (12)_8$. *Find* $(x - y)$ *by 7's complement and 8's complement method.*

(i) *By 7's complement method*:

$(45)_8$
$-(12)_8$
$\overline{(33)_8}$

7's complement of $y = (77 - 12) = 65$
Add x to the complement + 45
Carry 1 ← [1] $\overline{32}$
Add carry + 1
Result $\overline{(33)_8}$

Conventional decimal equivalent:

$(45)_8 = 4 \times 8^1 + 5 \times 8^0 = (37)_{10}$

$(12)_8 = 1 \times 8^1 + 2 \times 8^0 = (10)_{10}$

Thus, $(45)_8 - (12)_8 = (37)_{10} - (10)_{10} = (27)_{10}$.

Again, $(44)_8 = 3 \times 8^1 + 3 \times 8^0 = (27)_{10}$, hence confirmed.

Remember 8's complement of y = (7's complement of y) + 1

(ii) *By 8's complement method*: 8's complement of 12 = (7's complement of 12) + 1 = $(77 - 12) + 1 = 66$
+45
Carry 1 ← [1] 33

Discard carry, hence, we get the result = $(33)_8$.

Example 3.35. $x = (67)_8$, $y = (167)_8$. *Find* $(x - y)$ *by 7's complement and 8's complement method.*

(i) 7's *complemented method*: 7's complement of $167 = (777 - 167) = 610$
Add x to the complement +67
(No carry) $\overline{677}$

Re-complement $= (777 - 677) = (100)$.

Attach a negative sign and get the result: $-(100)_8 = -(64)_{10}$.

Conventional decimal equivalent: $(67)_8 = 6 \times 8^1 + 7 \times 8^0 = 55$

$(167)_8 = 1 \times 8^2 + 6 \times 8^1 + 7 \times 8^0 = 119$

Hence, $(55 - 119) = -64$.

(ii) 8's *complement method*:

8's complement of 167 = 7's complement of 167 + 1 = 610 + 1 = 611.

Add x to the complement = 611 + 67 = 700 (No carry).

Re-complement = $(777 - 700) = (077)$.

8's complement = 077 + 1 = 100.

Attach a negative sign and we get the result = −100.

3.4 How does a computer recognise the sign of a number?

In decimal number system, a positive number is indicated by a + sign before the number or no sign at all. But a negative number is denoted by a – sign before.

In a computer, however, there is no provision for such indication, because in a computer everything is expressed in terms of bits. So additional bits 0 is used to indicate a negative number. Thus, in 4 bits, +3 is indicated as 0011 and –3 is indicated as 1011, the prefix 0 and 1 represent respectively the positive and negative sign of the number 3.

The sign of any positive number is represented by 0 to the left of its binary expression and the magnitude by the binary expression itself. A negative number, on the other hand, is expressed by prefixing 1 to the binary expression but the magnitude of the number may be expressed in one of the following three ways: (a) Sign-magnitude (b) Sign 1's complement (c) Sign 2's complement.

Thus, +3 and –3 can be expressed in binary form as follows:

	+3	–3
Sign-magnitudc	0011	1011
Sign 1's complement	0100	1100
Sign 2's complement	0011	1101

A positive number is indicated by an additional bit 0 to the left of the MSD and a negative number is denoted by an additional bit 1 to the left of the MSD. Thus, binary 0100 denotes decimal value of + 4 whereas 1100 denotes a decimal value of –4.

A few positive and negative numbers are shown in the table below expressed in binary forms:

Decimal numbers	Sign-magnitude	Sign 1's complement	Sign 2's complement
0	0000	0000	0000
+1	0001	0001	0001
–1	1001	1110	1111
+2	0010	0010	0010
–2	1010	1101	1110
+3	0011	0011	0011
–3	1011	1100	1101
+4	0100	0100	0100
–4	1100	1011	1100

3.5 What is the representation of a real number in a computer?

There are two ways to represent a real number in a computer: (a) Fixed point representation (b) Floating point representation.

1. *Fixed point representation.* If the binary point is fixed at the left end or at the beginning of a binary number, the number is a fraction. Thus, 0.110 is a fraction.

If the binary point is fixed at the extreme right or at the end of a binary number, it is an integer. Thus, 110.0 is an integer.

If the binary point is fixed at some position of a binary number, the number has both integral part as well as fractional part. Thus, 1101.11 has integral part 1101 and fractional part 0.11.

3.6 Floating Point Representation

In this representation of a number, the position of the binary point can be shifted in position without altering the number.

Any number can be expressed as $N = mb^e$, where N is the number, m is called the mantissa, b is the base of the system and e is the exponent. By changing the mantissa and exponent, any number can be expressed in different equivalent forms. Thus, the binary number 0.1110 is a fraction, it can be put as:

$0.1110 = 11.10 \times 2^{-1}$, here $11.10 = m, -2 =$ exponent and 2 is the base.

$11.0001 = 0.11001 \times 2^2$, $0.11001 = m$, $2 =$ exponent, $2 =$ base.

Representation of a decimal number in floating point method:

$$14.162 = \underset{m}{0.14162} \times \underset{b=10}{10}^{\overset{}{2}} = \underset{m}{1.4162} \times \underset{b=10}{10}^{1}.$$

$e = 2$ (first), $e = 1$ (second)

Floating point addition:

1. Additional of two numbers having same exponents: $104.32 \times 10^2 + 110 \times 10^2 = 214.32 \times 10^2$.

Remarks: Same exponents, hence mantissa added.

2. Addition of two numbers having different exponents: By floating point method, the exponents are made equal and then to numbers are added as in (1).

$101.37 \times 10^2 + 10.0 \times 10^{-3} = 101.37 \times 10^2 + 1.00 \times 10^{-2} = 102.37 \times 10^{-2}$.

3. Two binary numbers to add: $110.01 \times 2^2 + 0.110 \times 2^{-1} = 110.01 \times 2^2 + 0.000110 \times 2^2 = 110.010110 \times 2^2$.

3.7 Data representation to a computer through different types of coding

Computers basically accept two types of data: (a) Numbers (b) Characters.

For feeding a numeric data to the computer, it must be transformed into its equivalent binary, because the computer knows only the binary language.

Character may be of two types: (a) Alphabetic (b) Alphanumeric.

1. The alphabetic characters refer to the 26 English alphabets in block letters A, B, C, etc. or small letters a, b, c, etc.

2. Alphanumeric data consists of both numeric character and alphabetic character.

For example, address of your residence.

Both the numeric data and alphabetic or alphanumeric data are to be transformed into a language compatible with the technology of the machine. To form of the data used inside a computer to perform a user's instruction is referred to as the internal data representation. For this purpose, we use different codes to form the internal data representation. After the instruction being executed by the machine, the output of the machine requires to be decoded so that the output is understood by the user.

There are different types of codes used for internal data representation:

BCD *code*. It is binary coded decimal code used for dealing with numeric data. If we use four bits, we can represent each digit of decimal number system from 0 to 9. So four bits are used in BCD code. This is the most popular BCD also called 8421 code, the MSB has a weight 8 and the LSB has a weight 1. This is shown in the table below:

Decimal	8 4 2 1 BCD	Decimal	8 4 2 1 BCD
0	0 0 0 0	5	0 1 0 1
1	0 0 0 1	6	0 1 1 0
2	0 0 1 0	7	0 1 1 1
3	0 0 1 1	8	1 0 0 0
4	0 1 0 0	9	1 0 0 1

Thus, BCD code of 47 in decimal system is 0100 0111

4 7

Remarks. In BCD code each digit of decimal number is converted into its four-bit binary form and not the entire number into its binary equivalent.

Thus, again $(11)_{10} = 0001\ 0001$.

If the decimal number is entirely converted to its binary form: $(11)_{10} = (1011)_2$.

So, in BCD code more bits are needed. But the computer recognises 4-bit representation as in BCD, and it is more convenient to remember and handle.

3.8 ASCII (ASKEE) Code

A computer is a binary machine; to communicate with it is alphabetic letters and decimal numbers, translation codes are necessary.

The BCD code speaks of numeric data representation but we require to represent alphabetic or alphanumeric data. So for comprehensive representation we have introduced two systems of codes, the first being *American Standard Code for Information Interchange* (ASCII), the second being an extension of this code, called *Extended Binary Coded Decimal Interchange Code* (EBCDIC pronounced as EB-SI-DIC).

In ASCII code a string of 7 bits is used thus leading to a generation of $2^7 = 126$ unique strings. So by ASCII code, we can easily represent 26 capital alphabets, 26 small alphabets, 10 digits (decimal system) and 32 other types of characters. In older computers, the number of characters was less than 64 and so 6-bit string was used to code the characters, but in our modern machines the characters have increased, hence, the 6-bit coding is inadequate. So, in ASCII coding using 7-bit string, characters numbering up to 94 or more can be represented. 6-bit coding or 7-bit coding was used as ASCII coding in older machine.

Extended ASCII, also called EBCDIC coding, used 8-bit, called a byte, so $2^8 = 256$ characters can be represented in this extended coding system.

3.9 Image Representation — Bitmap Representation

To represent an image, monochrome or coloured in digital form a matrix containing a specified number of rows and columns is used. Each element of this matrix is a pixel, i.e., a picture element. In a computer memory the image pixels are stored in an array. In case of monochrome image, a single bit is used to code the position of every pixel of the image but in colour images, bits are used to describe the position and colour of each pixel.

Bitmap is a generic term used for any type of image representation in a computer. Besides single bit, bitmap representation as in Monochrome images, there are 8-bit, 16-bit and 24-bit bitmap representations.

In 256 level, the bitmap initially contains a palette that displays brightness level for monochrome images, R-B-G (Red-Blue-Green) for colour images.

3.10 Short Theory Questions

1. **What is the base of an octal number system?**

   Ans. **The base of an octal number system is 8, with Arabic symbols 0, 1, 2, …, 7. Thus, base of this system is 2 raised to the power 3, hence, can be easily represented by three bits.**

2. **Is hexadecimal number system a positional number system?**

   Ans. **Yes, it is, because its base is 16 comprising 0, 1, 2, …, 9 Arabic symbols and A, B, C, D, E, F English alphabetic symbols representing numbers 10, 11, 12, 13, 14, 15 respectively. Since the base is $16 = 2^4$ any number in this system can be represented by 4 bits. Since the base is a positive integer, the value of a digit depends on the position of the digit, hence, the hexadecimal number system is positional number system.**

3. **How many types of data are possible?**

   Ans. **There are 4 types of possible data: (a) Numeric (b) Alphabetic (c) Alphanumeric (d) Logical.**

4. **Is it true that 1 KB equals to 1000 bytes?**

   Ans. **No, 1 byte consists of a string of 8 bits, so 1 KB = 125 bytes.**

5. **Are ASCII and EDCDIC codes interconvertible?**

   Ans. **Yes, in ASCII, the number of bits used in a string is 7 but in EBCDIC code it is 8, so EBCDIC code may be looked upon as an extension of ASCII code.**

6. **What do you mean by the following?**

   (a) 1's complement (b) 2's complement (d) 9's complement.

   Ans. **1's complement and 2's complement:**

   In binary number system, the radix or base is two; hence, we can talk about 1's complement or 2's complement of a binary number.

   1's complement of a binary number is obtained by subtracting each bit from 1, so complement of 1 and 0 are $1 - 1 = 0$ and $1 - 0 = 1$ respectively.

   1's complement of a binary number is obtained by replacing 1 by 0 and 0 by 1.

   To find the 2's complement of a binary number add 1 to the LSB of 1's complement of the number. 2's complement is obtained by adding 1 to the LSB of 1's complement.

Binary	1's complement	2's complement
1	0	0 + 1 = 1
0	1	1 + 1 = 10
110	001	001 + 1 = 010
010	101	101 + 1 = 110
110	001	001 + 1 = 010

   9's complement**: In decimal number system, the maximum digit's in unit's place, ten's place, hundred's place are 9,99,999, etc. So, 9's complement of a decimal number is $(9-x_1)$, $(99-x_2x_1)$, $(999-x_3x_2x_1)$, etc. Thus,**

Decimal	9's complement
4	$9-4=5$
69	$99-69=30$
830	$999-830=169$

10's complement: 10's complements are obtained as $(10-x_1), (100-x_2x_1), (100-x_3x_2x_1)$, etc. Thus,

Decimal	10's complement
4	$10-4=6$
69	$100-69=31$
830	$1000-830=170$

10's complement of a decimal number is thus obtained by adding 1 to the LSD of the 9's complement of the number.

7. Convert the following decimal numbers to octal form: (a) 2502 (b) 0.3125 (c) 332.

Ans. (a) By successive division by 8:

8 | 2502	Quotient	Remainder	
8 | 312	312	6	
8 | 39	39	0	
8 | 4	4	7	
0	0	4	(MSD)

So, $(2502)_{10}=(4706)_8$.

(b) By multiplication by 8:

	Integer	Fraction
0.3125×8	2 MSD	0.5
0.5×8	4 MSD	0

So, $(0.3125)_{10}=(0.24)_8$.

(c)

8 | 332	Quotient	Remainder
8 | 41	41	4
8 | 5	5	1
0	0	5

Thus, $(332)_{10}=(514)_8$.

8. Convert the following octal numbers to decimal equivalent: (a) 355 (b) 514 (c) 0.24.

Ans. (a) $355=3\times 8^2+5+5\times 8^1+5\times 8^0=237$.

Thus, $(355)_8=(237)_{10}$.

(b) $514=5\times 8^2+1\times 8^1+4\times 8^0=320+8+4=332$.

Thus, $(514)_8=(332)_{10}$.

(c) $0.24=2\times 8^{-1}+4\times 8^{-2}=0.25+0.0625=0.3125$.

Thus, $(0.24)_8=(0.3125)_{10}$.

9. Solve the following:

(a) $(124)_{10}=?_2$ (b) $(49)_8=?_2$ (c) $(1234F)_{16}=?_2$.

Ans. (a)

2 | 124	Quotient	Remainder	
2 | 62	62	0	LSD
2 | 31	31	0	
2 | 15	15	1	
2 | 7	7	1	
2 | 3	3	1	
2 | 1	1	1	
2 | 0	0	1	MSD

Thus, $(124)_{10} = (1111100)_2$.

(b) $(49)_8$ (Octal) → (Decimal) → (Binary).

$(49)_8 = 4 \times 8^1 + 9 \times 8^0 = (41)_{10}$.

$(41)_{10} =$	Divisor	Dividend	Quotient	Remainder	
	2	41			
	2	20	20	1	LSD
	2	10	10	0	
	2	5	5	0	
	2	2	2	1	
	2	1	1	0	
		0	0	1	MSD

Thus, $(49)_8 = (101001)_2$.

10. Find the binary difference by 1's complement method: (a) 100101 − 100011, (b) 11111 − 10000.

Ans. (a) Here $x = 100101$, $y = 100011$.

1's complement of $y = 011100$

Add x to the complement + 100101

Carry 1 [1] 000001

Adding carry to the result + 1

Hence, we have $(x - y) = 000010$.

(b) Here $x = 1111$, $y = 000010$.

1's complement of $y = 01111$

Add $x = 11111$

Carry 1 [1] 01110

Add Carry + 1

Hence, $(x - y) = 01111$.

3.11 Short Theory Questions with Answers

1. State the differences between octal and hexadecimal number systems.
2. Is hexadecimal number system a positional number system? Give reasons for your answer.
3. What do you mean by (a) 10's complement (b) 1's complement (c) 2's complement of a number?
4. What do you mean by BCD? Why is it at all necessary?
5. How many bits are there to form a string for representation of character in an ASCII code?
6. What is EDCDIC code? What is its advantage over older ASCII code?
7. Find the octal addition of the following: (a) $(37)_8 + (61)_8$ (b) $(43)_8 + (66)_8$. [(a) 120 (b) 131]
8. Find the hexadecimal addition of the following: (a) $(AB1)_{16} + (F58)_{16}$ (b) $(CA5)_{16} + (45A)_{16}$. [(a) 1A09, (b) 10FF]
9. What do you mean by sig ed numbers? How do you indicate the sign of a number to be accepted by a computer?
10. What do you mean by fixed and floating point representation?

3.12 Long Questions

1. Find the complements of a number in octal and hexadecimal system?
2. How do you code a data for representation in a computer?
3. Discuss how a signed number is represented in a positional number system?
4. What are the different types of complements of a number? Give examples.
5. What do you mean by bitmap representation? How is an image coded in a bitmap?

Chapter 4

Boolean Algebra — Digital Gates

4.1 Introduction

The basic concept of digital electronics is the outcome of the work of George Boole. In Boolean algebra the variables are allowed to assume only two values, mutually exclusive like YES or NO; ON or OFF; usually written as 1 or 0. This algebra is useful in designing digital systems where binary numbers and codes are used in data processing. It was Claude Shannon who first realised the importance of this algebra of logic and observed that electric and electronic circuits can be analysed and synthesised with the help of this algebra. In the world of electrical engineers, this algebra is popularly known as *switching algebra.*

A basic question in Boolean algebra may be asked: If Y be a function of two binary variables A, and B, then in how many ways Y can be defined? To answer the above question we see that A is allowed to take up two values 1 or 0, so also is B. So the possible combinations of A, B are

$$\begin{pmatrix} A & B \\ 0 & 0 \\ 1 & 0 \\ 0 & 1 \\ 1 & 1 \end{pmatrix} \quad \begin{array}{l} \text{Four possible combinations} \\ \{A, B\} = (\{0,0\}, \{1,0\}, \{0,1\}, \{1,1\}). \end{array}$$

For each of the combinations of A, B, we can define the function $Y = f(A, B)$ and so there will be $2^4 (= 16)$ possible Boolean functions as specified in the Table 4.1 below:

Table 4.1: Possible Boolean Functions

A	B	f_0	f_1	f_2	f_3	f_4	f_5	f_6	f_7	f_8	f_9	f_{10}	f_{11}	f_{12}	f_{13}	f_{14}	f_{15}	Remarks	
0	0	0	0	0	0	0	0	0	0	1	1	1	1	1	1	1	1	$f_{10} = \bar{B} = \bar{f}_5$	
0	1	0	0	0	0	1	1	1	1	0	0	0	0	1	1	1	1	$f_{12} = \bar{A} = \bar{f}_3$	
1	0	0	0	1	1	0	0	1	1	0	0	1	1	0	0	1	1	$f_8 = \bar{f}_7$	
1	1	0	1	0	1	0	1	0	1	0	1	0	1	0	1	0	1	$f_9 = \bar{f}_6$	$f_2 = \bar{f}_{13}$
																		$f_{14} = \bar{f}_1$	
																		$f_4 = \bar{f}_{11}$	

$\bar{f} \equiv f'$ is complement of f or NOT of f.

Gates defined in terms of Boolean function as in Table 4.1:

1. f_0 *and* f_{15} are not functions in real sense of the term, because they are independent of A and B.
2. f_1 is referred to as AND function and is characterised by saying $Y = 1$, if and only if $A = 1$ and $B = 1$. It is represented by

$$T = A \text{ AND } B = A \cdot B = AB. \tag{4.1}$$

In symbol,

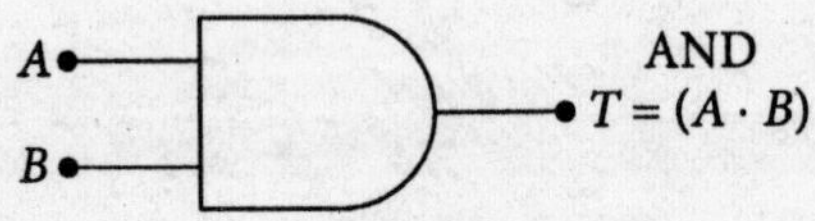

Fig. 4.1

3. f_3 is equal to A and do not care for B.
4. f_5 is equal to B and do not care for A.
5. f_7 is characterised by saying $Y = 1$, when $A = 1$ or, $B = 1$. So this function is referred to as OR-*operation.* It is also sometimes called inclusive OR, because it includes $Y = 1$, when both of A and B are 1.
   It is written as

$$Y = A \text{ OR } B = A + B. \tag{4.2}$$

In symbol,

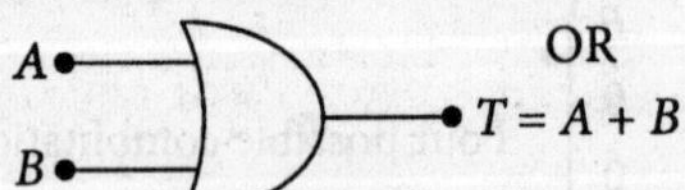

Fig. 4.2

6. f_6 — it is characterised by saying $Y = 1$ when $A = 1$ or, $B = 1$ and not when $A = 1$ and $B = 1$ simultaneously. It is, therefore, called also as an *Exclusive* OR or X-OR.
   It is written as

$$Y = A \oplus B = A\bar{B} + \bar{A}B. \tag{4.3}$$

In symbol,

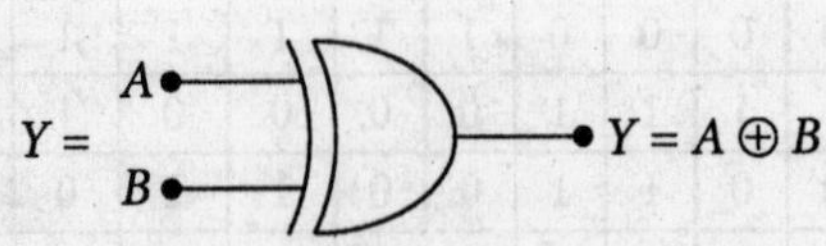

Fig. 4.3

7. $f_8 = \bar{f}_7$, it is complement of f_7 which is *inclusive* OR.
   It is OR-operation followed by NOT, called NOR-operation. It is written as

$$Y = \overline{A + B}. \tag{4.4}$$

In symbol,

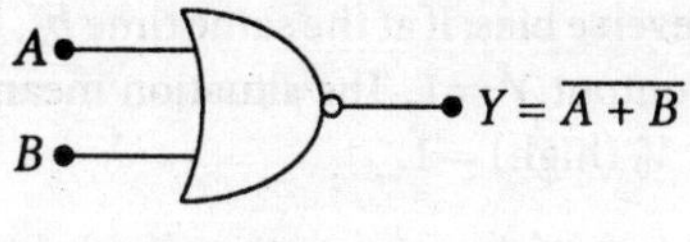

Fig. 4.4

8. $f_9 = \bar{f}_6$, it is complement of f_6 which is exclusive OR and hence, it is called X-NOR written as

$$Y = \overline{A \oplus B} = A \odot B. \tag{4.5}$$

In symbol,

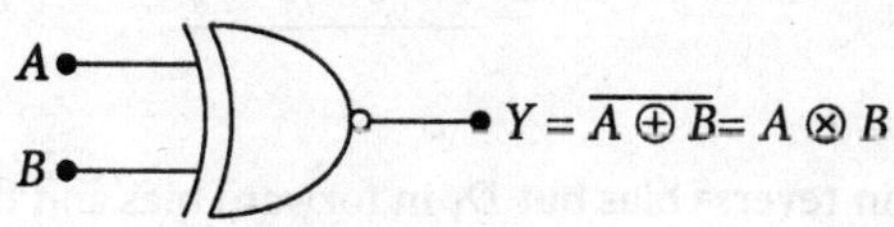

Fig. 4.5

9. $f_{14} = \bar{f}_1$, it is the complement of f_1, so it represents AND followed by NOT, called NAND and it is written as

$$Y = \overline{AB}. \tag{4.6}$$

In symbol,

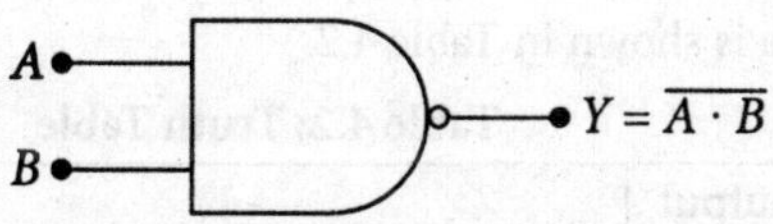

Fig. 4.6

10. $f_4 = \bar{f}_{11}$, complement of f_{11} and vice versa.
11. $f_2 = \bar{f}_{13}$, complement of f_{13} and vice versa.

4.2 More details about the basic gates

AND *gate.* It refers to the circuit which gives an output state 1 when all the inputs are simultaneously present, i.e., when all the inputs are in logical state 1. The standard symbol for a three-input AND gate is shown in Fig. 4.7(a) and its implementation circuit in Fig. 4.7(b).

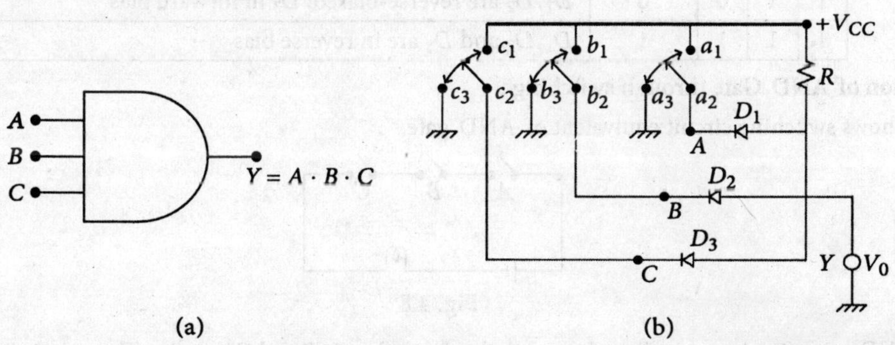

Fig. 4.7

A, *B* and *C* are the input terminals of the gate in Fig. 4.7(a).

Implementation circuit. The implementation circuit for the gate is shown in Fig. 4.7(b). How does it operate? Joining a_1, a_2, the diode D_1 is in reverse bias; if at the same time b_1, b_2 and c_1, c_2 are connected, all the three diodes are in reverse bias. So the output $Y = 1$. The situation means, $Y = 1$, when the inputs $A = 1, B = 1$ and $C = 1$. We take 0 volt = 0 and V_0 (high) = 1.

Performance of the circuit

1. Connect a_1, a_2; the input A is 1, the diode D_1 is in reverse bias and if b_2, b_3 and c_2, c_3 are connected D_2, D_3 are in forward bias. The output is in 0 state.

A	B	C	Y
1	0	0	0

2. When $A = 0, B = 1, C = 1$.
   The diodes D_2, D_3 are in reverse bias but D_1 in forward bias and the output $Y = 0$.

A	B	C	Y
0	1	1	0

In this way, we see that in this gate, only when $A = 1$, $B = 1$ and $C = 1$, simultaneously, the output is in state 1.

The truth table of the gate is shown in Table 4.2.

Table 4.2: Truth Table

A	B	C	Output Y	Remarks
0	0	0	0	All of D_1, D_2, D_3 are forward-biased
0	0	1	0	D_3 is in reverse bias while D_1 and D_2 are in forward bias
0	1	0	0	D_2 is in reverse bias while D_1, D_3 are in forward bias.
0	1	1	0	D_2 and D_3 are reverse-biased while D_1 is in forward bias
1	0	0	0	D_1 reverse-biased and D_2 and D_3 in forward bias
1	0	1	0	D_1, D_3 reverse-biased while D_2 forward-biased
1	1	0	0	D_1, D_2 are reverse-biased; D_3 in forward bias
1	1	1	1	D_1, D_2 and D_3 are in reverse bias

Realisation of AND Gate through switching

Fig. 4.8 shows switching circuit equivalent of AND gate.

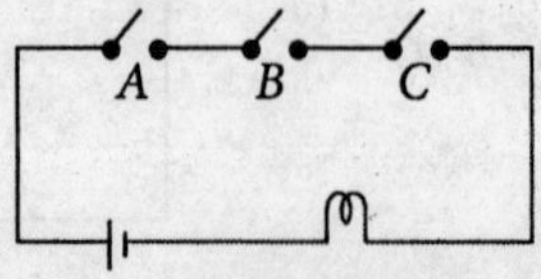

Fig. 4.8

The AND operation is accomplished through three switches A, B and C in series. The switch A represents a binary variable which can be in two states only, one ON and the other, OFF, respectively represented by logic symbol 1 and 0. There are eight possible states of the switches.

Table 4.3: Eight possible status of switches

Switch A	Switch B	Switch C	State of the lamp (output)	
Open (0)	Open	Open	OFF	0
Open (0)	Open	Closed	OFF	0
Open (0)	Closed	Open	OFF	0
Open (0)	Closed	Closed	OFF	0
Closed (0)	Open	Open	OFF	0
Closed (0)	Open	Closed	OFF	0
Closed (0)	Closed	Open	OFF	0
Closed (0)	Closed	Closed	ON	1
[00001111]	[00110011]	[01010101]		

OR *Gate.* It refers to the circuit which gives an output state when one of the inputs are in state 1. Consider the two-input OR gate shown in Fig. 4.9(a), symbol shown in Fig. 4.9(b) and the implementing circuit is drawn in Fig. 4.9(c).

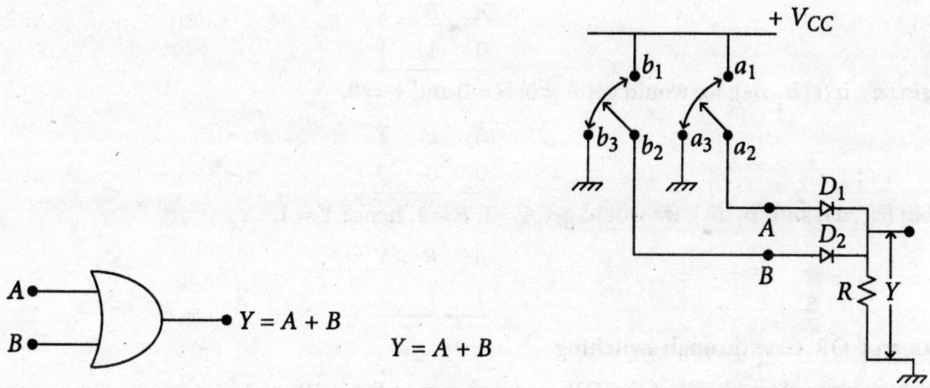

(a) (b) (c)

Fig. 4.9

A, *B* are the two inputs to the OR gate, Fig. 4.9(a). If the number of variables be *N*, then possible number of combinations is 2^N, so for two-input OR gate it is $2^2 = 4$.

Performance. The performing circuit is shown in Fig. 4.9(c). When input is supplied to the diode D_1 by joining (a_1, a_2), the diode is in conduction and current flows through *R* causing a high voltage drop almost equal to V_{CC} to appear. Similarly, closing (b_1, b_2) the diode D_2 is forward-biased so as to give a voltage drop across *R*.

The truth table for the performance of the circuit is shown in Table 4.4.

Table 4.4: Truth table (B for OR gate)

A	B	Y	Remarks
0	0	0	Both of D_1 and D_2 are not conducting
1	0	1	D_1 is in conducting, D_2 is in non-conducting states
0	1	1	Diode D_2 is conducting but D_1 is off
1	1	1	Both D_1 and D_2 are in conduction state

Performance of the OR Gate in short

1. Join (a_1, a_2) and we would get the input $A = 1$; D_1 is in forward bias and (b_2, b_3) joined together gives the input to $B = 0$. The voltage drop almost equal to V_{CC} occurs across R, so $Y = 1$.

A	B	Y
1	0	1

2. Join (a_2, a_3) and (b_1, b_2), so we would get input to $A = 0$, D_1 is off and input to B is high, the diode D_2 conducts to produce a high voltage output $Y = 1$.

A	B	Y
0	1	1

3. Join $(a_2, a_3), (b_2, b_3)$; we would get $A = 0$, $B = 0$ and $Y = 0$.

A	B	Y
0	0	0

4. Join (a_1, a_2) and (b_1, b_2); we would get $A = 1$, $B = 1$, hence $Y = 1$.

A	B	Y
1	1	1

Realisation of OR Gate through switching

The switching equivalent circuit of the OR gate is shown in Fig. 4.10.

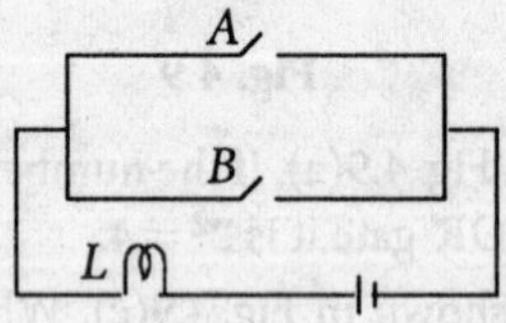

Fig. 4.10

When both switches A and B are open, the lamp does not glow. So, $A = 0$, $B = 0$, $Y = 0$.

When A is closed and B is open, the lamp glows.

So, $A = 1, B = 0, Y = 1$.

When A is open, B is closed, the lamp glows.

So, $A = 0, B = 1, Y = 1$.

When A is closed, B is also closed, the lamp glows.

So, $A = 1, B = 1, Y = 1$.

NOT Gate

This gate is used to negate a signal, to convert NO to YES or OFF to ON. Thus, NOT of A is $\bar{A}$ or A', which denotes negation of the signal A. The symbolic representation is shown in Fig. 4.11(a), its implementing circuit in Fig. 4.11(b), its truth table is shown in Fig. 4.11(c).

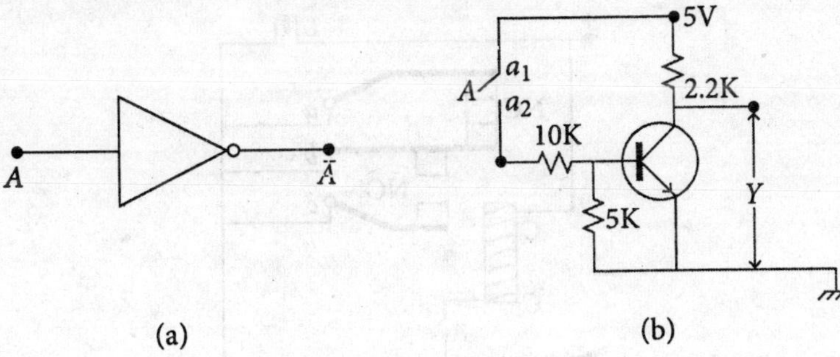

Fig. 4.11

Refer to Fig. 4.11(b). The base bias to the transistor is provided through resistive network arrangement consisting of 10k and 5k.

When (a_1, a_2) are connected, the transistor gets voltage across 5k as the base bias to strike. As the transistor strikes, the collector is thrown to ground potential and hence, output $Y = 0$. When (a_1, a_2) open, there is no base bias and the transistor does not strike. The output is high and so $Y = 1$. Thus, the truth table for the NOT gate is shown in Table 4.5.

Table 4.5: Truth Table (NOT Gate)

A	Y	Remarks
0	1	$Y = \bar{A}$
1	0	

The NOT gate is sometimes also called the *inverter gate.* This logic element gives output state opposite to the input state always.

NOT switching circuit

In order to make NOT operation electrically, we may think of an electromechanical switch which is a relay shown in Fig. 4.12.

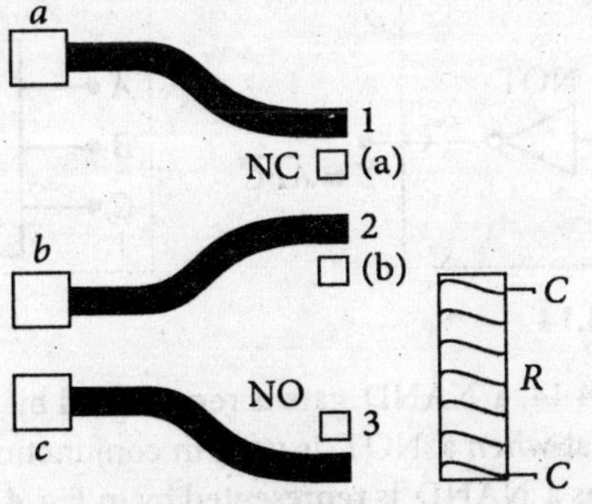

Fig. 4.12

The contacts a, b are normally closed (NC) and (b, c) are normally open (NO). The current passes through the relay coil CC, the rod R becomes magnetised and the NC contacts open and the normally open contacts close. Thus, the NC contact is identified with negation of the NO contact of the same relay.

The implementing circuit is shown in Fig. 4.13 as switching equivalent of NOT gate.

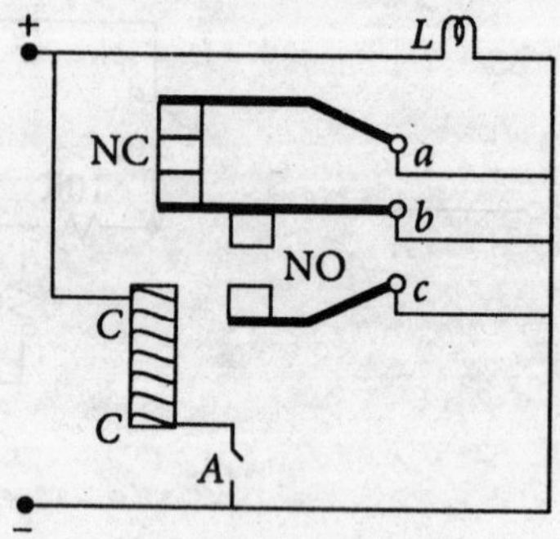

Fig. 4.13

When switch A is open, the relay contacts (a, b) remain closed corresponding to no-input state of the device, the lamp L glows, the output is high (i.e., 1).

When A is closed, the lever b is drawn below to close the contact between (b, c), the lamp is off. So there is no output. Thus, the truth table of the device is shown in Table 4.6.

Table 4.6: Truth Table

A	Binary state	Output (state of the lamp)		Remarks
Open	0	Lamp glows	Output state	$\bar{A}$
Closed	1	Lamp off	0	$\bar{A}$

Using these basic operations namely AND, OR and NOT, any Boolean function can be implemented by suitable electrical circuits.

NAND *Gate.* When a NOT gate is added to an AND gate, the resulting logical unit is known as a NAND gate. It provides a logic 0 when all inputs are in binary state 1 (AND followed by NOT = NAND).

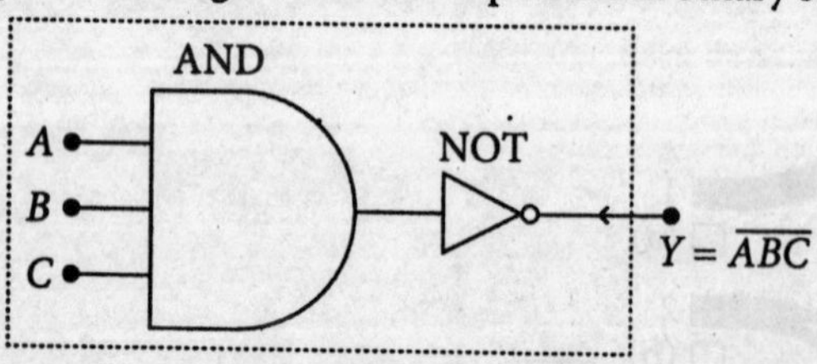

Fig. 4.14

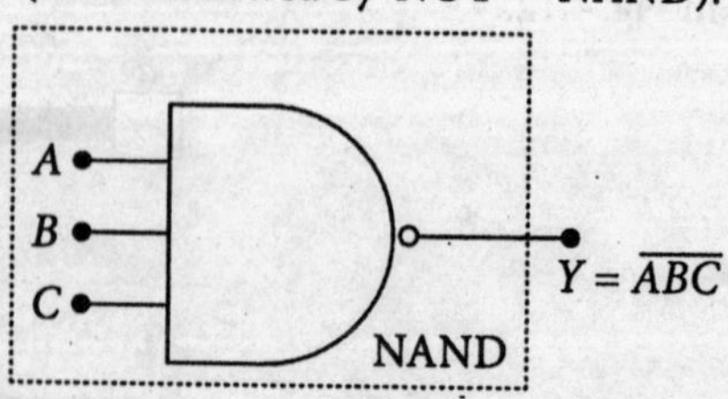

Fig. 4.15

Symbolically, as shown in Fig. 4.14, a NAND gate is represented by an AND gate followed by a NOT gate. It so happens sometimes that when a NOT is used in conjunction with other gates, only a circle is shown for the NOT gate and thus a NAND is represented by in Fig. 4.16.

The diode-transistor logic circuit (DTL) for implementation of a NAND gate is shown in Fig. 4.16.

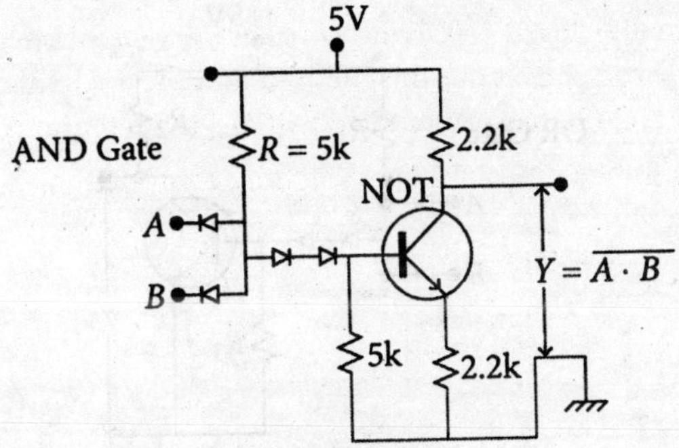

Fig. 4.16

The truth table is shown below for the composite gate, NAND.

Table 4.7: Truth Table

Input			Output	Remarks
A	B	AB	Y	
0	0	0	1	$\overline{AB}$
0	1	0	1	$\overline{AB}$
1	0	0	1	$\overline{AB}$
1	1	1	0	$\overline{AB}$

NOR *Gate.* When an OR is followed by an inverter (NOT) gate, the composite gate is known as NOR gate (OR followed by NOT = NOR).

The output of this gate is the binary 0 when any one or more inputs are in logic state 1; the symbolic representation is given in Fig. 4.17, truth table shown in Table 4.7 and its implementing circuit is given in Fig. 4.18.

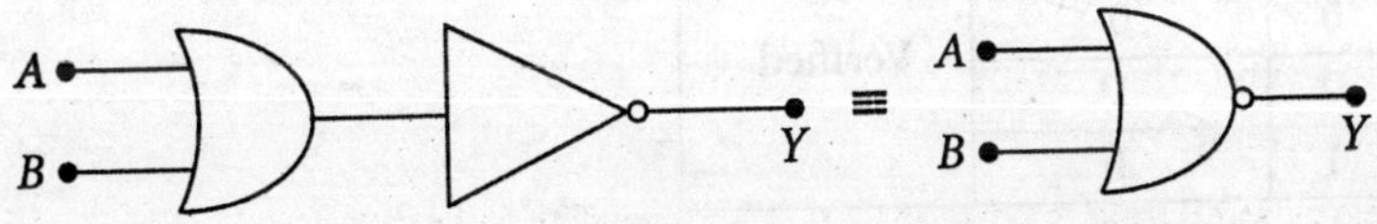

Fig. 4.17

In terms of notation, the gate operation may be written as $Y = \overline{A + B}$.

Truth table 4.8 for NOR gate.

Table 4.8: Truth Table

A	B	$(A+B)$	Y	Remarks
0	0	0	1	$\overline{(A+B)}$
0	1	1	0	$\overline{(A+B)}$
1	0	1	0	$\overline{(A+B)}$
1	1	1	0	$\overline{(A+B)}$

Implementing DTL circuit for NOR gate is shown in Fig. 4.18.

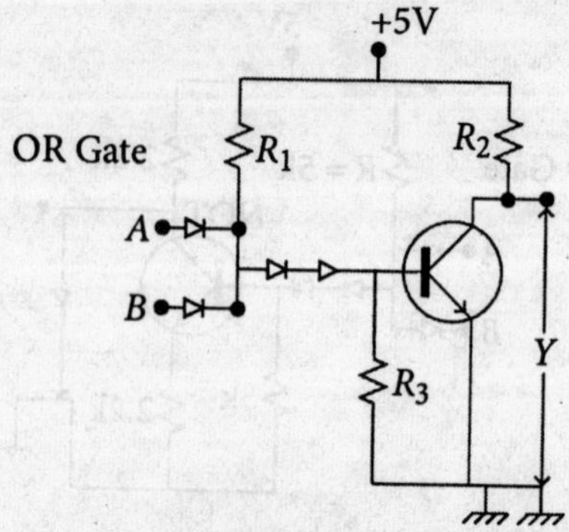

Fig. 4.18

4.3 A few Boolean theorems with their implementing circuits: De Morgan's theorems

Theorem 1 (Theorem of absorption). *This theorem is written as*

$$\boxed{A+AB=A}\,. \tag{4.7}$$

The theorem reads as: If A, B be two binary variables, then A OR (A AND B) equals A.

The truth table verification and the implementing circuit are shown in Table 4.9 and Fig. 4.19.

$A+AB=A(1+B)=A$, since $1+B=1$.

Table 4.9: Truth Table

(a)

A	B	AB	$Y=A+AB$	Remarks $A=A+AB$
0	0	0	0	
0	1	0	0	Verified
1	0	0	1	
1	1	1	1	

(b) *Verification circuit:*

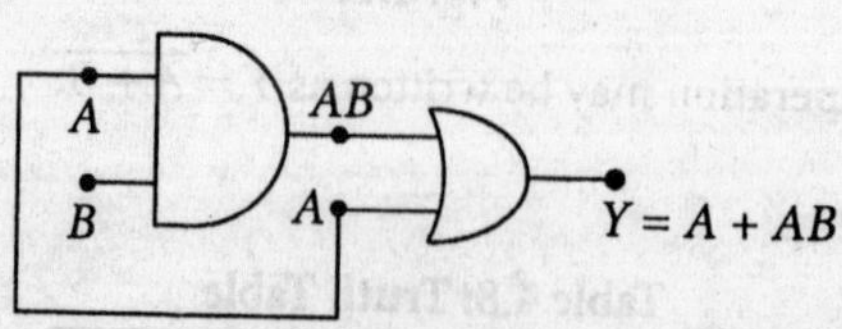

Fig. 4.19

Theorem 2.

$$\boxed{A+\bar{A}B=A+B}\,. \tag{4.8}$$

The theorem reads as: (A NOT AND B) OR A equals A OR B.

The truth table verification is shown in Table 4.10 and its implementing circuit is given in Fig. 4.20.

Table 4.10: Truth Table

A	B	$\bar{A}$	$\bar{A}B$	$A+\bar{A}B$	$A+B$	Remarks
0	0	1	0	0	0	
0	1	1	1	1	1	Theorem 2
1	0	0	0	1	1	is verified
1	1	0	0	1	1	

Verification circuit:

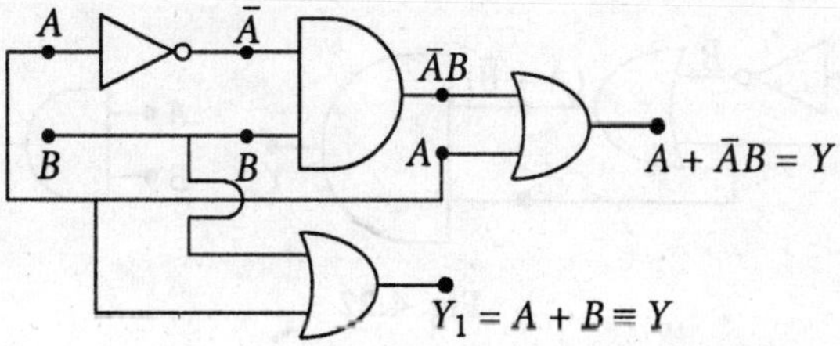

Fig. 4.20

Theorem 3.

$$\boxed{A(A+B)=A}\text{ Another absorption law.} \tag{4.9}$$

This theorem reads as: A AND (A OR B) equals A. This is also another form of absorption law. Truth table verification of the theorem is shown in Table 4.11.

Table 4.11: Truth Table

A	B	$A+B$	$A(A+B)$	Remarks
0	0	0	0	
0	1	1	0	Theorem 3
1	0	1	1	is verified
1	1	1	1	

Verification circuit:

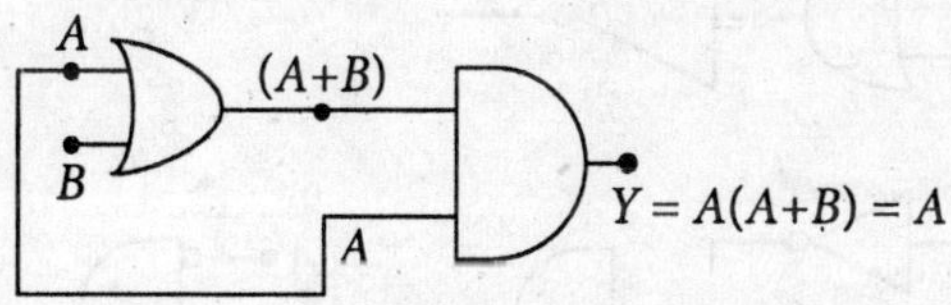

Fig. 4.21

Theorem 4.

$$\boxed{(A+\bar{B})B=AB}\,. \tag{4.10}$$

This can be read as: (A OR B NOT) and B equals A and B.
The truth table verification is shown in Table 4.12.

Table 4.12: Truth Table

A	B	$\bar{B}$	$A+\bar{B}$	AB	$(A+\bar{B})B$	Remarks
0	0	1	1	0	0	
0	1	0	0	0	0	Theorem 4
1	0	1	1	0	0	is verified
1	1	0	1	1	1	

Verification circuit:

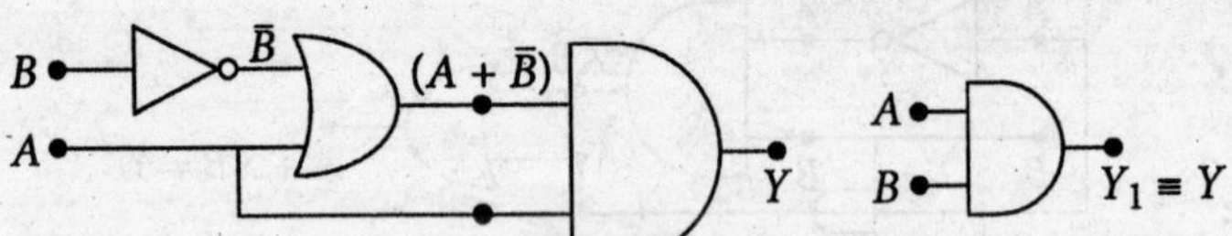

Fig. 4.22

Theorem 5 (De Morgan's theorem).

$$\overline{(A+B)} = (\bar{A}\bar{B}). \tag{4.11}$$

It reads as: OR gate followed by a NOT gate is equivalent to NAND gate. Truth table verification is given in table 4.13.

Table 4.13: Truth Table

A	B	$A+B$	$\overline{(A+B)}$	$\bar{A}\bar{B}$	Remarks
0	0	0	1	1	
0	1	1	0	0	Theorem 5
1	0	1	0	0	is verified
1	1	1	0	0	

Verification circuit:

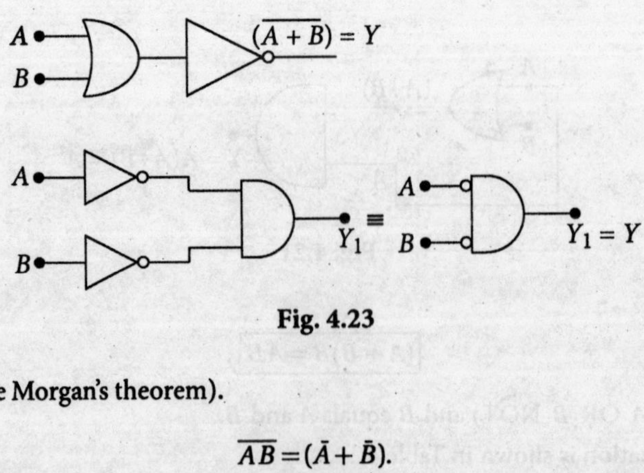

Fig. 4.23

Theorem 6 (Another De Morgan's theorem).

$$\overline{AB} = (\bar{A}+\bar{B}). \tag{4.12}$$

This states that (A AND B) NOT equals A NOT OR B NOT.

Truth table verification is given in Table 4.14.

Table 4.14: Truth Table

A	B	AB	$\overline{AB}$	$(\bar{A}+\bar{B})$	Remarks
0	0	0	1	1	
0	1	0	1	1	Theorem 6
1	0	0	1	1	is verified
1	1	1	0	0	

Verification circuit

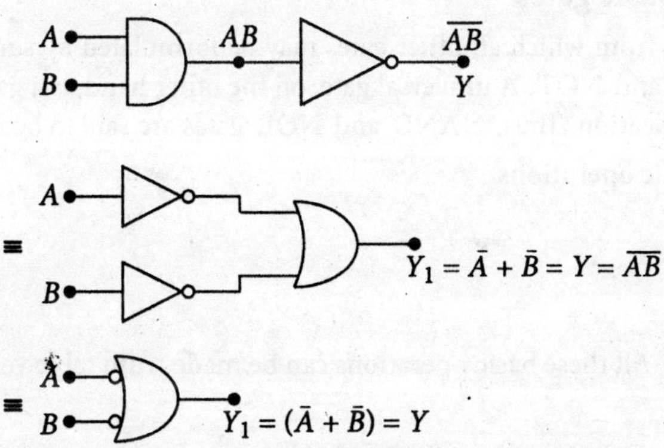

Fig. 4.24

Important point to note

Theorems 5 and 6 are called *De Morgan's* theorems, whose essence is to '*negate a Boolean* expression, *prime every unprimed symbol, unprime every primed symbol and replace* OR *by* AND *and* AND *by* OR.'

Theorem 7.

$$\boxed{AB + A\bar{B} = A\left(B + \bar{B}\right) = A}. \tag{4.13}$$

Table 4.15: Truth Table

A	B	B	AB	$A\bar{B}$	$\left(AB+A\bar{B}\right)$	Remarks
0	0	1	0	0	0	
0	1	0	0	0	0	Theorem 7
1	0	1	0	1	1	is verified
1	1	0	1	0	1	

Verification circuit:

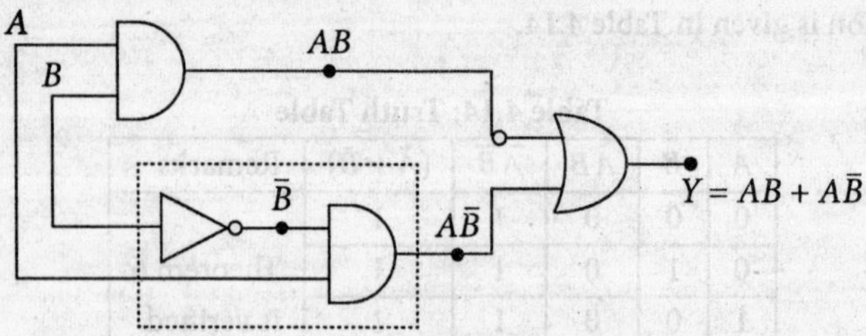

Fig. 4.25

4.4 Basic laws and basic gates

Basic gates are those gates from which all other gates may be formulated by suitable combination. These basic gates are AND, OR and NOT. A universal gate, on the other hand, is a gate which reduces to basic gates under suitable modification. Thus, NAND and NOR gates are said to be universal gates.

Basic laws satisfied by basic operations

AND *operations*:

$$\left.\begin{aligned} A\cdot 0 &= 0 \\ A\cdot 1 &= A \\ A\cdot A &= A \\ A\cdot \bar{A} &= 0 \end{aligned}\right\} \text{All these basic operations can be made truth table verification.}$$

OR *operations*:

$$\left.\begin{aligned} A+0 &= A \\ A+1 &= 1 \\ A+A &= A \\ A+\bar{A} &= 1 \end{aligned}\right\} \text{Can have truth table verification.}$$

NOT *operations*:

$$\bar{0} = 1$$
$$\bar{1} = 0.$$

4.5 Boolean algebra laws

$\bar{0} = 1$

$\bar{1} = 0$

If $A = 0, \bar{A} = 1$

If $A = 1, \bar{A} = 0$

$\bar{\bar{A}} = A$

$A\cdot 0 = 0$

$A\cdot 1 = 1$

$A\cdot A = A$

$A\cdot \bar{A} = 0$

$A+0 = A$

$A+1 = 1$

$A+A = A$, Identity law

$A+\bar{A} = 1$

$A+B = B+A$, Commutative law

$A\cdot B = B\cdot A$, Commutative law

$A+(B+C) = (A+B)+C$, Associative law

$A\cdot(B\cdot C)=(A\cdot B)\cdot C$, Associate law $\quad A+\bar{A}\cdot B=A+B$

$A\cdot(B+C)=(A\cdot B)+(A\cdot C)$, Distributive law $\quad \bar{A}\cdot\bar{B}=\overline{A+B}$.

$A+(B\cdot C)=(A+B)\cdot(A+C)$, Distributive law

To simplify Boolean expressions, we can use the above postulates and theorems as outlined below.

Step 1. Search for the term having a variable and its negation, e.g., $A\cdot\bar{A}\cdot B=0\cdot B=0$ and hence, it can be omitted.

Step 2. Collect the identical terms and using the corresponding law, one of them may be dropped.

Step 3. Search for the term having a variable and its negation, omit such terms such as $A\cdot\bar{A}\cdot B=0$ can be omitted.

Illustrative examples

1. *Simplify:* $A\cdot\left[B+C\cdot\left(\overline{A\cdot B+A\cdot C}\right)\right]=A\cdot\left[B+C\cdot\left(\overline{A\cdot B}\right)\cdot\left(\overline{A\cdot C}\right)\right]=A\cdot B+A\cdot C\left(\overline{A\cdot B}\right)\cdot\left(\overline{A\cdot C}\right)$
$$=A\cdot B+A\cdot C\left(\bar{A}\cdot\bar{B}\right)\cdot\left(\bar{A}+\bar{C}\right)=A\cdot B.$$

2. $\overline{\bar{A}+\bar{B}}+\overline{\overline{A+B}}=\overline{AB+\bar{A}\cdot\bar{B}}=\overline{AB}\cdot\left(\overline{\bar{A}\bar{B}}\right)=\left(\bar{A}+\bar{B}\right)\cdot(A+B)=\bar{A}B+A\bar{B}=A\oplus B.$

3. *Evaluate the following expression:* $\left(B\bar{C}+\bar{A}U\right)\cdot\left(A\bar{B}+C\bar{U}\right)$.

   Solution: Multiplying all variables to reduce the parenthesis, we have
$$\left(B\bar{C}+\bar{A}U\right)\cdot\left(A\bar{B}+C\bar{U}\right)=B\bar{C}\bar{B}+B\bar{C}C\bar{U}+\bar{A}AU\bar{B}+\bar{A}CU\bar{U}=0+0+0=0$$
$$\Rightarrow\left(B\bar{C}+\bar{A}U\right)\cdot\left(A\bar{B}+C\bar{U}\right)=\bar{0}=1.$$

4. *Simplify the Boolean expression:*
$$A=ab+abc+\bar{a}b+a\bar{b}c=(a+\bar{a})\,b+abc+a\bar{b}c=b+abc+a\bar{b}c=b+ac\left(b+\bar{b}\right)=b+ac.$$

5. *Verify the following Boolean identities:* (a) $A\cdot\left(\bar{A}+B\right)=AB$ (b) $A+\bar{A}B=A+B$ (c) $A+BC=(A+B)(A+C)$.

   Solution: (a) $A\cdot\left(\bar{A}+B\right)=A\cdot\bar{A}+A\cdot B=0+AB=AB.$

   (b) $A+\bar{A}B=A\cdot 1+\bar{A}B=A\cdot(1+B)+\bar{A}B,\ \because\ 1+B=1$
$$=A+A\cdot B+\bar{A}B=A+\left(A+\bar{A}\right)\cdot B,\ \because\ A+\bar{A}=1$$
$$=A+B.$$

   (c) $A+BC=(A+B)(A+C)$

   $Y=Y_1$.

A	B	C	BC	$A+BC$ (Y)	$(A+B)$	$A+C$	Y_1	Remarks
0	0	0	0	0	0	0	0	
0	0	1	0	0	0	1	0	
0	1	0	0	0	0	0	0	
0	1	1	1	1	1	1	1	$Y=Y_1$
1	0	0	0	1	1	1	1	
1	0	1	0	1	1	1	1	
1	1	0	0	1	1	1	1	
1	1	1	1	1	1	1	1	

6. (a) Complement and simplify the following expression: $(\bar{a}+b)\cdot\bar{c}\cdot\bar{d}$.

   Solution: Using De Morgan's theorem: $\overline{(\bar{a}+b)\cdot\bar{c}\cdot\bar{d}}=a\cdot\bar{d}+c+d.$

   (b) Solve the following Boolean expression: $ab+\overline{ab}+a\bar{b}c(ab+c)$.

   Solution: $ab+\overline{ac}+a\bar{b}c(ab+c)=ab+\overline{ac}+a\bar{b}cab+a\bar{b}cc=ab+\overline{ac}+0+a\bar{b}c$

$$=ab+\bar{a}+\bar{c}+a\bar{b}c=ab+\bar{c}+\bar{a}+a\bar{b}c$$
$$=ab+\bar{c}+\bar{a}+\bar{b}c=(\bar{a}+ab)+\bar{c}+\bar{b}c$$
$$=\bar{a}+b+\bar{c}+\bar{b}=\bar{a}+\left(b+\bar{b}\right)+\bar{c}.$$

4.6 Boolean algebra revisited

Based on several postulates or axioms in 1904, E.V. Huntington developed a definition of the two-value Boolean algebra, also called switching algebra.

A binary variable X, and three operators '(+)', '(·)' and '(−)' are said to form a set $S=(X,+,0,-)$ thus leading to six postulates as detailed below:

P1 Postulate 1: For all elements X and Y in set S : (i) $X+Y$ is an element of S and (ii) X, Y is also an element of S.

The above two parts (i) and (ii) are taken as the closure property w.r.t. operator '+' and '·'.

P2 Postulate 2:

1. *Zero axiom*: There exists a constant 0 in the set S such that for every element X in $S, X+0=X$.
2. *Unit axiom*: There exists a constant 1 in the sets such that for every element X in S, $X\cdot 1=X$.

P3 Postulate 3: For all elements X and Y in set S, (i) $X+Y=Y+X$ (ii) $X\cdot Y=Y\cdot X$. Commutative property.

P4 Postulate 4: For all elements X, Y and Z in set S

$$\left.\begin{aligned}&\text{(i) } X(Y+Z)=X\cdot Y+X\cdot Z\\&\text{(ii) } X+(Y\cdot Z)=(X+Y)\cdot(X+Z)\end{aligned}\right\}\text{ Distribution property.}$$

P5 Postulate 5: There exists an element $\bar{X}$ in sets for every element X in the same set such that (i) $X+\bar{X}=1$ (ii) $X\cdot\bar{X}=0$. They describe the inverse property.

P6 Postulate 6: There exist at least two distinct elements in the set S.

4.6.1 . What are the fundamental or basic gates?

Basic gates refer to those gates which are the building blocks of all gates. The three gates symbolised as '+', '•' and '–' (over bar) are the fundamental gates, because any gate can be constructed out of these three. They are called OR, AND and NOT operators or gates.

4.6.2 Give the implementing circuits of

(a) AND (b) OR (c) NOT using transistor as the switching elements.

(a) AND Gate using transistor

Earlier we have shown how diode logic can be used to implement an AND gate. Here we are using transistor logic for the implementation [Fig. 4.26].

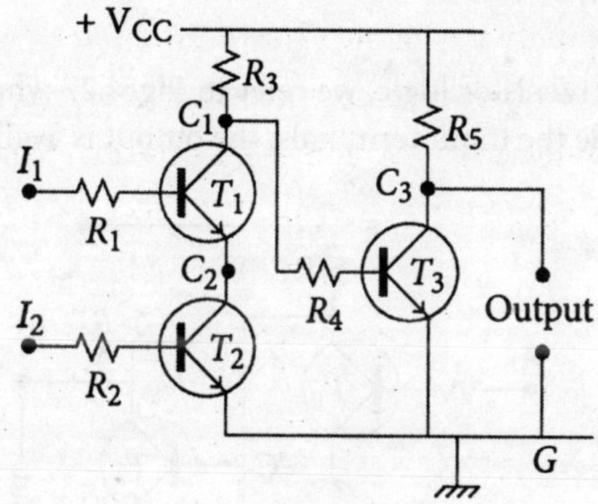

Fig. 4.26

Fig. 4.26 shows a two-transistor model for an AND gate having I_1 and I_2 as the input terminals to the bases of the two transistors T_1, T_2; C_1, C_2 forming the collector terminals. The third transistor T_3 provides the output between the collector terminal C_3 and ground.

When I_1 (one input) is high ($I_1 = 1$) and I_2 (the second input) is low ($I_2 = 0$), the first transistor T_1 is ON but T_2 is OFF so that the combined transistor system does not conduct, the collector of the first transistor C_1 is high supplying the base bias of the output transistor T_3. So the collector terminal C_3 is almost at ground potential.

So, $I_1 = 1$ Output
$I_2 = 0$ 0

Similarly, when $I_1 = 0$ Output
$I_2 = 1$ 0

When both inputs are low (= 0), both of T_1 and T_2 are OFF, C_2 is high which supplies the base bias to the output transistor, the output is zero (C_3 is almost at ground potential).

When $I_1 = 0$
$I_2 = 0$.

Both T_1 and T_2 are OFF, C_1 high, T_3 is in conduction, so C_3 is almost at ground potential.

When $I_1 = 1$
$I_2 = 1$.

Both T_1 and T_2 are conducting, reducing the potential of the point C_1 to almost zero value thus the base bias of the transistor $T_3 = 0$; hence, it is OFF and the output is high (= 1).

So to summarise the performance:

Input		Output	Remark
I_1	I_2	Y	
0	0	0	Output high when both inputs are high
1	0	0	
0	1	0	
1	1	1	

(b) OR Gate using transistor

To realise an OR gate through transistor logic, we refer to Fig. 4.27 where three transistors are used. The first two transistors T_1, T_2 provide the input terminals; the output is available across the third transistor T_3 (between C_0 and ground).

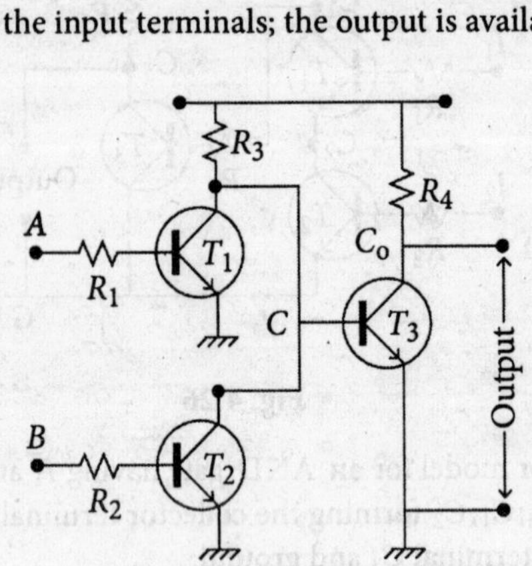

Fig. 4.27

Action: The bases of the transistors T_1 and T_2 provide input terminals and the output appears across the third transistor (T_3).

(i) When $A = 0$

$B = 0.$

Both the transistors T_1 and T_2 are OFF and so the point C is at high positive potential supplying the base bias to the transistor T_3. C_0 is thrown almost to the ground potential. The output is zero. So

A	B	Y
0	0	0

(ii) When $A = 1$

$B = 0.$

The transistor T_1 gets base bias; hence, ON but T_2 is OFF, as the potential at C will be almost equal to zero. The transistor T_3 is OFF, the output is high. Thus,

A	B	Y
1	0	1

(iii) When $A = 0$

$B = 1.$

It is similar to case (ii) with $A = 0$ and $B = 1$ and output Y will be high (= 1).

(iv) When $A = 1$

$B = 1.$

Both the transistors are ON, so the base bias to the transistor T_3 is zero, the output is high. So,

A	B	Y
1	1	1

So, the performance of the circuit is summarised as below:

A	B	Y	Remark
0	0	0	When one of the two inputs are ON, the output is high.
1	0	1	
0	1	1	
1	1	1	

(c) For NOT Gate, see the Fig. 4.11(b) and follow the description as given therein.

4.6.3 What are universal gates? Are NAND and NOR gates universal gates? Justify.

A universal gate is one by which we can construct the basic gates, namely AND, OR and NOT. From NAND and NOR gates we can construct all the basic gates; hence, different digital systems may be obtained through NAND and NOR gates. The convertibility is shown as given below:

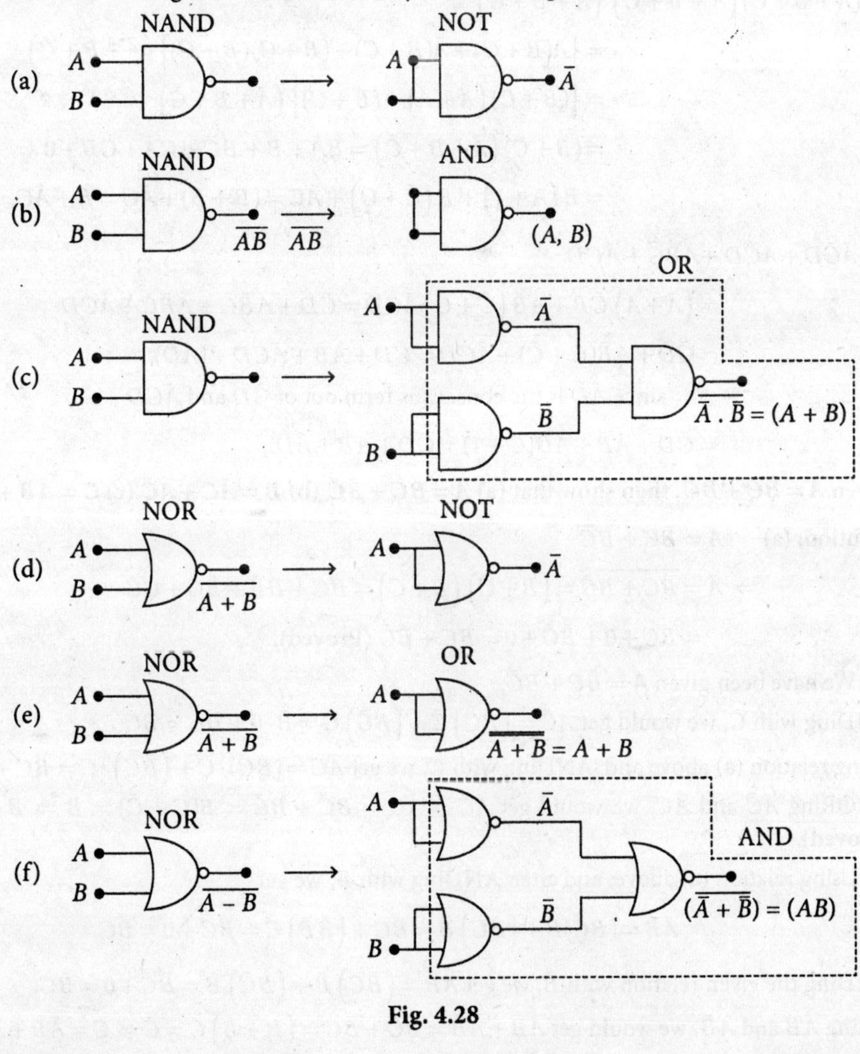

Fig. 4.28

Fig. 4.28 (a), (b) and (c) show conversion from NAND gate to the basic gates NOT, AND and OR respectively.

Fig. 4.28 (d), (e) and (f) show conversion from NOR gate to the basic gates NOT, OR and AND respectively. The representation of the process is self-explanatory.

1. **For all elements A in the set S: (a) $A+1=1$ (b) $A\cdot 0=0$. Prove the proposition.**

   Proof. (a) $A+1=(A+1)1=(A+1)\left(A+\bar{A}\right)=AA+A\bar{A}+A+\bar{A}=A+0+\bar{A}=A+\bar{A}=1$. [cf. P.5 (a)]

   (b) $A\cdot 0=A\cdot 0+A\bar{A}=A\cdot\left(\bar{A}+0\right)=A\bar{A}=0.$

2. **Prove the Boolean identity:**

   (a) $\boldsymbol{(A+B+C)\left(\bar{A}+B+C\right)\left(\bar{A}+B+\bar{C}\right)=B+\bar{A}C.}$

   (b) $\boldsymbol{\bar{A}\bar{C}D+A\bar{C}D+A\bar{B}\bar{C}+A\bar{B}C+\bar{A}CD=\bar{A}D+\bar{C}D+A\bar{B}.}$

   Proof.

   (a) $(A+B+C)\left(\bar{A}+B+C\right)\left(\bar{A}+B+\bar{C}\right)$

$$
\begin{aligned}
&=\left\{A(B+C)+\bar{A}(B+C)+(B+C)(B+C)\right\}\left(\bar{A}+B+\bar{C}\right)\\
&=\left\{(B+C)\left(A+\bar{A}\right)+(B+C)\right\}\left(\bar{A}+B+\bar{C}\right)\\
&=(B+C)\left(\bar{A}+B+\bar{C}\right)=B\bar{A}+B+B\bar{C}+C\bar{A}+CB+0\\
&=B\left(\bar{A}+1\right)+B\left(\bar{C}+C\right)+\bar{A}C=(B+B)+\bar{A}C=B+\bar{A}C.
\end{aligned}
$$

   (b) $\bar{A}\bar{C}D+A\bar{C}D+A\bar{B}\vec{C}+\bar{A}CD$

$$
\begin{aligned}
&=\left(\bar{A}+A\right)\bar{C}B+A\bar{B}\left(\bar{C}+C\right)\bar{A}CD=\bar{C}D+A\bar{B}\bar{C}+A\bar{B}C+\bar{A}CD\\
&=\bar{C}D+A\bar{B}(C+\bar{C})+\bar{A}CD=\bar{C}D+A\bar{B}+\bar{A}CD+(\bar{A}D),
\end{aligned}
$$

   since $\bar{A}D$ is the consensus term out of $\bar{C}D$ and $\bar{A}CD$

$$
=\bar{C}D+A\bar{B}+\bar{A}D(C+1)=\bar{C}D+A\bar{B}+\bar{A}D.
$$

3. **Given $A=\bar{B}C+B\bar{C}$, then show that (a) $\bar{A}=BC+\bar{B}\bar{C}$ (b) $B=\bar{A}C+A\bar{C}$ (c) $C=\bar{A}B+A\bar{B}$.**

   Solution: (a) $A=\bar{B}C+B\bar{C}$

$$
\begin{aligned}
\Rightarrow \bar{A}&=\overline{\bar{B}C+B\bar{C}}=\left(B+\bar{C}\right)\left(\bar{B}+C\right)=BC+B\bar{B}+\bar{B}\bar{C}+C\bar{C}\\
&=BC+0+\bar{B}\bar{C}+0=BC+\bar{B}\bar{C}\ \textbf{(Proved)}.
\end{aligned}
$$

   (b) We have been given $A=\bar{B}C+B\bar{C}$.

   ANDing with $\bar{C}$, we would get $A\bar{C}=\left(\bar{B}C\right)\bar{C}+\left(B\bar{C}\right)\bar{C}=\bar{B}\cdot 0+B\bar{C}=B\bar{C}$.

   Using relation (a) above and ANDing with C, we get $\bar{A}C=(BC)\cdot C+\left(\bar{B}\bar{C}\right)\cdot C=BC+0=BC$.

   So ORing $A\bar{C}$ and $\bar{A}C$, we would get $A\bar{C}=\bar{A}C=BC+B\bar{C}=B(C+\bar{C})=B\Rightarrow B=A\bar{C}+\bar{A}C$ **(Proved)**.

   (c) Using relation (a) above, and after ANDing with B, we get

$$
\bar{A}B=(BC)B+\left(\bar{B}\bar{C}\right)B=BC+\left(\bar{B}B\right)C=BC+0=BC.
$$

   ANDing the given relation with $\bar{B}$, we get $A\bar{B}=\left(\bar{B}C\right)\bar{B}+\left(B\bar{C}\right)\bar{B}=\bar{B}C+0=\bar{B}C$.

   ORing $\bar{A}B$ and $A\bar{B}$, we would get $\bar{A}B+A\bar{B}=BC+\bar{B}C=\left(B+\bar{B}\right)C=C\Rightarrow C=\bar{A}B+A\bar{B}$.

De Morgan's theorem:

4. For all elements $\bar{A}$ and $\bar{B}$ of the sets, verify that (a) $\overline{A+B}=\bar{A}\cdot\bar{B}$ (ii) $\overline{AB}=\bar{A}+\bar{B}$.

Verification:

(a)

A	B	$(A+B)$	$\bar{A}$	$\bar{B}$	$\overline{A+B}$	$\bar{A}\cdot\bar{B}$
0	0	0	1	1	1	1
0	1	1	1	0	0	0
1	0	1	0	1	0	0
1	1	1	0	0	0	0

The De Morgan's first relation verified.

(b)

A	B	$\bar{A}$	$\bar{B}$	$\overline{AB}$	$(\bar{A}+\bar{B})$
0	0	1	1	1	1
0	1	1	0	1	1
1	0	0	1	1	1
1	1	0	0	0	0

De Morgan's second relation verified.

4.7 From circuit representation to Boolean expression

(i)

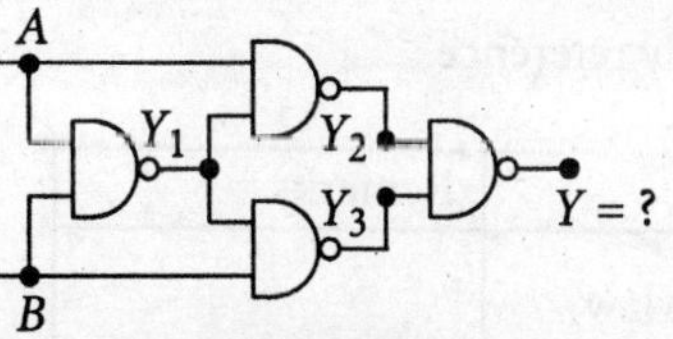

Fig. 4.29

Solution: $Y=\overline{Y_2Y_3}$.

$$Y_2=\overline{A\cdot\overline{AB}}=\bar{A}+\left(\overline{\overline{AB}}\right)=\bar{A}+AB$$

$$Y_3=\overline{B\cdot\overline{AB}}=\bar{B}+\left(\overline{\overline{AB}}\right)=\bar{B}+AB.$$

So, $Y=\overline{\left(\bar{A}+AB\right)\left(\bar{B}+AB\right)}=\overline{\left(\bar{A}+AB\right)}+\overline{\left(\bar{B}+AB\right)}=A\cdot\left(\bar{A}+\bar{B}\right)+\left[B\cdot\left(\bar{A}+\bar{B}\right)\right]$

$=A\bar{B}+B\bar{A}=A\oplus B$ XOR.

(ii)

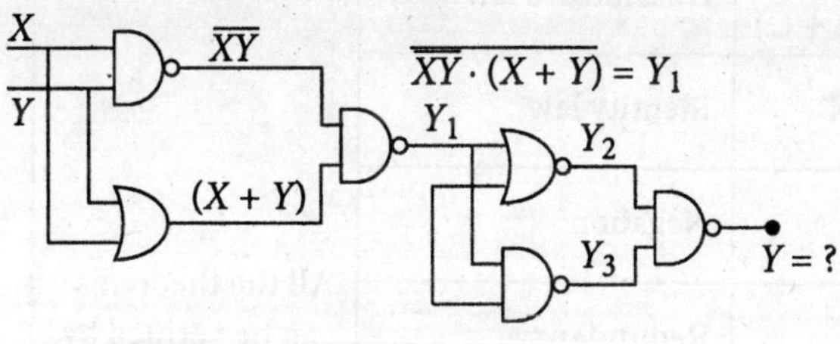

Fig. 4.30

Solution: $Y=\overline{Y_2\cdot Y_3}$, $Y_2=\overline{Y_1+Z}$, $Y_3=\overline{(Y_1\cdot Z)}$.

$$Y_1=\overline{\overline{XY}\cdot(X+Y)}=\overline{\overline{XY}}+\overline{(X+Y)}=XY+\left(\bar{X}\cdot\bar{Y}\right)$$

$$Y_2=Y_1+Z=XY+\left(\bar{X}\cdot\bar{Y}\right)\cdot Z$$

$$Y_3=\overline{XY+\left(\bar{X}\cdot\bar{Y}\right)\cdot Z}$$

$$Y=\overline{\left\{XY+\left(\bar{X}\cdot\bar{Y}\right)+Z\right\}\cdot\overline{\left(XY+\left(\bar{X}\cdot\bar{Y}\right)\right)Z}}=\overline{\left(XY+\bar{X}\cdot\bar{Y}+Z\right)}+\overline{\overline{\left(XY+\bar{X}\cdot\bar{Y}\right)Z}}$$

$$=\left(\left(\bar{X}+\bar{Y}\cdot(X+Y)\cdot\bar{Z}\right)\right)+\left(XY+\bar{X}\cdot\bar{Y}\right)Z=\left(\left(\bar{X}+\bar{Y}\right)\cdot\left(X\bar{Z}+Y\bar{Z}\right)\right)+XYZ+\bar{X}\bar{Y}Z$$

$$=\underbrace{X\bar{X}\bar{Z}}_{0}+\bar{X}Y\bar{Z}+X\bar{Y}\bar{Z}+\underbrace{Y\bar{Y}\bar{Z}}_{0}+XYZ+\bar{X}\bar{Y}Z$$

$$=XYZ+\bar{X}Y\bar{Z}+X\bar{Y}\bar{Z}+\bar{X}\bar{Y}Z.$$

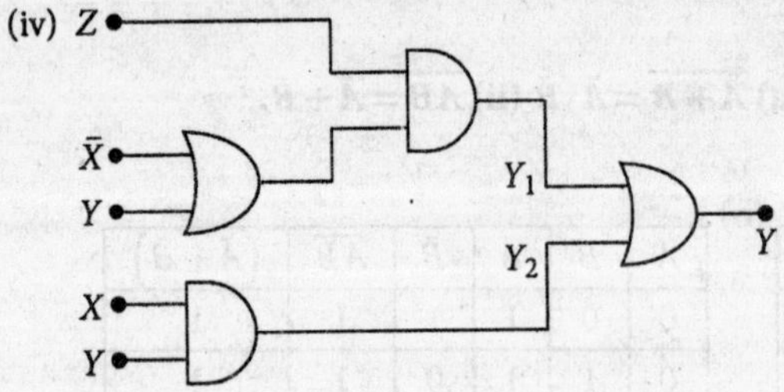

Fig. 4.31

Solution: $Y = Y_1 + Y_2$, $Y_1 = (\bar{X} + Y)Z$, $Y_2 = XY$.

$\therefore\ Y = (\bar{X} + Y)Z + XY.$

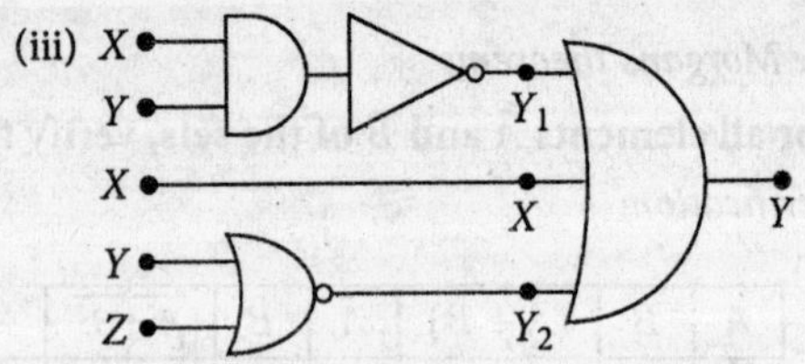

Fig. 4.32

Solution: $Y = Y_1 + X + Y_2$, $Y_1 = \overline{XY}$, $Y_2 = \overline{Y + Z}$.

$\therefore\ Y = \overline{XY} + X + \overline{Y + Z}.$

4.8 All the Boolean theorems at a glance for ready reference

We present below the table giving the Boolean theorems for ready reference.

No.	Theorem	Name	Remarks
1*a*	$A + B = B + A$	Commutative law	
1b	$A \cdot B = B \cdot A$		
2a	$(A + B) + C = A + (B + C)$	Associative law	
2b	$(A \cdot B) \cdot C = A \cdot (\bar{B} \cdot C)$		
3a	$A \cdot (B + C) \cdot = A \cdot B + A \cdot C$	Distributive law	
3b	$A + (B \cdot C) = (A + B) \cdot (A + C)$		
4a	$A + A = A$	Identity law	
4b	$A \cdot A = A$		
5a	$\bar{A} = \bar{A}$	Negation	
5b	$\bar{\bar{A}} = A$		
6a	$A + A \cdot B = A$	Redundancy	All the theorems can be verified by truth table
6b	$A \cdot (A + B) = A$		
7a	$0 + A = A$		
7b	$1 \cdot A = A$		
7c	$1 + A = A$		
7d	$0 \cdot A = 0$		
8a	$A + \bar{A} = 1$		
8b	$A \cdot \bar{A} = 0$		
9a	$A + \bar{A}B = A + B$		
9b	$A \cdot (\bar{A} + B) = A \cdot B$		
10a	$\overline{A + B} = \bar{A} \cdot \bar{B}$	De Morgan's first and second theorems	
10b	$\overline{A \cdot B} = (\bar{A} + \bar{B})$		

(a) IC AND Gate: 7408 quad two-input. AND gate.

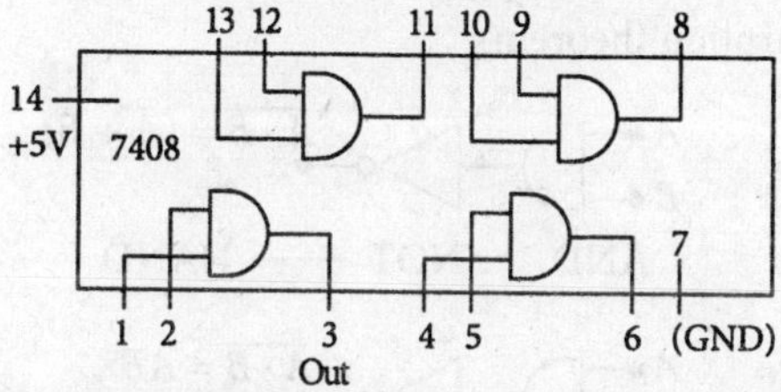

(b) IC OR Gate: 7432 quad two-input OR gate.

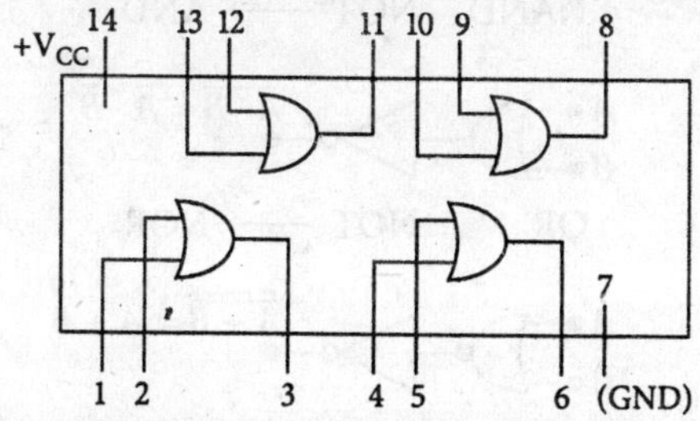

(c) IC NOT Gate: 7404 Hex-inverter.

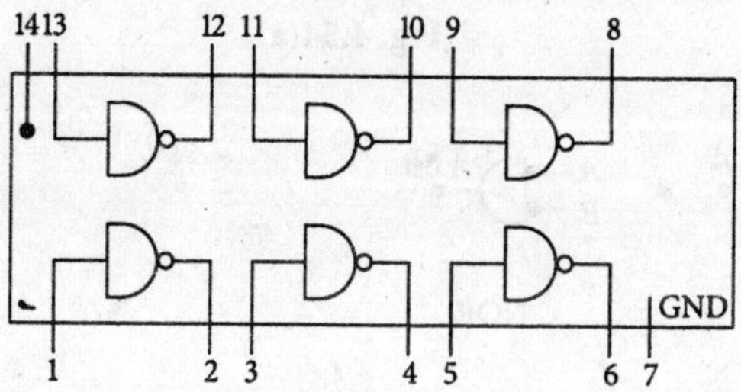

(d) IC NAND Gate: 7400 quad 2-input NAND.

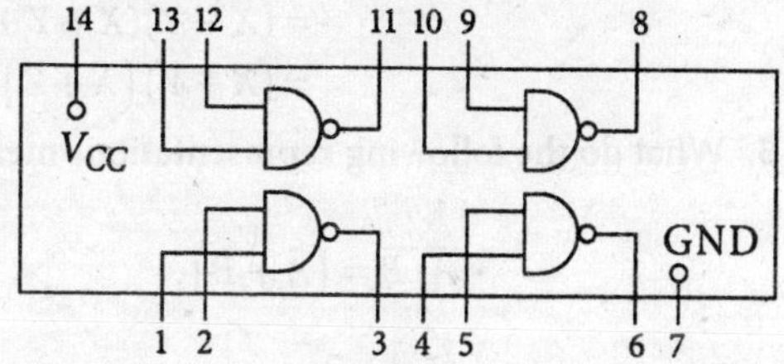

(e) IC NOR Gate: 7402.

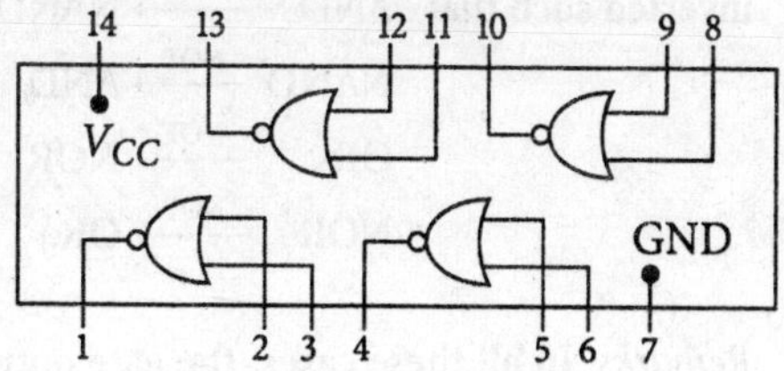

Fig. 4.33 (a), (b), (c), (d), (e).

4.9 Short Theory Questions with Answers

1. What is the number of inputs possible for a NOT gate?

   Ans. Only one input variable is possible for a NOT gate.

2. Under what condition/conditions, the output of a two-input AND gate is one?

   Ans. The output of a two-input AND gate is one when both the inputs are one.

3. What is known as consensus theorem?

   Ans. If a Boolean expression contains three terms in the variables, and if there is one term containing variables common to the other two terms, then ORing the three terms or ANDing the three terms, it is a consensus to drop the third term. Thus,

   (a) $X\cdot Y+\bar{X}\cdot Z+Y\cdot Z=X\cdot Y+\bar{X}\cdot Z$. (b) $(X+Y)(\bar{X}+Z)(Y+Z)=(X+Y)(\bar{X}+Z)$.

4. Prove the theorem as in Q.3.

   Solution: $X\cdot Y+\bar{X}\cdot Z+YZ=X\cdot Y+\bar{X}Z+YZ\cdot 1=X\cdot Y+\bar{X}Z+YZ\left(X+\bar{X}\right)$

   $=X\cdot Y+\bar{X}\cdot Z+XYZ+\bar{X}YZ=X\cdot Y+XYZ+\bar{X}Z(1+Y)$

   $=X\cdot Y+XYZ+\bar{X}Z=XY(1+Z)+\bar{X}Z=XY+\bar{X}Z.$

$$(X+Y)\left(\bar{X}+Z\right)(Y+Z)=(X+Y)\left(\bar{X}+Z\right)(Y+Z+0)=(X+Y)\left(\bar{X}+Z\right)\left(Y+Z+X+\bar{X}\right)$$
$$=(X+Y)(X+Y+Z)\left(\bar{X}+Z\right)\left(Y+Z+\bar{X}\right)$$
$$=(X+Y)\left(\bar{X}+Z\right), \text{ by Absorption theorem.}$$

5. What do the following representations mean?

$$\overline{A\cdot B}=\left(\bar{A}+\bar{B}\right).$$

Remarks. In all these cases, the gate outputs are inverted such that AND $\xrightarrow{\text{NOT}}$ NAND
NAND $\xrightarrow{\text{NOT}}$ AND
OR $\xrightarrow{\text{NOT}}$ NOR
NOR $\xrightarrow{\text{NOT}}$ OR.

Remarks. In all these cases, the gate outputs are inverted such that NOT $\xrightarrow{\text{AND}}$ NOR
NOT $\xrightarrow{\text{OR}}$ NAND
NOT $\xrightarrow{\text{NAND}}$ OR
NOT $\xrightarrow{\text{NOR}}$ AND.

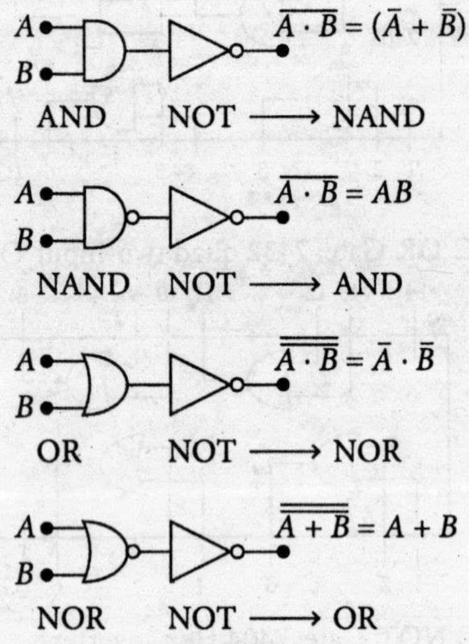

Fig. 4.34(a)

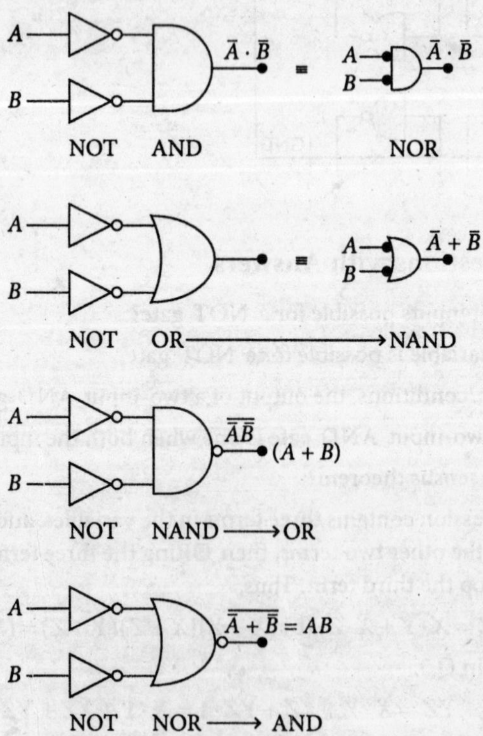

Fig. 4.34(b)

When both the inputs and outputs of a gate are inverted:

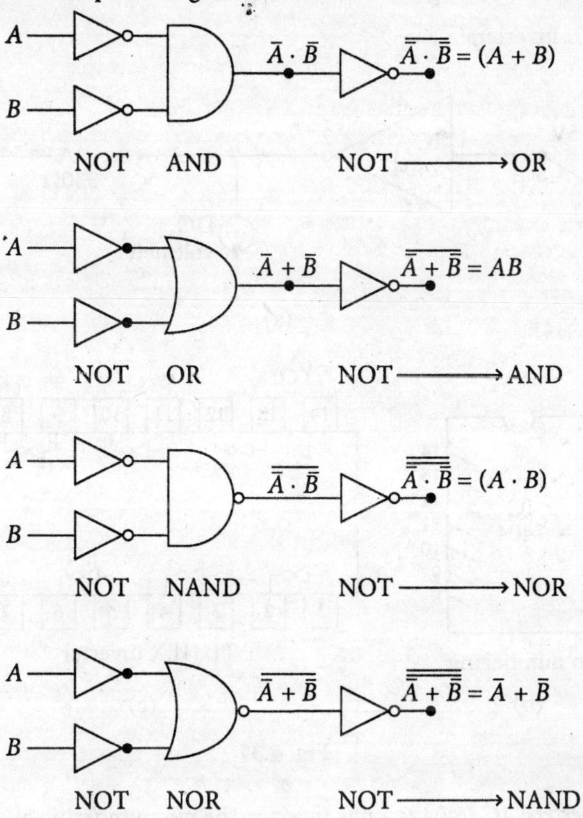

Fig. 4.35

6. How to recognise the gate from the outputs?

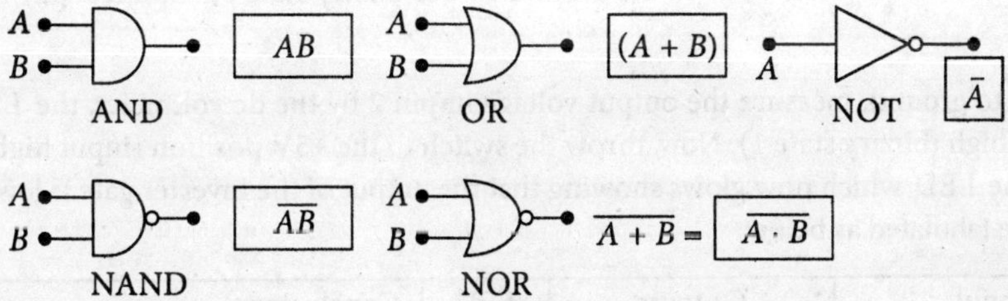

Fig. 4.36

7. Simplify the following expressions:

(a) $\overline{(\bar X+\bar Y)}\cdot X\bar YZ$ (b) $X\bar Y\bar Z+\bar X\bar Y\bar Z+\bar X\bar YZ$.

Solution: (a) $\overline{(\bar X+\bar Y)\cdot X\bar YZ}=\bar X, \overline{X\bar YZ+(\bar Y\cdot Y)XZ}=\overline{0+0}=1.$

8. Prove the following identities algebraically:

(a) $\bar A\bar CD+A\bar CD+AB\bar C+A\bar BC+\bar ACD=\bar AD+A\bar B+\bar CD$

(b) $(X+Y)\left(\bar X+\dot Z\right)(Y+Z)=(X+Y)\left(\bar X+Z\right)$.

9. A few practical experiments with gates:

Experiment. On TTL inverters:

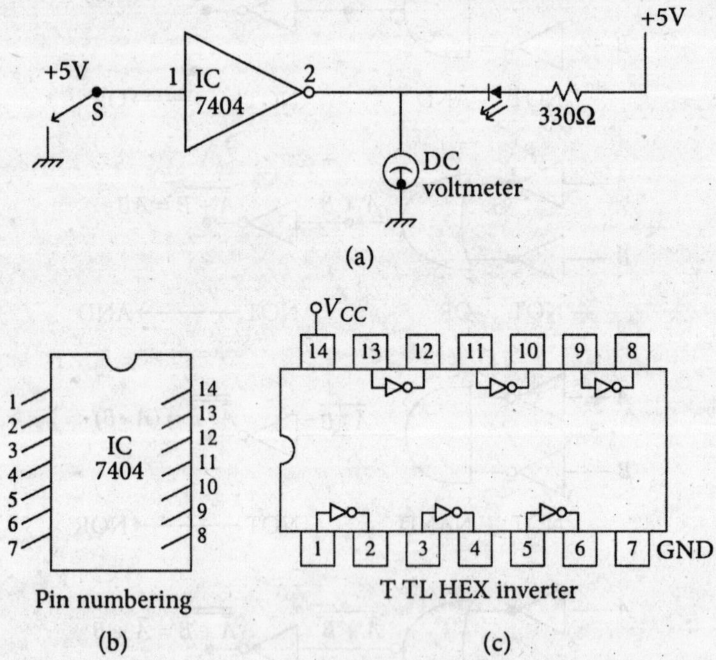

Fig. 4.37

Experiment 1. *On* TTL *inverters*: IC 7404 is a hex inverter, the pin numbering and pin-configuration are shown in Fig. 4.37 (b) and (c). Out of the six inverters you can use any inverter for the experiment. Connect the circuit as shown in Fig. 4.37 (a), the inverter between pins 1 and 2 being used. The switch *S* puts the input to the 5V level (Binary state 1) or to the ground 0V (or binary state 0). Connect pin 14 to +5V and pin 7 to ground.

Put the switch to ground, measure the output voltage in pin 2 by the dc voltmeter, the LED is off, the output voltage is high (binary state 1). Now, throw the switch to the +5V position (input high, binary state 1) and observe the LED which now glows showing that the output of the inverter gate is low (binary state 0). The results are tabulated as below:

Inputs		Outputs		LED	Conclusions
Voltage	Logic level	Voltage	Logic level		(i) Output is complement of the input.
0	0	High (+5V)	1	Off	(ii) If the input of the gate is left open, it behaves as if it has an input of binary 1.
+5V	1	Low (≃ 0)	0	Glows	
Nil (left open)	1	Low	0	Glows	(iii) LED glows when the output is low and vice versa.

Experiment 2. Study of a NAND (NOT AND) Gate. The NAND gate is an AND gate followed by an inverter. The output of a NAND gate is binary 0 if all its inputs are binary 1. Take a NAND gate IC 7400 and connect it in the circuit as shown in Fig. 4.39.

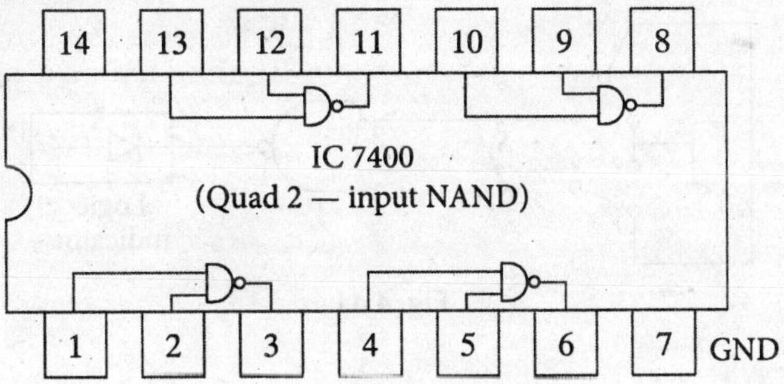

Fig. 4.38

Connect pin 14 to +5V and pin 7 to ground. The switches S_1 and S_2 are put to different positions and observe the output levels by watching LED. Disconnect the switches S_1 and S_2 and leave the pins 1 and 2 open. Observe the output level. Tabulate your results and make a truth table from your observations.

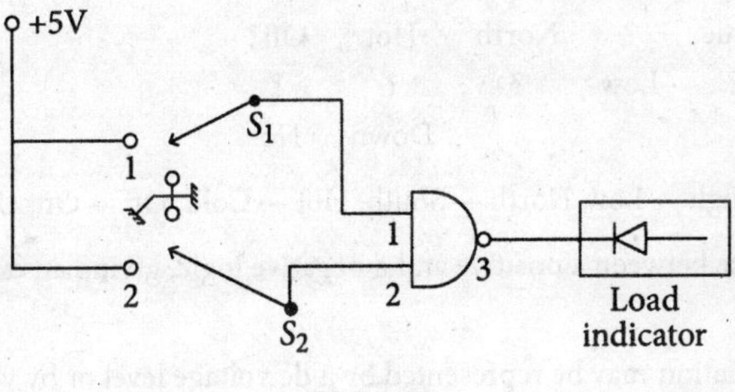

Fig. 4.39

Experiment 3. Study of a NOR Gate. The NOR gate, like a NAND gate, is also an improved logic element used for implementing decision making logic functions. A NOR gate is essentially a circuit containing the logic function of an OR gate and an inverter. We use IC 7402, which contains four identical NOR gates [Fig. 4.40].

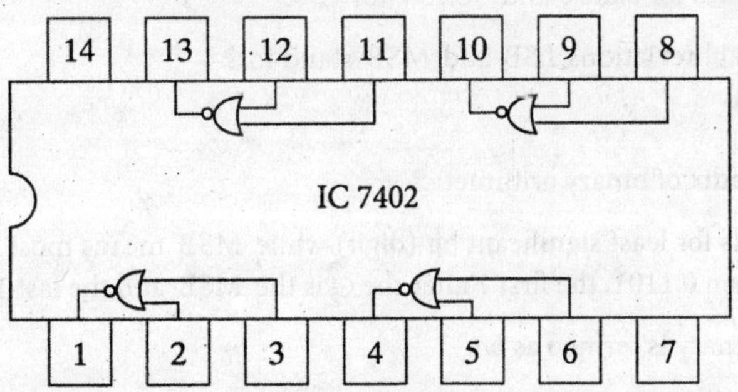

Fig. 4.40

The output of this gate is the binary 0 if one or more of the inputs are in binary state 1. Take the IC 7402 and connect as shown in Fig. 4.41, pin 14 to the +5V and pln 7 to the ground. Put switches S_1 and

S_4 to different positions and note the output level. Disconnect S_1 and leave the pin open. Put the pin S_2 to ground and note the output level. Tabulate your results and make a truth table or your observations and verify that it is an OR gate followed by inversion.

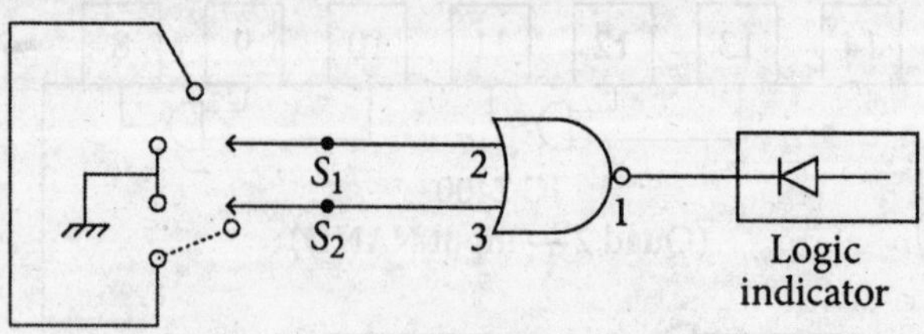

Fig. 4.41

4.10 Short Theory Questions with Answers

1. In binary language, we often use 1 and 0 to represent the two states. Given one state as in the following table, provide the other for each condition.

One state	1	True	?	North	Hot	Off?
Other state	0	?	Low	?	?	?
					Down	No.

Ans. True — False, High — Low, North — South, Hot — Cold, Off — On, Up — Down, Yes — No.

2. Explain the difference between a positive and a negative logic, giving an example of each as applied to voltage levels.

Ans. A digital information may be represented by a dc voltage level or by voltage pulse.

When voltage levels are used, two methods àre in use — positive logic and negative logic. In positive logic, the more positive of the two voltage levels represents binary state 1, while in negative logic, the more negative voltage represents binary state 1. In neither case 0 voltage is to represent 0 state. One system uses 0 to 1 volt as the lower level or binary 0 while 3 to 5 volts as binary 1. Another system uses -1.55V as typical for state 0 and -0.75V for 1.

3. (a) What do the abbreviations LSB and MSB stand for?

   (b) What is a *bit*?

   (c) What is the radix of binary arithmetic?

Ans. (a) LSB stands for least significant bit (digit) while MSB means most significant bit. Thus, in binary representation 0.1101, the first 1 after the ⊙ is the MSB and the last 1 is the LSB.

(b) The digit in a binary is termed as *bit*.

(c) The radix or base of the binary arithmetic is 2 whereas the base of decimal system is 10.

4. (a) Find the binary representation of the decimal number 176, 576, 47.6.

   (b) Find the decimal equivalent of 1111,10111.1.

Solution: (a) 176

	Remainder	
176/2 = 88	0	LSB
88/2 = 44	0	
44/2 = 22	0	
22/2 = 11	0	
11/2 = 5	1	
5/2 = 2	1	
2/2 = 1	0	
1/2 = 0	1	MSB

Therefore, binary equivalent of 176 is 10110000. To convert binary 10110000 into decimal system we proceed as under:

Binary: $1\times 2^7+0\times 2^6+1\times 2^5+1\times 2^4+0\times 2^3+0\times 2^2+0\times 2^1+0\times 2^0$.

Decimal: $128+0+32+16+0+0+0+0=176$.

576	Remainder	
$576\div 2=288$	0	LSB
$288\div 2=144$	0	
$144\div 2=72$	0	
$72\div 2=36$	0	
$36\div 2=18$	0	
$18\div 2=9$	0	
$9\div 2=4$	1	
$4\div 2=2$	0	
$2\div 2=1$	0	
$1\div 2=0$	1	MSB

$\therefore$ the binary equivalent of 576 is 1001000000.

47.6	Remainder	
47/2 = 23	1	LSB
23/2 = 11	1	
11/2 = 5	1	
5/2 − 2	1	
2/2 = 1	0	
1/2 = 0	1	MSB

For fraction	Binary	
$0.6\times 2=1.2$	1	MSB
$0.2\times 2=0.4$	0	
$0.4\times 2=0.8$	1	LSB

So the binary equivalent of 47.6 is 101111.101. Reverting to the decimal system.

Binary: $1\times 2^5+0\times 2^4+1\times 2^3+1\times 2^2+1\times 2^1+1\times 2^0+1\times 2^{-1}+0\times 2^{-2}+1\times 2^{-3}$.

Decimal: $32+0+8+4+2+1+0.5+0+0.1=47.6$.

1111:

Binary: $1\times 2^3+1\times 2^2+1\times 2^1+1\times 2^0$.

Decimal: $8+4+2+1=15$.

Try for the other binary numbers in the same way.

5. (a) What basic gates constitute English logic?
   (b) What is a truth table?
   (c) If a gate has 4 inputs, how may combinations of inputs are there?

Ans. (a) The basic gates the constitute English logic are AND, OR and NOT.

(b) A truth table shows the output for all possible combinations at the input. If a system of logic gates gives a truth table identical with the truth table of another system, then the two logical systems are identical.

(c) For a system of N inputs, the possible combinations are given by 2^n. So for 4 inputs, the possible number is $2^4=16$.

6. Use a truth table to show that $X_1=\bar{A}B+A\bar{B}$ and $X_2=\overline{AB}(A+B)$ are equivalent.

Solution:

X_1

A	B	$\bar{A}$	$\bar{B}$	$\bar{A}B$	$A\bar{B}$	$X_1=\bar{A}B+A\bar{B}$
0	0	1	1	0	0	0
0	1	1	0	1	0	1
1	0	0	1	0	1	1
1	1	0	0	0	0	0

X_2

A	B	AB	$\overline{AB}$	$A+B$	$X_2=\overline{AB}\,(A+B)$
0	0	0	1	0	0
0	1	0	1	1	1
1	0	0	1	1	1
1	1	1	0	1	0

7. (a) What does it mean to negate a signal?
   (b) What logic gate is used to do this?
   (c) If the input to this logic gate is left floating, what input does this represent and what is the output?

Ans. (a) *Negation* or *Complementation* means, absence of the signal at the input results in an output and presence of signal results in absence of the output. If $\left.\begin{matrix}A=1\\0\end{matrix}\right\}$, then negation of A denoted by $\bar{A}$ is given by $\left.\begin{matrix}\bar{A}=0\\1\end{matrix}\right\}$.

(b) The logic gate used to negate the signal is the NOT (or inverter) gate.

(c) If the input of the gate is left open, it behaves as if it has an input of binary 1, the output is binary 0.

8. Classify the following as positive or negative logic:

(a) 0 = 0V,	1 = 10V	Positive logic
(b) 0 = 0.2V,	1 = – 3.6V	Negative logic
(c) 0 = – 1.5V,	1 = – 0.75V	Positive logic
(d) 0 = – 0.8V,	1 = – 1.5V	Negative logic
(e) 0 = 5V,	1 = 0V	Negative logic
(f) 0 = 3.6V,	1 = 7.2V	Positive logic

9. Six inverters are connected in tandem as in Fig. 4.42. If the first input is A, what are the outputs at P, Q and R?

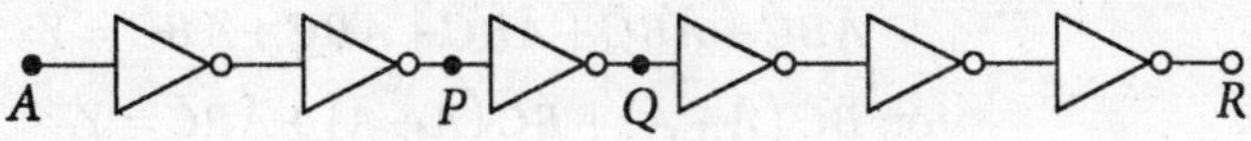

Fig. 4.42

Ans. The outputs at P, Q, R are $A, \bar{A}$ and A respectively.

4.11 Solved Problems

1. Verify $A+\overline{AB}=A+B$.

Solution:

A	B	$\bar{A}$	$\bar{A}B$	$A+\bar{A}B$	$A+B$
0	0	1	0	0	0
0	1	1	1	1	1
1	0	0	0	1	1
1	1	0	0	1	1

The truth table proves the relation.

2. Show that $A+\bar{A}=1$.

Solution:

A	$\bar{A}$	$Y=A+\bar{A}$
1	0	1
0	1	1

$\therefore A+\bar{A}=1.$

3. Solve the Boolean expression $C=A+B$ for the inputs $A=0$, $B=0$; $A=1$, $B=0$, $A=1$, $B=1$.

Solution:

A	B	$C=A+B$
0	0	0
1	0	1
1	1	1

So $C=0$

$=1$

$=1.$

4. Construct a logic gate design from the truth table given below.

Solution: From the truth table it appears that for any of the combinations of rows 2, 3, 4, 6 and 8, the output $Y=1$.

A	B	C	Y
0	0	0	0
0	0	1	1
0	1	0	1
0	1	1	1
1	0	0	0
1	0	1	1
1	1	0	1
1	1	1	1

So writing the Boolean equation, we get

$$\bar{A}\bar{B}C+\bar{A}B\bar{C}+\bar{A}BC+A\bar{B}C+ABC=Y \tag{4.14}$$

$$\text{or, } \bar{B}C\left(A+\bar{A}\right)+BC\left(A+\bar{A}\right)+\bar{A}B\bar{C}=Y.$$

Now, $A+\bar{A}=1$.

$\therefore\ \bar{B}C+BC+\bar{A}B\bar{C}=Y$ or, $C\left(B+\bar{B}\right)+\bar{A}B\bar{C}=Y$.

$$\therefore\ C+\bar{A}B\bar{C}=Y. \tag{4.15}$$

(4.15) can be further simplified with the help of De Morgan's law as below:

$$C+\bar{A}B\bar{C}=Y$$

$$\text{or, } C+\left(\overline{A+\bar{B}+C}\right)=Y$$

$$\text{or, } (C+C)+(A+B)=Y$$

$$\text{or, } C+\left(\bar{A}B\right)=Y. \tag{4.16}$$

Thus, the Boolean (4.14) is reduced to simplified (4.16), which can be utilised for gate circuit realisation of the truth table. The circuit design is given in Fig. 4.43.

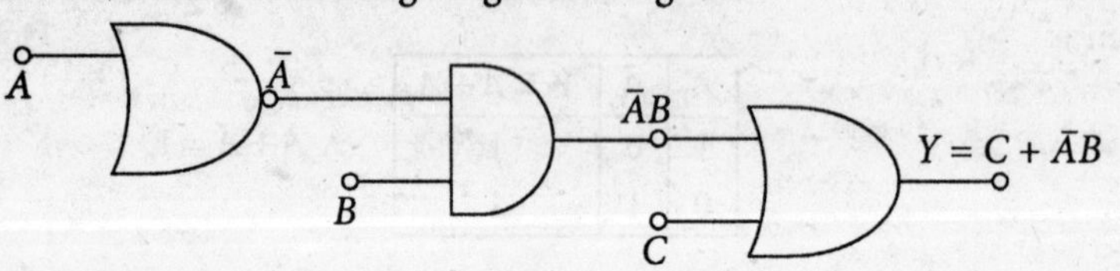

Fig. 4.43

5. In a certain company there are 4 shareholders, A owns 45 %, B owns 30 %, C owns 15 % and D owns 10 %. Each shareholder has a percentage vote equal to his share and a total vote greater than 50 % is required to pass a motion. Design an electronic voting system for the company. In the board meeting room each shareholder is to have a switch with which to indicate a Yes or No vote. A lamp is to light if the total vote cast is more than 50 %, indicating that the motion being voted is passed.

Solution: A NO vote is denoted by 0 and an YES vote by 1 (switch open and switch closed respectively). There are 4 inputs A, B, C, D and one output Y which is 1 when the vote is more than 50 %. The truth table is given below: All possible combinations are given by $2^4=16$.

45 % A	30 % B	15 % C	10 % D	Y	%
0	0	0	0	0	0
0	0	0	1	0	10 %
0	0	1	0	0	15 %
0	0	1	1	0	25 %
0	1	0	0	0	30 %
0	1	0	1	0	40 %
0	1	1	0	0	45 %

0	1	1	1	1	55 %
1	0	0	0	0	45 %
1	0	0	1	1	55 %
1	0	1	0	1	60 %
1	0	1	1	1	70 %
1	1	0	0	1	75 %
1	1	0	1	1	85 %
1	1	1	0	1	90 %
1	1	1	1	1	100 %

So from the truth table it is evident that the output $Y = 1$ for eight possible combinations, $\bar{A}BCD$, $A\bar{B}\bar{C}D$ or $A\bar{B}C\bar{D}$ or $A\bar{B}CD$ or $AB\bar{C}\bar{D}$ or $AB\bar{C}D$ or $ABC\bar{D}$ or $ABCD$. So the required Boolean equation is written as

$$Y = \bar{A}BCD + A\bar{B}\bar{C}D + A\bar{B}C\bar{D} + A\bar{B}CD + AB\bar{C}\bar{D} + AB\bar{C}D + ABC\bar{D} + ABCD. \tag{4.17}$$

(4.17) represents the truth table. When this is simplified with the help of Boolean theorems, we get

$$Y = BCD + A(B + C + D) \tag{4.18}$$

the gate implementation of (4.18) is shown in Fig. 4.44.

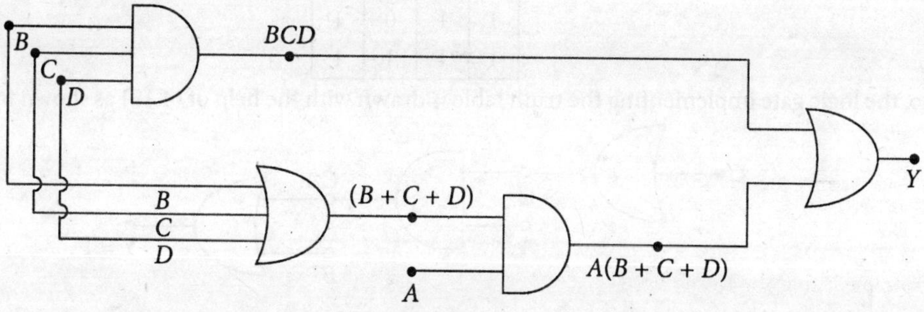

Fig. 4.44

6. Explain how an OR gate may be constructed with AND and NOT gates.

   Solution: By Boolean theorem we have $\left(\overline{\bar{A}\bar{B}}\right) = A + B$. This theorem can be applied to construct an OR gate from the combination of AND and NOT as shown in Fig. 4.45.

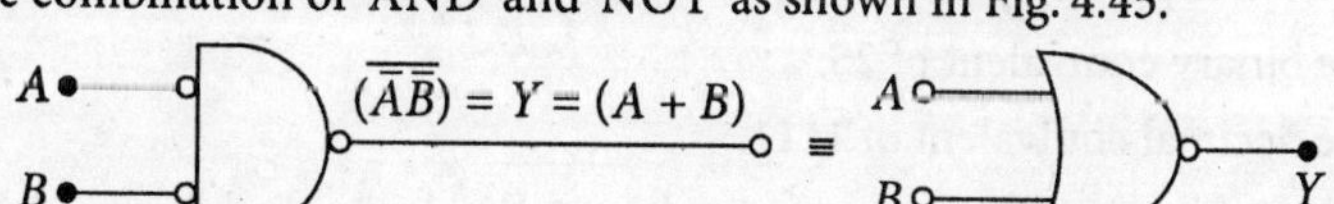

Fig. 4.45

7. Explain how a NOR circuit can be employed for AND.

   Solution: By Boolean theorem we have $\left(\overline{\bar{A} + \bar{B}}\right) = AB$, this relation can be applied to construct an AND from a combination of NOT and OR as shown in Fig. 4.46.

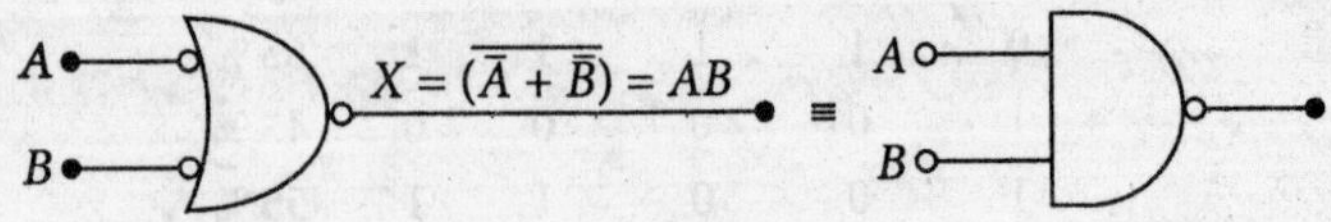

Fig. 4.46

8. With the following truth table, draw a simplified set of logic gates to provide a closed circuit when $Y = 1$.

Solution: The Boolean equation representing $Y = 1$ is given by

$$Y = \bar{A}B\bar{C} + \bar{A}BC + A\bar{B}\bar{C} + AB\bar{C} + ABC = BC\left(A + \bar{A}\right) + B\bar{C}\left(A + \bar{A}\right) + A\bar{B}\bar{C} = BC + B\bar{C} + A\bar{B}\bar{C}$$
$$= B(C + \bar{C}) + A\bar{B}\bar{C} = B + A\bar{B}\bar{C} = B + \left(\bar{A} + B + C\right) = (B + B) + \left(C + \bar{A}\right) = B + C + \bar{A} = B + \bar{C}A. \quad (4.19)$$

A	B	C	Y
0	0	0	0
0	0	1	0
0	1	0	1
0	1	1	1
1	0	0	1
1	0	1	0
1	1	0	1
1	1	1	1

So, the logic gate implementing the truth table is drawn with the help of (4.19) as shown in Fig. 4.47.

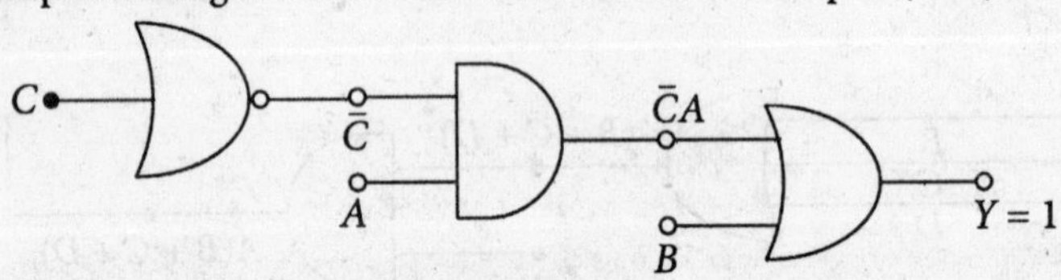

Fig. 4.47

4.12 Exercises

1. Draw a positive logic NOT gate and explain its operation. Give the truth table of a NOT gate.
2. Explain how an OR gate may be constructed with AND and NOT gates.
3. (a) Find the binary equivalent of 25.
   (b) Find the decimal equivalent of 1111.
   (c) Perform the binary addition corresponding to the decimal addition $2 + 2 = 4$.
   [*Hints.* $A = 10$
   $B = 10$
   $\overline{100}$]
4. Draw the logic diagrams to implement the following Boolean expression using (a) NAND (b) NOR gates. $X = \bar{A}(B + C) + D$.

5. Using a truth table, show that the term $(B+C)$ is redundant in the following: $(A+B)\left(\bar{A}+C\right)(B+C)$.
6. Negate the expression $A\bar{B}C+A\bar{B}\bar{C}+ABC$.
7. Implement $AB+\bar{C}\bar{D}=X$ with three NAND gates. Draw the logic circuit.
8. Change the OR gates to NOR, the AND gate to NAND.
9. Show how NAND gates can be used to produce OR operation $(X=A+B)$.
10. Simplify the expressions as far as possible: (a) $\bar{A}BC+\bar{A}\bar{C}$ (b) $\bar{A}\bar{C}+\bar{B}\bar{C}$ (c) $A\left(\bar{B}+\bar{C}\right)+AB$.
11. Give the truth table for (a) $A\bar{B}+AB$ (b) $AB\left(\bar{A}\bar{B}\bar{C}+\bar{B}C\right)$.
12. Prove that the NAND operation is not associative, i.e., $\overline{\left(\bar{A}\bar{B}\right)C}\neq\overline{A\left(\bar{B}\bar{C}\right)}$.

4.13 Problems Exercises

1. Find $A\cdot\bar{B}\cdot\bar{C}+A\cdot B$ under the given conditions as below:
   (a) $A=1,\ B=0,\ C=1$ [0] (c) $A=0,\ B=1,\ C=1.$ [0]
   (b) $A=0,\ B=0,\ C=0$ [0]
2. Simplify the Boolean functions as far as possible:
   (a) $xy+x\bar{y}$ [x] (d) $zx+z\bar{x}y$ [$z(x+y)$]
   (b) $(x+y)(x+\bar{y})$ [x] (e) $\left(\overline{x+y}\right)\left(\bar{x}+\bar{y}\right)$. [0]
   (c) $xyz+\bar{x}y+xy\bar{z}$ [y]
3. Reduce the following Boolean expressions:
   (a) $ABC+\bar{A}\bar{B}C+\bar{A}BC+AB\bar{C}+\bar{A}\bar{B}\bar{C}$ [$\bar{A}\bar{B}+B(A+C)$] (b) $BC+A\bar{C}+AB+BCD$. [$BC+A\bar{C}$]
4. Find the complement of the following and then reduce:
   (a) $\left(B\bar{C}+\bar{A}D\right)\left(A\bar{B}+C\bar{D}\right)$ [1] (c) $A\bar{B}+\bar{C}\bar{D}$. [$(\bar{A}+B)(C+D)$]
   (b) $\bar{B}D+\bar{A}B\bar{C}+ACD+\bar{A}BC[\bar{B}\bar{D}+A(\bar{D}+B\bar{C})]$
5. Given the Boolean function: $F=xy+\bar{x}\bar{y}+\bar{y}z$.
   (a) Implement it with the basic gates.
   (b) Implement it with only OR and NOT gates.
   (c) Implement it with AND and NOT gates.

Solution: (a) Using all the three basic gates, we draw the implementing circuit as shown below in Fig. 4.18.

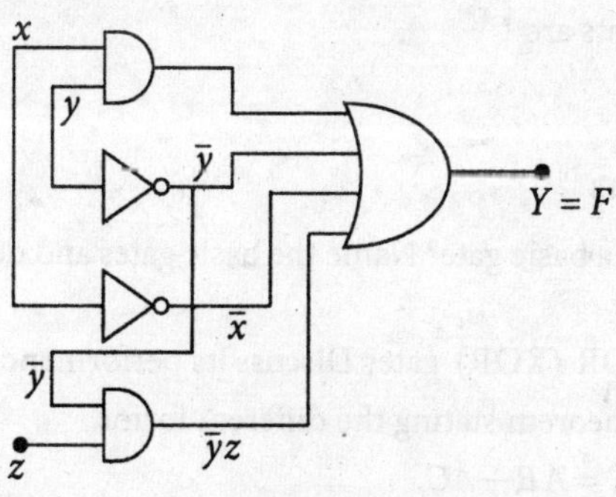

Fig. 4.48

(b) Complementing the function, we would get $\bar{F} = \overline{xy + \bar{x}\bar{y} + \bar{y}z} = (\bar{x}+\bar{y})(x+y)(y+\bar{z})$.
Using this relation and utilising OR and NOT gates, we would get $\bar{F}$ as the output and then re-complementing, we ultimately get F as shown in Fig. 4.49.

$$F = \overline{(\bar{x}+\bar{y})(x+y)(y+\bar{z})} = \overline{(\bar{x}+\bar{y})} + \overline{(x+y)} + \overline{(y+\bar{z})}.$$

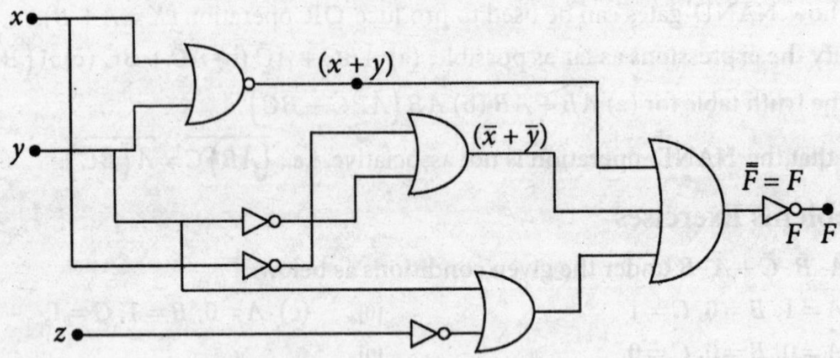

Fig. 4.49

(c) Scheme is $\bar{F} = \overline{xy + \bar{x}\bar{y} + y\bar{z}} = \left(\overline{\bar{x}\bar{y}}\right)\left(\overline{\bar{x}\bar{y}}\right) = \left(\overline{y\bar{z}}\right) \Rightarrow F = \overline{\bar{x}\bar{y}\left(\overline{\bar{x}y'}\right)(y\bar{z})}$.

6. Complete the following:

(a) $0 \cdot 0 = ?$ [0]
(b) $0 + 0 = ?$ [0]
(c) $0 \cdot 1 = ?$ [0]
(d) $A \cdot \bar{A} = ?$ [0]
(e) $A + \bar{A} = ?$ [1]
(f) $A + \bar{A}B = ?$ [$A + B$]
(g) $\overline{A + B} = ?$ [$\bar{A} \cdot \bar{B}$]

7. Determine the value of $L = \overline{XY + Z}$ for

(a) $X = Y = Z = 0$ [1]
(b) $X = 1, Y = Z = 0$ [1]
(c) $X = Y = Z = 1$. [0]

8. Reduce the following Boolean expression:

(a) $X \cdot 1 \cdot X$ [X]
(b) $X \cdot Y \cdot Y \cdot Y \cdot Z$ [XYZ]
(c) $YZ + Y\bar{Y}Z$. [YX]

9. An exclusive OR logic gate will have output as 1

(a) when all inputs are 0
(b) when all the inputs are 1
(c) when odd number of inputs are 1.

Which one is correct? [(c) is correct]

4.14 Theory Questions

1. What do you mean by a basic gate? Name the basic gates and describe any one of them with its truth table of performance.
2. What is an exclusive OR (XOR) gate? Discuss its performance giving its truth table.
3. Prove the consensus theorem stating the different forms.
[*Hints.* $AB + \bar{A}C + BC = AB + \bar{A}C$.
$(A+B)\left(\bar{A}+C\right)(B+C) = (A+B)\left(\bar{A}+C\right)$; these are the two forms.

$AB+\bar{A}C+BC=AB+\bar{A}C+BC(A+\bar{A})=AB+ABC+\bar{A}C+BC\bar{A}=AB(1+C)+\bar{A}C(1+B)=AB+\bar{A}C$, etc.]

4. What are universal gates? Name them. Discuss one of them with regard to its performance?
5. NOR and NAND gates are called universal gates. Explain.
6. Prove De Morgan's laws with the help of truth tables.
7. Reduce the Boolean expressions: (a) $\overline{\overline{ABC+\bar{A}\bar{B}}+BZ}$ (b) $C\left(\overline{ABC}+A\bar{B}C\right)$ (c) $\left(A+\bar{B}+C\right)\left(A\overline{BC}\right)$.
8. Show how NOR gates can be used to achieve both AND and OR functions.
9. Show that NOR operation is not associative.

   [*Hints.* Prove $\overline{\overline{A+B}+C}\neq\overline{A+\overline{B+C}}$.]
10. Find the truth table verification of the Boolean function: $F=xy+x\bar{y}+\bar{y}z$.
11. Show that NOR and NAND operators are not distributive.
12. Draw the logic diagram for the Boolean expression: $L=XY\bar{Z}+\bar{X}Y\bar{Z}+\bar{X}\bar{Y}$.
13. Draw a switching circuit representing the operation of NOT gate.

    [*Hints.*

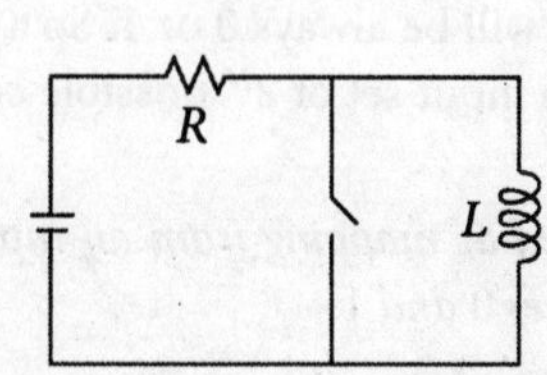

Fig. 4.50

14. Verify the following identities:

    (a) $A\bar{B}+\bar{A}B=(A+B)\left(\overline{AB}\right)=(A+B)\left(\bar{A}+\bar{B}\right)$

    (b) $AB+\bar{A}\bar{B}=\left(A+\bar{B}\right)\left(\bar{A}+B\right)=\left(\overline{A\bar{B}+\bar{A}B}\right)$

    (c) $AC+\bar{A}BC=AC+BC$

    (d) $AB+\bar{A}C=(A+C)\left(\bar{A}+B\right)$

    (e) $(A+B)\left(\bar{A}+C\right)=AC+\bar{A}B$.

Chapter 5

Boolean Function in Different Forms

5.1 Introduction

A Boolean function is a function of binary variables which can assume only two states. So, if f' be a function of N binary variables, then 2^N possible combinations of the variables can be generated. But the output of any combination by the digital systems will be always 0 or 1. So a Boolean function of N binary variables may be considered as mapping from an input set of 2^N possible combinations onto the output set of only two values 0 and 1.

So, *a Boolean function may be defined as mapping from an input set of all possible combinations of the binary variables to the output set of values* 0 *and* 1.

Thus, let

$$f(A) = \bar{A}. \tag{5.1}$$

Here, the function is a function of one binary variable A.

In this case,

$$f(0) = \bar{0} = 1$$
$$f(1) = \bar{1} = 0.$$

Let

$$f(A, B) = A + B. \tag{5.2}$$

Here the function is of two variables A and B leading to $2^2 (= 4)$ possible combinations

$$f(0,1) = 0 + 1 = 1$$
$$f(1,0) = 1 + 0 = 1$$
$$f(0,0) = 0 + 0 = 0$$
$$f(1,1) = 1 + 1 = 1.$$

Let

$$f(A, B) = \bar{A} + \bar{B}, g(A, B) = \overline{AB}. \tag{5.3}$$

$$f(0,0)=1+1=1,\ g(0\cdot 0)=1$$
$$f(0,1)=1+0=1,\ g(0\cdot 1)=1$$
$$f(1,0)=0+1=1,\ g(1\cdot 0)=1$$
$$f(1,1)=0+0=0,\ g(1\cdot 1)=0.$$

This example shows that the Boolean expressions $\bar{A}+\bar{B}$ and $\overline{AB}$ have got the same Boolean functional values.

So, a Boolean expression is an expression containing the Boolean variables but a Boolean function defines that for a particular value of the variables, there is one and only one value of the expression.

From the above example, Boolean expressions $\left(\bar{A}+\bar{B}\right)$ and $\overline{AB}$ are identical, because $f(A, B)$ and $g(A, B)$ assume same values for all possible values of A and B.

5.2 Maxterm and Minterm: Literals

For any two binary variables A and B, each may appear as $\left(A,\bar{A}\right),\left(B,\bar{B}\right)$; hence, each of $A, B, \bar{A}, \bar{B}$ is termed as *a literal.* So there are $2^2(=4)$ possible forms of their sum represented by $(A+B)$, $\left(\bar{A}+B\right)$, $\left(A+\bar{B}\right)$, $\left(\bar{A}+\bar{B}\right)$. *Such sum of the literals A, B is referred to as Maxterm.* In case of three variables A, B, C, there are $3^2(=9)$ maxterms.

Maxterms, also called standard sums of the refer to the forms sum containing all the variables, complemented or un-complemented.

Minterms: The minterms refer to the possible forms of the products of the literals.

Thus, minterms in case of two variables A, B are represented by $AB, \bar{A}B, A\bar{B}$ and $\bar{A}\bar{B}$. These terms are called the minterms or standard products of A and B.

A Boolean expression is one that contains either sums of minterms or products of maxterms. Thus, $f(A,B)=AB+\bar{A}B$ is an expression containing sum of minterms. Moreover, it contains all the variables either in the uncomplemented or complemented form. Such an expression is said to be in *canonical* form.

Consider,

$$f(A,B,C)=ABC+\bar{A}BC+A\bar{B}C+AB\bar{C}. \tag{5.4}$$

This Boolean expression is in canonical for...i, because it contains sum of minterms each containing all the variables either in complemented or uncompleted forms.

Note also that each minterm of the Boolean expression is the product of all the literals under consideration. This characteristic feature is the identification of the *disjunctive normal form* (DNF).

Disjunctive Normal Form (DNF) *refers to the Boolean expression containing a minterm or a sum of finite number of minterms of the literals.*

Conjunctive Normal Form (CNF)

A Boolean expression is said to be in conjunctive normal form (CNF) if it contains a single sum term or a sum of finite number of sum terms of the literals. Thus,

$$f(A,B,C)=(A+B)\left(B+\bar{C}\right)(A+B+C) \tag{5.5}$$

is a Boolean expression in CNF.

Full Disjunctive Normal Form (FDNF)

A Boolean expression is said to be in full disjunctive normal form (FDNF) if each term is a minterm, i.e., product of all the literals. Thus,

$$f(A,B,C)=ABC+\bar{A}BC+A\bar{B}C+AB\bar{C}+A\bar{B}\bar{C} \tag{5.6}$$

is an expression in FDNF.

Full Conjunctive Normal Form FCNF

Full conjunctive normal form (FCNF) refers to the Boolean expression which contains each sum term as a maxterm (i.e., sum of all literals).

Thus,

$$f(A,B)=(A+B)\left(\bar{A}+B\right)\left(A+\bar{B}\right) \tag{5.7}$$

is an expression in FCNF.

Complete Disjunctive Normal Form (CDNF)

FDNF expression may contain one minterm or a finite number of minterms, but complete disjunctive normal form expression contains all possible sum of the product terms of the literals.

Thus,

$$f(A,B)=AB+\bar{A}B+A\bar{B}+\bar{A}\bar{B}. \tag{5.8}$$

Complete Conjunctive Normal Form (CCNF)

In case of full conjunctive normal form, the expression contains a product term or finite number of product terms each of which is a maxterm. But in CCNF, the expression contains all possible products of maxterms.

Thus,

$$f(A,B)=(A+B)\left(\bar{A}+B\right)\left(A+\bar{B}\right)\left(\bar{A}+\bar{B}\right). \tag{5.9}$$

5.3 Development of Boolean expression from the truth table representation

A truth table lists the values of the Boolean function for each set of values assigned to the independent binary variables. Thus, suppose a Boolean function f is a function of three independent variables A, B, C. So we write, $Y = f(A, B, C)$. To form the truth table, a table is constructed with one column for each of the independent variables and one column for the dependent variables. The table below shows the constructions. In the present case, there are three independent variables A, B, C and the dependent variable is Y.

A	B	C	$Y=f(A,B,C)$
0	0	0	$=\ldots?$
0	0	1	
0	1	0	
0	1	1	
1	0	0	
1	0	1	
1	1	0	
1	1	1	

A truth table for a Boolean function is given as below:

A	B	C	Y
0	0	0	1
0	0	1	0
0	1	0	1
0	1	1	1
1	0	0	1
1	0	1	0
1	1	0	1
1	1	1	1

Construct the Boolean expression corresponding to this Boolean function Y.

Find the minterms containing variables A, B, C producing $Y=1$ and then ORing these terms, we get

$$\begin{aligned}Y &= \bar{A}\bar{B}\bar{C}+\bar{A}B\bar{C}+\bar{A}BC+A\bar{B}\bar{C}+AB\bar{C}+ABC\\ &= \bar{B}\bar{C}\left(A+\bar{A}\right)+B\bar{C}\left(A+\bar{A}\right)+BC\left(A+\bar{A}\right)\\ &= \bar{B}\bar{C}+B\bar{C}+BC=B\left(C+\bar{C}\right)+\bar{B}\bar{C}=B+\bar{B}\bar{C}.\end{aligned} \tag{5.10}$$

Thus, from the truth table we can develop the Boolean expression which can be simplified with the help of theorems in Boolean algebra to a minimum number of standard product sum for implementation.

In order to implement Y through minimum number of gates, we use the simplified version as in (5.10). The implementing circuit is shown in Fig. 5.1.

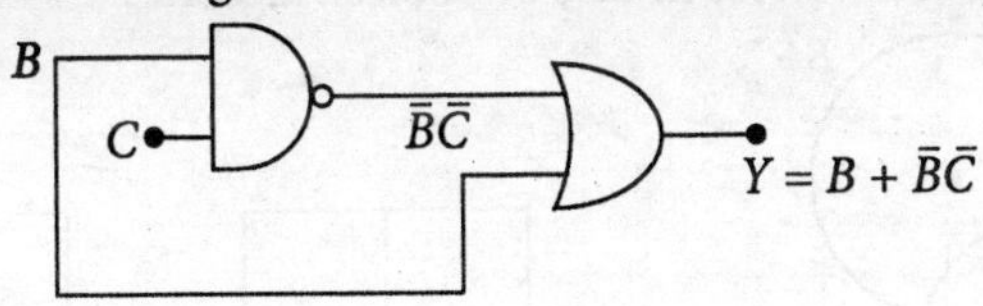

Fig. 5.1

5.4 Simplification of a Boolean expression by MAP method: Minimisation of Boolean function

An expression in binary variable representing some Boolean function requires to be simplified such as to contain minimal literals. This minimisation reduces the cost of implementation and the delay time, hence maximising variability.

Simplification technique may be applied through the use of the theorems of Boolean algebra but it is not used in practice. There are two standard methods for reduction. They are: (a) *Karnaugh map* (*K*-map) (b) *Quine-Mccluskey* (QM) *tabular method:*

K-map method is very suitable when the number of variables $n \leq 4$, whereas *Q-M* method is suitable for $n > 4$.

We will describe here only *K*-map method and not the tabular method.

Veitch-Karnaugh method. The essence of the method is similar to Venn diagram meant for pictorial representation and relation amongst the Boolean variables.

Venn diagram. In this diagram a number of circles are inscribed within a rectangle [Fig. 5.2] each circle representing a variable.

In case of two intersecting circles. In Fig. 5.2 we have shown two intersecting circles. The left circle represents the variable A and the right circle, the variable B. The region outside the left-hand circle is represented by $\bar{A}$, the complement of A. If $A=1$, it exists in the circle and $\bar{A}=0$ inside and $\bar{A}=1$ outside the circle. Similar consideration applies to the second circle for the variable B. The region of overlap represents the simultaneous existence of A and B, written as AB. The region of the rectangle outside the two circles is represented by $\bar{A}\bar{B}$.

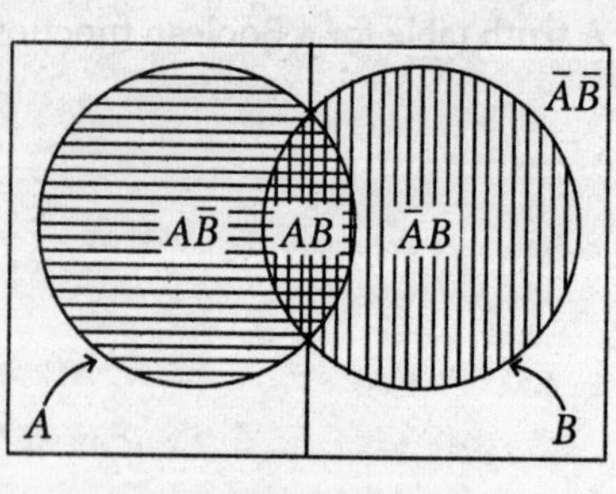

Fig. 5.2

1. Region belonging to neither A nor B is represented by $\bar{A}\bar{B}$.
2. Region inside the second circle and outside the first circle is represented by $\bar{A}B$.
3. Region inside the first circle but outside the second circle is denoted by $A\bar{B}$.
4. Region inside both the circles is represented by AB.

Consider the expression $A+AB$, this in Venn diagram would represent either the region of A or the region of overlap which also belongs to A.

$$\therefore\ A+AB=A \text{ (identity)}. \tag{5.11}$$

Three intersecting circles. In Fig. 5.3 we have shown three intersecting circles representing three variables A, B, C.

The blackened region belongs to A and B or C.

$\therefore\ A(B+C)=AB+AC.$

This shows that the distributive law holds in case of Boolean algebra.

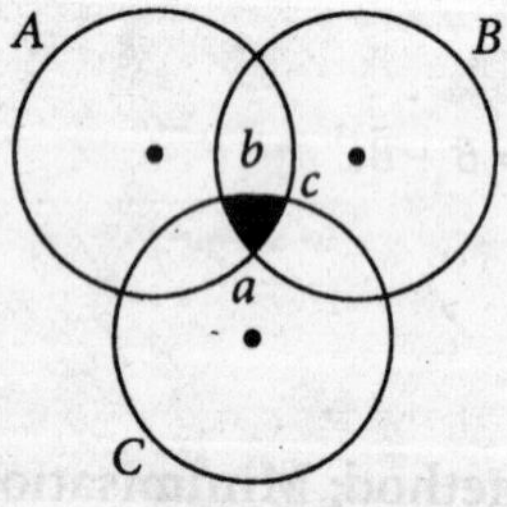

Fig. 5.3

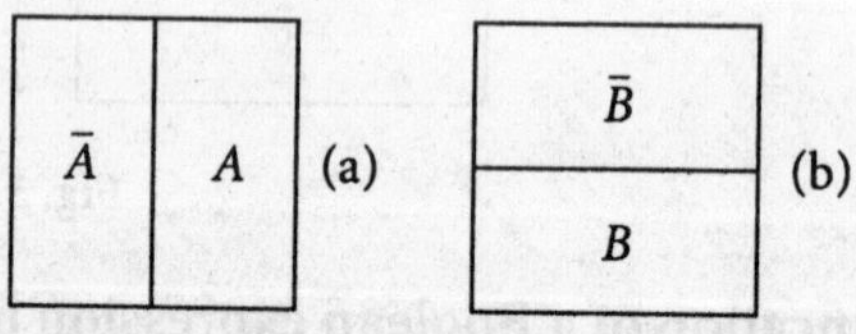

Fig. 5.4

Karnaugh map. Simplification of a Boolean expression can be more easily done through Karnaugh map technique if the number of Boolean variables is $N \leq 4$.

In this method, a square of unit area is divided into two equal parts by a vertical line [Fig. 5.4 (a)], the right part represents A and the left-part, its complement $\bar{A}$. Thus, $A+\bar{A}=1$.

Similarly, a square of unit area is divided into two equal parts by a horizontal [Fig. 5.4 (b)], the lower part represents B, the upper half, its complement $\bar{B}$ such that $B+\bar{B}=1$.

Combining these two ideas we draw the Fig. 5.5 to show the Karnaugh map for two variables A, B.

A unit square is divided into 4 square boxes representing four possible minterms $\bar{A}\bar{B}$, $A\bar{B}$, $\bar{A}B$, AB.

The bottom two squares combine to form the B strip thus giving

$$\bar{A}B+AB=B. \tag{5.12}$$

This is the essence of K-map approach to simplify Boolean expression. The second column of square combine to form the vertical strip A thus

$$A\bar{B}+AB=A \text{ (identity).} \tag{5.13}$$

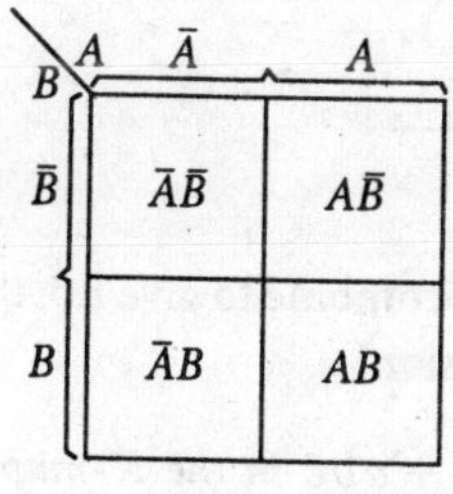

Fig. 5.5

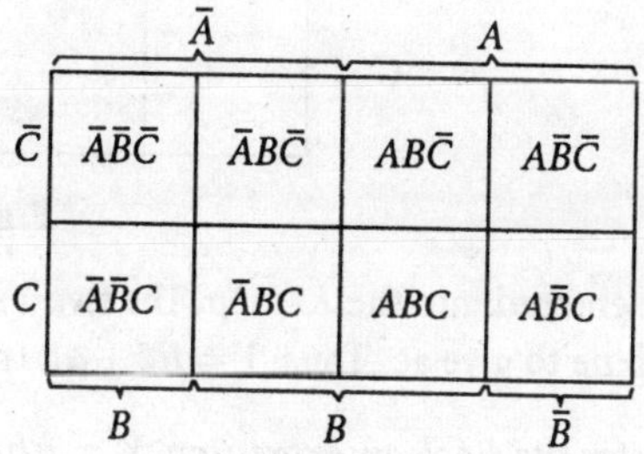

Fig. 5.6

In Karnaugh map, the adjacent squares in the horizontal or vertical directions combine. When they combine, one variable is eliminated. Please remember, diagonally adjacent squares do not combine.

In case of three variables A, B, C; the Karnaugh map is drawn in a rectangle of unit area divided into eight rectangular cells each of which represents a minterm [Fig. 5.6]. The Boolean expression in 3 variables consists of eight minterms given by the terms in the boxes [Fig. 5.6].

$$Y=\bar{A}\bar{B}\bar{C}+\bar{A}B\bar{C}+AB\bar{C}+A\bar{B}\bar{C}+\bar{A}\bar{B}C+\bar{A}BC+ABC+A\bar{B}C.$$

For further simplification of the K-map, consider Fig. 5.7.

The three-variable K-map is labelled as shown in the diagram.

Take the expression $Y=\bar{A}B\bar{C}+\bar{A}BC+AB\bar{C}+ABC.$

The first minterm is common in the column $\bar{A}B$ and row $\bar{C}$. The presence of this minterm is indicated by inserting 1 in the proper box. The names of the variables are written at the left-hand top corner; 0's and 1's used to indicate the co-ordinates. The four minterms are adjacent to each other and combine to give B. Thus, $Y=B$, minimal literals on simplification.

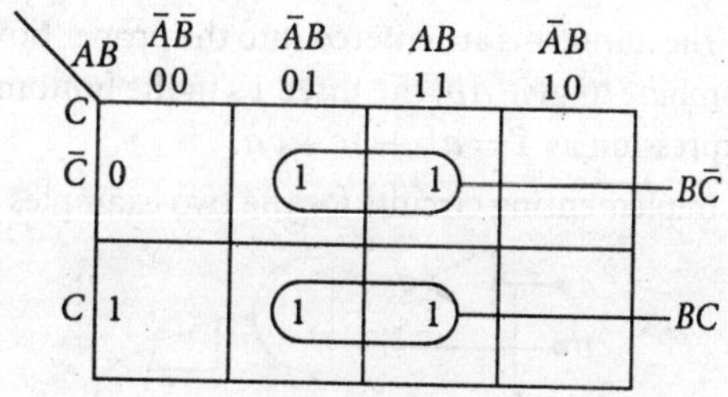

Fig. 5.7

Example 5.1. *From the truth table given below, construct the K-map and hence, simplify the Boolean function to obtain minimal expression.*

Find the combination of a, b, c giving $Y=1$. These combinations are 010, 101, 110, 111.

a	b	c	Y
0	0	0	0
0	0	1	0
0	1	0	1
0	1	1	0
1	0	0	0
1	0	1	1
1	1	0	1
1	1	1	1

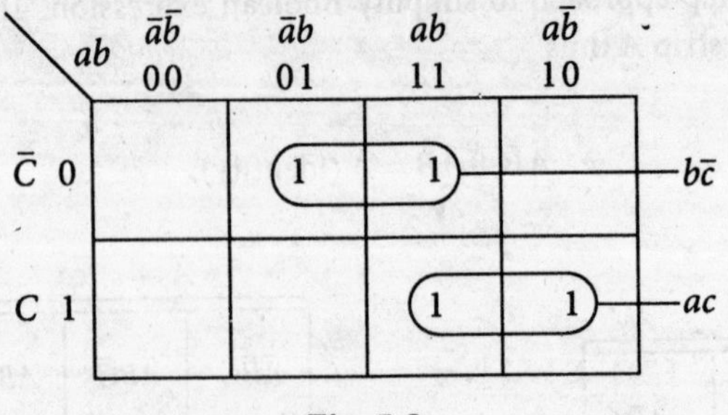

Fig. 5.8

The four 1's are entered into the *K*-map. The two 1's in the top row combine to give $b\bar{c}$; the two 1's in the bottom row combine to give ac. Thus, $Y = b\bar{c} + ac$ (minimal expression).

Example 5.2. *Enter the Boolean expression* $Y = abc + \bar{a}bc + a\bar{b}c + ab\bar{c}$ *in the K-map and obtain the minimal expression.*

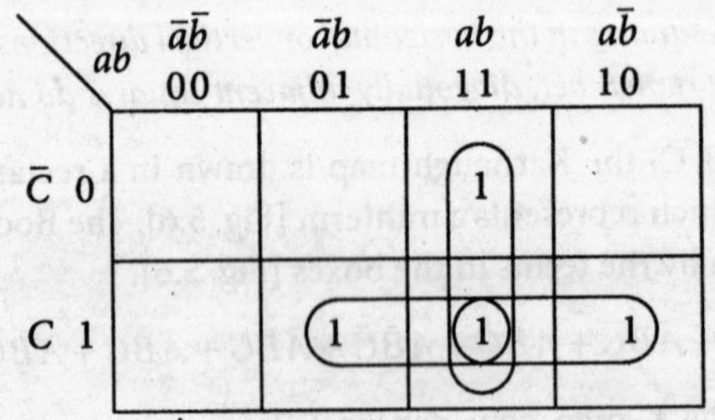

Fig. 5.9

The minterms are entered into the proper boxes indicated by 1's [Fig. 5.9]. The two 1's in the third column combine to give ab; the three 1's in the bottom row combine to give bc and ac thus *K*-map simplifies the expression as $Y = ab + bc + ca$.

Implementing circuits for the two examples are shown in Fig. 5.10 (a) and (b).

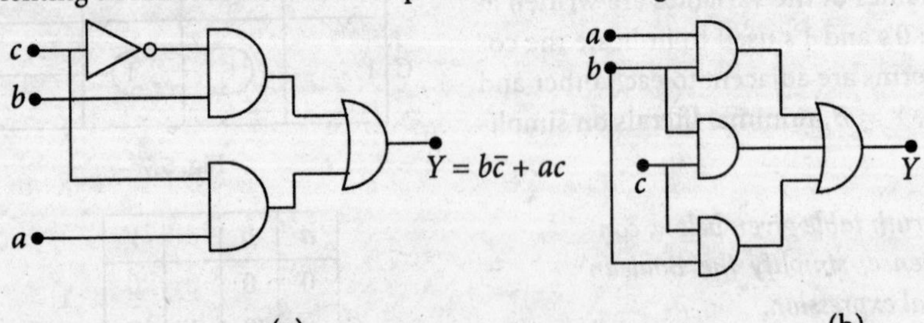

Fig. 5.10

Chapter 6

Switching Circuit Re-visited

Using the basic operations, namely, AND, OR and NOT, any Boolean function may be implemented by suitable electrical circuitry.

Example 6.1. *Draw the switching circuit to implement the Boolean function*

$$f(a,b,c)=a\bar{c}+ab+\bar{a}=\bar{a}(\bar{c}+b)+\bar{a}. \tag{6.1}$$

There are two (+) ORing operations and two AND operations, two NOT operations to be performed on a, c. So the implementing electrical circuit will be as shown in Fig. 6.1(a), (b) and (c) for the basic gates.

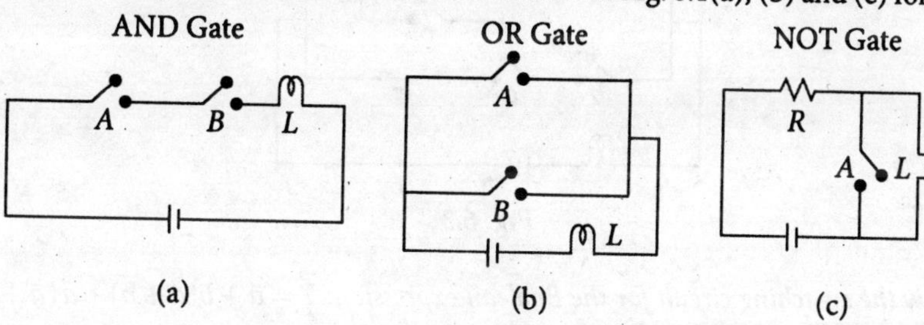

Fig. 6.1

We have shown three electrical circuits in Fig. 6.1.
(a) represents AND gate, (b) represents OR gate, (c) represents NOT gate.

(a)

A	B	L	Remarks
Open	Open	OFF	
Open	Closed	OFF (0)	AND
Closed	Open	OFF (0)	Gate
Closed	Closed	ON (1)	

(b)

A	B	L	Remarks
Open	Open	OFF	
Closed	Open	ON	OR gate
Open	Closed	ON	
Closed	Closed	ON	

(c)

A	L	Remarks
Open	ON	NOT gate
Closed	OFF	

Implementing (6.1) is utilised to implement the Boolean function $f(a,b,c)$ and is shown in Fig. 6.2.

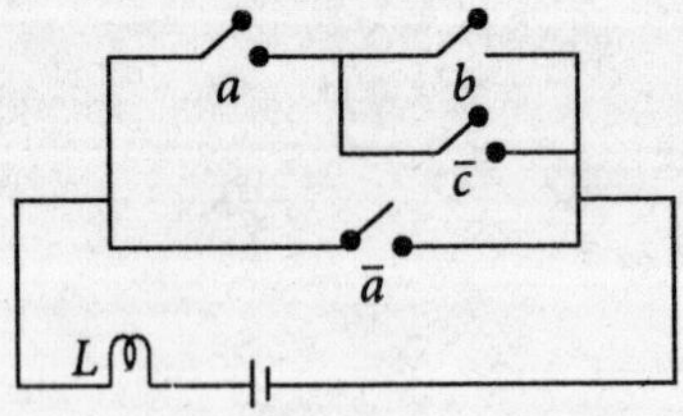

Fig. 6.2

Example 6.2. *Draw the switching circuit for implementation of the Boolean expression after simplification, if possible,* $Y = abc + \bar{a}bc + a\bar{b}c + ab\bar{c}$.

Solution: Simplification of the expression is to reduce the number of switches. So first simplify the expression with the help of standard formulae:

$$Y = abc + \bar{a}bc + a\bar{b}c + ab\bar{c} = (abc + \bar{a}bc) + \left(abc + a\bar{b}c\right) + (abc + ab\bar{c})$$
$$= bc + ac + ab = ab + bc + ca = a(b+c) + bc. \quad (6.2)$$

So this is the minimal expression. To implement by switching circuit, we refer to Fig. 6.3.

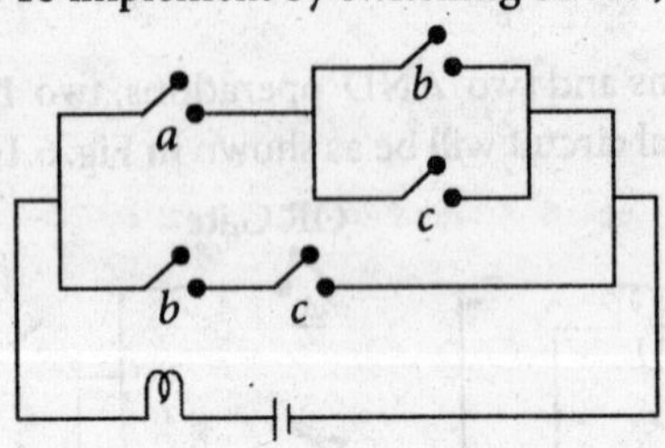

Fig. 6.3

Example 6.3. *Draw the switching circuit for the Boolean expression:* $Y = a + b(a+b) + a(\bar{a}+b)$.

Solution: Simplify the expression to a minimal expression by standard formulae:

$$Y = a + b(a+b) + a(\bar{a}+b) = a + ba + b + 0 + ab = a + ab + b = a(1+b) + b = a + b. \quad (6.3)$$

To implement the function, we use the minimal expression (6.3) and draw the circuit as in Fig. 6.4.

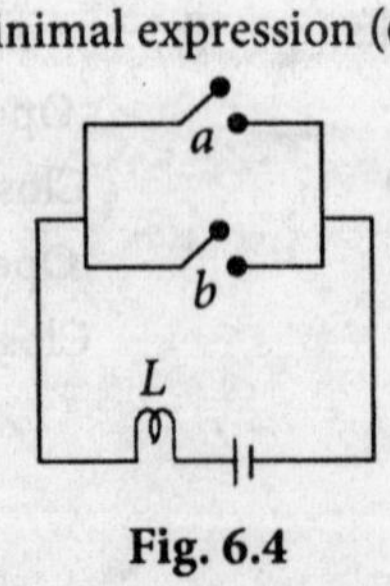

Fig. 6.4

Example 6.4. *In a committee consisting of three members decision is taken by majority votes. Design a simple switching circuit, which will indicate the acceptance of a proposal by lighting a bulb.*

Solution: Each member is provided with a switch denoted by a, b, c each of which is a 2-state Boolean variable. When at least two members are in favour, proposal is accepted and the lamp glows, i.e., the output $Y=1$. So, the truth table will be as shown in Table 6.1.

Table 6.1: Truth Table

a	b	c	Y	Remarks
0	0	0	0	There are 4 cases when acceptance of a proposal is indicated by lamp glow.
0	0	1	0	
0	1	0	0	
0	1	1	1	
1	0	0	0	
1	0	1	1	
1	1	0	1	
1	1	1	1	

So the Boolean expression is same as in Ex. 6.2; K-map representation is in Fig. 6.10 for simplification and the implementing switching circuit is shown in Fig. 6.3.

Example 6.5. *Examine the Boolean expression $f(x,y,z)=yz+zx$. Is it in* DNF? *Convert it into full* DNF. *Make the truth table of the function and draw the switching circuit to implement the function.*

Solution: It is in DNF, because it contains the sum of the product of the literals.

First make the truth table of the given function as shown in Table 6.2.

Table 6.2: Truth Table

x	y	z	yx	xz	f
0	0	0	0	0	0
0	0	1	0	0	0
0	1	0	0	0	0
0	1	1	1	0	1
1	0	0	0	0	0
1	0	1	0	1	1
1	1	0	0	0	0
1	1	1	1	1	1

To find the full DNF corresponding to the Truth Table 6.2, we follow the procedure below.

Step 1. Find the rows giving $f=1$.

Step 2. Complement the independent variables in each row giving value 0 and take product of the complemented variables and all other independent variables having value 1.

Step 3. Add all the terms ORing them.

The first minterm (fourth row) $=\bar{x}yz$.

The minterm corresponding to the sixth row $=x\bar{y}z$.

Eighth row $=xyz$.

So the full DNF of $f(x,y,z)$ is

$$f(x,y,z)=\bar{x}yz+x\bar{y}z+xyz=\bar{x}yz+x\bar{y}z+xyz+xyz$$
$$=(x+\bar{x})\,yz+xz\,(y+\bar{y})=yz+xz \text{ (minimal)}$$
$$=z(y+x). \quad (6.4)$$

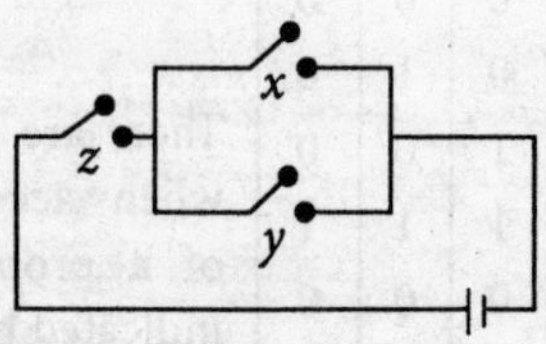

Fig. 6.5

Implementing switching circuit corresponding to (6.4) is shown in Fig. 6.5.

6.1 Short Theory Questions with Answers

1. Is it true that in Boolean algebra $= a+bc=(a+b)\cdot(a+c)$?

   Ans. Yes, because $(a+b)\cdot(a+c)=a\cdot a+a\cdot c+b\cdot a+b\cdot c=a+a\cdot c+b\cdot a+b\cdot c$
   $=a(1+c)+b\cdot a+b\cdot c=a+b\cdot a+b\cdot c$
   $=a(1+b)+b\cdot c=a+b\cdot c.$

2. $\left(\overline{a+b}\right)=?$

   Ans. $\left(\overline{a+b}\right)=\bar{a}\cdot\bar{b}$, according to De Morgan's law.

3. Does Boolean algebra deal with infinite number of elements like algebra of real numbers?

   Ans. No, it deals with finite number of elements.

4. What do you mean by a literal?

   Ans. A primed variable $\bar{A}$ or an unprimed variable A is called a *literal.*

5. What do you mean by monomial?

   Ans. A literal or a product of two or more literals is called a monomial.

   So, $x, \bar{x}y, \bar{y}, x\bar{y}\bar{z}$ are all monomials.

6. What do you mean by unary operation?

   Ans. If A be a binary variable, its complement $\bar{A}$ is obtained from A such that

A	$\bar{A}$
0	1
1	0

   This operation is called *unary operation.*

7. Two NOT gates connected in series produce output same as the input. — Explain.

   Ans. There is a common saying that two negations make an assertion. So two NOT gates in succession will exactly reproduce the input. Thus, the truth table of two NOT gates will be as shown in Fig. 6.6.

N_1	N_2	N_1
0	1	0
1	0	1

Fig. 6.6

8. Any switching circuit can be represented by the NOR and NAND gates. — Explain.

Ans. NOR and NAND are called universal gates, because all the basic gates, viz., AND, OR and NOT can be obtained through NOR and NAND gates. Since any switching circuit is a combination of basic gates, so it can be represented by NOR and NAND gates.

9. What do you mean by *Minterm* and *Maxterm* used in Boolean algebra?

Ans. A Boolean variable may appear in the complemented or uncomplemented form, called a literal. For a single Boolean variable a it may appear as a or $\bar{a}$, uncomplemented and complemented forms. Each refers to a literal. For two variables a and b, the literals are $a, b, \bar{a}, \bar{b}$ and there are four possible forms of the sum of literals, viz., $(a+b), (\bar{a}+b), (a+\bar{b}), (\bar{a}+\bar{b})$. Each of these terms is a maxterm of the literals.

Similarly, there are four possible forms of the product of the literals, viz., $ab, \bar{a}b, a\bar{b}, \bar{a}\bar{b}$. Each product term of the literals is called a *minterm of the literals.*

10. What do you mean by a truth table? What is its importance?

Ans. This table lists the values of the Boolean expression corresponding to all possible values of the independent Boolean variables contained in the expression. From the truth table representation, we can simplify the Boolean expression so as to implement the Boolean function with minimum number of switches.

11. What is the meaning of canonical form of Boolean expression?

Ans. A Boolean expression is expressed either as the sum of product (SOP) forms or product of the sum (POS) forms of the literals, i.e., sum of minterms or product of the maxterms.

So a Boolean expression is in canonical form if each term it contains is either a minterm or a maxterm of the variables $f(x,y) = x\bar{y} + \bar{x}y$.

12. What is the meaning of Disjunctive Normal Form (DNF)?

Ans. If a Boolean expression contains a single product term or sum of finite number of product terms of the literals, with no two terms identical, then the expression is said to be in Disjunctive Normal Form (DNF).

A Boolean expression, where each term is a minterm (i.e., product of all the literals) is said to be in full disjunctive form.

13. Prove De Morgan's theorems from theoretical reasoning: (a) $\overline{A \cdot B} = \bar{A} + \bar{B}$ (b) $\overline{A+B} = \bar{A} \cdot \bar{B}$.

Proof. (a) Take $A \cdot B = X$. According to postulates P5, there exists an element $\bar{X}$ in set S for every element X of the same set such that $X + \bar{X} = 1$ and $X \cdot \bar{X} = 0$.

Now, $A \cdot B + (\bar{A} + \bar{B}) = [A + (\bar{A} + \bar{B})] \cdot [B + (\bar{A} + \bar{B})] = [(A + \bar{A}) + \bar{B}] \cdot [(B + \bar{B}) + \bar{A}]$
$= (1 + \bar{B}) \cdot (1 + \bar{A}) = 1 \cdot 1 = 1.$

So, $(\bar{A} + \bar{B}) = \overline{A \cdot B} \Rightarrow \overline{A \cdot B} = (\bar{A} + \bar{B})$.

(b) $(A+B) \cdot (\bar{A} \cdot \bar{B}) = A \cdot (\bar{A} \cdot \bar{B}) + B \cdot (\bar{A} \cdot \bar{B}) = (A \cdot \bar{A})\bar{B} + (B \cdot \bar{B}) \cdot \bar{A} = 0 + 0 = 0.$

So $X = (A+B)$, $\bar{X} = (\bar{A} \cdot \bar{B}) \Rightarrow \overline{A+B} = \bar{A} \cdot \bar{B}$.

14. Draw diagram and explain how only NAND gates can be used to get the functions of AND and NOT gates. [CU 2006]

Ans. NAND gate is a universal gate in the sense that it leads to be used to get functions of all the basic gates.

NAND → NOT 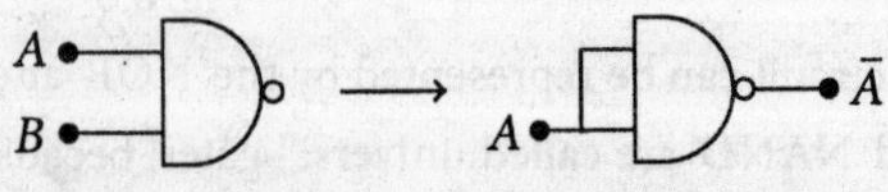

When the two-input terminals of a two-input NAND are tied together, it acts as a NOT gate.

NAND → AND 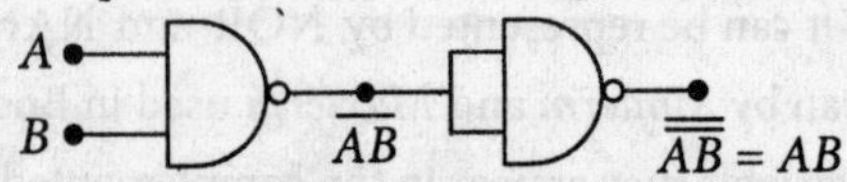

NAND + NOT → AND

15. Prove the Boolean identity: $AB + B\bar{C} + \bar{A}(B + C) + \bar{B}C + \bar{A}\bar{B}\bar{C} + ABC = \bar{A} + B + C$.

Solution: $L = AB + B\bar{C} + \bar{A}B + \bar{A}C + \bar{B}C + \bar{A}\bar{B}\bar{C} + ABC$

$$= AB(1 + C) + \left(B\bar{C} + \bar{B}C\right) + \bar{A}B + \bar{A}C + \bar{A}\bar{B}\bar{C}$$

$$= AB + B\bar{C} + \bar{B}C + \bar{A}B + \bar{A}C + \bar{A}\bar{B}\bar{C}$$

$$= B + B\bar{C} + \bar{B}C + \bar{A}C + \bar{A}\bar{B}\bar{C}$$

6.2 Questions Exercise

1. What are the conjunctive, disjunctive normal forms? Give example of each. Construct the truth table thereof.
2. Prove that EX-OR (exclusive OR) can be constructed by using NAND gates.

   [*Hints.* Scheme (i) $\overline{AB}$ (ii) $A \cdot \overline{AB} = \bar{A} + AB = \bar{A} + B$ (iii) $B \cdot \overline{AB} = \bar{B} + AB = \bar{B} + A$.

   $\overline{Y = \left(\bar{A} + B\right) \cdot \left(\bar{B} + A\right)} = A\bar{B} + \bar{A}B + A \oplus B]$

   Implementation (with all NANDs)

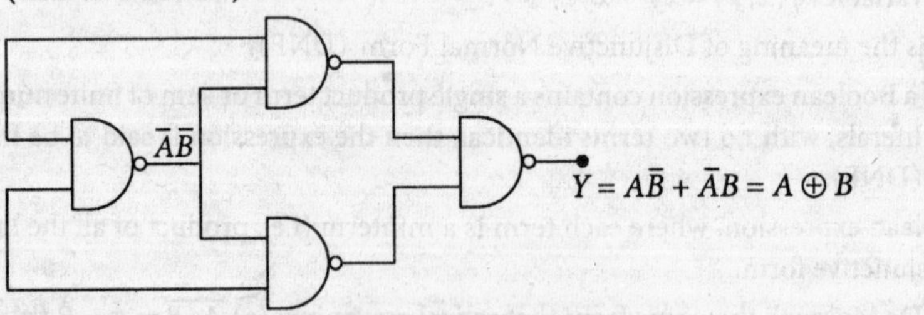

Fig. 6.7

3. What is Complete Conjunctive Normal Form (CCNF)?
4. What is Complete Disjunctive Normal Form (CDNF)?
5. Why are NAND and NOR gates called *universal gates*?
6. Simplify the Boolean expression $\bar{A}BC + A\bar{B}C + AB\bar{C} + ABC$. [$AB + BC + CA$] [BU 2002]
7. Simplify (a) $A + \overline{AB}$ (b) $(A + B)(B + C)(C + A)$, where A, B and C are Boolean variables.

Chapter 7

Combinational and Sequential Logical Circuits

7.1 Digital systems

Digital system refers to those electronic circuits where all the three basic gates are used at large with or without feedback. These systems are of two types: (a) Combinational digital (b) Sequential digital.

Combinational digital system. In this system, output at any instant depends on the instantaneous values of the inputs and without any reference to the previous history of the inputs. Even a ROM (Read only memory) is a combinational system, because it does not store bits but it memorises only the fundamental relationship between the input variables and the output variables.

Sequential digital system. In this system the output follows the same sequence as the input. A sequential circuit, therefore, possesses a memory. The output of the circuit remembers its previous logic state. Examples of such systems are flip-flops.

7.2 Combinational digital systems

Adder and subtractor

1. *Binary adder.* Addition of binary bits is done through adder circuit. Such a circuit is of two types: (i) Half-adder (ii) Full-adder. Digital computers perform a variety of information processing jobs. Among the basic functions performed are the various arithmetic operations, the most basic of which is the addition.

(i) *Half-adder.* Addition of two binary bits, one is called an *Addend,* the other an *Augend,* is referred to as half-adder.

Take as example of addition of two binary bits as shown in Table 7.1.

Table 7.1: Half-adder

Addend (A)	0	0	1	1
Augend (B)	+0	+1	+0	+1
S (Sum)	0	1	1	0
C (Carry)	0	0	0	1

The truth table for such addition is shown in Table 7.2.

Table 7.2: Truth table

A	B	C	D
0	0	0	0
0	1	1	0
1	0	1	0
1	1	0	1

So the S-digit is related to the Addend and Augend by

$$S=\bar{A}B+A\bar{B}, \tag{7.1}$$

hence, the relation is given by XOR operation,

$$C=AB. \tag{7.2}$$

Hence, C is given by an AND operation on the Addend and Augend.

The Fig. 7.1 represents the realisation of the two (7.1) and (7.2); Fig. 7.1(b) is the circuit symbol of a half-adder.

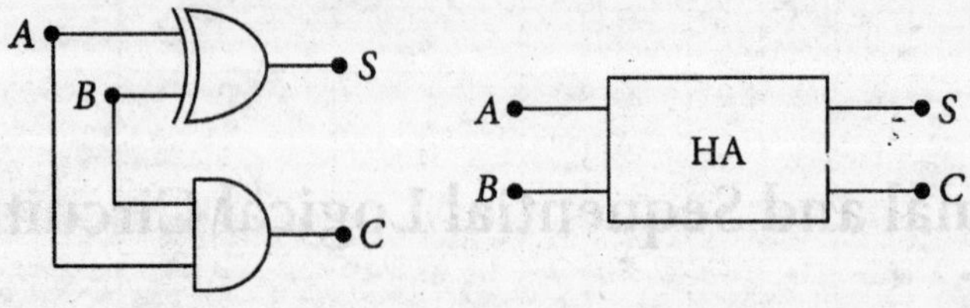

(a) (b)

Fig. 7.1

The implementing circuit for realisation of half-adder using NAND gate is shown in Fig. 7.2.

Operation of the circuit shown in Fig. 7.2 *explained.*

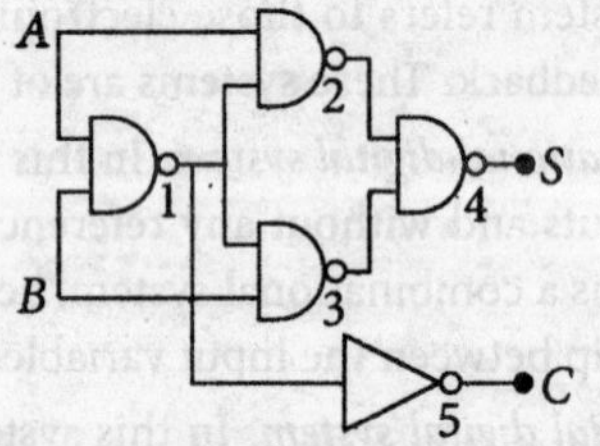

Fig. 7.2

Output

At point 1: $\overline{AB}$

At point 5: $\overline{\overline{AB}} = AB$ (Carry)

At point 2: $\overline{A\cdot\overline{AB}} = \bar{A} + AB = \bar{A} + B$

At point 3: $\overline{B\cdot\overline{AB}} = \bar{B} + AB = \bar{B} + A$

At point 4: $\overline{(\bar{A}+B)\cdot(\bar{B}+A)} = A\bar{B} + \bar{A}B = S$ (EX-OR).

Note. $\bar{A} + AB = \bar{A} + B$ is proved w.r.t. truth table as shown in Table 7.3.

Table 7.3: Truth table

A	B	AB	$\bar{A}$	$\bar{A}B$	$\bar{A}+AB$	$\bar{A}+B$
0	0	0	1	0	1	1
0	1	0	1	1	1	1
1	0	0	0	0	0	0
1	1	1	0	0	1	1

The table shows that $\bar{A} + AB = \bar{A} + B$.

Full-adder. It is a combinational circuit which performs the addition of three binary bits. So, in a full-adder, there are three inputs and two outputs. The symbolic representation is shown in Fig. 7.3 where x, y, z are the three binary inputs and S', C' are the two outputs representing respectively the sum and carry.

Fig. 7.3

The truth table for a full-adder is shown in Table 7.4.

Table 7.4: Truth table

x	y	z	Output	Sum (S')	Carry (C')
0	0	0	000	0	0
0	0	1	001	1	0
0	1	0	010	1	0
0	1	1	011	0	1
1	0	0	100	1	0
1	0	1	101	0	1
1	1	0	110	0	1
1	1	1	111	1	1

So the Boolean expression for

$$S' = \bar{x}\bar{y}z + \bar{x}y\bar{z} + x\bar{y}\bar{z} + xyz \tag{7.3}$$

$$C' = \bar{x}yz + x\bar{y}z + xy\bar{z} + xyz. \tag{7.4}$$

If we simplify the above two expressions, we would get

$$S' = \bar{x}\,(y\bar{z} + \bar{y}z) + x\,(yz + \bar{y}\bar{z}) = \bar{x}(y \oplus z) + x\,(yz + \bar{y}\bar{z}).$$

Now, $\overline{y \oplus z} = \left(\overline{y\bar{z} + \bar{y}z}\right) = \overline{y\bar{z}} \cdot \overline{\bar{y}z} = (\bar{y} + z) \cdot (y + \bar{z}) = yz + \bar{y}\bar{z}.$

$$\therefore\ S' = \bar{x}(y \oplus z) + x\left(\overline{y \oplus z}\right) = \bar{x} \oplus (y \oplus z). \tag{7.5}$$

C' can be written as

$$C' = \bar{x}yz + x\bar{y}z + xy\bar{z} + xyz = x\,(\bar{x}y + x\bar{y}) + xy\,(z + \bar{z}) = z(x \oplus y) + xy. \tag{7.6}$$

Implementation of the (7.5) and (7.6) is done through the circuit shown in Fig. 7.4.

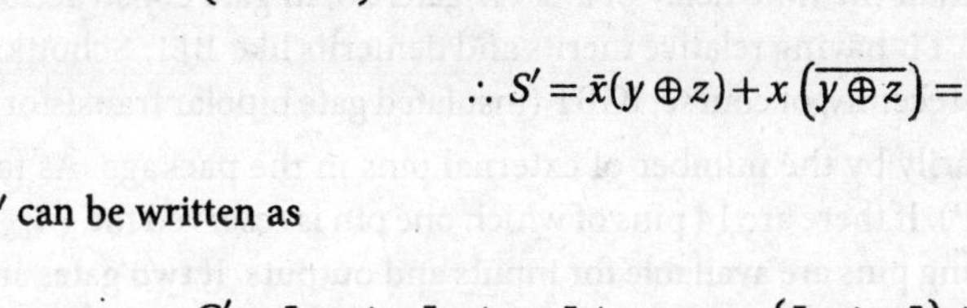

Fig. 7.4

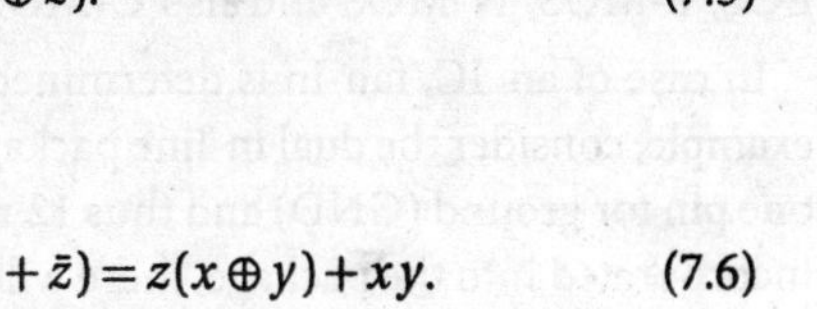

Fig. 7.5

A full-adder is essentially a half-adder with a carry input signal.

Fig. 7.5 shows the symbol of a full-adder and its truth table is shown in Table 7.4. z is to be taken as C_{i-1}.

7.3 Choice of implementing circuit for realisation of Boolean expression

In order to realise a Boolean expression, the implementing circuit should be such as to get the realisation through the use of minimum number of gates. This will minimise the hardware as well as the cost. With the introduction of IC gates, the cost minimisation is not a major problem. In this case, economy in use of a number of gates is w.r.t. a very important factor and that is the *delay time.*

What is delay time of an IC?

The *delay time* refers to the time difference between application of a signal at the input and its appearance at the output terminal. Every gate introduces a delay time and so use of minimum number of gates in the process of implementation reduces this delay time. This delay time is also related to the energy dissipation in a gate circuit.

Families of IC *gate circuits.* Fan-in and Fan-out of an IC circuit.

There are in general three classes of IC circuits, also called scales of integration of IC.

SSI: small scale integration, MSI: medium scale integration, LSI: large scale integration, VLSI: very large sale integration.

SSI: This type of IC consists of two to four gates.

MSI: MSI IC consists of 10–12 gates in a package.

LSI: LSI IC consists of about 100 gates.

VLSI: VLSI IC consists of more than 100 gates, it is a small system used in making a calculator, a telephone, a microprocessor, etc. on a single chip.

Gates may be constructed on DTL, i.e., diode-transistor logic or on TTL, called transistor transistor logic. The time delay of a DTL gate is greater than the time delay of a TTL gate. So, in gate construction TTL is preferred. There are several families of TTL having relative merits and demerits like BJT, Schottky, ECL, P-MOS, N-MOS and also C-MOS; very recently, of course, IGBT (insulated gate bipolar transistor).

In case of an IC, fan-in is determined primarily by the number of external pins in the package. As for example, consider the dual in-line package (DIP). If there are 14 pins of which one pin is reserved for $+V_{CC}$; one pin for ground (GND) and thus 12 remaining pins are available for inputs and outputs. If two gates are incorporated into the package, there will be two pins for the outputs, and thus 10 pins are left over. So each gate may be allowed 5 pins for inputs. We may say, fan-in per gates is 5; with four gate, fan-in becomes 2 per gates. So fan in is an important parameter of an IC.

Fan-out: Fan-out or loading specifies the number of standard loads that can be driven by the outputs of a gate without adversely affecting the logic levels. The fan-out limitation is set by the current value which can be drawn from the output of a gate. Consider Fig. 7.6 where we have shown the output of an IC gate used for LED glow, base bias of the transistor *Q* and also for driving the relay. Current output may not be enough for these three purposes, hence, overloading takes place.

Fig. 7.6

7.4 How to realise a Boolean expression by using universal gates?

The universal gates refer to the NAND and NOR gates. So how to realise a Boolean expression of the form $(AC+BD)$ in terms of NAND and NOR?

Take the function $Y = AC + BD$. This function can be realised through two ANDs and one OR gate [Fig. 7.7] or through all three NAND gates [Fig. 7.8].

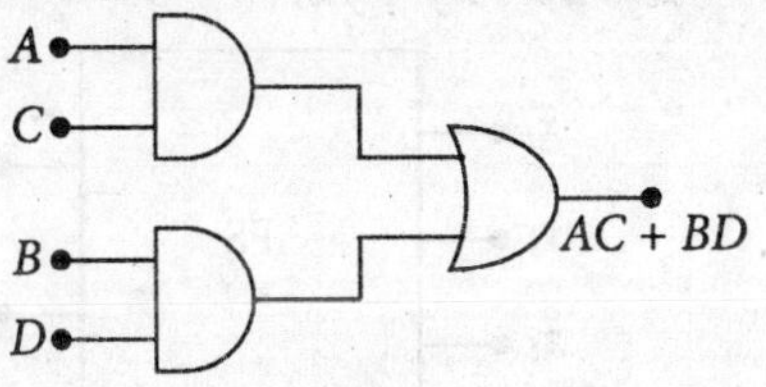

Fig. 7.7

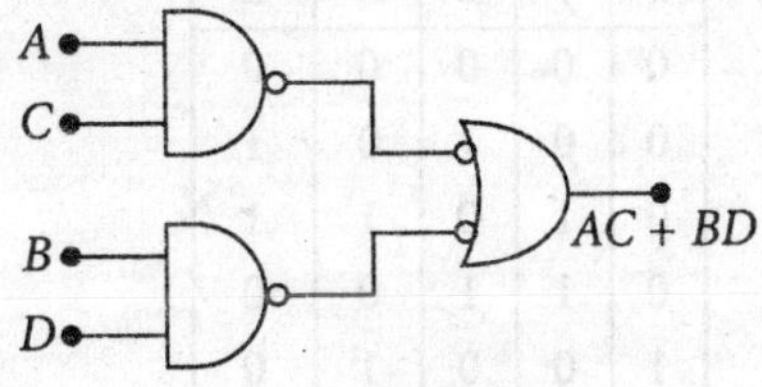

Fig. 7.8

7.5 Half-subtractor and full-subtractor

In a digital computer the subtraction is looked upon as addition of the complemented subtrahend to the minuend. But still to implement a circuit for subtraction of two binary numbers, the following consideration may be made.

Half-subtractor (HS). It is a combinational circuit which performs the subtraction of two single-bit binary numbers.

Table 7.5: Truth table

Inputs		Outputs	
x	y	D (difference)	B (borrow)
0	0	0	0
0	1	1	1
1	0	1	0
1	1	0	0

Let us consider the subtraction of y (subtrahend) from x (minuend) written as $(x - y)$. The truth table for such an operation is shown in Table 7.5.

The symbolic representation of HS is shown in Fig. 7.9.

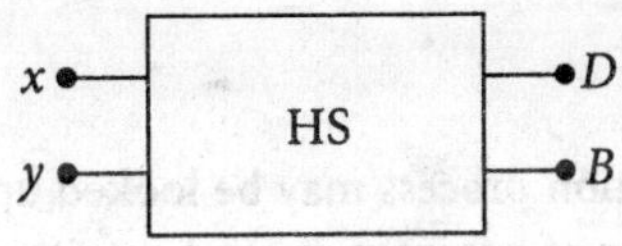

Fig. 7.9

The Boolean expression for D is

$$D = \bar{x}y + x\bar{y} = x \oplus y \qquad (7.7)$$

$$B = \bar{x}y. \qquad (7.8)$$

The implementing circuit is shown in Fig. 7.10.

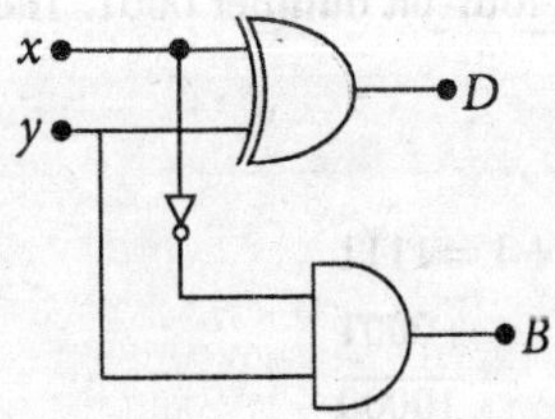

Fig. 7.10

So, the difference is same as sum (S) in case of half-adder.

Full-subtractor. It is the circuit that performs the subtraction between two bits along with the borrow (B) from the previous stage. So there are three inputs, one x (minuend), one y (subtrahend), the other z (borrow from the previous stage). There will be two outputs, one D, the other is borrow (B).

Thus, a full-subtractor truth table is shown in Table 7.6 and its symbolic representation is shown in Fig. 7.11.

Table 7.6: Truth table

x	y	z	D	B
0	0	0	0	0
0	0	1	0	1
0	1	0	1	1
0	1	1	0	0
1	0	0	1	0
1	0	1	0	0
1	1	0	0	0
1	1	1	1	1

So, $D = x \oplus y \oplus z$

$B = \bar{x}z + \bar{x}y + yz.$

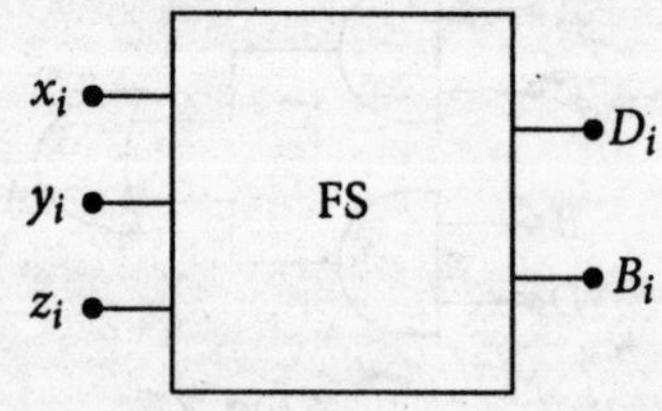

Fig. 7.11

The implementation circuit for the full-subtractor is shown in Fig. 7.12.

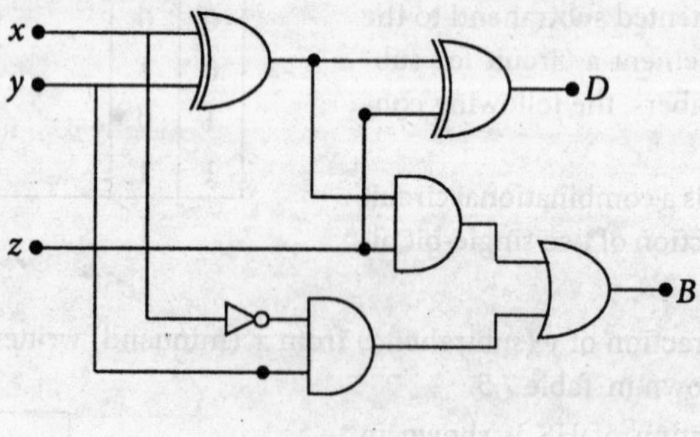

Fig. 7.12

Subtraction process may be looked upon as a process of summation also. Thus, if we want to perform subtraction of y from x, it can be written as $x - y = x + (-y)$, it can be looked upon as addition of x and complement of y. This way of looking at it is all the more useful if subtraction is applied to a 4-bit number.

Let y be a four-bit number 0001. Then $y + \bar{y} = 0001$

$$\begin{array}{r} 0001 \\ +1110 \\ \hline 1111 \end{array}$$

So, $y + \bar{y} + 1 = 1111$

$$\begin{array}{r} 1111 \\ +0001 \\ \hline 10000 \end{array}$$

$\therefore\ y = (10000 - \bar{y} - 1).$

Finally,

$$x \text{ minus } y = (x \text{ plus } \bar{y} \text{ plus } 1) - 10000. \tag{7.9}$$

(7.9) shows that for the subtraction of a 4-bit number y from a 4-bit number x, it is required to add $x, \bar{y}$ and 1. The term minus 10000 signifies that the addition of x, y and 1 may result in a fifth bit which must be ignored.

From the truth table of a full-subtractor, it is evident that when borrow z from the previous stage is zero, the outputs of a full-subtractor become identical with those of a half-subtractor.

7.6 Multiplexer (MUX)

The meaning of multiplexing is converting *many into one*, so a multiplexer performs the function of selecting one out of N input data sources and sending it to a single transmission channel with the help of control signals. In Fig. 7.13 we have shown a mechanical analog in the form of a single pole multi-throw switch and its symbolic representation is shown in Fig. 7.14.

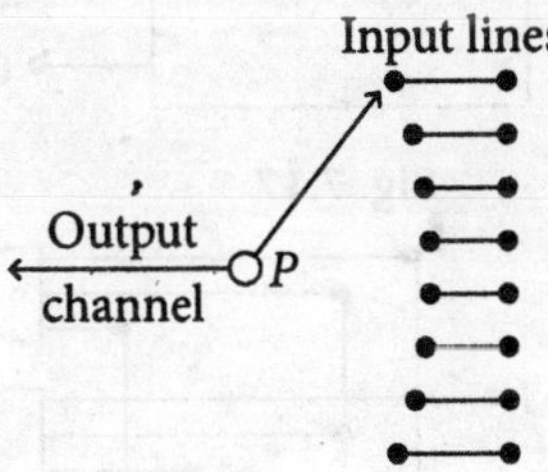

Fig. 7.13

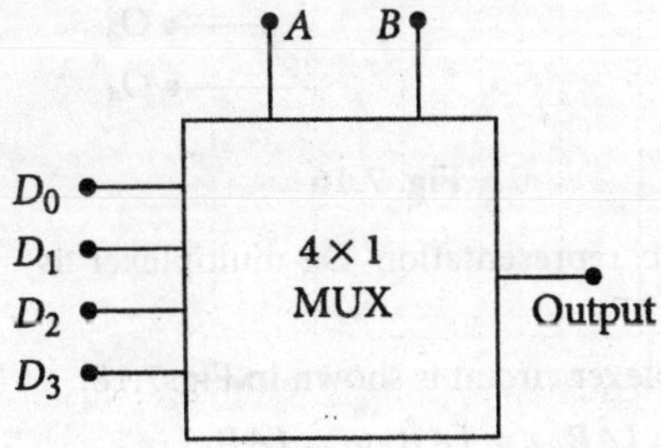

Fig. 7.14

The rotor spindle of the multi-throw switch connects at a time any of the input lines to the output. The pole P of the switch may be connected to the transmission line. The block diagram (Fig. 7.14) is the symbolic representation of 4 : 1 multiplexer meaning 4 input lines and 1 output line, the terminals A, B are the selector terminals. Such a multiplexer is popularly known as 4 to 1 multiplexer.

How is data selection actually made?

To understand the mechanism of selection of data in multiplexer, consider Fig. 7.15.

Input data bits are D_0, D_1, D_2, etc. Only one out of the 4-input data is to be transmitted to the output. The control bits are A, B and the choice of the input depends on the choice of A, B.

Thus, if $A = 0$, $B = 0$, $AB = 00$, then out of 4 ANDs, the first AND N_0 is enabled, all other ANDs are disabled. So the data bit D_0 is transmitted to the output giving

$$T = D_0. \tag{7.10}$$

If the control bit AB is 11, the bottom gate of Fig. 7.15 is enabled and all other gates are disabled.

Hence,

$$Y = D_3. \tag{7.11}$$

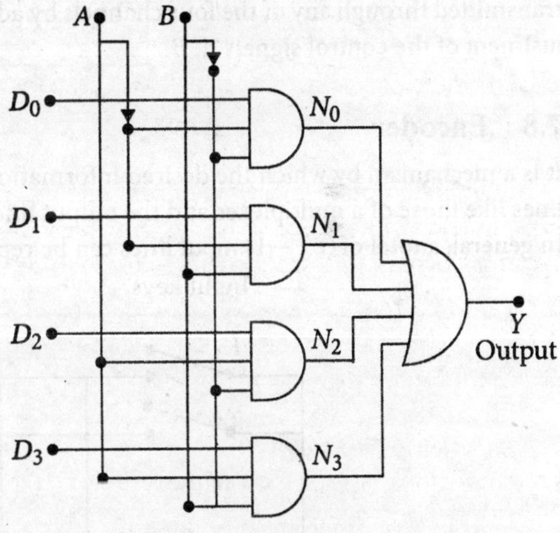

Fig. 7.15

So the control signal enables a particular AND disabling all the ANDs and the corresponding input data is transmitted to the output.

7.7 Demultiplexer

It is a device whose functioning is to produce *many outputs out of a single input* just reverse of the functioning of the device, called a *multiplexer.*

Its function is, therefore, to transmit an input signal to many output channels [Fig. 7.16].

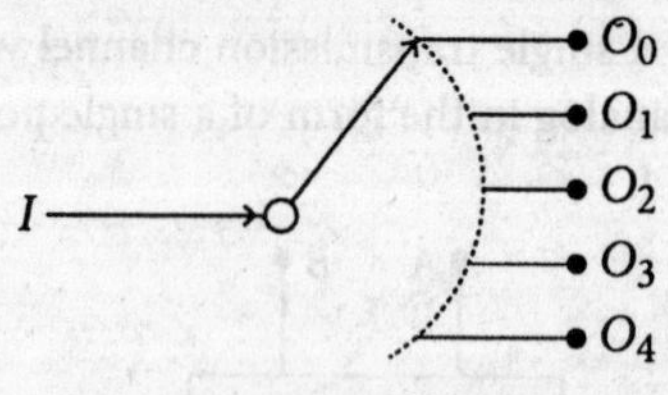

Fig. 7.16

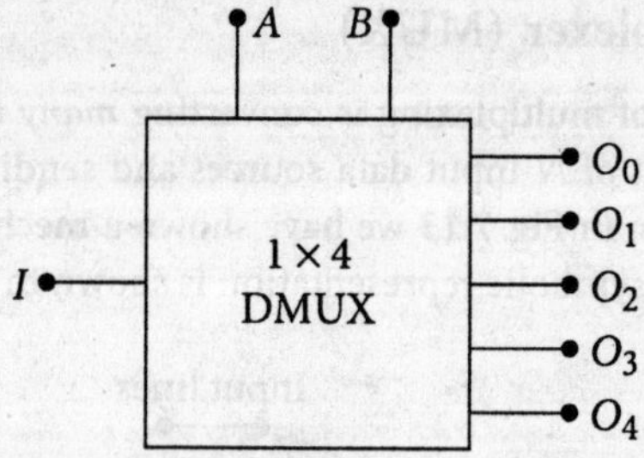

Fig. 7.17

The symbolic representation De-multiplexer is shown in Fig. 7.17.

The demultiplexer circuit is shown in Fig. 7.18.

$x = I\bar{A}\bar{B}$, $y = I\bar{A}B$, $z = IA\bar{B}$, $w = IAB$.

So, $x = I$, if $A = 0, B = 0$

$y = I$, if $A = 0, B = 1$

$z = I$, if $A = 1, B = 0$

$w = I$, if $A = 1, B = 1$.

We thus see that the given input signal may be transmitted through any of the four channels by adjustment of the control signals A, B.

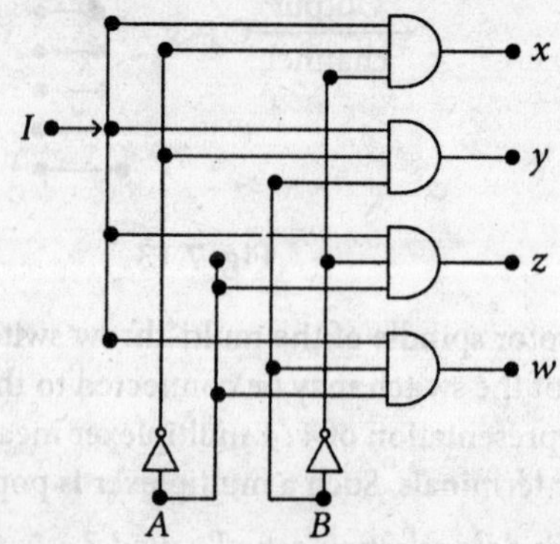

Fig. 7.18

7.8 Encoder

It is a mechanism by which the desired information is converted into binary codes. There are many input lines like those of a multiplexer and the output lines contain coded outputs identifying each of the inputs. In general, a total of $(2^N - 1)$ input lines can be represented by at best N-bit binary strings.

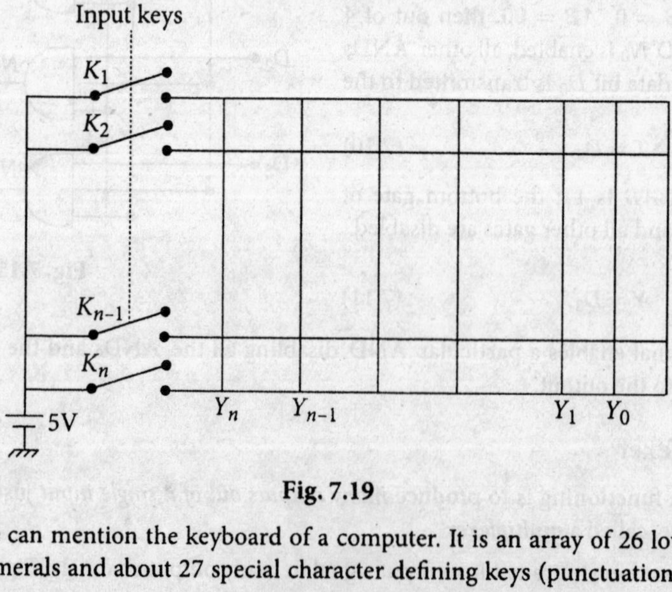

Fig. 7.19

As an example, we can mention the keyboard of a computer. It is an array of 26 lower case letters, 26 capital letters, 10 numerals and about 27 special character defining keys (punctuations and symbols and other special commands). Thus, a total of 84 codes are necessary for the output. So a minimum of 7 bits (because $2^6 < 84 < 2^7$) needed for output coding. The input keys have two states, pressed (logic 1) or not pressed (logic 0). The logic diagram, its implementing circuit and the encoder truth table are represented by Fig. 7.19, Table 7.7 and Fig. 7.20 respectively.

To illustrate the design procedure, consider the keyboard containing 10 keys corresponding to the numerals 0, 1, 2, ..., 9. A 4-bit output code is sufficient.

Table 7.7: Truth Table

Inputs										Outputs			
K_9	K_8	K_7	K_6	K_5	K_4	K_3	K_2	K_1	K_0	Y_3	Y_2	Y_1	Y_0
0	0	0	0	0	0	0	0	0	1	0	0	0	0
0	0	0	0	0	0	0	0	1	0	0	0	0	1
0	0	0	0	0	0	0	1	0	0	0	0	1	0
0	0	0	0	0	0	1	0	0	0	0	0	1	1
0	0	0	0	0	1	0	0	0	0	0	1	0	0
0	0	0	0	1	0	0	0	0	0	0	1	0	1
0	0	0	1	0	0	0	0	0	0	0	1	1	0
0	0	1	0	0	0	0	0	0	0	0	1	1	1
0	1	0	0	0	0	0	0	0	0	1	0	0	0
1	0	0	0	0	0	0	0	0	0	1	0	0	1

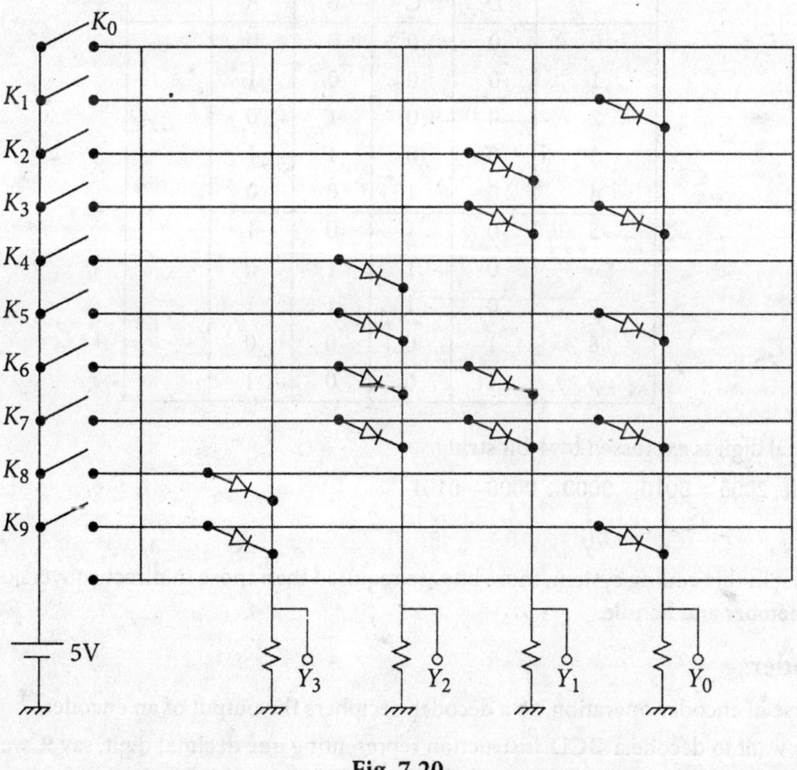

Fig. 7.20

The truth table for a 4-bit encoder is shown in Table 7.7. It will be assumed that not more than one key is pressed at a time, so that in any row, every input other than the key pressed is in 0-state. From the truth table, we get $Y_0 = 1$, if $K_1 = 1$ or, $K_3 = 1$ or, $K_5 = 1$ or, $K_7 = 1$ or, $K_9 = 1$. So the Boolean expression for Y_0 is $Y_0 = K_1 + K_3 + K_5 + K_7 + K_9$.

Similarly, $Y_1 = K_2 + K_3 + K_6 + K_7$

$Y_2 = K_4 + K_5 + K_6 + K_7$

$Y_3 = K_8 + K_9$.

All these expressions based on OR-logic are implemented by the diode matrix of Fig. 7.20. Thus, when K_5 is pressed, we get the output $Y_0 = 1, Y_1 = 0, Y_2 = 1, Y_3 = 0$, because K_5 line connecting the anodes of two diodes are in forward bias, hence, conduct producing voltage output across the resistances connected in the cathode part. So, the output is 0101, which defines decimal number 5.

Coding system: There are many schemes or codes for representing decimal number in binary but most common amongst them is the Binary Coded Decimal (BCD) system. When a decimal number is converted into binary form, it consists of a long string of bits which is difficult for a human being to remember. But in BCD, each digit is coded by four binary bits. Amongst many BCD systems, again, the 8421 code is most popular. In this code system the MSB has the weight 8, LSB has the weight 1. In tabular form this code is shown below [Table 7.8].

Table 7.8: 8421 code

Decimal	8 D	4 C	2 B	1 A	BCD
0	0	0	0	0	
1	0	0	0	1	
2	0	0	1	0	
3	0	0	1	1	
4	0	1	0	0	
5	0	1	0	1	
6	0	1	1	0	
7	0	1	1	1	
8	1	0	0	0	
9	1	0	0	1	

Each decimal digit is expressed by 4-bit string.

For example, 2005 = 0010 0000 0000 0101

↓ ↓ ↓ ↓

2 0 0 5

It is true that in this coding system, more bits are required than those in direct conversion scheme but it is easy to remember and handle.

7.9 Decoder

It is the reverse of encoder operation, so a decoder deciphers the output of an encoder.

Thus, if we want to decode a BCD instruction representing one decimal digit, say 9, we can use a four-input AND gate excited by 4-BCD bits as shown in Fig. 7.21(a) and (b).

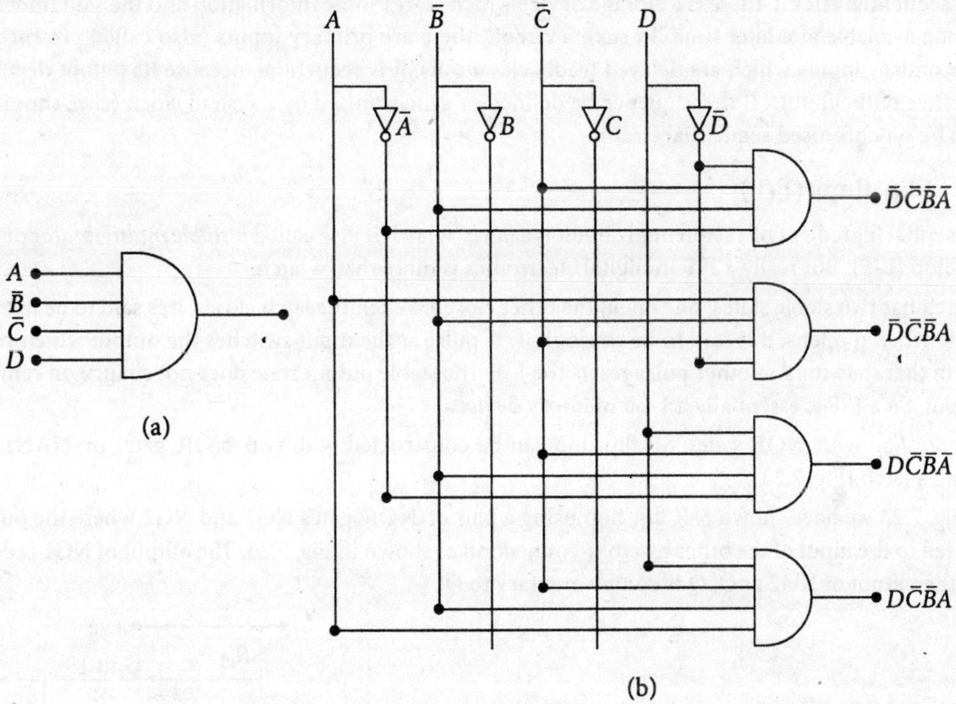

Fig. 7.21

Example: The output of the AND gate is 1, if and only if $D = 1, C = 0$ and $A = 1$. Since this code represents decimal number 9, the output is labelled line 9.

The logic circuit for BCD-decimal decoder is shown in Fig. 7.21(b). The logic for decimal 9 is

D	$\bar{C}$	$\bar{B}$	A	Y
1	0	0	1	9

7.10 Sequential digital systems

Combinational circuits are those where the outputs depend on the instantaneous values of the inputs excepting the small delay time. In such a circuit the output is independent of the history of the input.

There are again circuits where the outputs depend on the past inputs as well. So, in such a circuit the past input information are to be stored in some storage elements, called *memory elements.* This type of circuit is known as sequential circuits.

What basically is a sequential circuit?

Basically a sequential circuit may be looked upon as an ideal combinational circuit with feedback. An ideal combinational circuit is one in which the output depends on the instantaneous input and not on the history of the input. Of course, in a practical combinational circuit, there is time delay between application of an input and appearance of the corresponding output.

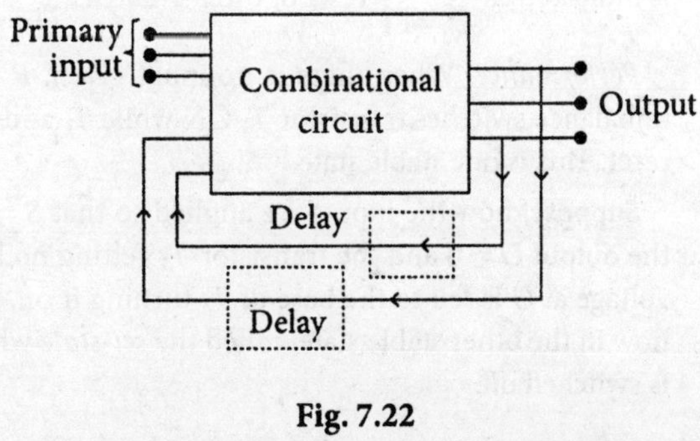

Fig. 7.22

In a sequential circuit, there is a storage device which stores some information and the said information are made available at a later time. In such a circuit, there are primary inputs (also called *present inputs*) and secondary inputs which are delayed feedback outputs. It is sequential, because its output depends on a sequence of the inputs. If the sequence be defined or synchronised by a system clock (clk), the circuit is said to be synchronised sequential circuit.

7.11 Flip-flops (F-F)

The essential ingredient of a sequential circuit is a *latch.* A latch is also called *bistable multivibrator* or simply a Flip-flop (F-F). But really a F-F in digital electronics is more than a latch.

A latch has two stable states, one open, the other closed. When the latch closes, it is said to be *reset,* logic 0; while when it opens, it is said to be *set,* logic 1. A pulse at the input switches the output which remains stable in that state until another pulse resets the F-F. The stable output state does not change on removal of the input. So a F-F is essentially a 1-bit memory device.

S-R *flip-flop with* NOR *gates.* SR flip-flop can be constructed with two NOR gates or NAND gates forming a latch [Fig. 7.23].

In Fig. 7.23 we have shown S-R flip-flop using a pair of NOR gates NG1 and NG2 where the output of one is fed to the input of the other by cross-connection as shown in Fig. 7.23. The output of NG1 is denoted by Q, the output of NG2 as $\bar{Q}$; $\bar{Q}$ is complementary to Q.

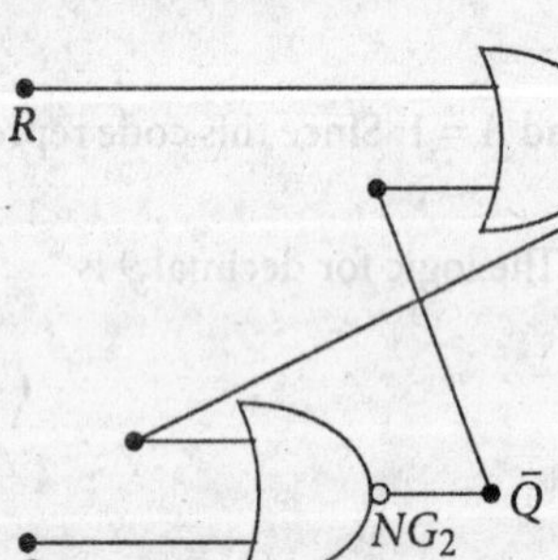

Fig. 7.23

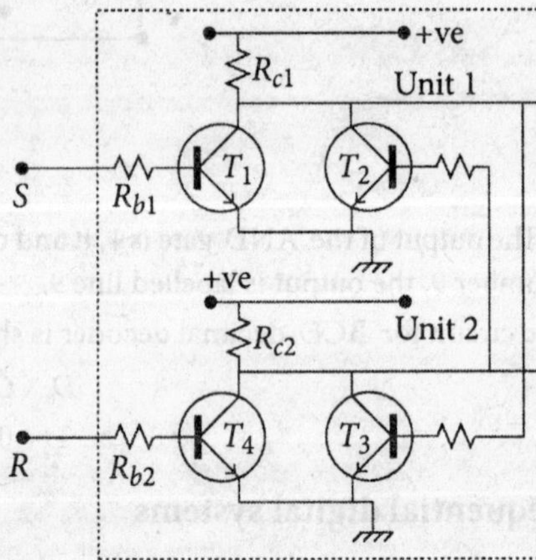

Fig. 7.24

Implementing circuit. The implementing circuit for R-S flip-flop is shown in Fig. 7.24 whose functioning is indicated as $\left.\begin{matrix} Q=0 \\ \bar{Q}=1 \end{matrix}\right\}$ reset of the F-F and as $\left.\begin{matrix} Q=1 \\ \bar{Q}=0 \end{matrix}\right\}$ when the condition for the F-F is to be set.

Performance. When there is no input given, $R=0, S=0$ and the power is switched on, some circuit imbalance switches transistor T_3 ON while T_1 and T_2 are OFF. So $Q=0$, $\bar{Q}=1$ and the F-F is said to be in reset. This is one stable state.

Suppose, now the inputs are applied so that $S=1, R=0$. This causes T_1 to go into conduction forcing the output $\bar{Q}=0$ and the transistor T_3 getting no base bias is held off and the output $Q=1$. This positive voltage at Q is fed to the base of T_2 turning it on, so that the input to S may now be removed. The F-F is now in the other stable state, called the *set-state* where it will stay if no further inputs are applied or power is switched off.

If $S=0, R=1$, the F-F will pass into reset condition.

Thus, an S-input makes Q high, $\bar{Q}$ low and R input makes Q low, $\bar{Q}$ high as is demanded by the truth table.

Table 7.9: Truth Table

R	S	Q	$\bar{Q}$
0	0	NC (No change)	NC (No change)
Set-state 0	1	1	0
Reset-state 1	0	0	1
1	1	0	0 ← Logically inconsistent

IC package MC 724 can provide two R-S F-Fs. In an R-S flip-flop, the last combination of inputs is not allowed and the F-F is said to have race or restriction on the inputs $S=1=R$, *not allowed.*

S-R *flip-flop with* NAND *gates.* Fig. 7.25 is the NAND gate realisation of the S-R flip-flop while Fig. 7.26 is the NOR-equivalent of the S-R flip-flop.

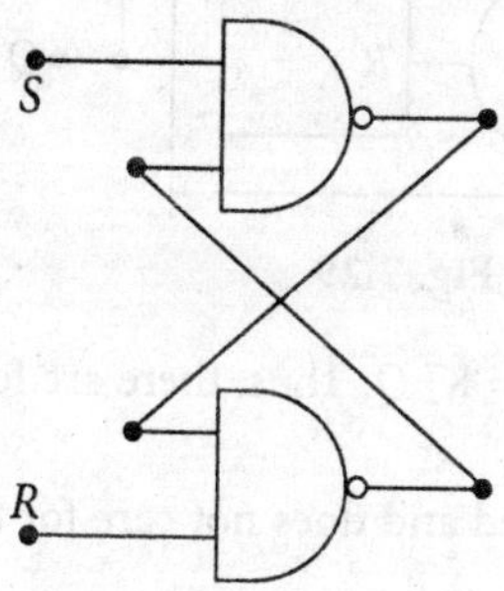

Fig. 7.25

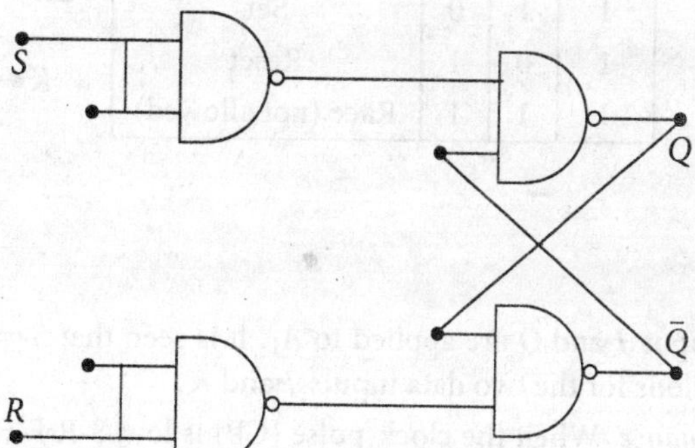

Fig. 7.26

The truth table for NAND S-R *flip-flop.* The truth table for NAND S-R flip-flop is shown in Table 7.10.

Table 7.10: Truth Table

Inputs		Outputs		Remarks
S	R	Q	$\bar{Q}$	
0	0	?	?	Not allowed
0	1	1	0	Set
1	0	0	1	Reset
1	1	Q	$\bar{Q}$	No change

Clocked S-R F-F. In digital system it is often necessary to set and reset a flip-flop in synchronisation with clock pulses. The clock pulse input decides whether the data is to be entered or ignored. Such a synchronised S-R F-F is shown in Fig. 7.27, its circuit symbol is shown in Fig. 7.28.

Performance. To understand the functioning of this F-F, we see that when the clock-pulse is absent or low (logic 0), then both the AND gates are disabled, thus S' and R' are in their low-state (zero) and Q and $\bar{Q}$ will not change. Therefore, as long as $CK=0$, the F-F will not change its state and there would not be any effect of S and R inputs on the outputs Q and $\bar{Q}$. When CK goes high (logic 1), then the AND gates are enabled and depending on the state of the inputs S-R, the F-F will set, reset or there can be no change in the state. The truth table is shown in Table 7.11.

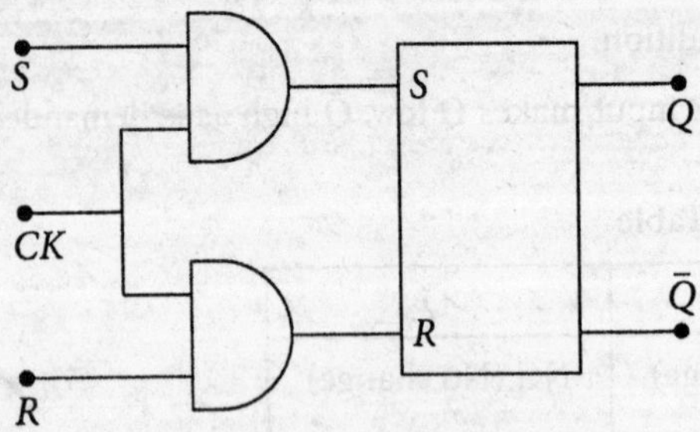

Fig. 7.27

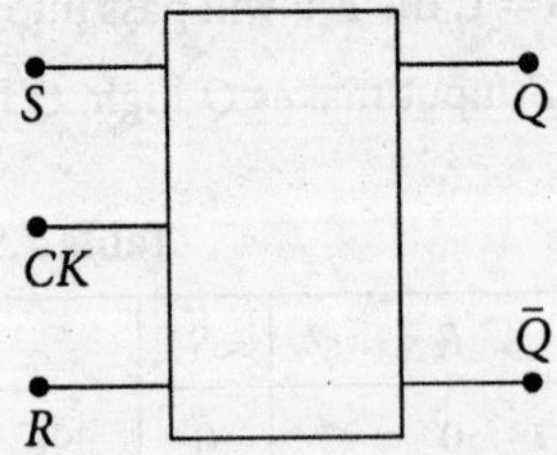

Fig. 7.28

J-K *flip-flop.* In order to avoid race in S-R flip-flop, J-K flip-flop is introduced.

It is a modified version of S-R flip-flop consisting of two AND-gates A_1 and A_2 [Fig. 7.29].

Table 7.11: Truth Table

CK	S	R	Q
0	×	×	No change (NC)
1	0	0	do
1	1	0	Set
1	0	1	Reset
1	1	1	Race (not allowed)

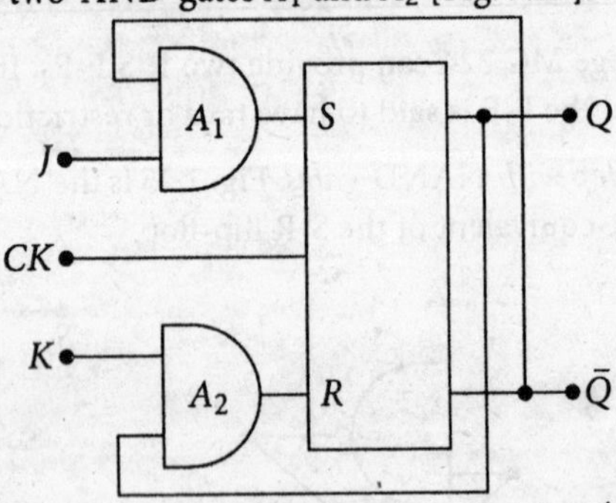

Fig. 7.29

Data input J and $\bar{Q}$ are applied to A_1. It is seen that $S = J, \bar{Q}$; $R = K, Q$. Thus, there are four possible combinations for the two data inputs J and K.

Performance. When the clock-pulse (CP) is low, S-R F-F is disabled and does not care for the inputs S and R. Consider the cases only when the CP is high, i.e., $CK = 1$.

If $J = 0$, $K = 0$, both of the AND gates are disabled and both S, R are low and there will be no change in outputs, Q and $\bar{Q}$ retain their earlier states.

If $J = 1$, $K = 0$, Q may be initially low or high.

If $Q = 0$ initially, $\bar{Q} = 1$.

For $A_1 : J = 1, \bar{Q} = 1 \Rightarrow S = 1$.

For $A_2 : K = 0, Q = 0 \Rightarrow R = 0$.

So Q will change to 1 from initial zero state.

If $Q = 1$ initially, $\bar{Q} = 0$, in this case:

For $A_1 : J = 1, \bar{Q} = 0 \Rightarrow S = 0$.

For $A_2 : K = 0, Q = 1 \Rightarrow R = 0$.

$S = 0 = R$ means there will be no change in the output, Q remains in the high state.

The two cases may now be summarised as below:

When $J = 1$, $K = 0$, then the output $Q = 1$, if it were not 1, or the output remains in state 1 if it were already in the state 1.

When $J = 0$, $K = 1$, we find that the output will reset to 0.

Now, scrutinise the critical case $J = 1 = K$ and

Initially $Q = 0, \bar{Q} = 1$.

For $A_1 : J = 1 = \bar{Q} \Rightarrow S = 1$.

For $A_2 : K = 1, Q = 0 \Rightarrow R = 0$.

Hence, Q should change to 1 (i.e., set).

Initially, $Q = 1, \bar{Q} = 0$.

For $A_1 : J = 1, \bar{Q} = 0 \Rightarrow S = 0$.

For $A_2 : K = 1, Q = 0 \Rightarrow R = 1$.

Hence, Q changes to 0 (reset).

So, whenever $J = 1 = K$, the output Q will change to 1 if it were initially at 0, and vice versa. Thus, *race* or ambiguity in the basic S-R flip-flop is removed. The truth table is shown in Table 7.12.

The change of state is called *toggling* and the F-F is said to toggle.

Table 7.12: Truth Table

CK	J	K	Q	$\bar{Q}$	Remarks
0	×	×	Q	$\bar{Q}$	NC (no change)
1	0	0	Q	$\bar{Q}$	NC
1	1	0	1	0	Set
1	-	1	0	1	Reset
1	1	1	$\bar{Q}$	Q	Toggle

Master-slave (MS) *flip-flop*. The clocked J-K flip-flop suffers from the restriction that the width of the clock pulse Δt_p should be smaller than propagation delay Δt_d. The propagation delay (Δt_d) is approximately the time interval between application of an input and its appearance at the output terminal. So, it requires $\Delta t_p < \Delta t_d < T$, where T is the period of the clock pulse.

In case of an IC F-F, $\Delta t_d \simeq 1$ ns, so $\Delta t_p < \Delta t_d$ is not satisfied and the output becomes indeterminate. The reason may be simply explained as below:

Start with $J = K = 1$ and $Q = 0$. If now a clock pulse be injected, Q will become 1 (toggling) after propagation delay Δt_d. At this instant $J = K = 1 = Q$, clock pulse is still ON, Q will change to 0. So during the lifetime of the clock pulse, the output oscillates between 1 and 0. So at the end of the pulse time, $CK = 0$, the value of Q is ambiguous. This ambiguous condition is known as *race around* condition.

The MS F-F is free from the race around condition. The Master-slave (MS) principle may be utilised in S-R or J-K flip-flip. However, J-K master-slave is more popular.

In a J-K M-S flip-flop, there are two units:

1. Master unit which contains a J-K clocked F-F.
2. Slave unit which also contains a clocked J-K F-F but its output is fedback to the input terminals of the master unit. Also the master clock pulse and slave clock pulse are mutually complemented, i.e., if the master has clock pulse CK, the clock pulse of the slave is $\overline{CK}$.

So what is there inside a J-K MS?

The schematic of the F-F is shown in Fig. 7.29.

Performance. Consider the clock pulse be ON, $CK = 1$ and $\overline{CK} = 0$. So the master is enabled and the slave unit disabled. On the arrival of the pulse at the master input and with $J = 1, K = 0$, the master output $Q_M = 1$ and the slave unit being disabled at this instant, the output Q_M of the master is stored in the slave in pulse. It is brought to the output of the slave when the clock pulse goes low $CK = 0 \Rightarrow \overline{CK} = 1$. Thus, with $J = 1, K = 0$ and $CK = 1$, the master output $Q_M = 1$. The ultimate output of the F-F, $Q = 1$ when $CK = 0 \Rightarrow \overline{CK} = 1$.

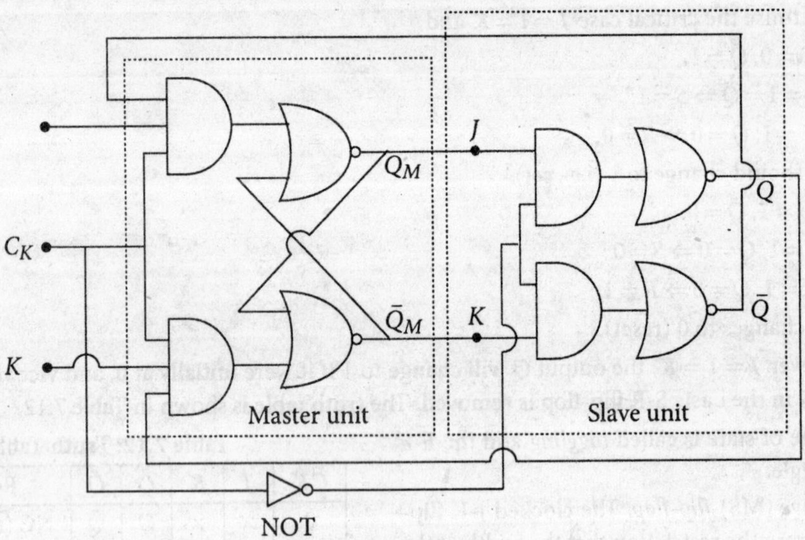

Fig. 7.29

Similarly, for $J=0, K=1$, the master is reset to 0, when $CK=1$ and the slave resets to 0 when $CK=0$. With $J=K=1$ and $CK=1$, the output Q_M toggles, i.e., $Q_M \rightarrow \bar{Q}_M$ and Q_M being the input to slave F-F, Q must follow Q_M. So, if Q_M toggles, Q also does it. Finally, with both J and K low, there is no change in the master latch and so also in slave latch.

To summarise — For a given J, K input, the master responds for $CK=1$ and the slave waits. But as the clock pulse goes low, the information stored in the master is transferred to the slave. Thus, the slave output follows the master output and there is no race around.

7.12 Registers

A register consists of several flip-flops and so it is a device for storing binary information. Thus, an n-bit register contains n flip-flops and can store n-bit binary information. For performing binary arithmetic in a computer, it is usual to store binary numbers as data words in a register. However, there would be means by which to shift the data in or out of the register.

In a calculator, a number is introduced by shifting the digit one by one.Thus, a shift register must have member in addition to the shifting action. The shifting is done by a pulse, called *command signa,l* and storing and shifting are done by a unit consisting of a chain of flip-flops. Such a register is called a *shift-register.*

How many classes of shift registers are in use?

Shift registers are classified in various ways: (1) Serial in and parallel out (2) Parallel to serial (3) Serial to serial (4) Parallel to parallel and so on.

Serial loading means loading a bit by means of one clock pulse.

The shifting or storing unit of a register consists of a chain of flip-flops. The flip-flops used are either S-R or J-K master-slave flip-flops to avoid occurrence of *race around* condition. One F-F is for one bit storage or shifting.

7.13 Counters and their classifications

A counter is a sequential circuit consisting of a set of flip-flops which go through a sequence of states on arrival of clock pulses. They form the important building blocks in many digital systems designed to count the occurrence of events, frequency divisions, and generate timing sequence.

There are variation in the design of an electronic counter but the heart of any counter is the bistable circuit or flip-flop of J-K M-S type.

Binary ripple counter. There is a chain of 4 flip-flops FF0(A), FF1(B), FF2(C) and FF3(D) in case of 4-bit binary ripple counter. The schematic of the counter is shown in Fig. 7.30. Flip-flops are clocked J-K type. With all the flip-flops $J = K = 1$, because they are connected to $+V_{CC}$ line and each Q output is connected to the CK input of the succeeding flip-flops. The effect of the input trigger pulse rippling down the counter sets and resets the flip-flops with each alternate input pulse. This is why this type of counter is called an *asynchronous* or *ripple counter.* The flip-flop FF0(A) will represent LSB and FF3(D) will represent MSB.

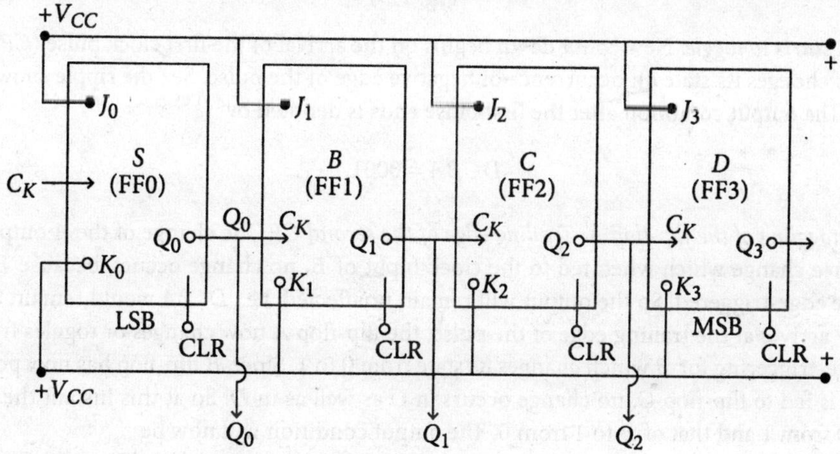

Fig. 7.30

Working. The pulses to be counted are applied to the clock input of FF0. The FFs are negative edge triggered, the master unit of the flip-flop will change, i.e., toggle from 1 to 0 and the new state of the master is transferred to the slave when the clock rises from 0 to 1. When $CK = 0$, all the FFs reset and start from $Q = 0000$. This can be obtained by applying clear signal (CLR) simultaneously to all the flip-flops, $DCBA$ = 0000.

Starting condition. $DCBA = 0000$ [See the clock pulse shown in Fig. 7.31].

Counting begins. Why called ripple counter? The output of one flip-flop drives another, the flip-flop A has to change its state before it can trigger the B flip-flop and B has to change for triggering C and so on. This motion of triggering from flip-flop to flip-flop is like ripple in water (see Fig. 7.44) and hence the name of the counter is ripple counter.

The pulses to be counted are applied to the clock input of the FF0(A). The FFs are negative edge triggered, the master unit of the flip-flop will change from 1 to 0 and the new state of the master is transferred to the slave when the clock rises from 0 to 1. When $CK = 0$, all the FFs reset and start from $Q = 0000$.

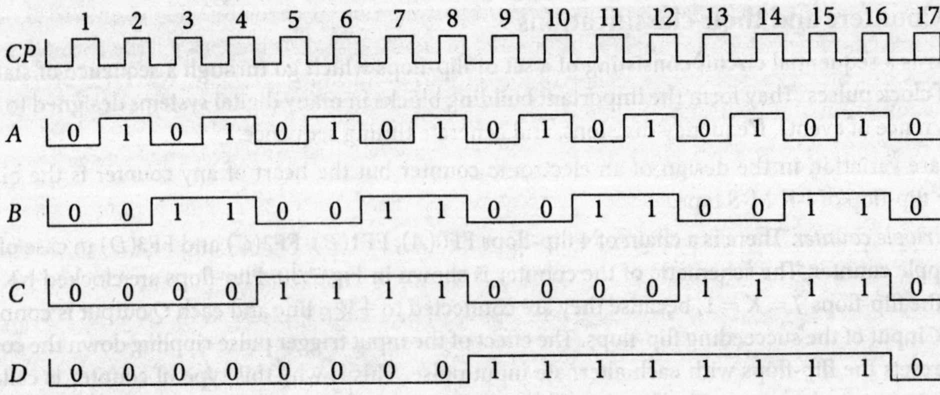

Fig. 7.31

Flip-flop A starts to toggle. New count down begins on the arrival of the first clock pulse (*CP*) at the flip-flop *A* which changes its state on occurrence of negative edge of the pulse. See the ripple shown against *A* in Fig. 7.31. The output condition after the first pulse ends is denoted by

$$DCBA = 0001. \tag{7.12}$$

After first toggling of the flip-flop A: Trailing edge of the second CP. The change of the *A* output is from 0 to 1, a positive change which when fed to the clock input of *B*, no change occurs, because *B* flip-flop is also negative edge-triggered. So the output will remain unaffected, i.e., *DCBA* would remain 0001. When a second CP arrives at the trailing edge of the pulse, the flip-flop *A* now changes or toggles from 1 to 0, a negative edge-triggering for *B* which changes its state from 0 to 1. Since *B* flip-flop has now positive going change, if it is fed to flip-flop *C*, no change occurs in *C* as well as in *D*. So at this instant the output of *A* changes to 0 from 1 and that of *B* to 1 from 0. The output condition will now be

$$DCBA = 0010. \tag{7.13}$$

On occurrence of negative edge of the third CP. When the third CP arrives on occurrence of the negative-edge, the state of the output of *A* changes and goes from 0 to 1. As this change is positive going when fed to *B*, the *B* flip-flop does not change its state. The flip-flops *C*, *D* or all other flip-flops remain unchanged. The output condition now becomes

$$DCBA = 0011. \tag{7.14}$$

On occurrence of negative trailing edge of the fourth CP. At the instant of occurrence of the negative trailing edge of the fourth *CP*, the FF *A* changes its state from 1 to 0; the FF *B* from 1 to 0 and the flip-flop *C* toggles from 0 to 1. Since this change in the state of *C* is positive going, it does not initiate any change in *D*. Thus, the output condition of the flip-flops will be

$$DCBA = 0100. \tag{7.15}$$

At the negative trailing edge of the fifth CP. The flip-flop *A* changes from 0 to 1, *B* remains unchanged, *C* also remains unchanged; the flip-flop *D* is not affected at this moment.

So the output now is

$$DCBA = 0101. \tag{7.16}$$

Conclusion. From the above discussion it is evident that the output of the flip-flop after arrival of each pulse is a binary number which is equal to the number of clock pulses received. When the 15th CP arrives on occurrence of the negative edge, the states of the outputs of flip-flops are $A=1, B=1, C=1, D=1$. So the output condition of the flip-flops will be

$$DCBA = 1111. \tag{7.17}$$

Remarks. From the waveforms as in Fig. 7.31 we can remark that the counter can work as a frequency divider. Waveforms at A, B, C, D are 1/2, 1/5, 1/8 and 1/16 of the clock frequency. Thus, a CP of frequency 16Hz will give a frequency of 8Hz, 4Hz, 2Hz and 1Hz at A, B, C and D respectively.

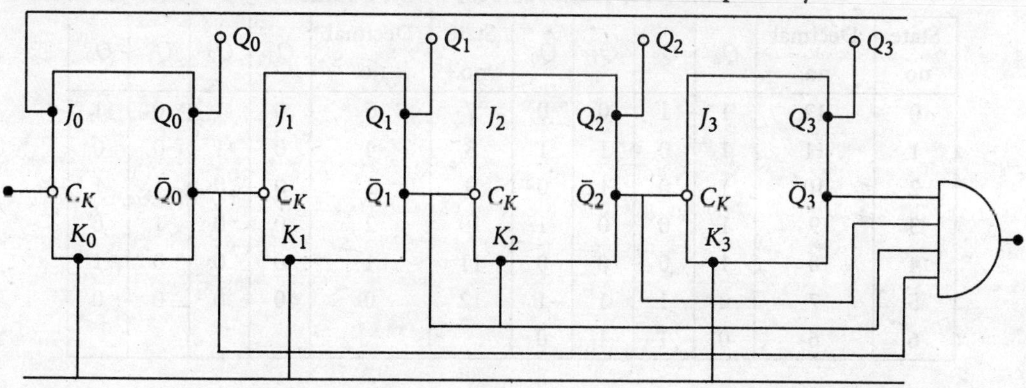

Fig. 7.32

The output conditions can be summarised as in Table 7.13.

Table 7.13: Output condition of Flip-Flop

Clock pulses (CP)	Output conditions of flip-flop				Clock pulses (CP)	Output conditions of flip-flop			
	D	C	B	A		D	C	B	A
0	0	0	0	0	9	1	0	0	1
1	0	0	0	1	10	1	0	1	0
2	0	0	1	0	11	1	0	1	1
3	0	0	1	1	12	1	1	0	0
4	0	1	0	0	13	1	1	0	1
5	0	1	0	1	14	1	1	1	0
6	0	1	1	0	15	1	1	1	1
7	0	1	1	1	16	0	0	0	0
8	1	0	0	0					

Four-bit binary counters are available in the market as MS1 circuits; for greater number of bits, several such counters may be cascaded.

7.14 Down-counter

The ripple counter described above is an up-counter, because it counts up. When an elevator goes up, the seven-segment display uses an up-counter to count the floor in the ascending order. Sometimes, it may also require counting in the decreasing order. Such a counter is called *down-counter,* which is used in various games and appliances like a micro-oven or a washing machine. They are examples of down-counter which counts pulses in decreasing order. A down-counter can be constructed from an up-counter by changing the trigger connection from Q to $\bar{Q}$ as shown in the schematic diagram [Fig. 7.32].

The output conditions may be summarised as shown in Table 7.14.

Table 7.14: Truth Table for Down Counter

State no.	Decimal no.	Q_3	Q_2	Q_1	Q_0	State no.	Decimal no.	Q_3	Q_2	Q_1	Q_0
0	12	1	1	0	0	7	5	0	1	0	1
1	11	1	0	1	1	8	4	0	1	0	0
2	10	1	0	1	0	9	3	0	0	1	1
3	9	1	0	0	1	10	2	0	0	1	0
4	8	1	0	0	0	11	1	0	0	0	1
5	7	0	1	1	1	12	0	0	0	0	0
6	6	0	1	1	0						

7.15 Short Theory Questions with Answers

1. At any instant, the output of a combinational circuit depends on which input?

   Ans. It depends on the input at that particular instant only.

2. Can you describe a combinational circuit by a truth table?

   Ans. Yes, every combinational circuit has its characteristic truth table.

3. How many types of inputs a multiplexer does have?

   Ans. A multiplexer has two types of inputs: (a) Data input (b) Control input.

4. What do you mean by a multiplexer (MUX) and a De-multiplexer (DMUX)?

   Ans. Multiplexing means *many into one*, so a multiplexer performs the function of selecting one out of N-input data sources and sending it to a single transmission channel with the help of control signals.

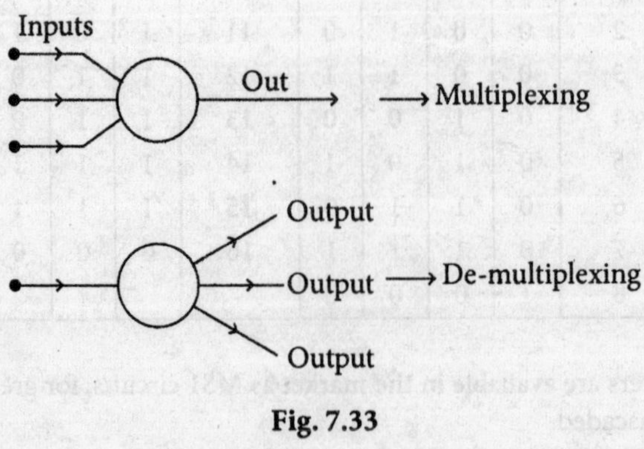

Fig. 7.33

De-multiplexing is the reverse of multiplexing operation, so its function is to transmit an input into many output channels.

5. What is a half-adder?

   Ans. A half-adder performs the binary addition of two bits.

6. How many input and output variables does a half-adder have?

   Ans. A half-adder has two-input variables and two output variables.

7. What is a combinational circuit?

   Ans. A combinational circuit is one which consists of a set of logic gates having inputs as well as outputs and the output depends on the instantaneous input.

   Example: Half-adder, multiplexer, etc.

8. What is a 2^n to n multiplexer?

   Ans. It means the multiplexer has 2^n data-input lines and control input lines. If $n = 2$, then the multiplexer has $2^2 = 4$ data input lines and 2 control lines as shown in Fig. 7.34.

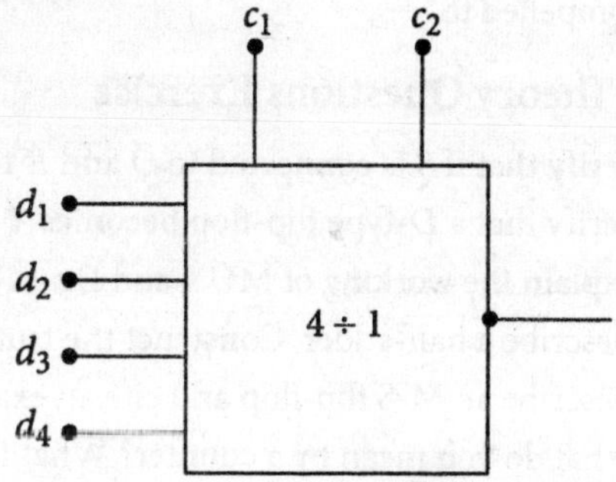

Fig. 7.34

9. Write the Boolean functions for the sum and carry of a half-adder.

   Ans. The truth table for half-adder is shown in Table 7.2.

10. Give the logic diagram for a *half-adder.*

    Ans. See discussion on half-adder and also Fig. 7.1.

11. Construct the Boolean function for a full-adder and draw the logic diagram for the same.

    Ans. A full-adder is a combinational circuit meant for summing three binary bits. The truth table for it is shown in Table 7.4. See the discussion on full-adder in the text.

12. Of all the flip-flops the M-S flip-flop is the best. Justify.

    Ans. It is free from *race around* ambiguity condition.

13. Write Boolean functions for the sum and carry of a half-adder.

    Ans. The truth table of a half-adder is given in Table 7.2, wherefrom we can write:

$$S = \bar{A}B + A\bar{B} = A \oplus B \ (XOR)$$
$$C - AB.$$

14. Give the logic diagram for realisation of the Boolean expression for the sum (S) and carry (C) of a half-adder.

    Ans.

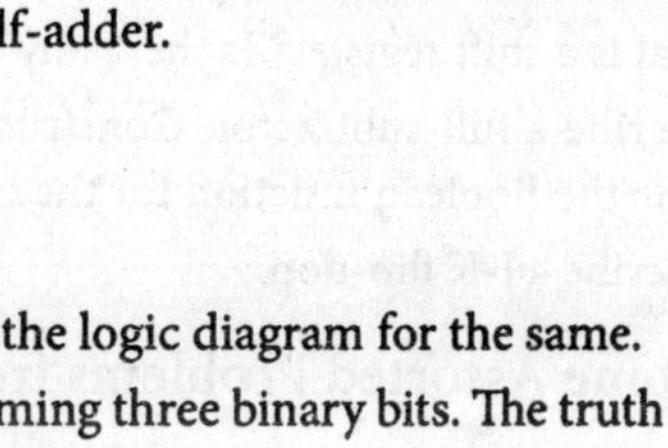

Fig. 7.35

15. What do you mean by synchronous and asynchronous circuits?

    Ans. Sequential circuits are of two types: (a) synchronous sequential (b) asynchronous sequential.

In synchronous sequential circuits, the memory elements change their state only at pre-assigned discrete time interval. Example: clocked F-F.

Asynchronous sequential circuit is one where the inputs and outputs do not change at pre-assigned times, because of propagation delay which cannot be strictly controlled. An asynchronous circuit works under *race-around* restriction.

16. Can you treat a flip-flop as memory element?

Ans. Yes; a flip-flop is a memory element in a sequential circuit. A flip-flop has two stable states — 1 and 0. Any information can be stored in the stable state of a flip-flop and it will not change unless compelled to.

7.16 Theory Questions Exercise

1. Verify that if S is connected to $\bar{Q}$ and R to Q, the S-R F-F is converted to a T-F-F.
2. Verify that a D-type flip-flop becomes T-type if D is connected to $\bar{Q}$.
3. Explain the working of MUX and De-MUX with examples.
4. Describe a half-adder. Construct the truth table and draw the logic diagram for the same.
5. Describe an M-S flip-flop and cite an example where it finds application.
6. What do you mean by a counter? What is a decade counter?
7. Is there any significance of the terms UP and DOWN attached to a counter?
8. What is the difference between a combinational and sequential circuit?
9. What is a shift register? Is there any register in a calculator?
10. Describe a full-subtractor. Construct the truth table and draw the logic diagram for three inputs. Write the Boolean function for the outputs.
11. Describe a J-K flip-flop.

7.17 Some Assorted Problems from Different Chapters

1. When an odd integer is represented in binary, the LSB is always 0. True or false? [True]
2. Hexadecimal is not a positional number system. — True or false? [False]
3. If two NOT-gates are connected in series in a circuit, then we can work without those two gates. True or false? [True]
4. Explain the NAND and XOR gates with suitable truth table. [See text]
5. What do you mean by basic duality of Boolean algebra.

   Ans. There is symmetry in the postulates of Boolean algebra w.r.t. (a) OR (+) operations (b) AND (•) operation (c) the identity element (1,0).

   If we interchange the operations and also the elements (1,0), we can deduce one set of axioms from the other and vice versa. This is called principle of duality. So we conclude that in Boolean algebra: (a) dual of any valid statement is also valid, (b) dual of any theorem is also a theorem, (c) dual of every identity remains an identity.
6. $(A+B)\cdot(A+C)=A+BC$, true or false? [True]
7. OR and AND gates are called basic gates. — Justify.
8. Realise through logic gates: $x+yz+xy=Q$.

   [*Hints.* $x+yz+xy = x(1+y)+yz = x+yz \Rightarrow x+yz=Q$, the basic diagram is shown in Fig. 7.36.

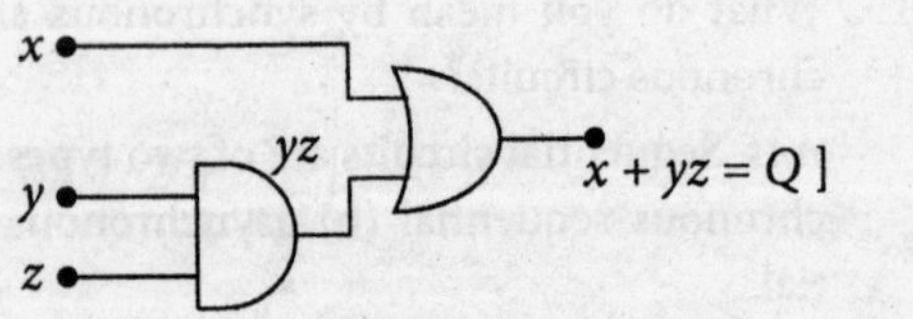

Fig. 7.36

9. For Boolean variable A, $A+A=0$. True or false? **[False]**

10. Simplify: $\bar{x}(x+y)(x+\bar{y})$.

[*Hints.* Complement of the given Boolean expression $=\overline{\bar{x}(x+y)(x+\bar{y})}=x+\overline{(x+y)}+\overline{x+\bar{y}}$

$$=x+\bar{x}\cdot\bar{y}+\bar{x}\cdot y=x+\bar{x}(y+\bar{y})=x+\bar{x}=1.$$

So, re-complementing, we get $\bar{x}(x+y)(x+\bar{y})=0$.]

11. Find the complement of $F=(x+y)(x+z)$.

[*Soln.* $\bar{F}=\overline{(x+y)(x+z)}=\overline{(x+y)}+\overline{(x+z)}=\bar{x}\cdot\bar{y}+\bar{x}\cdot\bar{z}=\bar{x}(\bar{y}+\bar{z})$.]

12. Using 2-input NAND gates, show how to design NOT gate.

[*Ans.* NAND gate is a universal gate, so any basic gate can be implemented with NAND. In Fig. 7.37 we have shown how a NAND produces a NOT gate

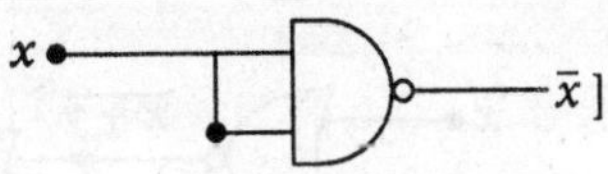

Fig. 7.37

13. What do you mean by universal gates? How do you get basic gates from the universal gates?

Ans. Basic gates are AND, OR and NOT, by combination of which we can realise any Boolean function. Two gates NAND and NOR are called *universal gates*, because each of them can lead to the basic gates, the transformation being shown in Fig. 7.38.

To get a NOT: Two terminals of a NAND tied together lead to a NOT.

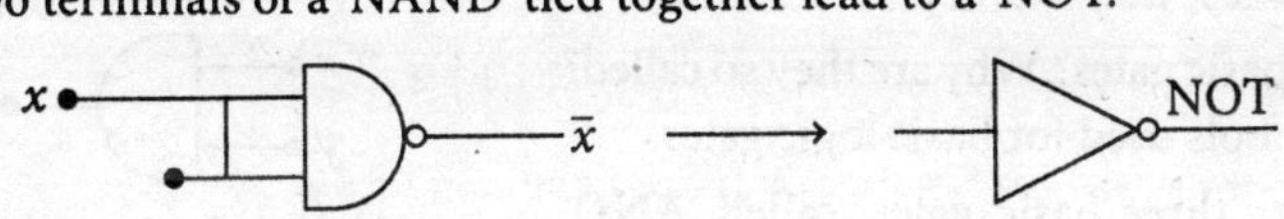

To get an AND: Two NANDs, one in the usual connection, the other connected in negation mode lead to the formation of an AND gate.

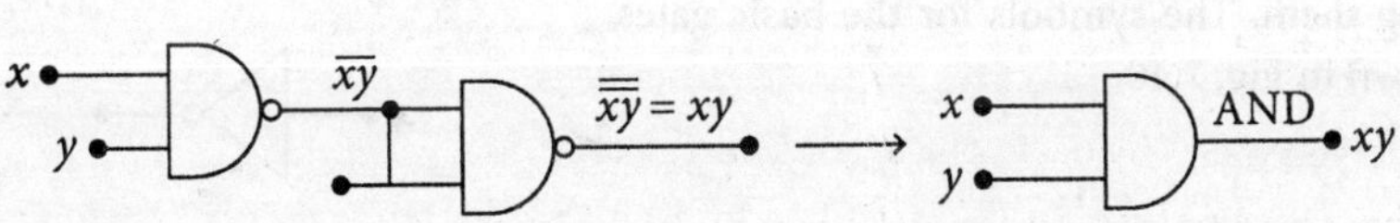

To get an OR: The outputs of two parallel NAND gates in negation mode are fed to another two-terminal NAND to form an OR.

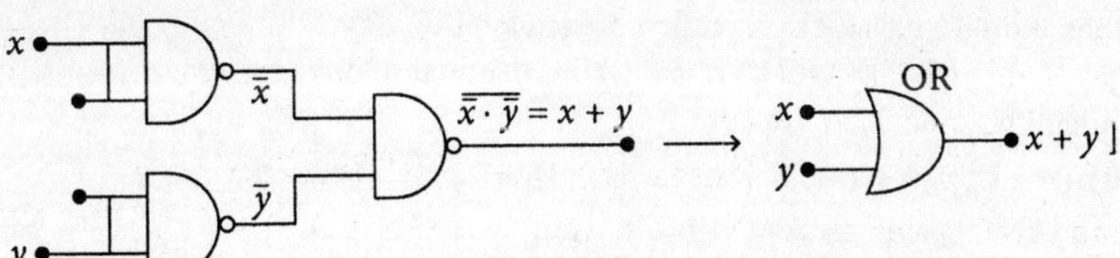

Fig. 7.38

NOR gate is also a universal gate. The scheme of conversion to the basic gates from NOR is shown in Fig. 7.39.

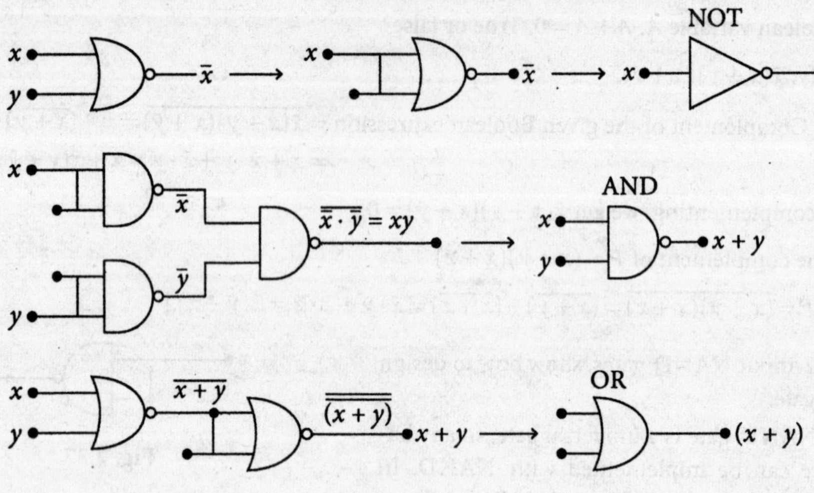

Fig. 7.39

14. What is an XOR gate? Develop its truth-table.

Ans. In such a gate, the output is in logic state 1 if either of the inputs is 1 and not both of them. Thus, the truth table for XOR gate will be as shown in Table.

Thus, the Boolean function for the output of the XOR gate $Q = \bar{x}y + x\bar{y} = x \oplus y$.

Table 7.15: Truth Table

x	y	Q
0	0	0
0	1	1
1	0	1
1	1	0

15. What are the basic gates? Why are they so called? Draw the symbols used for basic logic gates.

Ans. There are three basic gates, called AND, OR and NOT. They are called basic gates, because any Boolean function can be constructed by using them. The symbols for the basic gates are shown in Fig. 7.40.

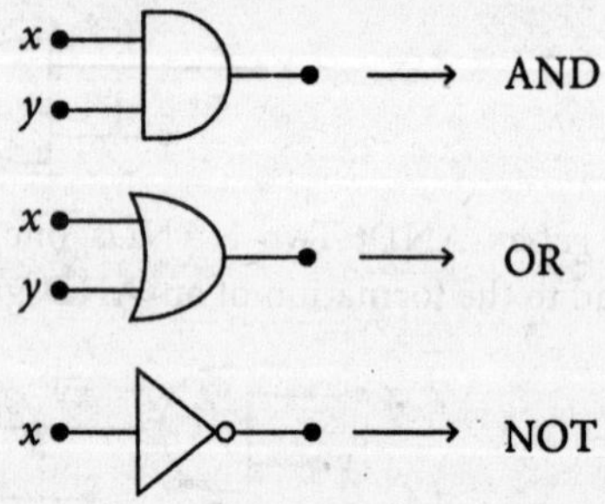

Fig. 7.40

Part IV

Signal Communication

Chapter 1

Analog Signal Communication

1.1 Communication Theory: Basic Theory of Signals and Its Communication

Communication is a very general term connoting *transfer* or *sharing* of experience or knowledge between a teacher and the taught; the information stored by a certain department disseminating to the different parts of a country; an alert from one country conveying to other different parts of the world, etc.

Now, what is a signal? Any data or information carrier is a signal. At present most of the signals are electromagnetic in nature, though in primitive days, signals were sent in the audio-form as through shouting over a short distance, as through beating of drums or blowing of trumpets over a comparatively longer distance in the crudest form.

Signals are electromagnetic waves: Most of the signals now-a-days are electromagnetic waves, first theoretically conceived and explored by Maxwell; the existence of the waves and their propagation was experimentally proved and verified by Hertz. It is now an established truth that all waves belong to a single family called *electromagnetic spectrum family* and light is a member belonging to it.

How does an electromagnetic wave originate? An electromagnetic wave contains two waves polarised at right angles to one another, one is the electric wave and the other, a magnetic wave, both mutually at right angles and also at right angles to the direction of propagation, hence a transverse wave.

An oscillating electric charge is the origin of a varying electric field, invariably accompanied by a magnetic field as well. Looking back into the atomistic theory of matter, every atom contains electrons, which move round the positive nucleus. An orbiting electron is equivalent to an oscillating charge, hence both electric and magnetic waves originate. An electromagnetic wave is born.

With the discovery of telegraph by Samuel F.B. Morse in 1818, there was an upheaval in the electric communication utilising more and more of the electromagnetic spectrum.

Table 1.1 shows the complete electromagnetic spectrum split into different regions depending on the frequency and wavelength. The application areas of the different regions are also indicated.

Table 1.1: Electromagnetic spectrum consists of the following constituents.

Wave region	Frequency (Hz)	Wavelength (cm)	Photon energy (eV)	Application
Long waves	$10^5 - 10^3$	$10^5 - 10^7$	$10^{-9} - 10^{-11}$	Long distance communication
Standard broadcasting waves	$10^7 - 10^5$	$10^3 - 10^5$	$10^{-7} - 10^{-9}$	Broadcasting
Microwaves	$10^{12} - 10^7$	$10^{-2} - 10^3$	$10^{-2} - 10^{-7}$	Microwave link
Infrared	$10^{14} - 10^{11}$	$10^{-4} - 10^{-1}$	$10^0 - 10^{-3}$	Object locator
Visible	$7.6 \times 10^{14} - 3.8 \times 10^{14}$	$3.9 \times 10^{-5} - 7.8 \times 10^{-5}$	$10^1 - 10^0$	
Ultraviolet	$10^{17} - 10^{15}$	$10^{-7} - 10^{-5}$	$10^3 - 10^1$	
X-ray	$10^{20} - 10^{16}$	$10^{-12} - 10^{-9}$	$10^6 - 10^2$	Medical Sciences
γ-rays	$10^{20} - 10^{19}$	$10^{-12} - 10^{-9}$	$10^8 - 10^5$	Medical Sciences

Most of the signals we come across at present are by and large analog signals where the signal strength at each and every instant is known.

There is another type of signal distinctly different from analog signal and this signal is termed as *digital signal* which exists only for discrete and desired time instants and quantised in amplitude. Whatever may be the initial form of energy in the signal, it is converted into electrical energy by some suitable transducer. In a signal two or more physical quantities speaking about the signal may be represented by writing one variable as a function of another or a number of other variables. These other variables which can be varied independently are called *independent variables.* The independent variable is usually time, but it may be space or both time and space.

Analog signal. If the dependent variable is a continuous function of time which means the signal exists for each and every instant of constant or variable amplitude. The signal is said to be *analog signal,* refer to Fig. 1.1(a) and (b).

Such a continuous time signal, also called analog signal, may be represented by

$$y = x_a(t). \tag{1.1}$$

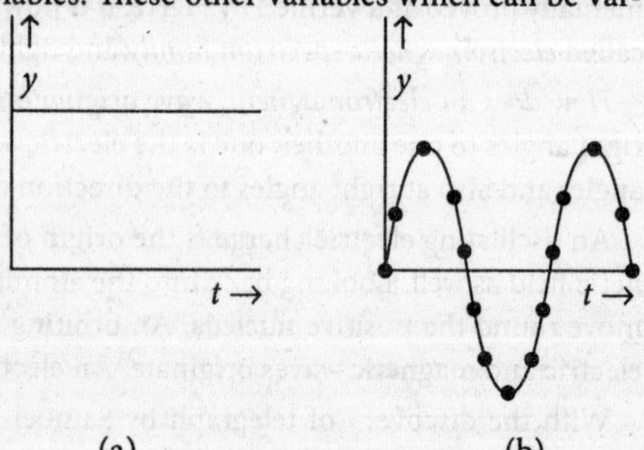

(a) (b)

Fig. 1.1

Digital signal. If the dependent variable in a signal is a discrete function of time, i.e., the signal exists for discrete time, the signal is said to be **discrete time signal.** It is denoted by $y = x_q(n)$, $n = 1,2,3,\ldots.$

The signals which are discretised w.r.t. time and quantised w.r.t. amplitude are called *digital signals.* Actual process by which we do the time discretisation and then amplitude quantisation will be discussed later. Here we will be making an important comment that analog signal processing and digital signal processing refer to the mechanism by which the two types of signals are processed and transmitted.

In Fig. 1.1 we have shown discrete time signal obtained from an analog signal by sampling it and it is represented by

$$x(nT) = x(t)\Big|_{t=nT}. \tag{1.2}$$

The instants at which the signal appears are called *sampling instants.* We write (1.2) for convenience as

$$x(nT)=\{x(n)\},\ n\in I_n,\ \text{+positive or negative including zero.} \tag{1.3}$$

Thus, a discrete time signal is represented by the sequence of numbers and is denoted by $\{x(n)\}$. So we can write

$$\{x(n)\}=x(-2),x(-1),x(0),x(1),x(2),\ldots. \tag{1.4}$$

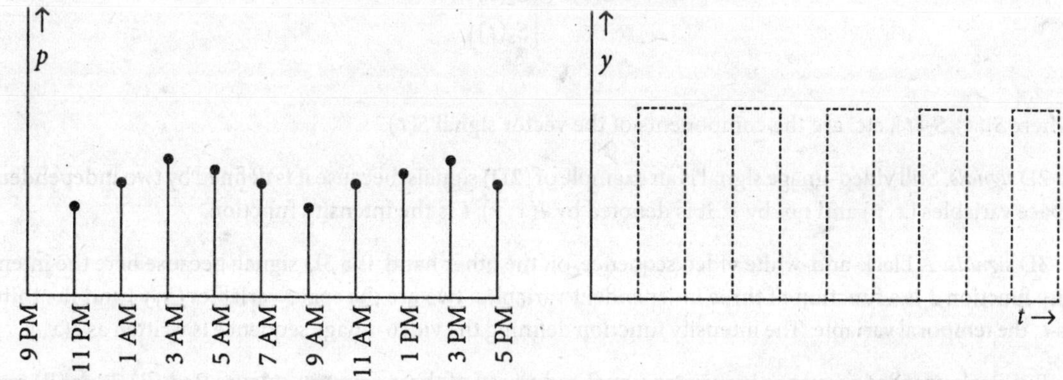

Fig. 1.2

1.2 Dimension of a Signal

In a signal two or more variables are involved. Of these variables, one variable may be written as a function of another or a number of other variables. These other variables may be varied independently and are called the independent variables. Depending on the number of independent variables defining the signal, a signal is referred to as:

1D (*one-dimensional*) *signal.* In this case, the number of independent variable defining the signal is just one. Such a signal is sometimes also called a scalar signal, it is a mono-source signal.

Practical examples are: Speech signal (1D). Mathematical representation of such a signal is

$$S_1(t)=A\cos\omega t. \tag{1.5}$$

It is a signal in real valued time domain. In complex representation such a signal is written as

$$S_2^*(t)=Ae^{j\omega t}. \tag{1.6}$$

Other examples of one-dimensional (1D) signal are:

(a) *Electro-cardiogram signal.* (ECG) representing the activity of heart. It is a periodic waveform and one period represents the time of transfer of blood from heart to arteries and forms an index of the performance of our heart.

(b) *Electro-encephalogram* (EEG) signal. The EEG signals do come from the brain cells (neurones). In this case, we use three electrodes, two being placed on the earlobes and the third, called the ground electrode, provides the base for producing differential voltages in the two electrodes.

When the brain cells are thus thrown into excitation, local current flows and signals from the cerebral cortex come out in 4 groups — beta-wave (7-13Hz), alpha-wave (8-13Hz), theta-wave (4-8Hz) and delta-wave (0.5-4Hz).

All (1D) signals as cited above are also mono-source, mono-channel and scalar signals.

Multi-space vector signals. We have signals produced by multi-sources, also called multi-channel signals, they are vector signals represented by

$$S(t) = \begin{bmatrix} S_1(t) \\ S_2(t) \\ S_3(t) \end{bmatrix}, \tag{1.7}$$

where $S_1(t), S_2(t)$, etc. are the components of the vector signal $S(t)$.

2D *signals.* Still video-image signal is an example of (2D) signals, because it is defined by two independent space variables (x, y) and not by t. It is denoted by $I(x, y)$, I is the intensity function.

3D *signals.* A black-and-white video sequence, on the other hand, is a 3D signal, because here the intensity function I is a function of three independent variables; two are the space variables (x, y) and the third is t, the temporal variable. The intensity function defining the video-image sequence is written as $I(x, y, t)$.

But a colour video sequence is a vector signal composed of three primary colours Red (R), Blue (B) and Green (G), each being a (3D) signal represented by

$$I(x, y, t) = \begin{bmatrix} I_r(x, y, t) \\ I_b(x, y, t) \\ I_g(x, y, t) \end{bmatrix}. \tag{1.8}$$

1.3 Essentials of Signal Processing

We have already made the remark that analog signal and its processing mechanism and transmission is an established truth and its origin dates back to ancient past. Compared to the modern development in the art of communication, called *digital signal processing, analog signal processing* and *transmission* is comparatively simpler; developed and established much, much earlier but digital signal processing began only in 1960s. Since then significant developments and breakthroughs in both theory and application took place. One point that we won't forget is that majority of the signals we come across are in reality analog signals and the processing of such a signal is already an established technique but in digital signal processing we have first to discretise an analog signal and then quantise it, thus getting the digital signal, it is then put to transmission.

Obvious question that arises in our mind, therefore, is why then digital signal processing when there is analog signal processing already available? We will turn to this question in later part of our discourse. We will follow the course of discussion according to chronology — starting with analog signal communication and ending with digital signal processing.

Analog signal communication. The essential parts of an analog signal transmission process are shown in block diagram (1.3).

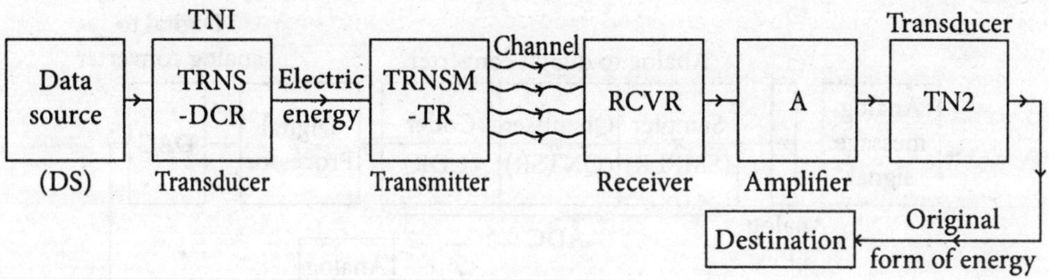

Fig. 1.3

There is a message or data source (DS) wherefrom the input whatever may be the energy form is passed through a transducer which converts the input into electric energy. After conversion the output of the transducer is fed to the transmitter coupled to the transmission channel which may be a guided transmission channel as in the form of a wire or waveguide or it may be an unguided one as in case of atmospheric or space channel. As the signal passes through the channel, there is progressive attenuation or distortion of the signal as in case of guided channel like telephone wire. When the signal is sent in the form of electric pulse, electrical power is lost through heat generation; in case of unguided channel like open air or atmosphere energy loss takes place, because of scattering and also due to absorption by the air molecules. So from time to time attempts have been made to improve the channel characteristics as regard fidelity and increase in data-rate of transfer so as to send more data or information over long distances.

The receiver (RCVR) at the end-point receives the signal and retrieves information from it, amplifies and then restores it to the original form before passing it on to the destination. Detailed discussion shortly follows.

Digital signal processing. A digital signal is not only discretised w.r.t. time (t) but its amplitude must also be quantised. A digital signal does not appear out of its own, we have to make a digital signal out of an analog signal which we frequently come across.

The essentials of digital signal processing are shown schematically in Fig. 1.4. So the initial step is how to get a digital signal out of an analog one. We have read in connection with analog communication that a few initial steps are common to both analog signal processing as well as digital signal communication, these few common steps are shown in schematics in Fig. 1.4.

Signal source	Initial form of energy	Transducer (TDCR)	Baseband signal energy output
Speech signal (Human voice)	Mechanical energy	Microphone	Electrical
TV picture (3D) image signal	Light energy	TV-camera (Image orthicon vidicon)	Electrical

Whatever may be the initial form of energy, it is transformed into electrical pulses by suitable transducer (TDCR) forming what is known as *message signal* or *baseband signal.* This baseband signal is analog in nature. In analog signal communication, the baseband signal is directly fed to the transmitter, then the sequence follows as depicted in Fig. 1.3. But in digital communication, the baseband signal is to be transformed into a digital signal (discrete in time as well as in amplitude) and the processing is schematically shown in Fig. 1.4.

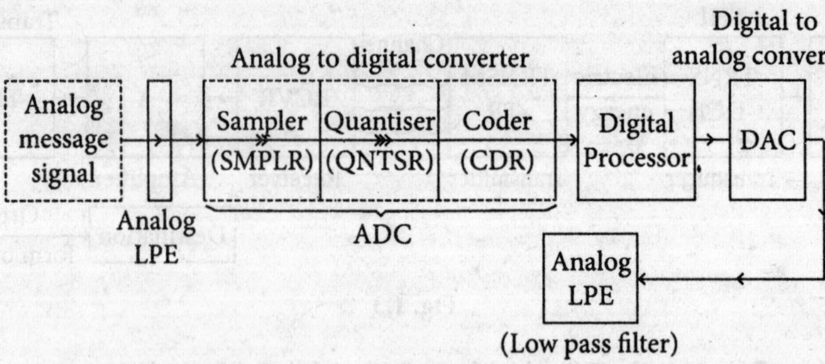

Fig. 1.4

It is evident from Fig. 1.4 that digital processing consists of three steps.

1. Conversion of analog signal to digital form by means of the device called ADC.
2. Processing the digital version through a digital processor according to desired algorithm.
3. Finally, conversion of the processed digital signal back to an analog form of desired frequency by DAC and low pass filter (LPF). The details will be explored later.

1.4 Review of Analog Signal Communication

Analog signal communication is a standard and established technique from a long time. But before discussion about the technique, we raise an important question. Why speech signal (low frequency audio-signal) cannot be radiated directly through an antenna? The answer is not far to seek.

For effective transmission of signal over a long distance by a radiating antenna, the dimension of the radiating system requires to be comparable to the wavelength of the signal wave.

If the transmission frequency is 1kHz, the corresponding wavelength of the electromagnetic wave (or baseband signal) is given by

$$c = \nu\lambda \Rightarrow \lambda = \frac{c}{\nu} = \frac{3\times 10^8}{1\times 10^3} = 3\times 10^5\text{m} = 300\text{km}. \tag{1.9}$$

This shows that the linear dimension of the radiating antenna system is unwieldly long and not feasible. So for transmission of low frequency audio-signal, it is to be mixed with high frequency carrier. The mixing of high frequency carrier with low frequency audio-signal wave is referred to as *modulation;* the high frequency component is the carrier wave; the low frequency audio-component is the modulating wave. The resultant wave after mixing is known as modulated wave.

Essence of modulation. The process by which to adjust some characteristics of the carrier wave by the instantaneous value of the low frequency modulating wave is called the process of modulation.

The carrier wave may be represented by the continuous sinusoid as

$$v_c = \hat{V}_c \sin(\omega_c t + \phi), \tag{1.10}$$

where v_c = instantaneous value of the carrier; $\hat{V}_c$ = peak value of the carrier; $\omega_c = 2\pi f_c$, f_c being the carrier frequency and ϕ = phase angle.

So three characteristics of the carrier may be modified giving rise to three types of modulation.

1. Amplitude $\hat{V}_c$ may be modified by the modulating wave — Amplitude Modulation (AM).

2. Frequency f_c may be modified — Frequency Modulation (FM).
3. Phase angle ϕ may be modified — Phase Modulation (PM).

   The first two types of modulation are very much in use while the third type of modulation is not so widespread.
4. *Amplitude modulation* (AM). We were may much familiar with this type of modulation in the past but with the advent of video-transmission we are now becoming accustomed with FM-transmission. In TV transmission, both audio and video-signals are communicated. We should always remember that the audio-signals are frequency-modulated while video-signals are amplitude-modulated. Obviously there are reasons behind this choice and they will be discussed at length later.

Mathematical discourse. The modulating voltage is represented by

$$v_s = \hat{V}_s \cos\omega_s t. \tag{1.11}$$

The carrier voltage wave is taken as

$$v_c = \hat{V}_c \cos\omega_c t. \tag{1.12}$$

The carrier voltage wave amplitude $\hat{V}_c$ is modified by the instantaneous value of the modulating voltage, hence, the amplitude of the modulated voltage is

$$\hat{V}_m = \left(\hat{V}_c + k\hat{V}_s \cos\omega_s t\right), \tag{1.13}$$

where k is some fraction

$$\Rightarrow \hat{V}_m = \hat{V}_c\left(1 + \frac{k\hat{V}_s}{\hat{V}_c}\cos\omega_s t\right) \Rightarrow \hat{V}_c(1 + m\cos\omega_s t), \tag{1.14}$$

m is a factor called *depth of modulation* or modulation index.

So we write $\hat{V}_m = \hat{V}_c(1 + m\cos\omega_s t)$.

Now, V_m = instantaneous value of the modulated voltage can be written as $v_m = \hat{V}_m \cos\omega_c t$, because in amplitude-modulation, frequency of the carrier remains unmodified.

$$\begin{aligned} \therefore\ v_m &= \hat{V}_m \cos\omega_c t = \hat{V}_c(1 + m\cos\omega_s t)\cos\omega_c t \\ &= \hat{V}_c\cos\omega_c t + \tfrac{1}{2}m\hat{V}_c\cos(\omega_c + \omega_s)\,t + \tfrac{1}{2}m\hat{V}_c\cos(\omega_c - \omega_c)\,t \qquad (1.15) \\ &= \hat{V}_c\cos 2\pi f_c t + \tfrac{1}{2}m\hat{V}_c\cos 2\pi\,(f_c + f_s)\,t + \tfrac{1}{2}m\hat{V}_c\cos\,(f_c - f_s)\,t, \qquad (1.16) \end{aligned}$$

where f_c and f_s are the frequencies of the carrier and the signal waves respectively.

Modulation index parameter (m). The factor m also called *index of modulation* in an important parameter related to the extent of modulation of the characteristic of the carrier by the low frequency modulating signal wave.

$$m = \frac{k\hat{V}_s}{\hat{V}_c}$$

from (1.14) and hence, can be defined as below:

$$\left(\hat{V}_m\right)_{\max} = \hat{V}_c(1 + m) \Rightarrow m = \frac{\left(\hat{V}_m\right)_{\max} - \hat{V}_c}{\hat{V}_c} \tag{1.17}$$

$$= \frac{\text{Maximum change in the amplitude of the carrier}}{\text{Original carrier wave amplitude}}$$

$$= \text{Maximum fractional change in the wave amplitude of the carrier.}$$

Depending on the value of m, we define the modulated wave as under-modulated or over-modulated. If $m = 100\%$, the carrier wave amplitude is maximum changed and that change is equal to carrier wave amplitude, we refer to this modulation as optimum modulation. If $m < 100\%$, it is referred to as *under-modulation* leading to low power transmission by the modulated wave; but if $m > 100\%$, modulation is said to be over-modulation leading to greater power transmission but with greater and greater distortion. In the process of amplitude-modulated transmission, optimum modulation is recommended.

Consider again (1.15). The equation shows that the AM-wave consists of three waves represented by the three terms of the equation.

1. The original carrier of frequency f_c.
2. The component wave of higher frequency $(f_c + f_s)$, called upper side frequency, having amplitude as $\frac{1}{2} m \hat{V}_c$.
3. Another component wave of lower frequency $(f_c - f_s)$, called lower side frequency, having amplitude as $\frac{1}{2} m \hat{V}_c$. The component waves of frequency lower and higher than the original carrier are the satellite frequencies. The two satellites have the same amplitude but with two different frequencies.

The graphical representations of

1. Modulating voltage: $v_s = \hat{V}_s \cos \omega_s t$.
2. Carrier wave: $v_c = \hat{V}_c \cos \omega_c t$.
3. Amplitude modulated wave: $v_m = \hat{V}_m \cos \omega_c t$

are shown in Fig. 1.5(a), (b) and (c) respectively.

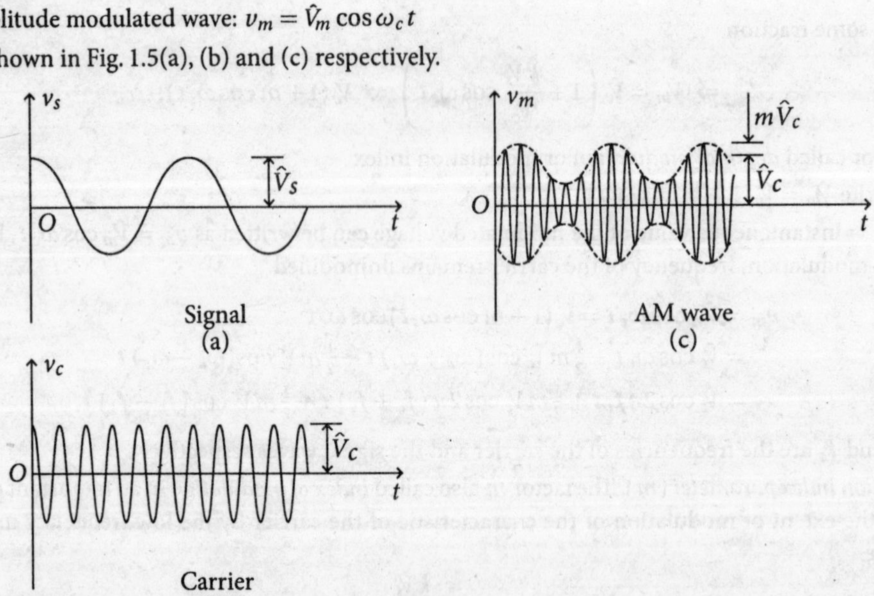

Fig. 1.5

1.5 Frequency Spectrum of AM-wave

There are three components in AM-wave if the modulating wave is a single harmonic wave of the frequency f_s and carrier wave is also single high frequency wave of frequency f_c.

These three waves are: (a) Original carrier of frequency f_c, (b) Upper side frequency $= (f_c + f_s)$ and (c) Lower side frequency $= (f_c - f_s)$.

The frequency spectrum is shown in Fig. 1.6.

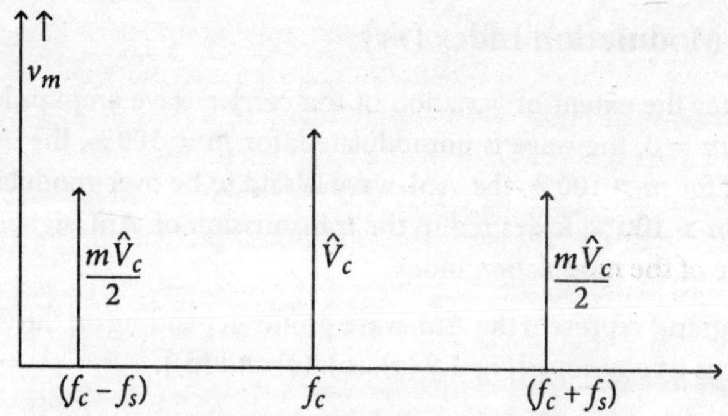

Fig. 1.6

Thus, the modulated wave has fundamental frequency f_c that of the carrier; the lower side frequency of value $(f_c - f_s)$ and the upper side frequency of value $(f_c + f_s)$. Since $f_c \gg f_s$, the side frequencies are very close to the fundamental frequency. The frequency bandwidth Δf is given by

$$\Delta f = (f_c + f_s) - (f_c - f_s) = 2f_s. \tag{1.18}$$

(1.18) shows that for quality audio-reception, the bandwidth required is 30kHz which means that for good audio-transmission, the maximum frequency that can be used is 15kHz. So we are allowed to utilise only a part of the frequency spectrum meant for quality audio-transmission.

The satellite bands — Lower and Upper. If the modulating voltage is of single frequency f_s, the AM-wave spectrum contains three frequencies — one fundamental and the other two symmetrically situated w.r.t. the fundamental one, is called lower side frequency and the other called *upper side frequency* forming a doublet. But if the modulating voltage has a frequency spectrum ranging from (f_{S1}) to (f_{SN}), the AM-wave spectrum consists of two symmetrically situated spectra, one on the lower side of frequencies, the other on the higher side w.r.t. the central fundamental frequency (f_c) (see Fig. 1.7).

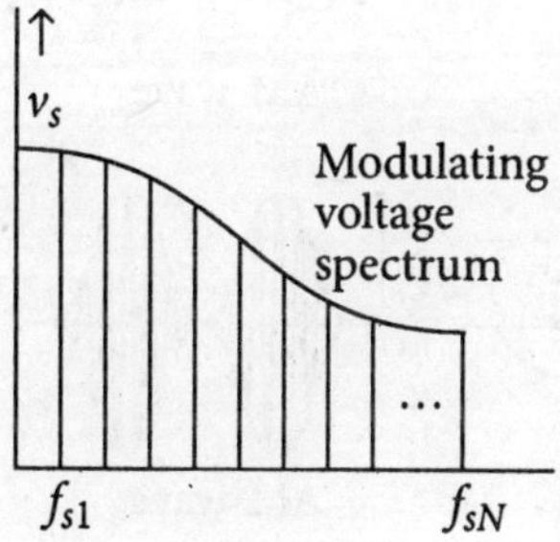

Fig. 1.7

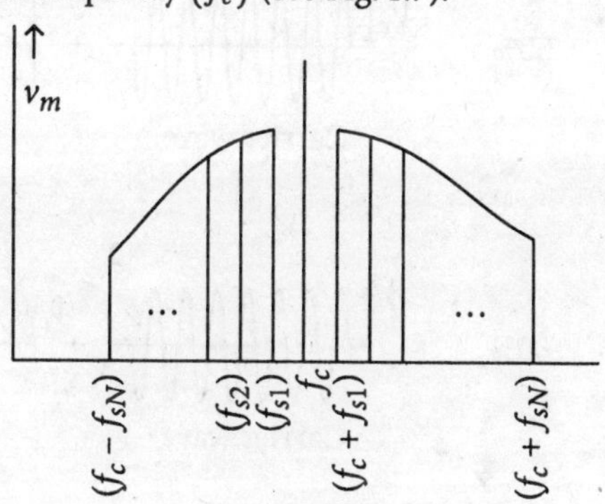

Fig. 1.8

Each of the frequencies contained in the modulating voltage spectrum, produces a symmetric doublet w.r.t. the fundamental frequency, thus producing in the AM-wave two frequency spectra, one having lower frequencies, called *lower band,* the other having higher frequencies, called *upper band* (Fig. 1.8). The modulating bandwidth or channel capacity for AM transmission is $(f_c + f_{SN}) - (f_c - f_{SN}) = 2f_{SN}$ = twice the maximum modulating band-frequency. It also represents channel capacity. The AM transmission has a channel capacity of 10kHz, so the maximum audio-frequency allowed for modulation is 5kHz only.

1.6 Significance of Modulation Index (m)

The parameter m indicates the extent of variation of the carrier wave amplitude due to modulation by the modulating wave. If $m = 0$, the wave is unmodulated; for $m = 100\%$, the AM-wave is said to have *optimum modulation* and for $m > 100\%$, the AM-wave is said to be over-modulated. Neither $m < 100\%$ (under-modulated) nor $m > 100\%$, is desired in the transmission of AM signal; the desired m value is 100 %, the optimum value of the modulation index.

The following set of diagrams represent the AM-wave profile depending on the value of m; the AM-wave has the modulating wave as its envelope [Fig. 1.9 (a), (b), (c) and (d)].

(a) When $m = 0$, the carrier wave remains unmodulated.

(b) When $m = 50\%$, the AM-wave amplitude is 1.5 times the amplitude of the carrier. The AM-wave profile is represented by the dashed curve which shows the modulating wave as the envelope.

(c) When $m = 100\%$, the AM-wave amplitude is twice the amplitude of carrier. The modulated wave profile is shown in Fig. 1.9(c) by the dashed curve. It is the optimum modulation used in AM transmission.

(d) When $m = 150\%$, the AM-wave amplitude is 2.5 times the amplitude of the carrier. AM-wave profile in this case has a special appearance in the sense that the two lobes are separated by a distance leading to distortion due to absence of any signal during this time interval.

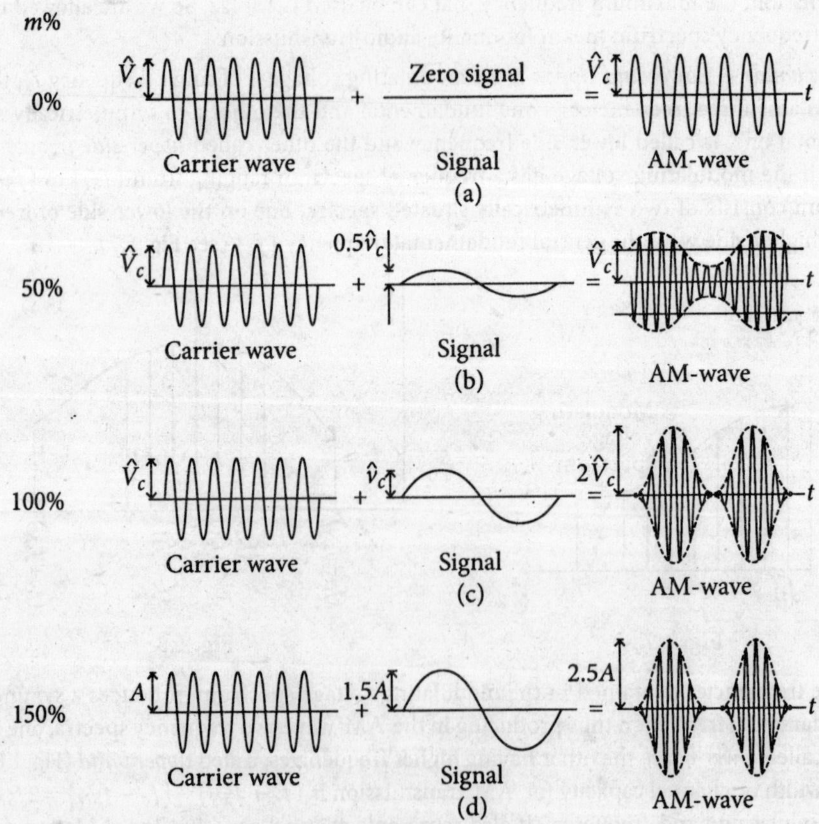

Fig. 1.9 (a), (b), (c) and (d)

1.7 Power Transmitted along with AM-wave

For different values of m, we get different shapes of the AM-wave profile and it is evident from the above discussion that for optimum modulation $m = 100\,\%$, it is the maximum value of m for which the AM-wave profile consists of two consecutive lobes just touching each other. It is also to remember that the information or message is contained in the two sidebands and not anything in the carrier. We will presently show the energy transmitted along with AM-wave very much depends on the value of m which, for no distortion, can go up to 100 % (optimum modulation per cent).

Average power in the AM-*wave.* The AM-wave is represented by

$$v_m = \sum_{l=1}^{N} \hat{V}_c(1 + m_l \cos\omega_{st}\, t)\cos\omega_c t$$

$$= V_c \cos\omega_c t + \sum_{l=1}^{N} \frac{\hat{V}_c m_l}{2}\cos(\omega_c + \omega_{sl})\, t + \sum_{l=1}^{N} \frac{1}{2} m_l \hat{V}_c \cos(\omega_c - \omega_{sl})\, t, \tag{1.19}$$

assuming signal wave to consist of a frequency spectrum ranging from $\omega_{s1}, \omega_{s2}, \ldots, \omega_{sN}$.

If the energy carried by the AM-wave is delivered on to a load of resistance R, then the instantaneous power P_i can be written as $P_i = v_m I_m = v_m^2/R$.

Taking help of (1.19) for v_m, we can write average power as

$$\langle P\rangle = \frac{1}{T}\int_{-T/2}^{T/2} \left(v_m^2/R\right)$$

$$= \frac{1}{TR}\int_{-T/2}^{T/2} \left[\hat{V}_c \cos\omega_c t + \sum \frac{1}{2} m_l \hat{V}_c \cos(\omega_c + \omega_{sl})\, t + \sum \frac{1}{2} m_l \hat{V}_c \cos(\omega_c - \omega_{sl})\, t\right]^2 dt. \tag{1.20}$$

Now

$$\hat{V}_c^2 \int_{-T/2}^{T/2} \cos^2\omega_c t = \hat{V}_c^2\left(\frac{T}{2}\right).$$

$$\int_{-T/2}^{T/2} \sum \left(\hat{V}_c^2/4\right) m_l^2 \cos^2\{(\omega_c + \omega_{sl})\, t\}\, dt = \frac{\hat{V}_c^2}{4}\sum m_l^2 \frac{T}{2} = \frac{T\hat{V}_c^2}{8}\sum m_l^2.$$

$$\int_{-T/2}^{T/2} \sum \frac{\hat{V}_c^2}{4} m_l^2 \cos^2\{(\omega_c - \omega_{sl})\, t\}\, dt = \frac{\hat{V}_c^2}{4}\sum m_l^2 \frac{T}{2} = \frac{TV_c^2}{8}\sum m_l^2.$$

So the average power carried by the AM-wave is

$$\langle P\rangle = \frac{1}{TR}\left[\hat{V}_c^2\left(\frac{T}{2}\right) + 2\frac{T\hat{V}_c^2}{8}\sum m_l^2\right] = \frac{\hat{V}_c^2}{2R} + \frac{\hat{V}_c^2}{4R}\sum m_l^2$$

$$= \left(\frac{\hat{V}_c}{\sqrt{2}}\right)^2 \Big/ R + \frac{1}{2}\left(\frac{\hat{V}_c}{\sqrt{2}}\right)^2 \times \frac{1}{R}\sum m_l^2 = \left(P_c + \frac{1}{2} P_c \times \sum m_l^2\right), \tag{1.21}$$

where P_c is the power of the carrier.

(1.21) clearly shows that $[\langle P\rangle] > P_c$.

If $\Sigma m_l = 100\,\%$, then we get

$$\langle P\rangle = P_c\left(1+\tfrac{1}{2}\right) = \tfrac{3}{2}P_c. \tag{1.22}$$

So if $\langle P\rangle = 150$ units (energy), $P_c = 100$ units; it means that out of 150 units of energy conveyed by AM-wave, 100 units will be in carrier component and remaining 50 units will be equi-partitioned between the two sidebands, so each sideband will contain 25 units. If Σm_l is lesser, lesser energy will be transported. It is to remember in this connection that the carrier does not carry any information or intelligence, the whole of it is equally contained in the two sidebands.

1.8 Neither Under-modulation nor Over-modulation is desired in AM-Transmission. Explain

If $\Sigma m_l = 100\,\%$, modulation is called the optimum modulation. It is the maximum value of depth of modulation for which maximum amount of energy conveyed by AM-wave without any distortion [cf. Fig. 1.9(c)].

If $\Sigma m_l < 100\,\%$, it is called under-modulation and lesser amount of energy will be transported with the AM-wave [cf. Fig. 1.9(b) with 50 % as (Σm_l)].

If $\Sigma m_l > 100\,\%$, called over-modulation, greater amount of energy is transported but with enough of distortion [cf. Fig. 1.9(d)].

1.9 Linear Amplitude Modulator

We have, after studying some details of the characteristics of AM-wave, now come to raise the question: *How to produce the* AM-*wave?*

There are two methods, one is called square-law modulation method; the second one is called the linear modulator method.

The first method uses a diode in proper bias through which the carrier voltage and modulating voltage pass in series. The second method uses a transistor as an amplifier. We will discuss here the second method, the linear modulator for producing the AM-wave.

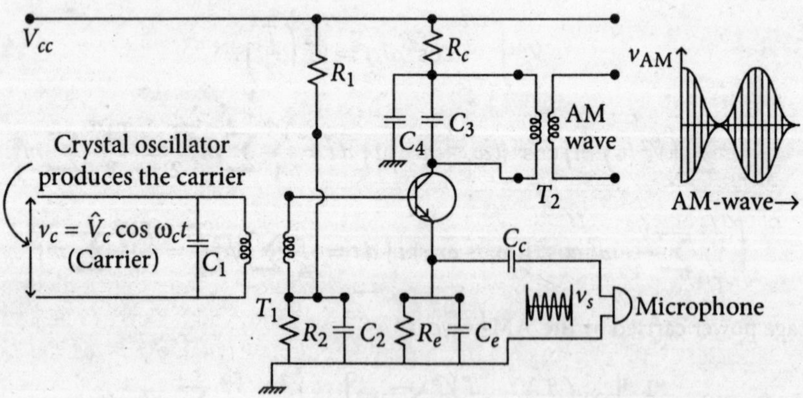

Fig. 1.10

Linear modulator. Linear modulator circuit is shown in Fig. 1.10. The biasing of the transistor is done through potential divider arrangement with emitter bias provided by R_1, R_2 and R_e; the parallel combination of R_e and C_e acts as a voltage stabiliser to keep the base bias line fixed against collector current variation due to the audio-signal being fed to emitter part from the microphone output. The carrier wave, produced by a crystal oscillator, is injected into the base of the transistor.

The output of the microphone is fed to the emitter of the transistor through the coupling capacitor C_c; the carrier voltage v_c as the input is fed through the transformer T_1 tuned by the capacitor C_1 to the base of the transistor.

The amplifier output is a high frequency (hf) voltage output transferred to the secondary of the transformer T_2 tuned by the capacitor C_3. Thus, we get AM-wave at the output port of the transformer T_2. The capacitor C_2 is used so that the dc base bias is not affected by the carrier voltage and the capacitor C_4 is used to isolate the dc supply against the AM voltage available across the primary of the transformer T_2.

Mathematical analysis. The modulating voltage is represented by $v_s = \hat{V}_s \cos\omega_s t$. It regulates the emitter current whose variation will be proportional to the modulating voltage. The collector current variation will be function of both modulating voltage and the carrier voltage. The instantaneous emitter current i_e can be written as $i_e = I_e + k\hat{V}_s \cos\omega_s t$, here k is termed as current conversion factor of voltage.

The voltage gain of the amplifier is A_v given as

$$A_v = k_1 i_e = k_1\left(I_e + k\hat{V}_s \cos\omega_s t\right) = k_1 I_e + k_1 k\hat{V}_s \cos\omega_s t. \tag{1.23}$$

The carrier voltage v_c which is the input to CE amplifier is amplified such that

$$\begin{aligned} v_0 = A_v v_c &= \left(k_1 I_e + k_1 k\hat{V}_s \cos\omega_s t\right) v_c = \left(k_1 I_e + k_1 k\hat{V}_s \cos\omega_s t\right)\hat{V}_c \cos\omega_c t \\ &= k_1 I_e \hat{V}_c\left(1 + \frac{k\hat{V}_s}{I_e}\cos\omega_s t\right)\cos\omega_c t = k_2\hat{V}_c(1 + m\cos\omega_s t)\cos\omega_c t, \end{aligned} \tag{1.24}$$

where $k_2 = k_1 I_e$; m = modulation depth $= \left(k\hat{V}_s/I_e\right)$.

$$\Rightarrow v_0 = k_2\hat{V}_c\cos\omega_c t + \frac{mk_2\hat{V}_c}{2}\cos(\omega_c+\omega_s)\,t + \frac{mk_2\hat{V}_c}{2}\cos(\omega_c-\omega_s)\,t. \tag{1.25}$$

So (1.25) suggests that output of the amplifier consists of a triplet.

1. $k_2\hat{V}_c\cos\omega_c t$ — Carrier wave component.
2. $\dfrac{mk_2\hat{V}_c}{2}\cos(\omega_c-\omega_s)\,t$ — Lower side frequency component.
3. $\dfrac{mk_2\hat{V}_c}{2}\cos(\omega_c+\omega_s)\,t$ — Upper side frequency component.

1.10 Merits and Demerits of AM-transmission

This process of transmission is very simple and very effective in sending low frequency modulating signal. But there is a number of demerits also as given below.

1. *Power of the signal in* AM *transmission is small.* It has been remarked earlier that the carrier wave does not contain the message but it is contained in the two sidebands. As the value of the modulation index Σm_l for a composite modulating wave decides the amount of power transmitted and to be contained in the two sidebands, if $\Sigma m_l < 100\%$ (under-modulation) and if $\Sigma m_l > 100\%$ (over-modulation), its value is restricted to be equal to the optimum modulation value (Σm_l is 100%).

   Thus, for $\Sigma m_l = 100\%$, we get from (1.21)

$$P_t = \langle P\rangle = P_c\left[1 + \left(\Sigma m_l^2/2\right)\right] = P_c\left(1 + \frac{\Sigma m_l^2}{2}\right)$$

$$\Rightarrow \frac{P_t}{P_c} = 1 + \frac{\Sigma m_l^2}{2} \Rightarrow P_s = (P_t - P_c) = \frac{P_t\left(1 + \frac{\Sigma m_l^2}{2}\right)}{\left(\frac{\Sigma m_l^2}{2}\right)} = P_t\left(\frac{\Sigma m_l^2}{2 + \Sigma m_l^2}\right).$$

$\therefore\ P_s = \frac{1}{3}P_t$. It means, in AM transmission only $\frac{1}{3}$ of the power transmitted is associated with the signal; so the maximum of the power is to be sacrificed as the carrying cost to the carrier wave. So amplitude-modulated analog signal processing is not cost efficient.

2. AM *transmission is susceptible to noise pick-up.* In AM transmission, the amplitude of the CW is modulated by the modulating signal voltage. There are so many other sources, both man-made and also natural (as in case of thunder, spark-plug firing of motor vehicle, etc.), which may affect the amplitude of the CW and thus produce spurious signals. It is not possible for the receiver to distinguish between the actual signal and the spurious signal, also called the noise. So AM transmission is noisy. To express how much of noise has crept in the process of transmission, we have introduced a ratio denoted by (S/N) which means signal to noise ratio. If it is high, it implies greater signal power and if it is low, noise rides over the actual signal.
3. *Poor audio-quality.* In AM broadcast, the allowed channel capacity is up to 10kHz implying maximum audio-frequency that can be used is only 5kHz. But for quality audio-reception, the required channel capacity is 30kHz. So the channel capacity allowed for AM transmission is a part of the channel capacity for quality audio-reception. For quality audio-reception, AM is not, therefore, very suitable. So, in TV analog signal transmission, video-signal is transmitted through AM but not the audio-signal. Audio signal is transmitted through a different modulation technique, called *frequency-modulated wave.*

1.11 Frequency Modulated Wave Transmission: FM-transmission

Because of several defects associated with AM, limitation being set by the modulation index; noisy transmission, etc., FM transmission is recommended for low frequency transmission.

In this modulation technique, the frequency of the carrier is modified by the modulating or signal voltage. The amplitude of the carrier here remains unaltered. The main objection against amplitude variation based transmission (AM transmission) is that the inanimate material agency which is the receiver, is unable to identify the actual signal and spurious signal, so that the signal effect and noise effect are mixed up.

FM *wave transmission — Theoretical basis.* The carrier voltage wave is represented by

$$v_c = \hat{V}_c \cos\omega_c t = \hat{V}_c \cos\phi_c,\ \phi_c = \omega_c t \tag{1.26}$$

$\Rightarrow \frac{d\phi_c}{dt} = \omega_c = 2\pi f_c$ giving carrier frequency.

In this processing of the signal, the frequency of the carrier is modified by the instantaneous value of the signal. So we write

$$\omega_{\text{FM}} = \omega_c + k_{fs}\hat{V}_s \cos\omega_s t, \tag{1.27}$$

where k_{fs} is termed as the frequency conversion factor for the signal voltage.

Now

$$\frac{d\phi_{\text{FM}}}{dt} = \omega_{\text{FM}} \Rightarrow \phi_{\text{FM}} = \int \omega_{\text{FM}} dt + \text{constant.}$$

Using (1.27), we would get

$$\phi_{\text{FM}} = \int \left(\omega_c + k_{fs}\hat{V}_s \cos\omega_s t\right) dt + \phi_0 = \omega_c t + \left(k_{fs}\hat{V}_s/\omega_s t\right)\sin\omega_s t + \phi_0,$$

where $\phi_0 = (\phi_{FM})_{t=0}$. If $\phi_0 = 0$, we get

$$\phi_{FM} = \omega_c t + \frac{k_{fs}\hat{V}_s}{\omega_s}\sin\omega_s t.$$

$$\therefore\ v_{FM} = \hat{V}_c \cos\phi_{FM} = \hat{V}_c \cos\left(.\omega_c t + \frac{k_{fc}\hat{V}_s}{\omega_s}\sin\omega_s t\right).$$

Now,

$$\omega_{FM} = \left(\omega_c + k_{fs}\hat{V}_s\cos\omega_s t\right) \Rightarrow (\omega_{FM})_{max} = \left(\omega_c + k_{fs}\hat{V}_s\right).$$

$$(\omega_{FS})_{min} = \left(\omega_c - k_{fs}\hat{V}_s\right).$$

Thus, we get

$$(\omega_{FM}) - \omega_c = k_{fs}\hat{V}_s = \omega_c - (\omega_{FM})_{min}. \qquad (1.28)$$

Hence

$$\Delta\omega = k_{fs}\hat{V}_s \Rightarrow 2\pi\Delta f = k_{fs}\hat{V}_s \Rightarrow \Delta f = \left(k_{fs}/2\pi\right)\hat{V}_s. \qquad (1.29)$$

$$\frac{\Delta f}{f_s} = \frac{k_{fs}\hat{V}_s}{2\pi f_s} = \frac{k_{fs}\hat{V}_s}{\omega_s} = m_F = \text{Frequency modulation index}$$

$$= \frac{\text{Maximum deviation of frequency from that of the carrier}}{\text{Signal frequency}} \qquad (1.30)$$

$$\therefore\ v_{FM} = \hat{V}_c \cos\left(\omega_c t + \frac{k_{fc}\hat{V}_s}{\omega_s}\sin\omega_s t\right) = \hat{V}_c\cos(\omega_c t + m_F\sin\omega_s t). \qquad (1.31)$$

(1.31) is the frequency-modulated wave and it shows that it has the same peak value as that of the carrier but the frequency of the FM has maxima and minima values at equal frequency interval w.r.t. the carrier frequency ω_c. In Fig. 1.11(a), (b) and (c), we have shown the signal voltage, the carrier wave and FM-wave respectively.

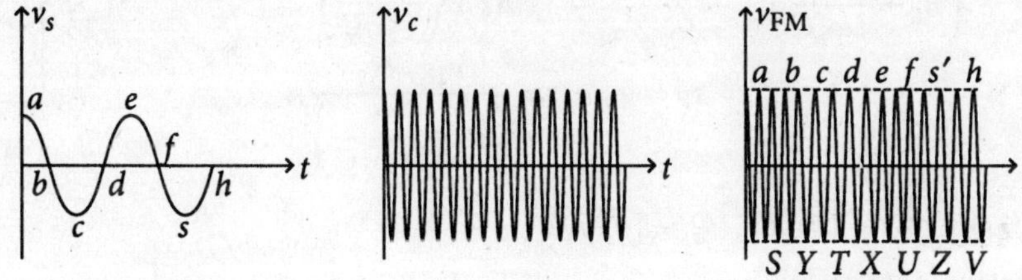

Signal voltage (a) | Carrier wave (b) | FM-wave (c)

Fig. 1.11

If Fig. 1.11(a), the signal voltage is shown as a cosine wave starting from a through b, c, d, e, f, g, h; the points a and e are the crests and c, g are the trough. In Fig. 1.11(b), the carrier wave of high frequency is shown. It is also a cosine wave. When the signal wave modulates the carrier, the frequency of the modulated wave changes from ω_c to $\omega_c + \Delta\omega$ (max) and from ω_c down to $(\Delta\omega - \omega_c)$ (min).

Starting from a peak value of the signal voltage, the voltage strength changes co-sinusoidally with time as represented in Fig. 1.11(a) down to c (negative peak) and rises again to e via the point d. In this way a wave is represented and its mathematical expression is

$$v_s = \hat{V}_s \cos \omega_s t. \tag{1.32}$$

The third Fig. (c) is the FM-wave which shows variation of frequency of the carrier by the modulating frequency which is written as

$$\omega_{\text{FM}} = \omega_c + k_{fs} \hat{V}_s \cos \omega_s t. \tag{1.33}$$

So from (1.33) lowest value of $\omega_{\text{FM}} = \omega_c$ but it increases or decreases which means $\omega_{\text{FM}} \geq \omega_c$ or $\leq \omega_c$ periodically as shown in Fig. 1.11(c).

1.12 Actual FM-wave Generation in Practice

The frequency modulator is more complex than the amplitude modulator.

The basic principle involved in this process of modulation is to vary the reactance of a circuit in accordance with the variation of the modulating voltage frequency.

Such a variable reactance circuit is placed across the tank circuit of an L-C tuned oscillator (either Hartley or Colpitts oscillator), the effective frequency of the oscillator will vary in accordance with that of the signal voltage, thus producing FM-wave. Such a modulator is called a *Reactance modulator.*

Theory of reactance modulator. In Fig. 1.12, we have shown a typical reactance modulator using an FET; it will be better if we use an FET instead of a BJT.

The modulator reactance appears across the output terminals A, B, the value being controlled by the modulating voltage applied across R_g between Gate and the source of the FET (N-channel).

Let us now see how this simple circuit produces a variable reactance across A, B.

The signal voltage is a cosine wave of frequency ω_s injected between the G and S of the FET. The resulting gate current i_G is given by

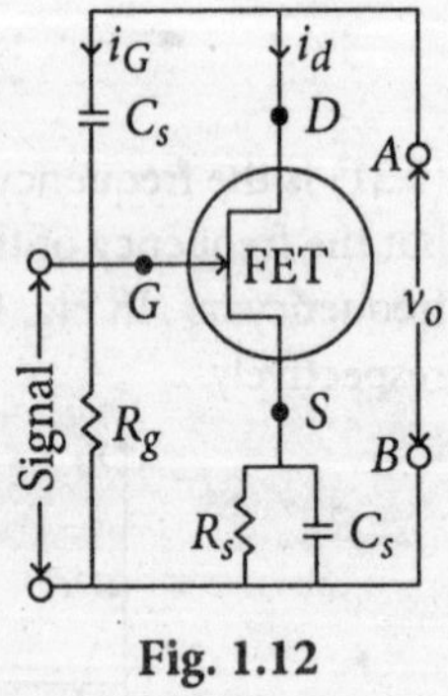

Fig. 1.12

$$i_G = \frac{V_0}{\left(R_g + \frac{1}{j\omega_s C}\right)} \Rightarrow v_{gs} = i_G R_g = \frac{v_0 R_g}{R_g + \left(\frac{1}{j\omega_s C}\right)} \tag{1.34}$$

$$\Rightarrow i_d = g_m v_{gs} = \frac{v_0 g_m R_g}{\left(R_g + \frac{1}{j\omega_s C}\right)}.$$

$\because \frac{1}{\omega_s C} \gg R_g$, we would get $i_d = j\omega_s C g_m R_g v_0$.

$\therefore$ output reactance is given by

$$z_0 = \left(\frac{v}{i_d}\right) = \left(\frac{1}{j\omega_s C R_g g_m}\right) = \frac{1}{j\omega_s C_e}, \tag{1.35}$$

where C_e in (1.35) is the capacitance $C_e = C R_g g_m$ and it appears across the terminals AB. The modulating voltage applied between the gate and the source varies g_m producing variation of C_e in accordance with the audio-signal. This varying C_e and hence, the varying reactance is placed across the tank circuit of Colpitts oscillator ultimately producing the frequency-modulated wave across L_2 in Fig. 1.13.

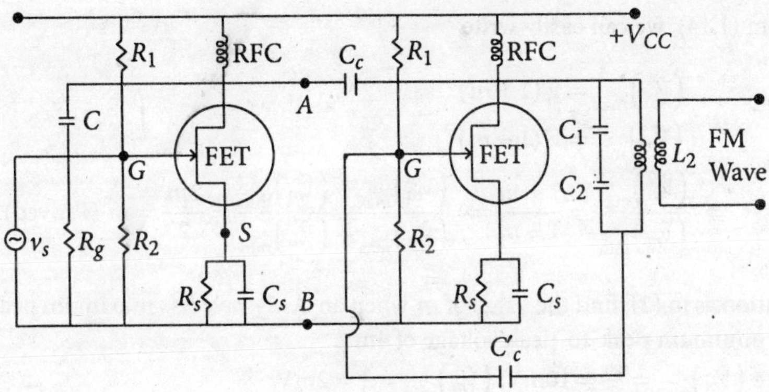

Fig. 1.13

Based on the above consideration, a complete circuit diagram is as shown in Fig. 1.13. We have used two FETs, both are N-channel FETs. The first FET is the reactance modulator whose varying reactance due to modulating voltage $v_s = \hat{V}_s \cos \omega_s t$ is controlled by v_{gs} and is placed across the tank of the Colpitts oscillator. The second FET acts as an amplifier meant for maintenance of oscillation.

Advantages of FM-wave

1. Power transmitted along with FM is not dependent on the depth of modulation m_F but is AM, it very much depends on m.
2. All the transmitted power in FM is associated with the wave itself, but in AM transmission, small amount of energy transmitted is contained in the two sidebands which actually carry the message or intelligence.
3. High transmission efficiency.
4. Noiseless reception, hence in TV, video-signal is transmitted through AM but audio-signal is transmitted through FM.
5. Better audio quality received.

1.13 Solved Numerical Problems

1. The modulating frequency is 15kHz, carrier frequency is 800kHz. What is the frequency spectrum of the AM-wave?

   Solution: The frequency spectrum of the AM-wave consists of three waves of frequencies f_c, $(f_c - f_s)$ and $(f_c + f_s)$. So the frequencies are $(800 - 15), 800, (800 + 15)$, i.e., 785kHz, 800kHz, 815kHz.

2. The modulation index m can be expressed as the fraction

$$m = \frac{(\hat{V}_m)_{\max} - (\hat{V}_m)_{\min}}{(\hat{V}_m)_{\max} + (\hat{V}_m)_{\min}}.$$

   Prove the proposition.

Solution: From (1.14), we can easily write

$$\left(\hat{V}_m\right)_{\max} = \hat{V}_c(1+m)$$
$$\left(\hat{V}_m\right)_{\min} = \hat{V}_c(1-m)$$
$$\Rightarrow \frac{\left(\hat{V}_m\right)_{\max}}{\left(\hat{V}_m\right)_{\min}} = \frac{1+m}{1-m} \Rightarrow \frac{\left(\hat{V}_m\right)_{\max} - \left(\hat{V}_m\right)_{\min}}{\left(\hat{V}_m\right)_{\max} + \left(\hat{V}_m\right)_{\min}} = \frac{2m}{2} = m \text{ (Proved).}$$

3. Using the relation as in (2), find the value of m when an AM-wave has maximum peak-to-peak voltage of 20mV and minimum peak-to-peak voltage of 4mV.
Solution: Here $\left(\hat{V}_m\right)_{\max} = \frac{20}{2} = 10\text{mV}; \left(\hat{V}_m\right)_{\min} = \frac{4}{2} = 2\text{mV}.$

$$\therefore\ m = \frac{(10-2)}{(10+2)} = \frac{2}{3} = 66.7\%.$$

4. Find fraction of the total power carried by the sidebands if the degree of modulation is 50 %.
Solution: P_s (power in the sidebands) $= P_t \dfrac{\Sigma m_l^2}{2+\Sigma m_l^2}$, here $m_l = 50\% = \frac{1}{2}$.

$$\therefore\ P_s = P_t \times \frac{1/4}{2+\frac{1}{4}} = P_t \times \frac{1}{9} \Rightarrow \frac{P_s}{P_t} = 11.1\%.$$

5. $\dfrac{P_c}{P_t} = \dfrac{I_c^2}{I_{\text{AM}}^2} = \dfrac{1}{\left(1+\frac{m^2}{2}\right)} \Rightarrow \dfrac{8^2}{9.1^2} = \dfrac{1}{\left(1+\frac{m^2}{2}\right)} \Rightarrow 1+\dfrac{m^2}{2} = \dfrac{9.1^2}{8^2} \Rightarrow m^2 = 2\left(\dfrac{9.1^2}{8^2} - 1\right) \Rightarrow m = 59\%.$

6. The antenna current of an AM transmitter for carrier wave only is 10A. What will be the antenna current when modulation is 50 %?
Solution: $\dfrac{P_t}{P_c} = \dfrac{I_t^2}{I_c^2} = \left(1+\dfrac{m^2}{2}\right) = \left(1+\dfrac{1}{8}\right) \Rightarrow I_t^2 = 10^2 \times \dfrac{9}{2} \Rightarrow I_t = 10.60.$

7. An AM transmitter supplies 8kW to the antenna when unmodulated. Determine the total power radiated by the transmitter when modulated to 60%.
Solution: $\dfrac{P_t}{P_c} = 1+\dfrac{m^2}{2} \Rightarrow P_t = 8\left(1+\frac{0.6^2}{2}\right) = 9.44\text{kW}.$

8. An FM transmitter sends out a carrier of frequency 100×10^6Hz modulated by a 15kHz sinusoidal signal. The maximum deviation of frequency is 30kHz. Find (a) m_F (b) three significant pairs of side frequencies (c) the channel width required.
Solution: (a) $m_F = \dfrac{30 \times 10^3}{15 \times 10^3} = 2.$
(b) $\omega_{\text{FM}} = (\omega_c + m_F \omega_s \cos\omega_s t) \Rightarrow f_{\text{FM}} = (f_c + m_F f_s \cos\omega_s t)$.
The three pairs of significant side frequencies are $(f_c + f_s)$, $(f_c + 2f_s)$ and $(f_c + 3f_s)$.
(c) The channel width required to accommodate the significant third pair $= 2 \times 3 \times 15 = 90 \times 10^3$Hz.

1.14 Short Theory Questions with Answers

1. What is frequency modulation?
Ans. It is the modification of the frequency of the carrier by the modulating signal voltage.

2. Amplitude modulated transmission is beset with noise. Why?

Ans. In amplitude-modulation, the amplitude of the carrier is modified by the signal. This variation of amplitude may be due to actual signal or spurious signal, also called noise due to man-made cause or natural cause. The receiver fails to recognise which variation in amplitude in AM-wave is due to actual signal and which is not. So AM transmission is noisy.

3. Define modulation and name the different types of modulation.

Ans. The modification of some characteristics of the carrier by the audio-signal wave is known as modulation.

A carrier wave is represented by $v_c = \hat{V}_c \cos(\omega_c t + \phi)$.

So there are three characteristics of the carrier which can be modified by the signal voltage and accordingly we have three types of modulations: (a) $\hat{V}_c$ is varied — amplitude-modulation (AM). (b) ω_c is varied — frequency modulation (FM). (c) ϕ is varied — phase modulation (PM).

4. What is under-modulation and over-modulation?

Ans. There is a term called modulation index defined as

$$m = \frac{\text{Maximum change in the amplitude of the carrier}}{\text{Original carrier amplitude}}.$$

Suppose, carrier wave amplitude is A. It is modulated by audio wave of amplitude $\frac{1}{2}A$, A and $\frac{3}{2}A$. Then

$m\,\%$ is $\frac{1.5A-A}{A} \times 100 = 50\,\%$, under-modulation. $m\,\%$ is $\frac{2A-A}{A} \times 100 = 100\,\%$, optimum modulation.

$m\,\%$ is $\frac{(5/2)A-A}{A} \times 100 = 150\,\%$, over-modulation.

Neither under-modulation nor over-modulation is desired in AM transmission.

5. Show that the total power for a fully amplitude-modulated wave is 1.5 times the unmodulated carrier power.

Ans. Prove the power relation as in (1.21) $\langle P\rangle = P_c\left(1+\frac{m^2}{2}\right)$.

For $m = 100\,\%$, $\langle P\rangle = P_c\left(1+\frac{1}{2}\right) = 1.5P_c$.

6. An AM broadcast transmitter radiates 10kW of unmodulated carrier and 12.5kW of total power when the carrier is modulated by a sinusoidal message signal. Find the modulation index. If another sinusoidal message signal modulates the carrier simultaneously to a depth of 60 %, what will be the total power radiated?

Ans. From (1.21), we get

$$\frac{P_t}{P_c} = \frac{\langle P\rangle}{P_c} = \left(1+\frac{\Sigma m_l^2}{2}\right) \Rightarrow \frac{12.5}{10} = \left(1+\frac{m^2}{2}\right) \Rightarrow m^2 = 2\left(\frac{12.5}{10}-1\right) \Rightarrow m = 0.707.$$

When there are two message signals simultaneously acting on the carrier, then

$$\frac{\langle P\rangle}{P_c} = \frac{P_t}{P_c} = \left(1+\frac{\Sigma m_l^2}{2}\right) \Rightarrow P_t = P_c\left[1+\tfrac{1}{2}\left(m_1^2+m_2^2\right)\right] = 10\left(1+\frac{0.707^2+0.6^2}{2}\right) = 14.3\text{kW}.$$

7. The peak-to-peak value of an AM voltage has maximum value of 5V and minimum value of 1V. Find the depth of modulation and the amplitude of the unmodulated carrier.

Solution: $m = \dfrac{(\hat{V}_m)_{max} - (\hat{V}_m)_{min}}{(\hat{V}_m)_{max} + (\hat{V}_m)_{min}} = \dfrac{(5/2)-(1/2)}{(5/2)+(1/2)} = 0.66.$

$$(V_m)_{max} = \frac{5}{2} = \hat{V}_c(1+m), \quad \hat{V}_c = \frac{2.5}{1+0.66} = 1.51\text{V}.$$

8. 1kW carrier is amplitude-modulated by an audio-signal. If the modulation is 60 %, find total power with the modulated wave.

   Solution: $\dfrac{P_t}{P_c} = \left(1+\dfrac{m^2}{2}\right) \Rightarrow P_t = 1\left(1+\dfrac{0.6^2}{2}\right) = 1.18\text{kW}.$

9. What is an amplitude modulator?

   Ans. An electronic circuit that produces an AM-wave is called an amplitude modulator.

10. What are the limitations of AM transmissions?

    Ans. (a) Poor efficiency (b) noisy reception (c) smaller operating range and (d) poor audio-quality.

1.15 Review Questions

1. High frequency waves are employed as carrier. Explain.
2. In television communication audio-signal is transmitted through FM but video-signal is transmitted through AM. Why?
3. In AM transmission modulation index requires to be 100 % (optimum). Why?
4. AM gives noisy reception. Why?
5. FM gives noiseless reception. Explain.
6. Though FM gives noiseless reception, in radio communication FM is rarely used. Why?
7. In the message signal, the carrier wave and the modulated wave are represented by

   $$v_s = \hat{V}_s \cos\omega_s t \qquad \text{signal wave}$$
   $$v_c = \hat{V}_c \cos(\omega_c t + \phi) \qquad \text{carrier wave}$$
   $$v_m = \hat{V}_c \cos(\omega_c t + \phi_s \sin\omega_s t) \qquad \text{modulated wave.}$$

   What is the process of modulation utilised?
8. What are the disadvantages of FM transmission?

   [*Hints.* (a) Wider channel needed (b) Processors more complex and expensive (c) Smaller area coverage.]

1.16 MCQs

1. In India mostly we use for radio communication (a) AM transmission (b) FM transmission (c) PM transmission.
2. In India video-signal is transmitted by (a) FM (b) AM (c) PM.
3. A carrier of peak voltage 0.1V and frequency 150kHz is amplitude-modulated by a message signal of peak voltage 10V and frequency 2kHz.

   A. The frequencies in the output spectrum are: (i) 148kHz (ii) 150kHz (iii) 152kHz

   (i) 4kHz (ii) 150kHz (iii) 154kHz

   (i) 146kHz (ii) 150kHz (iii) 156kHz.

B. The peak values of output components if $m = 50\%$ and voltage gain $A = 50$.
(i) 1.25V (ii) 5V (iii) 1.25V
(i) 2.5V (ii) 5V (iii) 2.5V.

[*Hints.* Output Frequencies
A: $(f_c - f_s)$ f_c $(f_c + f_s)$.
B: $V_{out} = A \times 10 = \frac{1}{2} \times 10 = 5V = \hat{V}_c$.
Peak values of output components (i) $\left(m\hat{V}_c/2\right)$ (ii) $\hat{V}_c$ (iii) $\left(m\hat{V}_c/2\right)$.

4. The power carried by the modulated P_t for greater value of m is (a) greater (b) lesser and (c) no change.
5. An AM signal has a maximum peak-to-peak voltage of 10V and minimum peak-to-peak voltage is 0. The modulation index percentage is (a) 20 % (b) 100 % and (c) 50 %.
6. The minimum length for the transmitting antenna of a radio station whose frequency is 1080kHz will be (a) 277.7m (b) 27.77m (c) 300.7m.
7. Modulation is necessary because: (a) Sound waves require a medium for propagation (b) Audio waves require large antenna for their transmission (c) Message signal is contained in the sidebands in the output.
8. If f_c be the frequency of the carrier, f_s be the frequency of the modulating wave, then (a) $f_c = f_s$ (b) $f_c < f_s$ (c) $f_c > f_s$.
9. In an AM, the characteristic of the carrier is modified: (a) w.r.t. its frequency (b) w.r.t. its amplitude (c) w.r.t. the frequency of the modulating signal.
10. If m represents the modulation index in AM, then its value is kept at: (a) 50 %, (b) 60 % (c) 100 % for distortionless transmission.
11. If P_c be the power of the carrier, P_s be the power in the sidebands, then (a) $P_s > P_c$ (b) $P_s = P_c$ (c) $P_s < P_c$.
12. In an FM, there is (a) no existence of any sideband (b) a number of pairs of sideband (c) one sideband.
13. A 80MHz carrier is frequency-modulated, the modulation index being 4. The message signal has a frequency of 10kHz. The maximum frequency deviator is (a) 40kHz (b) 45kHz (c) 50kHz.
14. If m_F be the modulation index in an FM, then (a) $0 < m_F < 1$ (b) $0 < m_f < \infty$ (c) $m_F > 1$.

1.17 Random Signals, Random Variables

We have so long discussed signals or processes all deterministic in nature; there are also signals or processes which are indeterministic like speech signal, temperature of air, air pressure, thermal noise generated in a resistor, etc. They are all referred to as Random phenomena and deserve special attention, Random signals are generated in random fashions and cannot be predicted ahead of time. They originate from man-made or natural sources, temporal or spatial, can be described statistically to characterise one's degree of knowledge or ignorance about the randomness. Detailed discussion is given in statistical description of random process (also called *stochastic processes*).

1.18 Demodulation of Radio Signal

In previous articles of the present chapter, we discussed about what is modulation in both AM and FM transmission, its need in the process of transmission, etc. Now, it is time for taking up discussion on *demodulation* and we will wind up our discussion on analog signal processing and transmission with the introduction of superheat receiving set.

What is demodulation? We have learnt that modulation is the process of mixing a modulating wave (audio) with a high frequency carrier wave. Why? To transmit the audio effectively over a long distance as otherwise an unwisely long antenna system needs be used in the transmission process. The process of de-mixing or separation of audio from the carrier is called *demodulation*. There are three correlated terms in this connection: (a) modulating wave (b) carrier wave (c) modulated wave.

(a) Modulating wave refers to the low frequency audio wave (30Hz–30kHz).

(b) Carrier wave high frequency 3-30MHz for all types of communication; very high frequency for TV, FM, radar, etc., frequency range 30-300MHz.

(c) Modulated wave refers to the resultant wave after the audio is mixed with the carrier. This modulated wave of high frequency is transmitted by the different broadcasting stations in the form of audio wave enveloping the carrier wave.

1.19 Radio Communication System: Radio Transmitter and Receiver

The transmitter sending radio signals and the receiver receiving them are combined to form what is known as a *transceiver* or *transreceiver.*

The complete radio communication system is schematically represented by shown in block diagram [Fig. 1.14].

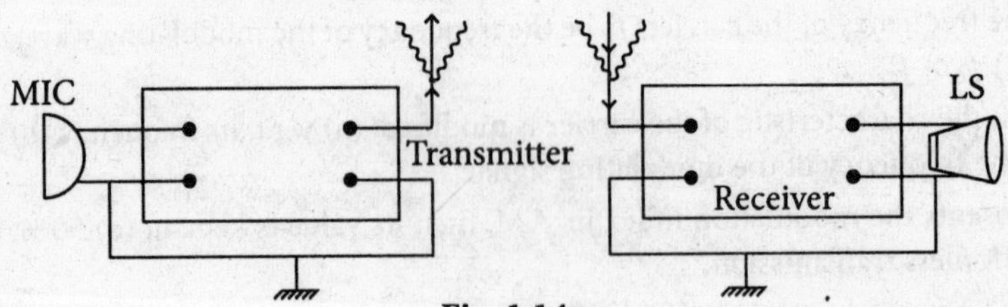

Fig. 1.14

The left-hand portion of the communication system, called the transmitting portion for either amplitude-modulated wave or frequency-modulated wave is described in Fig. 1.10 and Fig. 1.13 respectively.

Here we will be describing the receiving section first for amplitude-modulated signal and then for FM signal.

Amplitude-modulated signal receiver. The first element to receive the amplitude-modulated signal is the antenna system of the receiver. There are two sets of coils in the receiver, one set is referred to as antenna coil, the other set is called the oscillator, both of which include a variable capacitor, also called a gang-condenser — one for selective section of the input signal and the other for producing the local oscillation frequency. The complete receiving components of the radio-receiver are as shown in block diagram (1.15).

Before the introduction of superhet (superb) receiver, there was the straight receiver which had several limitations of which the most important one is that in the straight receiver set, different modulated waves sent by different radio stations would require separate tuning paths and the audio amplifier to get the required frequency response is costly and bulky. This difficulty is overcome in the new design of the receiver first suggested by Major A. Armstrong and it is the superheterodyne receiver, whose block diagram representation is shown in Fig. 1.15. The various stages are described as given below:

Antenna system. The first element at the entry end of the receiver is the antenna or aerial, external or indoor connected to a set of coils, called antenna coils. Each of these coils is connected at one end to one half of a variable capacitor C_1, the other half of the capacitor C_2, connected to one end of another set of coils, called oscillator coil; the capacitors C_1 and C_2 are mounted on the same shaft.

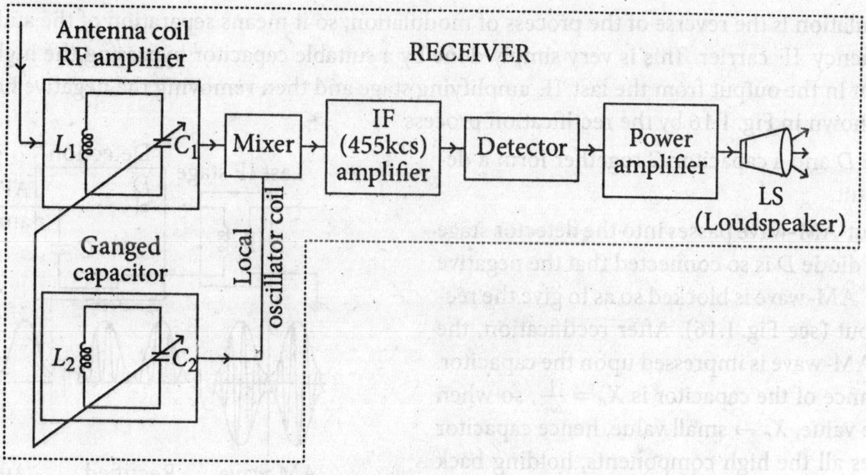

Fig. 1.15

Selector stage and RF ***amplifier.*** The AM-waves transmitted by the radio stations are first intercepted by the antenna and they pass through the primary of the antenna coil, the secondary being tuned by the capacitor C_1. Due to the phenomenon of resonance of different frequencies, one frequency is selectively transferred from the primary to the secondary. This stage is followed by RF amplifier stage.

Local oscillator stage. It is basically an LC oscillator which produces 455kc/s higher than the selected radio frequency. One half of the ganged condenser C_2 is connected to one end of the oscillator coil L_2 (L_2-C_2 form the tank circuit of the local oscillator).

RF amplifier output and the local oscillator output are made to beat so as to always produce a resultant frequency 455kc/s, called the IF (intermediate frequency)

IF = local oscillator frequency – selected radio frequency = 455kc/s (always).

Thus, in the mixer oscillator stage, as a result of beating between the selected radio frequency and local oscillator frequency, we always get the IF equal to 455kc/s whatever may be the incoming wave frequency. This phenomenon of conversion of local oscillator frequency beating with the selected radio frequency always producing IF as 455kc/s is known as super heterodyning.

A radio receiver is usually required to tune over 550kc/s to 1600kc/s; so the frequency of the local oscillator must vary from (550 + 455) = 1005 to (1600 + 455) = 2055kc/s, because of IF requirement being equal to 455kc/s always.

Table 1.1: gives the required frequencies of the local oscillator to give the IF

Frequency of selected radio frequency kHz	Local oscillator frequency kHz	IF frequency kHz
550	1005	455
940	1395	455
1200	1655	455
1600	2055	455

455kHz is internationally accepted intermediate frequency.

IF ***amplifier stages.*** There are usually three IF amplifier stages containing IF transformers all tuned to a single frequency of 455kc/s which is the internationally accepted intermediate frequency. Intermediate frequency though of lower frequency is also a radio frequency and it contains all the audio information.

Detector stage. Detection refers to the combined process of demodulation followed by rectification.

Demodulation is the reverse of the process of modulation, so it means separation of the audio from the high frequency IF carrier. This is very simply done by a suitable capacitor bypassing the high frequency component in the output from the last IF amplifying stage and then removing the negative half of the IF signal as shown in Fig. 1.16 by the rectification process.

A diode D and a capacitor C together form a detector circuit.

The input AM-wave passes into the detector stage where the diode D is so connected that the negative half of the AM-wave is blocked so as to give the rectified output (see Fig. 1.16). After rectification, the rectified AM-wave is impressed upon the capacitor. The reactance of the capacitor is $X_c = \frac{1}{\omega c}$, so when $\omega \to$ large value, $X_c \to$ small value, hence capacitor C bypasses all the high components, holding back the low frequency component. So the audio wave comes out of the detector stage.

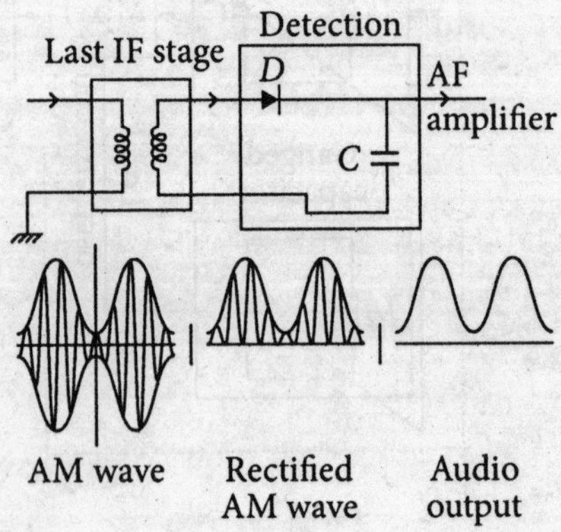

Fig. 1.16

Chapter 2

Digital Signal and Its Communication

2.1 Introduction

In any scientific discussion we confront with three basic questions — *what, why* and *how.* The discussion is said to be complete if we can provide answers to all these queries. The present discussion also centres round answering these three queries: *What* is digital signal? *Why* to consider the digital signal? *How* to get the digital signal?

2.2 What is a Digital Signal?

We often make the statement as: All digital signals are *discrete* but *all discrete signals* are not digital. A signal may be discrete w.r.t. time but it necessarily does not mean a digital signal. A signal may be discrete w.r.t. amplitude but in this case also it does not necessarily mean a digital signal. So *what is it, then?*

A signal which *is discrete in time and also discrete in amplitude as well* is referred to as a *digital signal.*

It is to remember that most of the signals we come across are in reality analog by nature and there is standard and established technique of transmission of analog signals. We have discussed the processing of the analog signals in the previous chapter. Of course, the data or information stored in a computer are digital in form.

Since analog signals are very much there and mechanism of its communication is well known, why then digital signal and its processing?

2.3 Digital Signal Processing and Its Usefulness

Actually, however, the digital processing and transmission was recognised as a separate field by itself towards the beginning of 1960s, and since then significant developments and breakthroughs in both theory and applications took place. A very important point to remember is that an analog signal exists for each and very time instant requiring energy consumption as a continuous function of time but in digital signal, it exists only for discrete and desired instants; hence, here energy consumption function is also discrete time function.

Digital communication system has been introduced as replacement of analog communication system, because of several reasons:

1. The digital communication system is very much immune to channel noise which offsets the relative simplicity of analog communication system.
2. The digital communication system is highly reliable, has high fidelity, security and low error rate.

3. The digital communication system permits the use of computers and LSI circuits; it can accommodate various sources of information in a flexible and compatible manner.
4. The most important point in favour of the digital communication system is that the rate of data transfer volume is much higher than that in analog communication.
5. In case of digital signal, it can be stored in magnetic tape or disk without any further loss of fidelity beyond that introduced by A/D conversion process. As a consequence, digital signals become transportable and can be processed off-line in a remote laboratory.
6. It is usually very difficult to perform some mathematical operations on signals in analog form but these operations can be done very comfortably on digital signals.
7. Digital signal processing in some cases becomes cheaper than corresponding counterpart, maybe due to the fact that the digital hardware is cheaper.

As a result of these advantages, digital signal processing has been applied in practical systems covering a wide range of disciplines as in speech processing; transmission in telephone channel; in image processing and transmission; in seismology and geo-physical prospecting.

2.4 Essential Schematics of Digital Communication System

Input signal is analog in nature, designated as message signal and stored in a space, called *data source* (DS) [Fig. 2.1].

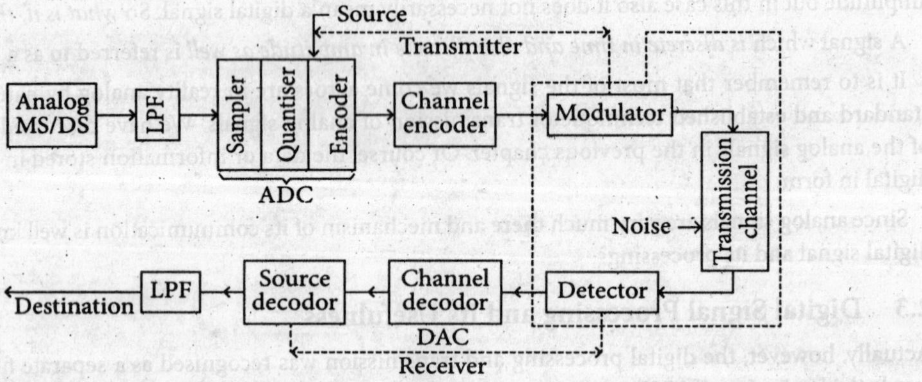

Fig. 2.1

The schematics of the digital communication system have been drawn in Fig. 2.1, the details of the functioning of each unit of the system will be given later.

The digital signal processing starts with analog message source. This analog message signal is passed through ADC for conversion of analog signal to digital signal. The digitisation process starts with SAMPLER, sampling the analog signal followed by quantisation of the sample values and ends with the ENCODER. The whole story of conversion is written in a device, called ADC (analog to digital converter).

The output of ADC consists of data-stream of binary numbers giving rise to signal digital in form. So the digital conversion of analog input signal consists of three steps [Fig. 2.1].

1. Conversion of analog input into a digital form as it passes through ADC.
2. Processing of the digital version through a digital processor according to desired algorithm.
3. Reconversion of the processed digital signal back into an analog form of desired frequency by DAC and low pass filter (LPF) to make the message ready for despatch to the destination.

ADC *device.*

This compartment has three parts.

1. *Sampler.* The sampler samples out the analog signal and produces the discrete time signal.
2. *Quantiser.* It produces discreteness in amplitude of the signal. So the sampling followed by quantisation produces signal discrete in time as well as discrete in amplitude. *Such a signal is called the digital version of the input analog signal.*
3. Finally, *the encoder* which codes the digital output into streams of binary numbers that come out as the finished product of the device, is called ADC. So, here again we are raising the basic question: What is a digital signal?

The answer is: **Any signal which when converted into one having discreteness w.r.t. amplitude and time, is referred to as a digital signal.**

SAMPLER samples out the input analog signal

The SAMPLER does the function of sampling the input analog signal to produce a discrete time signal, mathematically represented by

$$x(n) = x_a(nT), \; -\infty < n < \infty, \quad (2.1)$$

where $x(n)$ is the discrete time signal obtained by taking samples of the signal $x_a(t)$ for every $t = T$ sec, T being the sampling time. The resolution in the AD conversion is decided by the sampling period T or by its frequency $f_s = 1/T$.

How to decide the sampling rate? It is decided w.r.t. a theorem called *Nyquist theorem* which states that for accurate A/D conversion, the sampling frequency f_s must satisfy the inequality

$$f_s \geq 2f_m, \quad (2.2)$$

where f_m is the maximum frequency in the input analog signal. So the minimum sampling rate is $(f_s)_{\min} = 2f_m$ and this rate of sampling is known as *Nyquist sampling rate.* So, for precision of sampling we must know what is the maximum frequency present and hence, determine the Nyquist sampling frequency. For example, in speech highest frequency used is 3kHz; whereas in TV signal, the highest frequency is 5MHz.

Let us take an analog input in the time-domain $x_a(t)$ represented by in Fig. 2.2.

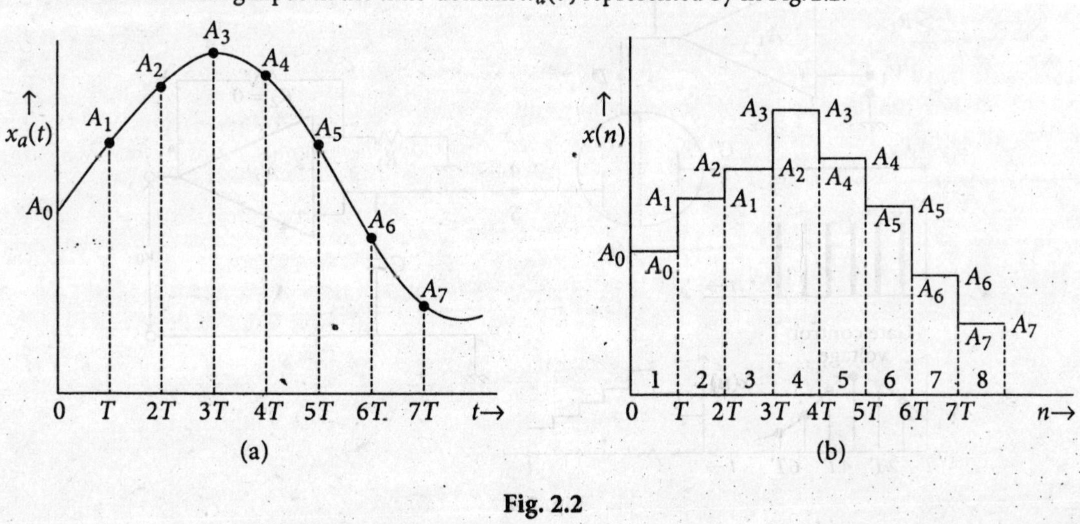

Fig. 2.2

To satisfy the Nyquist condition for accurate A/D conversion, the analog signal is to be passed through a low pass filter before sampling so that required range of frequency is allowed by the filter to enter into the sampler. The process of sampling gives rise to discrete time signal. Sampling is followed by quantisation in the quantizer, the result is a digital signal which is discrete in time as well as in amplitude represented by the staircase type signal shown in Fig. 2.2(b).

The discrete form of the input analog signal is accomplished through an electronic circuit in the ADC device. This electronic circuit is known as SAMPLE-HOLD-Circuit (S/H circuit) explained as below.

SAMPLE-HOLD Circuit (S/H circuit)

The basic principle of the S/H circuit is explained w.r.t. Fig. 2.3 producing the staircase type of signal out of the input analog signal. The implementing circuit details are shown in Fig. 2.4.

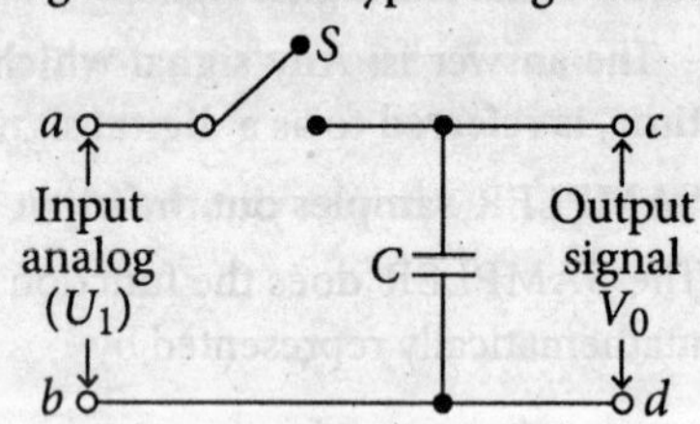

Fig. 2.3

In Fig. 2.3 S is the switch in series with the capacitor C. The switch S closes and opens at intervals of time T, called sampling interval $T = 1/f_s$. The terminals a, b are the input ports where the input analog signal of Fig. 2.2 shown as a continuous curve (a). This analog signal is sampled out at regular intervals of T secs, the sample points being A_0, A_1, A_2, A_3, etc.

The analog signal is an electric wave usually and the capacitor receives the charging voltage when the switch S closes. Almost instantaneously the capacitor charges up to the voltage corresponding to the point A_0 of the input signal. The switch now opens and it remains open until the next sampling is done thus getting the horizontal part of the staircase. The voltage in between sampling is held constant across the capacitor represented by the horizontal portions A_0A_0', A_1A_1', A_2A_2', etc. So for periodic sampling, we get at the output a staircase type of signal $A_0A_0'\ A_1A_1'\ A_2A_2' \ldots$.

Actual implementing circuit

MOSFET is a very suitable semiconducting element having switching characteristics better than those of BJT.

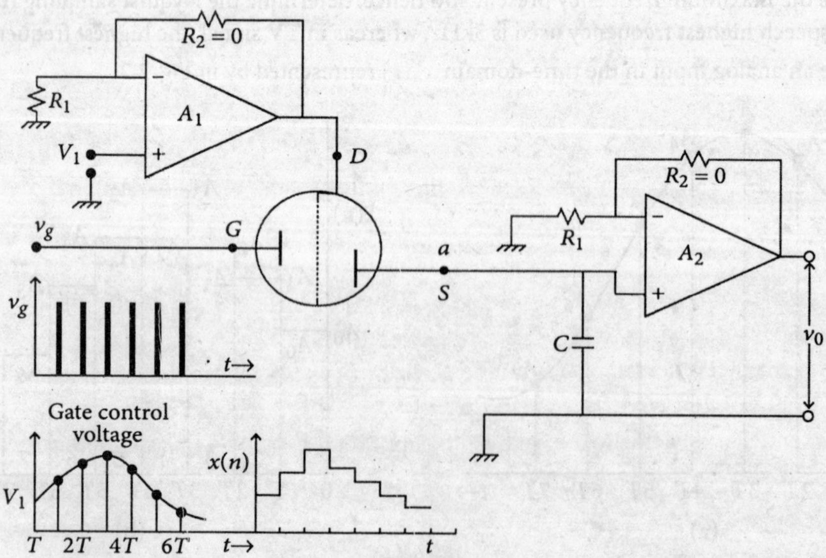

Fig. 2.4

Op-Amps A_1 and A_2 act as unit gain buffer. When a positive pulse is applied on the gate, the MOS A_1 is switched on and the capacitor is charged immediately to the applied input voltage, because the time constant of the charging circuit having the resistance as the sum of R_a (Op-Amp A_1 output resistance) and R_{DS} (Drain-source resistance) is very small. In absence of the positive pulse on the gate, the MOS-switch is thrown off; the capacitor C remains in isolation and holds the charge until the reception of positive pulse again by the gate G of the MOSFET from the gate-control voltage v_g. If the period of the control voltage is same as T, the sampling period, the capacitor will faithfully track down the input signal voltage v_i.

2.5 Discrete Time Signal — It is the Result of Sampling

The sampling process produces discrete time signal represented by sequence of numbers, called samples. Thus, $x_a(b)$ converts into $x(n)$, where $x(n)$ is the discrete time signal obtained by taking samples at every T sec. So we write: $x(n) = x_a(nT)$, where $n \in I_n$, all integers positive or negative including zero, and $x(n)$ is the nth sample value of the sequence denoted by $[x(n)]$.

Basic operations on sequence $[x(n)]$

A single-input, a single-output discrete time system operates on a sequence, called as input sequence, according to some rules to generate another sequence, called output sequence, with more desirable properties. For example, the input may be a signal corrupted by additive noise. The discrete time system is designed to remove the noise content from the input. The designing is done w.r.t. some basic operations on sequences.

Operation 1: Modulation

Two sequences $[x(n)]$ and $[y(n)]$ operate such as to produce a sequence $[w_1(n)]$ defined by

$$T\{[x(n)\ [y(n)]]\} = [w_1(n)] = [x(n)y(n)].$$

Such an operation is referred to as *modulation* and the device doing it is called a *modulator.* This operation is schematically represented by in Fig. 2.5.

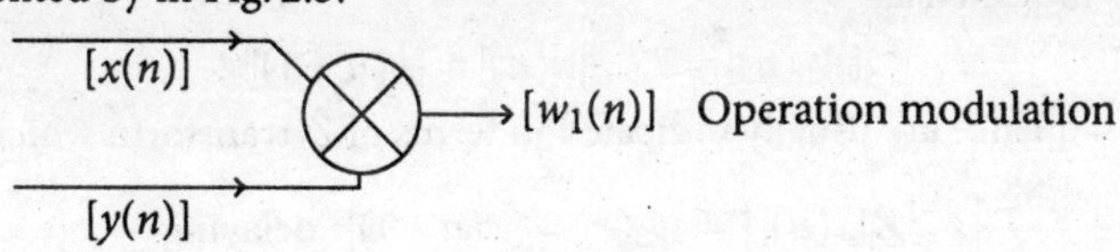

Fig. 2.5

Example 2.1. $[x(n)] = [\cos(w_1 n)]$

$[y(n)] = [\cos(w_2 n)]$.

The modulated signal sequence $[w_1(n)]$ is given by $[w_1(n)] = T\{[x(n)] \cdot [y(n)]\} = [\cos(w_1 n) \cdot \cos(w_2 n)]$

$$= \left[\tfrac{1}{2}\cos(w_1 + w_2)\, n + \tfrac{1}{2}\cos(w_1 - w_2)\, n\right].$$

Operation 2: Scalar multiplication

When each sample of a sequence $[x(n)]$ is multiplied by a scalar constant A, we get another sequence as a product represented mathematically as $[w_2(n)] = T\{A[x(n)]\} = [Ax(n)]$. Operation: scalar multiplication.

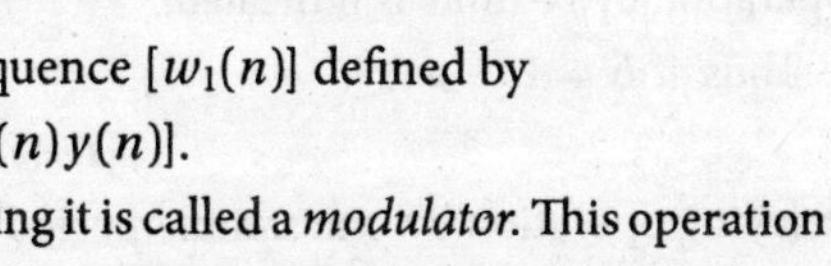

Fig. 2.6

Schematic representation is shown in Fig. 2.6.

Operation 3: Addition

This basic operation, called *addition*, refers to addition of two sequences $[x(n)]$ and $[y(n)]$ to produce a new sequence represented by $[w_3(n)]$, shown as below

$$[w_3(n)] = T\{[x(n)] + [y(n)]\} = [x(n) + y(n)].$$

It states that the sum of two sequences is another sequence $w_3(n)$ formed by the addition of the corresponding sampled values of the two sequences. The device implementing this operation is called an *adder* [Fig. 2.7].

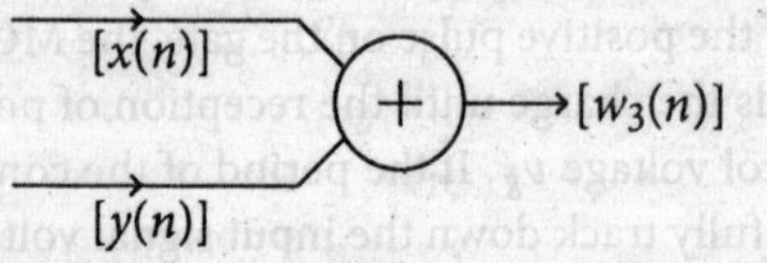

Fig. 2.7

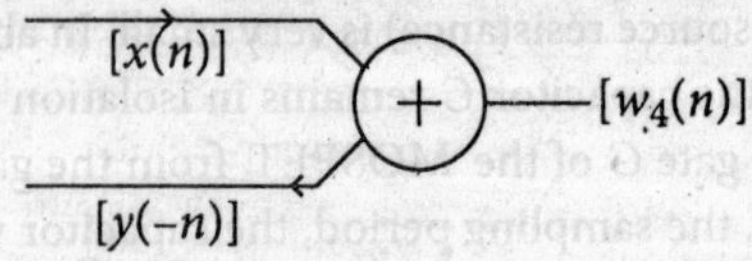

Fig. 2.8

Operation 4: Subtraction

Subtraction is equivalent to addition of $[x(n)]$ with reverse of $[y(n)]$ denoted by $[y(-n)]$. Thus,

$$\{[x(n)] - [y(n)]\} = [x(n) - y(n)] = [x(n) + y(-n)] = w_4(n).$$

This operation is schematically represented by in Fig. 2.8.

Two more basic operations are there and they depend on Z-transform which we will discuss later. Here, we simply mention the essence of the two operations.

Time shifting operation on sequence

$[w_5(n)] = T_{D-N}\{[x(n)]\} = [x(n-N)]$, it means for $N > 0$, delaying operation, but for $N < 0$, advancing operation by N-units is indicated.

Thus, if $N = 1$,

$$[w_5(n)] = T_{D-1}\{[x(n)]\} = [x(n-1)],$$

delaying operation by one sample period is indicated and it is *unit delay operation*. Advancing operation by one sample period is indicated as

$$[w_5(n)] = T_{A+1}[x(n)] = [x(n+1)].$$

These operations on sequence are usually indicated in terms of Z-transform which we will see later.

$$T\left\{z^{-1}Z[x(n)]\right\} = w_6(n) = [x(n-1)];\ \text{delaying}$$
$$T\{z\,Z[x(n)]\} = w_7(n) = [x(n+1)];\ \text{advancing.}$$

2.6 Illustrative Examples

1. Discrete-time system is represented in block diagram as shown in Fig. 2.9. Write the input-output relationship.

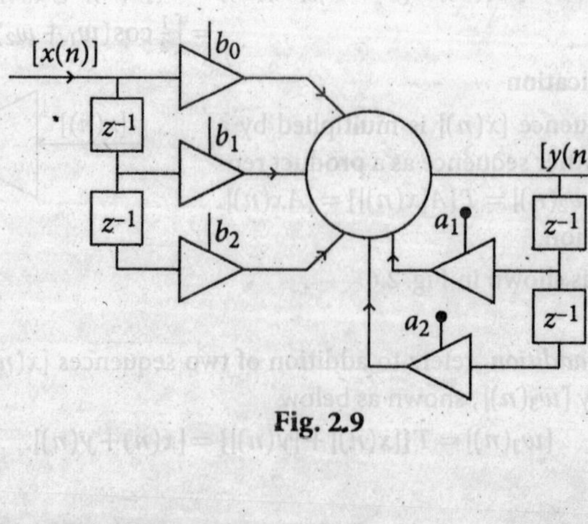

Fig. 2.9

Solution: $[y(n)] - [a_1y(n-1)] - [a_2y(n-2)] = [b_0x(n)] + [b_1x(n-1)] + [b_2x(n-2)]$
$\Rightarrow [y(n)] = b_0[x(n)] + b_1[x(n-1)] + b_2[x(n-2)] + a_1[y(n-1)] + a_2[y(n-2)]$.

2. The input signal $[x(n)]$ is defined for $-3 \leq n \leq 3$ as

$$x(n) = \begin{cases} |n|, & -3 \leq n \leq 3 \\ 0, & \text{otherwise.} \end{cases}$$

Determine the response of the following systems:

(a) $[y(n)] = [x(n)]$, identity system.

(b) $[y(n)] = \frac{1}{3}([x(n-1)] + [x(n)] + [x(n+1)])$, averaging system.

(c) $[y(n)] = [x(n+1)]$, 1 unit advancing system.

Solution: The input signal is represented by

$$[x(n)] = \begin{cases} |n|, & -3 \leq n \leq 3 \\ 0, & \text{otherwise.} \end{cases}$$

So the input sequence is represented by $[x(n)] = [\ldots, 0, 3, 2, 1, \underset{\uparrow}{0}, 1, 2, 3, 0, \ldots]$
(for $-n$) $n = 0$ (for $+n$).

(a) Output sequence is exactly same as the input sequence. The system in this case is known as the *identity system.*

(b) The sequence $[x(n)]$ represents the input signal, $x(n)$ represents the nth sample value; similarly, $[y(n)]$ is the output sequence and hence, the output discrete signal; $y(n)$ represents the nth sample value.

The input-output relation shows that the output signal is the average of three inputs at the present moment, 1 sec earlier and 1 sec later, such a system is the *averaging operation system.*

For $n = 0$, $y(0) = \frac{1}{3}[x(-1) + x(0) + x(1)] = \frac{1}{3}[1 + 0 + 1] = \frac{2}{3}$

$n = 1$, $y(1) = \frac{1}{3}[x(0) + x(1) + x(2)] = \frac{1}{3}[0 + 1 + 2] = 1$

$y(-1) = \frac{1}{3}[x(-2) + x(-1) + x(0)] = \frac{1}{3}[2 + 1 + 0] = 1$

$n = 2$, $y(2) = \frac{1}{3}[x(1) + x(2) + x(3)] = \frac{1}{3}[1 + 2 + 3] = 2$

$y(-2) = \frac{1}{3}[x(-3) + x(-2) + x(-1)] = \frac{1}{3}[3 + 2 + 1] = 2$

$n = 3$, $y(3) = \frac{1}{3}[x(2) + x(3) + x(4)] = \frac{1}{3}[2 + 3 + 0] = \frac{5}{3}$

$y(-3) = \frac{1}{3}[x(-4) + x(-3) + x(-2)] = \frac{1}{3}[0 + 3 + 2] = \frac{5}{3}$.

So, $[y(n)] = \left[\ldots, 0, \frac{5}{3}, 2, 1, \underset{\uparrow}{\frac{2}{3}}, 1, 2, \frac{5}{3}, 0, \ldots\right]$.
(n-negative) (n=0) (n-positive)

3. The system produces unit advancement of the samples; hence, we write:

$$[y(n)] = [\ldots, 0, 3, 2, 1, 0, \underset{\uparrow}{1}, 2, 3, 0, \ldots].$$

The block diagram representation of the input-output relations as in 2(a), (b) and (c) is shown below.

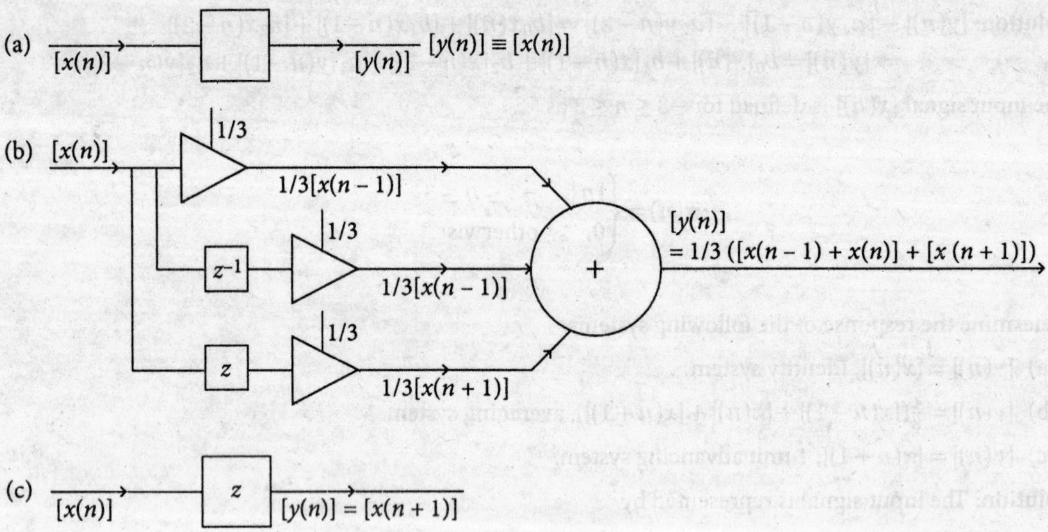

Fig. 2.10

2.7 Quantising and Encoding

With the sample and hold circuit (S/H), just discussed earlier, the sampling produces discrete time signal and quantisation finally produces signal discrete in amplitude as well.

Let us illustrate the process of quantisation with an example.

Example 2.1. *Sample an analog signal given as $x_a(f) = 0.9^t$, $t \geq 0$; every $T = 1$ sec, so that the sampling frequency, $f_s = 1$Hz. The discrete time signal is written as*

$$x(n) = \begin{cases} 0.9^n, & n \geq 0 \\ 0, & n < 0. \end{cases} \tag{2.3}$$

In Table I, we have given sample values $x(n)$ for different values of n, truncated values $x_t(n)$ and round-off values of $x_t(n)$= quantised values.

Table I

n	Notation of the sample	Sample value $x(n)$	$x_t(n)$ truncated to 1 decimal place	$x_q(n)$ (rounded-off)
0	$x(0)$	1.0	1.0	1.0
1	$x(1)$	0.9	0.9	0.9
2	$x(2)$	0.81	0.8	0.8
3	$x(3)$	0.729	0.7	0.7
4	$x(4)$	0.6561	0.7	0.7
5	$x(5)$	0.59049	0.6	0.6
6	$x(6)$	0.531441	0.5	0.5
7	$x(7)$	0.4782969	0.5	0.5
8	$x(8)$	0.43046721	0.4	0.4

Taking sample values $x(n)$ for different values of n and taking the values only to 1 significant digit, they are plotted against different n values. We get the curve as shown in Fig. 2.11. Analog signal sampled at every $T = 1$ sec, the sampling frequency is 1 Hz. Sampling simply produces signal discrete in time. It is followed by the process of quantisation which converts discrete time signal into discrete-amplitude signal.

Quantisation process. The process of quantisation will be explained w.r.t. the above diagram.

First step towards quantisation is to find the step size Δ or quantile interval. In order to find it, we follow the definition:

$$\Delta = \frac{[\{x(n)\}_{max} - \{x(n)\}_{min}]}{(L-1)}, \qquad (2.4)$$

where L is the number of quantisation levels.

$\{x(n)\}_{max}$ in the present case = 1.0.

$\{x(n)\}_{min}$ in the present case = 0.0.

If $L = 11$, we would get $\Delta = \frac{1-0}{10} = 0.1$.

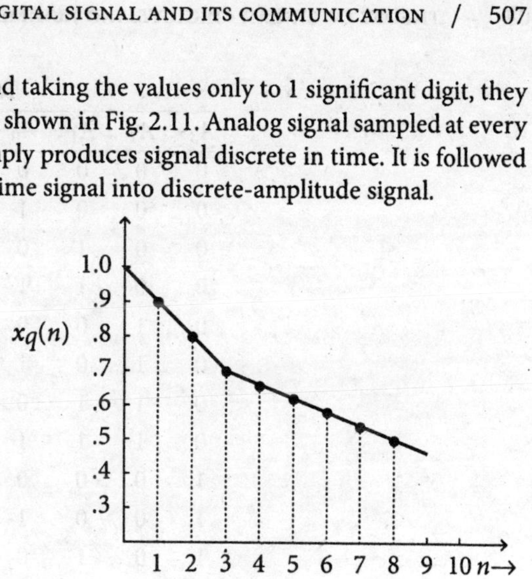

Fig. 2.11

So, the sampled value $x(n)$ lies between the range $\left\{x(n) - \frac{\Delta}{2}\right\}$ and $\left\{x(n) + \frac{\Delta}{2}\right\}$, it is taken to have quantised value $= \left\{x(n) \pm \frac{\Delta}{2}\right\}$. Thus, the quantisation error $e_q(n)$ is given by $e_q(n) = x_q(n) = \left\{x(n) \pm \frac{\Delta}{2}\right\}$.

So, the instantaneous error cannot exceed $(\Delta/2)$, half the quantisation step length.

We have then:

$n =$	0	1	2	3	4	5	6	7	8
$x_q(n) =$	1.0	0.9	0.8	0.7	0.7	0.6	0.5	0.5	0.4

The values $x_q(n)$ are the quantised values of $x(n)$ and thus the process of sampling followed by quantisation converting an analog signal $x_q(t)$ into digital signal. The result of the combined process (sampling + quantisation) is a digital signal expressed as a coding of binary bits in the compartment, called *encoder* in the ADC device.

To appreciate the coding process of $x_q(n)$, we take $\left\{x_q(n)\right\}_{max} = V_0$ (volt) = 1 volt in the present problem. The binary equivalent of the voltage as an N-bit binary word is given by

$$x_q(n) = v = k\left(2^{N-1}A_{N-1} + 2^{N-2}A_{N-2} + \cdots + 2^3A_3 + 2^2A_2 + 2^1A_1 + 2^0A_0\right),$$

where

$$k = V_0/\left(2^N - 1\right). \qquad (2.5)$$

For a 4-bit encoder

$$k = \frac{V_0}{(2^N - 1)} = \frac{1}{(2^4 - 1)} = \frac{1}{15} = 0.06$$

is the resolution of the encoder.

So, 4-bit binary equivalent of the ADC output will be expressed as shown in Table II.

Table II

A_3	A_2	A_1	A_0	$v = x(n)$	$v_q = x_q(n)$
0	0	0	0	0	0
0	0	0	1	0.06	0.1
0	0	1	0	0.12	0.1
0	0	1	1	0.18	0.2
0	1	0	0	0.24	0.2
0	1	0	1	0.30	0.3
0	1	1	0	0.36	0.3
0	1	1	1	0.42	0.4
1	0	0	0	0.48	0.5
1	0	0	1	0.54	0.5
1	0	1	0	0.60	0.6
1	0	1	1	0.66	0.7
1	1	0	0	0.72	0.7
1	1	0	1	0.78	0.8
1	1	1	0	0.84	0.8
1	1	1	1	0.90	0.9

Example 2.2. *Consider another example, where an analog signal $x_a(t)$ has been shown by a graphical plot as in* Fig. 2.12. *The analog signal has maximum sampled value $\{x(n)\}_{max} = 10V$ $\{x(n)\}_{min} = 3V$. Show graphically the sampling process; and show also in tabular form the quantised output voltage and finally the binary output of* ADC *device.*

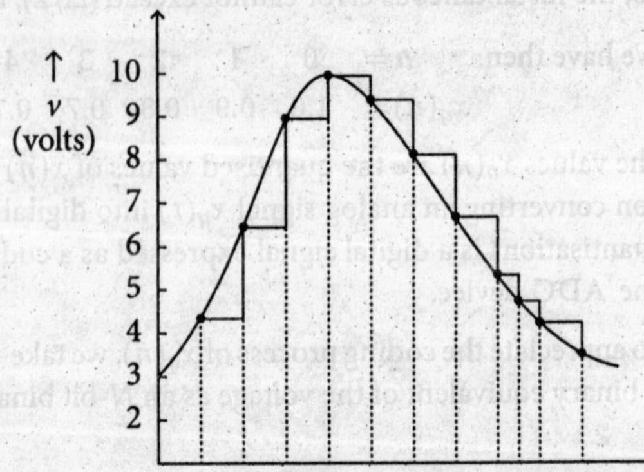

Fig. 2.12

$$\{x(n)\}_{max} = 10V = V_0$$

$$\{x(n)\}_{min} = 3V$$

$$\therefore \{x(n)\}_{max} - \{x(n)\}_{min} = 10 - 3 = 7V.$$

Let us take the number of quantisation levels $L = 8$, then the step length Δ is given by $\Delta = 7/(L-1) = 7/7 = 1V$ per level. If the sampled value be $x(n)$, then it lies within the range $\{x(n) - \Delta/2\}$ to $\{x(n) + \Delta/2\}$. Hence, $x_q(n) = \{x(n) \pm \Delta/2\}$. So, instantaneous error cannot exceed $(\Delta/2)$, the quantisation step length.

Take a 4-bit binary encoder. The value $x(n)$ can be expressed as $x(n) = v = L(2^3 A_3 + 2^2 A_2 + 2^1 A_1 + 2^0 A_0)$, where $k = V_0/(2^4 - 1) = 10/15 = 0.7$.

$\therefore x(n) = 0.7(2^3 A_3 + 2^2 A_2 + 2^1 A_1 + 2^0 A_0)$.

LSB = 0.7.

Table III

A_3	A_2	A_1	A_0	Coded value (v) $x(n)$	$x_q(n)$	Digital output
0	0	0	0	0	0	0000
0	0	0	1	0.7	0.7	0001
0	0	1	0	1.34	1.0	0010
0	0	1	1	2.01	2.0	0011
0	1	0	0	2.8	3.0	0100
0	1	0	1	3.35	4.0	0101
0	1	1	0	4.02	4.0	0110
0	1	1	1	4.69	5.0	0111
1	0	0	0	5.36	6.0	1000
1	0	0	1	6.03	6.0	1001
1	0	1	0	6.7	7.0	1010
1	0	1	1	7.77	8.0	1011
1	1	0	0	8.04	8.0	1100
1	1	0	1	8.71	9.0	1101
1	1	1	1	9.38	10.0	1111

Example 2.3 (Sampling and quantisation of sinusoidal analog signal). *Consider a sinusoidal analog signal represented by $x_a(t) = A\sin\omega_0 t$. This consideration is a useful case, because sinusoids are used test signals in A/D converter. Here, the step length Δ is given by*

$$\Delta = \frac{x(n)_{\max} - x(n)_{\min}}{2^b},$$

where the quantizer has b bits of accuracy. If the sampling rate f_s satisfies Nyquist theorem, the error in A/D conservation is only due to quantisation.

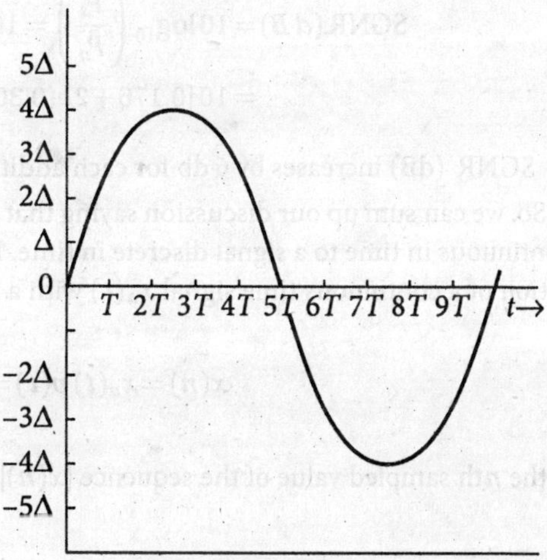

Fig. 2.13

The step length here is given by

$$\Delta = \frac{\{x(n)\}_{\max} - \{x(n)\}_{\min}}{2^b} = \frac{A-(-A)}{2^b} = \frac{2A}{2^b},$$

where the quantizer has b bits of accuracy.

In order to find the quantisation error, we can quantise analog signal $x_a(t)$ instead of discrete time signal $x(n) = x_a(nT)$. We note that the signal $x_a(t)$ is almost linear between two quantisation levels as shown in Fig. 2.13(a).

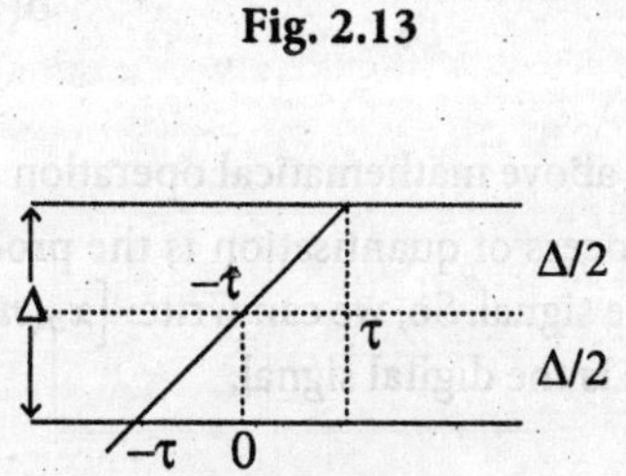

Fig. 2.13(a)

The quantisation error is $e_q(t) = x_a(t) - x_q(t)$. The signal stays in a quantisation level of width $(\Delta/2)$ for time $t = \tau$. So, the mean-square error power P_q is given by

$$P_q = \frac{1}{2\tau}\int_{-\tau}^{\tau} e_q(t)^2\,dt = \frac{1}{\tau}\int_0^{\tau} e_q^2(t)\,dt.$$

Now,

$$e_q(t) = \frac{\Delta/2}{\tau}t,\ -q \le t \le \tau.$$

$$\therefore\ P_q = \frac{1}{\tau}\int_0^{\tau}\left(\frac{\Delta}{2\tau}\right)^2 t^2\,dt = \frac{\Delta^2}{12} = \frac{1}{12}\left(\frac{2A}{2^b}\right)^2 = \frac{A^2/3}{2^{2b}}. \tag{2.6}$$

The mean-square power of the signal $x_a(t)$,

$$P_s = \frac{1}{T}\int_0^{T} A^2\sin^2\omega_0 t\,dt = \left(\frac{A^2}{2}\right).$$

So, SGNR (signal to quantisation noise ratio) $= \left(\dfrac{P_s}{P_q}\right) = \dfrac{A^2}{2} \times \dfrac{2^{2b}\times 3}{A^2} = \dfrac{3}{2}\cdot 2^{2b}$.

Expressed in decibel (dB)

$$\begin{aligned}\text{SGNR}\,(dB) &= 10\log_{10}\left(\frac{P_s}{P_q}\right) = 10\log_{10}\left(\frac{3}{2}\times 2^{2b}\right) = 10\left[\log_{10}\frac{3}{2} + 2b\log_{10}2\right]\\ &= 10\,[0.176 + 2b(0.30)] = 1.76 + 6b\end{aligned} \tag{2.7}$$

⇒ SGNR (dB) increases by 6 db for each addition of bit to the word length.

So, we can sum up our discussion saying that sampling can be defined as a process of going from a signal continuous in time to a signal discrete in time. It can be thought mathematically as equivalent to multiplication of a continuous time signal $x_a(t)$ with a train of unit pulses $p(t)$. Sampling can be represented as

$$x(n) = x_a(t)\,p(t) = \sum_{-\infty}^{\infty} x(nT)\,\delta(t - nT),\quad x(n)$$

is the nth sampled value of the sequence $[x(n)]$. We note that

$$\delta(t - nT) = \begin{cases}1, & t = nT\\ 0, & \text{otherwise.}\end{cases}$$

So, the above mathematical operation in the summation notation really represents switching operation.

The process of quantisation is the process of conversion of a continuous amplitude signal into discrete amplitude signal. So, we can write: $\left[x_q(n)\right] = Q\{[x(n)]\}$, Q is called the *quantisation operator.* This $\left[x_q(n)\right]$ sequence is the digital signal.

2.8 The Practical AD Converter (ADC)

There are different types of analog-to-digital converters: (a) Simultaneous A/D converters (b) Stair-step-ramp A/D converter (c) Dual slope A/D converter (d) Successive approximation A/D converter (e) IC A/D converter (ADC 0804, 8-bit).

Of these several types, the simplest design is the simultaneous A/D converter which uses a number of comparator circuits as parallel differential comparators which compare reference voltage with analog input voltage.

Most accurate and widely used A/D converter is, however, the dual-slope A/D converter used extensively in digital voltmeter, etc. Finally, IC version of A/D converter will be introduced in brief.

Simultaneous A/D *converter.* This is the simplest type of A/D converter using a number of comparators. Analog signal is one input to a comparator and the other input to it is a reference voltage V_R provided by the voltage divider resistive network shown in Fig. 2.14.

Suppose, it is a 3-bit converter, so it requires $(2^3 - 1)$, i.e., 7 comparators as shown in Fig. 2.14.

$V(+)$ is a fixed reference voltage and inputs to the non-inverting terminals are supplied by the voltage divider resistive network producing a voltage drop equal to $(V/8)$ on each of the resistors in the resistive network. Analog signal voltage v_i is serving as the second input to the comparators. Thus, the comparator C_1 has two inputs, one due to the analog voltage v_i and the other due to the reference voltage $V_R = V/8$, the former being injected into the inverting terminal (negative) of the comparator, the latter into the non-inverting terminal (positive) of the comparator.

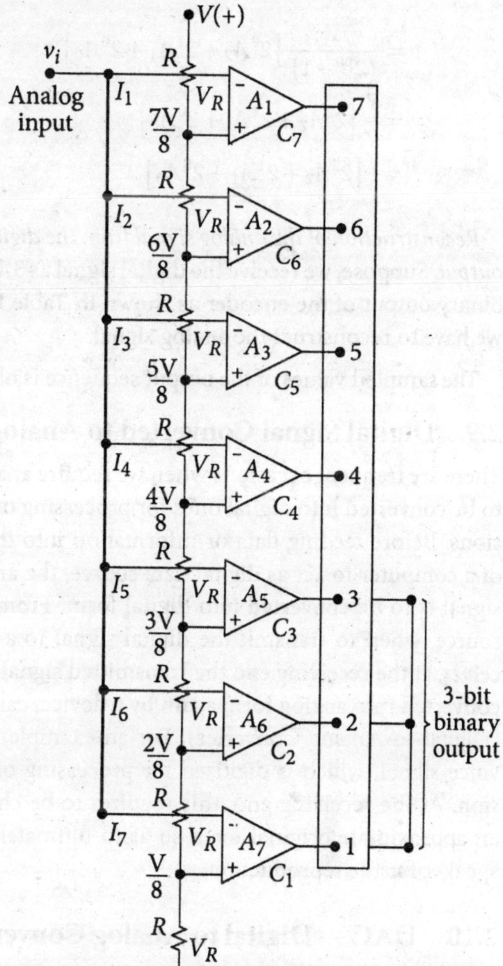

Fig. 2.14 Simultaneous A/D counter.

When v_i becomes slightly greater than the reference voltage V_R for a particular comparator, that particular comparator has output *high*. When the comparator C_1 has high output, the analog input $v_i = x(n)$ is such that $\left(\frac{V}{8}\right) < x(n) < \frac{2}{8}V$.

If the outputs of C_1 and C_2 are high, then the input $x(n)$ is to satisfy $\frac{2}{8}V < x(n) < \frac{3}{8}V$ and so on. If all comparators have low output, it would imply $x(n) < \frac{V}{8}$.

When the outputs of all the comparators are high, the analog input $x(n)$ is to satisfy $\frac{7V}{8} < x(n) < V$.

Suppose, $V(+) = 7$ volt.

$$v = \frac{V}{(2^N - 1)}\left[2^2 A_2 + 2^1 A_1 + 2^0 A_0\right]$$
$$= \frac{7}{7}\left[2^2 A_2 + 2^1 A_1 + 2^0 A_0\right]$$
$$= 1\left[2^2 A_2 + 2^1 A_1 + 2^0 A_0\right].$$

Table IV

Digital output				v
0	0	0		0
0	0	1	C_1	1
0	1	0	C_2	2
0	1	1	C_3	3
1	0	0	C_4	4
1	0	1	C_5	5
1	1	0	C_6	6
1	1	1	C_7	7

Reconstruction of the analog signal from the digital output. Suppose, we receive the digital signal as 3-bit binary output of the encoder as shown in Table IV, we have to reconstruct the analog signal.

The sampled values of the output sequence is obtained from the formula $[x(n)] = v = [0, 1, 2, 3, 4, 5, 6, 7]$.

2.9 Digital Signal Converted to Analog Signal (DAC)

There are frequent occasions when we require analog signals to be converted into digital ones for processing or transmissions. Before feeding data or information into the registers of a computer to act as digital data source, the analog input signal is to be converted into digital form. From such data source, when to transmit the digital signal to a distant receiver, at the receiving end the transmitted signal is to be reconverted into analog form again by a device, called a DAC (Digital-to-Analog Converter). For an example, consider a voice-signal, which is digitised for processing or transmission. At the receiving end, this requires to be changed into an approximate original form so as to ultimately actuate a speaker for the reproduction.

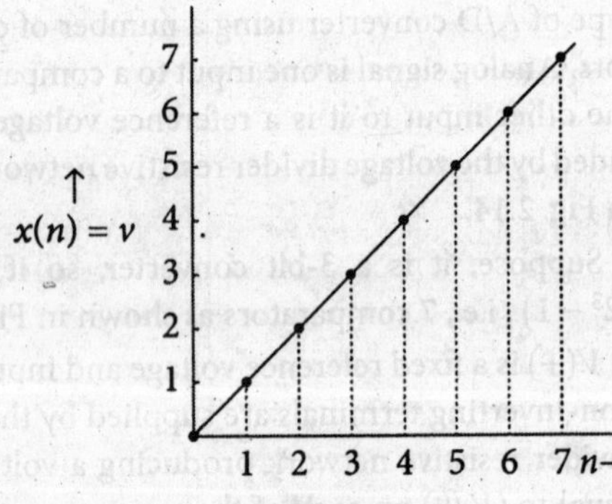

Fig. 2.15

2.10 DAC — Digital to Analog Converter

In this process of conversion, reverse of A/D conversion takes place and for this purpose also there are several devices, either discrete or IC version, both are available. However, we will describe the simplest and straightforward method in the name of weighted-resistor D/A converter to explain the basic principle and finally we will conclude our discussion on the process of conversion with the introduction of IC ADC-DAC (converter).

Weighted-register DAC. We have shown a resistive network for a 4-bit converter in the Fig. 2.16 for explaining the basic principle, the IC versions of both ADC and DAC have been given elsewhere.

In Fig. 2.16 we have shown a resistive network in the 4 parallel rails having resistances taken as ascending or descending powers in 2 of a minimum value equal to R. Thus, we have resistances $2^3 R$, $2^2 R$, $2^1 R$, $2^0 R$ connected to a common voltage reference V_R or to different voltage references V_0, V_1, V_2, V_3, etc. The bits from the registers are fed to the bases of transistors S_0, S_1, S_2 and S_3 acting as electronic switches. When a bit pulse from D_0 enters the transistor (S_0), the switch S_0 closes, the first transistor is ON and other transistors are OFF, because of absence of bit pulse to the bases. When switch S_0 closes, the digital input to the DAC device is 0001, current through resistor $8R$ flows while other resistors do not pass any current, so $I = I_0 = V_0/8R$; similarly, if S_0 switch remaining closed and also S_1 switches on, but all other switches are open, then the digital input to the DAC is represented by (0011), in this case current flows through $8R$ and $4R$, so

$$I = I_0 + I_1 = \frac{V_0}{8R} + \frac{V_1}{4R}.$$

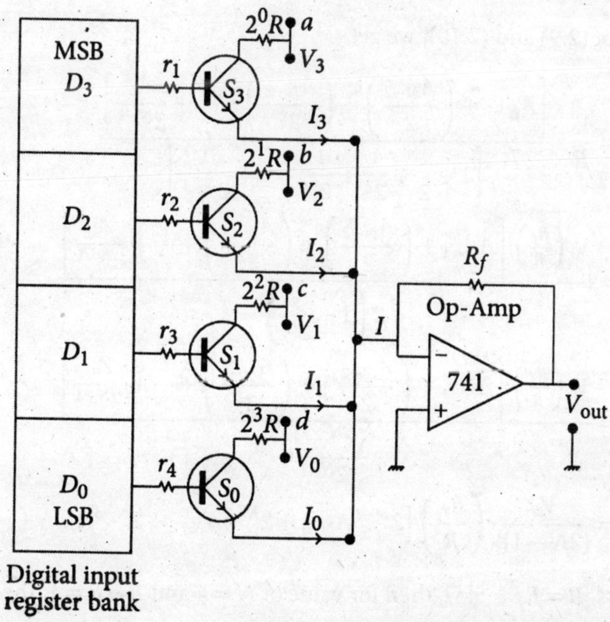

Fig. 2.16

When all the switches are closed, the digital inputs is (1111), the currents allowed by all the transistors combine to give the total current

$$I = \underset{\text{MSB}}{\frac{V_3}{R}} + \frac{V_2}{2^1 R} + \frac{V_1}{2^2 R} + \underset{\text{LSB}}{\frac{V_0}{2^3 R}} = I_3 + I_2 + I_1 + I_0, \tag{2.8}$$

I_3 corresponds to the maximum value of current flowing through minimum value of resistance in the resistive network; hence, it represents MSB.

If $V_0 = V_1 = V_2 = V_3 = V_c = 5$ volts, $R = 5\text{k}\Omega$, then

$$I = \frac{5}{5 \times 10^3}\left(\frac{1}{8} + \frac{1}{4} + \frac{1}{2} + 1\right) = 1.875\text{mA}.$$

We thus see that the current flowing through a resistor is proportional inversely to the resistance which is taken as multiples in 2. So, in general, the output current I can be written for a N-bit DAC device as $I = I_{N-1} + I_{N-2} + I_{N-3} + \cdots + I_0$ corresponding to the resistances $2^{N-1}R, 2^{N-2}R, 2^{N-3}R, \ldots, 2^0 R$.

So for an N-bit device we can write w.r.t (2.8):

$$I = V_c\left[\frac{A_{N-1}}{R} + \frac{A_{N-2}}{2R} + \frac{A_{N-3}}{2^2 R} + \cdots + \frac{A_0}{2^{N-1}R}\right] = \frac{V_c}{R}\left[A_{N-1} + \frac{A_{N-2}}{2^1} + \frac{A_{N-3}}{2^2} + \cdots + \frac{A_0}{2^{N-1}}\right], \tag{2.9}$$

where V_c is the common input voltage. If V_{out} be the voltage output (analog) of the DAC, we can also write using the basic principle of an Op-Amp .

$$I=\left(\frac{V_{out}}{R_f}\right)\left[1+\frac{1}{2}+\frac{1}{2^2}+\cdots+\frac{1}{2^{N-1}}\right]. \tag{2.10}$$

Equating the RHS of (2.9) and (2.10), we get

$$\frac{V_{out}}{R_f}=\frac{V_c}{R}\cdot\frac{\left[A_{N-1}+\left(\frac{A_{N-2}}{2}\right)+\left(\frac{A_{N-3}}{2^2}\right)+\cdots+\frac{A_0}{2^{N-1}}\right]}{\left[1+\frac{1}{2}+\frac{1}{2^2}+\frac{1}{2^3}+\cdots+\frac{1}{2^{N-1}}\right]}$$

$$\Rightarrow V_{out}=\frac{V_0\left(\frac{R_f}{R}\right)\left[A_{N-1}+\left(\frac{A_{N-2}}{2}\right)+\left(\frac{A_{N-3}}{2^2}\right)+\cdots+\frac{A_0}{2^{N-1}}\right]}{2\left(1-\frac{1}{2^N}\right)}$$

$$\Rightarrow V_{out}=\frac{V_c\left(\frac{R_f}{R}\right)\left[A_{N-1}+\left(\frac{A_{N-2}}{2}\right)+\left(\frac{A_{N-3}}{2^2}\right)+\cdots+\frac{A_0}{2^{N-1}}\right]}{\frac{(2^N-1)}{2^{N-1}}}$$

$$=\frac{V_c}{(2N-1)}\cdot\left(\frac{R_f}{R}\right)\left[2^{N-1}A_{N-1}+2^{N-2}A_{N-2}+2^{N-3}A_{N-3}+\cdots+2^0A_0\right]. \tag{2.11}$$

Suppose, $V_c=5$ volts, $R=R_f=5\text{k}\Omega$, then for value of $N=4$ and $R_f/R=1$, then (2.11) becomes

$$V_{out}=\frac{5}{(16-1)}[8A_3+4A_2+2A_1+A_0].$$

The coefficients A_3, A_2, A_1, A_0 are 1 if the switch connects a resistor to V_c, otherwise it is 0. Thus, if $V_c=5\text{V}$ and $A_3=1$, $A_2=A_1=A_0=0$, so, for digital input 1000, $V_{out}=\frac{5}{15}(8\times1+0+0+0)=\frac{8}{3}$ volts.

Similarly, for $A_3\ A_2\ A_1\ A_0$, V_{out}

0 0 0 1 $\frac{5}{15}\times(0+0+1) \quad =\frac{5}{15} \quad =0.33$ volt

0 0 1 0 $\frac{5}{15}\times(0+0+2+0)=\frac{2}{15}\times5=0.66$ volt

0 1 0 0 $\frac{5}{15}\times(0+4+0+0)=\frac{5}{15}\times4=1.33$ volts.

Example 2.1. *For 4-bit DAC device, find the analog voltage output for all possible digital inputs, assume $V_c=5\text{V}$, $R=R_f=5\text{k}\Omega$. What is the resolution of the device?*

From (2.11), we get

$$V_{out}=\frac{V_c}{(2^N-1)}\left[2^{N-1}A_{N-1}+2^{N-2}A_{N-2}+\cdots+2^0A_0\right]$$

$$=\frac{5}{(2^4-1)}\left[2^3A_3+2^2A_2+2^1A_1+2^0A_0\right] \text{ for a 4-bit DAC}$$

$$=\frac{5}{15}[8A_3+4A_2+2A_1+A_0]. \tag{2.12}$$

The resolution of the device is defined as the value of 1 bit as a fraction of the total input voltage. This is given by a formula:

$$\text{Resolution } R = \frac{1}{(2^N - 1)}.$$

Thus, for a 4-bit device, the resolution is

$$R\% = \frac{1}{(2^N - 1)} \times 100 = \frac{1}{15} \times 100\% = 6.67\%.$$

For an 8-bit DAC, $R\% = \dfrac{1}{(2^8 - 1)} \times 100 = 0.39\%$.

For a 6-bit DAC, $R\% = \dfrac{1}{(2^6 - 1)} \times 100 = 1.59\%$.

The possible digital inputs for a 4-bit DAC device and the corresponding analog voltage outputs are shown in Table V using the formula (2.12): $V_{out} = \frac{5}{15}[8A_3 + 4A_2 + 2A_1 + A_0]$.

Table V

Digital inputs				V_{out} (volt)
A_3	A_2	A_1	A_0	
0	0	0	0	0.00
0	0	0	1	0.06
0	0	1	0	0.66
0	0	1	1	0.99
0	1	0	0	1.33
0	1	0	1	1.67
0	1	1	0	2.00
0	1	1	1	2.33
1	0	0	0	2.67
1	0	0	1	3.00
1	0	1	0	3.33
1	0	1	1	3.67
1	1	0	0	4.00
1	1	0	1	4.33
1	1	1	0	4.67
1	1	1	1	5.00

2.11 IC Version of ADC and DAC

We have already discussed the basic principle of ADC and also DAC devices and described the implementing circuit using discrete semiconductor elements. Now, we will describe the IC versions of the same.

ADC IC: IC 0804. Nowadays, a large number of such ICs are available in the market, of which IC 0804 is a C-MOS 8-bit successive approximation A/D converter and also designed to be used with 8080A microprocessor (shown in pin diagram in Fig. 2.17). A simplified block diagram is shown in Fig. 2.18.

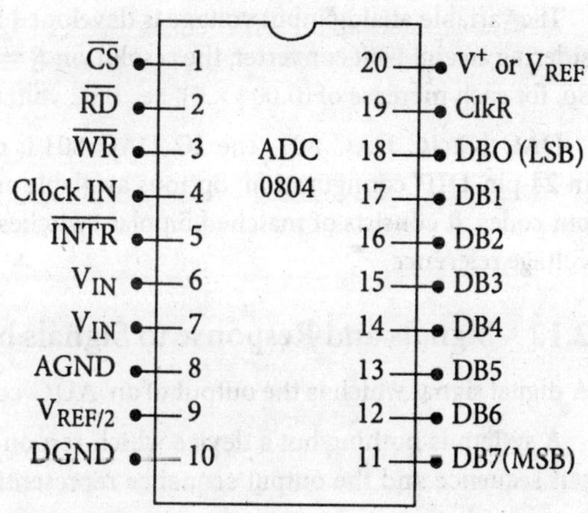

Fig. 2.17

The control lines activate the A/D to first sample out and digitise the input analog voltage, also direct the A/D to generate 8-bit binary outputs. If the input analog voltage is 0, the binary output will be (00000000); if the input is 5 volts (say), the output will be (11111111).

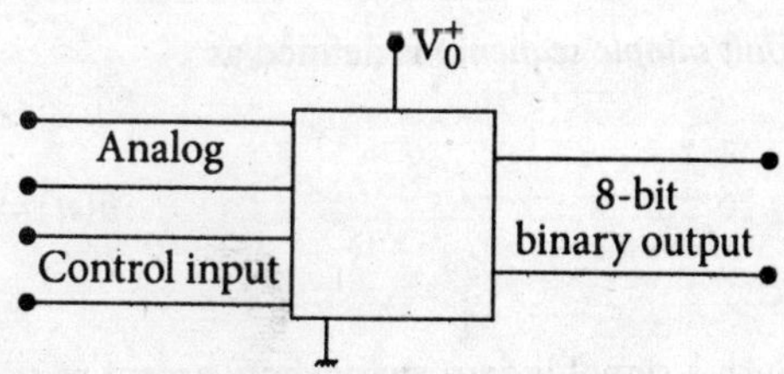

Fig. 2.18

The implementing circuit for ADC 0804 is shown in Fig. 2.19.

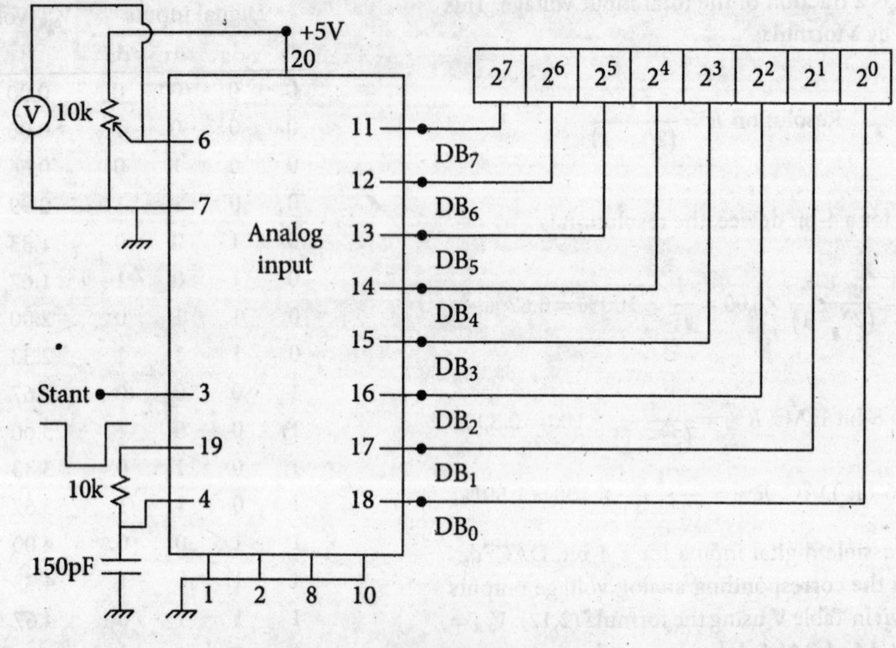

Fig. 2.19

The variable analog input voltage is developed between pins 6 and 7 across 10kΩ pot. Since we are considering an eight-bit converter, the resolution $R = 1/(2^8 - 1) = 0.004$ of the full-scale voltage (here 5 volts). So, for each increase of (0.004×5), i.e., 0.02 volt, there is increase of one binary count.

DAC IC: IC DAC 801. The IC DAC 801 is a twelve-bit (12-bit) digital to analog converter available in 24-pin DIP configuration; options available are CB1 (complementary binary) or three digit BCD input codes. It consists of matched bipolar switches, a precision resistive network, a low drift, high-stability voltage reference.

2.12 Signals and Response to Signals by Different Systems

A digital signal which is the output of an ADC consists of a series of binary numbers forming a sequence.

A system is nothing but a device which responds and establishes a certain relationship between the input sequence and the output sequence representing the input digital signal and the output digital signal respectively.

Depending on the nature of the sequence, the digital signals are classified as below:

Unit sample sequence is defined as

$$\delta(n) = \begin{cases} 1, & \text{for } n = 0 \\ 0, & \text{for } n \neq 0. \end{cases}$$

Such a signal is zero everywhere except at $n = 0$, where its value is unity. Such a signal has a property like that of Dirac δ-function represented by diagram as in Fig. 2.20.

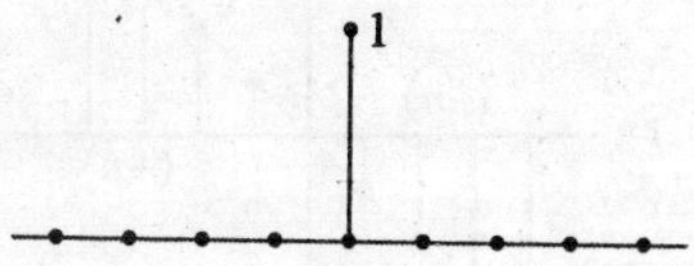

Fig. 2.20

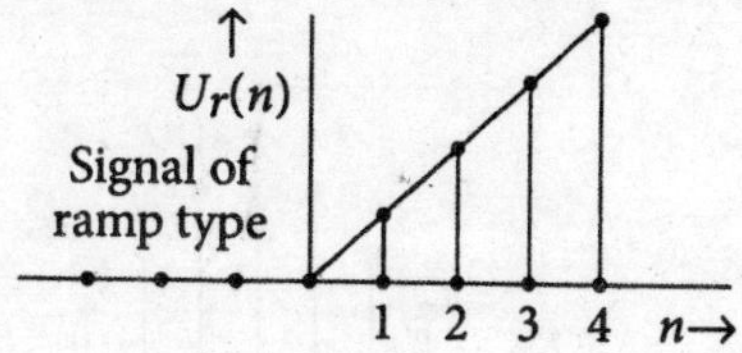

Fig. 2.21

Unit ramp signal. Denoted by $u_r(n)$ and defined as

$$u_r(n) = \begin{cases} n, & \text{for } n \geq 0 \\ 0, & \text{for } n < 0. \end{cases}$$

This signal is shown in Fig. 2.21.

Unit step signal. [Fig. 2.22] This signal is denoted by $u(n)$ and defined as

$$u(n) = \begin{cases} 1, & \text{for } n \geq 0 \\ 0, & \text{for } n < 0. \end{cases}$$

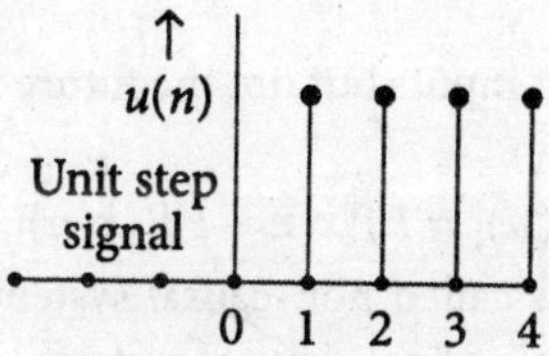

Fig. 2.22

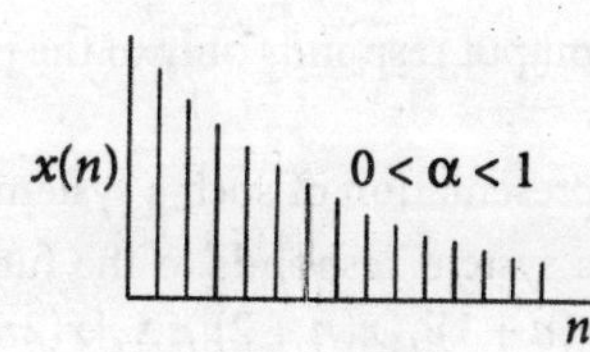

Fig. 2.23

Exponential signal. It is a sequence of the form given as $[x(n)] = \alpha^n$, for all n, α is a parameter. If α is real, then $[x(n)]$ is a real signal. Its diagrammatic representation is shown in Fig. 2.23.

LTI (Linear time invariant) system. LTI system is one whose response does not depend on time. It means that if a system has an output $[y(n)]$ as response to the input signal represented by the sequence $[x(n)]$, and that if the input sequence $[x(n)]$ is time-delayed but the output sequence $[y(n)]$ remains unaffected, the system is said to be time invariant. Thus, if the input and output sequences are such that $y(n,k) = T_{-k}[x(n)] = [x(n-k)] = [y(n)]$, time invariance is indicated.

The linearity of the system is indicated if it satisfies the principle of superposition. It is a system that for any arbitrary input $[x_1(n)]$ and $[x_2(n)]$, the output sequence is given by

$$[y(n)] = T\{[a_1 x_1(n)] + [a_2 x_2(n)]\} = a_1 T[x_1(n)] + a_2 T[x_2(n)].$$

Such a system is called *linear system.* So a system having the property of time invariance and also linearity is called LTI system.

Symmetric signal and anti-symmetric signal. Symmetric signal, also called even signal, is one such that $[x(n)] = [x(-n)]$.

On the other hand, a signal with $[x(n)] = -[x(-n)]$ is called an *antisymmetric* or *odd signal.*

An even signal is shown in Fig. 2.24.

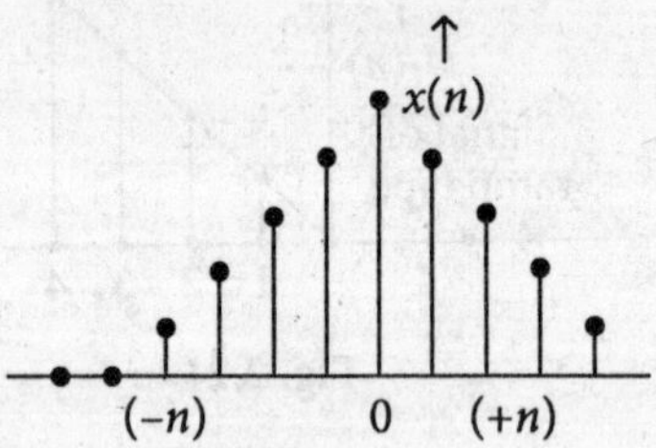

Fig. 2.24

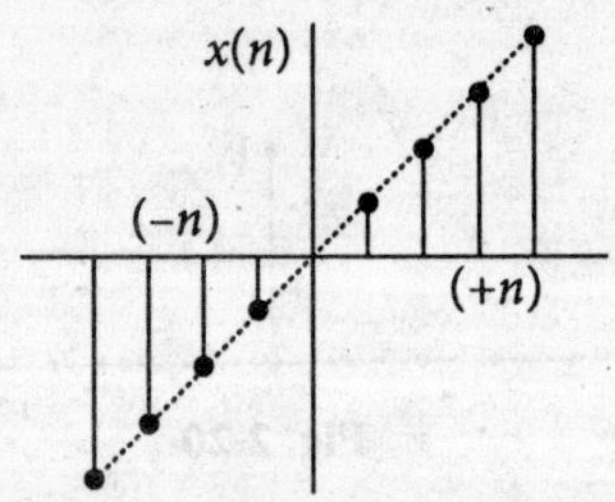

Fig. 2.25

Anti-symmetric signal is shown in Fig. 2.25.

Periodic and aperiodic signals. A signal is periodic with a period N if $[x(n)] = [x(n+N)]$ for all n. The smallest value of N for which the above relation holds is called the fundamental period. Thus, if $[x(n)] = a\sin 2\pi f_0 n$, it represents a periodic signal if $f_0 = s/N$, where s, N are both integers. If s and N are not integers, the signal is not periodic. It is said to be aperiodic.

Causal and non-causal signals. A signal in which the value of a member $x(n)$ belonging to the sequence $[x(n)]$ representing a signal is equal to 0 for all $n < 0$ and $[x(n)] = a^n u(n)$, where $u(n)$ is a unit step sequence, is a causal signal.

But if $x(n) \neq 0$ for $n < 0$, the sequence is non-causal. Thus, $x(n) = a^n u^n(-n+1)$ is an example of non-causal signal.

A system whose output responds only to the present and past inputs but not the future inputs is referred to as *causal systems.*

Mathematical representation of such a system is given by $[y(n)] = F\{[x(n-1)], [x(n)], [x(n-2)] \cdots\}$.

If the output of a system responds to the future inputs, it is called non-causal system or anticipatory. Thus, $[y(n)] = F\{[x(n+1)], [x(n+2)], \ldots, [x(2n)]\}$ is an example of non-causal system.

Deterministic and random signal. A signal having predictable characteristics is referred to as deterministic signal whereas a signal whose characteristics cannot be predicted is called a *random signal* or *noise.*

A deterministic signal is uniquely described by an explicit mathematical expression, a table of data or a set of well-defined rules. In case of such a signal, all the past, present and future values are precisely known.

In case of random signal, however, the signal is generated in a random fashion and cannot be predicted. So it is not futuristic. Examples of the random signals are speech signals, music and seismic signals; they include also quantisation noise a_0 in A/D conversion. The motion of pollen grains suspended in water or the motion of micro-particles in a solvent, called Brownian motion, is an example of random motion.

The random signals do not have finite energy and hence, they do not have Z-transform or Fourier transform. However, for mathematical description of such processes, we can replace the infinite energy sequence characteristic of random signals by a finite energy sequence through the introduction of auto-correlation and cross-correlation for two random variables.

FIR *and* IIR *systems.* FIR refers to finite impulse response system and IIR refers to infinite impulse response system. LTI systems are classified according to the type of impulse response as FIR and IIR.

FIR *system.* An impulse is represented by a Dirac delta function $\delta(n)$, so $[x(n)] = \delta(n)$ is an impulse sequence. If system response $h(n)$ is of finite length, i.e., $h(n) = 0$ for $n < N_1$ and $n > N_2$, then it is known as a finite impulsive response discrete time system for which the convolution sum reduces to

$$[y(n)] = \sum_{k=N_1}^{N_2} h(k)x(n-k).$$

The above convolution sum being finite, can be calculated directly for $[y(n)]$.

IIR *system.* If $h(n)$ is of infinite length, the system is referred to as infinite impulse response discrete time system.

For such a system, the convolution sum can be expressed as

$$[y(n)] = \sum x(k)\, h(n-k).$$

The above expression can be used to compute the output sequence. For increasing n, the computational complexity to compute the output samples increases as the number of products to be summed increases.

Correlation length. The correlation length of a signal is a measure of the length over which similarity exists. To know the extent of similarity in case of one signal, we would require to know the auto-correlation F_{xx} of the same signal or to know the cross-correlation F_{xy} of two different signals.

Cross-correlation of two signals $x(n)$, $y(n)$ is a measure of similarity of the two signals and is defined by the mathematical relation:

$$F_{xy} = \sum_{-\infty}^{\infty} x(n)y(n-k) \quad F_{x_1x_2} = \sum_{-\infty}^{\infty} x_1(n)x_2(n-k), \; k = 0, \pm 1, \pm 2, \text{ etc.,}$$

$x(n)$ is the reference sequence and $y(n-k)$ is the shifted sequence by k units w.r.t. $x(n)$.

$$F_{xy} = \sum_{-\infty}^{\infty} x(n-k)y(n),$$

where $[y(n)]$ is the reference sequence and $[x(n-k)]$ is the shifted sequence.

The auto-correlation sequence F_{xx} is available from F_{xy} by putting $x = y$.

∴ auto-correlation sequence:

$$F_{xx} = F_{xy}\Big|_{x=y} = \sum_{-\infty}^{\infty} x(n)x(n-k).$$

If $k = 0$,

$$F_{xx} = \sum_{-\infty}^{\infty} x^2(n) = \text{energy of the signal.}$$

2.13 Short Theory Questions with Answers

1. What is the meaning of a deterministic signal? Give an example.

   Ans. A signal whose characteristics can be predicted with a high degree of accuracy is referred to as a deterministic signal.

   Example. $v(t) = a \cos \omega t$.

2. What is a random signal?

   Ans. A signal whose characteristics cannot be predicted with a high degree of accuracy and which cannot be predicted before its occurrence, is called a random signal.

   Example. Noise.

3. What do you mean by (a) energy signal (b) power signal? Find (E) and (P) for a signal given in the form of $x(n)=\left(\frac{1}{3}\right)^n u(n)$, $u(n)$ is a unit step function:

$$u(n)=\begin{cases}1, & n\geq 0\\ 0, & n<0.\end{cases}$$

Ans. The energy signal is defined by

$$E=\sum_{-\infty}^{\infty}[x(n)]^2,$$

if and only if the right-hand sum tends to a finite quantity.

The power signal, on the other hand, is defined by the relation

$$P=\lim_{N\to\infty}\cdot\frac{1}{(2N+1)}\sum[x(n)]^2,$$

if and only if the average value as represented by the right-hand sum tends to a finite quantity. The condition for an energy signal is that here $P=0$ and for a power signal $E=\infty$. If a signal does not satisfy either of them, it is not an energy signal or a power signal.

Solution: $[x(n)]=\left(\frac{1}{3}\right)u(n)$.

$$E=\sum_{-\infty}^{\infty}[x(n)]^2=\sum_{0}^{\infty}\left(\frac{1}{9}\right)^n\{u(n)\}^2=\sum_{0}^{\infty}\left(\frac{1}{9}\right)^n=1+\frac{1}{9}+\left(\frac{1}{9}\right)^2+\ldots$$

$$=\frac{1}{1-\frac{1}{9}}=\frac{9}{8},\text{ finite.}$$

$$P=\lim_{N\to\infty}\cdot\frac{1}{(2N+1)}\sum_{-\infty}^{\infty}[x(n)]^2=\lim_{N-\infty}\cdot\frac{1}{(2N+1)}\sum_{0}^{\infty}\left(\frac{1}{9}\right)^2=\lim_{N\to\infty}\cdot\frac{1}{(2N+1)}\cdot\frac{1-\left(\frac{1}{9}\right)^{N+1}}{1-\frac{1}{9}}=0.$$

So the given signal is an energy signal.

4. What are the operations on signals?

Ans. A number of operations can be performed on a signal to get a desired signal according to the designer's choice. They are:

Modulation: $[w(n)]=[x(n)]\cdot[y(n)]=[x(n)\cdot y(n)]$.

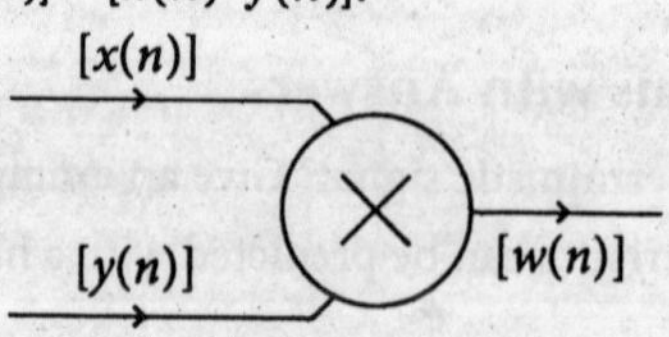

Scalar multiplication: $[w(n)]=A[x(n)]=[Ax(n)]$.

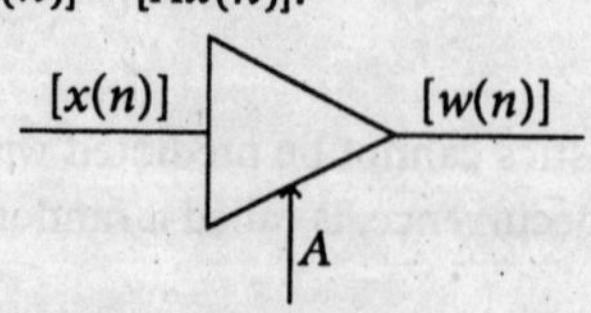

Time shifting: $T\{[x(n)]\}=[x(n-k)]$.

Time reversal: $T\{[x(n)]\}=x(-n)$.

Addition: $[w(n)]=[x(n)]+[y(n)]=[x(n)+y(n)]$.

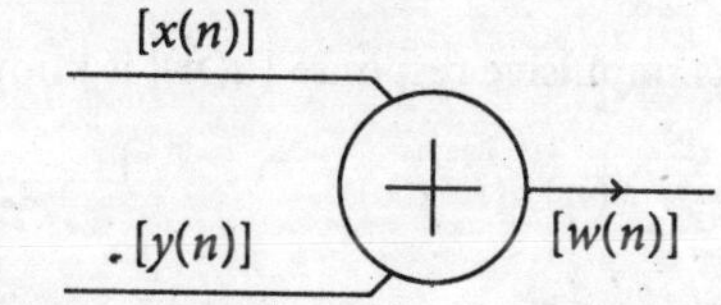

Scaling: $[w(n)]=T\{a^n[x(n)]\}=[a^n x(n)]$.

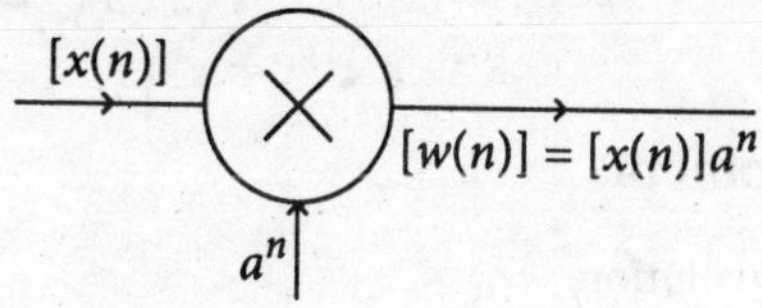

5. Define: (a) system (b) discrete time system.

   Ans. (a) A system is defined as a physical device that performs an operation on signal.

   (b) A discrete time system is a device or algorithm that operates on a discrete time input signal $[x(n)]$ according to certain rules to produce another discrete time sequence $[y(n)]$, called the output signal.

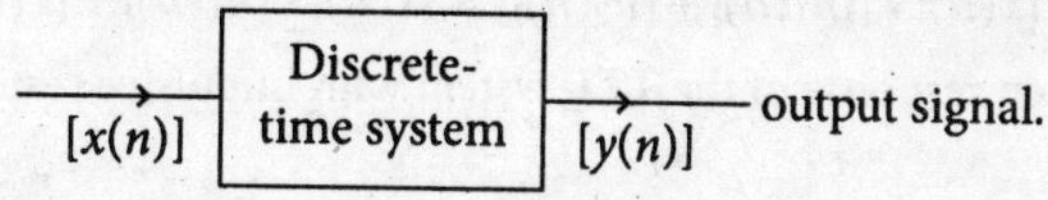

6. Represent the finite duration sequence $x(n)=(2;4,-1,-2)$ as a sum of weighted impulse response.

   Solution: Given $[x(n)]=(2,\underset{\uparrow}{4},-1,-2)$. $x(0)=4$, $x(-1)=2$, $x(1)=-1$, $x(2)=-2$.

$$[x(n)]=\sum_{k=-1}^{2}x(k)\delta(n-k)=x(-1)\delta(n+1)+x(0)\delta(n)+x(1)\delta(n-1)+x(2)\delta(n-2)$$

$$=2\delta(n+1)+4\delta(n)+(-1)\delta(n-1)-(2)\delta(n-2)=2\delta(n+1)+4\delta(n)-\delta(n-1)-2\delta(n-2).$$

7. What is an LTI system?

   Ans. An LTI is a linear time invariant device. Linearity satisfies the principle of superposition, which states that the effect of superposition of a number of signals is the linear combination of the effects due to the individual signal. Time invariance of the system refers to the fact that the input-output relation of the system does not change with time.

8. What do you mean by impulsive response or unit sample response of a system?

   Ans. Response of a system to a unit sample sequence is known as impulsive response, denoted by $h(n)$. So the output $[y(n)]$ of the device for an input signal $[x(n)]$ as an impulsive response is given as the convolution of $h(n)$ and $x(n)$:

$$[y(n)]=[x(n)]*[h(n)]=\sum_{k=-\infty}^{\infty}x(k)\,h(n-k).$$

9. What do you mean by discrete convolution?

   Ans. Convolution of discrete time signals is known as discrete convolution.

Consider an input signal $[x(n)]$ to an LTI system and let $[y(n)]$ be the output of the system.

So, $[y(n)] = \sum_{-\infty}^{\infty} x(k)\,h(n-k) = \sum_{-\infty}^{\infty} h(k)\,x(n-k)$.

10. Find the step response when its impulsive response $[h(n)]$ is known.

Ans. $[y(n)] = [x(n)] * [h(n)] = \sum_{-\infty}^{\infty} x(n)\,h(n-k)$.

Now, when $[x(n)] = [u(n)]$, we have $[y(n)] = [u(n)] * [h(k)] = \sum_{k=-\infty}^{n} u(n-k)\,h(k)$.

2.14 Solved Numerical Problems

1. Enumerate properties of convolution.

Ans. The properties of convolutions are:

Commutative property: $[x(n)] * [h(n)] = [h(n)] * [x(n)]$.

Associative property: $\{[x(n)] * [h_1(n)]\} * [h_2(n)] = [x(n)] * \{[h_1(n)] * [h_2(n)]\}$.

Distributive property: $[x(n)] * \{[h_1(n)] + [h_2(n)]\} = [x(n)] * [h_1(n)] + [x(n)] * [h_2(n)]$.

2. Determine the unit step response of the LTI system with impulsive response $[h(n)] = [a^n u(n)]$.

Ans. $y(n) = \sum_{-\infty}^{\infty} h(k) = \sum_{0}^{n} a^k = \dfrac{1-a^{n+1}}{(1-a)}$.

3. Examine whether the system is time invariant or time variant: $[y(n)] = [x(n)] + [x(n-1)]$.

Solution: $[y(n)] = [x(n)] + [x(n-1)] \Rightarrow [y(n,k)] = [x(n-k)] + [x(n-k-1)]$.

If the output is delayed by k-units, $[y(n-k)] = [x(n-k)] + [x(n-k-1)]$.

So, $[y(n,k)] = [y(n-1)]$ and, therefore, the system is time invariant.

4. If the input to a system is unit impulsive $[x(n)] = \delta(n)$, the output of the system is called an impulsive response $h(n)$ given as $T\{\delta(n)\} = h(n)$. For linear time-invariant system, if the input sequence $x(n)$ and impulse response $h(n)$ are given, we can find the output $y(n)$ by using the equation:

$$[y(n)] = \sum_{h=-\infty}^{\infty} x(k)\,h(n-k),$$

the right-hand side is known as convolution sum of $\{x(n)\}$ and $\{h(n)\}$. The convolution of $\{[x(n)]\}$ and $\{[h(n)]\}$ is represented by $\{[x(n)]\} * \{[h(n)]\}$.

5. Determine the convolution sum of two finite sequences $[x(n)] = (3,2,1,2)$ and $h(n) = (1,\underset{\uparrow}{2},1,2)$.

Solution: $x(n) = (3,2,1,2)$

$h(n) = (1,\underset{\uparrow}{2},1,2)$.

Strip-wise addition gives us $\{3,(6+2),(3+4+1),(6+2+2+2),(4+1+4),(2+2),4\} = (3,8,8,12,9,4,4)$;

$y(n) = \{3,\underset{\uparrow}{8},8,12,9,4,4\}$.

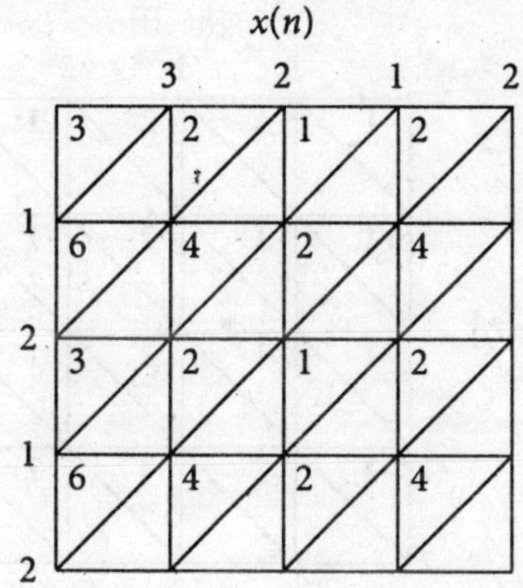

6. Find the cross-correlation between a pair of signals of finite lengths given as $[x(n)] = \{1,2,1,1\}$, $[y(n)] = \{1,1,2,1\}$.

Solution: $[x(l)] = \{1,2,1,1\}$
$[y(-l)] = \{1,2,1,1\}$.

Cross-correlation $F_{xy}(l)$ is given by $F_{xy}(l) = [x(l)] * [y(-l)]$.

Folding $y(l)$, we would get $y(-l) = \{1,2,1,1\}$; $F_{xy}(l) = [x(l)] * [y(-l)]$.

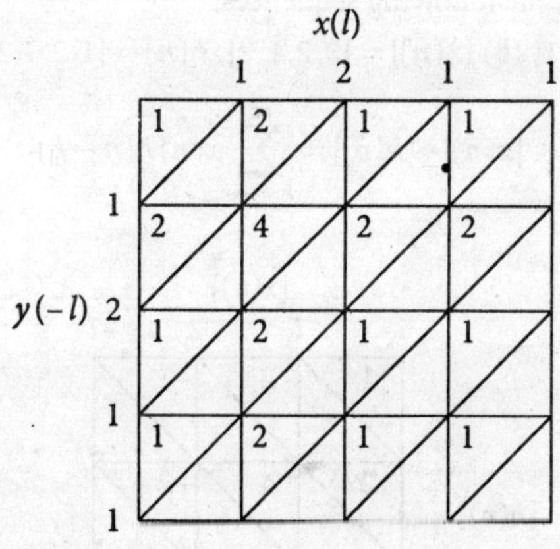

Strip sum = (1, 4, 6, 6, 5, 2, 1). $F_{xy}(l) = \{1,4,6,6,5,2,1\}$.

7. Find the cross-correlation of two finite length sequences $[x(n)] = \{1,-1,1,2\}$; $h(n) = \{1,2,1,-1\}$.

Solution: $[x(l)] = \{1,-1,1,2\}$
$[h(l)] = \{1,2,1,-1\}$.

Cross-correlation $F_{xy}(l) = [x(l)] * [h(-l)]$.

Folding $h(l)$, we would get $h(-l) = \{-1,1,2,1\}$.

Strip sum = (−1, 2, 0, −2, 3, 5, 2). So, $F_{xy}(l) = \{-1,2,0,-2,3,5,2\}$.

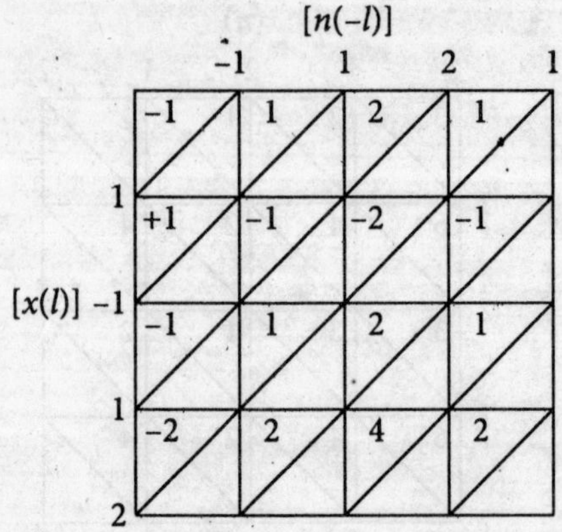

8. *Auto-correlation:* Auto-correlation of a sequence is the correlation of a sequence with itself. The auto-correlation of $[x(n)]$ is defined as

$$F_{xx}(l)=\sum_{-\infty}^{\infty}[x(n)]*[x(n-l)]=\sum_{-\infty}^{\infty}x(n)x(n-l)=\sum_{-\infty}^{\infty}x(n+l)x(n).$$

If $l=0, F_{xx}(0)=\sum_{-\infty}^{\infty}x^2(n)$.

9. Compute the convolution of the following sequences:
(a) $[x(n)]=[h(n)]=\{\underset{\uparrow}{1},2,-1\}$ (b) $[x(n)]=\{3,2,1,2\}$, $h(n)=\{1,\underset{\uparrow}{2},1,2\}$.

Solution: $[x(n)]=\{\underset{\uparrow}{1},2,-1\}$ $\quad [x(n)]*[h(n)]=\sum_{k=-\infty}^{\infty}x(n)\,h(n-k)$.

$[h(n)]=\{\underset{\uparrow}{1},2,-1\}$.

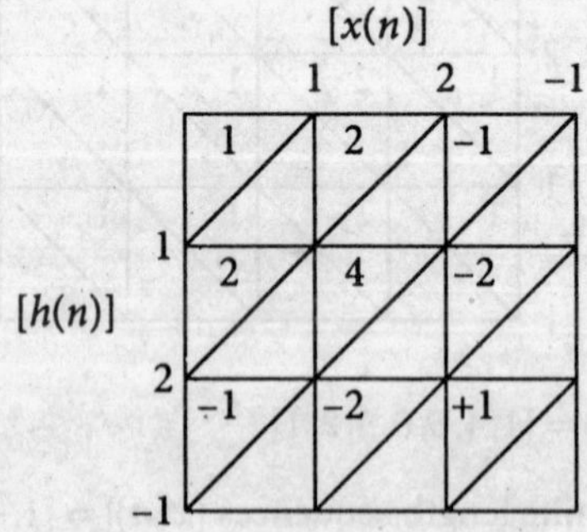

Strip sum = (1, 4, +2, −4, +1). $y(n)=\{1,4,+2,-4,1\}$.
(b) Same as in Example 5.

2.15 Theory Questions Exercises

1. What is the causality condition for an LTI system? [$h(n)=0$ for $n<0$]

2. What is the meaning of discrete convolution. $[y(n)=\sum_{-\infty}^{\infty}x(n)h(n-k)]$
3. What are the convolution properties?
4. Is it true that a causal system is one whose impulsive response $h(n)=0$ for all negative values of n?
5. Define a linear system with example.
6. What are the different types of operations which can be performed on discrete time signals?
7. What is a time-invariant system. Cite an example.
8. Represent the following finite duration sequence $[x(n)]=\{2,\underset{\uparrow}{4},-1,-2\}$ as a sum of weighted impulse sequences.

   [*Hints.* $[x(n)]=\{2,\underset{\uparrow}{4},-1,-2\}$. $x(-1)=2$, $x(0)=4$, $x(1)=-1$, $x(2)=-2$, so we can write:

$$
\begin{aligned}
[x(n)] = \sum_{k=-1}^{k=\infty} x(k)\delta(n-k) &= 2\delta(n+1)+4\delta(n)+(-1)\delta(n-1)+(-2)\delta(n-2) \\
&= 2\delta(n+1)+4\delta(n)-\delta(n-1)-2\delta(n-2)].
\end{aligned}
$$

9. What is the characteristic of a causal system?

   [*Hints.* Its output at any time depends only on the present and past inputs only but not on the future.]
10. Define: (a) system (b) discrete time system.
11. (a) How many classes of discrete time system are there?

    [*Hints.* (i) static and dynamic (ii) time variant and time invariant (iii) linear and non-linear (iv) stable and unstable (v) causal and non-causal (vi) IIR and FIR.]

    (b) Define static system and dynamic system.
12. What do you mean by an impulsive system?
13. Define convolution sum.
14. Define: (a) finite impulsive response (FIR) (b) infinite impulse response (IIR) systems.
15. Define symmetric and antisymmetric signals?
16. Distinguish between an energy signal and power signal.
17. What is (a) analog signal (b) discrete time signal (c) digital signal?
18. What are the different ways of classification of signals?

    [*Hints.* Graphical, tabular, functional, sequence representation.]
19. Define: (a) unit sample sequence (b) unit step signal (c) unit ramp signal. Also represent them graphically.
20. Find the convolution of two signals as given below:

    (a) $[x(n)]=3^n u(-n), [h(n)]=\left(\frac{1}{3}\right)^n u(n-2)$

    (b) $[x(n)]=\{\underset{\uparrow}{1},0,2,5,4\}, [h(n)]=\{1,-1,1,-1\}$.

Chapter 3

Transformation Techniques for Signal Analysis

3.1 What are the Types of Transformations?

Many times we find that time-domain representation of a signal is not very convenient for its analysis, rather the transformation in some other domain (viz., frequency) makes the analysis and design of LTI system more convincing and simple.

LTI system (linear time invariant) is one whose response does not depend on time. It means that if a system has an output $[y(n)]$ as response to an input $[x(n)]$ and that if the input $[x(n)]$ is time-delayed by k-units, the output is not affected, the system is said to *be time invariant,* i.e., $[y(n)] = T[x(n)]$, T is an operator and if $[y(n,k)] = T[x(n-k)] = [[y(n)]]$, the time invariancy is represented.

A system is *linear* if it satisfies the principle of superposition. It is a system that for any arbitrary input sequence $[x_1(n)]$ and $[x_2(n)]$, the output sequence is given by $[y(n)] = T\{[a_1x_1(n)] + [a_2x_2(n)]\} = a_1T[x_1(n)] + a_2T[x_2(n)]$. Such a system is called linear system.

The schematic diagram, showing the linearity property, is illustrated as shown below:

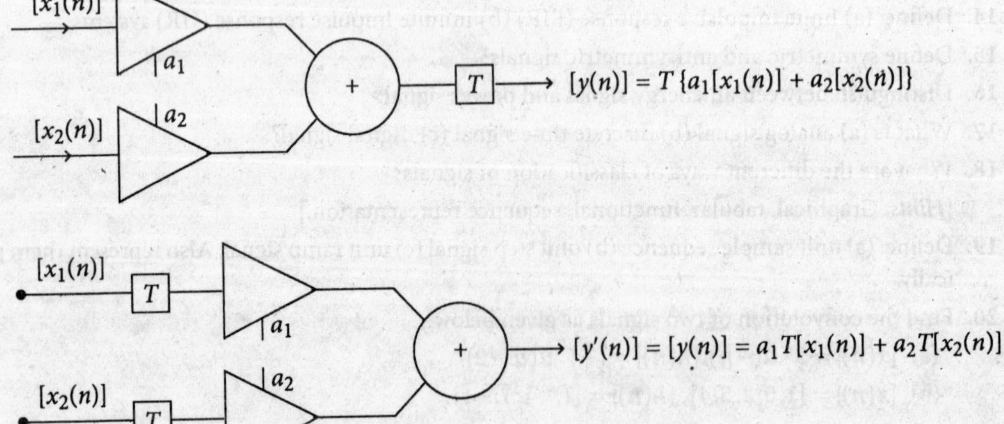

Fig. 3.1

Of the several transformations, the frequently occurring transformations are the following.

Z-transform by which time-domain signal is transformed into the complex z-plane, $z = x + jy$. The transformed function in the z-plane is denoted by $G(z)$. Once we get $G(z)$, its inversion leads to its time-domain representation once again. However, we should remember that Z-transformation applies only to the discrete time signal and *not to the analog signal* $x_a(t)$.

Integral transform. There are in general four types of transform, called *integral transform,* whose generic mathematical representation is

$$F(\alpha) = \int_a^b f(x)\,k(\alpha,x)\,dx, \tag{3.1}$$

where the function $F(\alpha)$ is said to be the integral transform of the function $f(x)$ by the kernel $k(\alpha,x)$. Depending on the choice of the kernels we have different types of transforms as described below.

Laplace transform (LT):

$$F(\alpha) = \int_0^\infty f(x)\,e^{-\alpha x}\,dx. \tag{3.2}$$

Fourier transform (FT):

$$F(\alpha) = \frac{1}{\sqrt{2\pi}}\int_{-\infty}^{\infty} f(x)\,e^{j\alpha x}\,dx. \tag{3.3}$$

Fourier-Bessel transform (FBT):

$$F(\alpha) = \int_0^\infty f(x)\,x\,J_n(\alpha x)\,dx. \tag{3.4}$$

Mellin transform (MT):

$$F(\alpha) = \int_0^\infty f(x)\,x^{\alpha-1}\,dx. \tag{3.5}$$

Of them the most oft-applied transforms are LT and FT. It is to remember that LT technique is applicable to continuous time signal while FT applies to both of continuous and discrete time signal. FT for discrete signal is called *Discrete Fourier Transform,* whereas that for continuous signal, is called *Continuous Fourier Transform* (CFT). Both of LT and FT are transformations from time-domain to frequency domain.

Another representation of a signal is in terms of series sum of sines and cosines which is popularly known as *Fourier Series* representation. This is a representation in terms of harmonic components. Thus, if we know the harmonic components hidden in the signal, we can design suitable LTI system, design different types of filters according to our need and purpose. Fourier series expansion is for a continuous or discrete *periodic signal* while an aperiodic but continuous or discrete signal admits of FT. We will begin our discussion with Z-transform, its properties and applications and then make analysis through FS and finally through Fourier Transform representation of a periodic and aperiodic function respectively.

3.2 *Z*-transform and Its Application

Transformation techniques are important tools in the analysis of signals and LTI systems. One such transform is the Z-transform which plays the same role in the analysis of discrete time signals as Laplace transform does in continuous time signal. While Laplace transform is restricted to real domain, the Fourier transform applies to complex domain.

Z-transform is important, because it has the advantage that the Z-transform of *convolution of two time-domain signals is equivalent to multiplication of their respective Z-transforms.* This property of the Z-transform has greatly simplified the analysis of the response of an LTI system to various types of discrete time signal.

3.3 Z-transform of Some Standard Time-domain Functions

Functions: Z-transform defined. The Z-transform of a discrete time signal $[x(n)]$ is defined as

$$Z\{[x(n)]\} = G(z) = \sum_{-\infty}^{\infty} [x(n)]\, z^{-n}, \tag{3.6}$$

where z is a complex variable and the domain of z is the complex domain.

The inverse of Z-transform, if it exists, gives us the signal back in the original domain. Thus,

$$Z^{-1}\{[G(z)]\} = [x(n)] = \frac{1}{2\pi j} \oint G(z) z^{n-1}\, dz. \tag{3.7}$$

Of course, we will not use this formula frequently to find out the inverse Z-transform. We will consider only those signals and systems whose Z-transforms are expressible in the form of the ratio of two polynomials for our convenience.

Properties of Z-transform. Z-transform technique is a very simple and convincing technique for the analysis of discrete time signals and design of LTI systems. Its strength is due to several properties possessed by the transform.

Property 1: *Linearity property.* If a function $f(x)$ is expressible in terms of two other functions $\phi(x)$ and $\psi(x)$ as a linear combination of the two, the function is said to possess property of linearity. Because of this property we can write:

$$f(x) = a_1\phi(x) + a_2\psi(x). \tag{3.8}$$

Similarly, suppose, an operator L is called a linear operator if we can write:

$$L(au + bv) = aL(u) + bL(v). \tag{3.9}$$

Now, suppose, we have two discrete signals $[x_1(n)]$ and $[x_2(n)]$ such that $Z\{[x_1(n)]\} = G_1(z)$ and $Z\{[x_2(n)]\} = G_2(z)$. These two signals on superposition gives rise to a signal $[x(n)]$ such that

$$[x(n)] = a_1[x_1(n)] + a_2[x_2(n)]. \tag{3.10}$$

If (3.10) is describing a signal $[x(n)]$, we say that $[x(n)]$ is a linear combination of two other functions $[x_1(n)]$ and $[x_2(n)]$.

Taking Z-transform of (3.10), we would get

$$\begin{aligned} Z\{[x(n)]\} = Z\{a_1[x_1(n)] + a_2[x_2(n)]\} &= \sum_{-\infty}^{\infty} \{a_1[x_1(n)] + a_2[x_2(n)]\} z^{-n} \\ &= \sum_{-\infty}^{\infty} \{a_1[x_1(n)]\} z^{-n} + \sum_{-\infty}^{\infty} \{a_2[x_2(n)]\} z^{-n} \\ &= a_1 \sum_{-\infty}^{\infty} [x_1(n)]\, z^{-n} + a_2 \sum_{-\infty}^{\infty} [x_2(n)]\, z^{-n} \end{aligned} \tag{3.11}$$

$$\Rightarrow G(z) = a_1G_1(z) + a_2G_2(z). \tag{3.12}$$

Convergence ROC of $G(z)$ is the intersection of ROCs of $[x_1(n)]$ and $[x_2(n)]$. Thus, the Z-transform operator is a linear operator.

Property 2: Time shifting property. If the input sequence $[x(n)]$ is time shifted (delayed) by k units, it is denoted by $[x(n-k)]$, $k>0$, it is delayed and for $k<0$, it is advanced by k-units and is denoted by $[x(n-k)]$ and $[x(n+k)]$ respectively.

So if $Z\{[x(n)]\}=G(z)$, then

$$Z\{[x(n-k)]\}=\sum_{-\infty}^{\infty}\{[x(n-k)]\}z^{-n}=\sum_{-\infty}^{\infty}[x(s)]\,z^{-(k+s)},\ [\text{putting } n-k=s]$$

$$=z^{-k}\sum_{-\infty}^{\infty}[x(s)]\,z^{-s}=z^{-k}G(z). \tag{3.13}$$

Thus,

$$Z\{[x(n-1)]\}=z^{-1}G(z) \tag{3.14}$$

$$Z\{[x(n-2)]\}=z^{-2}G(z) \tag{3.15}$$

$$Z\{[x(n+1)]\}=zG(z) \tag{3.16}$$

$$Z\{[x(n+2)]\}=z^{2}G(z). \tag{3.17}$$

Example 3.1. *Find the Z-transform of the signals of finite duration as given:* (a) $[x_1(n)]=\{1,2,3,5,7,0,1\}$ (b) $[x_2(n)]=\{1,2,\underset{\uparrow}{5},7,0,1\}$ (c) $[x_3(n)]=\{0,0,1,2,5,7,0,1\}$ (d) $[x_4(n)]=\{2,4,\underset{\uparrow}{5},7,0,1\}$.

Solution: (a) $[x_1(n)]=\{1,2,3,5,7,0,1\}$.

$$Z\{[x_1(n)]\}=\sum_{-\infty}^{\infty}[x_1(n)]\,z^{-n}=1+2z^{-1}+3z^{-2}+5z^{-3}+7z^{-4}+0z^{-5}+1z^{-6}$$

$$=\left(1+2z^{-1}+3z^{-2}+5z^{-3}+7z^{-4}+z^{-6}\right). \tag{3.18}$$

(b) $[x_2(n)]=\{\underset{(-2)}{1},\ \underset{(-1)}{2},\ \underset{n\uparrow n=0}{5},\ \underset{1}{7},\ \underset{2}{0},\ \underset{3}{1}\}$.

$$Z\{[x_2(n)]\}=\sum_{-\infty}^{\infty}[x_2(n)]\,z^{-n}=\left(1\times z^{2}+2z+5+7z^{-1}+z^{-3}\right). \tag{3.19}$$

(c) $[x_3(n)]=\{0,0,1,2,5,7,0,1\}$.

$$Z\{[x_3(n)]\}=\sum_{-\infty}^{\infty}[x_3(n)]\,z^{-n}=1\times z^{-2}+2\times z^{-3}+5\times z^{-4}+7\times z^{-5}+1\times z^{-7}$$

$$=\left(z^{-2}+2z^{-3}+5z^{-4}+7\times z^{-5}+1\times z^{-7}\right). \tag{3.20}$$

(d) $[x_4(n)]=\{2,4,5,7,0,1\}$.

$$Z\{[x_4(n)]\}=\sum_{-\infty}^{\infty}[x_4(n)]\,z^{-n}=\left(2z^{2}+4z^{1}+5+7z^{-1}+z^{-5}\right). \tag{3.21}$$

Example 3.2. *Take the signal $[x_1(n)]$ as described above. Find $Z[x_1(n+2)]$, $Z[x_1(n-2)]$.*

Solution: $Z[x_1(n+2)]=z^2G(z)$, where $G(z)$ is $Z[x_1(n)]$.

Now,

$$Z\{[x_1(n)]\} = \left(1+2z^{-1}+5z^{-2}+7z^{-3}+z^{-5}\right)$$

$$Z\{[x_1(n+2)]\} = z^2\left(1+2z^{-1}+5z^{-2}+7z^{-3}+z^{-5}\right) = \left(z^2+2z+5+7z^{-1}+z^{-3}\right) \tag{3.22}$$

$$Z\{[x_1(n-2)]\} = z^{-2}G(z) = z^{-2}\left(1+2z^{-1}+5z^{-2}+7z^{-3}+z^{-5}\right)$$

$$= \left(z^{-2}+2z^{-3}+5z^{-4}+7z^{-5}+z^{-7}\right). \tag{3.23}$$

Property 3: Scaling property. Suppose, a signal is represented by $[x(n)]$. If we write $a^n[x(n)] = [a^n x(n)]$, each element of the sequence $[x(n)]$ is multiplied by factor a^n, called the scale factor. We are to find $Z\{[a^n x(n)]\}$.

$$Z\{[a^n x(n)]\} = \sum_{-\infty}^{\infty}[a^n x(n)]\, z^{-n} = \sum_{-\infty}^{\infty}[x(n)]\left(a^{-1}z\right)^{-n} = G\left(a^{-1}z\right), \tag{3.24}$$

where $G(z) = Z\{[x(n)]\}$.

Example 3.3. *Find $\{Z[a^n u(n)]\}$, $[u(n)]$ is the unit step function sequence, defined as*

$$u(n) = \begin{cases} 1, & n \geq 0 \\ 0, & n < 0. \end{cases}$$

Solution: $Z[u(n)] = \sum_{-\infty}^{\infty}[u(n)]\, z^{-n} = \sum_{0}^{\infty} z^{-n}$

$$= \left(1+z^{-1}+z^{-2}+z^{-3}+\ldots\right)$$

$$= \frac{1}{(1-z^{-1})} = G(z).$$

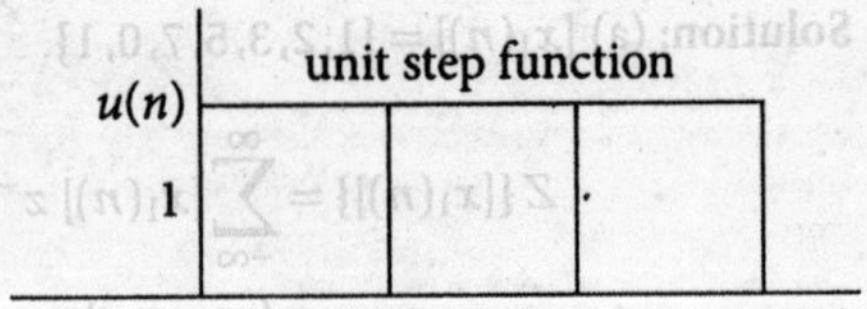

Fig. 3.2

So,

$$Z\{a^n[u(n)]\} = G\left(a^{-1}z\right) = \frac{1}{1-(a^{-1}z)^{-1}} = \left(\frac{1}{1-az^{-1}}\right).$$

$$\boxed{Z\{a^n[u(n)]\} = \frac{1}{(1-az^{-1})}}. \tag{3.25}$$

Property 4: Differentiating property in the z domain. We take a signal $[x(n)]$ and its Z-transform is $G(z)$. We get

$$Z\{[x(n)]\} = G(z) = \sum_{-\infty}^{\infty}(-n)[x(n)]\, z^{-n}.$$

Differentiating the above w.r.t. z, we will have

$$\frac{d}{dz}\{G(z)\} = \sum_{-\infty}^{\infty}(-n)[x(n)]\, z^{-n-1} = -z^{-1}\sum_{-\infty}^{\infty}[nx(n)]\, z^{-n} = -z^{-1}Z\{[nx(n)]\}$$

$$\Rightarrow Z\{nx(n)\} = -z\frac{d}{dz}[G(z)] = -zG'(z). \tag{3.26}$$

Example 3.4. *A signal is represented by $[x(n)]$ given by $[x(n)] = [na^n u(n)]$. Find $Z\{[x(n)]\}$.*

Take $x_1(n) = [u(n)] \Rightarrow Z\{x_1(n)\} = Z[u(n)] = 1/\left(1-z^{-1}\right) = G(z)$.

Solution: Let us take $[x_2(n)] = [a^n u(n)] \Rightarrow Z\{[x_2(n)]\} = Z\{[a^n u(n)]\} = G_1(z) = G(a^{-1}z)$

$$= 1/(1 - az^{-1}).$$

$$[x(n)] = [nx_2(n)]$$

$$\Rightarrow Z\{[x(n)]\} = Z\{[nx_2(n)]\} = -z\frac{d}{dz}\{G_1(z)\} = -z(-1)\frac{az^{-2}}{(1-az^{-1})^2} = \frac{az^{-1}}{(1-az^{-1})^2}. \quad (3.27)$$

Property 5: *Convolution and connected theorem.* Convolution is a very important property of Z-transform and because of this property, analysis and designing of discrete time signal and LTI systems become very much simplified.

What do we actually mean by convolution?

Two signals $[x_1(n)]$ and $[x_2(n)]$ are really two sequences; the convolution of the two signals is denoted by $[x_1(n)] * [x_2(n)]$ generates a new signal $[x(n)]$ generated such that $[x(n)] = [x_1(n)] * [x_2(n)]$.

The convolution of $[x_1(n)]$ and $[x_2(n)]$ is defined as

$$[x(n)] = \sum_{k=-\infty}^{k=\infty} [x_1(k)]\,[x_2(n-k)] = \sum_{-\infty}^{\infty} [x_2(k)]\,[x_1(n-k)].$$

$$\therefore\ Z\{[x(n)]\} = \sum_{n=-\infty}^{n=\infty} [x(n)]\,z^{-n} = \sum_{n=-\infty}^{n=\infty}\left\{\sum_{k=-\infty}^{\infty} [x_1(k)]\,[x_2(n-k)]\,z^{-n}\right\}$$

$$= \sum_{k=-\infty}^{\infty}\left\{\sum_{n=-\infty}^{n=\infty} [x_1(k)]\,[x_2(n-k)]\,z^{-n}\right\}$$

changing the order of summation. We can write:

$$Z\{[x(n)]\} = \sum_{k=-\infty}^{k=\infty} [x_1(k)]\left\{\sum_{n=-\infty}^{n=\infty} [x_2(n-k)]\,z^{-(n-k)}\cdot z^{-k}\right\}$$

$$= \sum_{k=-\infty}^{k=\infty} [x_1(k)]\,z^{-k}\left\{\sum_{n=-\infty}^{n=\infty} [x_2(n-k)]\,z^{n-k}\right\} = G_1(z)G_2(z), \quad (3.28)$$

where $G_1(z)$ is $Z\{[x_1(n)]\}$ and $G_2(z)$ is $Z\{[x_2(n)]\}$. Thus, the convolution of two signals $[x_1(n)]$ and $[x_2(n)]$ denoted by $[x(n)] = [x_1(n)] * [x_2(n)]$ and defined by

$$[x(n)] = \sum_{n=-\infty}^{\infty} x_1(n)x_2(n-k),$$

so simply we can identify the convolution of two sequences as the sum of the products of the nth sample value of the first sequence and $(n-k)$th sample value of the second sequence or equivalently, as the product sum of $(n-k)$th sample value of the first sequence and the kth sample value of the second sequence represented by

$$\sum_{-\infty}^{\infty} x_1(n-k)x_2(k).$$

(3.28) embodies a theorem, called convolution theorem, which states that Z-transform of the convolution of the two sequences is the product of the Z-transforms of the individual sequences. Thus,

$$Z\{[x(n)]\} = Z\{[x_1(n)] * [x_2(n)]\} = Z\{[x_1(n)]\}\cdot Z\{[x_2(n)]\} \Rightarrow G(z) = G_1(z)G_2(z). \quad (3.29)$$

Some properties of convolution sum operation

1. *Convolution sum operation is commutative:*

$$[x_1(n)] * [x_2(n)] = [x_2(n)] * [x_1(n)].$$

2. *Convolution sum operation has associative property:*

$$\{[x_1(n)] * [x_2(n)]\} * [x_3(n)] = [x_1(n)] * \{([x_2(n)] * x_3(n))\}.$$

3. *Convolution sum operation satisfies distributive law:*

$$[x_1(n)] * ([x_2(n)] + [x_3(n)]) = [x_1(n)] * [x_2(n)] + [x_1(n)] * [x_3(n)].$$

3.4 Worked-out Examples

1. If $[x_2(n)] = [x_1(n-2)]$, $[x_3(n)] = [x_1(n+2)]$, where $[x_1(n)]$ is the input sequence of finite length given as $[x_1(n)] = \{1, 2, 5, 7, 0, 1\}$. Find their Z-transform.

Solution: $Z\{[x_1(n)]\} = \sum_{n=0}^{n=6} [x_1(n)]\, z^{-n} = \left(1 + 2z^{-1} + 5z^{-2} + 7z^{-3} + 0 \times z^{-4} + 1 \times z^{-5}\right).$

$$\therefore\ G_1(z) = \left(1 + 2z^{-1} + 5z^{-2} + 7z^{-3} + z^{-5}\right).$$

$$\therefore\ Z\{[x_1(n+2)]\} = z^2 G_1(z) = \left(z^2 + 2z + 5 + 7z^{-1} + z^{-3}\right)$$

$$Z\{[x_1(n-2)]\} = z^{-2} G_1(z) = \left(z^{-2} + 2z^{-3} + 5z^{-4} + 7z^{-5} + z^{-7}\right).$$

2. If $Z\{[x(n)]\} = G(z)$, find $Z\{[x(-n)]\}$. Time reversal.

Solution: $Z\{[x(n)]\} = \sum_{n=-\infty}^{\infty} [x(n)]\, z^{-n}$, by definition.

If we set $l = -n$, we would get $Z\{[x(-n)]\} = \sum_{-\infty}^{\infty} [x(l)]\, z^{l} = \sum_{l=-\infty}^{\infty} \left\{[x(l)] \left(z^{-1}\right)\right\}^{-l} = G\left(z^{-1}\right).$

This is the time reversal property of Z-transform.

3. Find the Z-transform of the sequence $[x(n)]$ as defined below:

$$[x(n)] = \left(\tfrac{1}{3}\right)^{n-1} u(n-1).$$

Solution: $[x_1(n)] = \left(\dfrac{1}{3}\right)^n u(n) = a^n u(n).$

$$\therefore\ Z\{[x_1(n)]\} = \frac{1}{1 - az^{-1}} = \frac{1}{1 - \frac{1}{3}z^{-1}} = \frac{z}{\left(z - \frac{1}{3}\right)} = G(z).$$

$$Z\{[x_1(n-1)]\} = z^{-1}\left(\frac{z}{z - \frac{1}{3}}\right) = \frac{1}{\left(z - \frac{1}{3}\right)}.$$

4. Find the Z-transform of the sequence $[x(n)] = na^n u(n)$.

Solution: We know, $Z\{[x_1(n)]\} = Z\{[a^n u(n)]\} = \dfrac{1}{(1 - az^{-1})} = G_1(z).$

$$\therefore\ Z\{[nx_1(n)]\} = Z\{[na^n u(n)]\} = -z\frac{d}{dz}(G_1(z)) = -z\frac{d}{dz}\left(\frac{1}{1 - az^{-1}}\right)$$

$$= -z\left((-1)\frac{az^{-2}}{(1 - az^{-1})^2}\right) = \frac{az^{-1}}{(1 - az^{-1})^2}.$$

5. Find the Z-transform of $[x(n)] = \sin \omega n u(n)$.

Solution: $Z\{[x(n)]\} = \sum_{-\infty}^{\infty} [\sin \omega n u(n)]\, z^{-n} = \sum_{0}^{\infty} \left(\frac{e^{jn\theta} - e^{-jn\theta}}{2j} \right) z^{-n}$

$$= \frac{1}{2j} \left[\sum e^{jn\theta} z^{-n} - \sum e^{-jn\theta} z^{-n} \right]$$

$$= \frac{1}{2j} \left[\sum \left(e^{+j\theta} z^{-1}\right)^n - \sum \left(e^{-j\theta} z^{-1}\right)^n \right]$$

$$= \frac{1}{2j} \left[\frac{1}{1 - e^{j\theta} z^{-1}} - \frac{1}{1 - e^{-j\theta} z^{-1}} \right]$$

$$= \frac{1}{2j} \left[\frac{z^{-1}(\cos\theta + j\sin\theta - \cos\theta + j\sin\theta)}{1 - z^{-1}\left(e^{j\theta} + e^{-j\theta}\right) + z^{-2}} \right]$$

$$= \frac{z^{-1}\sin\theta}{\left(2 - 2z^{-1}\cos\theta + z^{-2}\right)}.$$

6. $[x(n)] = \cos \omega n u(n)$, find $G(z)$.

Solution: $G(z) = Z\{[x(n)]\} = \sum_{-\infty}^{\infty} \{\cos \omega n u(n)\} z^{-n} = \sum_{0}^{\infty} \left(\frac{e^{j\omega n} + e^{-j\omega n}}{2} \right) z^{-n}$

$$= \frac{1}{2} \left(\sum_{0}^{\infty} \left(e^{j\omega} z^{-1}\right)^n + \sum_{0}^{\infty} \left(e^{-j\omega} z^{-1}\right)^n \right) = \frac{1}{2} \left(\frac{1}{\left(1 - e^{j\omega} z^{-1}\right)} + \frac{1}{\left(1 - e^{-j\omega} z^{-1}\right)} \right)$$

$$= \frac{1}{2} \left\{ \frac{2 - \left(e^{j\omega} + e^{-j\omega}\right) z^{-1}}{1 - \left(e^{j\omega} + e^{j\omega}\right) z^{-1} + z^{-2}} \right\} = \frac{1 - z^{-1}\cos\omega}{\left(1 - 2z^{-1}\cos\omega + z^{-2}\right)}.$$

3.5 Inverse Z-transform: $Z^{-1}\{G(z)\}$

If discrete time signal $[x(n)]$ is having its Z-transform as $G(z)$, and $G(z)$ is available, to retrieve the signal from $G(z)$, we have to make the inverse transformation. Mathematically expressed the process stands as

$$[x(n)] = Z^{-1}\{G(z)\} = \frac{1}{2\pi j} \oint G(z) z^{n-1}\, dz.$$

Of course, we do not use this formula too often for retrieval of the signal from the Z-transform. We make use of the following methods for inverse transformation.

(a) *Table look-up method.* Often the inverse Z-transform can be obtained by inspection of Z-transform table.

Illustration 1. Find $Z^{-1}\{G(z)\}$, where $G(z) = \frac{0.5z}{(z^2 - z + 0.25)}$.

Solution: $G(z) = \frac{0.5z}{(z^2 - z + 0.25)} = \frac{0.5z}{(z - 0.5)^2} = \frac{0.5z^{-1}}{(1 - 0.5z^{-1})^2}$.

From the table we see, $Z\{na^n u(n)\} = \frac{az^{-1}}{(1 - az^{-1})^2}$.

Here, $a = 0.5$, so, $Z^{-1}\{G(z)\} = n(0.5)^n u(n)$.

(b) *Long division method.* Suppose, we are given $G(z)$ as $G(z)=\dfrac{1+2z^{-1}}{1+0.4z^{-1}-0.12z^{-2}}$.

$$
\begin{array}{r|l}
 & 1+1.6z^{-1}-0.52z^{-2}+0.4z^{-3}-0.2224z^{-4} \\
\hline
1+0.4z^{-1}-0.12z^{-2} & 1+2z^{-1} \\
 & 1+0.4z^{-1}-0.12z^{-2} \\
 & -\quad -\qquad + \\
\hline
 & 1.6z^{-1}+0.12z^{-2} \\
 & 1.6z^{-1}+0.64z^{-2}-0.192z^{-3} \\
 & -\qquad -\qquad + \\
\hline
 & -0.52z^{-2}+0.192z^{-3} \\
 & -0.52z^{-2}-0.208z^{-3}+0.0624z^{-4} \\
 & +\qquad +\qquad - \\
\hline
 & 0.400z^{-3}-0.0624z^{-4} \\
 & 0.400z^{-3}+0.16z^{-4}-0.048z^{-5} \\
 & -\qquad -\qquad + \\
\hline
 & -0.2224z^{-4}+0.048z^{-5}
\end{array}
$$

$\therefore\ G(z)=1+1.6z^{-1}-0.52z^{-2}+0.4z^{-3}-0.2224z^{-4} \Rightarrow [x(n)]=(1,1.6,-0.52,0.4,-0.2224,\ldots)$.

(c) *Partial fraction expansion method.* $\dfrac{G(z)}{z}=\dfrac{C_1}{(z-p_1)}+\dfrac{C_2}{(z-p_2)}+\ldots\dfrac{C_N}{(z-p_N)}$, $C_1, C_2, \cdots, C_N$, etc. are determined by using the formula:

$$C_k=\frac{(z-p_k)\,G(z)}{z}\bigg|_{z=p_k},\qquad k=1,2,\ldots,N.$$

If $G(z)$ has a pole of multiplicity l, there occurs a factor $(z-p_k)^l$ in the denominator which now can be expanded and expressed as (assuming $l=2$):

$$\frac{G(z)}{z}=\frac{1}{(z-p_1)(z-p_2)^2}. \tag{3.30}$$

(3.30) can be expanded in terms of the partial fractions as

$$\frac{G(z)}{z}=\frac{C_1}{(z-p_1)}+\frac{C_2}{(z-p_2)}+\frac{C_3}{(z-p_2)^2},$$

where C_1, C_2, C_3 are evaluated by using

$$C_1=(z-p_1)\frac{G(z)}{z}\bigg|_{z=p_1} \tag{3.31}$$

$$C_2=\frac{d}{dz}(z-p_2)^2\frac{G(z)}{z}\bigg|_{z=p_2} \tag{3.32}$$

$$C_3=(z-p_2)^2\frac{G(z)}{z}\bigg|_{z=p_2}. \tag{3.33}$$

Illustration 1. Find the inverse of $G(z)$, where $G(z)$ is given as

$$G(z)=\frac{1+3z^{-1}}{1+3z^{-1}+2z^{-2}}.$$

Solution: To remove the negative powers, multiply the numerator and denominator by z^2, thus we get

$$G(z)=\frac{z^2+3z}{z^2+3z+2}\Rightarrow\frac{G(z)}{z}=\frac{z+3}{(z^2+3z+2)}=\frac{z+3}{(z+1)(z+2)}=\frac{C_1}{(z+1)}+\frac{C_2}{z+2}. \quad (3.34)$$

$$C_1=(z+1)\frac{G(z)}{z}\bigg|_{z=-1}=(z+1)\cdot\frac{z+3}{(z+1)(z+2)}\bigg|_{z=-1}\Rightarrow C_1=\frac{2}{1}=2.$$

$$C_2=(z+2)\frac{G(z)}{z}\bigg|_{z=-2}=(z+2)\frac{z+3}{(z+1)(z+2)}\bigg|_{z=-2}=-1.$$

$$\therefore\quad\frac{G(z)}{z}=\frac{2}{(2+1)}+\frac{(-1)}{(z+2)}\Rightarrow G(z)=\frac{2z}{(z+1)}-\frac{z}{z+2}.$$

On inversion, we would get

$$[x(n)]=Z^{-1}\left(\frac{2z}{z+1}\right)-Z^{-1}\left(\frac{z}{z+2}\right)=2Z^{-1}\left(\frac{1}{1+z^{-1}}\right)-Z^{-1}\left(\frac{1}{1+2z^{-1}}\right)=2(-1)^n u(n)-(-2)^n u(n).$$

(d) *Convolution method.* $G(z)$ is split into parts $G_1(z)$ and $G_2(z)$ such that $G(z)=G_1(z)G_2(z)$.

Now, we find $[x_1(n)]$ and $[x_2(n)]$ by taking inversion of $G_1(z)$ and $G_2(z)$ respectively. Then using the convolution property of Z-transform, we would get

$$Z\{[x_1(n)]*[x_2(n)]\}=G_1(z)G_2(z)=G(z)\Rightarrow[x_1(n)]*[x_2(n)]=Z^{-1}\{G(z)\}$$

$$\Rightarrow\sum_{k=-\infty}^{\infty}x_1(k)x_2(n-k)=Z^{-1}\{G(z)\}. \quad (3.35)$$

Illustration: Find $Z^{-1}\left\{\frac{z(z+1)}{(z-1)(z-3)}\right\}$.

Solution: $G(z)=\frac{z(z+1)}{(z-1)(z-3)}=\frac{z}{(z-1)}\cdot\frac{z+1}{(z-3)}=G_1(z)\cdot G_2(z).$

$$G_1(z)=\frac{z}{(z-1)}=\frac{1}{1-z^{-1}}.$$

$$[x_1(n)]=Z^{-1}\{G_1(z)\}=Z^{-1}\left\{\frac{1}{1-z^{-1}}\right\}=u(n).$$

See the Z-transform Table 1.

$$G_2(z)=\frac{z+1}{(z-3)}=\frac{z}{z-3}+\frac{1}{(z-3)}.$$

$$Z^{-1}\{G_2(z)\}=Z^{-1}\left\{\frac{z}{z-3}\right\}+Z^{-1}\left\{\frac{1}{z-3}\right\}=Z^{-1}\left(\frac{1}{1-3z^{-1}}\right)+Z^{-1}\left(\frac{1}{z-3}\right)$$

$$=(3)^n u(n)+Z^{-1}\left(z^{-1}\frac{z}{z-3}\right)=(3)^n u(n)+3^{n-1}u(n-1)=[x_2(n)].$$

$$[x(n)] = [x_1(n)] * [x_2(n)] = u(n) * \left\{3^n u(n) + 3^{n-1} u(n-1)\right\}$$
$$= u(n) * 3^n u(n) + u(n) * 3^{n-1} u(n-1)$$
$$= \sum_{k=0}^{\infty} 3^{n-k} + \sum_{k=0}^{\infty} 3^{n-1-k} = 3^n \sum_{0}^{n} 3^{-k} + 3^{n-1} \sum_{0}^{n-1} 3^{-k}$$
$$= 3^n \left(\frac{1-3^{-n-1}}{2/3}\right) + 3^{n-1}\left(\frac{1-3^{-n}}{2/3}\right) = \frac{3^{n+1}}{2}\left(1-3^{-n-1}\right) + \frac{3^n}{2}(1-3^{-n})$$
$$= 2(3)^n u(n) - u(n).$$

(e) *Using complex contour integration.* The usual formula for finding the inverse Z-transform is the Cauchy residue theorem in terms of the contour integral defined by

$$[x(n)] = \frac{1}{2\pi j} \oint_C G(z) z^{n-1} \, dz.$$

Cauchy residue theorem

$$Z[x(n)] = G(z) = \sum_{n=-\infty}^{\infty} [x(n)] \, z^{-n}. \tag{3.36}$$

Multiplying both sides of (3.36) by z^{k-1} and integrating over a closed contour C (circle) in anticlockwise sense inside the ROC of $G(z)$, encircling the point $z = 0$, we get

$$\oint_C G(z) z^{k-1} \, dz = \oint_C \sum_{k=-\infty}^{\infty} [x(n)] \, z^{-n} \cdot z^{k-1} \, dz = \sum_{k=-\infty}^{\infty} \oint_C [x(n)] \, z^{-n} \cdot z^{k-1} \, dz, \tag{3.37}$$

changing the order of operation in (3.37).

Multiplying both sides of (3.37) by $1/2\pi j$, we would get

$$\frac{1}{2\pi j} \oint_C G(z) z^{k-1} \, dz = \frac{1}{2\pi j} \sum_{k=-\infty}^{k=\infty} \oint_C [x(n)] \, z^{-n+k-1} \, dz = \frac{1}{2\pi j} \sum_{-\infty}^{\infty} [x(n)] \oint_C z^{-n+k-1} \, dz.$$

Now, according to Cauchy's residue theorem,

$$\oint_C z^{-n+k-1} = 2\pi j \, \delta_{kn}, \quad \delta_{kn} = \begin{cases} 1, & k = n \\ 0, & k \neq n. \end{cases}$$

$$\therefore \frac{1}{2\pi j} \oint_C G(z) z^{k-1} \, dz = \frac{1}{2\pi j} \sum_{-\infty}^{\infty} [x(k)] \times 2\pi j \, \delta_{kn} \Rightarrow \frac{1}{2\pi j} \oint_C G(z) z^{n-1} \, dz = [x(n)]$$

$$\Rightarrow [x(n)] = \frac{1}{2\pi j} \oint_C G(z) z^{n-1} \, dz, \tag{3.38}$$

where C is a circle in the z-plane in the ROC of $G(z)$. (3.38) is the general expression for finding the inverse transform of $G(z)$. (3.38) can be written as

$$[x(n)] = \sum \{\text{Residues of } G(z) z^{n-1} \text{ at the poles inside } C\}$$
$$= \sum (z - z_0) G(z) z^{n-1} \Big|_{z=z_0}. \tag{3.39}$$

If $G(z)z^{n-1}$ has no poles inside C for one or more values of n, then $x(n)=0$ for these values.

Residues of the poles can be determined by using Cauchy residue theorem.

Residue theorem of Cauchy. If $f(z)$ be a function of the complex variable z, C be a closed path in the z-plane and $\frac{d}{dz}\{f(z)\}$ exists on and inside the contour C encircling the origin $z=0$ and if $f(z)$ has no poles at $z=z_0$, then

$$\frac{1}{2\pi j}\oint_C \frac{f(z)}{(z-z_0)}dz = \begin{cases} f(z_0), & \text{if } z_0 \text{ is inside } C \\ 0, & \text{if } z_0 \text{ is outside } C. \end{cases} \tag{3.40}$$

If $(k+1)$ order derivative of $f(z)$ exists and $f(z)$ has no poles at $z=z_0$, then

$$\frac{1}{2\pi j}\oint_C \frac{f(z)}{(z-z_0)}dz = \frac{1}{(k-1)}\cdot\frac{d^{k-1}}{z^{k-1}}f(z)\Big|_{z=z_0}, \text{ if } z_0 \text{ is outside } C$$
$$=0. \tag{3.41}$$

The right-hand side value of (3.40) and (3.41) are called residues of the poles $z=z_0$. For multipoles inside the contour, the integral can be evaluated by adding up the results from each pole. For example, if there are n distinct poles $z_1, z_2, \cdots, z_n$ and $f(z)$ has no poles inside C, then

$$\frac{1}{2\pi j}\oint_C \frac{f(z)dz}{(z-z_1)(z-z_2)\cdots(z-z_n)} = \sum_{i=1}^{n}\lim_{z\to z_i}\{(z-z_i)\,\psi(z)\},$$

$$\psi(z) = \frac{f(z)}{(z-z_1)(z-z_2)\cdots(z-z_n)}. \tag{3.42}$$

For multiplicity of any pole, (3.42) is used to determine their contribution to the integral while simple pole, $\psi(z)$, is used.

3.6 Solved Numerical Problems

1. Find Z of $[x(n)] = [u_r(n)]$.

   Solution: Sequence $[u_r(n)]$ is called unit ramp function $[u_r(n)] = \begin{cases} n, & n \geq 0 \\ 0, & n < 0. \end{cases}$

$$Z\{[x(n)]\} = G(z) = \sum_{-\infty}^{\infty}[x(n)]\,z^{-n} = \sum_{0}^{\infty}u_r(n)z^{-n} = \sum_{0}^{\infty}z^{-n} = 0+z^{-1}+z^{-2}+z^{-3}+\ldots$$
$$= \left(1+z^{-1}+z^{-2}+\ldots\right)-1 = \frac{1}{(1-z^{-1})},\quad |z|>1$$
$$= \frac{z}{z-1}.$$

2. Find Z of $[x(n)] = a^n u(n)$.

   Solution: $Z\{[x(n)]\} = \sum[a^n u(n)]\,z^{-n} = \sum[u(n)]\left(a^{-1}z\right)^{-n} = \frac{1}{1-(a^{-1}z)^{-}} = \frac{1}{1-az^{-1}}, |z|>|a|.$

3. Find the Z-transform of (a) $\delta(n)$ (b) $\delta(n-2)$ and (c) $\delta(n+2)$.

   Solution: (a) $[x(n)] = [\delta(n)]$, $\delta(n)$ unit impulse sequence.

$$Z\{\delta[n]\} = \sum_{-\infty}^{\infty}\{[\delta(n)]\}z^{-n} = 1 = G(z), \text{ for all } z.$$

(b) $Z\{[\delta(n-2)]\}=\{[\delta(n-2)]\}z^{-n}=\{[\delta(s)]\}z^{-(s+2)}=z^{-2}[\delta(s)]z^{-s}=z^{-2}Z\{[\delta(n)]\}=z^{-2}G(z).$

$$Z\{[\delta(n+2)]\}=\sum_{-\infty}^{\infty}\{[\delta(s)]\}z^{-(s-2)}=z^{2}\sum_{-\infty}^{\infty}\{[\delta(s)]\}z^{-s}=z^{2}G(z).$$

4. **Consider a two-sided sequence defined as $[x(n)]=a^{n}u(n)+b^{n}u(-n-1)$. Find $G(z)$.**

Solution: Signal is two-sided and exists in both the domains of n.

$$[x(n)]=a^{n}u(n)+b^{n}u(-n-1).$$

$$\Rightarrow Z\{[x(n)]\}=Z\{[a^{n}u(n)]\}+Z\{[b^{n}u(-n-1)]\}=\sum_{-\infty}^{\infty}\{[a^{n}u(n)]\}z^{-n}+\sum_{-\infty}^{\infty}\{[b^{n}u(-n-1)]\}z^{-n}.$$

$$u(n)=\begin{cases}0, & n<0\\ 1, & n>0\end{cases}\quad\Bigg|\quad u(-n-1)=\begin{cases}0, & \text{for } n\geq 0\\ 1, & \text{for } n\leq 1.\end{cases}$$

So, the two Z-transforms as given above reduce to

$$Z\{[x(n)]\}=\sum_{0}^{\infty}[a^{n}u(n)]\,z^{-n}+\sum_{-\infty}^{-1}b^{n}u(-n-1)z^{-n}=\frac{1}{1-az^{-1}}+\sum_{1}^{\infty}b^{-n}1\cdot z^{n}$$

$$=\frac{1}{1-az^{-1}}+\sum_{1}^{\infty}\left(b^{-1}z\right)^{n}=\frac{1}{1-az^{-1}}+\left[\left(b^{-1}z\right)+\left(b^{-1}z\right)^{2}+\left(b^{-1}z\right)^{3}+\ldots\right].$$

The first power series converges, if $\left|az^{-1}\right|<1$ or $|z|>|a|$. The second series converges, if $\left|b^{-1}z\right|<1$ or, $|z|<|b|$.

So, $Z\{x(n)\}=\dfrac{1}{1-az^{-1}}+\left\{\dfrac{1}{1-b^{-1}z}-1\right\}=\dfrac{1}{(1-az^{-1})}\quad\dfrac{1}{(1-bz^{-1})}.$

5. **Find the Z-transform of the following signals:**

(a) $[x(n)]=\left(\frac{1}{2}\right)^{n}u(n)$

(b) $[x(n)]=\left(\frac{1}{3}\right)^{n}u(-n-1)$

(c) $[x(n)]=\left(\frac{1}{2}\right)^{n}u(n)+\left(\frac{1}{3}\right)^{n}u(-n-1).$

Solution: (a) $Z\{x(n)\}=Z\{[a^{n}u(n)]\}=\sum_{0}^{\infty}\{[a^{n}u(n)]\}z^{-n}=\sum_{0}^{\infty}\left(az^{-1}\right)^{n}$

$$=1+az^{-1}+(az)^{2}+\cdots+\infty=\frac{1}{1-az^{-1}}=\frac{z}{(z-a)},\ |z|>a$$

$$=\frac{z}{z-\left(\frac{1}{2}\right)}.$$

(b) $$Z\{[x(n)]\}=Z\left\{\left(\frac{1}{3}\right)^{n}u(-n-1)\right\}=\sum_{-\infty}^{\infty}\left\{\left[\left(\frac{1}{3}\right)^{n}u(-n-1)\right]\right\}z^{-n}=\sum_{-\infty}^{-1}\left(\frac{1}{3}\right)^{n}z^{-n}$$

$$=\sum_{1}^{\infty}\left(\frac{1}{3}\right)^{-n}z^{n}=\left\{\left(\frac{1}{3}\right)^{-1}z\right\}+\left\{\left(\frac{1}{3}\right)^{-1}z\right\}^{2}+\cdots\infty$$

$$=-\frac{1}{1-3z}-1=\frac{3z}{(1-3z)}=\frac{1}{\left(\frac{1}{3z}-1\right)}$$

$$=-\frac{1}{\left(1-\frac{1}{3}z^{-1}\right)},\ \text{ROC: } z\leq\frac{1}{3}.$$

(c) Similar to solved numerical problem (4), where $a = 1/2$, $b = 1/3$.

We will find, $Z\{x(n)\} = \dfrac{1}{\left(1-\frac{1}{2}z^{-1}\right)\left(1-\frac{1}{3}z^{-1}\right)}$.

6. Determine the Z-transform of the signal $[x(n)] = \begin{cases} a^n, & 0 \le n \le M-1 \\ 0, & \text{elsewhere.} \end{cases}$

Solution: $Z\{[x(n)]\} = \sum_0^{M-1} (a^n)\, z^{-n} = \sum_0^{M-1} \left(az^{-1}\right)^n = 1 + az^{-1} + \left(az^{-1}\right)^2 + \cdots + (M-1)$

$$= \frac{1-(az^{-1})^M}{(1-az^{-1})} = \frac{\left(1-\frac{a^M}{z^M}\right)z}{(z-a)} = \frac{z^M - a^M}{z^{M-1}(z-a)}.$$

Table I: Z-transforms

Sequence	Transform	Region of convergence [ROC]
$\delta(n)$	1	All z
$u(n)$	$\dfrac{1}{(1-z^{-1})}$	$\lvert z\rvert < 1$
$-u(-n-1)$	$\dfrac{1}{(1-z^{-1})}$	$\lvert z\rvert > 1$
$\delta(n-m)$	z^{-m}	All z except 0 (if $m > 0$) All z except 0 (if $m < 0$)
$a^n u(n)$	$\dfrac{1}{(1-az^{-1})}$	$\lvert z\rvert > \lvert a\rvert$
$-a^n u(-n-1)$	$\dfrac{1}{(1-az^{-1})}$	$\lvert z\rvert < \lvert a\rvert$
$na^n u(n)$	$\dfrac{az^{-1}}{(1-az^{-1})^2}$	$\lvert z\rvert > \lvert a\rvert$
$-na^n u(-n-1)$	$\dfrac{az^{-1}}{(1-z^{-1})^2}$	$\lvert z\rvert < \lvert a\rvert$
$[\cos n\omega]\, u(n)$	$\dfrac{1-(\cos\omega)z^{-1}}{1-(2\cos\omega)z^{-1}+z^{-2}}$	$\lvert z\rvert > \lvert a\rvert$
$[\sin n\omega]\, u(n)$	$\dfrac{(\sin\omega)z^{-1}}{1-(2\cos\omega)z^{-1}+z^{-2}}$	$\lvert z\rvert > \lvert 1\rvert$
$[r^n \cos\omega n]\, u(n)$	$\dfrac{(1-r\cos\omega)z^{-1}}{1-(2r\cos\omega)z^{-1}+z^{-2}r^2}$	$\lvert z\rvert > \lvert r\rvert$
$[r^n \sin\omega n]\, u(n)$	$\dfrac{(r\sin\omega)z^{-1}}{1-(2r\cos\omega)z^{-1}+r^2z^{-2}}$	$\lvert z\rvert > \lvert r\rvert$

Table II: *Z*-transforms inverse

	$G(z)$	$[x(n)]=Z^{-1}\{G(z)\}$
1.	$\frac{z}{(z-a)}$	$a^n u(n)$
2.	$\frac{z}{(z-a)^2}$	$na^{n-1}u(n)$
3.	$\frac{z}{(z-a)^3}$	$\frac{n(n-1)a^{n-2}}{2!}u(n)$
4.	$\frac{z}{(z-a)^4}$	$\frac{n(n-1)(n-2)a^{n-3}}{3!}u(n)$
5.	$\frac{1}{(1-2z^{-1})(1-z^{-1})^2}$	$4(2)^n u(n)-3u(n)-nu(n)$
6.	$\frac{z^3+z^2}{(z-1)(z-3)}$	$\delta(n+1)-u(n)+6(3)^n u(n)$
7.	$\frac{z}{(z-2)(z-3)}, \lvert z\rvert<2$	$[2^n-3^n]\,u(-n-1)$
8.	$\frac{1}{(1-z^{-1})(1-2z^{-1})}$	$\left[2^{n+1}-1\right]u(n)$
9.	$\frac{z}{\left(z^2+\frac{1}{12}z-\frac{1}{12}\right)}$	$-\frac{12}{7}\left(-\frac{1}{4}\right)^n u(n)+\frac{12}{7}\left(\frac{1}{3}\right)^n u(n)$
10.	$\frac{z^2+3z}{z^2+3z+2}$	$2(-1)^n u(n)-(-2)^n u(n)$

3.7 Solved Numerical Problems

1. Determine the *Z*-transform of the signal $[x(n)]=\{3,0,0,0,0,\underset{\uparrow}{6},1,-4\}$.

Solution: $Z\{[x(n)]\}=\sum_{-\infty}^{\infty}x(n)z^{-n}=3z^5+6+1\times z^{-1}-4z^{-2}\Rightarrow G(z)=3z^5+6+z^{-1}-4z^{-2}$.

2. Determine the *Z*-transform of the signal represented by the sequence $[x(n)]=(1+n)u(n)$.

Solution: $Z\{[x(n)]\}=\sum[x(n)]z^{-n}=\sum\{(1+n)u(n)\}z^{-n}=\sum u(n)z^{-n}+\sum nu(n)z^{-n}$

$$=\left(1+z^{-1}+z^{-2}+\cdots\right)+Z\{nu(n)\}=\frac{1}{(1-z^{-1})}+\frac{z^{-1}}{(1-z^{-1})^2}$$

$$=\frac{(1-z^{-1})+z^{-1}}{(1-z^{-1})^2}=\frac{1}{(1-z^{-1})^2}.$$

3. Find the Z-transform of the signal represented by $[x(n)] = [(a^n + a^{-n})\, u(n)]$.

Solution: $Z\{[x(n)]\} = Z\{[a^n u]\} + Z\{[a^{-n} u]\} = \sum_{-\infty}^{\infty}[a^n u(n)]\, z^{-n} + \sum_{-\infty}^{\infty}[a^{-n} u(n)]\, z^{-n}$

$$= \sum_0^{\infty} a^n z^{-n} + \sum_0^{\infty} a^{-n} z^{-n} = \frac{z}{(z-a)} + \frac{z}{\left(z-\frac{1}{a}\right)}.$$

4. $[x(n)] = \{\underset{\uparrow}{2}, -1, 3, 2, 5\}$, find $G(z)$.

Solution: $G(z) = \sum_{n=0}^{5}[x(n)]\, z^{-n} = 2 - 1z^{-1} + 3z^{-2} + 2z^{-3} + 5z^{-4} = \left(2 - z^{-1} + 3z^{-2} + 2z^{-3} + 5z^{-4}\right).$

5. $[x(n)] = \{2, -1, 3, 2, \underset{\uparrow}{1}, 0, 2, 3, -1\}$, find $G(z)$.

Solution: $G(z) = \sum_{-4}^{4}[x(n)]\, z^{-n} = \left(2z^4 + (-1)z^3 + 3z^2 + 2z^1 + 1 + 0z^{-1} + 2z^{-2} + 3z^{-3} - 1z^{-4}\right)$

$$= \left(2z^4 - z^3 + 3z^2 + 2z + 1 + 2z^{-2} + 3z^{-3} - z^{-4}\right).$$

6. $x(n) = a^n u(n) + b^n u(-n-1)$, find $G(z)$.

Solution: $G(z) = \sum_0^{\infty} a^n z^{-n} + \sum_{-\infty}^{-1} b^n z^{-n}$

$$= \sum_0^{\infty}\left(az^{-1}\right)^n + \sum_1^{\infty} b^{-n} z^n \qquad u(n) = \begin{cases} 1, & n \ge 0 \\ 0, & n < 0 \end{cases}$$

$$= \sum_0^{\infty}\left(az^{-1}\right)^n + \sum_1^{\infty}\left(b^{-1}z\right)^n \qquad u(-n-1) = 0,\ n \ge 0$$
$$= 1,\ n \le 1$$

$$= \frac{1}{(1-az^{-1})} + \frac{1}{(1-b^{-1}z)}$$

$$= \frac{1}{(1-az^{-1})} + \frac{1}{(1-b^{-1}z)} \qquad \text{for } n \to \infty, \left(b^{-1}z\right)^n \to 0$$

$$= \frac{z}{(z-a)} - \frac{z}{(z-b)}.$$

7. $[x(n)] = \left(\frac{1}{3}\right)^n u(n) - \left(\frac{1}{2}\right)^n u(-n-1)$, find $G(z)$.

Solution: Take $\frac{1}{3} = a$, $\frac{1}{2} = b$. $\therefore$ $[x(n)] = a^n u(n) - (b)^n u(-n-1)$.

$$\Rightarrow Z\{[x(n)]\} = Z\{a^n u(n)\} - Z\{(b)^n u(-n-1)\}$$

$$Z\{a^n u(n)\} = \sum_0^{\infty} a^n z^{-n} \text{ [by definition]}$$

$$= \left(1 + az^{-1} + \left(az^{-1}\right)^2 + \left(az^{-1}\right)^2 + \cdots\right)$$

$$= \frac{1}{1-az^{-1}} = \frac{1}{1-\frac{a}{z}} = \frac{z}{\left(z-\frac{1}{3}\right)}.$$

$$Z\{b^n u(-n-1)\} = \sum_{-\infty}^{-1} b^n z^{-n} = \sum_{1}^{\infty} \left(b^{-1}z\right)^{-n} \qquad \left[u(-n-1) = \begin{cases} 0, & n \geq 0 \\ 1, & n \leq 0 \end{cases}\right]$$

$$= \sum_{0}^{\infty} \left(b^{-1}z\right)^{-n} - 1 = \left(1 + \left(b^{-1}z\right)^{-1} + \left(b^{-1}z\right)^{-2} + \left(b^{-1}z\right)^{-3} + \cdots\right) - 1$$

$$= \frac{1}{(1-b^{-1}z)} - 1 = \frac{b^{-1}z}{(1-b^{-1}z)} = \frac{(z/b)}{\left(1-\frac{z}{b}\right)} = -\frac{z}{\left(z-\frac{1}{2}\right)}.$$

$$\therefore\ Z\{[x(n)]\} = \frac{z}{\left(z-\frac{1}{3}\right)} + \frac{z}{\left(z-\frac{1}{2}\right)}.$$

8. $[x(n)] = \left(-\frac{1}{5}\right)^n u(n) + 5 \times \left(\frac{1}{2}\right)^{-n} u(-n-1)$, find $G(z)$.

Solution: Here, $a = -1/5, b = 2$. Putting these values, we would get from Prob. (7),

$$Z\{[x(n)]\} = \frac{z}{\left(z+\frac{1}{5}\right)} - \frac{5z}{(z-2)} = \frac{1}{(1+0.2z^{-1})} - \frac{5}{(1-2z^{-1})}.$$

9. $[x(n)] = \frac{1}{2}\delta(n) + \delta(n-1) - \frac{1}{3}\delta(n-2)$, find $G(z)$.

Solution: $G(z) = Z\{[x(n)]\} = \frac{1}{2}Z\{\delta(n)\} + Z\{[\delta(n-1)]\} - \frac{1}{3}Z\{[\delta(n-2)]\}$

$$= \tfrac{1}{2}(1) + z^{-1}(1) - \tfrac{1}{3}z^{-2}(1) = \left(\tfrac{1}{2} + z^{-1} - \tfrac{1}{3}z^{-2}\right).$$

10. $[x(n)] = u(n-2)$, find $G(z)$.

Solution: $G(z) = Z\{[x(n)]\} = Z\{[x(n-2)]\} = z^{-2}Z\{[u(n)]\} = z^{-2} \cdot \dfrac{z}{(z-1)} = \dfrac{z^{-1}}{(z-1)}.$

11. $[x(n)] = \cos n\theta\, u(n)$, find $G(z)$.

Solution: $G(z) = Z\{[x(n)]\} = \sum_{-\infty}^{\infty} [x(n)]\, z^{n} = \sum_{-\infty}^{\infty} (\cos n\theta) z^{-n} u(n) = \sum_{0}^{\infty} \frac{1}{2}\left\{e^{jn\theta} + e^{-jn\theta}\right\} z^{-n}$

$$= \frac{1}{2}\left\{\sum_{0}^{\infty} e^{jn\theta} z^{-n} + \sum_{0}^{\infty} e^{-jn\theta} z^{-n}\right\} = \frac{1}{2}\left\{\sum_{0}^{\infty} \left(e^{j\theta}z^{-1}\right)^n + \sum_{0}^{\infty} \left(e^{-j\theta}z^{-1}\right)^n\right\}$$

$$\Rightarrow G(z) = \frac{1}{2}\left(\frac{1}{1-e^{j\theta}z^{-1}} + \frac{1}{1-e^{-j\theta}z^{-1}}\right) = \frac{1}{2}\left(\frac{1-e^{-j\theta}z^{-1} + 1 - e^{j\theta}z^{-1}}{(1-z^{-1}(e^{j\theta}+e^{-j\theta}) + z^{-2})}\right)$$

$$= \frac{1}{2}\left(\frac{2-(e^{j\theta}+e^{-j\theta})\, z^{-1}}{1-(2\cos\theta)z^{-1} + z^{-2}}\right) = \left(\frac{1-z^{-1}\cos\theta}{1-2z^{-1}\cos\theta + z^{-2}}\right).$$

12. $[x(n)] = \{3(3)^n - 4(2)^n\}\, u(n)$, find $G(z)$.

Solution: $Z\{[x(n)]\} = \sum_{-\infty}^{\infty}\{(3(3)^n - 4(2)^n)\, u(n)\} z^{-n} = \sum_{0}^{\infty}\{3(3)^n - 4(2)^n\} z^{-n}$

$$= \sum_{0}^{\infty} 3(3)^n z^{-n} - \sum_{0}^{\infty} 4(2)^n z^{-n} = 3\sum_{0}^{\infty}\left(3z^{-1}\right)^n - 4\sum_{0}^{\infty}\left(2z^{-1}\right)^n$$

$$= 3 \times \frac{1}{(1-3z^{-1})} - 4 \times \frac{1}{(1-2z^{-1})} = \frac{3z}{(z-3)} - \frac{4z}{(z-2)},\ \text{ROC } |z| > 3.$$

13. $[x(n)] = \left(\frac{1}{3}\right)^{n-1} u(n-1)$, find $G(z)$.

Solution: Take $[x_1(n)] = \left(\frac{1}{3}\right)^n u(n) = a^n u(n)$.

$$G_1(z) = \frac{z}{\left(z - \frac{1}{3}\right)}.$$

$$\therefore\ G(z) = z^{-1} \times G_1(z) = z^{-1} \cdot \frac{z}{\left(1 - \frac{1}{3}\right)} = \frac{1}{\left(z - \frac{1}{3}\right)} \Rightarrow Z\left\{\left(\frac{1}{3}\right)^{n-1} u(n-1)\right\} = G(z) = \frac{1}{\left(z - \frac{1}{3}\right)}.$$

14. $[x(n)] = (n+1)\, a^n u(n)$, find $G(z)$.

Solution: Take $[x_1(n)] = a^n u(n) \Rightarrow Z\{[x_1(n)]\} = G_1(z)$.

$$[x(n)] = (n+1)\, a^n u(n) = [n x_1(n)] + [x_1(n)].$$

$$x(z) = Z\{[y(n)]\} = Z\{[y_1(n)]\} + (-z)\frac{dG_1(z)}{dz} = G_1(z) - z\frac{dG_1(z)}{dz}$$

$$= \frac{1}{(1-az^{-1})} - z \times (-1)\frac{(+a)z^{-2}}{(1-az^{-1})^2} = \frac{1}{a - az^{-1}} + \frac{az^{-1}}{(1-az^{-1})^2}$$

$$= \frac{(1-az^{-1}) + az^{-1}}{(1-az^{-1})^2} = \frac{1}{(1-az^{-1})^2}.$$

15. $[x(n)] = n^2 a^n u(n)$, find $G(z)$.

Solution: Take $x_1(n) = a^n u(n)$.

$$G_1(z) = Z\{[x_1(n)]\} = \frac{1}{(1-az^{-1})}.$$

$$Z\{[n x_1(n)]\} = -z\frac{d}{dz}\{G_1(z)\} = -z \times (-1)\frac{(a/z^2)}{(1-az^{-1})^2} \Rightarrow Z\{[n x_1(n)]\} = \frac{az^{-1}}{(1-az^{-1})^2}.$$

$$Z\{[n x_2(n)]\} = Z\left\{\left[n^2 a^n u(n)\right]\right\} = -z\frac{d}{dz}\left(\frac{az^{-1}}{(1-az^{-1})^2}\right)$$

$$= -Z\left\{\frac{-(a/z^2)}{(1-az^{-1})^2} + az^{-1} \times (-2)\frac{(a/z^2)}{(1-az^{-1})^3}\right\} = \frac{(a/z)}{(1-az^{-1})^2} + \frac{2a^2/z^2}{(1-az^{-1})^3}$$

$$= \frac{az^{-1}}{(1-az^{-1})^2} + \frac{2a^2z^{-2}}{(1-az^{-1})^3} = \frac{az^{-1} + a^2z^{-2}}{(1-az^{-1})^3}.$$

3.8 Short Theory Questions with Answers

1. Find the Z-transform of the following sequences with their ROCs:

(a) $[x_1(n)] = a^n u(n-2)$ (c) $[x_3(n)] = a^n u(n+4)$

(b) $[x_2(n)] = -a^n u(-n-1)$ (d) $[x_4(n)] = a^n u(-n)$.

Solution: (a) $Z\{a^n u(n-2)\} = z^{-2} Z\{[a^n u(n)]\} = z^{-2} \cdot \dfrac{1}{(1-az^{-1})} = \dfrac{+z^{-2}}{(1-az^{-1})}$ $|z| > a$ (ROC).

(b) $Z\{[-a^n u(-n-1)]\} = \dfrac{z}{(z-a)}$, $|z| < a$ (ROC).

(c) $Z\{[x_3(n)]\} = Z\{[a^n u(n+4)]\} = z^4 \dfrac{1}{(1-az^{-1})} = \dfrac{z^4}{(1-az^{-1})}$, $|z| < a$.

(d) $Z\{[x_4(n)]\} = Z\{[a^n u(-n)]\} = \sum_{-\infty}^{\infty} a^n u(-n) z^{-n} = \sum_{-\infty}^{0} a^n z^{-n}$, put $l = -n$

$$= \sum_{0}^{\infty} a^{-n} z^n = \sum_{0}^{\infty} \left(a^{-1}z\right)^n = 1 + \left(a^{-1}z\right) + \left(a^{-1}z\right)^2 + \cdots = \frac{1}{(1-a^{-1}z)}, \ \left|a^{-1}z\right| < 1.$$

2. What is a Z-transform? What is its importance. What are the different types of sequences?

Ans. The Z-transform of a sequence $[x(n)]$ representing a discrete time signal is defined as

$$G(z) = Z\{[x(n)]\} = \sum_{-\infty}^{\infty} [x(n)]\, z^{-n}.$$

Since Z-transform is an infinite power series, it exists for only those values of z, which make the series converge and so the ROC (Region of convergence) refers to those set of values of z for which $G(z)$ is finite.

Now, z is a complex quantity which can be expressed also as $z = re^{jwn}$.

$$\therefore\ G(z) = \sum_{-\infty}^{\infty} [x(n)]\, r^{-n} e^{-j\omega n}.$$

Thus, we see that for $r = 1$,

$$G(z) = \sum_{\infty}^{\infty} [x(n)]\, e^{-j\omega n} = G(j\omega) \rightarrow \text{Fourier transform.}$$

It shows that the Z-transform evaluated on the unit circle corresponds to the Fourier transform.

3. What is the meaning of ROC?

Ans. ROC (region of convergence) of $G(z)$ is the set of z values for which $G(z)$ remains finite.

4. What is meant by a causal signal and anti-causal signal? Also indicate their ROCs.

Ans. A causal signal is one for which $[x(n)] = 0$ for $n < 0$ and for such a signal the Z-transform is indicated as $G_+(z)$.

An anti-causal signal is one for which $[x(n) = 0]$ for $n > 0$ and for such a signal the Z-transform is indicated as $G_-(z)$.

A causal signal the bridges past and the present and not the future, whereas an anti-causal signal also connects the future, it has anticipatory aspect.

5. Discuss the existence of Z-transform of a signal represented by

$$[x(n)] = \left(\frac{1}{2}\right)^n u(n) + \left(\frac{1}{3}\right)^n u(-n-1).$$

Solution:

$$Z\{[x(n)]\} = \sum_{-\infty}^{\infty} [x(n)]\, z^{-n}$$

$$\Rightarrow Z\{[x(n)]\} = \sum_{-\infty}^{\infty} \{[a^n u(n) + b^n u(-n-1)]\} z^{-n} = \sum_{-\infty}^{\infty} \{a^n u(n)\} z^{-n} + \sum_{-\infty}^{\infty} \{b^n u(-n-1)\} z^{-n}.$$

Now

$$u(n) = \begin{cases} 1, & n \geq 0 \\ 0, & n < 0 \end{cases} \qquad u(-n-1) = \begin{cases} 0, & n > 0 \\ 1, & n \leq 0. \end{cases}$$

$$\therefore\ G(z) = Z\{[x(n)]\} = \sum_{0}^{\infty} (az^{-1})^n + \sum_{-\infty}^{-1} b^n z^{-n} = \left(1 + (az^{-1}) + (az^{-1})^2 + \ldots\infty\right) + \sum_{1}^{\infty} b^{-n} z^n$$

$$= \underset{\text{ROC } |z|>a}{\frac{1}{(1-az^{-1})}} + \underset{\text{ROC } |z|<b}{\left(b^{-1}z + b^{-2}z^2 + \ldots\infty\right)} = \underset{\text{ROC } |z|>a}{\frac{1}{(1-az^{-1})}} + \underset{\text{ROC } |z|<b}{\left(1 + b^{-1}z + b^{-2}z^2 + \ldots\infty\right)}$$

$$= \frac{1}{(1-az^{-1})} + \left(\frac{1}{1-b^{-1}z} - 1\right) = \frac{1}{(1-az^{-1})} + \underset{\text{ROC } |z|<b}{\frac{b^{-1}z}{(1-b^{-1}z)}} = \frac{z}{(z-a)} + \frac{(z/b)}{\left(1-\frac{z}{b}\right)}$$

$$= \underset{\text{ROC } |z|>a}{\frac{z}{(z-a)}} - \underset{\text{ROC } |z|<b}{\frac{z}{(z-b)}} \qquad \begin{array}{l|l} az^{-1} < 1 & b^{-1}z < 1 \\ \text{or, } \frac{a}{z} < 1 & \text{or, } \frac{z}{b} < 1 \\ \therefore\ z > a & \therefore\ z < b. \end{array}$$

The region of overlap between two circles of radii a and b represents the ROC of the signal $[x(n)]$. So the condition for the existence of the Z-transform of the signal is $b < |z| < a$. But in our case, the values of a and b are $\frac{1}{2}$ and $\frac{1}{3}$ respectively. So we have here $b < a$. So the Z-transform of the signal exists.

6. Does the signal in the given form $[x(n)]$ exist?

$$[x(n)] = \left(-\frac{1}{5}\right)^n u(n) + 5\left(\frac{1}{2}\right)^{-n} u(-n-1).$$

Solution:

$$Z\{[x(n)]\} = Z\left\{\left[\left(-\frac{1}{5}\right)^n u(n) + 5\left(\frac{1}{2}\right)^{-n} u(n-1)\right]\right\}$$

$$= \sum_{-\infty}^{\infty} \left\{\left(-\frac{1}{5}\right)^n u(n)\right\} z^{-n} + 5\sum_{-\infty}^{\infty} \left[\left(\frac{1}{2}\right)^{-n} u(-n-1)\right] z^{-n}$$

$$= \sum_{0}^{\infty} \left[\left(-\frac{1}{5}\right)^{-1} z^{+1}\right]^{-n} + 5\sum_{-\infty}^{-1} (2^n z^{-n})$$

$$=1+\left(-\frac{1}{5}\right)^{-1}z+\left[\left(-\frac{1}{5}\right)^{-1}z\right]^{-2}+\left[\left(-\frac{1}{5}\right)^{1}z\right]^{-3}+\ldots$$
$$+5\left[2^{-1}z+\left(2^{-1}z\right)^{2}+\left(2^{-1}z\right)^{3}+\ldots\right]$$
$$=\frac{1}{(1+0.2z^{-1})}+5\left[\frac{1}{\left(1-\frac{1}{2}z\right)}-1\right]=\frac{1}{(1+0.2z^{-1})}+\frac{\frac{5}{2}z}{\left(1-\frac{1}{2}z\right)}$$
$$=\frac{1}{(1+0.2z^{-1})}+\frac{5}{\left(\frac{1}{2}z-1\right)}=\frac{1}{(1+0.2z^{-1})}-\frac{5}{(1-0.2z^{-1})}\ \left[\text{ROC } \tfrac{1}{5}<|z|<2\right].$$

7. Discuss the existence of Z-transform of the following signal:

$$[x(n)]=\left(\frac{1}{2}\right)^{n}u(n)+\left(\frac{1}{3}\right)^{n}u(-n-1).$$

Solution: $a=1/2$, $b=1/3$, so $b \not> a$, and hence, the Z-transform does not exist.

8. Find the Z-transform of the signal:

$$[x(n)]=\left(\frac{1}{3}\right)^{n}u(n)-\left(\frac{1}{2}\right)^{n}u(-n-1).$$

Solution: $b=1/2>a(=1/3)$, hence, the Z-transform exists.

$$Z\{[x(n)]\}=\sum_{-\infty}^{\infty}\{[x(n)]\}z^{-n}=\sum_{-\infty}^{\infty}\left[\left(\frac{1}{3}\right)^{n}u(n)\right]z^{-n}-\sum_{-\infty}^{\infty}\left[\left(\frac{1}{2}\right)^{n}u(-n-1)\right]z^{-n}$$
$$=\sum_{0}^{\infty}\left\{\left(\frac{1}{3}\right)z^{-1}\right\}^{n}-\sum_{-\infty}^{-1}\left(\frac{1}{2}\right)^{n}z^{-n}=\left[1+\frac{1}{3}z^{-1}+\left(\frac{1}{3}z^{-1}\right)^{2}+\ldots\infty\right]-\sum_{1}^{\infty}\left(\frac{1}{2}\right)^{-n}z^{n}$$
$$=\frac{1}{\left(1-\frac{1}{3}z^{-1}\right)}-\sum_{1}^{\infty}\left\{\left(\frac{1}{2}\right)^{-1}z\right\}^{n}=\frac{1}{\left(1-\frac{1}{3}z^{-1}\right)}-\left\{\left(\frac{1}{2}\right)^{-1}z+\left(\frac{1}{2}\right)^{-2}z^{2}+\ldots\infty\right\}$$
$$=\frac{1}{\left(1-\frac{1}{3}z^{-1}\right)}-\left\{\left(\frac{1}{\left(1-\frac{1}{2}z\right)}-1\right)\right\}=\frac{1}{\left(1-\frac{1}{3}z^{-1}\right)}-\frac{\frac{1}{2}z}{\left(1-\frac{1}{2}z\right)}$$
$$=\frac{1}{\left(1-\frac{1}{3}z^{-1}\right)}-\frac{1}{1\left(\frac{1}{2}z^{-1}-1\right)}=\frac{1}{\left(1-\frac{1}{2}z^{-1}\right)}+\frac{1}{\left(1-\frac{1}{3}z^{-1}\right)}\ \left[\text{ROC: } \tfrac{1}{3}\le z\le\tfrac{1}{2}\right].$$

9. $Z\left\{\left(\frac{1}{3}\right)^{n}u(-n-1)\right\}=?$

Solution:

$$Z\left\{\left(\frac{1}{3}\right)^{n}u(-n-1)\right\}=\sum_{-\infty}^{\infty}\left\{\left(\frac{1}{3}\right)^{n}u(-n-1)\right\}z^{-n}=\sum_{-\infty}^{-1}\left(\frac{1}{3}\right)^{n}z^{-n}=\sum_{1}^{\infty}\left(\frac{1}{3}\right)^{-n}z^{n}$$
$$=\left\{\left(\frac{1}{3}\right)^{-1}z\right\}^{n}=\left(\frac{1}{3}\right)^{-1}z+\left\{\left(\frac{1}{3}\right)^{-1}z\right\}^{2}+\cdots=\frac{1}{\left[1-\left(\frac{1}{3}\right)^{-1}z\right]}-1$$
$$=\frac{\left(\frac{1}{3}\right)^{-1}z}{\left(1-\left(\frac{1}{3}\right)^{-1}z\right)}=\frac{\left(\frac{1}{3}\right)^{-1}z}{\left(\frac{1}{3}\right)^{-1}z\left(\frac{1}{3}z^{-1}-1\right)}=-\frac{1}{\left(1-\frac{1}{3}z^{-1}\right)}\ \left[\text{ROC: } |z|<\tfrac{1}{3}\right].$$

10. Determine the impulse response of the system described by the difference equation

$$y(n)-\frac{1}{6}y(n-1)-\frac{1}{6}y(n-2)=x(n),$$

where $x(n)=nu(n)$.

Solution: Z-transforming both sides, we get

$$Z\{x(n)\}=Z\{y(n)\}-\frac{1}{6}Z\{y(n-1)\}-\frac{1}{6}Z\{y(n-2)\}=Y(z)-\frac{1}{6}z^{-1}Y(z)-\frac{1}{6}z^{-2}Y(z)$$

$$\Rightarrow Y(z)=\frac{G(z)}{1-\frac{1}{6}z^{-1}-\frac{1}{6}z^{-2}}.$$

$$G(z)=Z\{nu(n)\}=\frac{z}{1-z}$$

$$Y(z)=\frac{\frac{z}{1-z}}{1-\frac{1}{6}z^{-1}-\frac{1}{6}z^{-2}}=\frac{z^2\left(\frac{z}{1-z}\right)}{(\delta z^2-z-1)}.$$

$$\frac{Y(z)}{z}=\frac{z^2}{(1-z)(2z-1)(3z+1)}=\frac{C_1}{(1-z)}+\frac{C_2}{(2z-1)}+\frac{C_3}{(1+3z)}$$

$$C_1=(1-z)\frac{Y(z)}{z}\bigg|_{z=1}=\frac{1^2}{1\cdot 4}=\frac{1}{4}$$

$$C_2=(2z-1)\frac{Y(z)}{z}\bigg|_{z=1/2}=\frac{1/4}{\left(1-\frac{1}{2}\right)\left(1+3\cdot\frac{1}{2}\right)}=\frac{1/4}{\frac{1}{2}\times\frac{5}{2}}=\frac{1}{5}$$

$$C_3=(1+3z)\frac{Y(z)}{z}\bigg|_{z=-1/3}=\frac{\left(-\frac{1}{3}\right)^z}{\left(1+\frac{1}{3}\right)\left(-\frac{2}{3}-1\right)}=\frac{1/9}{-20/9}=-\frac{1}{20}.$$

$$\therefore \frac{Y(x)}{z}=\frac{1}{4(1-z)}+\frac{1}{5(2z-1)}+\frac{1}{20(1+3z)}$$

$$\Rightarrow Y(z)=\frac{z}{4(1-z)}+\frac{z}{5(2z-1)}+\frac{z}{20(1+3z)}=-\frac{z}{4(z-1)}+\frac{z}{10\left(z-\frac{1}{2}\right)}+\frac{z}{60\left[z-\left(-\frac{1}{3}\right)\right]}$$

$$\Rightarrow Y(b)=-\frac{1}{4}u(n)+\frac{1}{10}\left(\frac{1}{2}\right)^n u(n)+\frac{1}{60}\left(-\frac{1}{3}\right)^n u(n)=\left[\frac{1}{20}\left(\frac{1}{2}\right)^n+\frac{1}{60}\left(\frac{1}{3}\right)^n-\frac{1}{8}\right]u(n).$$

11. Find the system function $h(n)$, when the input and output satisfy the difference equation $y(n)=x(n)+3x(n-1)+2y(n-1)-y(n-2)$.

Solution: $H(z)=\dfrac{Z\{y(n)\}}{Z\{x(n)\}}=\dfrac{Y(z)}{X(z)};\ Z^{-1}\{H(z)\}=h(n).$

The difference equation is given by $y(n)=x(n)+3x(n-1)+2y(n-1)-y(n-2)$.

Z-transforming both sides and denoting $Z\{y(n)\}=Y(z), Z\{x(n)\}=X(z)$, we would get

$$Y(z)=X(z)+3z^{-1}x(z)+2z^{-1}Y(z)=z^{-2}Y(z)$$

$$\Rightarrow Y(z)\left[1-2z^{-1}+z^{-2}\right]=X(z)\left[1+3z^{-1}\right]$$

$$\Rightarrow H(z)=\frac{Y(z)}{X(z)}=\frac{1+3z^{-1}}{\left[1-2z^{1}+z^{-2}\right]}=\frac{(1+3z^{-1})\,z^2}{(1-2z+z^2)}=\frac{z(z+3)}{(z-1)^2}$$

Using the method of partial fraction, we can write:

$$H(z)=\frac{z}{z-1}+\frac{4z}{(z-1)^2}$$

$$\Rightarrow Z^{-1}\{H(z)\}=h(n)=Z^{-1}\left\{\frac{z}{z-1}\right\}+Z^{-1}\left\{\frac{4z}{(z-1)^2}\right\}=u(n)+4nu(n).$$

3.9 Theory Questions Exercise

1. Define Z-transform of a discrete time signal. Explain the meaning of ROC of Z-transformed signal.
2. Explain the linearity property of Z-transform. What do you mean by the time shifting property of Z-transform.
3. Explain the importance of the Z-transform in designing the LTI system.
   [*Hints.* State the convolution theorem applied to discrete time signals.]
4. State the properties of ROC.
   [*Hints.* (i) ROC is a ring or disc in the Z-plate centred on the origin (ii) ROC cannot contain any pole (iii) ROC must be a connected region (iv) ROC of an LTI stable system contains the unit circle.]
5. Determine the Z-transform and ROC of the finite duration signal: $\{x(n)\}=\{2,4,\underset{\uparrow}{5},7,0,1\}$.
   $[x(z)=2z^2+4z+5+7z^{-1}+z^{-3}]$
6. Find Z-transform of (a) digital impulse (b) digital step signal.
7. If Z-transform of a signal $[x(n)]$ is $G(z)$, what is the Z-transform of $[nx(n)]$? $[-z\frac{d}{dz}[G(z)]]$
8. What are the different methods available for evaluation of $z^{-1}\{G(z)\}$?
   [*Hints.* (i) Long division method (ii) Partial fraction expansion method (iii) Residue method (iv) Convolution method.]
9. Find the convolution of the two signals using Z-transform: $[x(n)]=\{\underset{\uparrow}{1},2,1\}$.
10. Determine the system function and unit sample response of the system described by the difference equation: $y(n)=\frac{1}{2}y(n-1)+2x(n)$. $[h(n)=2\left(\frac{1}{2}\right)^n u(n)]$

3.10 Numerical Problems Exercise

1. Find the convolution of the following using Z-transform: $[x(n)]=\{1,\underset{\uparrow}{2},1\}$; $h(n)=(\underset{\uparrow}{1},1,1)$.
   [*Hints.* $Z\{x(n)\}=G(z), Z\{h(n)\}=H(z)$, then $Z\{[x(n)]*[h(n)]\}=G(z)H(z), G(z)=(1+2z^{-1}+z^{-2})$ and $H(z)=(1+z^{-1}+z^{-2})$; so conv $\{[x(n)]\}*\{[h(n)]\}=\{1,3,4,3,1\}$.]
2. Find the Z-transform of the following signals:
   (i) $(-1)^n u(n)$ (ii) $\frac{1}{2}(n^2+n)\left(\frac{1}{3}\right)^{n-1}u(n-1)$ (iii) $[x(n)]=\{-1,0,\underset{\uparrow}{1},0,1\}$.
3. If $[x(n)]$ is a causal sequence, find the Z-transform:
   (i) $[x(n)]=\{\underset{\uparrow}{1},-2,1,3,4\}$ (ii) $[x(n)]=\{\underset{\uparrow}{1},0,3,-1,2\}$. $[1-2z+z^{-2}+3z^{-3}+4z^{-4}]$
4. If $[x(n)]$ is an anticlockwise sequence find its Z-transform:
   (i) $[x(n)]=\{-3,-2,-2,-1,0,\underset{\uparrow}{1}\}$ (ii) $[x(n)]=\{\ldots,11,8,5,2,\underset{\uparrow}{0}\}$.
5. If the output sequence is $[y(n)]$ and the response of the system is $[h(n)]$ as given, find the input sequence.

[*Hints.* Find $Y(z)$ and $H(z)$ and then find $X(z)=Y(z)/H(z)$, hence, $[x(n)]$ by long division method.]

6. Using convolution method find the input sequence $[x(n)]$, whose Z-transform is given as

$$G(z)=\frac{z^2}{\left(1-\frac{1}{z}\right)^2}.$$

7. Solve the problem of finding, the input sequence when the Z-transform is $X(z)$ given as

$$X(z)=\frac{1-\frac{1}{4}z^{-1}}{\left(1-\frac{1}{9}z^{-2}\right)}.$$

8. Find the inverse Z-transform of
   (a) $G(z)=\dfrac{1}{1-3z^{-1}+2z^{-2}}$, use convolution method.
   (b) $G(z)=\dfrac{z}{z^2+\frac{1}{12}z-\frac{1}{12}}$.
   (c) $G(z)=\dfrac{z^2+z}{(z-1)(z-3)}$.
9. Find the impulse response of the system described by the difference equation:
   $$y(n)-3y(n-1)-4y(n-2)=x(n)+2x(n-1).$$ $[\frac{6}{5}(4)^n u(n)-\frac{1}{5}(-1)^n u(n)]$
10. A causal LTI system is described by the difference equation:
   $$y(n)=y(n-1)+y(n-2)+x(n-1);$$
   $x(n)$ is the input and $y(n)$ is the output. Find response of the system and also indicate its ROC.
11. Determine the signal $[x(n)]$ whose Z-transform is given by $G(z)=\log(1+az^{-1}), |z|>a$.
12. A Fibnacci sequence which is described by the difference equation $y(n)=y(n-1)+y(n-2)$, with initial conditions:
   $$y(0)=y(-1)+y(-2)=1$$
   $$y(1)=y(0)+y(-1)=1.$$
13. Compute the convolution of the following signals by means of the Z-transform:
   $$x_1(n)=\begin{cases}\left(\frac{1}{3}\right)^n, & n\geq 0\\ \left(\frac{1}{2}\right)^{-n}, & n<0.\end{cases}$$
   $$x_2(n)=\left(\frac{1}{2}\right)^n u(n).$$
14. Using long division, determine the inverse Z-transform of
   $$G(z)=\frac{1+2z^{-1}}{1-2z^{-1}+z^{-2}},$$
   if (a) $[x(n)]$ is causal (b) $[x(n)]$ is anti-causal.
15. Determine the Z-transforms of the following finite duration signals:

(a) $[x(n)] = \{1,2,5,7,0,1\}$.

(b) $[x(n)] = \{1,2,\underset{\uparrow}{5},7,0,1\}$.

(c) $[x(n)] = \{1,0,-1,0,1,-1,\ldots\}$.

(d) $[x(n)] = \delta(n+k),\ k>0$.

(e) $[x(n)] = \begin{cases} a^n, & n \geq 0 \\ 0, & n<0. \end{cases}$

16. A linear time invariant system is characterised by the system function $H(z)$ given as

$$H(z) = \frac{3-4z^{-1}}{1-3.5z^{-1}+1.5z^{-2}}.$$

17. Determine the response of the system described by the difference equation:

$$y(n) = \tfrac{5}{6}y(n-1) - \tfrac{1}{6}y(n-2) + x(n)$$

to the input signal $[x(n)] = \delta(n) - \frac{1}{3}\delta(n-1)$.

[*Hints.* $Z\{y(n)\} = \frac{5}{6}Z\{[y(n-1)]\} - \frac{1}{6}Z\{[y(n-2)]\} + Z\{[x(n)]\}$

$$\Rightarrow Y(z) = \frac{5}{6}z^{-1}Y(z) - \frac{1}{6}z^{-2}Y(z) + X(z) \Rightarrow Y(z)\left[\frac{1}{6}z^{-2} - \frac{5}{6}z^{-1} + 1\right] = X(z).$$

$$H(z) = \frac{Y(z)}{X(z)} = \frac{1}{\left[1-\frac{5}{6}z^{-1}+\frac{1}{6}z^{-2}\right]} = \frac{1}{\left(1-\frac{1}{2}z^{-1}\right)\left(1-\frac{1}{3}z^{-1}\right)}.$$

$$Z\{x(n)\} = X(z) = \left(1-\frac{1}{2}z^{-1}\right) \Rightarrow Y(z) = \frac{1}{\left[1-\frac{1}{2}z^{-1}\right]} \Rightarrow y(n) = \left(\frac{1}{2}\right)^n u(n).]$$

18. Determine the causal signal $[x(n)]$ if the Z-transform $X(z)$ is given by

(a) $X(z) = \dfrac{1+2z^{-1}}{1+z^{-2}}$

(b) $X(z) = \dfrac{1+2z^{-1}+z^{-2}}{1+4x^{-1}+4z^{-2}}$

(c) $X(z) = \dfrac{1-\frac{1}{2}z^{-1}}{1+\frac{1}{2}z^{-1}}$

(d) $X(z) = \dfrac{1-az^{-1}}{z^{-1}-a}$.

19. Find the causal signal $[x(n)]$ whose Z-transform is

$$X(z) = \frac{1}{\left(1+z^{-1}\right)\left(1-z^{-1}\right)^2}.$$

20. The system function is described by

$$H(z) = \frac{1}{\left(1-\frac{5}{6}z^{-1}+\frac{1}{6}z^{-2}\right)}.$$

The Z-transform of the input signal $X(z) = \left(1-\frac{1}{3}z^{-1}\right)$. Find the response of the system.

21. Find the Z-transform of the following signals:

(a) $[x(n)] = n(-1)^n u(n)$

(b) $[x(n)] = -na^n u(-n-1)$

(c) $[x(n)] = (-1)^n u(n)$

(d) $[x(n)] = (n+1)u(n)$.

22. Determine the convolution of the following pairs of signals:

(a) $[x_1(n)] = \left(\dfrac{1}{4}\right)^n u(n-1), [x_2(n)] = \left[1+\left(\dfrac{1}{2}\right)^n\right] u(n)$

(b) $[x_1(n)] = \left(\dfrac{1}{2}\right)^n u(n), [x_2(n)] = \cos n\pi u(n)$

(c) $[x_1(n)] = nu(n), [x_2(n)] = 2^n u(n-1)$.

23. Determine $[x(n)]$ whose Z-transforms are given as:

(a) $X(z) = \log(1-2z), |z| < \frac{1}{2}$ (b) $X(z) = \log(1-z^{-1}), |z| > \frac{1}{2}$.

24. Determine signals that can have following Z-transforms:

(a) $X(z) = \dfrac{1}{1-1.5z^{-1}+0.5z^{-2}}$ (c) $X(z) = \dfrac{3}{\left(1-\frac{10}{3}z^{-1}+z^{-2}\right)}$

(b) $X(z) = \dfrac{1}{\left(1-\frac{1}{2}z^{-1}+\frac{1}{4}z^{-2}\right)}$

25. Consider the system function $H(z) = \dfrac{1-2z^{-1}+2z^{-2}-z^{-3}}{(1-z^{-1})(1-0.5z^{-1})(1-0.2z^{-1})}$ with ROC: $0.5 < |z| < 1$.

26. Using Cauchy integral method, find the inverse Z-transform of

$$X(z) = \frac{z}{(z-1)(z-2)}, \; 1 < |z| < 2.$$

27. Find inverse Z-transform of $X(z)$ described as below using residue method:

(a) $X(z) = \dfrac{1-\frac{1}{4}z^{-1}}{1-\frac{1}{9}z^{-2}}$, ROC: $|z| > \dfrac{1}{3}$ (c) $X(z) = \dfrac{z+1}{(z+0.2)(z-1)}$, ROC: $|z| > 1$

(b) $X(z) = \dfrac{z}{(z-2)(z-3)}$, ROC: $|z| < 2$ (d) $X(z) = \dfrac{z}{Z\left(z-\frac{1}{2}\right)}$, ROC $|z| > \dfrac{1}{2}$.

28. Using convolution method, find the inverse Z-transform of

(a) $X(z) = \dfrac{1}{1-3z^{-1}+2z^{-2}}$ (c) $X(z) = \dfrac{z^2+z}{(z-1)(z-3)}, |z| > 3$

(b) $X(z) = \dfrac{z^2}{\left(z-\frac{1}{4}\right)^2}$ (d) $X(z) = \dfrac{1-\frac{1}{2}z^{-1}}{\left(1+\frac{1}{2}z^{-1}\right)(1-z^{-1}+z^{-2})}$.

29. Given $x_1(n) = \left(\dfrac{1}{3}\right)^n u(n), x_2(n) = \left(\dfrac{1}{5}\right)^n u(n)$, find $X(z)$ by convolution method.

30. Find the Z-transform of the following sequences:

(a) $[x(n)] = \left(-\dfrac{1}{5}\right)^n u(n) + 5\left(\dfrac{1}{2}\right)^{-n} u(-n-1)$ (b) $[x(n)] = u(n-2)$

(c) $[x(n)] = n^2 a^n u(n)$.

31. Find the system function $H(z)$ for a system described by a difference equation

$$y(n) = \sum_{1}^{N} a_k y(n-k) + \sum_{0}^{N} b_k x(n-k).$$

32. Determine the Z-transform of

(a) $2^n u(-n)$ (d) $[x(n)] = \left(\dfrac{1}{4}\right)^n u(n) + \left(\dfrac{1}{2}\right)^n u(n)$

(b) $\dfrac{1}{2}(n^2+n)\left(\dfrac{1}{3}\right)^{n-1} u(n-1)$ (e) $[x(n)] = n(n-1)(n-2)u(n)$.

(c) $[x(n)] = \{-1, 0, \underset{\uparrow}{1}, 0, 1\}$

33. Find $H(z)$ for the system described by

(a) $y(n) = x(n) + \dot{x}(n-1) - 2x(n-2) + 3x(n-3)$

(b) $y(n) = \frac{3}{2}y(n-1) - \frac{1}{2}y(n-2) + x(n) + x(n-1)$

(c) $y(n)-\frac{3}{4}y(n-1)+\frac{1}{8}y(n-2)=x(n)$

(d) $y(n)=x(n)+2x(n-1)-4x(n-2)+x(n-3)$.

3.11 Objective Type Questions

1. Explain the meaning of causal and anti-causal sequence.
2. What do you mean by ROC?
3. Explain the time-shifting property of the Z-transform.
4. Explain the following properties of Z-transform: (a) Scaling property (b) Time reversal property (c) Convolution property (d) Inversion property.
5. What are the different methods for the determination of inverse Z-transform?
6. The Z-transform of $[x(n)]$ is $G(z)$, then what is the Z-transform of $[nx(n)]$?
7. Find $H(z)$ for a system described by the difference equation:

   (a) $8y(n)-6y(n-1)+y(n-2)=x(n)+x(n-1)$

   (b) $y(n)=x(n)+2x(n-1)-4x(n-2)+x(n-3)$.
8. What are the properties of DFT?

   [*Ans.* The properties of DFT are:

   (a) *Periodicity.* If $X(k)$ be N-point DFT of finite duration sequence $[x(n)]$, the periodicity property says: $[x(n+N)]=[x(n)]$ for all n

   $X(k+N)=X(k)$ for all k.

   (b) *Linearity property.* If $X_1(k)=\text{DFT}[x_1(n)]$ and $X_2(k)=\text{DFT}[x_2(n)]$, then

   $$\text{DFT}\{a_1[x_1(n)]+a_2[x_2(n)]\}=a_1\text{DFT}[x_1(n)]+a_2\text{DFT}[x_2(n)]=a_1X_1(k)+a_2X_2(k).$$

   (c) *Time reversibility of a sequence.* Let $\text{DFT}[x(n)]=X(k)$, then

   $$\text{DFT}\{[x(-n)]_N\}=\text{DFT}\{[x(N-n)]\}=X(-k)_N=X(N-K).$$

   (d) *Circular shifting of a sequence.* If $\text{DFT}[x(n)]=X(k)$, then $\text{DFT}\,[x(n-l)_N]=X(k)\,e^{-2j\pi kl/N}$.]
9. Find the DFT of the sequence $[x(n)]=\{1,1,0,0\}$.

Solution: $X(k)=\sum_{n=0}^{N-1}[x(n)]\,e^{-j2\pi kn/N},\ k=0,1,\ldots,(N-1)$.

Here, $N=4$.

$$X(k)=\sum_{0}^{3}[x(n)]\,e^{-j2\pi kn/4}$$

$$X(0)=\sum_{0}^{3}[x(n)]=[1+1+0+0]=2$$

$$X(1)=\sum_{0}^{3}[x(n)]\,e^{-j2\pi n/4}=\left[1+1e^{-j2\pi/4}+0+0\right]=(1-j)$$

$$X(2)=\sum_{0}^{3}[x(n)]\,e^{-j4\pi n/4}=\left(1e^{-j0}+1e^{-2\pi}+0+0\right)=(1-1)=0$$

$$X(3) = \sum_{0}^{3} [x(n)]\, e^{-j6\pi n/4} = (1+j)$$

$$\therefore\ X(k) = \{2, 1-j, 0, 1+j\}.$$

10. Define discrete Fourier series.
11. What is the relationship between Fourier series coefficients of a periodic sequence and DFT?
12. Distinguish between DFT and DTFT.
13. Distinguish between Fourier series and Fourier transform.
14. Find the DFT of a sequence $[x(n)] = \{1,1,0,0\}$ and find the IDFT of $X(k) = \{1,0,1,0\}$.
15. Determine the Fourier transform $X_1(k)$, $X_2(k)$ and $X_3(k)$ of the following sequences:
    (a) $[x_1(n)] = \{1,1,\underset{\uparrow}{1},1,1\}$
    (b) $[x_2(n)] = \{1,0,1,0,1,0,1,0,1\}$
    (c) $[x_3(n)] = \{1,0,0,1,0,0,\underset{\uparrow}{1},0,0,1,0,0,1\}$.
16. What do you mean by convolution of two sequences. Does convolution satisfy commutative property?
17. Enumerate the properties satisfied by convolution of two sequences.

3.12 Numerical Problems

1. Find the FT of a sequence:

$$[x(n)] = \begin{cases} 1, & \text{for } -2 \le n \le 2 \\ 0, & \text{otherwise.} \end{cases}$$

2. Definite FT of a sequence and give its symmetry properties.
3. Examine whether the FT of $[x(n)] = (0.8)^n$ exists for negative values of n.
4. Explain the frequency shifting property of discrete time Fourier system.
5. Explain the time-shifting property of DTFT.
6. Complete the FT of the following signals:
    (a) $[x(n)] = u(n) + u(n-6)$
    (b) $[x(n)] = 2^n u(-n)$
    (c) $[x(n)] = \left(\frac{1}{4}\right)^n u(n+4)$.
7. Retrieve the signal from the given FTs:

    (a) $X(\omega) = \begin{cases} 0, & 0 \le \omega \le \omega_0 \\ 1, & \omega_0 \le |\omega| \le \pi. \end{cases}$

    (b) $X(\omega) = \begin{cases} 1, & \omega_0 - \delta\omega/2 \le |\omega| \le \omega_0 + \delta\omega/2 \\ 0, & \text{otherwise.} \end{cases}$

8. A signal $[x(n)]$ has the following FT:

$$X(\omega) = \frac{1}{(1 - ae^{-j\omega})}.$$

Determine FT of the following signals: (a) $[x(2n+1)]$ (b) $[x(-2n)]$ (c) $x(n) * x(n-1)$.

9. What do you mean by correlation of signals?

10. Prove the following theorems of the DTFT: (a) Linearity (b) Time shifting (c) Time reversal (d) Frequency shifting theorem.
11. Determine the DTFT of each of the following sequences:
    (a) $[x_1(n)] = a^n u(n-1),\ a < 1$
    (b) $[x_2(n)] = na^n u(n),\ a < 1$
    (c) $[x_3(n)] = a^n u(n+1),\ a < 1$
    (d) $[x_4(n)] = a^n u(-n-1),\ a < 1.$
12. Determine the DTFT of each of the following sequences:
    (a) $[x_1(n)] = u(n) - u(n-5)$
    (b) $[x_2(n)] = a^n\{u(n) - u(n-8),\ a < 1$
    (c) $[x_3(n)] = (n+1)a^n u(n),\ a < 1.$
13. Represent the finite duration sequence $[x(n)] = \{3, 4, -1, -2\}$ as a sum of weighted impulse sequences.
14. Find the step response of a system if the impulse response $h(n)$ is known.
15. Determine the unit step response of the LTI system with impulse response $h(n) = a^n u(n), |a| < 1$.
16. Enumerate the important theorems of DFT that are useful in digital signal processing applications.
17. What are the different theorems on DTFT?
18. Define frequency shifting theorem related to the Fourier transform.
19. Define convolution theorem related to Fourier transform.
20. State and mathematically formulate the modulation theorem or windowing theorem related to the convolution of two sequences.
21. Show that the inverse DTFT of

$$X(\omega) = \frac{1}{(1 - ae^{-j\omega})^m},\ a < 1$$

is given by

$$[x(n)] = \frac{(n+m-1)!}{n!(m-1)!} a^n u(n).$$

Part V

Spectral Analysis

Chapter 1

Fourier Series and Its Application

1.1 Introduction

Spectral analysis of signals means decomposition of the signal into its constituents; in this respect it is similar to the functioning of an *analysing prism* which disperses white light into its constituent colours. Composition of a signal from its decomposed components is called the synthesis resembling the functioning of a recombination prism.

There are in general five different mathematical artifices to solve problems of spectral analysis of signals:

1. *Z-transform.* Exclusively meant for solving problems on signals discrete in time.
2. *Fourier series* (CFS) *for continuous* time periodic signals.
3. *Fourier transform* (CFT) *for continuous* time aperiodic signals (CTFT).
4. *Fourier series for discrete signals* (DFS).
5. *Fourier transform for discrete time aperiodic* signals (FTDT).

1.2 Fourier Series Representation of a Continuous Time Periodic Signals

In this whole realm of physical science, particularly in the analysis of vibrational problems, the FS plays a very important role. In the analysis and also in the synthesis of a complex type of periodic function, Fourier series is of immense importance.

A simple harmonic function is the simplest type of periodic function but any periodic function is not simple harmonic, because it may be synthesised by superposition of a number of simple harmonic functions. Fourier's theorem is a mathematical artifice by which a periodic function can be analysed into its harmonic components.

Formal statement of the theorem

Any function $f(t)$, provided it satisfies a set of conditions, can be expanded in a trigonometric series, i.e., a series of sines and cosines as

$$f(t)=a_0+\sum_{n=1}^{\infty}a_n\cos n\omega t+\sum_{n=1}^{\infty}b_n\sin n\omega t \text{ or, } f(x)=a_0+\sum_{1}^{\infty}a_n\cos nx+\sum_{1}^{\infty}b_n\sin nx. \tag{1.1}$$

The requirements for conditional representation of $f(t)$ by (1.1) are:

1. $f(t)$ is single valued in the interval $(-T/2, T/2)$.
2. $f(t)$ is bounded.

3. $f(t)$ has at most a finite number of maxima and minima.
4. $f(t)$ has finite number of discontinuities, if any; piecewise continuous.
5. $f(t+T)=f(t)$ for values of t outside of $(-T/2, T/2)$, it implies that it is periodic and continuous.

This set of conditions laid down in (1) through (5) is referred to as Dirichlet's conditions, where a_0 is a constant defining the average value of the function and a_n, b_n are the Fourier coefficients.

The Fourier coefficients a_n and b_n are defined by

$$a_n = \frac{1}{2\pi}\int_{-\pi}^{\pi} f(x)\cos nx\,dx,\ b_n = \frac{1}{2\pi}\int_{-\pi}^{\pi} f(x)\sin nx\,dx.$$

(1.1) may be expressed in terms of complex exponentials in view of the fact that

$$\cos\theta = \frac{e^{j\theta}+e^{-j\theta}}{2},\ \sin\theta = \frac{e^{j\theta}-e^{-j\theta}}{2j}.$$

$$\therefore f(t)=a_0+\sum_1^{\infty}\frac{a_n}{2}\left(e^{jn\omega t}+e^{-jn\omega t}\right)+\sum_1^{\infty}\frac{b_n}{2j}\left(e^{jn\omega t}+e^{-jn\omega t}\right)=\sum_{-\infty}^{\infty}c_n e^{jn\omega t}, \tag{1.2}$$

where $c_0=a_0, c_n=\frac{1}{2}(a_n-jb_n), c_{-n}=\frac{1}{2}(a_n+jb_n)$.

Evaluation of coefficient c_n

To find c_n, we multiply both sides of (1.2) by $e^{-jm\omega t}$ and integrating over a complete period T, we get

$$\int_0^T f(t)e^{-jm\omega t}\,dt=\int_0^T e^{-jm\omega t}\left(\sum_{-\infty}^{\infty}c_n e^{jn\omega t}\right)dt$$

$$\Rightarrow \int_0^T f(t)e^{-jm\omega t}\,dt=\sum_{-\infty}^{\infty}c_n\int_0^T e^{j(n-m)\omega t}\,dt=\sum_{-\infty}^{\infty}c_n T\delta_{nm},\ \delta_{nm}=\begin{cases}1, & n=m\\ 0, & n\neq m\end{cases}$$

$$\Rightarrow c_n=\frac{1}{T}\int_0^T f(t)e^{-jn\omega t}\,dt. \tag{1.3}$$

In general, the Fourier coefficients c_n are complex valued. It is to remember that if the periodic signal is real, c_n and c_{-n} are complex conjugates.

Thus, $c_0=\frac{1}{T}\int_0^T f(t)\,dt$ = average value of $f(t)$ over a periodic interval.

$$c_n=\frac{1}{T}\int_0^T f(t)e^{-jn\omega t}\,dt \tag{1.4}$$

$$c_{-n}=\frac{1}{T}\int_0^T f(t)e^{jn\omega t}\,dt \tag{1.5}$$

$$f(t)=c_0+\sum(c_n+c_{-n})\cos n\omega t+\sum\{(c_n-c_{-n})/j\}(\sin\omega t). \tag{1.6}$$

1.3 FS for Continuous Time Periodic Function: Application to Continuous Time Signal

1. *Analysis of square-wave signal.* The continuous time signal of square-wave type is given by

$$y = f(t) = c_0 + \sum_{-\infty}^{\infty} c_n\, e^{jn\omega t}. \tag{1.7}$$

Synthesis equation	$f(t) = \sum_{-\infty}^{\infty} c_n\, e^{jn\omega t}$
Analysis equation	$c_n = \frac{1}{T}\int_T f(t) e^{-jn\omega t}\, dt$

$$y = \begin{cases} k \text{ (constant)}, & \text{for } 0 \le t \le T/2 \\ 0, & \text{for } T/2 \le t \le T. \end{cases}$$

Such a function is diagrammatically represented by shown in Fig. 1.1:

$$c_0 = \frac{1}{T}\int_{-T/2}^{T/2} f(t)dt = \frac{1}{T}\int_{-T/2}^{T/2} f(t)dt = \frac{1}{T}\left[\int_{-T/2}^{0} 0\,dt + \int_0^{T/2} k\,dt\right]$$
$$= \frac{k}{2} \text{ (position of the axis or the dashed line).}$$

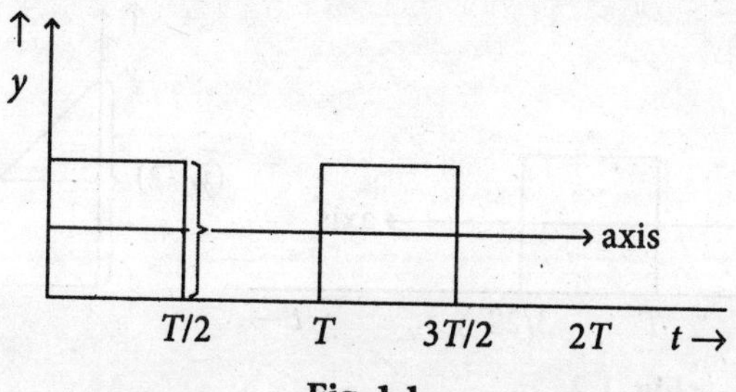

Fig. 1.1

$$c_n = \frac{1}{T}\int_{-T/2}^{T/2} f(t)e^{-jn\omega t}\, dt = \frac{1}{T}\left[\int_{-T/2}^{0} f(t)e^{-j\omega t}\, dt + \int_0^{T/2} ke^{-jn\omega t}\, dt\right]$$
$$= \frac{k}{T}\left[-\frac{1}{jn\omega}\left\{e^{-jn\omega t}\Big|_0^{T/2}\right\}\right] = -\frac{k}{jn\omega t}\left[e^{-jn\omega T/2} - 1\right] = \frac{k}{j\omega nT}\left[1 - e^{-jn\omega T/2}\right].$$

$$c_{-n} = \frac{k}{j\omega(-n)T}\left[1 - e^{j\omega nT/2}\right] = \frac{k}{j\omega nT}\left[e^{j\omega nT/2}\right].$$

$$a_n = (c_n + c_{-n}) = \frac{k}{j\omega nT}\left[1 - e^{-jn\omega T/2}\right] + \frac{k}{j\omega nT}\left[e^{j\omega nT/2} - 1\right] = \frac{k}{j\omega nT}\left[e^{j\omega nT/2} - e^{-j\omega nT/2}\right]$$
$$= \frac{k}{\omega nT}\left[\frac{e^{jn\pi} - e^{-jn\pi}}{j}\right] = \frac{2k}{\omega nT}\sin n\pi = 0 \text{ for all values of } n.$$

$$b_n = -\frac{1}{j}(c_n - c_{-n}) = -\frac{k}{j\omega nT}\left[2-\left(e^{jn\omega T/2}+e^{-jn\omega T/2}\right)\right] = \frac{k}{\omega nT}[2-2\cos n\pi]$$

$$= \frac{2k}{2n\pi}[1-\cos n\pi] = \frac{k}{n\pi}[1-(-1)^n]$$

$$\Rightarrow b_n = \begin{cases} \frac{2k}{n\pi}, & \text{for } n \text{ odd} \\ 0, & \text{for } n \text{ even.} \end{cases}$$

So only the odd terms of the sine series appear in the expansion.

Hence,

$$y = f(t) = \frac{k}{2} + \frac{2k}{\pi}\left(\sin\omega t + \frac{1}{3}\sin 3\omega t + \frac{1}{5}\sin 5\omega t + \ldots\right). \tag{1.8}$$

The series when plotted gives rise to square-waveform.

(a) $y = \frac{k}{2}$, first term defining the axis.

(b) $y = \frac{k}{2} + \frac{2k}{\pi}\sin\omega t$, the first two terms.

(c) $y = \frac{k}{2} + \frac{2k}{\pi}\sin\omega t + \frac{2k}{3\pi}\sin 3\omega t$,
the first three terms and so on.

(1.8) is plotted with y vs. t and it is seen that as more and more terms are added, the curve displays the form of square-wave [Fig. 1.2].

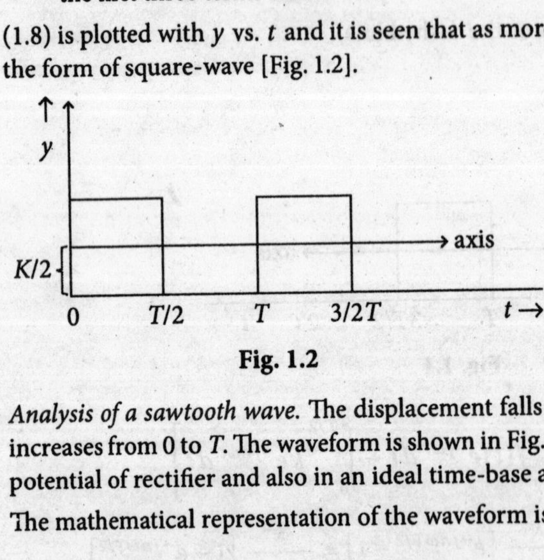

Fig. 1.2

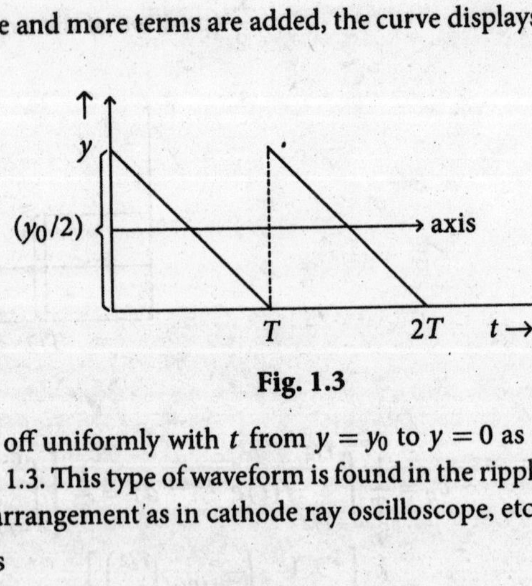

Fig. 1.3

2. *Analysis of a sawtooth wave.* The displacement falls off uniformly with t from $y = y_0$ to $y = 0$ as t increases from 0 to T. The waveform is shown in Fig. 1.3. This type of waveform is found in the ripple potential of rectifier and also in an ideal time-base arrangement as in cathode ray oscilloscope, etc.

The mathematical representation of the waveform is

$$\frac{y}{y_0} + \frac{t}{T} = 1 \Rightarrow y = y_0\left(1-\frac{t}{T}\right),\ 0 \le t \le T.$$

$$c_n = \frac{1}{T}\int_{-T/2}^{T/2} f(t)e^{-jn\omega t}\,dt.$$

$$a_0 = c_0 = \frac{1}{T}\int_{-T/2}^{T/2} f(t)\,dt = \frac{1}{T}\int_{-T/2}^{T/2} y_0\left(1-\frac{t}{T}\right)dt = \frac{y_0}{T}\left[T - \frac{T^2}{2T}\right] = \left(\frac{y_0}{2}\right).$$

$$a_n = (c_n - c_{-n}) = \frac{y_0}{T}\left[\int_0^T e^{-j\omega nt}\left(1-\frac{t}{T}\right)dt + \int_0^T e^{j\omega nt}\left(1-\frac{t}{T}\right)dt\right]$$

$$= \frac{y_0}{T}\left[\int_0^T \left(e^{-j\omega nt}+e^{j\omega nt}\right)dt - \frac{1}{T}\int_0^T t\left(e^{j\omega nt}+e^{-j\omega nt}\right)dt\right]$$

$$= \frac{y_0}{T}\left[2\int_0^T \cos\omega nt\,dt - \frac{1}{T}\left\{\int_0^T te^{-j\omega nt}\,dt + \int_0^T te^{j\omega nt}\,dt\right\}\right].$$

Now,

$$\int_0^T te^{-j\omega nt}\,dt = -\frac{t}{j\omega n}e^{-j\omega nt}\Big|_0^T + \frac{1}{j\omega n}\int_0^T e^{-j\omega n}\,dt$$

$$= -\frac{T}{j\omega n}e^{-j2\pi n} + \frac{1}{j\omega n}\left(-\frac{1}{j\omega n}\right)\left(e^{-j2\pi n}-1\right)$$

$$\Rightarrow \frac{1}{T}\int_0^T te^{-j\omega nt}\,dt = -\frac{1}{j\omega n}e^{-j2\pi n} + \frac{1}{\omega^2 nT}\left(e^{-j2\pi n}-1\right).$$

Similarly,

$$\frac{1}{T}\int_0^T te^{j\omega nt}\,dt = \frac{1}{j\omega n}e^{j2\pi n} + \frac{1}{\omega^2 n^2 T}\left(e^{j2\pi n}-1\right).$$

$$a_n = (c_n + c_{-n}) = \frac{y_0}{T}\left[-\frac{1}{j\omega n}e^{j2\pi n} + \frac{1}{\omega^2 n^2 T}\left(e^{-j2\pi n}-1\right) + \frac{1}{j\omega n}e^{j2\pi n} + \frac{1}{\omega^2 n^2 T}\left(e^{j2\pi n}-1\right)\right]$$

$$= \frac{y_0}{T}\left[\frac{1}{\omega^2 n^2 T}\left\{\left(e^{-j2\pi n}+e^{j2\pi n}\right)-2\right\}\right] = y_0\left[\frac{1}{(2\pi n)^2}\cdot 2\{\cos 2n\pi - 1\}\right] = y_0 \times 0 = 0.$$

So the FS expansion consists of sine terms only, no cosine terms.

$$b_n = j(c_n - c_{-n}) = \frac{j}{T}\int_0^T f(t)\left(e^{-j\omega nt}-e^{j\omega nt}\right)dt = -\frac{j}{T}\int_0^T f(t)\left(e^{j\omega nt}-e^{-j\omega nt}\right)dt$$

$$= -\frac{j}{T}\int_0^T y_0\left(1-\frac{t}{T}\right)2j\sin\omega nt\,dt = \frac{2y_0}{T}\left[\frac{-(1-t/T)}{\omega n}\cos\omega nt\Big|_0^T + \int_0^T \frac{1}{T\omega n}\cos\omega nt\,dt\right]$$

$$= \frac{2y_0}{T}\left[\frac{1}{\omega n}\right] = \frac{2y_0}{2\pi n} = \frac{y_0}{\pi n}.$$

So the sawtooth periodic wave admits of expansion as

$$y = \frac{y_0}{2} + \sum_1^{\infty}\frac{y_0}{n\pi}\sin n\omega t = \frac{y_0}{2} + \frac{y_0}{\pi}\sum\frac{1}{n}\sin n\omega t. \tag{1.9}$$

Graphical representation. (1.9) when plotted and as more and more terms in the expansion are considered, the plot tends to assume a definite shape which is a sawtooth type [Fig. 1.3].

$y = \frac{y_0}{2}$, axis of curve.

$y = \frac{y_0}{2} + \frac{y_0}{\pi}\sin\omega t$, first two terms.

$$y = \frac{y_0}{2} + \frac{y_0}{\pi}\sin\omega t + \frac{y_0}{2\pi}\sin 2\omega t,$$

the first three terms. Thus, we see that as more and more terms are added, the graph tends to assume the form as that of the sawtooth type.

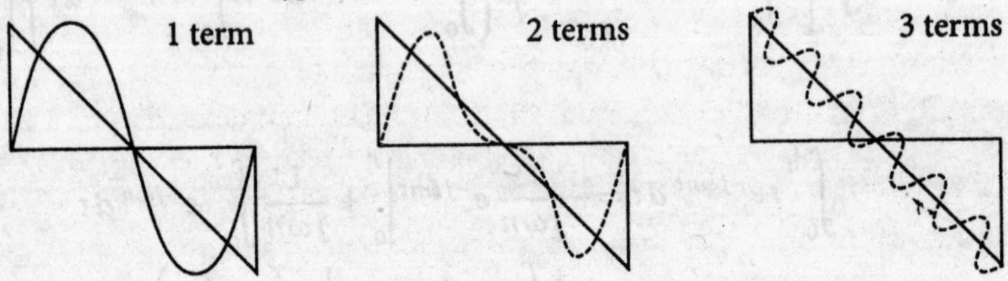

Fig. 1.4

3. *Half-wave rectification.* The output of the single-phase half-wave rectifier may be analytically expressed in terms of Fourier's expansion.

When an alternating source of emf is fed to a diode, the current flows in the circuit only when the collector (anode) becomes positive w.r.t. the emitter (cathode), but the diode refuses to conduct when the anode is negative w.r.t. the cathode.

The mathematical counterpart of this statement is to define a function such that

$$y = \hat{V}\sin\omega t = \begin{cases} \hat{V}\sin\theta, & \text{when } 0 \le \omega t \le \pi \\ 0, & \text{when } \pi \le \omega t \le 2\pi \end{cases}$$

$$A_0 = \frac{1}{2\pi}\int_0^{2\pi} y\,d\theta = \frac{1}{2\pi}\int_0^{\pi} \hat{V}\sin\theta\,d\theta + \frac{1}{2\pi}\int_{\pi}^{2\pi} y\,d\theta,\ \theta = \omega t$$

$$= \frac{\hat{V}}{\pi},$$

which is the axis of the curve.

$$B_n = \frac{1}{2\pi}\int_0^{2\pi} y\sin n\theta\,d\theta = \frac{1}{\pi}\int_0^{\pi} \hat{V}\sin\theta\sin n\theta\,d\theta + \frac{1}{\pi}\int_{\pi}^{2\pi} y\sin n\theta\,d\theta$$

$$= \frac{\hat{V}}{2\pi}\left\{\int_0^{\pi}[\cos(n-1)\theta - \cos(n+1)\theta]\,d\theta\right\} + 0$$

$$= \frac{\hat{V}}{2\pi}\left\{\left[\frac{\sin(n-1)\theta}{(n-1)}\bigg|_0^{\pi}\right] - \left[\frac{\sin(n+1)\theta}{(n+1)}\bigg|_0^{\pi}\right]\right\}$$

$$= \frac{\hat{V}}{2\pi(n-1)}[\sin(n-1)\pi] - \frac{\hat{V}}{2\pi(n+1)}\sin(n+1)\pi.$$

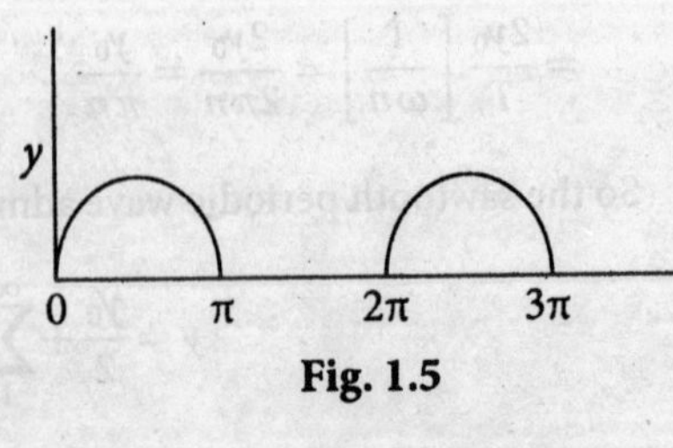

Fig. 1.5

$\therefore$ $B_n = 0$, except $n = 1$ for which we get

$$B_1 = \frac{1}{2\pi}\int_0^{2\pi} y\sin\theta\,d\theta = \frac{1}{2\pi}\left[\int_0^{\pi} \hat{V}\sin^2\theta\,d\theta + \int_{\pi}^{2\pi} y\sin\theta\,d\theta\right] = \frac{\hat{V}}{z}.$$

Calculating for A_n in the same way, we get

$$A_n = -\frac{\vec{V}}{2\pi(n+1)(n-1)}[(n-1)\cos(n+1)\pi-(n+1)\cos(n-1)\pi+2] = -\frac{2\hat{V}}{\pi}\sum_{2,4,6}\frac{1}{(n+1)(n-1)}.$$

$$\therefore\ y = \vec{V}\left[\frac{1}{\pi}+\frac{1}{2}\sin\omega t-\frac{2}{\pi}\sum_{2,4,6}\frac{\cos n\omega t}{(n+1)(n-1)}\right].$$

This equation when plotted gives the graph shown in Fig. 1.5.

4. *Full-wave rectifier.* In the full-wave rectification, our function is defined as

$$y = \begin{cases} \hat{V}\sin\omega t, & \text{for } 0\le\omega t\le\pi \\ -\hat{V}\sin\omega t, & \text{for } \pi\le\omega t\le 2\pi. \end{cases}$$

Hence, Fourier's series expansion is obtained as

$$y = \hat{V}\left[\frac{2}{\pi}-\frac{4}{\pi}\sum_{2,4,6}\frac{\cos n\omega t}{(n+1)(n-1)}\right],$$

which is represented in the Fig. 1.6.

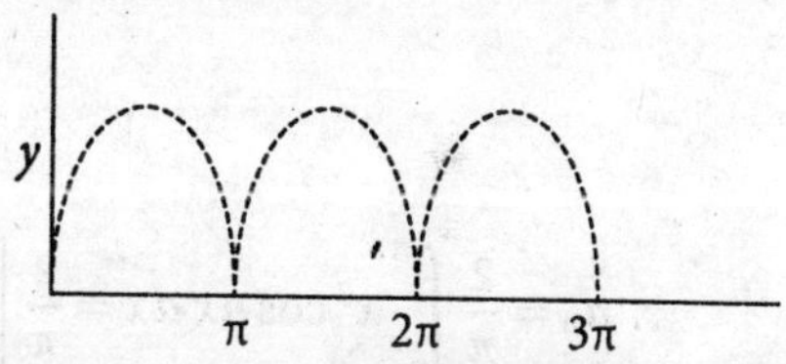

Fig. 1.6

1.4 Importance of Fourier's Theorem

In the analysis and synthesis of vibration, this theorem finds important application. If we are to find out the harmonic components in a given vibration, we can take help of Fourier's theorem. Again, if we are to produce a vibration of given period, this theorem will tell us which harmonic components must be chosen. The harmonic component having the lowest frequency is called the *fundamental*, the components having frequencies different from that of fundamental are called *overtones* and the overtones whose frequencies are multiplies of fundamental frequency are termed as harmonics.

The quality of a musical sound depends on the number of overtones and their relative positions. So the quality of a musical note can be determined by applying Fourier's theorem after proper analysis of the note.

Fourier's theorem is of help also in the analysis of the output of rectifier circuit and hence, also in the designing of suitable filter circuits.

1.5 Evaluation of Riemann Zeta Function: Evaluation of the Sum of Some Standard Series

The Riemann Zeta function denoted by $\xi(2)$ and $\xi(4)$ and in general denoted by $\xi(k)$ are defined as

$$\xi(2) = \sum_{1}^{\infty}\frac{1}{n^2} = \frac{1}{1^2}+\frac{1}{2^2}+\frac{1}{3^2}+\ldots\infty$$

$$\xi(4) = \sum_{1}^{\infty}\frac{1}{n^4} = \frac{1}{1^4}+\frac{1}{2^4}+\frac{1}{3^4}+\ldots\infty.$$

In general,

$$\xi(k) = \sum_{1}^{\infty}\frac{1}{n^k} = \frac{1}{1^k}+\frac{1}{2^k}+\frac{1}{3^k}+\ldots\infty.$$

The above series sum can be evaluated by the method of Fourier expansion.

Let us first take the function $f(x)$ defined as $f(x)=x^2, -\pi \le x \le \pi$.

Since $f(x)$ is an even function $b_n=0$. $\therefore\ x^2=a_0+\Sigma a_n \cos nx$ and the coefficient a_n is given by

$$a_n=\frac{1}{\pi}\int_{-\pi}^{\pi} f(x)\cos nx\,dx=\frac{1}{\pi}\int_{-\pi}^{\pi} x^2\cos nx\,dx.$$

Now, $F(x)=x^2\cos nx$.

$$\therefore\ \int_{-\pi}^{\pi} F(x)\,dx=\int_0^{\pi}[F(x)+F(-x)]\,dx.$$

Since $F(x)$ is an even function $F(-x)=F(x)$.

$$\therefore\ \int_{-\pi}^{\pi} F(x)\,dx=2\int_0^{\pi} F(x)\,dx.$$

$$\therefore\ a_n=\frac{2}{\pi}\int_0^{\pi} x^2\cos nx\,dx=\frac{2}{\pi}\left[\frac{x^2}{n}\sin nx\bigg|_0^{\pi}-2\int_0^{\pi} x\cdot\frac{1}{n}\sin nx\,dx\right]=\frac{2}{\pi}\left[0-\frac{2}{n}\int_0^{\pi} x\sin nx\,dx\right]$$

$$=\frac{4}{n\pi}\int_0^{\pi} x\sin nx\,dx=-\frac{4}{n\pi}\left[-\frac{x}{n}\cos nx\bigg|_0^{\pi}+\int_0^{\pi}\frac{1}{n}\cos nx\,dx\right]$$

$$=-\frac{4}{n\pi}\left[-\frac{\pi}{n}\cos n\pi+\frac{1}{n}\cdot\frac{1}{n}\sin nx\bigg|_0^{\pi}\right]=\frac{4}{n^2}\cos n\pi=\frac{4}{n^2}(-1)^n.$$

Also,

$$a_0=\frac{1}{2\pi}\int_{-\pi}^{\pi} f(x)\,dx=\frac{1}{2\pi}\int_0^{\pi}[f(x)+f(-x)]\,dx.$$

Here, $f(x)=x^2$ is an even function, hence, $f(-x)=f(x)$.

$$\therefore\ a_0=\frac{1}{2\pi}\cdot 2\int_0^{\pi} f(x)\,dx=\frac{1}{\pi}\int_0^{\pi} x^2\,dx=\frac{\pi^3}{3\pi}=\frac{\pi^2}{3}.$$

$$\therefore\ f(x)=x^2=\frac{\pi^2}{3}+\sum a_n\cos nx=\frac{\pi^2}{3}+\sum\frac{4}{n^2}(-1)^n\cos nx.$$

Putting $x=\pi$, we get

$$\pi^2=\frac{\pi^2}{3}+\sum\frac{4}{n^2}(-1)^n\cos n\pi=\frac{\pi^2}{3}+\sum\frac{4}{n^2}(-1)^n\cdot(-1)^n$$

$$=\frac{\pi^2}{3}+4\sum\frac{1}{n^2}=\frac{\pi^2}{3}+4\xi(2)\Rightarrow\xi(2)=\sum\frac{1}{n^2}=\frac{\pi^2}{6}.$$

To find $\xi(4)$:

Take $f(x)=x^4$.

$$\therefore\ a_0=\frac{1}{2\pi}\int_{-\pi}^{\pi} f(x)\,dx=\frac{1}{2\pi}\cdot 2\int_0 x^4\,dx=\frac{\pi^4}{5}.$$

Since $f(x)$ is an even function, we have $b_n = 0$.

$$f(x) = x^4 = a_0 + \sum a_n \cos nx,$$

where $a_0 = \pi^4/5$.

$$a_n = \frac{1}{\pi}\int_{-\pi}^{\pi} f(x)\cos nx\,dx = \frac{2}{\pi}\int_0^{\pi} x^4 \cos nx\,dx = \frac{2}{\pi}(-1)^n\left(\frac{4\pi^3}{n^2} - \frac{24\pi}{n^4}\right) = (-1)^n\left(\frac{8\pi^2}{n^2} - \frac{48}{n^4}\right).$$

$$\therefore\ f(x) = x^4 = \frac{\pi^4}{5} + \sum(-1)^n\left(\frac{8\pi^2}{n^2} - \frac{48}{n^4}\right)\cos nx.$$

Putting $x = \pi$, we get

$$\pi^4 = x^4 = \frac{\pi^4}{5} + \sum(-1)^n\left(\frac{8\pi^2}{n^2} - \frac{48}{n^4}\right)\cos n\pi = \frac{\pi^4}{5} + \sum(-1)^n\left(\frac{8\pi^2}{n^2} - \frac{48}{n^4}\right)\times(-1)^n$$

$$= \frac{\pi^4}{5} + \sum\left(\frac{8\pi^2}{n^2} - \frac{48}{n^4}\right) \Rightarrow \frac{4\pi^4}{5} = 8\pi^2\sum\frac{1}{n^2} - 48\sum\frac{1}{n^4} = 8\pi^2\xi(2) - 48\xi(4)$$

$$= 8\pi^2\frac{\pi^2}{6} - 48\xi(4) \Rightarrow \sum\frac{1}{n^4} = \frac{\pi^4}{90}.$$

1.6 Short Theory Questions with Answers

1. What is a Fourier series?

   Ans. Fourier's series is a trigonometric series into which a function can be expressed provided the function satisfies a set of conditions known as Dirichlet's conditions.

2. What is the advantage of Fourier's expansion over Taylor's expansion?

   Ans. The Fourier's series can be used to represent discontinuous functions where all order of derivatives need not exist whereas Taylor's expansion is possible in case all orders of derivatives do exist.

3. Show that if a real function $f(x)$ be expanded in the complex exponential Fourier series $\sum_{n=-\infty}^{\infty} c_n e^{inx}$, then $c_{-n} = c_n^*$, where c_n^* is the conjugate of c_n.

   Solution: $c_n = \frac{1}{2\pi}\int_{-\pi}^{\pi} f(x)e^{inx}\,dx;\ c_n^* = \frac{1}{2\pi}\int_{-\pi}^{\pi} f(x)e^{-inx}\,dx = c_{-n}.$

4. How can the Fourier series expansion be applicable to the periodic function?

   Ans. Fourier's series expansion applies to the periodic function since outside of the closed interval under consideration there is periodic extension of the function.

5. What are the conditions under which a function $f(x)$ can be Fourier expanded into a convergent trigonometric series?

   Ans. The necessary conditions which the function $f(x)$ should satisfy so that it can be Fourier expanded are those of Dirichlet's conditions mentioned in the statement of the theorem.

6. Assuming that the Fourier expansion of $f(x)$ is uniformly convergent, find the average value of $[f(x)]^2$.

Solution: Average value

$$= \frac{1}{2\pi}\int_{-\pi}^{\pi}[f(x)]^2\,dx = \frac{1}{2\pi}\int_{-\pi}^{\pi}\left[a_0 + \sum_1^{\infty} a_n \cos nx + \sum_1^{\infty} b_n \sin nx\right]^2 dx$$

$$= \frac{1}{2\pi}\left[a_0^2\int_{-\pi}^{\pi} dx + \sum_1^{\infty} a_n^2 \int_{-\pi}^{\pi}\cos^2 nx\,dx + \sum_1^{\infty} b_n^2\int_{-\pi}^{\pi}\sin^2 nx\,dx + \sum 2a_0a_n\int_{-\pi}^{\pi}\cos nx\,dx\right.$$

$$\left. + 2\sum\sum a_\alpha a_\beta\int_{-\pi}^{\pi}\cos\alpha x\sin\beta x\,dx + 2\sum a_0 b_n\int_{-\pi}^{\pi}\sin nx\,dx\right]$$

$$= \frac{1}{2\pi}\left[2\pi a_0^2 + \sum \pi a_n^2 + \sum \pi b_n^2\right] = a_0^2 + \frac{1}{2}\sum_1^{\infty}\left(a_n^2 + b_n^2\right).$$

This is known as Parseval's relation.

1.7 A Few More Solved Problems

1. Obtain the Fourier series for a function of $f(x)$, where

$$f(x) = \begin{cases} \cos x, & 0 \le x \le \pi \\ -\cos x, & -\pi \le x \le 0. \end{cases}$$

Solution: $f(x) = a_0 + \sum_1^{\infty} a_n \cos nx + \sum_1^{\infty} b_n \sin nx$

$$a_0 = \frac{1}{2\pi}\int_{-\pi}^{\pi} f(x)\,dx = \frac{1}{2\pi}\int_{-\pi}^{0} f(x)\,dx + \frac{1}{2\pi}\int_0^{\pi} f(x)\,dx$$

$$= \frac{1}{2\pi}\int_{-\pi}^{0} -\cos x\,dx + \frac{1}{2\pi}\cos x\,dx = \frac{1}{2\pi}\left(-\sin x\Big|_{-\pi}^{0}\right) + \frac{1}{2\pi}\left(\sin x\Big|_0^{\pi}\right) = 0.$$

$$a_n = \frac{1}{\pi}\int_{-\pi}^{\pi} f(x)\cos nx\,dx = 0.$$

$$b_n = \frac{1}{\pi}\int_{-\pi}^{\pi} f(x)\sin nx\,dx = \frac{4n}{\pi(n^2-1)}\left[1 + (-1)^n\right]$$

$$\Rightarrow b_n = \begin{cases} 0, & \text{for } n \text{ odd} \\ \dfrac{8n}{\pi(n^2-1)}, & \text{for } n \text{ even.} \end{cases}$$

2. Find the Fourier series for the function $f(x)$ defined as

$$f(x) = \begin{cases} -k, & -\pi < x < 0 \\ k, & 0 < x < \pi. \end{cases}$$

Hence, show that $\frac{\pi}{4} = 1 - \frac{1}{3} + \frac{1}{5} - \frac{1}{7} + \ldots$.

Solution: $f(x)=a_0+\sum a_n\cos nx+\sum b_n\sin nx$

$$a_0=\frac{1}{2\pi}\int_{-\pi}^{\pi}f(x)\,dx=\frac{1}{2\pi}\int_{-\pi}^{0}-k\,dx+\frac{1}{2\pi}\int_{0}^{\pi}k\,dx=0$$

$$a_n=\frac{1}{\pi}\int_{-\pi}^{\pi}f(x)\cos nx\,dx=\frac{1}{\pi}\int_{-\pi}^{0}-k\cos nx\,dx+\frac{1}{\pi}\int_{0}^{\pi}k\cos nx\,dx=0$$

$$b_n=\frac{1}{\pi}\left[\int_{-\pi}^{0}-k\sin nx\,dx+\int_{0}^{\pi}k\sin nx\,dx\right]=\frac{2k}{n\pi}(1-\cos nx)$$

$$\Rightarrow b_n=\frac{2k}{n\pi}[1-(-1)^n]\Rightarrow b_n=\begin{cases}0, & \text{for } n \text{ even}\\ \frac{4k}{n\pi}, & \text{for } n \text{ odd.}\end{cases}$$

$$\therefore f(x)=\sum_{1,3,5}^{\infty}\frac{4k}{n\pi}\sin nx=\frac{4k}{\pi}\left[\sin x+\frac{\sin 3x}{3}+\frac{\sin 5x}{5}+\ldots\right].$$

Putting $x=\pi/2$,

$$f\left(\frac{\pi}{2}\right)=k=\frac{4k}{\pi}\left[1-\frac{1}{3}+\frac{1}{5}+\frac{1}{7}+\ldots\right]\Rightarrow\frac{\pi}{4}=1-\frac{1}{3}+\frac{1}{5}-\frac{1}{7}+\ldots.$$

3. Expand the function $f(x)$ in a Fourier series in the interval $(-\pi,\pi)$,

$$f(x)=\begin{cases}2x, & 0\le x\le\pi\\ x, & -\pi\le x\le 0.\end{cases}$$

Solution: $$a_0=\frac{1}{2\pi}\int_{-\pi}^{\pi}f(x)\,dx=\frac{1}{2\pi}\left[\int_{-\pi}^{0}x\,dx+\int_{0}^{\pi}2x\,dx\right]=\frac{1}{2\pi}\left[\frac{x^2}{2}\Big|_{-\pi}^{0}+2\frac{x^2}{2}\Big|_{0}^{\pi}\right]=\frac{\pi}{4}.$$

$$a_n=\frac{1}{\pi}\int_{-\pi}^{\pi}f(x)\cos nx\,dx=\frac{1}{\pi}\left[\int_{-\pi}^{0}x\cos nx\,dx+\int_{0}^{\pi}2x\cos nx\,dx\right]$$

$$=\frac{1}{\pi}\left[\frac{x}{\pi}\sin nx\Big|_{-\pi}^{0}-\frac{1}{n}-\frac{1}{n}\int_{-\pi}^{0}\sin nx\,dx+\frac{2x}{n}n.x\Big|_{0}^{\pi}-\frac{2}{n}\int_{0}^{\pi}\sin nx\,dx\right]$$

$$=\frac{1}{\pi}\left[0+\frac{1}{n^2}\cos nx\Big|_{-\pi}^{0}+\frac{2}{n^2}\cos nx\Big|_{0}^{\pi}\right]=\frac{1}{\pi}\left[\frac{1}{n^2}-\frac{1}{n^2}\cos n\pi-\frac{2}{n^2}+\frac{2}{n^2}\cos n\pi\right]$$

$$=\frac{1}{\pi}\left[-\frac{1}{n^2}+\frac{1}{n^2}\cos n\pi\right]=\frac{1}{n^2\pi}[-1+\cos n\pi]=\frac{1}{n^2\pi}[-1+(-1)^n]$$

$$\Rightarrow a_n=\begin{cases}0, & \text{for } n \text{ even}\\ -\frac{2}{n^2\pi}, & \text{for } n \text{ odd.}\end{cases}$$

$$b_n = \frac{1}{\pi}\int_{-\pi}^{\pi} f(x)\sin nx\,dx = \frac{1}{\pi}\left[\int_{-\pi}^{0} x\sin nx\,dx + \int_{0}^{\pi} 2x\sin nx\,dx\right]$$

$$= \frac{1}{\pi}\left[-\frac{x}{n}\cos nx\Big|_{-\pi}^{0} + \frac{1}{n}\int_{-\pi}^{0}\cos nx\,dx - \left(\frac{2x}{n}\cos nx\right)\Big|_{0}^{\pi} + \frac{2}{n}\int_{0}^{\pi}\cos nx\,dx\right]$$

$$= \frac{1}{\pi}\left[-\frac{\pi}{n}\cos n\pi + \frac{1}{n^2}\sin nx\Big|_{-\pi}^{0} - \frac{2\pi}{n}\cos n\pi + \frac{2}{n^2}\sin nx\Big|_{0}^{\pi}\right]$$

$$= \frac{1}{\pi}\left[-\frac{\pi}{n}\cos n\pi - \frac{2\pi}{n}\cos n\pi\right] = -\frac{3}{n}\cos n\pi = -\frac{3}{n}(-1)^n.$$

$$\therefore\ f(x) = a_0 + a_1\cos x + a_2\cos 2x + a_3\cos 3x + \cdots + b_1\sin x + b_2\sin 2x + b_3\sin 3x + \ldots$$

$$= \frac{\pi}{4} - \frac{2}{1^2\pi}\cos x - \frac{2}{3^2\pi}\cos 3x + \cdots + \frac{3}{1}\sin x - \frac{3}{2}\sin 2x + \frac{3}{3}\sin 3x + \ldots$$

$$= \frac{\pi}{4} - \frac{2}{\pi}\left(\frac{\cos x}{1^2} + \frac{\cos 3x}{3^2} + \frac{\cos 5x}{5^2} + \ldots\right) + 3\left(\frac{\sin x}{1} - \frac{\sin 2x}{2} + \frac{\sin 3x}{3} - \ldots\right).$$

4. Expand in a Fourier series the periodic function $f(x)$ with period $2l$ which on the interval $(-l, l)$ is given by the equation $f(x) = |x|$.

Solution: $f(x) = |x|$ means $f(x) = \begin{cases} -x, & -l \le x \le 0 \\ x, & 0 \le x \le l. \end{cases}$

$$a_0 = \frac{1}{2l}\int_{-l}^{l} f(x)\,dx = \frac{1}{2l}\left[\int_{-l}^{0} -x\,dx + \int_{0}^{l} x\,dx\right] = \frac{1}{2l}\left[-\frac{x^2}{2}\Big|_{-l}^{0} + \frac{x^2}{2}\Big|_{0}^{l}\right]$$

$$= \frac{1}{2l}\left[\frac{l^2}{2} + \frac{l^2}{2}\right] = \frac{l^2}{2l} = \frac{l}{2}.$$

$$a_n = \frac{1}{l}\int_{-l}^{l} f(x)\cos\frac{n\pi x}{l}\,dx = \frac{1}{l}\left[\int_{-l}^{0} -x\cos\frac{n\pi x}{l}\,dx + \int_{0}^{l} x\cos\frac{n\pi x}{l}\,dx\right]$$

$$= \frac{1}{l}\left[-x\frac{l}{n\pi}\sin\frac{n\pi x}{l}\Big|_{-l}^{0} - \int_{-l}^{0}(-1)\cdot\frac{l}{n\pi}\sin\frac{n\pi x}{l}\,dx + x\cdot\frac{l}{n\pi}\sin\frac{n\pi x}{l}\Big|_{0}^{l} - \int_{0}^{l}\left(\frac{l}{n\pi}\right)\sin\frac{n\pi x}{l}\,dx\right]$$

$$= \frac{1}{l}\left[0 + \frac{l}{n\pi}\cdot\frac{l}{n\pi}\left(-\cos\frac{n\pi x}{l}\Big|_{-l}^{0}\right) + 0 + \left(\frac{l}{n\pi}\right)^2\cos\frac{n\pi x}{l}\Big|_{0}^{l}\right]$$

$$= \frac{1}{l}\left(\frac{l}{n\pi}\right)^2\left[\cos\frac{n\pi x}{l}\Big|_{0}^{l} - \cos\frac{n\pi x}{l}\Big|_{-l}^{0}\right]$$

$$= \frac{1}{l}\left(\frac{l}{n\pi}\right)^2[(\cos n\pi - 1) - (1 - \cos n\pi)] = \frac{1}{l}\cdot\left(\frac{l}{n\pi}\right)^2[2\cos n\pi - 2] = \frac{2l}{n^2\pi^2}[(-1)^n - 1]$$

$$\Rightarrow a_n = \begin{cases} 0, & \text{for } n \text{ even} \\ -\frac{4l}{n^2\pi^2}, & \text{for } n \text{ odd.} \end{cases}$$

$$b_n = \frac{1}{l}\int_{-l}^{l} f(x)\sin\frac{n\pi x}{l}\,dx = \frac{1}{l}\Big|_{-l}^{0} -x\sin\frac{n\pi x}{l}\,dx + \frac{1}{l}\int_0^l x\sin\frac{n\pi x}{l}\,dx.$$

$$\int_{-l}^{0} -x\sin\frac{n\pi x}{l}\,dx = x\cdot\frac{l}{n\pi}\cos\frac{n\pi x}{l}\Big|_{-l}^{0} - \int_{-l}^{0}(-1)\times\frac{l}{n\pi}\left(-\cos\frac{n\pi x}{l}\right)dx$$

$$= \frac{l^2}{n\pi}\cos n\pi - \frac{l}{n\pi}\cdot\frac{l}{n\pi}\sin\frac{n\pi x}{l}\Big|_{-l}^{0} = \frac{l^2}{n\pi}\cos n\pi - \frac{l^2}{n^2\pi^2}\times 0 = \frac{l^2}{n\pi}\cos n\pi.$$

$$\int_0^l x\sin\frac{n\pi x}{l}\,dx = -x\frac{l}{n\pi}\cos\frac{n\pi x}{l}\Big|_0^l + \int_0^l \frac{l}{n\pi}\cos\frac{n\pi x}{l}\,dx$$

$$= -\frac{l^2}{n\pi}\cos n\pi + \frac{l^2}{n^2\pi^2}\sin\frac{n\pi x}{l}\Big|_0^l = -\frac{l^2}{n\pi}\cos n\pi.$$

So the expansion is $f(x) = |x| = a_0 + \sum a_n \cos n\frac{\pi x}{l}$, n odd

$$= \frac{l}{2} + a_1\cos\frac{\pi x}{l} + a_3\cos\frac{3\pi x}{l} + a_5\cos\frac{5\pi x}{l} + \cdots = \frac{l}{2} - \frac{4l}{\pi^2}\left[\frac{\cos\frac{\pi x}{l}}{1} + \frac{\cos\frac{3\pi x}{l}}{3^2} + \frac{\cos\frac{5\pi x}{l}}{5^2} + \ldots\right].$$

1.8 Exercises

1. Explain the importance of Fourier's series in acoustics. Show how it may be used to quantitatively compare the qualities of two musical notes.
2. An alternating emf of sine form is applied to a full-wave rectifier. Obtain in an expression for the rectified current in harmonic series.
3. Find a harmonic series which will represent the current in a circuit consisting of a diode value of impedance R to which an alternating emf $E\sin\omega t$ is applied.

[*Hints.* Here, $a_0 = \frac{E}{\pi R}$, hence, y = current $= \frac{E}{R}\left[\frac{1}{\pi} + \frac{1}{2}\sin\omega t - \frac{2}{\pi}\sum_{2,4,6}\frac{\cos n\omega t}{(n+1)(n-1)}\right]$

$$= \frac{E}{2\pi R}\left[2 + \pi\sin\omega t - 4\sum_{2,4,6\ldots}\frac{\cos n\omega t}{(n+1)(n-1)}\right]]$$

4. A displacement curve is defined as $y = \frac{4at}{T}$, as t varies from 0 to $T/4$

$$= 2a - \frac{4at}{T}, \text{ as } t \text{ varies from } T/4 \text{ to } 3T/4$$

$$= \frac{4at}{T} - 4a \text{ for } t = \frac{3T}{4} \text{ to } T.$$

Find the amplitude of the fundamental and of the first two harmonics.

[*Hints.* $a_0 = \frac{1}{T}\int_0^T y\,dt = \frac{1}{T}\left[\int_0^{T/4}\frac{4at}{T}dt + \int_{T/4}^{3T/4}\left(2a - \frac{4t}{T}\right)dt + \int_{3T/4}^{T}\left(\frac{4at}{T} - 4a\right)dt\right]$.]

Similarly, find out a_n and b_n and hence, obtain the Fourier series expansion. [$\frac{8a}{\pi^2}, \frac{8a}{9\pi^2}, \frac{8a}{25\pi^2}$]

5. Expand the function $f(x) = 1$ in the interval $(0, \pi)$ and calculate the sum of the series

$$1-\tfrac{1}{3}+\tfrac{1}{5}-\tfrac{1}{7}+\ldots\infty. \qquad [(\tfrac{\pi}{4})]$$

6. Expand $f(x)=x^2$ in Fourier series, compute the sum of the series

$$\tfrac{1}{1^2}-\tfrac{1}{2^2}+\tfrac{1}{3^2}-\tfrac{1}{4^2}+\infty. \qquad [(\tfrac{\pi^2}{12})]$$

7. Expand the function in a Fourier series in the interval $(-\pi,\pi)$ when the function is defined as below:

$$f(x)=\begin{cases}-\dfrac{(\pi+x)}{2}, & -\pi\le x\le 0\\ \dfrac{\pi-x}{2}, & 0\le x\le\pi.\end{cases} \qquad [\sin x+\tfrac{1}{2}\sin 2x+\tfrac{1}{3}\sin 3x+\ldots]$$

8. Expand $f(x)=x+x^2$ in Fourier series in the interval $(-\pi,\pi)$.

$$\left[\frac{\pi^2}{3}-\sum(-1)\frac{4}{n^2}\cos nx-\sum(-1)^n\frac{2}{n}\sin nx\right].$$

9. Expand $f(x)$ in a Fourier series in the interval $(-\pi,\pi)$ when the function is defined as

$$f(x)=\begin{cases}(\pi-x)x, & 0<x<\pi\\ -(\pi+x)x, & -\pi<x<0.\end{cases} \qquad [\tfrac{8}{\pi}\textstyle\sum_{1,3,5}\frac{\sin nx}{n^2}]$$

1.9 Fourier Analysis of a Rectangular Pulse of length A_0 and width T

Determine the Fourier series and power density spectrum of a rectangular pulse train as shown in Fig. 1.7 and defined as $y=\begin{cases}A_0, & 0\le t\le T\\ 0, & t>T.\end{cases}$

$$c_0=\frac{1}{T}\int_0^T A_0\,dt=A_0.$$

$$c_{\pm n}=\frac{1}{T}\left[\int_0^T f(t)e^{-j\omega nt}\,dt+\int_0^T f(t)e^{j\omega nt}\,dt\right]$$

$$=\frac{A_0}{T}\int_0^T\left[e^{-j\omega nt}+e^{j\omega nt}\right]dt$$

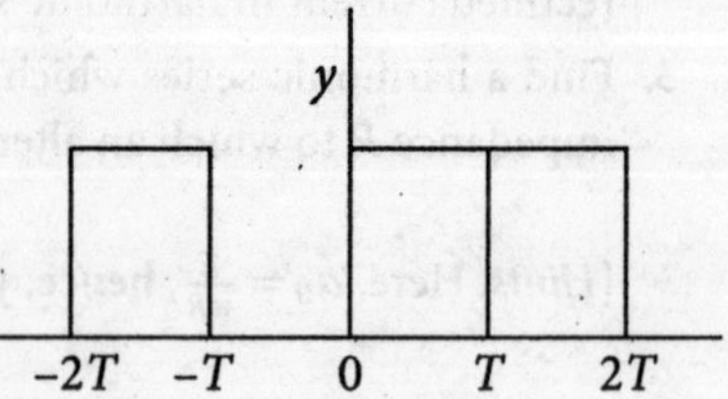

Fig. 1.7

$$=\frac{A_0}{T}\frac{1}{j\omega n}\left[\left(e^{j\omega nt}-e^{-j\omega nt}\right)\right]_0^T=\frac{A_0}{2\pi n}\sin 2\pi n=A_0\frac{\sin\phi}{\phi},\ \pi=2\pi n. \qquad (1.10)$$

$$y=A_0+\sum c_{\pm n}\left(e^{-j\omega nt}-e^{j\omega nt}\right)=A_0+\sum A_0\frac{\sin\phi}{\phi}\left(e^{-j\omega nt}+e^{-j\omega nt}\right)=A_0+\sum A_0\frac{\sin\phi}{\phi}\cos\pi nt.$$

Discussion on the function defined as

$$I=Y^2-A_0^2\frac{\sin^2\phi}{\phi^2}\Rightarrow\frac{dI}{d\phi}=A_0^2\left[\frac{2\sin\phi\cos\phi}{\phi^2}-\frac{2\sin^2\phi}{\phi^3}\right].$$

For maximum or minimum,

$$\frac{dI}{d\phi}=0\Rightarrow\sin\phi(\phi\cos\phi-\sin\phi)=0.$$

Solutions of the equation $\sin\phi=0$ or, $\tan\phi=\phi$.

$\sin\phi = 0$ is the condition for minimum.

So $\phi = m\pi, m = \pm1, \pm2, \pm3$, etc.

For max: $\phi = \tan\phi$.

This is a transcendental equation which can be solved numerically or can be solved graphically. We show here graphically.

Let $y = \phi$, y vs. ϕ is a straight line, then $y = \tan\phi$, y vs. $\tan\phi$ is the tan-graph. The points of intersection of these two graphs give the points of maxima shown in Fig. 1.8.

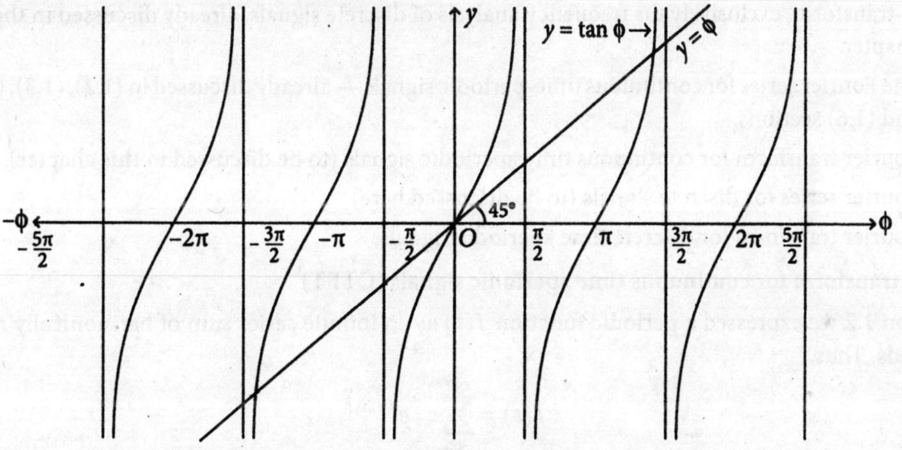

Fig. 1.8

The points of intersection and hence maxima points are located at $\phi = 0, \pm1.43\pi, \pm2.462\pi$, etc.

So the graphical plot is shown in Fig. 1.9.

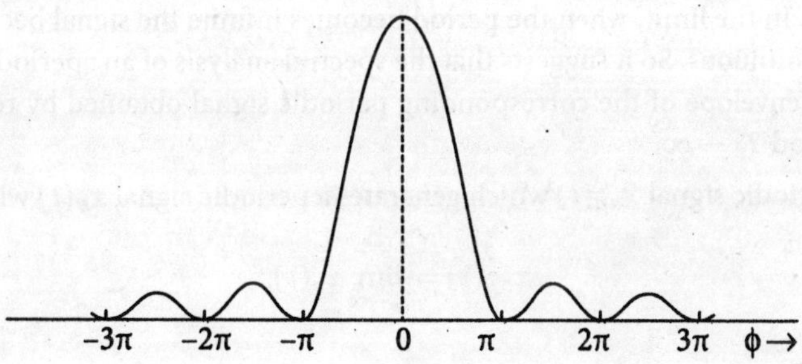

Fig. 1.9

1.10 Fourier Transform for Continuous Time Function (CTFT)

Fourier transform is another tool for frequency analysis of a signal like Fourier series but the basic difference is that FS is applicable to periodic signals whereas FT is having its use in the *frequency spectrum analysis* for *periodic functions* of discrete or continuous nature. Fourier transform for continuous function is known as CFT and for discrete function, known as DFT like CFS and DFS in case of Fourier series analysis.

The essential difference features for FS and FT are illustrated in the form of a table below:

Fourier series	Fourier transform
1. Continuous or discrete function but periodic in nature.	1. Continuous or discrete function but aperiodic in nature.
2. Discrete frequency spectrum.	2. Continuous frequency spectrum.

In general, there are five frequency analysing techniques, some of which have already been discussed earlier, the remaining ones will be seen in the present sections.

The methods are:

1. Z-transform, exclusively for frequency analysis of discrete signals, already discussed in the previous chapter.
2. The Fourier series for continuous time-periodic signals — already discussed in (1.2), (1.3), (1.4), (1.5) and (1.6) sections.
3. Fourier transform for continuous time aperiodic signals (to be discussed in this chapter).
4. Fourier series for discrete signals (to be discussed here).
5. Fourier transform for discrete time aperiodic signals.

Fourier transform for continuous time aperiodic signals (CTFT)

In section 1.2 we expressed a periodic function $f(t)$ as an infinite series sum of harmonically related exponentials. Thus,

$$f(t)=\sum_{-\infty}^{\infty} c_n\, e^{jn\omega_0 t},$$

where ω_0 is the fundamental frequency value ($\omega_0=2\pi f_0$) related to the fundamental period T_0 as an inverse $f_0=1/T_0$. As a consequence of the periodicity possessed by the Fourier series representation, the frequency spectrum consists of a series of equidistant lines with spacing equal to the fundamental period T_0. It will be remembered that if we allow the period to increase without limit, the line spacing tends to become equal to zero. In the limit, when the period becomes infinite the signal becomes aperiodic but its spectrum becomes continuous. So it suggests that the spectral analysis of an aperiodic signal may be taken as the analysis of the envelope of the corresponding periodic signal obtained by repeating the aperiodic signal with some period $T_p \to \infty$.

Let there be an aperiodic signal $x_{ap}(t)$ which generates a periodic signal $x_p(t)$ with period T_p such that

$$x_{ap}(t)=\lim_{T_p\to\infty} x_p(t). \tag{1.11}$$

(1.11) implies that the spectrum of $x_{ap}(t)$ leads to the spectrum of $x_p(t)$ when the period $T_p \to \infty$.

From the FS for continuous periodic signal developed in section (1.3), we can write

$$x_\phi(t)=\sum_{-\infty}^{\infty} c_n\, e^{j\omega n t},\ \omega=\frac{2\pi}{T_p},$$

where $c_n=\dfrac{1}{T_p}\displaystyle\int_{-T_p/2}^{T_p/2} x_\phi(t)\, e^{-j\omega n t}\, dt.$

Since $x_{ap}(t) = x_p(t)$ for $T_p \to \infty$, we get

$$(c_n)_{ap} = \frac{1}{T_p}\int_{-\infty}^{\infty} x_{ap}(t)e^{-j\omega n t}\,dt \tag{1.12}$$

$$\Rightarrow \left(x_{ap}(t)\right) = \frac{1}{T_p}\int_{-\infty}^{\infty} (c_n)_{ap} e^{j\omega n t}\,dt. \tag{1.13}$$

We now define $\tilde{X}(\omega)$ as

$$F\left\{x_{ap}\right\} = \tilde{X}(\omega) = \int_{-\infty}^{\infty} x_{ap}(t)e^{-j\omega t}\,dt. \tag{1.14}$$

(1.14) is called the Fourier transform (FT) of $x_{ap}(t)$ in the continuous variable ω. Its inversion gives rise to $F^{-1}\left\{\tilde{X}(\omega)\right\}$ obtained as

$$x_{ap}(t) = \int_{-\infty}^{\infty} \tilde{X}(\omega)e^{j\omega t}\,d\omega. \tag{1.15}$$

$\tilde{X}(\omega)$, called the Fourier transform and $x_{ap}(t)$, the continuous aperiodic function, form a pair of analysis and synthesis equation pair, shown in the table below:

Synthesis equation — Transform inverse	Analysis equation, Direct transform
$x_{ap}(t) = \int_{-\infty}^{\infty} \tilde{X}(\omega)e^{j\omega t}\,d\omega$	$\tilde{X}(\omega) = \int_{-\infty}^{\infty} x_{ap}(t)e^{-j\omega t}\,dt$

Definition of Fourier transform of continuous time signal. The frequency domain representation of a continuous time signal $x_p(t)$ is given by the continuous time Fourier transform (CTFT) defined as

$$\tilde{X}(\omega) = \int_{-\infty}^{\infty} x_p(t)e^{-j\omega t}\,dt.$$

Example 1.1. *Find the* FT *of a rectangular pulse signal defined as*

$$x_p(t) = \begin{cases} A, & |t| \le \tau/2 \\ 0, & |t| > \tau/2. \end{cases}$$

This pulse is shown in Fig. 1.10.

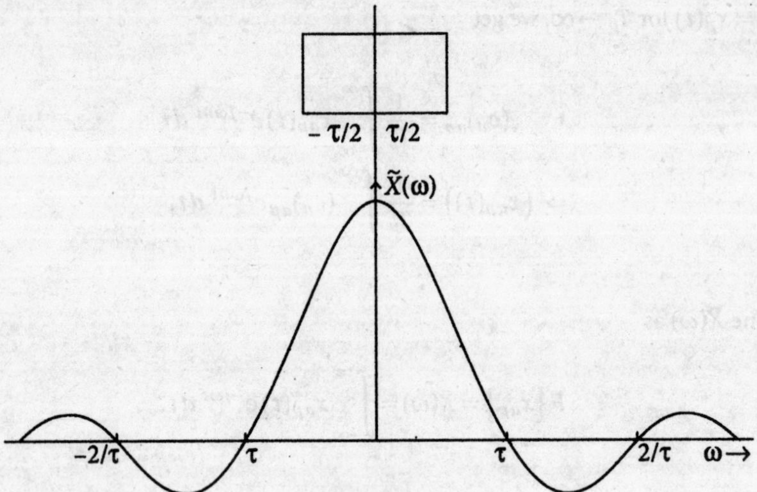

Fig. 1.10

Solution: The signal is defined by

$$x_{ap}(t) = \begin{cases} A, & |t| \le \tau/2 \\ 0, & |t| > \tau/2. \end{cases}$$

The signal is aperiodic and the Dirichlet conditions, hence amenable to Fourier transform.

Taking FT of $x_p(t)$ with the help of (1.14), we get

$$\tilde{X}(\omega) = \int_{-\infty}^{\infty} x_{ap}(t) e^{-j\omega t}\, dt = \int_{-\tau/2}^{\tau/2} A e^{-j\omega t}\, dt = -\frac{A}{j\omega} e^{-j\omega t}\Big|_{-\tau/2}^{\tau/2} = -\frac{A}{j\omega}\left[e^{-j\omega\tau/2} - e^{j\omega\tau/2}\right]$$

$$= \frac{2A}{\omega}\left[\frac{e^{j\omega\tau/2} - e^{-j\omega\tau/2}}{2j}\right] = \frac{2A}{\omega}\sin(\omega\tau/2) = A\tau\frac{\sin(\omega\tau/2)}{(\omega\tau/2)}.$$

$$\therefore\ \tilde{X}(\omega) = A\tau\frac{\sin(\omega\tau/2)}{(\omega\tau/2)}. \tag{1.16}$$

$\tilde{X}(\omega)$ has the shape of $\frac{\sin\phi}{\phi}$ as shown in Fig. 1.10.

Chapter 2

Fourier Transform and Its Application

2.1 Introduction: What is Fourier Transform?

If $f(x)$ be a function defined in $(-\infty,\infty)$ and be piece-wise continuous in each finite partial interval and integrable in $(-\infty,\infty)$, then

$$F[f(x)]=\frac{1}{\sqrt{2\pi}}\int_{-\infty}^{\infty} e^{+ikx}\,dx=g(k)$$

is known as the Fourier transform and its inverse is written as

$$F^{-1}\{g(k)\}=\frac{1}{\sqrt{2\pi}}\int_{-\infty}^{\infty} g(k)\,e^{-ikx}\,dx=f(x).$$

Sometimes $F[f(x)]$ is also written as $\tilde{f}(x)$.

***Fourier sine and cosine transforms*: The cosines and sine transformation pairs are defined as:**

$$g_c(k)=\sqrt{\frac{2}{\pi}}\int_0^{\infty} f(x)\cos kx\,dx;\quad f(x)=\sqrt{\frac{2}{\pi}}\int_0^{\infty} g_c(k)\cos kx\,dk;$$

$$g_s(k)=\sqrt{\frac{2}{\pi}}\int_0^{\infty} f(x)\sin kx\,dx;\quad f(x)=\sqrt{\frac{2}{\pi}}\int_0^{\infty} g_s(k)\sin kx\,dk.$$

The Fourier transform method is also a very useful method of solving differential equation involving variables defined in the whole domain ($-\infty$ to ∞).

2.2 Some Basic Properties of this Transform

Property 1 (Linearity property). ***If $f_1(k)$ and $f_2(k)$ be the Fourier transform of two functions $f_1(x)$ and $f_2(x)$ respectively, then $F\,[a f_1(x)+b f_2(x)]=a f_1(k)+b f_2(k)$.***

Proof. $F\,[f_1(x)]=f_1(k)=\dfrac{1}{\sqrt{2\pi}}\displaystyle\int_{-\infty}^{\infty} f_1(x)\,e^{+ikx}\,dx.\quad F\,[f_2(ax)]=f_2(k)=\dfrac{1}{\sqrt{2\pi}}\displaystyle\int_{-\infty}^{\infty} f_2(x)\,e^{ikx}\,dx.$

$$F\,[a f_1(x)+b f_2(x)]=\frac{1}{\sqrt{2\pi}}\int_{-\infty}^{\infty}[a f_1(x)+b f_2(x)]\,e^{ikx}\,dx$$

$$=\frac{1}{\sqrt{2\pi}}\int_{-\infty}^{\infty} a f_1(x)\,e^{ikx}\,dx+\frac{1}{\sqrt{2\pi}}\int_{-\infty}^{\infty} f_2(x)\,e^{ikx}\,dx=a f_1(k)+b f_2(k).$$

Property 2 (Change of scale property). *If $f(k)$ be the FT of $f(x)$, then $F[f(ax)] = \frac{1}{a} f(k/a)$.*

Proof. $f(k) = F[f(x)] = \dfrac{1}{\sqrt{2\pi}} \displaystyle\int_{-\infty}^{\infty} f(x) e^{jkx}\, dx. \quad F[f(ax)] = \dfrac{1}{\sqrt{2\pi}} \displaystyle\int_{-\infty}^{\infty} f(ax) e^{ikx}\, dx.$

Put $ax = p \Rightarrow a\,dx = dp \Rightarrow dx = \frac{1}{a}\, dp$.

$$\therefore\ F[f(ax)] = \frac{1}{\sqrt{2\pi}} = \int_{-\infty}^{\infty} \frac{1}{a} f(p) e^{i(k/a)p}\, dp = \frac{1}{a} \cdot \frac{1}{\sqrt{2\pi}} \int_{-\infty}^{\infty} f(p) e^{ik'p}\, dp$$

$$= \frac{1}{a} F[f(p)] = \frac{1}{a} g(k') = \frac{1}{a} g(k/a).$$

Property 3 (Shifting property). *If $f(k)$ be the complex Fourier transform of the function $f(x)$, then the Fourier transform of $f(x-a)$ is given by $F[f(x-a)] = e^{ika} f(k)$.*

Proof. $f(k) = \dfrac{1}{\sqrt{2\pi}} \displaystyle\int_{-\infty}^{\infty} e^{ikx} f(x)\, dx. \quad F[f(x-a)] = \dfrac{1}{\sqrt{2\pi}} \displaystyle\int_{-\infty}^{\infty} e^{ikx} f(x-a)\, dx.$

Put $x - a = p \Rightarrow dx = dp$.

$$\therefore\ \frac{1}{\sqrt{2\pi}} \int_{-\infty}^{\infty} e^{ikx} f(x-a)\, dx = \frac{1}{\sqrt{2\pi}} \int_{-\infty}^{\infty} f(p) e^{ik(p+a)}\, dp = e^{ika} \frac{1}{\sqrt{2\pi}} \int_{-\infty}^{\infty} f(p) e^{ikp}\, dp = e^{ika} f(k).$$

Property 4. *If $f(k)$ be the complex Fourier transform of $f(x)$, then the Fourier transform of $f(x)\cos bx$ is $\frac{1}{2}[g(k-b) + g(k+b)]$.*

Proof. $f(k) = \dfrac{1}{\sqrt{2\pi}} \displaystyle\int_{-\infty}^{\infty} f(x) e^{ikx}\, dx.$

$$F[f(x)\cos bx] = \frac{1}{\sqrt{2\pi}} \int_{-\infty}^{\infty} f(x) \cos bx\, e^{ikx}\, dx = \frac{1}{\sqrt{2\pi}} \left[\int_{-\infty}^{\infty} f(x) \cdot \frac{e^{ibx} + e^{-ibx}}{2} \cdot e^{ikx}\, dx \right]$$

$$= \frac{1}{\sqrt{2\pi}} \left[\frac{1}{2} \int_{-\infty}^{\infty} f(x) e^{i(k+b)x}\, dx + \frac{1}{2} \int_{-\infty}^{\infty} f(x) e^{i(k-b)x}\, dx \right] = \frac{1}{2}[g(k+b) + g(k-b)].$$

Property 5 (Convolution property). *If $f(k)$ and $g(k)$ be the* FTs *of $f(x)$ and $g(x)$ respectively, then the* FT *of the convolution of $f(x)$ and $g(x)$ is the product of their Fourier transforms.*

The convolution of $f(x)$ and $g(x)$ is usually denoted by $f * g$ and defined as

$$f * g = \frac{1}{2\pi} \int_{-\infty}^{\infty} f(u) g(x-u)\, du.$$

Proof. $f(k) = \dfrac{1}{\sqrt{2\pi}} \displaystyle\int_{-\infty}^{\infty} f(x) e^{ikx}\, dx. \quad g(k) = \dfrac{1}{\sqrt{2\pi}} \displaystyle\int_{-\infty}^{\infty} g(x) e^{ikx}\, dx.$

$$F[f * g] = F\left[\frac{1}{2\pi} \int_{-\infty}^{\infty} f(u) g(x-u)\, du \right] = \frac{1}{\sqrt{2\pi}} \int_{-\infty}^{\infty} \left[\frac{1}{\sqrt{2\pi}} \int_{-\infty}^{\infty} f(u) g(x-u)\, du \right] e^{ikx}\, du$$

$$= \frac{1}{\sqrt{2\pi}} \int_{-\infty}^{\infty} f(u) \left[\frac{1}{\sqrt{2\pi}} \int_{-\infty}^{\infty} g(x-u) e^{ikx}\, dx \right] du = \frac{1}{\sqrt{2\pi}} \int_{-\infty}^{\infty} f(u) \left[e^{iku} \right] du \times g(k)$$

$$= g(k) \cdot \frac{1}{\sqrt{2\pi}} \int_{-\infty}^{\infty} f(u) e^{iku}\, du = g(k) f(k).$$

Property 6 (Fourier transform of the derivative). *If $f(k)$ is the* FT *of $f(x)$, first derivative of $f(x)$ is $-ikf(k)$.*

Proof. $f(k)=\frac{1}{\sqrt{2\pi}}\int_{-\infty}^{\infty} f'(x)e^{ikx}\,dx=\frac{1}{\sqrt{2\pi}}\int_{-\infty}^{\infty}\lim_{h\to 0}\frac{f(x+h)-f(x)}{h}e^{ikx}\,dx$

$$=\frac{1}{\sqrt{2\pi}}\lim_{h\to 0}\int_{-\infty}^{\infty}\frac{f(x+h)}{h}e^{ikx}\,dx-\frac{1}{\sqrt{2\pi}}\lim_{h\to 0}\int_{-\infty}^{\infty}\frac{f(x)}{h}e^{ikx}\,dx$$

$$=\lim_{h\to 0}\frac{1}{\sqrt{2\pi}}\int_{-\infty}^{\infty}\frac{f(x+h)}{h}e^{ik(x+h)}\cdot d(x+h)\cdot e^{-ikh}-\lim_{h\to 0}\frac{1}{\sqrt{2\pi}}\int_{-\infty}^{\infty}\frac{f(x)}{h}e^{ikx}\,dx$$

$$=\lim_{h\to 0}\frac{f(k)\cdot e^{-ikh}}{h}-\lim_{h\to 0}\frac{f(k)}{h}=\lim_{h\to 0}\left[\frac{e^{-ikh}}{h}-\frac{1}{h}\right]f(k)=\lim_{h\to 0}\left[\frac{e^{-ikh}-1}{h}\right]$$

$$=f(k)\lim_{h\to 0}\left[\frac{1-ikh+\frac{1}{2!}(ikh)^2-\cdots-1}{h}\right].$$

Note. You can use L'Hôspital's theorem to evaluate the above limit

$$=f(k)(-ik)=-ikf(k).$$

For nth derivative, the Fourier transform is equal to $(-ik)^n$ times $f(k)$ if $f(k)$ is the FT of $f(x)$ and if the first $(n-1)$ derivatives of $f(x)$ vanish as $x\to\pm\infty$, i.e., $F[f^n(x)]=(-ik)^n f(k)$.

2.3 Some Properties of sine and cosine FT

Property 7. *If $f_s(k)$ be the sine transforms of $f(x)$, then FT of $f(ax)$ is $\frac{1}{a}f_s(k/a)$. Similar property is also possessed by cosine transform.*

Property 8. *If $f_s(k)$ and $f_c(k)$ are the sine and cosine transform of $f(x)$, then*

$$f_x[f(x)\cos bx]=\tfrac{1}{2}[f_s(k+b)+f_s(k-b)]$$
$$f_c[f(x)\sin bx]=\tfrac{1}{2}[f_s(k+b)+f_s(k-b)]$$
$$f_s[f(x)\cos bx]=\tfrac{1}{2}[f_c(k-b)+f_c(k+b)].$$

2.4 Solved Problems

1. (a) Find the FT if $f(x)$ defined by $f(x)=\begin{cases}1, & -a\le x<a\\ 0, & |x|>a.\end{cases}$

Solution: $f(k)=\frac{1}{\sqrt{2\pi}}\int_{-\infty}^{\infty} f(x)e^{ikx}\,dx$

$$=\frac{1}{\sqrt{2\pi}}\int_{-\infty}^{-a} f(x)e^{ikx}\,dx+\frac{1}{\sqrt{2\pi}}\int_{-a}^{a} f(x)e^{ikx}\,dx+\frac{1}{\sqrt{2\pi}}\int_{a}^{\infty} f(x)e^{ikx}\,dx$$

$$=\frac{1}{\sqrt{2\pi}}\int_{-a}^{a} f(x)e^{ikx}\,dx=\frac{1}{\sqrt{2\pi}}\int_{-a}^{a} e^{ikx}\,dx=\frac{1}{\sqrt{2\pi}}\cdot\frac{1}{ik}e^{ikx}\Big|_{-a}^{a}$$

$$=\frac{1}{\sqrt{2\pi}}\cdot\frac{1}{ik}\left(e^{ia}-e^{-ika}\right)=\frac{1}{\sqrt{2\pi}}\cdot\frac{2\sin ka}{k},\ k\neq 0.$$

For $k=0$,

$$\frac{1}{\sqrt{2\pi}}2a\frac{\sin ka}{ka}=\frac{2a}{\sqrt{2\pi}}.$$

(b) Find the Fourier transform of $e^{-x^2/2}$.

Solution: $F\left[e^{-x^2/2}\right]=\dfrac{1}{\sqrt{2\pi}}\displaystyle\int_{-\infty}^{\infty}e^{-x^2/2}.e^{ikx}\,dx=\dfrac{1}{\sqrt{2\pi}}\displaystyle\int_{-\infty}^{\infty}e^{-\frac{1}{2}(x-ik)^2}.e^{-k^2/2}\,dx$

$$=\frac{e^{-k^2/2}}{\sqrt{2\pi}}\int_{-\infty}^{\infty}e^{-\frac{1}{2}(x-ik)^2}\,dx.$$

Put $\dfrac{1}{\sqrt{2}}(x-ik)=y \Rightarrow dx=\sqrt{2}\,dy.$

$$\therefore F\left[e^{-x^2/2}\right]=\frac{e^{-k^2/2}}{\sqrt{2\pi}}\int_{-\infty}^{\infty}e^{-y^2}\sqrt{2}\,dy=\frac{e^{-k^2/2}}{\sqrt{\pi}}\sqrt{\pi}=e^{-k^2/2}.$$

2. Find the FT of $f(x)$ defined by $f(x)=\begin{cases}e^{i\omega x}, & a<x<b\\ 0, & x<a,\ x>b.\end{cases}$

Solution: $F[f(x)]=f(k)=\dfrac{1}{\sqrt{2\pi}}\displaystyle\int_{-\infty}^{\infty}e^{ikx}\,dx$

$$=\frac{1}{\sqrt{2\pi}}\int_{-\infty}^{a}f(x)e^{ikx}\,dx+\frac{1}{\sqrt{2\pi}}\int_{a}^{b}f(x)e^{ikx}\,dx+\frac{1}{\sqrt{2\pi}}\int_{b}^{\infty}f(x)e^{ikx}\,dx$$

$$=\frac{1}{\sqrt{2\pi}}\int_{a}^{b}f(x)e^{ikx}\,dx=\frac{1}{\sqrt{2\pi}}\int_{a}^{b}e^{i\omega x}\,e^{ikx}\,dx=\frac{1}{\sqrt{2\pi}}\int_{a}^{b}e^{i(\omega+k)x}\,dx$$

$$=\frac{1}{\sqrt{2\pi}}\cdot\frac{1}{i(\omega+k)}e^{i(\omega+k)x}\Big|_{a}^{b}=\frac{1}{\sqrt{2\pi}}\cdot\frac{1}{i(\omega+k)}\left[e^{i(\omega+k)b}-e^{i(\omega+k)a}\right]$$

$$=\frac{i}{\sqrt{2\pi}}\cdot\frac{1}{(\omega+k)}\left[e^{i(\omega+k)a}-e^{-i(\omega+k)b}\right]$$

3. Find the Fourier sine transform of $f(x)$ given as $f(x)=1/x$.

Solution: $f_s[f(x)]=f(k)=\sqrt{\dfrac{2}{\pi}}\displaystyle\int_{0}^{\infty}f(x)\sin kx\,dx=\sqrt{\dfrac{2}{\pi}}\displaystyle\int_{0}^{\infty}\dfrac{1}{x}\sin kx\,dx.$

Now, $\displaystyle\int_{0}^{\infty}e^{-\alpha x}\sin kx\,dx=\frac{k}{\alpha^2+k^2}.$

Integrating w.r.t. α between α_1 and α_2, we get

$$\int_{0}^{\infty}\left[\int_{\alpha_2}^{\alpha_1}e^{-\alpha x}\,d\alpha\right]\sin kx\,dx=\int_{\alpha_2}^{\alpha_1}\frac{k}{(\alpha^2+k^2)}d\alpha$$

$$\Rightarrow\int_{0}^{\infty}\frac{e^{-\alpha_1 x}-e^{\alpha_2 x}}{x}\sin kx\,dx=\tan^{-1}\frac{\alpha_1}{k}-\tan^{-1}\frac{\alpha_2}{k}.$$

As $\alpha_1 \to \infty$ and $\alpha_2 \to 0$, we get

$$\int_0^\infty \frac{\sin kx}{x}\,dx = \frac{\pi}{2}.$$

$$\therefore\ f_s(k) = \sqrt{\frac{2}{\pi}}\int_0^\infty \frac{\sin kx}{x}\,dx = \sqrt{\frac{2}{\pi}} \times \frac{\pi}{2} = \sqrt{\frac{\pi}{2}}.$$

4. Find Fourier cosine transform of $f(x) = \dfrac{1}{1+x^2}$.

Solution: $F_c[f(x)] = f_c(k) = \sqrt{\dfrac{2}{\pi}}\displaystyle\int_0^\infty f(x)\cos kx\,dx = \sqrt{\dfrac{2}{\pi}}\int_0^\infty \dfrac{1}{1+x^2}\cos kx\,dx.$

Differentiating both sides w.r.t. k, we get

$$\frac{d}{dk}[f_c(k)] = -\sqrt{\frac{2}{\pi}}\int_0^\infty \frac{x\sin kx}{1+x^2}\,dx = -\sqrt{\frac{2}{\pi}}\int_0^\infty \frac{(1+x^2-1)\sin kx}{x(1+x^2)}\,dx$$

$$= -\sqrt{\frac{2}{\pi}}\left[\int_0^\infty \frac{\sin kx}{x}\,dx - \int_0^\infty \frac{1}{x(1+x^2)}\sin kx\,dx\right]$$

$$= -\sqrt{\frac{2}{\pi}} \times \frac{\pi}{2} + \sqrt{\frac{2}{\pi}}\int_0^\infty \frac{\sin kx}{x(1+x^2)}\,dx.$$

Differentiating again w.r.t. k, we get

$$\frac{d^2}{dk^2} f_c(k) = \sqrt{\frac{2}{\pi}} \times \sqrt{\frac{\pi}{2}}\int_0^\infty \frac{x\cos kx}{x(1+x^2)}\,dx = f_c(k).$$

So the solution is given as $f_c(k) = c_1 e^k + c_2 e^{-k}$.

Now, when $k = 0$,

$$f_c(k) = \sqrt{\frac{2}{\pi}}\int_0^\infty \frac{\cos kx}{1+x^2}\,dx$$

$$\Rightarrow f_c(k)\Big|_{k=0} = \sqrt{\frac{2}{\pi}}\int_0^\infty \frac{1}{1+x^2}\,dx = \sqrt{\frac{2}{\pi}}\tan^{-1}x\Big|_0^\infty = \sqrt{\frac{2}{\pi}} \times \frac{\pi}{2} = \sqrt{\frac{\pi}{2}}.$$

$$\therefore\ c_1 + c_2 = \sqrt{\frac{\pi}{2}}.$$

Again, $\dfrac{d}{dk} f_c(k) = -\sqrt{\dfrac{\pi}{2}}$ at $k = 0$. $\therefore\ c_1 - c_2 = -\sqrt{\dfrac{\pi}{2}}$.

Solving for c_1 and c_2, we get $c_1 = 0$, $c_2 = \sqrt{\dfrac{\pi}{2}}$. $\therefore\ f_c(k) = \sqrt{\dfrac{\pi}{2}}\,e^{-k}$.

5. Find the cosine transform of $f(x) = e^{-x^2}$.

Solution: $F_c[f(x)] = f_c(k) = \sqrt{\dfrac{2}{\pi}}\displaystyle\int_0^\infty f(x)\cos kx\,dx = \sqrt{\dfrac{2}{\pi}}\int_0^\infty e^{-x^2}\cos kx\,dx = I$ (suppose).

Differentiating w.r.t. k, we get

$$\frac{dI}{dk}=\sqrt{\frac{2}{\pi}}\int_0^\infty -xe^{-x^2}\sin kx\,dx=\frac{1}{2}\sqrt{\frac{2}{\pi}}\int_0^\infty \left(-2xe^{-x^2}\right)\sin kx\,dx$$

$$=\frac{1}{2}\sqrt{\frac{2}{\pi}}\left[\int_0^\infty d\left(e^{-x^2}\right)\sin kx\right]=\frac{1}{2}\sqrt{\frac{2}{\pi}}\left[\sin kx\,e^{-x^2}\Big|_0^\infty-\int_0^\infty k\cos kx\,e^{-x^2}\,dx\right]$$

$$=-\frac{1}{2}\sqrt{\frac{2}{\pi}}k\int_0^\infty e^{-x^2}\cos kx\,dx=-\frac{k}{2}I$$

$$\Rightarrow \frac{dI}{I}=-\frac{k}{2}dk \Rightarrow \ln I=-\frac{k^2}{4}+\ln A \Rightarrow I=Ae^{-k^2/4}.$$

Now,

$$I=\sqrt{\frac{2}{\pi}}\int_0^\infty e^{-x^2}\cos kx\,dx.$$

When $k=0$,

$$I=\sqrt{\frac{2}{\pi}}\int_0^\infty e^{-x^2}\,dx=\sqrt{\frac{2}{\pi}}\times\frac{\sqrt{\pi}}{2}=\frac{1}{\sqrt{2}}. \quad \therefore A=\frac{1}{\sqrt{2}}.$$

Hence,

$$f_c(k)=I=\frac{1}{\sqrt{2}}e^{-k^2/4}.$$

2.5 Evaluation of Some Integration by Fourier Transform

Evaluate the integrals: (a) $\int_0^\infty \frac{\cos kx}{a^2+k^2}dk$ (b) $\int_0^\infty \frac{k\sin kx}{a^2+k^2}dk.$

Solution: Let $I_1=\sqrt{\frac{2}{\pi}}\int_0^\infty e^{-ax}\cos kx\,dx$ and $I_2=\sqrt{\frac{2}{\pi}}\int_0^\infty e^{-ax}\sin kx\,dx.$

Now,

$$I_1=\sqrt{\frac{2}{\pi}}\int_0^\infty e^{-ax}\cos kx\,dx=\sqrt{\frac{2}{\pi}}\left[-\frac{1}{a}\cos kx\,e^{-ax}\Big|_0^\infty+\int_0^\infty k\sin kx\left(-\frac{1}{a}\right)e^{-ax}\,dx\right]$$

$$=\left(\sqrt{\frac{2}{\pi}}\cdot\frac{1}{a}-\frac{k}{a}I_2\right).$$

Similarly, $I_2=\frac{k}{a}I_1$.

Solving, $I_1=\sqrt{\frac{2}{\pi}}\cdot\frac{a}{a^2+k^2}$, $I_2=\sqrt{\frac{2}{\pi}}\cdot\frac{k}{a^2+k^2}$.

If we take $f(x)=e^{-ax}$, then

$$I_1=F_c[f(x)]=f_c(k)=\sqrt{\frac{2}{\pi}}\cdot\frac{a}{a^2+k^2}=\sqrt{\frac{2}{\pi}}\int_0^\infty e^{-ax}\cos kx\,dx$$

$$I_2=F_s[f(x)]=f_s(k)=\sqrt{\frac{2}{\pi}}\cdot\frac{a}{a^2+k^2}=\sqrt{\frac{2}{\pi}}\int_0^\infty e^{-ax}\sin kx\,dx.$$

By inversion we get

$$e^{-ax}=\sqrt{\frac{2}{\pi}}\int_0^{\infty} f_c(k)\cos kx\,dx=\sqrt{\frac{2}{\pi}}\int_0^{\infty}\frac{a}{k^2+a^2}\cos kx\,dk$$

$$\Rightarrow \int_0^{\infty}\frac{a}{a^2+k^2}\cos kx\,dk=\frac{\pi}{2}e^{-ax}\Rightarrow \int_0^{\infty}\frac{\cos kx}{a^2+k^2}dk=\frac{\pi}{2a}e^{-ax}.$$

Similarly,

$$\int_0^{\infty}\frac{k}{k^2+a^2}\sin kx\,dk=\frac{\pi}{2}e^{-}.$$

Parseval's relation and evaluation of integral $\int_0^{\infty}\frac{\sin^2 ax}{x^2}dx$

The Parseval's relation or identity as it is sometimes called is defined as

(i) If $f(k)$ be the FT of $f(x)$, then $\int_{-\infty}^{\infty}|f(x)|^2\,dx=\int_{-\infty}^{\infty}|f(k)|^2\,dk$

(ii) $\int_{-\infty}^{\infty}f(x)\,g^*(x)\,dx=\int_{-\infty}^{\infty}f(k)\,g^*(k)\,dk,$

where $g^*(x)$ is the complex conjugate of $g(x)$.

Example 2.1. *Evaluate* $\int_0^{\infty}\frac{\sin^2 ax}{x^2}dx.$

Solution: Let $f(x)=\begin{cases}1, & |x|<a\\ 0, & |x|>a.\end{cases}$

$$f(k)=\frac{1}{\sqrt{2\pi}}\cdot\frac{2\sin ka}{k}.$$

According to Parseval's identity

$$\int_{-\infty}^{\infty}|f(x)|^2\,dx=\int_{-\infty}^{\infty}|f(k)|^2\,dk=\frac{1}{2\pi}\int_{-\infty}^{\infty}\frac{4\sin^2 ka}{k^2}\Rightarrow\int_{-\infty}^{\infty}\frac{\sin^2 ka}{k^2}dk=\frac{\pi}{2}\int_{-\infty}^{\infty}|f(x)|^2\,dx.$$

Now, the meaning of $|x|$ is given as $|x|=\begin{cases}-x, & \text{when } x \text{ is negative}\\ x, & \text{when } x \text{ is positive.}\end{cases}$

So

$$\int_{-\infty}^{\infty}|f(x)|^2\,dx=\int_{-\infty}^{-a}|f(x)|^2\,dx+\int_{-a}^{0}|f(x)|^2\,dx+\int_{-a}^{a}|f(x)|^2\,dx=\int_{-a}^{a}|f(x)|^2\,dx=\int_{-a}^{a}1^2\,dx=2a.$$

$$\therefore\int_{-\infty}^{\infty}\frac{\sin^2 ka}{k^2}dx=\frac{\pi}{2}\times 2a=\pi a\Rightarrow 2\int_0^{\infty}\frac{\sin^2 ka}{k^2}dk=\pi a\Rightarrow\int_0^{\infty}\frac{\sin^2 ax}{x^2}dx=\frac{\pi a}{2}.$$

Example 2.2. *Evaluate* $\int_0^{\infty} \frac{\cos\lambda x}{\lambda^2+1}\, d\lambda$.

Solution: This is similar to $\int_0^{\infty} \frac{\cos kx}{a^2+k^2}\, dk$, here $\lambda = k$ and $a = 1$.

$$\therefore \int_0^{\infty} \frac{\cos\lambda x}{1+\lambda^2}\, d\lambda = \frac{\pi}{2} e^{-x}.$$

2.6 Application of Fourier Transform to Solve Differential Equation

2.6.1 Solution of ordinary differential equation

Consider the problem of damped harmonic oscillator to be solved by FT method.

The equation of motion of such an oscillator is given as

$$\frac{d^2x}{dt^2} + 2b\frac{dx}{dt} + \omega_0^2 x = f(t), \tag{2.1}$$

where $f(t)$ is any function such that

$$f(t) = \frac{1}{\sqrt{2\pi}} \int_{-\infty}^{\infty} f(\omega) e^{-i\omega t}\, d\omega \text{ and } f(\omega) = F[f(t)] = \frac{1}{\sqrt{2\pi}} \int_{-\infty}^{\infty} e^{i\omega t}\, dt.$$

Taking FT of both sides of the equation, we get

$$F\left[\frac{d^2x}{dt^2}\right] + 2bF\left[\frac{dx}{dt}\right] + \omega_0^2 F[x] = F[f(t)]$$

$$\Rightarrow \frac{1}{\sqrt{2\pi}} \int_{-\infty}^{\infty} \frac{d^2x}{dt^2} e^{i\omega t}\, dt + \frac{2b}{\sqrt{2\pi}} \int_{-\infty}^{\infty} \frac{dx}{dt} e^{i\omega t}\, dt + \frac{\omega_0^2}{\sqrt{2\pi}} \int_{-\infty}^{\infty} x e^{i\omega t}\, dt = \frac{1}{\sqrt{2\pi}} \int_{-\infty}^{\infty} f(t) e^{i\omega t}\, dt$$

$$\Rightarrow (-i\omega)^2 X(\omega) + 2b(-i\omega)^1 X(\omega) + \omega_0^2 X(\omega) = f(\omega)$$

$$\Rightarrow \left(\omega^2 - i2b\omega + \omega_0^2\right) X(\omega) = f(x) \Rightarrow X(\omega) = \frac{f(x)}{\omega_0^2 - \omega^2 - 2ib\omega}.$$

By inversion we get

$$x(t) = \frac{1}{\sqrt{2\pi}} \int_{-\infty}^{\infty} \frac{f(\omega) e^{-i\omega t}}{\omega_0^2 - \omega^2 - 2bi\omega}\, d\omega.$$

2.6.2 Solution of partial differential equation

Example 2.1 (Heat conduction equation). $\frac{\partial u}{\partial t} = h^2 \frac{\partial^2 u}{\partial x^2}$, *given* $u(x,0) = f(x)$.

Solution: Let us define the following Fourier transforms:

$$u(k,t) = \frac{1}{\sqrt{2\pi}} \int_{-\infty}^{\infty} u(x,t) e^{ikx}\, dx. \tag{2.2}$$

$$f(k) = \frac{1}{\sqrt{2\pi}} \int_{-\infty}^{\infty} f(x) e^{ikx}\, dx. \tag{2.3}$$

Taking $F-T$ of the heat conduction equation, we get

$$h^2\frac{1}{\sqrt{2\pi}}\int_{-\infty}^{\infty}\frac{\partial^2 u}{\partial x^2}e^{ikx}\,du=\frac{1}{\sqrt{2\pi}}\int_{-\infty}^{\infty}\frac{\partial u}{\partial t}e^{ikx}\,dx$$

$$\Rightarrow h^2(-ik)^2u(k,t)=\frac{1}{\sqrt{2\pi}}\cdot\frac{\partial}{\partial t}\int_{-\infty}^{\infty}ue^{ikx}\,dx=\frac{\partial}{\partial t}u(k,t)\Rightarrow -h^2k^2u(k,t)=\frac{\partial}{\partial t}u(k,t).$$

$$\frac{du}{u}=-h^2k^2\,dt\Rightarrow \ln u(k,t)=-h^2k^2t+\ln A\Rightarrow u(k,t)=Ae^{-h^2k^2t}.\quad \therefore u(k,0)=A.$$

From (2.2), we get

$$u(k,0)=\frac{1}{\sqrt{2\pi}}\int_{-\infty}^{\infty}u(x,0)e^{ix}\,dx=\frac{1}{\sqrt{2\pi}}\int_{-\infty}^{\infty}f(x)e^{ix}\,dx=f(k).$$

$$\therefore A=f(k).\quad \therefore u(k,t)=f(k)e^{-h^2k^2t}.$$

By inversion, we get

$$u(x,t)=\frac{1}{\sqrt{2\pi}}\int_{-\infty}^{\infty}f(k)e^{-h^2k^2t}e^{-ikx}\,dk=\frac{1}{\sqrt{2\pi}}\int_{-\infty}^{\infty}\left(f(k)e^{-h^2k^2t}\right)e^{-ikx}\,dk$$
$$=\frac{1}{\sqrt{2\pi}}\int_{-\infty}^{\infty}[f(k)g(k)]\,e^{-ikx}\,dk.$$

By convolution theorem

$$\frac{1}{\sqrt{2\pi}}\int_{-\infty}^{\infty}f(k)g(k)e^{ikx}\,dx=\frac{1}{\sqrt{2\pi}}\int_{-\infty}^{\infty}f(\xi)g(x-\xi)\,d\xi.$$

$g(x-\xi)$ is the FT inverse of $e^{-h^2k^2t}$, i.e.,

$$g(x)=\frac{1}{\sqrt{2\pi}}\int_{-\infty}^{\infty}e^{-h^2k^2t}\cdot e^{-ikx}\,dk.$$

Put $u^2=2k^2h^2t\Rightarrow u\,du=2h^2t\,k\,dk\Rightarrow dk=\dfrac{u\,du}{2h^2tk}=\dfrac{\sqrt{2}kht^{1/2}du}{2h^2tk}=\dfrac{1}{\sqrt{2}}\cdot\dfrac{du}{h\sqrt{t}}=\dfrac{du}{h\sqrt{2t}}.$

$$\therefore g(x)=\frac{1}{\sqrt{2\pi}}\int_{-\infty}^{\infty}e^{-u^2/2}e^{-ikx}\,dk=\frac{1}{\sqrt{2\pi}}\int_{-\infty}^{\infty}e^{-u^2/2}\cdot e^{-i(u/h\sqrt{2t})x}\cdot\frac{du}{h\sqrt{2t}}$$
$$=\frac{1}{h\sqrt{4\pi t}}\int_{\infty}^{\infty}e^{-u^2/2}\cdot e^{-ik'u}\,du$$
$$=\frac{du}{h\sqrt{2t}}=\frac{1}{h\sqrt{2t}}\cdot\frac{1}{\sqrt{2\pi}}\int_{-\infty}^{\infty}e^{-u^2/2}e^{-ik'u}\,du,\ k'=\frac{x}{h\sqrt{2t}}$$
$$=\frac{1}{h\sqrt{2t}}e^{-k'2/2}=\frac{1}{h\sqrt{2t}}e^{-x^2/4h^2t}.$$
$$\therefore u(x,t)=\frac{1}{\sqrt{2\pi}}\int_{-\infty}^{\infty}f(\xi)\frac{1}{h\sqrt{2t}}e^{-(x-\xi)^2/4h^2t}\,d\xi=\frac{1}{2h\sqrt{\pi t}}\int_{-\infty}^{\infty}f(\xi)e^{-(x-\xi)^2/4h^2t}\,d\xi.$$

2.7 Wave Equation

Consider the variation of a string fixed at both ends, satisfying the wave equation

$$\frac{\partial^2 u}{\partial t^2}=c^2\frac{\partial^2 u}{\partial x^2},$$

under the condition that at $t=0$, the string is at rest but it has a displacement $f(x)$ for $-\infty<x<\infty$.

Taking $F-T$ of the wave equation we can write

$$\frac{1}{\sqrt{2\pi}}\int_{-\infty}^{\infty}\left(\frac{\partial^2 u}{\partial t^2}\right)e^{ikx}\,dx=\frac{c^2}{\sqrt{2\pi}}\int_{-\infty}^{\infty}\left(\frac{\partial^2 u}{\partial x^2}\right)e^{ikx}\,dx.$$

$$\frac{d^2u(k,t)}{dt^2}=c^2(-ik)^2u(k,t)=-c^2k^2u(k,t).$$

The general solution to the above equation is given as

$$u(k,t)=c_1\,e^{ickt}+c_2\,e^{-ickt}. \tag{2.4}$$

Initially the string is at rest, so $\frac{\partial u}{\partial t}=0$ at $t=0$.

$$\therefore\ F\left[\frac{\partial u}{\partial t}\right]=0\Rightarrow\frac{1}{\sqrt{2\pi}}\int_{-\infty}^{\infty}\left(\frac{\partial u}{\partial t}\right)e^{ikx}\,dx=0\Rightarrow\frac{1}{\sqrt{2\pi}}\cdot\frac{\partial}{\partial t}\int_{-\infty}^{\infty}u\,e^{ikx}\,dx=0\Rightarrow\frac{\partial}{\partial t}u(k,0)=0.$$

Differentiating (2.4) w.r.t. t, we get

$$\frac{\partial u}{\partial t}=ick\left(c_1\,e^{ickt}-c_2\,e^{-ickt}\right),\quad\therefore\ \left.\frac{\partial u}{\partial t}\right|_{t=0}=0=(c_1-c_2).$$

Also, at $t=0$, $u(x,0)=f(x)$.

$$u(k,0)=\frac{1}{\sqrt{2\pi}}\int_{-\infty}^{\infty}f(x)e^{ikx}\,dx=c_1+c_2.$$

$$\therefore\ c_1-c_2=0,\quad c_1+c_2=u(k,0).$$

On solution we get

$$c_1=\frac{u}{2}(k,0),\quad c_2=\frac{u}{2}(k,0).\quad\therefore\ u(k,t)=\frac{u}{2}(k,0)\left(e^{ickt}+e^{-ickt}\right).$$

Inverting the above equation, we get

$$u(x,t)=\frac{1}{\sqrt{2\pi}}\int_{-\infty}^{\infty}u(k,t)e^{-ikx}\,dk=\frac{1}{\sqrt{2\pi}}\left[\int_{-\infty}^{\infty}\frac{u}{2}(k,0)\left(e^{ickt}+e^{-ickt}\right)e^{-ikx}\,dk\right]$$

$$=\frac{1}{\sqrt{2\pi}}\left[\frac{1}{2}\int_{-\infty}^{\infty}u(k,0)\,e^{ickt}\,e^{-ikx}\,dk+\frac{1}{2}\int_{-\infty}^{\infty}u(k,0)\,e^{-ickt}\,e^{-ikt}\,dk\right]$$

$$=\frac{1}{\sqrt{2\pi}}\left[\frac{1}{2}\int_{-\infty}^{\infty}u(k,0)\,e^{ik(ct-x)}\,dk+\frac{1}{2}\int_{-\infty}^{\infty}u(k,0)\,e^{-ik(ct+x)}\,dk\right]$$

$$=\frac{1}{2}\left[\frac{1}{\sqrt{2\pi}}u(k,0)\,e^{-ik\sigma}\,dk+\frac{1}{\sqrt{2\pi}}\int_{-\infty}^{\infty}u(k,0)\,e^{-ik\sigma'}\,dk\right]$$

$$=\frac{1}{2}[f(x-ct)+f(c+ct)]=\frac{1}{2}f(x+ct)+\frac{1}{2}f(x-ct).$$

2.8 Solution of Poisson equation

Let us take Poisson's equation represented by

$$\nabla^2 u(\bar{r}) = -\frac{\rho(\bar{r})}{\epsilon},$$

where $u(\bar{r})$ is the potential function and $\rho(\bar{r})$ is the charge density.

Let us write in three dimensions, the FT of $u(\bar{r})$ as

$$u(\bar{k}) = \frac{1}{(2\pi)^{1/2}} \int_{-\infty}^{\infty} u(\bar{k})\, e^{i\bar{k}\bar{r}}\, d^3 r,$$

where $d^3 r = dx\, dy\, dz$.

The inverse transform is given by

$$u(\bar{r}) = \frac{1}{(2\pi)^{3/2}} \int_{-\infty}^{\infty} u(\bar{k})\, e^{-i\bar{k}\bar{r}}\, d^3 k,$$

where $d^3 k$ is the elementary volume in k-space.

Taking FT of both sides of the above equation, we get

$$-k^2 u(\bar{k}) = -\frac{\rho(\bar{k})}{\epsilon} \Rightarrow u(\bar{k}) = \frac{\rho(\bar{k})}{\epsilon k^2}, \tag{2.5}$$

where $\rho(\bar{k}) = \frac{1}{(2\pi)^{3/2}} \int_{-\infty}^{\infty} \rho(\bar{r}')\, e^{i\bar{k}\bar{r}}\, d^3 r'$.

So, Fourier inversion of the above (2.5) gives

$$u(\bar{r}) = \frac{1}{\epsilon(2\pi)^{1/2}} \int_{-\infty}^{\infty} \frac{\rho(\bar{k})}{k^2}\, e^{-i\bar{k}\bar{r}}\, d^3 k = \frac{1}{(2\pi)^{3/2}\epsilon} \int_{-\infty}^{\infty} \left[\frac{1}{(2\pi)^{3/2}} \int_{-\infty}^{\infty} \frac{\rho(\bar{r}')}{k^2}\, e^{i\bar{k}\bar{r}}\, d^3 r' \right] e^{-i\bar{k}\bar{r}}\, d^3 k$$

$$= \frac{1}{(2\pi)^3 \epsilon} \int_{-\infty}^{\infty} \int_{-\infty}^{\infty} \rho(\bar{r}')\, \frac{e^{-\bar{k}(\bar{r}-\bar{r}')}}{k^2}\, d^3 k\, d^3 r'.$$

Now

$$\int_{-\infty}^{\infty} \frac{e^{i\bar{k}(\bar{r}-\bar{r}')}}{k^2}\, d^3 k,$$

$d^3 k$ is the elementary volume in k-space and if spherical polar coordinates are chosen in k-space, then $d^3 k = k^2 \sin\theta\, d\theta\, d\phi\, dk$.

$$\therefore \int_{-\infty}^{\infty} \frac{e^{ik(\bar{r}-\bar{r}')}}{k^2}\, d^3 k = \int_0^{\infty} \int_0^{\pi/2} \int_0^{2\pi} \frac{e^{-ik|\bar{r}-\bar{r}'|\cos\theta}}{k^2} \sin\theta\, d\theta\, d\phi\, dk$$

$$= \int_0^{\infty} \left[\int_0^{\pi} \left(e^{-ik\,|\bar{r}-\bar{r}'|\cos\theta} \sin\theta\, d\theta \right) \right] dk \times 2\pi$$

$$= 2\pi \int_0^{\pi} \left[\int_0^{\pi} e^{-ik|\bar{r}-\bar{r}'|\cos\theta} \sin\theta\, d\theta \right] dk. \tag{2.6}$$

Take the integral

$$\int_0^{\pi} e^{-ik|\bar{r}-\bar{r}'|\cos\theta}\sin\theta\, d\theta.$$

Put $-ik|\bar{r}-\bar{r}'|\cos\theta = x \Rightarrow ik|\bar{r}-\bar{r}'|\sin\theta\, d\theta = dx.$

$$\therefore \int e^{-ik|\bar{r}-\bar{r}'|\cos\theta}\sin\theta\, d\theta = \int e^x \frac{dx}{ik|\bar{r}-\bar{r}'|} = \frac{e^x}{ik|\bar{r}-\bar{r}'|}.$$

$$\therefore \int_0^{\pi} e^{-ik|\bar{r}-\bar{r}'|\cos\theta}\sin\theta\, d\theta = \left.\frac{e^{-ik|\bar{r}-\bar{r}'|\cos\theta}}{ik|\bar{r}-\bar{r}'|}\right|_0^{\pi} = \frac{e^{ik|\bar{r}-\bar{r}'|} - e^{-ik|\bar{r}-\bar{r}'|}}{ik|\bar{r}-\bar{r}'|}. \tag{2.7}$$

So the integral in (2.7) can be written as

$$\int_0^{\infty} \frac{e^{-ik|\bar{r}-\bar{r}'|} - e^{-ik|\bar{r}-\bar{r}'|}}{ik|\bar{r}-\bar{r}'|}\, dk = 2\int_0^{\infty} \frac{\sin k|\bar{r}-\bar{r}'|}{k|\bar{r}-\bar{r}'|}\, dk.$$

$$\therefore \int_{-\infty}^{\infty} \frac{e^{-ik|\bar{r}-\bar{r}'|}}{k^2}\, d^3k = 4\pi\int_0^{\infty} \frac{\sin|\bar{r}-\bar{r}'|}{k|\bar{r}-\bar{r}'|}\, dk = \frac{4\pi}{|\bar{r}-\bar{r}'|}\int_0^{\infty} \frac{\sin k|\bar{r}-\bar{r}'|}{k}\, dk = \frac{4\pi}{|\bar{r}-\bar{r}'|}\cdot\frac{\pi}{2} = \frac{2\pi^2}{|\bar{r}-\bar{r}'|}.$$

$$\therefore u(\bar{r}) = \frac{1}{(2\pi)^3\epsilon}\int_{-\infty}^{\infty} \rho(\bar{r}')\, d^3r' \frac{2\pi^2}{|\bar{r}-\bar{r}'|} = \frac{1}{4\pi\epsilon}\int_{-\infty}^{\infty} \frac{\rho(\bar{r}')}{|\bar{r}-\bar{r}'|}\, d^3r'. \tag{2.8}$$

The solution as in (2.7) is in accordance with the definition of potential. $\rho(\bar{r}')$ is the charge density at $P'(\bar{r}')$, then $\rho(\bar{r}')\, d^3r'$ defines charge associated with the volume element d^3r', hence, the potential due to this charge at the point $P(\bar{r})$ distant $|\bar{r}-\bar{r}'|$ w.r.t. P' is given by

$$\frac{1}{4\pi\epsilon}\cdot\frac{\rho(\bar{r}')\, d^34'}{|\bar{r}-\bar{r}'|},$$

hence, the total potential at P is given by

$$u(r) = \frac{1}{4\pi\epsilon}\int_{-\infty}^{\infty} \frac{\rho(\bar{r}')\, d^3r'}{|\bar{r}-\bar{r}'|}.$$

2.9 Short Theory Questions with Answers

1. What is the condition that a function can be Fourier transformed?
   Ans. A function $f(x)$ that satisfies the following conditions, called *Dirichlet's condition* can be Fourier transformed. (a) $f(x)$ is defined and is single-valued in some interval (a, b); (b) $f(x)$ is bounded; (c) $f(x)$ has at most a finite number of maxima and minima; (d) $f(x)$ has only a finite number of discontinuities — piecewise continuous.
2. What is Fourier transform of a function? What is its importance in practical problems?
   Ans. If $f(x)$ be a function in ordinary space, then its $F-T$ in any space (k-space) is defined as

$$f(k) = \frac{1}{\sqrt{2\pi}}\int_{-\infty}^{\infty} f(x)e^{ikx}\, dx.$$

Its inversion gives the function in x-space and is given by

$$f(x)=F^{-1}[f(k)]=\frac{1}{\sqrt{2\pi}}\int_{-\infty}^{\infty} f(k)e^{-ikx}\,dk.$$

The Fourier transform may be used to evaluate integrals, to solve differential equations both ordinary and partial.

3. Show that the convolution of $f(x)$ and $g(x)$ is commutative.
Solution: The convolution of the two functions $f(x)$ and $g(x)$ written as $f^* g$ is written as

$$f^* g=\frac{1}{\sqrt{2\pi}}\int_{-\infty}^{\infty} f(x-u)g(u)\,du.$$

So

$$g^* f=\frac{1}{\sqrt{2\pi}}\int_{-\infty}^{\infty} g(x-u)f(u)\,du.$$

Now put $x-u=\xi \Rightarrow -du=d\xi$.

$$\therefore\ g^* f=\frac{1}{\sqrt{2\pi}}\int_{-\infty}^{\infty} g(x-u)f(u)\,du=\frac{1}{\sqrt{2\pi}}\int_{-\infty}^{\infty} -g(\xi)f(x-\xi)\,d\xi$$

$$=\frac{1}{\sqrt{2\pi}}\int_{-\infty}^{\infty} f(x-\xi)g(\xi)\,d\xi=f^* g.$$

4. If a function is defined as $f(t)=\begin{cases} e^{-xt}\,g(t), & t>0 \\ 0, & t<0. \end{cases}$
Show that Fourier transform (FT) of $f(t)$ is related to the Laplace transform (LT) of $g(t)$.
Solution: The Fourier transform of $f(t)$ is given by

$$F[f(t)]=f(k)=\frac{1}{\sqrt{2\pi}}\int_{-\infty}^{\infty} f(t)e^{ikt}\,dt=\frac{1}{\sqrt{2\pi}}\int_{-\infty}^{0} 0\cdot e^{ikt}\,dt+\frac{1}{\sqrt{2\pi}}\int_{0}^{\infty} e^{-xt}\,g(t)\,e^{ikt}\,dt$$

$$=\frac{1}{\sqrt{2\pi}}\int_{0}^{\infty} e^{(ik-x)t}\,g(t)\,dt.$$

Put $ik-x=-p$.

$$\therefore\ F[f(t)]=\frac{1}{\sqrt{2\pi}}\int_{0}^{\infty} e^{-pt}g(t)\,dt=\frac{1}{\sqrt{2\pi}}L\{g(t)\}.$$

2.10 Exercises

1. Calculate the Fourier transform of $f(x)$ defined as $f(x)=\begin{cases} 1, & -a\le x\le a \\ 0, & |x|>a. \end{cases}$
2. By use of Fourier transform solve the equation

$$\frac{\partial\theta}{\partial t}=h^2\frac{\partial^2\theta}{\partial x^2},$$

under the condition $\theta(x,t) \to 0$ and $\dfrac{\partial\theta}{\partial x} \to 0$ as $x \to \pm\infty$; $\theta(x,0) = e^{-a^2x^2}$, where α is a positive constant.

3. Find the Fourier transform of $f(x)$ defined by $f(x) = \begin{cases} e^{i\omega x}, & a < x < b \\ 0, & x < a,\ x > b. \end{cases}$
4. Find the Fourier transform of $f(x)$ defined by $f(x) = \begin{cases} (1-x^2), & |x| < 1 \\ 0, & |x| > 1. \end{cases}$

   and hence find
$$\int_0^\infty \left(\frac{x\cos x - \sin x}{x^3}\right)\cos\frac{x}{2}\,dx.$$
5. Find the Fourier sine transform of $f(x) = 1/x$.
6. Find the sine and cosine transform of $f(x) = \frac{e^{ax}+e^{-ax}}{e^{\pi x}-e^{-\pi x}}$.
7. Find the Fourier transform of $f(x) = e^{-x^2/2}$ and $f(x) = \frac{\sin ax}{x}$.
8. Using the method of Fourier transform, find the displacement $y(x,t)$ of an infinite string, given that the string is initially at rest and that the initial displacement is $f(x)$ for $-\infty < x \leq \infty$. Show that the solution can be put as $y(x,t) = \frac{1}{2}[f(x+ct) + f(x-ct)]$.
9. If $F(k) = \frac{N}{\sqrt{2a}} e^{-k^2/4a}$, find $F(x)$.

2.11 Integral Equations

Many a time in Physics we have an unknown function occurring under integral sign. An equation containing the unknown function under an integral cover is known as an *integral equation.* The solution of the integral equation furnishes a method for the study of eigenvalue problems in mathematical physics.

The most general type of integral equation is written as

$$g(x)f(x) = \phi(x) + \lambda \int_a^b k(x,z)\phi(z)\,dz. \tag{2.9}$$

The known functions are $g(x)$, $f(x)$ and $k(x,z)$, the last one being called the *Kernel.* The limits of integration a, b are either known functions of x or are constants. We will consider solutions of special type integral equation which is obtained from the general type (1) by choosing the known functions $g(x)$, $k(x,z)$ and a, b, etc. Thus, putting $g(x) = 0$, we get

1. $0 = \phi(x) + \lambda \int_a^b k(x,z)\phi(z)\,dz$, called *Fredholm's equation* of first kind.
2. Putting $g(x) = 1$, we get Fredholm's equation of second kind written as
$$f(x) = \phi(x) + \lambda \int_a^b k(x,z)\phi(z)\,dz. \tag{2.10}$$
3. *Volterra's equation of the second kind.* The integral equation is written as
$$f(x) = \phi(x) - \int_0^x k(x-u)\phi(u)\,du. \tag{2.11}$$

4. *Volterra's equation of the first kind.* This integral equation is given as

$$f(x)=\int_a^b k(x-u)\phi(u)\,du. \tag{2.12}$$

5. *Abel's integral equation.* This is written as

$$f(x)=\int_0^t \frac{\phi(u)\,du}{(x-u)^n},\ 0<n<1. \tag{2.13}$$

If $f(x)=0$ in (1), (2) and (3), the resulting equations becomes homogeneous.

Solution of integral equations: By Laplace transform

Take Volterra's equation of the second kind

$$f(x)=\phi(x)-\int_0^x k(x-u)\phi(u)\,du. \tag{2.14}$$

Here, $f(x)$ and $k(x-u)$ are known and $\phi(x)$ is unknown. Taking LT of both sides, we get

$$L\{f(x)\}=L\{\phi(x)\}-L\left\{\int_0^x k(x-u)\phi(u)\,du\right\}.$$

Now let $L\{\phi(x)\}=g(s)$, $L\{f(x)\}=f(s)$, $L\{k(x)\}=k(s)$.

$$\therefore L\int_0^x k(x-u)\phi(u)\,du=k(s)\,g(s) \text{ by Faultung theorem.}$$

$$\therefore \tilde{X}(\omega)=A\tau\frac{\sin(\omega\tau/2)}{(\omega\tau/2)}. \tag{2.15}$$

$\tilde{X}(\omega)$ has the shape of $\frac{\sin\phi}{\phi}$ as shown in Fig. 1.10.

2.12 Discrete Fourier Series (DFS)

Let us consider a sequence $x_p(n)$ with period N such that $x_p(n)=x_p(n+lN)$, l is an integer. $x_p(n)$ being periodic, it can be expressed as a weighted sum of complex exponentials whose frequencies are integral multiples of the fundamental frequency $\frac{2\pi}{N}$. They are of the form

$$e^{j2\pi k(n/N)}=e^{j2\pi k(n+lN)/N}. \tag{2.16}$$

Since the discrete-time FS of a periodic sequence is periodic, when sampled, it consists of a finite number of frequency components. If they are N in number, the harmonics are given by $\frac{2\pi k}{N}$, $k=0,1,2,\ldots,N-1$.

Any periodic sequence according to FS representation can be expressed as

$$x_p(n)=\sum_{k=0}^{N-1} C_k e^{i2\pi kn/N}, \tag{2.17}$$

where C_k, $k=0,1,2,\ldots,N-1$, are called discrete Fourier series coefficients.

In order to find the discrete FS coefficient, let us multiply both sides of (2.17) by $e^{-j2\pi l n/N}$ and sum the product for $n=0$ to $(N-1)$, we will get

$$\sum_{n=0}^{N-1} x_p(n)\,e^{-j2\pi l n/N} = \sum_{n=0}^{N-1}\sum_{k=0}^{N-1} C_k\, e^{j2\pi(n-l)n/N}. \tag{2.18}$$

Changing the order let us find the summing first w.r.t. n, for $k-l=x=0,\pm N,\pm 2N\pm,\ldots$, etc.

Let

$$\sum e^{j2\pi(k-l)/N} = \sum e^{j2\pi x/N},\ x=k-l;$$

then the r.h.s. of the (2.17) can be written as

$$\sum_{n=0}^{N-1} x_p(n)\,e^{-j2\pi l n/N} = \sum_{k=0}^{N-1} C_k \sum_{n=0}^{N-1} e^{j2\pi x n/N}. \tag{2.19}$$

Then

$$\sum_{n=0}^{N-1} e^{j2\pi x n/N} = \sum_{0}^{N-1} e^{(j2\pi x/N)n} = \sum_{n=0}^{N-1} a^n = \left(1+a+a^2+\cdots+a^{N-1}\right) = \frac{(1-a^N)}{(1-a)} = \frac{1-e^{j2\pi x}}{1-e^{j2\pi x/N}}. \tag{2.20}$$

The right-hand side of (2.19) is of the form $\frac{0}{0}$ as $x\to 0$, so we apply L'Hôspital's theorem to get the ratio value as

$$\lim_{x\to 0}\frac{1-e^{j2\pi x}}{1-e^{j2\pi x/N}} = \frac{\sum_{x\to 0}\frac{d}{dx}\left(1-e^{j2\pi x}\right)}{\sum_{x\to 0}\frac{d}{dx}\left(1-e^{j2\pi x/N}\right)} = N.$$

So, we can write

$$\sum x_p(n)\,e^{j2\pi l/N} = \sum_{k=0}^{N-1} C_k \sum e^{j2\pi(k-l)n/N} = \begin{cases} C_k N, & \text{for } x=0\\ 0, & \text{for } x\neq 0.\end{cases}$$

$$\therefore\ C_k = C_l = \frac{1}{N}\sum x_p(n)\,e^{-j2\pi nk/N}. \tag{2.21}$$

This equation evaluates the Fourier's coefficients of the discrete time FS (DTFS) for $k=0,1,2,\ldots,(N-1)$.

2.13 Fourier Transform for Discrete Time Aperiodic Signal (DTFT)

The present analysis is very much similar to that of continuous time aperiodic sequence. The procedure is to divide sequence into N-lengths in the time domain. Then evaluate the Fourier transform of length-N time domain sequence.

The DFT of a finite length time-domain sequence $[x(n)]$ is defined as

$$\boxed{X(k) = \sum_{k=0}^{N-1} [x(n)]\,e^{-j2\pi kn/N},\ 0\le k\le (N-1)}, \tag{2.22}$$

where $X(k)$ is the DFT coefficients.

Multiplying both sides by $e^{j2\pi ln/N}$, we get after summing over n first, we would get

$$\sum_{n=0}^{N-1} X(k)\,e^{2j\pi ln/N} = \sum_{n=0}^{N-1}\sum_{k=0}^{N-1} X(n)\,e^{j2\pi(l-k)n/N} \tag{2.23}$$

Proceeding exactly in the same way as in the previous article to build (2.20) through (2.17), (2.18) and (2.19), we can get

$$\boxed{[x(n)] = \frac{1}{N}\sum X(k)\,e^{j2\pi l/N}}. \tag{2.24}$$

2.14 Short Theory Questions with Answers

1. What do you mean by DFT and DTFT?

   Ans. DFT refers to discrete Fourier transform obtained by performing sampling operation in both time as well as frequency domains; its spectrum is discrete.

   But DTFT refers to time discrete Fourier transform obtained by sampling in time domain only; its spectrum is continuous function of frequency (ω).

2. What is the difference between FS and FT?

   Ans. FS gives harmonic content of a periodic time function and the frequency spectrum is discrete whereas FT gives frequency information for an aperiodic signal and here the frequency spectrum is continuous.

3. Define discrete Fourier series (DFS).

   Ans. It refers to a sequence $\left[x_p(n)\right]$ with a period of N samples so that $\left[x_p(n)\right] = \left[x_p(n+N)\right]$. Then DFS of the sequence $\left[x_p(n)\right]$ is defined as

$$X_p(k) = \sum_{n=0}^{N-1} \left[x_p(n)\right]\, e^{-j2\pi kA/N}.$$

4. Establish the relation between DFT and Z-transform.

   Ans. Z-transform of a sequence $[x(n)]$ is

$$X(z) = Z\{[x(n)]\} = \sum_{n=-\infty}^{\infty} [x(n)]\,z^{-n}$$

   with ROC that includes the unit circle.

   If $X(z)$ is sampled at N equally spaced points on the unit circle $z_k = e^{j2\pi k/N}$, $k = 0, 1, 2, \ldots, (N-1)$ we obtain

$$\begin{aligned} X(k) &= X(z)\Big|_{z=e^{j2\pi k/N}}, \quad k = 0, 1, 2, \ldots, (N-1) \\ &= \sum [x(n)]\,z^{-j2\pi nk/N}. \end{aligned} \tag{2.25}$$

   If the sequence $[x(n)]$ has a finite duration of length N, then

$$X(z) = \sum_{0}^{N-1} [x(n)]\,z^{-n} = \sum_{0}^{N-1} \left\{\frac{1}{N}\left[X(k)\,e^{-j2\pi ln/N}\right\} z^{-n}\right] = \frac{1}{N}\sum_{0}^{N-1} X(k) \sum_{l=0}^{N-1} \left(e^{j\pi l/Nz^{-1}}\right)^n.$$

5. What is the N-point DFT of a sequence $[x(n)]$?

Ans. This is defined by the relation

$$X(k) = \sum_{n=0}^{N-1} [x(n)]\, e^{-j2\pi kn/N}, \; k = 0, 1, \ldots, (N-1).$$

6. The N-point DFT of a sequence $[x(n)]$ is $C(k)$. Find the IDFT.

Ans. IDFT $C(k) = [x(n)] = \sum_{n=0}^{N-1} C(k)\, e^{j2\pi kn/N}, \; n = 0, 1, \ldots, (N-1).$

Part VI

Microelectronic Devices

Chapter 1

Microcontroller

1.1 Introduction

It is said that necessity is the mother of invention. It is true to a very large extent. Majority of the micro-electronic devices have come through different types of project-works assigned to solve different types of practical problems.

Let us refer to the problem of counting. The most ancient method of solving counting problem was through the introduction of a machine in the name of *Abacus* — a hand-operated counting machine which consisted of beads on wires. But this machine is so old that its origin has now almost melted into oblivion. We felt the need of a better computing machine and the necessity ushered in an era of calculating machine, both electric and electronic types. The first large scale electronic calculating machine, called ENIAC developed around 1946 with the basic hardware made of valves. About 18,000 valves were used to form the heart of the system requiring a large room for its installation and also requiring a power supply system almost half the size of the machine. The cost of running the system is prohibitive.

Since then drastic changes were taking place in the arena of electronic technology, the valve technology being replaced by semiconductor discrete technology, the discrete technology being soon replaced by IC technology, etc.

In IC technology, electronic elements are fabricated and inter-connected on a chip of silicon as substrate. Depending on the number of units fabricated on a chip, there are four classes of integration: (a) SSI (small scale integration) (b) MSI (medium scale integration) LSI (large scale integration) and VLSI (very large scale integration). In the first three fabrication processes, one of the standard techniques is the process of diffusion. But in VLSI, a special diffusion technique, called 3D diffusion is adopted. With the advent of these new technologies, there was a complete revolution in the manufacturing of computing machines. Thus, there was the birth of computer machines of the following three types:

Main-frame computer. It is the most expensive, most powerful and extensive hardware based computer capable of handling a large volume of data and solving variety of problems.

Minicomputer. Less expensive than the main-frame computer and also less powerful than the main-frame one.

Microcomputer. Less expensive and less powerful than the mini-computer.

Of course, the differences among these classifications and the point where one class leaves off and the other begins are not well defined. In general, the cost and packaging differentiates the three classes.

1.2 Microcontroller: In quest of RD work with a microcontroller

Let there be a project proposal for a digital timer having the following specifications:

1. Timing range between 1 s to 1 hour.
2. It is to be programmable with a resolution of 1 s; to be operated with two push-bottom switches.
3. To be powered by 1.5 volts dc source.
4. Cost not to exceed ₹ 300 or so.

Design consideration w.r.t. the above specifications requires that: (a) Discrete logical circuit is of no use here, because the device is to be programmable, small in size and to have low power consumption. (b) Ambitious microprocessor like 8085 or Z 80 will not be a proper choice, because of cost and size consideration. (c) Naturally, therefore, the straightforward choice is a microcontroller. But the question is which microcontroller to be chosen?

In order to get the answer, we must have a fair knowledge about a microcontroller, its construction, working and various types of applications.

1.3 Learning about a microcontroller

Today the commercial market is flooded with great variety of microcontrollers. To study all of them is next to impossible, nor it is absolutely necessary. It will suffice to have recourse to a general approach to study the salient features of a microcontroller.

Our study begins with two basic questions: (a) What is a microcontroller? (b) What is its difference with a microprocessor?

Microcontroller defined. A microcontroller is looked upon as a microcomputer on a single chip, because it contains all the essential basic ingredients of a full-fledged microcomputer.

It contains CPU (central processing unit), ROM (read only memory), RAM (read access memory), I/O (input-output) ports and some smart peripherals like serial communication devices, inbuilt timers, etc.

Though a microcontroller is usually referred to as a microcomputer it is to be borne in mind that a microcontroller is aimed at fixed and relatively light applications whereas a microcomputer is not dedicated to a fixed task but its use is restricted by user's imagination which can direct its activity by loading application program in its RAM.

1.4 A microcontroller and a microprocessor compared

A microprocessor is basically a CPU but a microcontroller contains a microprocessor within it just like an atom containing the nucleus within it. A nucleus is a part of an atom but it is not the atom as a whole. So a microcontroller is a microprocessor but a microprocessor is not a microcontroller.

A microcontroller is a single chip microcomputer having all the features of a microcomputer but it is dedicated in most cases for a particular task. The microcomputer, on the other hand, can be used for different programs by the user by loading the programs in its RAM. The striking difference between a microcomputer and a microcontroller, it is that in a general purpose microcomputer, the RAM/ROM ratio is pretty high while that in a microcontroller is low.

What is the significance of high value and low value of RAM/ROM *ratio?*

$\tau =$ RAM/ROM is the ratio between RAM space and ROM space.

RAM space is meant for storing stack and temporary variables while ROM space contains the control program. If τ is large, RAM space is greater than the ROM space and τ low would mean RAM space is less than the ROM space.

In case of a microprocessor, τ is large, so in this case RAM space > ROM space, but in case of a microcontroller ROM space > RAM space. So we can write:

$S_{RA} > S_{RO}$, microprocessor

$S_{RA} < S_{RO}$, microcontroller.

Thus, a microprocessor can store temporary variables much more than a microcontroller while a microcontroller contains control programs much more than that in a microprocessor.

1.5 CPU is the brain of a microcontroller

In any microcomputer machine and hence, also in a microcontroller, there is a central processing unit, called a CPU which is supposed to be the brain of the machine. This CPU is shown in block diagram as in Fig. 1.1.

Fig. 1.1

This unit consists of circuits meant for arithmetic operation such as addition, subtraction and also logical functions like ANDING and ORING. There is another component, called control unit (CU). A CPU is not in its elf self-sufficient, because in order to process data and do arithmetic operation on them, there must be a source of the data (DS) in the so-called memory space which stores the data from where it can be taken by the CPU when needed. Further, when the CPU does some calculation, how would we know the result? So we need an output device. So to make a CPU complete in itself, additional components are added thus composing the characteristic computer system schematically shown in Fig. 1.2.

The central processing unit is an electronic circuit to perform arithmetic operation and also logic functions. This CPU which processes and controls data is popularly known as *microprocessor.* So what is a microprocessor?

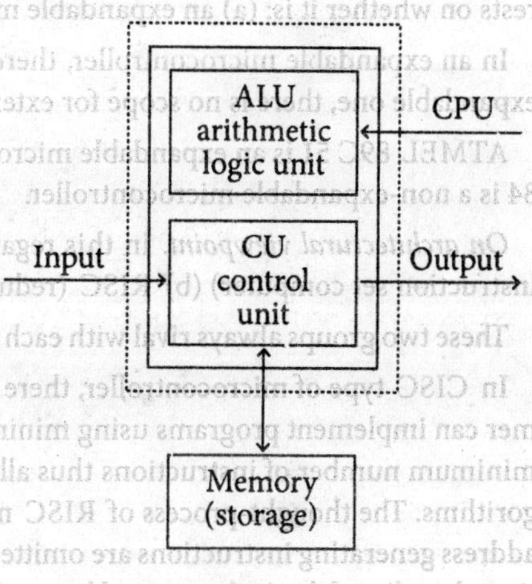

Fig. 1.2

A microprocessor is a digital *computer* CPU but it is in no sense a complete digital computer.

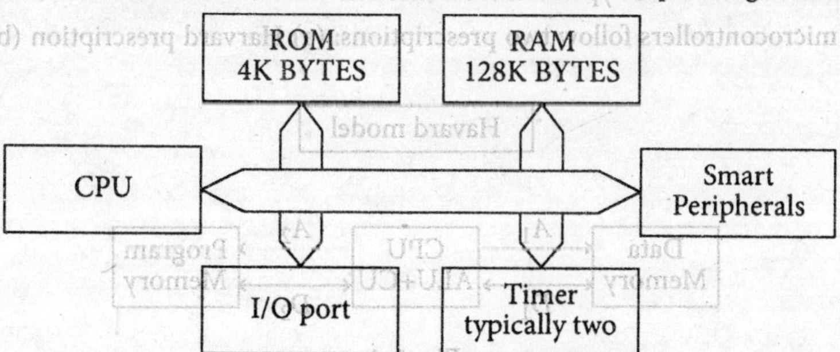

Fig. 1.3

Born with the same idea and by the same group of people who invented microprocessor, microcontrollers were also designed and fabricated. But microcontrollers do not enjoy the same amount of popularity as the microprocessor, although the former is more a micro-computer than a *microprocessor.*

A look into the block diagrams of a hypothetical microcontroller will reveal what has been depicted in Fig. 1.3.

1.6 Classification of microcontroller

There are three ways in which microcontrollers are usually classified depending on: (a) Design engineering viewpoint (b) Practical viewpoint (c) Architectural viewpoint.

Classification depending on engineering viewpoint. This classification depends on the available hardware features of the chip. On this view we have: (a) Low-end microcontroller (b) Mid-range microcontroller (c) High-end microcontroller.

Thus, the microchip PIC 12C 508 is a low-end microcontroller while INTEL 80C 196 is a high-end microcontroller, it is because the former contains bare minimum hardware resources and the latter contains all possible conceivable hardware resources.

On practical viewpoint. Another way of classification of a microcontroller as suggested by some engineers rests on whether it is: (a) an expandable microcontroller (b) closed or non-expandable microcontroller.

In an expandable microcontroller, there is scope for addition of external memory; but in case of non-expandable one, there is no scope for external attachment.

ATMEL 89C 51 is an expandable microcontroller, so also is INTEL 80C 196 but 89C 2051 or PIC 16F 84 is a non-expandable microcontroller.

On architectural viewpoint. In this regard, there are two types of microcontroller: (a) CISC (complex instruction set computer) (b) RISC (reduced instruction set computer).

These two groups always rival with each other thus they are under continuous revision and modification.

In CISC type of microcontroller, there is a large number of available instructions so that the programmer can implement programs using minimum of them. In an RISC type, on the other hand, there is bare minimum number of instructions thus allowing the programmers the freedom of designing their own algorithms. The thought-process of RISC microcontroller user is wider. The complex arithmetic, logical or address generating instructions are omitted in RISC microcontroller; there is drastic reduction in instruction, decoding delay is shorter and hence the increased speed of this type of microcontroller. In this respect this type excels over CISC type but the bad thing about this microcontroller is that the program tends to be longer than that in case of CISC type of microcontroller.

Architecturally microcontrollers follow two prescriptions: (a) Harvard prescription (b) Princeton prescription.

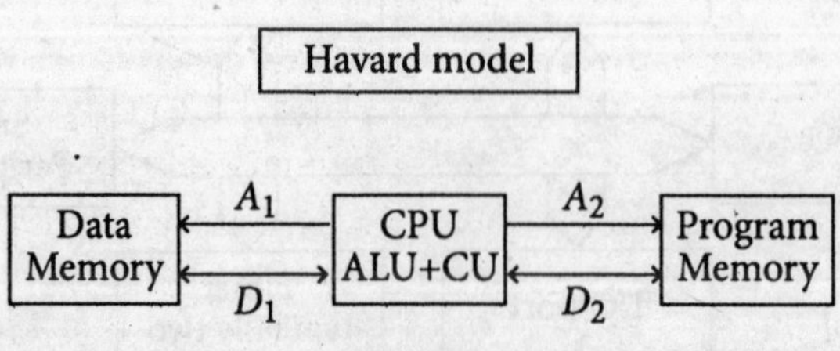

Fig. 1.4

About Harvard prescription. This design of the microcomputer was developed in Harvard University under the leadership of Howard Aiken (1900–72) in the early twentieth century.

Microcontroller 8051 is built on the Harvard model. This model is schematically represented by in Fig. 1.4.

It shows that here there is separate memory space for data and program with separate buses for each; A_1, A_2 are *address buses* and D_1, D_2 are *data buses*. This mechanism gives greater speed improvement.

About Princeton design or von Neumann model. This microcomputer device was designed in Princeton University under the leadership of Dr John von Neumann (1903–1957), a brilliant American mathematician of Hungarian origin.

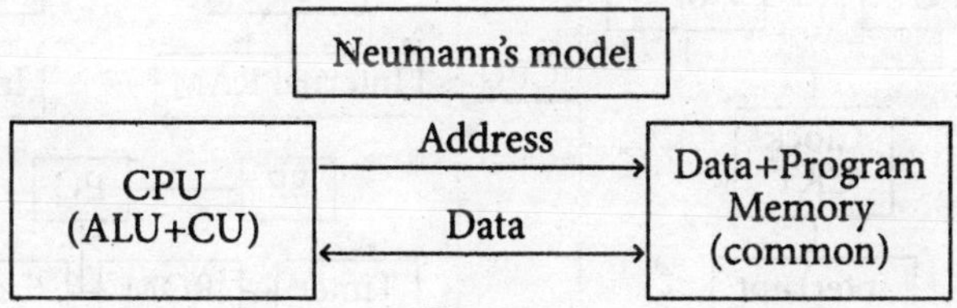

Fig. 1.5

This model had ready acceptance, because it is simpler and very well suited to the then available technology (before the advent of solid state technology). The model is represented by in Fig. 1.5. But after late 1970s, single-chip micro-computer manufacturer realised that the Harvard architecture had the merits for embedded real time controllers and gained popularity.

In Harvard design [Fig. 1.4], separate memory space is used for data and program but in Princeton design there is single memory space for both data and program and other variables. There is a single bus for transferring data into and out of CPU. Because of separate buses for data and program in Harvard model, it has greater speed improvement since data and program instructions can be fetched separately and independently. No such independence and simultaneous operations on data and program instruction exist in Neumann's model. Simplicity and cost is the primary advantage of von Neumann's model over Harvard architecture.

The above two models compared

Neumann's model	*Harvard model*
1. Common memory space for data and program.	1. Separate memory for data and program.
2. There is a single bus for transferring data into and out of CPU.	2. Separate bus for each of data and program.
3. Data and program instructions cannot be fetched simultaneously.	3. Since there are separate memory spaces for data and program and also there are two buses, they can be fetched; simultaneously, hence greater speed of operation.
4. von Neumann's design is comparatively cheaper and simpler.	4. This design is relatively complex and more expensive. It needs twice as many address and data pins on a chip compared to the Neumann's design.
5. Most computers today use Neumann's architecture.	5. Harvard architecture is used in specialised microprocessor intended for real time embedded applications, this is most suited for digital signal processors.

1.7 Microcontroller vs. Microprocessor (Architectural differences)

A general purpose microprocessor has a CPU consisting of ALU, a program counter (PC), a stack pointer (SP), some registers, clock-timers and interrupt circuit [Fig. 1.6].

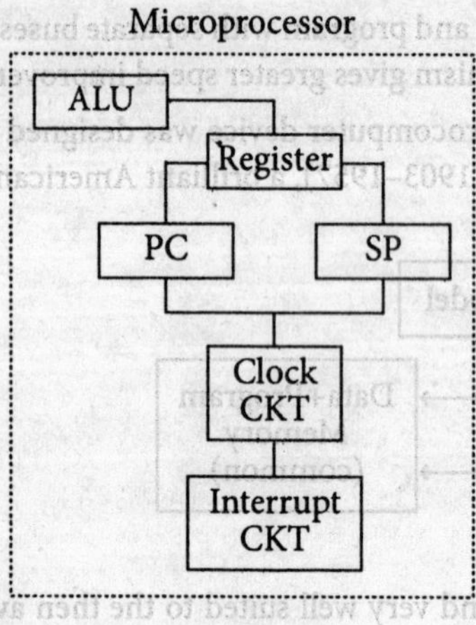

Fig. 1.6

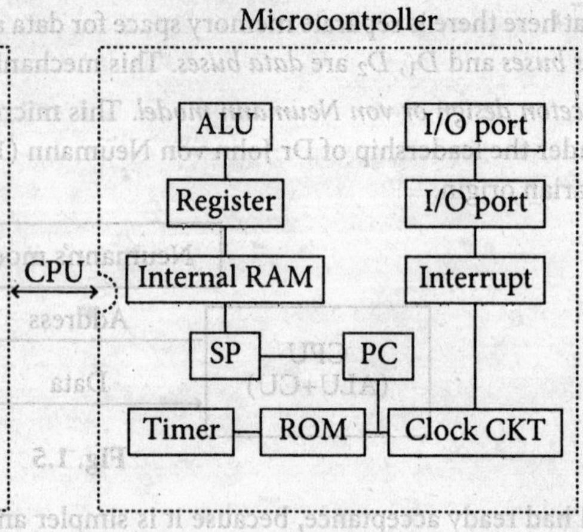

Fig. 1.7

A microcontroller, on the other hand, is a microprocessor plus something. This something refers to addition to memory (ROM and RAM), memory decoders and a number of I/O devices, such as a keyboard and CRT or LCD display [Fig. 1.7].

1.8 Microcontroller 8051 — a member of MCS 51 family

8051 is the generic representation of a family of microcontrollers MCS 51 whose members range from 8031–8051. After its introduction in 1980 it became very popular with the circuit designers and it so happens that it can be taken as a representative member of the whole family.

8051 is a CISC type, it follows from Harvard design. It is housed in a dual-in-line package (DIP) having forty-pin-configuration as shown in Fig. 1.8.

The basic hardware features of microcontroller 8051 are:

- 8-bit CPU (control processing unit).
- 64k PMS (program memory space).
- 64k DMS (data memory space).
- 4k on chip memory (ROM space).
- 128 bytes on chip memory (RAM space).
- 32 I/O ports.
- Two 16-bit Timers/counters.
- UART (universal asynchronous trans-receiver).
- Built-in clock-oscillator.

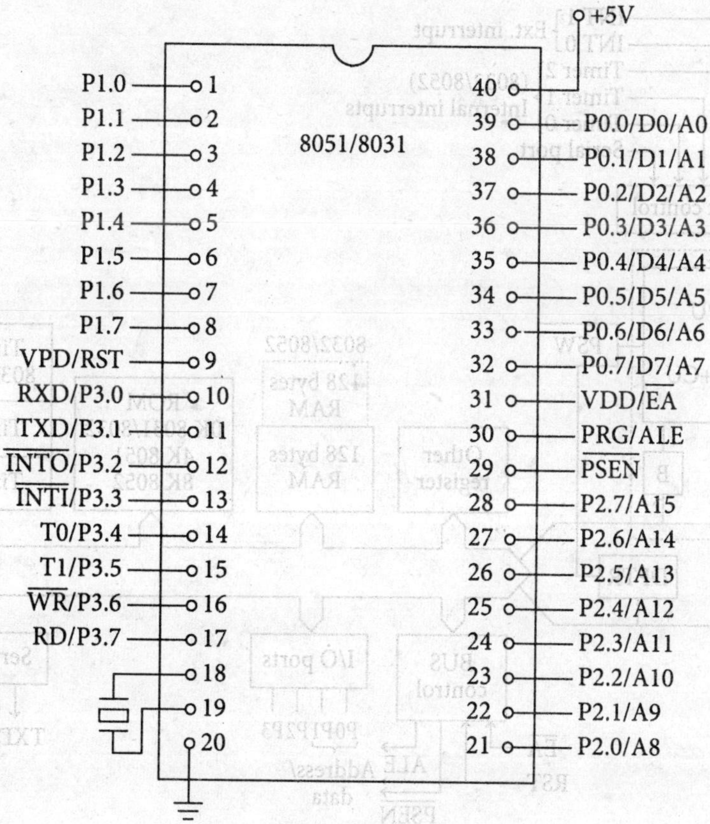

Fig. 1.8 Dual In-line Package (DIP).

1.9 Learning about the pin-configuration

In order to use a microcontroller, one has to learn about the internal hardware design, also called the architecture of the device, to determine the type, the number and size of the registers and other circuits.

Knowledge about the microcontroller would not be complete if we cannot manipulate the hardware by an accompanying set of instructions or what is termed as the software. Once familiar with both of hardware and software, the designer can apply the microcontroller to problems at hand. It is almost similar to one's desire to write a poem in a foreign language. To do it one must have sufficient vocabulary in that language equivalent to gathering knowledge about the hardware and also he must know the rules of grammar equivalent to familiarisation with the software of the device.

So we will study first the architecture or hardware contents of 8051, the representative member of the generic microcontroller family having members ranging from 8031 to 8051.

Architecture of 8051. The pin-configuration of this microcontroller is shown in Fig. 1.8, where it is shown to have different functions. In Fig. 1.9 we have shown the architecture in block diagram and subsequently the different parts are described.

Block diagram as shown in Fig. 1.9 will be of immense help to us. The exploded view of the interior of the microcontroller 8051 is exhibited in Fig. 1.9; corresponding pin-configurations as in Fig. 1.8 has the descriptions as mentioned overleaf:

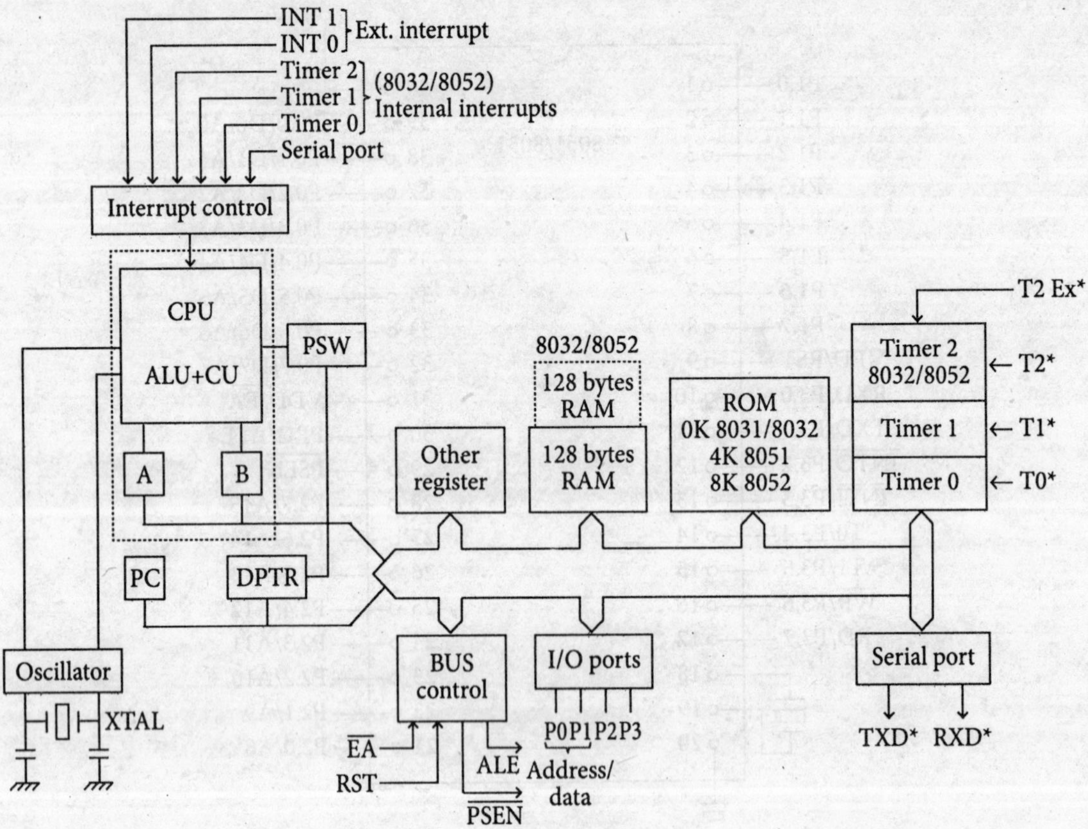

Fig. 1.9

Power supply points: Pins 40 *and* 20. V_{CC} positive is connected to pin 40 while its negative is connected to pin 20 which is grounded.

Usually for 8051 V_{CC} is of value equal to 5 volts but ATMEL 89C 51 and 89C 2051 operate with wider voltage range 2.7V to 6V.

Oscillator on chip: Pins 18 *and* 19. These two pins are the inputs to the external crystal oscillator. Usually, the inbuilt oscillator is driven by an external crystal oscillator connected between 18 and 19. The normal crystal frequency in 8051 is 12 MHz but its ATMEL, Philips derivatives run with 24 MHz oscillation frequency.

Pin 9: RST (*resetting*) *pin.* This is the resetting pin which is activated by a circuit shown in Fig. 1.10. High input signal at this pin resets the microcontroller. Typically one RC network is hooked up to this pin.

Pin 10 RXD: P3.0. This pin is the external serial input pin.

Pin 11: TXD: P3.1. This pin is the external serial output.

Pin 12: $\overline{\mathrm{INTO}}$: P3.2 }
Pin 12: $\overline{\mathrm{INTI}}$: P3.3 } Interrupt pins.

Pin 14: P3.4 (T0) Timer 0 input pin.

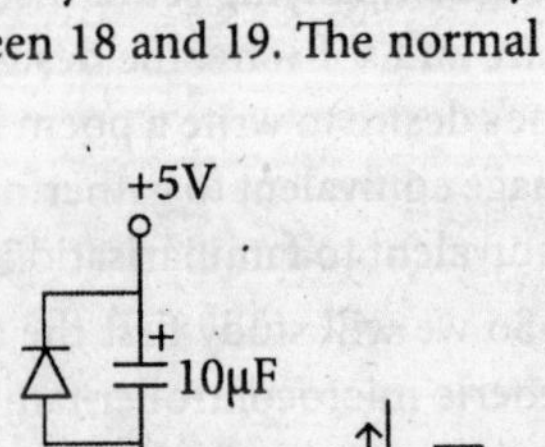

Fig. 1.10

Pin 15: P3.5 (T1) Timer 1 input pin.

Pin 16: P3.6 ($\overline{WR}$) Write strobe.

Pin 17: P3.7 ($\overline{RD}$) Read strobe.

Pin 21: P2.0 (A8) Address 8
(Port 2 Bit 0)

Pin 22: P2.1 (A9) Address 9
(Port 2 Bit 1)

Pin 23: P2.2 (A10) Address 10
(Port 2 Bit 2)

Pin 24: Port 2 Bit 3 Address 11

Pin 25: Port 2 Bit 4 Address 12

Pin 26: Port 2 Bit 5 Address 13

Pin 27: Port 2 Bit 6 Address 14

Pin 28: Port 2 Bit 7 Address 15

Pin 29: $\overline{PSEN}$ Program store enable

Pin 30: Address latch enable (EPROM program pulse)

Pin 31: External enable (EPROM programming voltage)

Pin 32: Port 0 Bit 7 (Address/data 7)

Pin 33: Port 0 Bit 6 (Address/data 6)

Pin 34: Port 0 Bit 5 (Address/data 5)

Pin 35: Port 0 Bit 4 (Address/data 4)

Pin 36: Port 0 Bit 3 (Address/data 3)

Pin 37: Port 0 Bit 2 (Address/data 2)

Pin 38: Port 0 Bit 1 (Address/data 1)

Pin 39: Port 0 Bit 0 (Address/data 0)

Pin 40: $V_{CC} = 5V$

1.10 Describing the functions of different pins

Group 1

1. *Pins* 1–8 (*dedicated* I/O *port* — *Port* 1). These pins are designated as P1.0, P1.1, P1.2, etc. They are not dual functions pins, unlike pins 10–17 of port 3 or pins 21–28 of port 1, both of which group of pins are dual purpose pins. This group from 1–8 are used for interfacing external devices. P1.0 and P1.1 are used as analog comparator inputs in 89C 2051 microcontroller and in 8052/89C52/8032 they are used as inputs to the third timer.
2. *Pin* 9 (*resetting input*): *Active high.* This pin on getting logic high signal resets the microcontroller.
3. Pins 10–17 (*dual purpose group, port* 3). This is a double purpose serving group of pins for (a) use as general purpose I/O (b) implementation of some special features of 8051.
4. *Pins* 18–19 (*oscillator inputs*). The inbuilt oscillator is driven by external crystal connected between pins 18 and 19. The capacitor of value equal to 30pF at both pins is invariably connected for frequency stabilisation.
5. *Pin* 20: *Power supply negative* (GND).

6. *Pins* 21–28: (*dual purpose, port* 2). They serve as general purpose I/O or as high address lines for external memory.
7. *Pin* 29 (PSEN — *program store enable*). This output signal enables external program or code memory and appears as a consequence of Harvard architecture. The PSEN signal goes low during fetch cycle.
8. *Pin* 30 (ALE, *address latch enable*). This output pin is used for demultiplexing the address and data bus. In 8051 when port 0 is used in alternate mode, i.e., as the data and lower address bus, the ALE latches the address into an external register during first half of a memory cycle. In the other half of the memory cycle, the port 0 lines are available for data.

   An ALE signal generates pulse 1/6th of the rate of the on-chip oscillator, so it can be used as a general purpose clock source.
9. *Pin* 31: (EA — *external access*). If this pin is tied to logic high (1), this microcontroller executes program from internal ROM but if it is tied to logic low (0), it executes program from external ROM (internal ROM is ignored). EPROM version of 8051, the EA pin is used as V_{PP} (21 or 12V) as programming voltage.
10. *Pins* 32–39. These group of pins also serves double purpose, as I/O port or data/address bus. In the latter case EA pin must be made low.

Application areas of microcontrollers. Day by day, the application area of the microcontroller is expanding and the working mechanism is becoming more well defined and fine.

In our discussion here we will divide the area into zones to show the vast expanse of applicability of the microcontroller systems.

Automobile industry. Most of the new generation cars are fitted with electronic gadgets with microcontrollers embedded in them. One such a gadget is meant for adjusting the air-fuel mixture to a correct ratio; another for monitoring fuel economy. In some cars again some talking annunciators have been introduced who proclaim the expected time of arrival at a particular place or to announce if the seat-belt is unfastened or a door is ajar.

Industrial zone. Microcontrollers are used in instruments meant for: (a) power line monitoring (b) ac/dc — motor control (c) UPS — uninterrupted power supply (d) electronic energy meters (e) temperature control and monitoring (f) fluid-flow measurement and control.

As smart sensors. Analog sensors converted to digital form by a microcontroller; it is also used in practically all the electronic security gadgets ranging from burglar alarm to highly complex access control systems.

Telecommunication area. Microcontroller is used in the making of telephone caller ID, telephone monitoring system with print-out facilities.

Chapter 2

Microprocessor

2.1 Introduction

Born with same idea and by the same group of people and company who invented microcontrollers, microprocessors were also designed and fabricated by them. The first microprocessor, a 4-bit microprocessor 4004 came in the year 1971 through Intel Corporation, USA and Busicom of Japan.

A microprocessor is an LSI (large scale integration) circuit in metal oxide semiconductor (MOS) technology. Intel corporation after introduction of 4-bit 4004 microprocessor, introduced 8-bit and 16-bit microprocessors in the name of Intel 8080, 8085 and 8085A of 8-bit type and Motorola 68000, Z800, TMS 9900 of 16-bit type. Nowadays even 32-bit microprocessors like Intel 80386, Z8000, MC 68030 are also available.

A microprocessor is a physical system which receives an input signal and delivers an output signal whose characteristic would be decided by the internal behaviour of the system. A microprocessor accepts only binary inputs. The signal is fed through a data bus consisting of 8 conductors for an eight-bit processor, each bit of the signal being carried by one conductor.

So the data bus can convey $2^8 (= 256)$ possible combinations of input data. So, before feeding data to the microprocessor, the data if initially be analog require to be converted into digital signal by a suitable ADC. The digital signal so obtained passes through the microprocessor which performs its task according to the software and the results obtained are brought out through the data bus to output ports. A suitable microprocessor acting as DAC converts the digital signal to the analog signal.

A microprocessor functions not only to receive input signal but produce an output one. For the purpose of processing signals by a microprocessor, several input data are to be received by it. So there must be a data source, called the microprocessor *data memory, read/write memory* or popularly coined as RAM (read access memory).

Besides RAM memory space, there is another memory space, called ROM, which means *read only memory*. The program stored in ROM using special devices cannot be changed by the user.

In a microprocessor system in addition to data bus and address bus, one more bus is there, known as *control bus*.

The essential components of a microprocessor. There are four essential ingredients of microprocessor system. They are: (a) CPU (b) The memory units (MU) (c) Input unit (IU) (d) Output unit (OU).

A CPU: World of Inside CPU. A typical microprocessor CPU is shown in Fig. 2.1 w.r.t. our discussion held previously. The schematic diagram of the CPU is redrawn here for ready reference [Fig. 2.1].

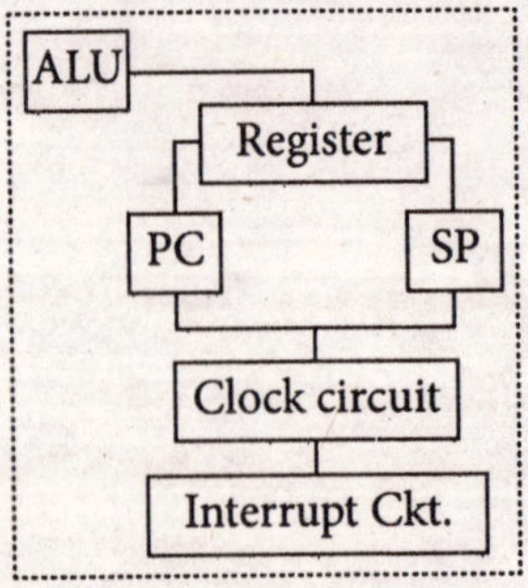

CPU

ALU—Arithmetic logic unit
PC—Program counter
SP—Stack pointer

Fig. 2.1

The CPU world may be divided into five functional blocks.

1. ALU: *Arithmetic and logical unit.* This is where the arithmetic and logic operations are carried out. This family consists of three family members (a) Accumulator (Ace) (b) Program stable register (c) Arithmetic logic unit. Accumulator is an 8-bit special register which is used in almost all arithmetic and logic operations.

Status register is also a special register where the condition and state of the latest arithmetic or logic operation is stored. In different bit positions of this register, information like *parity, zero, sign, overflow, interrupt mask, interrupt request, carry, auxiliary carry,* etc. are stored.

The contents of the status register is called PSW (program status word) and the register itself is called PSW register.

What is a microprocessor?

So the question that may be posed is: what exactly is a microprocessor? The name itself is self-explanatory, it means that it is small (micro) and it is able to process data (processor). *A microprocessor is a* CPU, the central processing unit. What is it, the CPU? It is an electronic device which can interpret and execute instructions and control input and output. It may be said to be the brain of a computer.

2.2 A wonder gift from the micro-wonderland — a computer, a counting machine

A humble beginning in the process of counting with the device known as *Abacus,* the counting machine has undergone many thorough changes in the design and applications. Evolutionary phases of the computers are referred to as *generations.* Till date computers are classified into five generations.

1. *First generation.* The first generation computers were built with valves as its hardware.
2. *Second generation.* In this type, valves were replaced by semiconducting elements.
3. *Third generation.* In this type, the integrated circuits (IC) replaced the transistor semiconductor circuits.
4. *Fourth generation.* Computers were built with VLSI IC circuits in place of ordinary integrated circuits.
5. *Fifth generation computers* refers to fourth generation computers with the addition of intelligence. These generation computers are not yet fully developed.

Up to the fourth generation, computers work on *data/logic information processing system* (DIPS/LIPS) but the fifth generation computers are to be made to work on *knowledge information processing system* (KIPS).

Analog computer and digital computer. An analog computer works on analog input (continuously variable electrical voltages) while a digital computer works on digital input signal.

We will confine our discussion only to digital computers. The schematic diagram of a computer system is shown in Fig. 2.2.

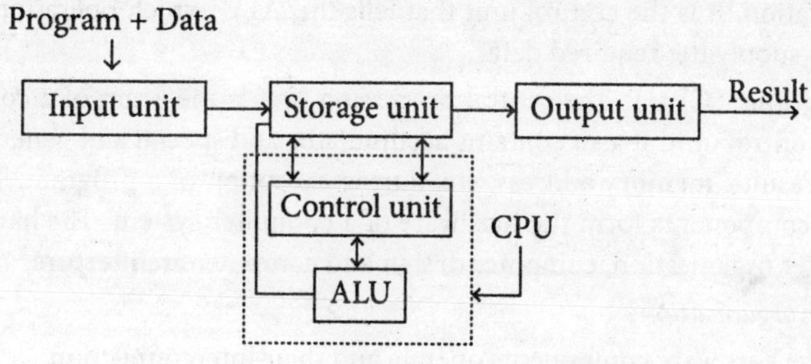

Fig. 2.2

The digital computers are of the following four types: (a) Microcomputer — smallest in size and speed (b) Minicomputer (c) Mainframe computer (d) Supercomputer.

The microcomputer developed by IBM corporation is named as PC, personal computer. Micro-computers are of three types: (a) Personal computer (PC) (b) Personal computer with extended technology (PC/XT) (c) Personal computer with advanced technology (PC/AT).

In a PC we use floppy diskettes and no hard disk is present. Both PC/XT and PC/AT possess hard disk. The number of microprocessor used in a PC go to name it such as 286, 386, 486 P, PH, P HI, etc.

Essential building blocks of a digital computer. There are five major building blocks as shown in Fig. 2.1. The blocks are: (a) Input unit (b) Output unit (c) Storage unit (d) Arithmetic and logic unit (ALU) (e) Control unit.

Arithmetic and logic unit combined with control unit constitute what is known as central processing unit (CPU).

Main functional blocks described:

1. *Input unit.* It contains a device which transfers data and instruction at the entry point before any computation is performed on the inputs. The input unit connects the computer system with external environment. The inputs must be converted into binary bits which the primary memory of a computer can accept. This is called interfacing the computer with the inputs.
2. *Output unit.* The output unit refers to the device which accepts the processed data by the computer and thus links the computer to the external world. The output devices are monitors, printers, plotter, floppy disks, etc.
3. *Storage unit.* It refers to the device which stores the input data after processing for the purpose of computation. It is the memory section of the computer.
4. *Arithmetic logic unit* (ALU). Herein all the basic arithmetical operations like addition, subtraction, multiplication, division and also logic operations are performed. The data and instructions are fetched from the storage device to the ALU as and when required. Data may flow back and forth between memory section and ALU several times before the processing is over.

The information handled by a computer is generally divided into words, each containing a group of bits. The word length of the ALU is that of the microprocessor used. If the μP used be of word length of 16 bits, the ALU is capable of performing arithmetic operations on 16-bit words; if μP is of word length of 32 bits, the ALU can handle operation on 32-bit words, etc.

Control unit. This unit coordinates the operation of all sections of the computer and sends control signals. The control circuitry interprets the instruction that constitute the program and directs the rest of the machine in its operation. It is the control unit that tells the ALU which operation to execute, and issues necessary signals to supply the required data.

Central processing unit (CPU). The central processing unit is the brain of a computer and it includes the ALU plus the control unit. It also contains accumulator and special and general registers which store data, intermediate results, memory address, etc. during execution of a program. The electronic circuitry and other physical components form the hardware of a computer system. The hardware is built on three concepts — computer organisation, computer design and computer architecture.

What is computer organisation?

It refers to how the hardware components operate and their interconnection.

What is computer design?

It is the design of the computer w.r.t. hardware viewpoint. The design starts after the specifications being formulated. Computer design refers to the type of hardware to be used and its connection. Computer design is also known as system implementation.

Computer architecture. Computer architecture refers to the structure and behaviour of the computer as seen by the user. It is concerned with the information formats, the instruction set and memory addressing techniques. Architectural design means specification of various functional blocks and structuring them together so as to form the computer system.

2.3 What are the characteristics of a computer storage system?

Storage system of a computer is a device for storing data and instructions. Following are the characteristics of the storage unit of a computer.

1. *Access time* refers to the time in which a required data is located and retrieved from the storage unit in response to the call of the program.
2. *Storage capacity* refers to the volume of data that can be stored in the storage unit.
3. *Cost/stored bit* refers to the cost effectiveness of storage unit.

2.4 Primary and secondary storages

On the basis of the three characteristics as mentioned above, the storage is primarily of two types (a) Primary storage (b) Secondary storage.

Primary storage. It is the main memory storage unit made of small areas, called locations or cells. Each location can store a fixed number of bits, which denote the word length.

The primary memory is measured in terms of $2^{10} = 1024$ bits or 1 Kb of word length. Each location is assigned a unique number, called the address of the location. A location can contain a data or instruction and whatever may be the contents of a location, its address remains unaltered. The organisation of the primary memory is shown in Fig. 2.3.

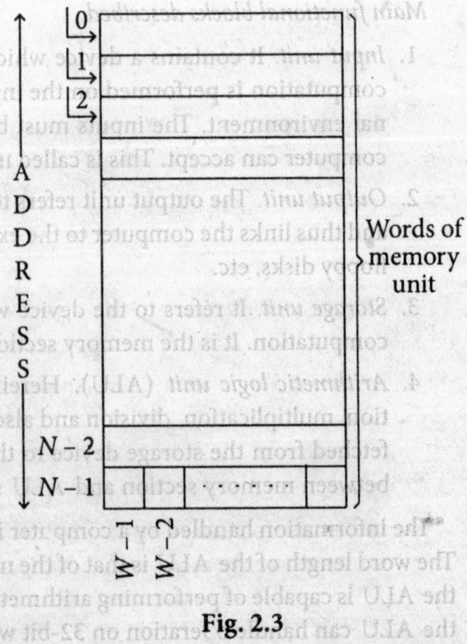

Fig. 2.3

Address starts at zero and the highest is $(N - 1)$, where N is the number of words that can be stored in the memory. A word read at an address stays on until a new word is written or stored at that address.

Majority of primary memory are of two types: (a) Random access memory (RAM) (b) Read only memory (ROM).

Random access memory. In this case, the memory cells can be accessed from any randomly located cell such that the process of locating a word is the same and requires an equal amount of time no matter where the cells are located physically in memory organisation.

The two operations that RAM can perform are the READ and WRITE operations. RAM is a volatile memory, because its contents can be altered during computational process and it is lost when the main power is switched off.

ROM: It is a permanent memory and only one operation READ is possible. The information stored in ROM are made permanent during hardware preparation and any information stored stays within the unit even when the power is made ON or OFF.

Secondary storage. In a computer a large volume of data is to be stored and storing of a word in the main memory is not cost-effective. Moreover, the RAM memory is not permanent. So there is provision for secondary storage devices which are less expensive and permanent. The access time is large and so billions of bytes of data are stored in the secondary storage devices. The data can be accessed in two ways: (a) Sequential or serial access (b) Direct or random access.

1. *Sequential access.* In the sequential access devices, stored data are accessed serially one after another, starting from the first address to the last desired address. So, in sequential access, the time of access depends on the position of the address. Sequential processing is utilised in the preparation of electric bills, monthly pay bills, etc.

2. *Random access.* Here stored data is randomly accessed as opposed to sequential accession.

What are the examples of sequential storage device and random storage device?

1. Magnetic tape is an example of sequential storage device.
2. *Magnetic disks* — direct storage device, are of two types, hard disk, floppy disk, etc.

Index

Acceptor impurities 121
Alpha 149
AM transmitter 488
Amplification
  Concept 158-178
  Types of gains 149
Amplifier frequency response 196
Amplifiers
  CB 159-161
  CE 166-169
  CC 171-172
  Cascaded 191-217
Analog to digital converter 478
Analog signal 478
Audio frequency
  Amplifier 191
  Oscillator 218
Avalance
  Breakdown 141

Backward diode 146
Barkhausen criterion 219
Base
  Band signal 481
  Bias 155
  Spreading resistance 193
Binary
  Adders 451
  Addition 299
  Division 300
  Numbers 299
Bipolar transistor 146
Bit 375
  Byte 375
Boolean algebra 401
Canonical form 439
Carrier
  Wave 482
Channel
  Width 485
Chip 261
Clamping circuit 134
Clipping circuit 133
Closed loop gain 285
CMOS inverter 263
CMRR 296
Common
  Base mode 183
  Collector mode 183
  Source drain characteristics 242
  Source FET amplifier 246
Complement method 119
Counter
  Decade 467
  Ripple 467
  Synchronous 471
Crystal
  Oscillator 222
Current source
  Equivalent circuit 4
Cut-off region 152

Darlington pair 315
De Morgan's theorem 430
Decimal system 370
Decoder, Encoder 458-460
Demodulation 497
Demultiplexer 457
Depletion MOSFET 254
Depth of modulation 483-484
Diac 350

Differential amplifier 275
Differentiator 289
Digital to analog converter 514
Diode, different types 141-143
  as clamper 134
  as clipper 133
  for rectification 137
  as voltage multiplier 135
Distribution function
  Fermi-Dirac 121

Electron ballistics 111
Enhancement MOS 258

Fermi level in semiconductor 121
Field effect transistor 239
Fixed bias 153
Flip flop 330
FM-transmitter 492
Frequency modulation 490

Gain bandwidth product 203
Gate 401

h-equivalent circuit 176
Half adder 452
Hartley oscillator 219
Hexadecimal number system 381
Hybrid parameters 177

Ideal current source 3-4
Ideal voltage source 3
Impedance 42
Input
  Offset current 283
  Voltage 284
Integrated circuit (IC) 266
Inverting amplifier 283

Junction field effect 239

K-map 440
Kirchhoff's law 4

L, R; LCR circuits 44-46
Light emitting diode 142
Linear element 3
Lithography 261
Logarithmic amplifier 288

Maxterm 439
Mesh analysis 14
Metal-oxide-semiconductor FET 254
Micro
  Controller 597
  Processor 607
Millman's theorem 7
Min term 439
Modulated wave 482
Modulating wave 483
Modulation index 483
Modulator
  Linear 488
  Variable reactance 492
Multi
  Vibrator 237
  Astable 233
  Monostable 235
Multiplexer 457

n-channel
  JFET 239
  MOSFET 241
NAND gate 419
Negation 407
NMOS inverter 260
Nodal analysis 9
Noise 490
Non-inverting amplifier 285
NOR gate 419
NOT circuit 407
npn-transistor 147
Nyquist criterion 503

Octal number system 158
Offset
  Current 283
  Voltage 283
Operational amplifier 273
Output characteristics 152

p-channel
  JFET 239

MOSFET 255
Parallel combination 48
Pass transistor 309
Peak inverse voltage 132
Phase modulation 483
Phasor diagram 44
Photo lithographic process 261
Piezoelectric effect 229
Pinchoff
Current 241
Point 241
Voltage 241
pn-junction 126
pnp-transistor 147

Quantiser 503

Race-around condition 462
Radio
Communication 498
Receiver 498
Transmitter 498
RAM 599
RC-coupled amplifier 191
Reactance modulator 493
Reciprocity theorem 6
Rectifier 130, 132
Register 466
Regulated power supply 305
Reverse
Bias 129
rotating magnetic field 86
ROM 599
RS-flip flop 331

Sampling 503
Schottky diode effect 142
Semiconductor 119
controlled rectifier 341
doped 120
extrinsic 120
n-type 119
p-type 120
Sequential circuit 460
Series voltage regulator 309
Shunt regulator 308
Side
band 485
frequency 392
Sign 1's complement 392
Sign 2's complement 392
Source transformation theorem 18
Summing amplifier 286
Super heterodyne receiver 498
Supernode analysis 13
Switching mode power supply (SMPS) 324

Tank circuit 225
Thévenin's theorem 4
Transformer
coupled amplifier 67
transistor 146-173
Triac 349
Tuned
collector 223

Uni-junction transistor 353
Unity gain buffer 286
Universal gate 419

Valence band 119
Veitch diagram 441
Virtual ground 277
Voltage
divider bias 155
multiplier 135
source equivalent circuit 4

Wien bridge oscillator 228

XNOR gate 403
XOR gate 402

Zener
diode 139
breakdown 140

Some exciting gifts are up for grabs with this book!!

THEORY OF LINEAR CIRCUITS AND RUDIMENTS OF ELECTRONICS

Dr P K Chakrabarti

FREE Gifts Rs. 1000/-

DEAR READERS

By now you must have gone through this book in the entirety. So we invite your assessment on this title with absolute impartiality and candid observations. After all, we publishers faithfully depend on your benchmarking. Friends, please fill up the open spaces below with CAPITAL LETTERS. You may add some extra sheets of paper if the space provided is not sufficient for you.

i. What are the important topics of your syllabus which are not adequately covered in this book?

ii. In which chapters and on what topics the treatment of the subject-matter by the author is not satisfactory and why?

iii. Have you come across the typos, factual inaccuracies, inconsistencies, and other forms of mistakes in the book? If yes, then please specify the mistakes and page numbers.

iv. Name top three books on the same subject in your consideration. Could you provide some reasons?

Your 1st Preference: ....

Your 2nd Preference: ....

Your 3rd Preference: ....

v. Any suggestion to improve the quality of illustrations, diagrams, etc.?

vi. How did you like the cover design? Any suggestion on colours?

vii. Suggestions for further improvement of production-standard:

viii. Who recommended you this book:

☐ Teacher ☐ Friend ☐ Parents ☐ Bookseller ☐ Librarian

ix. Name of the person who has recommended:

Designation: ....

Address: ....

Phone No.: ....

e-mail: ....

x. Name and address of the bookseller from whom you have purchased this book:

....

....

....

xi. Name and address of the library where you would like to find the book:

....

....

xii. Name and address of your institution (Please mention the University or Board):

....

....

xiii. Your name, e-mail, and complete postal address:

....

....

....

xiv. A few other titles published by us that you would like to read:

....

....

The best assessment will be rewarded half-yearly. The award will be in the form of our publications, as decided by the Editorial and Marketing Operations Division, amounting to Rs. 1000/- (Rupees one thousand) only.

....

Please mail the filled up coupon at your earliest to :

Editorial & Marketing Operations Division

NEW CENTRAL BOOK AGENCY (P) Ltd

8/1 Chintamoni Das Lane, Kolkata 700 009

West Bengal, INDIA

Phone + Fax: 033 + 2241 8411, 2241 8573, 2241 9094

e-mail: ncbapvtltd@eth.net, ncba@cal2.vsnl.net.in

Website: www.centralgroup.com